Noble gases

			IIIA	IVA	VA	VIA	VIIA	2 4.0026 **He** s² Helium

| | | | 5   10.81 **B** s²2s²2p¹ Boron | 6   12.011 **C** s²2s²2p² Carbon | 7   14.0067 **N** s²2s²2p³ Nitrogen | 8   15.9994 **O** s²2s²2p⁴ Oxygen | 9   18.9984 **F** s²2s²2p⁵ Fluorine | 10   20.179 **Ne** s²2s²2p⁶ Neon |

| | | | 13   26.9815 **Al** [Ne]3s²3p¹ Aluminum | 14   28.086 **Si** [Ne]3s²3p² Silicon | 15   30.9738 **P** [Ne]3s²3p³ Phosphorus | 16   32.064 **S** [Ne]3s²3p⁴ Sulfur | 17   35.453 **Cl** [Ne]3s²3p⁵ Chlorine | 18   39.948 **Ar** [Ne]3s²3p⁶ Argon |

VIII	IB	IIB						
28 58.71 **Ni** [Ar]3d⁸4s² Nickel	29 63.54 **Cu** [Ar]3d¹⁰4s¹ Copper	30 65.37 **Zn** [Ar]3d¹⁰4s² Zinc	31 69.72 **Ga** [Ar]3d¹⁰4s²4p¹ Gallium	32 72.59 **Ge** [Ar]3d¹⁰4s²4p² Germanium	33 74.992 **As** [Ar]3d¹⁰4s²4p³ Arsenic	34 78.96 **Se** [Ar]3d¹⁰4s²4p⁴ Selenium	35 79.904 **Br** [Ar]3d¹⁰4s²4p⁵ Bromine	36 83.80 **Kr** [Ar]3d¹⁰4s²4p⁶ Krypton
46 106.4 **Pd** Palladium	47 107.870 **Ag** Silver	48 112.40 **Cd** Cadmium	49 114.82 **In** [Kr]4d¹⁰5s² Indium	50 118.69 **Sn** [Kr]4d¹⁰5s²5p² Tin	51 121.75 **Sb** [Kr]4d¹⁰5s²5p³ Antimony	52 127.60 **Te** [Kr]4d¹⁰5s²5p⁴ Tellurium	53 126.904 **I** [Kr]4d¹⁰5s²5p⁵ Iodine	54 131.30 **Xe** [Kr]4d¹⁰5s²5p⁶ Xenon
78 195.09 **Pt** Platinum	79 196.967 **Au** [Xe]4f¹⁴5d¹⁰6s¹ Gold	80 200.59 **Hg** [Xe]4f¹⁴5d¹⁰6s² Mercury	81 204.37 **Tl** [Xe]4f¹⁴5d¹⁰6s²6p¹ Thallium	82 207.19 **Pb** [Xe]4f¹⁴5d¹⁰6s²6p² Lead	83 208.981 **Bi** [Xe]4f¹⁴5d¹⁰6s²6p³ Bismuth	84 (210) **Po** [Xe]4f¹⁴5d¹⁰6s²6p⁴ Polonium	85 (210) **At** [Xe]4f¹⁴5d¹⁰6s²6p⁵ Astatine	86 (222) **Rn** [Xe]4f¹⁴5d¹⁰6s²6p⁶ Radon

63 151.96 **Eu** [Xe]4f⁷5d⁰6s² Europium	64 157.25 **Gd** [Xe]4f⁷5d¹6s² Gadolinium	65 158.925 **Tb** [Xe]4f⁹5d⁰6s² Terbium	66 162.50 **Dy** [Xe]4f¹⁰5d⁰6s² Dysprosium	67 164.930 **Ho** [Xe]4f¹¹5d⁰6s² Holmium	68 167.26 **Er** [Xe]4f¹²5d⁰6s² Erbium	69 168.934 **Tm** [Xe]4f¹³5d⁰6s² Thulium	70 173.04 **Yb** [Xe]4f¹⁴5d⁰6s² Ytterbium	71 174.97 **Lu** [Xe]4f¹⁴5d¹6s² Lutetium
95 (243) **Am** [Rn]5f⁷6d⁰7s² Americium	96 (247) **Cm** [Rn]5f⁷6d¹7s² Curium	97 (247) **Bk** [Rn]5f⁹6d⁰7s² Berkelium	98 (251) **Cf** [Rn]5f¹⁰6d⁰7s² Californium	99 (254) **Es** [Rn]5f¹¹6d⁰7s² Einsteinium	100 (253) **Fm** [Rn]5f¹²6d⁰7s² Fermium	101 (256) **Md** [Rn]5f¹³6d⁰7s² Mendelevium	102 (254) **No** [Rn]5f¹⁴6d⁰7s² Nobelium	103 (257) **Lr** [Rn]5f¹⁴6d¹7s² Lawrencium

Chemistry: A Study of Matter

 nd. edition

Alfred B. Garrett

W. T. Lippincott

Frank Henry Verhoek

The Ohio State University

CHEMISTRY

A Study of Matter

XEROX   Lexington,

COLLEGE   Massachusetts

PUBLISHING Toronto

SECOND EDITION

ISB Number: 0-536-00181-2   (Domestic)
ISB Number: 0-536-00700-4   (International)
Library of Congress Catalog Card Number 73-190288
Printed in the United States of America.

PREFACE

The past few years have seen a decided upswing in student desire to understand the major forces that control their lives. More and more, students are recognizing the importance of comprehending and appreciating the forces of nature and the nature of the material world. With this has come an impatience with learning that has no obvious use outside the learning situation and a willingness to struggle with difficult concepts that have important bearing on life and living. The general chemistry course, dealing as it does with the fabric and the dynamics of the material world, holds great fascination and promise for large numbers of students. Most expect to be challenged by this course; most will work diligently to learn, provided they see how this learning will benefit them.

In writing this text, and in revising it for the second edition, we have sought to provide both a reliable guide and resource for helping students understand the chemistry that is taking place around them and within them, and a vehicle to illustrate how such knowledge is obtained, how it can be used, how new ideas evolve and are examined, and how all this is connected to decision-making and ultimately to consequences.

To avoid overwhelming students by presenting most or all of the great unifying concepts in the first weeks of the course without providing adequate time to assimilate and use these ideas in examining actual chemical systems, we have organized the text for the early introduction of only one or two major themes. These are studied in detail, and this knowledge illustrated and applied in appropriate chemical systems. The next major theme is then presented, studied in detail, and once again illustrated and applied; however, this time the applications require students to use not only the major concepts currently under study but all that were studied previously.

In addition, each chapter is concluded with a two or three page essay giving specific examples of how the knowledge presented in the chapter is used or becomes important in the life of the individual or the society.

The major themes in the first seven chapters are stoichiometry, atomic and molecular structure, and some principles of reactivity. These are illustrated and applied repeatedly in the early chapters and used extensively to explain and correlate the behavior of families of representative metals and a family of nonmetals in Chapters 8 and 9. In Chapters 10–13, the major themes are molecules in motion, energy absorption by molecules, the nature of solutions and of solids. These concepts along with those developed in the first seven chapters are then applied in examining the chemistry of p-type elements. In Chapters 18–22 chemical equilibrium, introductory thermodynamics and kinetics are presented. This is followed by chapters on nuclear chemistry, transition elements and coordination chemistry where previously developed concepts are used repeatedly. Three chapters on organic chemistry, emphasizing principles and practical applications, including polymers, drugs and foods, and a chapter on biochemistry, giving an overview of the chemistry of living cells, complete the text.

The chapters are written as essays on each topic building from the simple to the complex. Certain enriching material supplementing the text discussion appears throughout in a special type face; this material includes historical developments, elaboration and refinement of definitions, extended descriptions of experiments and equipment, and some illustrative examples. Epigraphs at the beginning of the chapters commemorate important discoveries and people in the history of chemistry. Each chapter concludes with a summary, a list of important terms, a reading list, and a variety of questions and problems.

The essays at the ends of the chapters discuss such diverse topics as "Odor and Molecular Geometry," "Measurement Standards," "Sources of Energy," "Modern Ceramics," "Chemistry of Plant Growth," "Atmospheric Stability and Air Quality," "Respiration," and many others. Each of these attempts to lead the student into a realm of wider application for the principles discussed in the body of the text.

The text is arranged so that the teacher can select any of a variety of sequences of topics. For teachers wishing to emphasize introductory physical chemistry, the sequence of Chapters 1–13 and 18–23 is recommended; for those favoring a principles-oriented course in descriptive inorganic chemistry, the sequence of Chapters 1–17 and 24–25 is suggested; for those wishing to prepare students for more biologically directed curricula, the sequence of Chapters 1–12, 18–21, and 26–29 might be appropriate. Rarely would the entire book be covered in a single course.

The text is designed for the student in the first course in college chemistry. A background in high school chemistry is assumed, but the diligent student should succeed without it. The book is written for all, including the best in the class. The authors have made a special effort, in developing the difficult areas, to keep the background and the intellectual maturity of the student in mind. The mathematics required is a good understanding of algebra and a sound, thorough comprehension of arithmetic.

We acknowledge with grateful appreciation assistance received from past and present colleagues in the Department of Chemistry at The Ohio State University.

We invite you to explore with us this venture in understanding and interpreting the universe.

A.B.G.

W.T.L.

F.H.V.

CONTENTS

Chemistry:
An Approach to Interpreting
the Universe
1

1

Some of the Procedures and the Language of Chemistry   4
Relevance of Science: Scientific Literacy   12
Important Terms   13
Questions and Problems   13
Special Problems   15
References   16

PART ONE   ATOMS, MOLECULES, AND IONS

Atoms:
Their Composition and Size
19

2

Development of the Theory of Atomic Structure   19
The Nature of the Nucleus   24
Chemical Atomic Weights   29
The Atom as a Source of Power   33
Summary   34
Important Terms   35
Questions and Problems   35
Special Problems   38
References   39

Moles:
Symbols, Formulas, and Equations
40

3

Chemical Symbols   41
Chemical Formulas   42
Chemical Equations   46
The Importance of Measurement Standards   56
Summary   57
Important Terms   58
Questions and Problems   58
Special Problems   60
References   61

The Electron
Configuration of Atoms
62

4

A Theory for the Arrangement of Electrons in Atoms   62
The Wave Nature of Electrons (The Electron-Cloud
    Model of the Atom)   71
Electron Configuration of Atoms   76
Other Evidence for Electron Configurations   80
How We See Nonluminous Objects   85
Summary   87
Important Terms   87
Questions and Problems   88
Special Problems   89
References   91

Chemical Bonds:
What Holds Atoms to Each Other
in Compounds?
92

5

Materials and the Quality of Life   110
Summary   111
Important Terms   112
Questions and Problems   112
Special Problems   114
References   115

Molecular Structure and Intermolecular Forces of Attraction
116

6

Molecular Structure   118
Intermolecular Forces of Attraction   133
Odor and Molecular Geometry   136
Summary   138
Important Terms   138
Questions and Problems   138
Special Problems   140
References   141

PART TWO   STRUCTURE AND REACTIVITY

Principles of Reactions;
Patterns of Reactivity;
The Chemistry of Hydrogen
145

7

An Introduction to Chemical Equilibrium   146
Equilibria Involving Acids and Bases   148
Oxidation-Reduction Reactions   154
Examples of the Calculation of Oxidation State   156
The Chemistry of Hydrogen   158
Some Facts about Hydrogen   158
Chemical Processes in the Biosphere   164
Summary   165
Important Terms   166
Questions and Problems   166
Special Problems   167
References   168

The Alkali and Alkaline Earth Elements: Two Families of Representative Metals 169

8

Comparison of the Two Families 169
Some Important Compounds of Group IA and Group IIA Elements 181
The Periodic Table and Trends Observed in Group IA and Group IIA 184
Simple Ions in Human Health 187
Summary 189
Important Terms 189
Questions and Problems 189
Special Problems 191
References 192

The Halogen Family: A Family of Nonmetals 193

9

A Broad Overview of the Halogen Family 193
Properties of the Halogen Elements 196
Preparation of the Halogen Elements 201
Some Compounds of the Halogens 202
Chemical Solutions to Some Every Day Health Problems 207
Summary 209
Important Terms 209
Questions and Problems 209
Special Problems 211
References 212

PART THREE   ENERGY AND THE STATES OF MATTER

Energy and Molecules of Gases 215

10

The Gas Laws 218
The Kinetic-Molecular Theory 225
Internal Energy of Molecules 230
Heat Capacity 235
Plasma 235
Atmospheric Stability and Air Quality 235
Summary 237
Important Terms 238
Questions and Problems 238
Special Problems 240
References 241

Attractive Forces: Three States of Aggregation 242

11

Fogs, Clouds, and Precipitation 252
Summary 253
Important Terms 254
Questions and Problems 254
Special Problems 255
References 257

Solutions—Homogeneous Mixtures 258

12

The Process of Dissolving   258
Other Types of Solutions   266
Ionization   268
Body Fluids: Precious Solutions   273
Summary   275
Important Terms   275
Questions and Problems   275
Special Problems   277
References   278

The Metallic State: Architecture of Solids 279

13

Use of X Rays to Determine Crystal Structure   279
Metals: The Arrangement of Atoms in Metals   283
Band Theory of Metals   290
Arrangement of Ions in Crystals   294
Arrangement of Molecules in Crystals   297
Lattice Energy and Crystal Stability   299
Application of Solid State Models   299
Important Terms   301
Questions and Problems   301
Special Problems   303
References   304

PART FOUR   FAMILIES OF ELEMENTS

The Oxygen Family 307

14

An Overview of Oxygen Family Chemistry   307
Properties of the Elements   313
Important Compounds of Oxygen Family Elements   315
Respiration   317
Summary   319
Important Terms   319
Questions and Problems   320
Special Problems   321
References   321

The Nitrogen Family 322

15

An Overview of Nitrogen Family Chemistry   322
The Elements   329
Important Compounds of Nitrogen Family Elements   331
Compounds in Positive Oxidation States   333
Some Chemistry of Plant Growth   334
Summary   335
Important Terms   336
Questions and Problems   336

Special Problems   337
References   338

The Carbon Family   339

16

A Broad Overview of the Carbon Family   339
The Elements   345
Some Inorganic Compounds of the Carbon Family   347
Modern Ceramics   351
Summary   353
Important Terms   353
Questions and Problems   353
Special Problems   354
References   355

The Boron Family   356

17

A Broad Overview of the Boron Family   356
The Elements   361
Electronic Ceramics—You Use Them Every Day   363
Summary   364
Important Terms   364
Questions and Problems   364
References   365

PART FIVE   ENERGY AND CHEMICAL CHANGE

Energy   369

18

Heats of Reaction   369
Calculation of Bond Energies   375
Applications of Thermochemical Measurements   378
Change in Enthalpy with Change in Temperature   381
The Sun as a Main Source of Energy   382
Summary   384
Important Terms   384
Questions and Problems   385
Special Problems   387
References   388

Chemical Equilibria—Measurement of Extent of Reactions   389

19

The Equilibrium Constant   389
Shifting the Position of Equilibrium   395
Equilibria in Ionic Reactions in Aqueous Solution   398
Exact Solutions for Ionization Processes   403
Ionization of Diprotic Acids   406
Equilibria Involving Ionic Compounds of Low
    Solubility   408
Reinforced Buffering in Blood   413

Summary   415
Important Terms   415
Questions and Problems   415
Special Problems   418
References   419

Free Energy
420

20

Free Energy and Equilibrium   427
Free Energy and the Oxidation of Glucose   435
Summary   437
Important Terms   437
Questions and Problems   437
Special Problems   440
References   441

Electrochemical Cells—
The Measurement of Free Energy
442

21

Fuel Cells: Infants with Unusual Promise   457
Summary   458
Important Terms   459
Questions and Problems   459
Special Problems   461
References   463

Chemical Kinetics; Reaction Pathways
464

22

Rate and Mechanism of Reaction   465
Theories of Reaction Rates   475
Chain Reactions and Explosions   481
Are Reaction Rates Important to Life?   482
Summary   483
Important Terms   484
Questions and Problems   484
Special Problems   487
References   489

PART SIX   NUCLEAR CHEMISTRY

Nuclear Chemistry
493

23

The Nature and Properties of the Rays   494
Some Properties of the Nucleus   496
Some Radiochemistry   499
Applications of Nuclear Science   505
Nuclear Power: Is It Safe?   507
Summary   509
Important Terms   509
Questions and Problems   510
Special Problems   511
References   512

PART SEVEN THE TRANSITION ELEMENTS

Overview of the Transition Elements—Coordination Chemistry **515**

24

Properties of Transition Elements   516
Properties and Bonding in Complex Ions   517
Effect of Ligands on the Properties of Ions   529
Transition Elements and Human Health   530
Summary   531
Important Terms   531
Questions and Problems   532
Special Problems   533
References   533

The Transition Elements **534**

25

First-Series Transition Elements   535
The Chemistry of the Elements   538
Second- and Third-Row Transition Elements   548
Lanthanides and Actinides   549
Zinc, Cadmium, and Mercury   550
Magnets from Transition Metals   551
Summary   552
Important Terms   552
Questions and Problems   552
Special Problems   554
References   555

PART EIGHT ORGANIC CHEMISTRY

Organic Chemistry I: Structures and Nomenclature **559**

26

Classification of Organic Substances   560
Some Important Homologous Series   562
Theory: The Alkanes, RH   562
Alkenes, $RCH=CHR$ 568
Alkynes, $R-C\equiv C-R$ 570
Alkyl Halides, RX   571
Alkanols or Alcohols, ROH   573

Aldehydes (Alkanals), $R\overset{O}{\overset{\|}{C}}-H$; Ketones (Alkanones), $R-\overset{O}{\overset{\|}{C}}-R$ 575

Carboxylic Acids, $R\overset{O}{\overset{\|}{C}}-OH$ 577
Derivatives of Carboxylic Acids   579
Amines   580
Aromatic Compounds   580
Optical Isomerism   583

How Organic Chemistry Helps Us All   588
Summary   590
Important Terms   590
Questions and Problems   590
Special Problems   593
References   594

**Organic Chemistry II:
Reactivity of
Some Organic Structures
595**

27

Classes of Organic Reactions   595
Oxidation and Reduction Reactions   596
Condensation Reactions   601
Addition Reactions   603
Elimination Reactions   608
Substitution Reactions   610
The Need and Basis for Safe Insecticides   617
Summary   620
Important Terms   620
Questions and Problems   620
Special Problems   623
References   625

**Polymers and
Other Complex Molecules
626**

28

Polymers   626
Other Important Complex Molecules   639
A Challenge for the Plastics Industry   647
Summary   648
Important Terms   649
Questions and Problems   649
Special Problems   650
References   651

**Some Chemistry of Living Cells
652**

29

The Chemistry of Muscle Activity   670
Summary   672
Important Terms   672
Questions and Problems   672
Special Problems   673
References   674

Appendix A. Values of Important Constants   i
Appendix B. Some Rules for Chemical Nomenclature   iv
Appendix C. Balancing Oxidation-Reduction Equations   vii
Appendix D. Wave Motion and Wave Functions   x
Appendix E. Appendix Tables   xiii

INDEX   xix

Chemistry: A Study of Matter

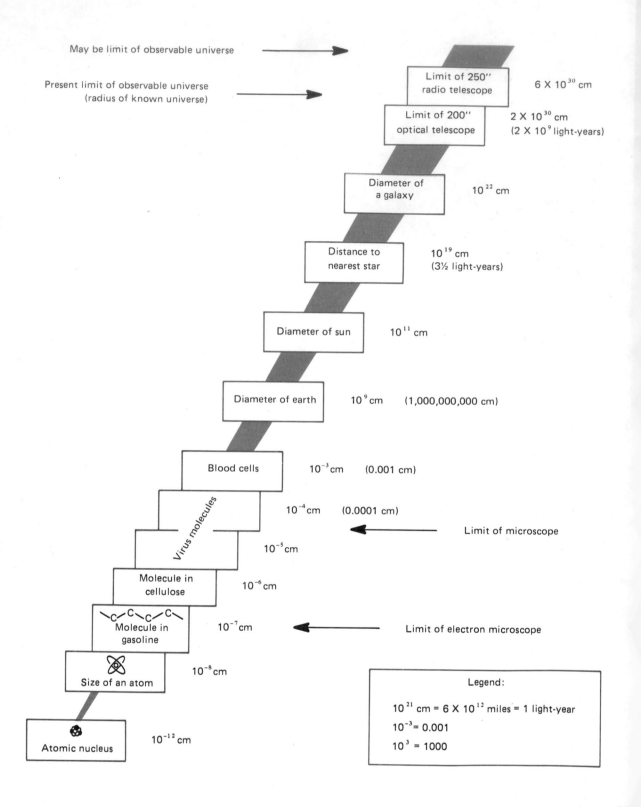

May be limit of observable universe →

Present limit of observable universe
(radius of known universe) →

Limit of 250''
radio telescope 6×10^{30} cm

Limit of 200''
optical telescope 2×10^{30} cm
(2×10^{9} light-years)

Diameter of
a galaxy 10^{22} cm

Distance to
nearest star 10^{19} cm
($3\frac{1}{2}$ light-years)

Diameter of sun 10^{11} cm

Diameter of earth 10^{9} cm   (1,000,000,000 cm)

Blood cells 10^{-3} cm   (0.001 cm)

Virus molecules 10^{-4} cm   (0.0001 cm)   ← Limit of microscope

10^{-5} cm

Molecule in
cellulose 10^{-6} cm

Molecule in
gasoline 10^{-7} cm   ← Limit of electron microscope

Size of an atom 10^{-8} cm

Atomic nucleus 10^{-12} cm

Legend:

10^{21} cm = 6×10^{12} miles = 1 light-year

10^{-3} = 0.001

10^{3} = 1000

Figure 1.1   Range in sizes and distances of objects in the universe.

CHEMISTRY: AN APPROACH TO INTERPRETING THE UNIVERSE

Today we can describe with considerable confidence the composition and structure not only of the very tiny objects in the universe—molecules, atoms, and atomic nuclei—but also of the great massive bodies such as the moons, the planets, the stars, and even the galaxies. We can describe as well the kinds of changes or reactions that occur not only in the laboratory or on the earth's surface but also in the sun and in the farthest star in the farthest galaxy. Man's information about the structure and composition of the stuff in the universe, his understanding of the changes in energy, structure, and composition of this stuff, and his studies of the types of forces and motion of objects in the universe comprise an impressive part of our cultural heritage.

Man's Early Attempts Earliest history records man's interest in the many natural phenomena such as fire, lightning, and thunder as well as his concern with common things such as air, water, and soil and with the moon, sun, and stars. For thousands of years he struggled to find a way to describe them in terms of simpler substances. In his early attempts, superstitions and black magic played a major role. Folklore and religious dictums often dominated. For hundreds of years many people lived under the beliefs that—

The earth was the center of the universe.
The four elements were fire, air, earth, and water.
A philosopher's stone could change base metals to gold.
Life developed by spontaneous generation.

As knowledge about the universe gradually was distilled and refined, the quest to know the universe gradually became known as *science*.

Science The collection of facts and theories about natural phenomena and objects in the universe is the body of knowledge now called science. The feature that distinguishes today's search for knowledge of the universe from the procedure of the early days is the toughminded attitude of the scientist—his

1

requirement that conclusions or theories be subjected to experimentation or testing before they are accepted. This unique process establishes the validity of a theory. It is one of the important steps in the scientific method, which is often summarized in the following terms: observation, classification, theorizing, and testing.

The Science of Chemistry

The body of information that describes the structure and composition of the material in the universe together with the changes in structure and composition this material undergoes and the energy relations involved in these changes is known as chemistry.

Man's concepts of the material world have been fashioned from his own experiences, refined by his dedication to reason, and perfected by a disciplined imagination kept honest by the skepticism of his peers. The beginnings of chemistry grew from two tap roots—one philosophical, the other utilitarian. Greek philosophers sought to account for the material diversity of the world in orderly and intellectually satisfying terms. The practical men of the several ancient civilizations expanded the arts and crafts into useful technologies for smelting and alloying metals, making glass, producing charcoal and lime, etc.

Inheriting the rudimentary concepts of four elements—earth, air, fire and water—and seeking an orderly explanation for the observation that one thing can be transformed into another—wood to ash, grass to milk, seed to growing plant—Greek philosophers attempted to explain these phenomena in terms of fundamental principles. They developed the continuistic and atomistic views of nature and provided the logic patterns for using these views in solving problems. They demanded reasonable, even mathematical explanations for observed behavior of matter. They posed the great problems of science, many of which are still unsolved. However, the Greeks failed to understand the necessary connection between thought and practice—between the life of the mind and the needs of the living.

Figure 1.3 A modern chemistry laboratory requires facilities for exact measurements, analysis of small quantities of material, and determination of properties by physical methods, and is equipped with an array of scientific instruments to make these tasks more precise and more rapid. (*Photograph courtesy of Fisher Scientific Company.*)

Meanwhile, association of the metallic arts with the Egyptian priesthood gave rise to the mysticism and obscurity of the period of alchemy. Alchemy was developed by Arabic cultures between A.D. 600 and 1000 and transmitted to the then Western world toward the end of that time. Its practitioners were largely concerned with the transmutation of base metals into gold, and they flourished until the early 18th century, having in the meantime given rise to a school that emphasized the role of chemistry in the cure of disease (the iatrochemists). The alchemists discovered a great many chemical facts in their search for transmutation processes.

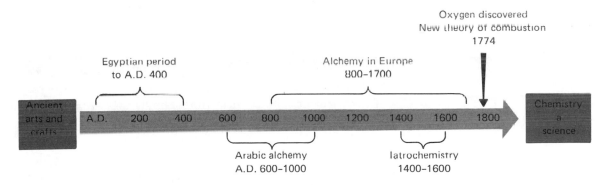

Figure 1.4 Time scale for the development of chemistry.

OBJECTIVES OF MAN THE INTELLECTUAL

In the Sciences: An interpretation of the material universe.

In the Social Sciences and Humanities: The way man can live most effectively in the universe he learns to interpret.

In the Communicative Arts (mathematics, oral and written word, art, music, poetry, drama, etc.): An effective method of communicating to other men information about the reality he discovers in his searches.

Not until the Renaissance and Francis Bacon's contention that "knowledge is power" did emphasis on application of knowledge become popular among the well-educated. Also during this period, dependence upon the immutable results of experiments rather than upon the mystical pronouncements of authorities became the accepted basis for discussion.

An important milestone was the discovery of oxygen (in 1774) by Joseph Priestley, followed by elucidation of the true nature of combustion by Antoine Lavoisier.

Building on these foundations, early 19th century chemistry was in the process of replacing its alchemical and empiricist traditions with simple but rigorous numerical laws which brought order to the seeming chaos of understanding the nature of matter. The chemistry of the modern industrial age— dyes, bleaches, soaps, detergents, drugs, antiseptics, anesthetics, fertilizers— rests upon a base of only a few general principles. The search for, and study of, these principles is what chemistry is all about.

The knowledge that is chemistry has been created by men seeking to understand and possibly to control the world of materials. Like many human enterprises that have survived the ravages of time, that knowledge is at once magnificent, imperfect, and incomplete. There remains for future contributors to chemistry the task of understanding what is now known, so they can correct its errors and advance its frontiers.

SOME OF THE PROCEDURES AND LANGUAGE OF CHEMISTRY

How to Organize the Search for the Truth about the Universe

The problem of understanding the universe is a huge one. Many years of toil and study were required before the method of tackling the problem became clear. The method finally chosen is that used in approaching any big problem, that is, *simplification by classification.*

The Two Entities of the Universe

Careful observation of the world around us indicates the two great entities, matter and energy. Matter is anything that has mass (or weight) and occupies space. Matter includes things that can be observed, such as rocks, water, soil, machines, trees, as well as things that cannot be observed directly but can be detected by other means, such as air and other gases, or particles too small to see such as atoms, ions, molecules, electrons, protons, and neutrons.

Energy is the ability to do work. It also may be defined as heat or anything transformable into heat. Some familiar forms are mechanical, electrical, radiant, and chemical energy.

Some Further Classifications in the Study of Matter

The study of matter can be further organized by using several different classifications: homogeneous and heterogeneous substances, for example, or solids, liquids, and gases. The classification chosen depends upon the convenience desired or the particular phenomenon being studied. Several different classifications are indicated in Figure 1.5. Each of these classifications is used in the study of chemistry.

Homogeneous and Heterogeneous Substances. The term *substance* is used for any form of matter. If the substance appears to be uniform throughout, it is called a *homogeneous* substance; for example, sugar, water, alcohol, or

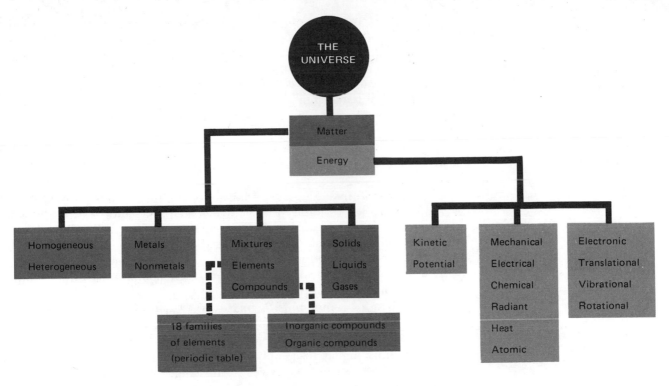

Figure 1.5 Simplification by classification.

a solution. Any homogeneous system of material is called a *phase*. If a substance is not uniform throughout but is of such a nature that we can distinguish two or more parts (phases) which are themselves homogeneous, it is called *heterogeneous*; for example, salt and pepper mixture, sandstone, and muddy water.

Metals and Nonmetals. Metals are substances that have a metallic luster, are malleable and ductile, and are good conductors of heat and electricity. Nonmetals are poor conductors of heat and electricity, and are neither malleable nor ductile.

	Metals	Nonmetals
Examples	Iron, copper, gold, tin	Sulfur, carbon, iodine, phosphorus

Solids, Liquids, and Gases. These are called the physical states of matter. They are also called the solid, liquid, and gaseous phases. Theoretically, every substance can exist in these three states. Actually, all substances can be solidified, but some substances decompose when they are heated in an attempt to change them to a liquid or to a gas. For example, sugar decomposes before it vaporizes; limestone decomposes at ordinary pressures when the temperature is raised.

Mixtures, Elements, and Compounds. For the chemist, these are perhaps the most useful classifications of substances. Mixtures contain two or more substances; they may be homogeneous or heterogeneous, but they can always be separated by physical processes. For example, a solution of sugar and water is a homogeneous mixture; the sugar can be separated from the water simply by evaporating the water. Iron and sulfur is a heterogeneous mixture; the sulfur can be dissolved with carbon disulfide, leaving the iron, or the iron can be removed from sulfur with a magnet.

Careful study has shown that the universe is made of about 100 substances called elements , from which all other substances (compounds and mixtures) are made. Examples of the elements are copper, iron, lead, hydrogen, oxygen, and chlorine. A considerable part of the subject matter of chemistry involves the study of elements and the compounds and mixtures they form.

Elements combine or react with each other to form new substances called compounds ; for example, the elements hydrogen and oxygen combine to form the compound water. The elements combined in compounds cannot be separated by physical means; thus they differ from mixtures. Furthermore, each compound always contains the same elements in the same ratio by weight. There are about 2 million different compounds known today and millions more probably exist.

Are solutions compounds or mixtures? Much of the time chemists work with solutions rather than pure elements or compounds. Air is a solution of gases; ocean water is a solution of salts in water. Solutions are homogeneous and are thus similar to compounds in this respect, but the composition of solutions can be varied often over a wide range. Therefore, a solution cannot be considered to be a compound. A solution is defined as a homogeneous mixture. The composition of a solution is described in terms of the amount of dissolved substance (called the solute) and the amount of the solvent . (See Chapter 12.)

Classification of the Elements

The problem of learning about the elements is further simplified by classifying them into groups of elements on the basis of their properties, so that each member element of a particular group has properties similar to the properties of the other elements in the group. Such groups of elements with similar properties are called families. An arrangement of elements into families produces a periodic table (Figure 1.6 and inside front cover) where the families appear in vertical columns marked by a distinguishing Roman numeral at the top of the column. Thus the study of the elementary substances in the entire universe resolves itself into the much more simple problem of learning about the various families of elements and the compounds they form.

Chemical and Physical Properties

The term property , as used in the preceding paragraph, means a mark or characteristic that enables us to identify or describe a substance. For convenience, a distinction is sometimes made between physical properties and chemical properties, although a definitive distinction between these is somewhat difficult to establish. Among physical properties are included such characteristics as color, hardness, solubility, odor, boiling point, freezing point, and vapor pressure. Among chemical properties are included such characteristics as heat of combustion, stability (toward decomposition), or reactivity. Some of these properties are illustrated in Table 1.1.

Figure 1.6 — Periodic table

	IA	IIA	IIIB	IVB	VB	VIB	VIIB		VIII		IB	IIB	IIIA	IVA	VA	VIA	VIIA	O
1	1 Hydrogen H																	2 Helium He
2	3 Lithium Li	4 Beryllium Be											5 Boron B	6 Carbon C	7 Nitrogen N	8 Oxygen O	9 Fluorine F	10 Neon Ne
3	11 Sodium Na	12 Magnesium Mg											13 Aluminum Al	14 Silicon Si	15 Phosphorus P	16 Sulfur S	17 Chlorine Cl	18 Argon Ar
4	19 Potassium K	20 Calcium Ca	21 Scandium Sc	22 Titanium Ti	23 Vanadium V	24 Chromium Cr	25 Manganese Mn	26 Iron Fe	27 Cobalt Co	28 Nickel Ni	29 Copper Cu	30 Zinc Zn	31 Gallium Ga	32 Germanium Ge	33 Arsenic As	34 Selenium Se	35 Bromine Br	36 Krypton Kr
5	37 Rubidium Rb	38 Strontium Sr	39 Yttrium Y	40 Zirconium Zr	41 Niobium Nb	42 Molybdenum Mo	43 Technetium Tc	44 Ruthenium Ru	45 Rhodium Rh	46 Palladium Pd	47 Silver Ag	48 Cadmium Cd	49 Indium In	50 Tin Sn	51 Antimony Sb	52 Tellurium Te	53 Iodine I	54 Xenon Xe
6	55 Cesium Cs	56 Barium Ba	57-71 Series of Lanthanide Elements	72 Hafnium Hf	73 Tantalum Ta	74 Tungsten W	75 Rhenium Re	76 Osmium Os	77 Iridium Ir	78 Platinum Pt	79 Gold Au	80 Mercury Hg	81 Thallium Tl	82 Lead Pb	83 Bismuth Bi	84 Polonium Po	85 Astatine At	86 Radon Rn
7	87 Francium Fr	88 Radium Ra	89-103 Series of Actinide Elements	104	105 Hahnium Hn													

Series of Lanthanide Elements

57 Lanthanum La	58 Cerium Ce	59 Praseodymium Pr	60 Neodymium Nd	61 Promethium Pm	62 Samarium Sm	63 Europium Eu	64 Gadolinium Gd	65 Terbium Tb	66 Dysprosium Dy	67 Holmium Ho	68 Erbium Er	69 Thulium Tm	70 Ytterbium Yb	71 Lutetium Lu

Series of Actinide Elements

89 Actinium Ac	90 Thorium Th	91 Protactinium Pa	92 Uranium U	93 Neptunium Np	94 Plutonium Pu	95 Americium Am	96 Curium Cm	97 Berkelium Bk	98 Californium Cf	99 Einsteinium Es	100 Fermium Fm	101 Mendelevium Md	102 Nobelium No	103 Lawrentium Lr

Figure 1.6 Periodic table.

Table 1.1 Typical Physical and Chemical Properties of Several Substances

Substance	Physical Properties	Chemical Properties
Helium	Gas, low boiling point	Inert
Water	Liquid at 25°C, colorless	Stable at 25°C; reacts with some metals
Sulfur	Solid, yellow, brittle	Stable at 25°C; burns in oxygen, releasing energy
Iron	Solid, metallic luster, high density	Rusts readily; burns in oxygen, releasing energy

Chemical Changes

Those changes in which a new substance or new substances are formed are called chemical changes. For example, iron rusts; in this change a new substance, rust or iron oxide, is formed. Carbon burns; in this change a new substance, carbon dioxide, is formed. Nitroglycerine explodes; in this reaction new substances, nitrogen, carbon dioxide, water, and other gases, are formed. Water is electrolyzed; in this change oxygen and hydrogen are formed. An energy change also accompanies a chemical change. As a result, there is a difference in the chemical energy of the reactants and products.

Physical Changes

Those changes which involve a change in the form or state of a substance but do not involve production of a new substance are called physical changes For example, water boils; in this change no new substance is formed, but water changes from the liquid state to the gaseous state. Sulfur melts; solid sulfur is changed to liquid sulfur.

Atoms Are Building Blocks of Elements

For several hundred years philosophers debated the question of whether matter is continuous or whether it is discontinuous and therefore made of particles. As early as 400 B.C., the Greek philosopher Democritus found it helpful to assume that matter is made of particles (which he called atoms). But it remained just a philosophical question until 1803, when John Dalton gathered enough data from experiments on the solubility of different gases in liquids and the definite amounts of elements that combine with each other to lead him to suggest that matter was definitely made of particles or atoms. Further evidence confirming that suggestion is available today, and we can even describe with confidence the structure and composition of atoms. Each of the 100-odd elements has atoms of one particular type different from the atoms of all other elements. We define an element as a substance made of one type of atom only.

A summary of the main points of the atomic theory as accepted today is as follows:

1. All matter is made up of unit particles called atoms
2. The atoms of a particular element have the same mass, or at least an average mass characteristic of the element; atoms of different elements have different average masses.
3. In chemical reactions whole atoms, never fractions of atoms, combine or separate, or change places.
4. When atoms combine with other atoms, they do so in one or more whole-number ratios which are usually small.

Names and Symbols for Elements

The privilege of naming an element is that of the discoverer. Most of the elements are given English names, but some were given names in Latin; for example, silver was called *argentum* and iron *ferrum*. Abbreviations of the names of elements are called *chemical symbols*; for example, hydrogen is H, oxygen is O, silver is Ag, and iron is Fe. In the atomic theory each symbol represents one atom of the element. Thus, H represents one atom of hydrogen, 2H represents two atoms of hydrogen, H_2 represents a molecule of hydrogen containing two atoms, and H_2O represents a molecule of water containing two atoms of hydrogen and one atom of oxygen.

Three Basic Laws Explained by Assuming the Existence of Atoms

1. Law of Conservation of Matter and Energy. One of the questions that bothered the early scientists was whether matter is lost or destroyed when substances combine or decompose. What happens to wood when it burns? What happens to sugar when it dissolves? What happens to the energy in coal when the coal burns?

Many careful experiments in which the weights of the reactants were compared with the weights of the products of reactions showed no detectable loss of weight when the reactions occurred. These results are expressed in a formal law of conservation of mass: *Matter is neither created nor destroyed in chemical process.*

Other similar experiments on the transformations of energy led scientists to believe that a similar conservation law exists for energy: *Energy is neither lost nor destroyed, but may be transformed from one form to another.*

Today it is known that in nuclear changes, such as those which occur in radioactive transformations, matter can be converted to energy and energy to matter, but neither is lost. In chemical changes, however, our interpretation is that atoms combine with each other, or rearrange in different combinations; this means they are not destroyed but appear in new forms and arrangements without measurable change in mass.

2. Law of Definite Composition. Another observation of the early scientists was that a given weight of an element always combines with definite weights of other elements. For example:

1.000 g of oxygen combine with 1.125 g of aluminum.
1.008 g of hydrogen combine with 16.03 g of sulfur.
1.008 g of hydrogen combine with 35.45 g of chlorine.

Finding that results of this sort are generally true, scientists expressed the law of definite composition: *In chemical compounds, the several elements are always combined in definite proportions by weight* (Figure 1 7).

How can this be explained? John Dalton's theory, that elements are made of atoms, provides an answer. Dalton reasoned in this manner: The atoms of a given element all have the same mass, but this mass is different from the masses of atoms of other elements. Reaction of one element with another takes place between the atoms of each element. Hence, if a specified number of atoms of the one element always combine with a specified number of atoms of the other element, then a specified mass of the first element must always combine with a specified mass of the second. Therefore, the elements should have definite combining weights, and the product formed will always have a definite composition.

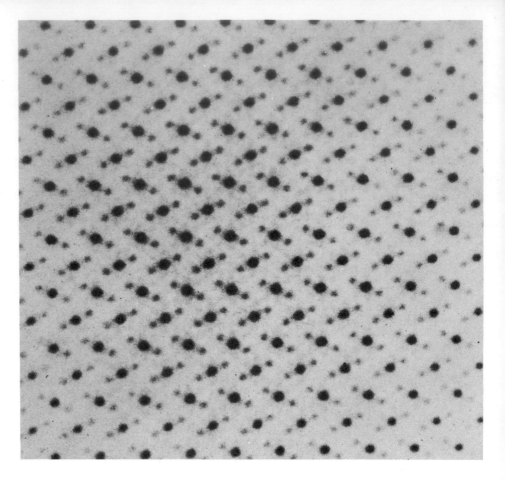

3. Law of Multiple Proportions. It was observed that some elements have more than one combining weight. For example:

1.008 g of hydrogen combine with 8.000 g of oxygen in the compound water.
1.008 g of hydrogen combine with 16.000 g of oxygen in hydrogen peroxide.

It was further noted that *the ratios of the weights of the second element* (oxygen in the example given) *combined with a fixed weight of the first* (hydrogen) *can always be expressed as a ratio of small whole numbers* (8:16 or 1:2 in the example given). The preceding sentence is, in fact, a statement of the formal law of multiple proportions

Again, Dalton's theory of atoms was helpful in explaining this observation. We need only to assume that the same atoms can combine in more than one ratio. For example, in water, one atom of oxygen combines with two atoms of hydrogen, while in the hydrogen peroxide, two atoms of oxygen combine with two of hydrogen.

$$
\begin{array}{cc}
\text{H} & \text{H} \\
\quad \diagdown & \qquad \diagup \\
\quad\text{O} & \quad\text{O—O} \\
\quad \diagup & \qquad\qquad \diagdown \\
\text{H} & \qquad\qquad \text{H}
\end{array}
$$

Twice as much oxygen combines with the same weight of hydrogen in hydrogen peroxide as in water, and the two-to-one ratio of the combining weights of oxygen is explained.

Units of Measurement

By agreement, scientists make and record measurements either in the cgs (centimeter-gram-second) or the mks (meter-kilogram-second) system of units. Both systems arise from the metric system which, in turn, is based on the length of the meter now defined as 1,650,763.73 times the wavelength of the orange-red line in the spectrum of krypton isotope 86. The meter is 39.37 inches. Appendix A lists commonly used units and some relations among them.

Forms and Units of Energy

Energy must be supplied if a piece of matter initially at rest is to be placed in motion. Once in motion we say it possesses *kinetic energy*, the amount of which is given by $\frac{1}{2}mv^2$, where m is the mass of the object and v is its velocity. If the mass is given in kilograms and the velocity in meters per second, the energy is in *joules*. The energy supplied in changing the position of an object against a force—such as in lifting a weight against the force of gravity—is said to be stored in the object in its new position as *potential energy*. From the physical laws that force equals mass times acceleration, and that energy equals force times distance, it follows that the amount of potential energy stored in an object lifted a distance h is mgh, where m is the mass of the object and g the acceleration of gravity. Kinetic and potential energy are interconvertible—an object falling to earth decreases in potential energy and increases in kinetic energy.

Heat and radiation are other forms of energy. Radiation, which may exist in a vacuum, includes light, X rays, gamma rays, and radio waves (see Chapter 4). Heat is a form of energy associated with the motion of atoms and molecules, and capable of being transferred through fluid media by convection and through empty space by radiation. Heat is usually measured in units of *calories*—the energy needed to increase the temperature of 1 g of water from 14.5°C to 15.5°C—or in kilocalories, thousands of calories. (The kilocalorie also is known as the large calorie and it is this unit that is used in dietary considerations of calories.) The heat capacity of a substance is the quantity of heat needed to raise the temperature of a given mass of the substance by 1°C.

Temperature is a measure of the relative "hotness" of an object or system. To exist at a certain temperature, the object must contain a definite quantity of heat. When objects at different temperatures are brought into contact, heat will flow from the hotter to the cooler body until the temperatures are equal.

To illustrate the relationship between temperature and heat, we might compare the heat present in a 1 g sample of water at 25°C with that present in 1 kg of water at the same temperature. Obviously, the larger sample contains 1000 times as much heat as the smaller sample even though both are at the same temperature. We say that heat measures the quantity of energy present, and temperature describes the intensity of the heat.

Usually, temperature is measured using a thermometer. Convenient and arbitrary temperature scales have been devised. The *Celsius scale* is used in scientific work. This scale assigns a temperature of 0° to the freezing point and 100° to the boiling point of water and divides the distance between these points into 100 equal units, each known as a degree Celsius (°C). On the

Fahrenheit scale, the freezing and boiling points of water are assigned values of 32°F and 212°F, respectively. Thus a Fahrenheit degree is 100/180 or 5/9 as large as a Celsius degree. The *Kelvin scale* (K) is defined as °C + 273.15. Here the size of the degree remains the same as that on the Celsius scale, but the frame of reference is changed from the freezing and boiling points of water to the lower absolute limit of attainable temperatures.

Chemistry, Other Sciences, and Technology

Chemistry has a close relationship to other sciences because of the application of the study of chemical change and the structure of matter to every other science. Consequently, it is important to remember that the division between the sciences is not sharp. Today all sciences overlap each other, and oftentimes the same subject matter is to be found in several sciences. Science and technology are related in that one (science) is the process of study, search, and discovery, and the other (technology) is the application of the results of that study or that discovery for some practical purpose.

Chemistry and an Enlightened World

The results of the search for an understanding of the universe have enlightened man in a satisfying and exciting manner as to the nature of his world. Many of his old fears and superstitions have been removed, and a new dimension has been added: a new method of learning about the universe—the method of science.

Relevance of Science: Scientific Literacy. To analyze chemically a sample of moon dust, to do the same for Mars and the other planets, and to know what reactions occur in the sun and the stars—such is the chemistry of the cosmos. To learn of the reactions in the heart of the living cell where replication of each species is achieved, or to know the process that produces energy from matter in the heart of the atom—such is chemistry of the microcosmos. To make available through synthesis or by extraction from natural materials items that take much of the misery and drudgery out of living—such is the chemistry of humanity.

The chemical industry, built on fundamental research done in universities, foundations, government, and in its own laboratories, is our largest industry. The food we eat, the clothes we wear, our medicines, antiseptics, vitamins, anesthetics, pesticides, and germicides, as well as the cars we drive, the jets we fly, our plastics, rubbers, fibers, fertilizers, and fuels are examples of the contribution of applied chemistry to a more comfortable and safer life. These and other scientific developments help provide for all who would contribute to building a better world a greater opportunity through a longer and potentially more productive life span.

Such contributions show the power of science but they cannot assure a happier world. One of the limitations of science is that while these contributions make such a world possible, man must, to attain it, couple the objective of science with a search to find how he can live most effectively in the world he learns to interpret.

Needed: A Scientifically Literate Generation. Cognizance of these facts helps us appreciate the importance of scientific literacy in all professions today—not only to know science but to have developed a perspective on the power and limitations of science. Both are important in developing a generation that is scientifically literate and able to use the products of science to advantage.

Matter
homogeneous
heterogeneous
states of matter
solids
liquids
gases

Energy
kinetic energy
potential energy
heat
light
conservation of energy
temperature

Properties of matter
physical
chemical

Elements
metals
nonmetals

Laws of chemical combination
law of conservation of mass
law of definite composition
law of multiple proportions

Atoms
atomic theory
atomic masses
ions
molecules

Equations
symbols
formulas
units
cgs
mks

QUESTIONS AND PROBLEMS

1. Give examples of each of the following: (a) a substance, (b) several physical properties, (c) several chemical properties, (d) a property that might be considered either chemical or physical, (e) several physical and several chemical changes, (f) a substance of low density, (g) a substance of high density.

2. Give two specific examples to illustrate each of the following terms: element, compound, phase, homogenous mixture, heterogeneous mixture, solution, metal, nonmetal.

3. Describe the interconversion of kinetic and potential energy in: (a) a waterfall, (b) a burning torch, (c) a bouncing ball, (d) the condensing of steam, (e) the melting of ice.

4. (a) Why should chemistry be concerned with energy changes? (b) Show by examples how energy may be obtained more easily from some materials than from others. (c) What commonplace substances are important sources of energy?

5. Distinguish between temperature and heat, using an example.

6. (a) Find the number of calories of heat needed to raise the temperature of 1 liter of water from 30°C to 40°C. (b) How many grams of water can be heated from 10°C to 90°C by the application of 800 cal of heat energy?

7. Do the readings on the Fahrenheit and Celsius temperature scales ever coincide? If so, at what temperature? Draw a diagram of the two scales and show this common point.

8. Account for each of the following facts in terms of the law of conservation of mass: (a) A candle burns and the material composing it entirely disappears. (b) A piece of iron burns and the resulting product weighs more than the original iron. (c) The material of an atomic bomb "explodes" and the resulting material weighs slightly less than the original.

9. Classify each of the following as verifiable or nonverifiable information. Justify your answer in each case.
(a) A pound is 454 g.
(b) An atom exists.
(c) Energy is the ability to do work.

(d) The sun is a star.

(e) The Kelvin temperature scale is more significant than the Fahrenheit temperature scale.

10. Design an experiment to: (a) Test any (or all) postulates in the atomic theory. (b) Illustrate the law of definite composition. (c) Measure the heat evolved in the burning of a match.

11. Design an experiment or experiments that will reveal whether the following are mixtures or single substances: air, water, a nickel, natural gas, gasoline, ink, dry ice.

12. State how you would separate into its components: (a) a mixture of sugar and sand, (b) a mixture of sand and water, (c) a mixture of ethyl alcohol and water, (d) a mixture of iron filings and sand, (e) a mixture of salt and sugar.

13. Describe an experiment in which the mass of a golf ball could be determined without weighing it on a balance or scale.

14. (a) How could the weight of a cannon ball be changed without changing its mass? (b) Could its mass be changed without changing its weight?

15. Find the names of the elements whose symbols are Ti, V, Fe, As, Ar, Ag, Y, W, Tc, Nd, No.

16. Using this text as a source, find the formulas for the following compounds: sulfur hexafluoride, tetraphosphorus decaoxide, metaphosphoric acid, iron(II) chloride, nitrogen dioxide, calcium carbonate, potassium superoxide, sodium hypochlorite, iodine trifluoride, diborane.

17. What properties distinguish water from other colorless liquids?

18. What properties distinguish solid white salts such as sodium chloride and potassium chloride from one another?

19. State the units in which each of the following is measured: (a) mass, (b) length, (c) density, (d) heat, (e) light, (f) concentration, (g) temperature.

20. Calculate your weight in (a) kilograms, (b) grams.

21. Calculate the length of your arm in (a) meters, (b) centimeters.

22. Convert:
    (a) 250 cc to quarts
    (b) 90 mg to kg
    (c) 10 oz to g
    (d) 10 cm to mm
    (e) 400 ml to liters
    (f) 400 cm to m
    (g) 10 in to cm
    (h) 3 Angstroms (the diameter of an atom of gold) to in
    (i) A cubic decimeter to liters

23. The visible spectrum of light ranges in wavelength from approximately 4000 A to 7000 A. What are these wavelengths in centimeters?

24. (a) A cube of aluminum weighs 21.6 g. The density of aluminum at 20°C is 2.7 g/cc. What is the length of the cube edge? (b) How much would a cube of gold (density at 20°C = 19.3 g/cc) of the same size weigh?

25. The cube of aluminum in Problem 24 contains 5.06×10^{22} atoms. What is the diameter of a single aluminum atom, assuming that the atoms are touching each other?

26. One gram of matter, if completely transformed into energy, would produce 2.15×10^{13} cal. How many tons of coal would be required to furnish this amount of heat if 1 g of the coal, when burned, produced 7400 cal?

27. (a) The density of osmium is 22.5 g/cc; that of lead is 11.3 g/cc. What volume would 1 kg of each metal occupy? (b) A piece of nickel having a volume of 10.0 cc weighs 89.00 g. Calculate its density.

28. (a) A tank measures 5.0 m $\times$ 40 cm $\times$ 120 mm and is filled with water at 32°F. Calculate the heat required to heat the water to 212°F (the boiling point). (b) The burning of 1 kg of coal results in the evolution of 7600 kcal of heat. How much coal must be burned to supply the heat needed in (a) above?

29. Consider the following data based on experimental evidence that element x and element y unite to form one or more compounds:

	Grams of x	Grams of y	Grams of compound
Exp. 1:	12.16	8.00	20.16
Exp. 2:	9.12	6.00	15.12
Exp. 3:	6.08	4.00	10.08

(a) Which laws of chemical combination are illustrated by these data? Explain. (b) If 40.0 g of element y will combine with 177.5 g of a third element z, what weight of element x will combine with 71 g of element z?

SPECIAL PROBLEMS

1. Does science provide absolute truth? Why or why not?

2. When a TV commercial claims a product has been tested "scientifically," what might be meant by "scientifically"?

3. Should a scientist be responsible for the uses made by society of the discoveries he makes?

4. A film of oil is spread on the surface of water in such a way that 1 cc of oil covers 100 m². Express the thickness of the film in Angstrom units.

5. A graduated cylinder is filled with copper shot to the 25-cc mark, and water is added to the same level to fill the empty space. The following data are recorded at 20°C:

Weight of graduate alone	52.0 g
Weight of graduate and copper	185.5 g
Weight of graduate, copper, and water	195.5 g

(a) Calculate the packing density of the copper at 20°C.
(b) Calculate the absolute density of the copper at 20°C.
(c) Calculate the per cent of empty space.

6. A dump truck delivers 8.10×10^3 kg of gravel. The capacity of the truck is 7.0 m³. A piece of gravel of representative size is found to weigh 19.2 g and displace 6.98 cc of water (at 4°C).
   (a) Find the absolute density of the gravel expressed in g/cc.
   (b) Find the packing density of the gravel in g/cc.
   (c) Find the per cent void space in the packed gravel.

7. A saline solution is made up containing 10 g of salt in 1 liter of solution. What weight of salt is contained in a 25-ml sample of the solution?

8. An analysis of five samples containing elements A and B yielded the following results:

Sample No.	Weight of A (g)	Weight of B (g)
1	28	50
2	14	25
3	14	12.5
4	14	37.5
5	50	28

One of the samples is a mixture, while four are single compounds.
   (a) What sample is the mixture?
   (b) Which samples are the same compound?
   (c) How can you determine your answer in (b)? Which laws of chemistry did you use in determining your answers?

REFERENCES   AUSTIN, A. V., "Standards of Measurement," *Sci. Amer.*, 218, 50 (June 1968).
   BICKEL, C. L., HOGG, J. C., LIPPINCOTT, W. T., and NICHOLSON, M., *Chemistry: Patterns and Properties*, American Book, New York, 1971.
   SCHNEER, C. J., *Mind and Matter*, Grove Press, New York, 1969.
   WEEKS, M.E., and LEICESTER, H. M., *Discovery of the Elements*, third edition, Chemical Education, Easton, Pa., 1968.

PART ONE

ATOMS, MOLECULES, AND IONS

ATOMS: THEIR COMPOSITION AND SIZE

I have chosen the word atom to signify those ultimate particles, in preference to particle, molecule, or any other diminutive term, because I believe it to be more expressive; it includes in itself the notion of indivisible, which the other terms do not. . . . All the changes we can produce, consist in separating particles that are in a state of cohesion or combination, and joining those that were previously at a distance.

John Dalton (1766–1844)

The atomic age really began with Dalton's formulation of the atomic theory at the beginning of the 19th century (Chapter 1). For the next hundred years, chemists were busy testing the theory. In convincing themselves of its validity, they further explored the important laws of chemical combination, they determined the combining weights of the elements and the formulas for many compounds, and they developed a scale of relative atomic weights.

Nearly all of the early tests of the atomic theory involved experiments with matter in bulk, and required some very clever and ingenious methods of reasoning, at which chemists became especially skilled. After the concept of atoms was firmly established, the techniques of the physicists, involving spectroscopes and electrical instruments such as cathode-ray tubes, proved to be extremely useful in showing that atoms were not indivisible particles, as had originally been supposed, but composite structures made up of still smaller units. These smaller units differ in number and in arrangement in the atoms of different elements, and the properties of each element are a consequence of the particular structures of its atoms.

DEVELOPMENT OF THE THEORY OF ATOMIC STRUCTURE

The problem of determining the structures of atoms, however, was an enormous one. Even the largest atoms are about a thousand times too small to be seen with the most powerful microscope. This means, of course, that we cannot make direct observations of the structure of atoms and so must resort to inference and deduction from data obtained from many indirect experiments. Such inference and deduction lead us to imagine a particular internal structure for an atom and, if predictions on the basis of our imagined structure agree with experiment, we dignify our imagined structure by calling it a theory. Working out an acceptable theory of the structure of the atom has been one of the most absorbing problems of physics and chemistry during the present century. The task is by no means completed, and our

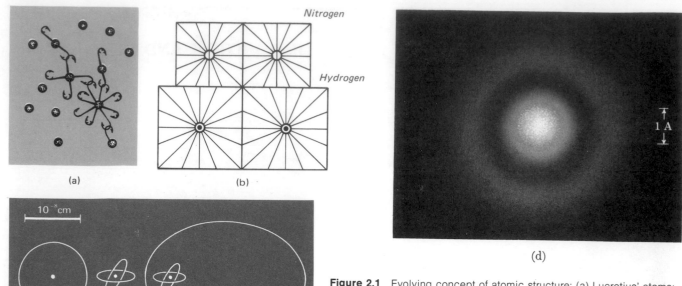

Figure 2.1 Evolving concept of atomic structure: (a) Lucretius' atoms; (b) Dalton's concept of nitrogen and hydrogen; (c) Bohr's atoms; (d) modern construct of a lithium atom showing regions of high and low negative charge.

present theory is constantly growing and improving. However, the experimental facts of atomic behavior are clearly consistent with our modern concepts of atomic structure and, in that sense, the theory is correct (Figure 2.1).

Electrons in Atoms

By 1900 several experiments had been performed that indicated that atoms or combinations of atoms acquired a negative or positive electrical charge under certain conditions. Michael Faraday, in 1833, had observed that various forms of matter in the molten state or in aqueous solution conduct an electric current and undergo a chemical change as a result of the passage of current through them. He reasoned that, to explain these results, he would have to assume that atoms or molecules could be electrically charged, and he coined the word ion for these charged particles.

As a result of Faraday's work with electricity, George Stoney concluded, in 1881, that there must be an *elementary unit of electricity*, and he introduced the term electron to designate this unit.

In 1879, Sir William Crookes performed a series of experiments in which electricity was passed through gases in tubes at low pressure. He noticed unusual radiation (or rays) from the negative electrode, which became known as cathode rays

Later, Joseph J. Thomson studied extensively the nature of the electrical discharge which could be made to occur in gases at low pressure in a cathode-ray tube such as that shown in Figure 2.2. The observed deflections of the cathode ray were just those to be expected for a beam of negatively charged particles. Thomson found that these particles always appeared, no matter what residual gas was in the tube or what metal was used in the electrodes. They thus seemed to be components of all matter.

Figure 2.2 Experiments with cathode-ray tubes: (a) A beam of cathode rays is produced with the help of a slit in the disc. (b) The beam is deflected by an electric field; the direction of deflection indicates that negative electrical charges are present in the beam.

The tube is made of glass with several metal electrodes sealed into the glass. A source of high voltage is connected to the electrodes A and C such that A is charged negatively with respect to C. The disc B contains a small slit. When the tube is partially evacuated and the voltage applied between A and C, the characteristic cathode radiation is observed on the screen at the end of the tube opposite the slit. The screen is fashioned like that of a television picture tube: it is coated with a compound such as zinc sulfide which fluoresces when struck by the cathode ray. Thus the passage of the ray through the tube can be followed by the position of the streak of light at the end of the tube. Normally the ray causes a bright streak at position I. When a potential difference is produced between the two plates E which parallel the normal direction of flow of the cathode rays, the rays are attracted toward the positive plate as indicated by the new position of the streak at point II.

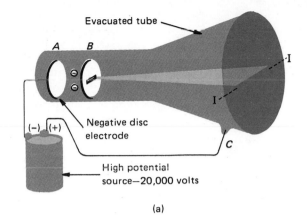

(a)

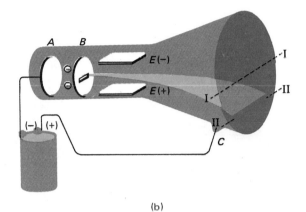

(b)

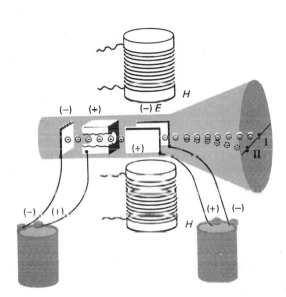

Figure 2.3 J. J. Thomson's apparatus for measuring the ratio of charge-to-mass of cathode-ray particles (electrons). The narrow cathode beam was first deflected by the electrostatic field at E and then brought back to its original position by a magnetic field between H-H. Knowing the magnetic and electrostatic field strengths, the ratio of e/m was calculated.

If these particles carry a negative charge, how big is this charge? Thomson was not able to determine this, but he was able to measure the ratio of charge to mass, e/m. He accomplished this by measuring the magnetic field strength necessary to return the spot to position I after deflection to II by a measured electric field (Figure 2.3). The value he obtained was close to the recent accurate value of $e/m = 1.76 \times 10^8$ coul*/g. Thomson found, as he changed the nature of the metal electrodes (Al, Fe, Pt) or the type of residual gas left in the partially evacuated tube (CO_2, H_2, air), that the value of the e/m ratio for the particles produced remained unchanged. This was a severe test of

*A coulomb (coul) is a unit of electric charge.

his suggestion that these rays were all alike even though they were produced in different tubes with different electrodes and different gases. He concluded that he was dealing with a common fundamental particle, obtainable from all matter. This negatively charged particle, originally called the cathode ray, must be the unit of electricity, the existence of which had already been postulated by Stoney and named by him the electron.

Evidence for Positive Particles

If negatively charged particles exist in all matter, surely there must be positive particles, since matter as a whole is electrically neutral. A test of this assumption was obtained by placing a small hole in the negative electrode of a cathode-ray tube, behind which was placed a fluorescent screen. Positive particles, if present, should move toward this negative electrode. It was observed that particles evidently did move toward this electrode, pass through the hole, and strike the screen. From this experiment it was concluded that there were positively charged particles present in a discharge tube. The positively charged particle of smallest mass found was called the proton.

Further experiments showed these positive rays or particles to have these characteristics:

1. The ratio of charge-to-mass (e/m) was found to be several thousand times *smaller* than the value of e/m for electrons (note that the larger the value of m, the smaller the value of the ratio e/m).
2. The value of e/m was *dependent* on the nature of the residual gas in the tube, in contrast to the behavior of the negatively charged electrons, which always had the same value of e/m no matter what was in the tube.

To explain these data it was necessary to assume that the particles, unlike the cathode-ray particles, varied in mass depending on the nature of the gas. The reasonable hypothesis was made that these positively charged particles were formed in the tube from neutral atoms or molecules, probably as a result of collisions between the electrons and atoms or molecules of the gas in the tube. If such a collision could knock an electron from the neutral substance, a positive particle would be left behind. This would then be drawn toward the negative electrode and, if moving in the right direction, would pass through the hole and hit the fluorescent screen.

Evidence for Positive and Negative Particles from Radioactivity

While physicists were studying cathode rays, some chemists were studying the interesting phenomenon of fluorescence. Henri Becquerel, a French scientist, was attempting in 1896 to convert sunlight into a deeply penetrating radiation by causing the sunlight to produce fluorescence of uranium compounds. His procedure was to lay a piece of uranium ore on a wrapped photographic plate and expose the uranium to sunlight. When the photographic plate under the uranium was developed, he found it streaked. But when he repeated the experiment without exposing the uranium to sunlight, he found the plate was again streaked. The sunlight evidently had nothing to do with it at all! Further work on this astonishing result by Marie Sklodowska Curie and her husband Pierre Curie led to the identification of a new element in uranium ore, *polonium*, as the source of the radiation affecting the photographic plate. This work by the Curies was followed shortly by their discovery of *radium* as a similar radioactive substance.

Table 2.1 Types of Radiation.

Rays	Symbol	Characteristics
Alpha rays—positive particles, with $e/m = 4.82 \times 10^4$ coul/g	α	He^{+2} ions
Beta rays—negative particles, with $e/m = 1.76 \times 10^8$ coul/g	β^-	Electrons
Gamma rays—electromagnetic rays of shorter wavelength than light rays, able to penetrate matter even more deeply than X rays	γ	Similar to light but very deeply penetrating, even more so than most X rays

These two elements emitted radiations that had a much greater effect on photographic plates than the original uranium ore. The radiations were made of three different rays (Figure 2.4 and Table 2.1).

Thus, we have further evidence that matter is made of positive and negative particles. Additional studies showed that the positive alpha particles are emitted by many of the heavy elements from uranium to lead.

Figure 2.4 The relative penetrating power of alpha, beta, and gamma radiations.

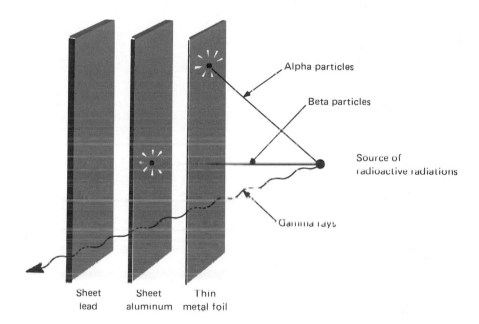

Alpha particles

Beta particles

Source of radioactive radiations

Gamma rays

Sheet lead   Sheet aluminum   Thin metal foil

Determining the Charge and the Mass of Electrons

Thomson could determine from his experiments neither the mass nor the charge of the electron alone, but only the ratio e/m. It remained for an American scientist, Robert Millikan, to devise an ingenious experiment to measure the charge on the electron. He studied the rate of fall of electrically charged oil drops (that is, drops which had acquired one or more electrons or lost one or more electrons) suspended between charged plates (Figure 2.5). From these data he calculated the charge on the oil drops; he found that this charge was always equal to, or was some multiple of, a definite unit charge (1.6020×10^{-19} coul). He concluded that the unit charge was the charge on the electron, and he explained his observations by assuming that the droplets picked up one or more electrons or lost one or more electrons. After

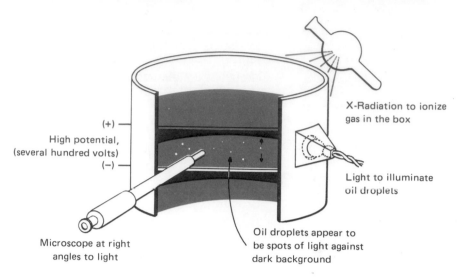

X-Radiation to ionize
gas in the box

Light to illuminate
oil droplets

Oil droplets appear to
be spots of light against
dark background

Microscope at right
angles to light

Figure 2.5 Diagram of apparatus used by Millikan to determine the charge on an electron. Oil was sprayed into a chamber between two electrically charged plates, and the rate at which the drops fell toward the lower plate under the influence of gravity was observed with a microscope. X rays then were passed through the box, knocking electrons off some atoms in the gas space between the box. Some of these electrons were captured by the oil droplets and traveled toward the positively charged plate. From the rate at which the drops moved against the force of gravity (when the upper plate was positively charged) or the faster rate at which they fell (when the lower plate was positively charged), it was possible to calculate how large a charge the droplets carried. The charge varied, but the differing charges did not have a random value; they differed by a unit charge or by some integral multiple of that unit charge. Thus the charges on different oil drops depended upon how many electrons a drop had captured.

Millikan's success in determining the charge on the electron, he used Thomson's values for the ratio of the charge to the mass, $e/m = 1.76 \times 10^8$ coul/g, and calculated the mass:

$$\frac{e}{m} = 1.76 \times 10^8 \text{ coul/g}$$

$$e = 1.60 \times 10^{-19} \text{ coul}$$

$$m = \frac{e}{\frac{e}{m}} = \frac{1.60 \times 10^{-19} \text{ coul}}{1.76 \times 10^8 \text{ coul/g}} = 9.11 \times 10^{-28} \text{ g, mass of the electron}$$

The electron, then, is assumed to be a *particle* of negative electricity of definite charge and mass that is a universal constituent of matter and, in suitable multiples, makes up all negative charges.

To balance the negative electrons there must be positive entities, as already indicated by the positive-ray experiments. The next paragraphs discuss the experiments which provided information about the positive part of the atom.

THE NATURE
OF THE NUCLEUS

Rutherford's Metal Foil
Experiment

In 1911, Ernest Rutherford and his students, Hans Geiger and Ernest Marsden, performed a classic experiment that showed that the positive portions of an atom were concentrated in a very small, dense particle. They bombarded atoms in gases as well as in metal films with alpha particles in an experiment similar to that described in Figure 2.6. They used very thin films of metal foil fabricated from very malleable metals, such as gold, silver, copper, and platinum; the films had a thickness of about 1000 atoms or about 1×10^{-5} cm. This film was placed in the path of a beam of alpha particles. Behind the foil was placed a screen, C, coated with zinc sulfide, which would scintillate when struck with an alpha particle. Another zinc sulfide screen, D, was placed at the side of the path to detect alpha particles that might be highly deflected.

It was found that most of the alpha particles passed through the foil and struck screen C near P, undeflected in their paths. This result showed that the atoms of the metals were mostly free space, not occupied by massive particles that might hinder the passage of the alpha particles.

On careful study, they discovered that a few of the particles were deflected on passage through the film and were seen as bright spots of light as they impinged at other places on C. Furthermore, they found that one alpha particle in 10,000 or so seemed to undergo a collision with some very dense solid matter and be deflected backwards from the metal foil, striking screen D.

Rutherford interpreted these results to indicate that the atom contains a small, heavy particle, which must be positively charged. It must be *small* because only a few alpha particles struck it compared to the number passing through the foil; it must be *heavy* because the alpha particle bounced off it, rather than having it pushed aside by the alpha particle; and since the negatively charged particles (electrons) were light in weight, it seemed likely that this *heavy* particle was the seat of *positive* charge. With these assumptions, Rutherford calculated how large the charge on such a small, heavy particle would have to be to produce the number of particles at various angles of deflection that was observed in the experiment. The data were not reliable enough to do this exactly, but he estimated that a positive charge about 100 times the charge on an electron was required to give the observed effects with gold foils, for example; or, for other metals in general, the charge must be approximately $+\frac{1}{2} Ae$, where e was the charge on an electron and A was a known number representing the weight of the metal atom relative to the weight of other atoms. Since the negatively charged electrons must be situated more or less symmetrically around this particle, it was appropriate to call it the nucleus of the atom

If the atom is mostly free space and contains a small nucleus, what are the relative sizes of the two? The amount of space required for an atom is readily

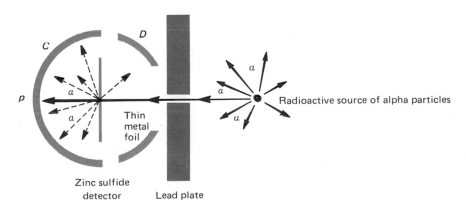

Figure 2.6 Ernest Rutherford's metal foil experiment. Approximately one in 10,000 alpha (α) particles bounced back from the metal foil, providing evidence for the existence of a heavy, compact, positively charged atomic nucleus.

calculated from Avogadro's number and the volume occupied by one mole of an element; the results show that atoms have a diameter of about 10^{-8} cm. From Rutherford's experiments it was possible to calculate the diameter of the nucleus to be about 10^{-12} cm, or 1/10,000th the diameter of the atom.

Atomic Numbers

About the time that these scattering experiments were in progress, several groups of physicists discovered methods of measuring the wavelengths of X rays, the new type of radiation discovered by Roentgen only several years earlier (1895). Rutherford knew that X rays could be produced by bombarding metals with rapidly moving electrons, so he proposed to a member of his research group, H. G. J. Moseley, that they plan a series of experiments with X rays to try to determine the charge on the nucleus. They proposed to bombard a series of metal targets in X-ray tubes and to examine the wavelengths of the X rays emitted.

Moseley found that the X-ray wavelengths of different metals were reproducible and characteristic of the particular metal used. But more than this, he found that a simple mathematical relationship existed between the wavelength of the X ray and a whole number that is equal to about one-half the relative atomic weight of the metal. Moseley noted that these whole numbers (or atomic numbers) were nearly identical to the estimated nuclear charges in Rutherford's experiments with metal foils. He concluded that the atomic number was the charge on the nucleus, in units of the electronic charge, 1.60×10^{-19} coul (Figure 2.7). For scientists, this was a major breakthrough in the study of atomic structure, for now not only did they know the relative size of the nucleus but they had an exact measure of the nuclear charge. The picture unfolded rapidly after this.

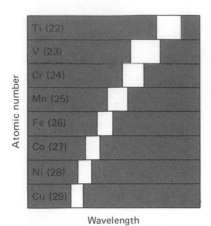

Figure 2.7 Results of H. G. J. Moseley's experiments. The wavelengths of certain X rays emitted from metals decreased systematically as the nuclear charge increased. Moseley obtained values for the charge on atomic nuclei from these experiments. These values are known as atomic numbers.

Building Blocks of the Nucleus

Millikan's work showed that there existed a minimum negative charge on a particle called the electron; experiments with positive rays indicated that heavy positive particles could be obtained from atoms. Moseley's work led to the conclusion that the nuclear charge varied from element to element. Since atoms are neutral and since hydrogen is the lightest element, the presumption was that hydrogen atoms consisted of a heavy nucleus containing one positive particle (named the proton) and one electron. Atoms of other elements were imagined to contain a number of protons comparable to their nuclear charge and an equal number of electrons.

Measurements of e/m for hydrogen in positive-ray tubes give 9.4×10^4 coul/g. The mass of the proton then is calculated to be 1.67×10^{-24} g, about 1837 times the mass of the electron (9.1×10^{-28} g). Other atomic nuclei should have masses equal to their nuclear charge times the mass of the proton.

Rutherford said when considering the results of his experiments: "It was quite the most incredible event that has ever happened to me in my life. It was almost as incredible as if you fired a 15-inch shell at a piece of tissue paper and it came back and hit you. On consideration I realized that this scattering backwards must be the result of a single collision, and when I made calculations I saw that it was impossible to get anything of that order of magnitude unless you took a system in which the greater part of the mass of the atom was concentrated in a minute nucleus. *It was then that I had the idea of an atom with a minute massive centre carrying a charge.*"

That this picture is incomplete became evident when the e/m value for the helium nucleus was measured in an improved positive-ray tube. Moseley's work indicated that the nuclear charge on helium was +2; but the mass of the helium nucleus calculated from its e/m value was 6.68×10^{-24} g, or *four* times that of the hydrogen nucleus instead of *two* times. How could this be? Could there be more than two protons in the helium nucleus? If not, how could the larger mass be accounted for? This spurred a search for the existence of a third particle.

Evidence for a Third Fundamental Particle— the Neutron

One avenue of approach was to extend the Rutherford experiments on the bombardment of elements with alpha particles. High-velocity alpha particles should have energies high enough to penetrate the nuclei of some elements, be captured by these nuclei, and combine with them to form excited unstable nuclei.

Experiments in which alpha particles were used to bombard beryllium, boron, or other light elements produced a radiation that did not have the set of unique characteristics of any of the then-known radiation. Alpha, beta, and gamma rays have varying efficiencies of penetration of matter as shown in Figure 2.4. Alpha particles and beta particles can penetrate only thin films of matter. Most gamma rays penetrate considerably farther, but at most only a few inches of solid objects. However, it was found that the radiation from the alpha-particle bombardment of light elements passed readily through a thickness of several feet of solid matter. The origin and nature of the radiation was not clear for several years. For a time it was considered to be extremely short-wavelength electromagnetic radiation, similar to gamma rays.

An English scientist, James Chadwick, in 1932, made an important observation that led to a better understanding of the nature of the "rays." He observed that hydrogen-containing compounds (water, paraffin, and the like) were particularly effective in reducing the energy until the radiation stopped. Application of the simple principles of colliding particles shows that the most efficient interchange of energy between colliding particles occurs when the particles have the same mass. In view of this fact, Chadwick suggested that the "very penetrating radiation" was not really radiation but neutral solid particles which had a mass about the same as the proton; he called the particle the neutron. Its high penetration of matter could be understood since the neutral particle would not be affected by the strong electric forces on approach to the regions of the atom rich in electrons or protons. Only "hits" on the massive center of the atoms could cause the particle to be deflected. Chadwick was able to calculate the mass of the neutron from the energy and direction of the recoil of protons produced on impact of neutrons with hydrogen atoms (Figure 2.8). Accurate data from recent experiments show that the mass of the neutron is 1.676×10^{-24} g, or about equal to the mass of the proton.

Today, the nuclear reactor is direct evidence of the existence of neutrons, for its operation depends upon producing and using neutrons (Chapter 24). Neutrons bombard uranium atoms and, being uncharged, are able to penetrate the nuclei, causing the excited nucleus to split into smaller nuclei and some neutrons, with a net loss of mass and evolution of much energy. The neutrons formed serve to continue the process by reacting with other uranium nuclei, and the energy is available for power production.

Figure 2.8 Chadwick's experiments leading to the discovery of the neutron. The alpha (α) particles striking the beryllium give rise "to unidentified particles which are allowed to bombard paraffin. Protons are then ejected from the paraffin. Knowing the momentum of the ejected proton, it is possible to estimate the masses of the unidentified particles, the neutrons.

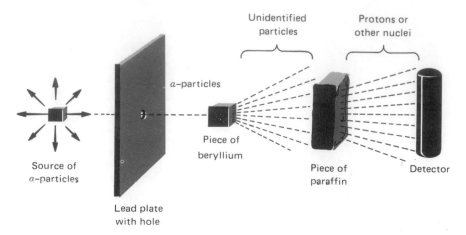

The existence of neutrons explains the unexpected mass of the helium nucleus mentioned in the last section. The greater mass of the helium nucleus, four times that of the hydrogen nucleus, but only twice the charge is explained if it contains two protons and two neutrons. The two protons give it the charge of two positive units and its two neutrons add weight but no charge. The discrepancy in the observed weight, 6.65×10^{-24} g, compared to 6.69×10^{-24} g calculated from the sum of the masses of two protons and two neutrons, represents one of the examples of the conversion of mass to energy in nuclear changes mentioned in Chapter 1. The loss of mass appears as the energy released as the four particles fuse to become the compact helium nucleus (Chapter 24).

The neutron is extremely small about 10^{-12} cm in diameter or 1/10,000 the diameter of the hydrogen atom, the smallest of the atoms. Since it has the mass of a hydrogen atom concentrated in a volume about 1/1,000,000,000,000 the volume of the hydrogen atom, it has an exceptionally high density. If a thimbleful of neutrons could be collected and packed like marbles in a box, it would be found that the thimbleful of neutrons would weigh approximately 200,000,000 tons.

Other Particles in the Nucleus

It is an oversimplification to describe the nucleus only in terms of protons and neutrons. Today, over 100 different particles have been identified among the products resulting when atoms are bombarded by very-high-energy particles. They range in mass from the neutrino mass (zero) to 2400 times the mass of an electron. Their lifetimes are very short. However, our present purposes are aimed at an understanding of the chemical reactions of atoms, ions, and molecules; particles in the nucleus other than protons and neutrons need not concern us further here.

The Components of the Atom

The material discussed in this chapter concerning the structure of the atom is summarized in Figure 2.9.

The model of the atom that emerged after Rutherford's work on bombarding metal foil and Chadwick's work on identifying the neutron consisted of a nucleus, composed of protons and neutrons, surrounded by electrons. Atoms of various elements were pictured as differing from one another in the number of protons, neutrons, and electrons present. The number of protons in the nucleus of an atom of a particular element is equal to the atomic number

Figure 2.9 Development of the concept of the nuclear atom.

Fundamental electrical nature of matter	Conduction in solution – Faraday Conduction in gases – Crookes

Electrons as particles	Cathode ray experiments – Thomson
	  Electrons exist in all matter; e/m determined.
	Beta particles in radioactivity
	  Electrons are produced by processes other than
	    those in cathode ray tubes.
	Unit electron charge – Millikan
	  Charge and mass (from known e/m) determined.

Positive entities in atoms	Positive rays–Thomson
	  Positive particles the counterpart of negative electrons.
	Alpha particles in radioactivity
	  Identification as He^{+2} confirms concept that electrons
	  can be removed from and added to the positive com-
	  ponents of the atom.
	Alpha-particle bombardment of metal foils– Rutherford
	  An atom has a nucleus, heavy, positively charged, and
	  small.
	Nuclear charge identified with atomic number– Moseley
	Neutrons are present in the nucleus– Chadwick

To come	Arrangement and energies of electrons in the atom (Chapter 4) Energy levels in the nucleus (Chapter 23)

of the element. The atom is neutral; therefore, the number of electrons in the atom is equal to the number of protons in the nucleus (Figure 2.10). Since protons and neutrons are about 1837 times heavier than electrons, the mass of the atom is very nearly equal to the sum of the masses of the neutrons and protons present.

CHEMICAL ATOMIC WEIGHTS

As indicated in Figure 2.9, further discussion of atomic structure will appear in Chapter 4. The remainder of this chapter will be devoted to the methods currently used for determining atomic weights, preparing the way for a discussion of quantitative relations in chemical combination in Chapter 3.

Exact Atomic Masses

By 1860 chemists had developed a generally accepted scale of *relative* atomic weights based on very careful determinations of the combining weights of elements and the formulas of compounds.* Thus it was known that a carbon atom weighed 12 times as much as a hydrogen atom, an oxygen atom $1\frac{1}{3}$ times that of a carbon atom, a sulfur atom twice as much as an oxygen atom, and so on. But once the essential electrical nature of the atoms had been established, it became evident that experiments similar to those which had been used to establish this could be used to determine the actual masses of atoms. Thomson's positive-ray experiments had shown that the masses of the positive-ray particles were characteristic for each gas in the positive-ray tube, and this ability of the positive-ray apparatus to determine masses was put to use by F. W. Aston in his design of a mass spectrograph. This original instrument has been modified until today the masses of positive ions can be determined with great precision (Figure 2.11).

The existence of atoms of the same element with different masses had been discovered earlier in the identification of the elements formed in radioactive disintegrations. Frederick Soddy, in 1913, was the first to suggest that these were in fact atoms of the same element, even though they had different atomic weights, and he coined the name isotopes for these different-weight atoms.

Figure 2.10 An early model of a beryllium atom showing four protons and five neutrons in the nucleus and four electrons orbiting the nucleus. This model with its orbiting electrons is inconsistent with much experimental evidence.

*For a discussion of the methods and reasoning used, see a recent history of chemistry.

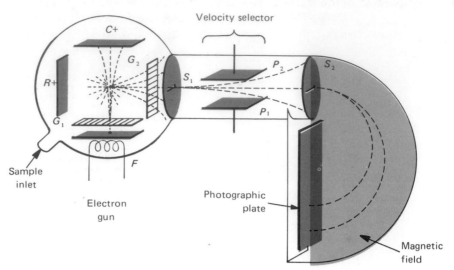

Velocity selector

Sample
inlet

Electron
gun

Photographic
plate

Magnetic
field

Figure 2.11 Diagram showing features of a mass spectrometer. Electrons, produced at the hot filament F, pass through a slit and are accelerated by an electric field between grid G_1 and the collector plate C^+. Positive ions, produced by collisions of electrons with the sample gas, are accelerated by an electric field between the repulsion plate R^+ and G_2. Passing through slit S_1, they enter the velocity selector, passing through an electric field between plates P_1 and P_2 and a magnetic field perpendicular to the plane of the paper. The two fields are arranged so that deflection of ions by the electric field is opposed by the magnetic field; for adjusted field strengths, this means that only ions having identical velocities pass through slit S_2. All others miss the opening. The selected ions then pass into another magnetic field, where they undergo a circular deflection of radius proportional to their ratio of mass to charge.

For two ions of the same charge, the heavier will experience the smaller amount of bending in the magnetic field and will strike a detector at a point different from that struck by the lighter one. The distance between the two positions is a measure of the difference in the masses of the two ions. The intensity of the current produced when the detector is at the point where the heavier ions strike, compared to its intensity at the point where the lighter ions strike, measures the relative number of ions of the two different masses. In the mass spectrometer, therefore, we may not only measure exactly the relative masses of the positive ions in the beam, but also their relative numbers.

Aston's work confirmed the existence of isotopes for nonradioactive elements as well. Later work has shown that nearly all of the naturally occurring elements consist of two or more isotopes (Appendix Table E4). Thus, neon has isotopic atoms of masses 33.2×10^{-24} g, 34.9×10^{-24} g, and 36.5×10^{-24} g.

Table 2.2 Masses of Some Atoms (in Grams)

Atom	Mass (g)
^{1}H	1.67×10^{-24}
^{4}He	6.65×10^{-24}
^{9}Be	14.98×10^{-24}
^{12}C	19.93×10^{-24}
^{19}F	31.55×10^{-24}
^{23}Na	38.17×10^{-24}
^{27}Al	44.80×10^{-24}

Composition of the Nuclei of Isotopes

How can atoms of the same element have different masses? Each of these atoms must have the same number of electrons, since the number (and arrangement) of electrons, as we shall see in Chapter 5, determine the chemical behavior of atoms. Each must contain the same number of protons, since the atoms are neutral. The mass difference must be accounted for by different numbers of neutrons in the nucleus. Thus each neon atom contains 10 protons, but 10, 11, or 12 neutrons. The sum of the number of protons plus the number of neutrons gives the mass number of the isotope. To specify the particular isotope of an element, we write the atomic number (= number of protons) as a subscript before the symbol of the atom, and the mass number (= number of protons *plus* number of neutrons) as a superscript at the left. For instance, the symbol $^{21}_{10}$Ne indicates that particular isotope of neon which has 11 neutrons. An atom of this isotope has a mass of 34.9×10^{-24} g.

A More Useful Unit for Atomic Weights

Table 2.2 shows that the mass of an atom when expressed in grams is a very small number. To avoid using the cumbersome exponentials needed to express such small numbers and to express atomic weights in numbers approximating the mass number of the isotope, chemists have devised another unit called the atomic mass unit (amu). This is defined as $\frac{1}{12}$ the mass of that carbon isotope that contains six protons, six neutrons, and six electrons. The measured mass of this isotope is 19.93×10^{-24} g; thus, 1 amu has the mass $19.93 \times 10^{-24}/12 = 1.66 \times 10^{-24}$ g (measuring masses in grams), just as the conversion of lengths is that 1 yard (measuring lengths in yards) corresponds to 3 feet (measuring lengths in feet). When the atomic masses in grams of Table 2.2 are converted to atomic mass units, the masses of atoms expressed in these units are as given in Table 2.3. That table shows that the mass number of an atom is equal to its atomic weight in atomic mass units, rounded off to the nearest whole number. This follows from the fact that the proton and the neutron each have masses of about 1.0 amu (neutron mass = 1.008665 amu; proton mass = 1.007277 amu).

Table 2.3 Masses of Some Atoms (in Atomic Mass Units)

Atom	Mass (amu)
^{9}Be	9.012
^{12}C	12.000
^{19}F	18.998
^{23}Na	22.990
^{27}Al	26.982
^{31}P	30.974
^{45}Sc	44.956
^{55}Mn	54.938
^{59}Co	58.933

The chemical atomic weight of an element is a weighted average of the masses of the isotopes, expressed in atomic mass units. Table 2.4 gives the mass spectrometric data on atomic masses and abundances for neon as an example. The calculation of the last column shows that the weighted average atomic weight of neon is 20.18 amu. This is in close agreement with the value obtained from measurement of properties of the naturally occurring mixture of isotopes, 20.183.

Table 2.4 Masses and Relative Abundance of Neon Isotopes

Mass Number of Isotope	Mass (g)	Mass (amu)	Abundance (per cent)	Contribution to Atomic Weight (amu)
20	33.2×10^{-24}	19.992	90.92	18.18
21	34.9×10^{-24}	20.994	0.26	0.055
22	36.5×10^{-24}	21.991	8.82	1.94
			Weighted Atomic Weight	20.18

In general, the atomic weight values obtained by mass spectrometry are more precise than those obtained by chemical methods. We may use with confidence the values of the chemical atomic weights listed in the table on the inside back cover. These represent the average atomic weights, in atomic mass units, of the isotopic atoms of each element as fixed by the relative abundance of their isotopes in nature (Figure 2.12).

Atoms, Molecules, and Ions
The types of combinations of protons, neutrons, and electrons in atoms and molecules will be described in Chapters 4, 5, and 6, but we shall point out here the information that an atom has the same number of electrons as protons and, therefore, is electrically neutral; the same is also true of molecules. However, some atoms and molecules can gain one or more extra electrons, or lose one or more, and thus become electrically charged. Such atoms and molecules are called ions. Examples of some ions are

Ag^+	silver ion	NO_3^-	nitrate ion	S^{-2}	sulfide ion
Cl^-	chloride ion	NH_4^+	ammonium ion	CO_3^{-2}	carbonate ion

The charges given in the formulas for the ions are in units of the charge on the electron. Thus Ag^+ indicates that the neutral silver atom has lost one electron, leaving a positive ion of one unit charge. S^{-2} indicates that the neutral sulfur atom has gained two electrons, thus acquiring two electron units of negative charge.

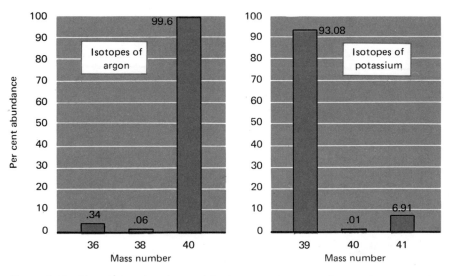

Figure 2.12 The relative abundance of the isotopes of argon and potassium.

The Atom as a Source of Power. All of us clearly recognize that our total environment is composed of both the natural world and the technological superstructures we have built. While expansion of technology often is associated with convenience, comfort, and affluence, it is in fact vitally linked to jobs and income and to health and security.

An important illustration is our harnessing and utilizing energy. Without this, human civilization simply could not have advanced beyond the stage where all but the most favored are manual laborers with short and pain-filled lives. In North America today, the average energy output per capita is about 320 horsepower-hours per day. This is roughly 800 times the daily work output of the average strong laborer. The current total electrical generating capacity in the U.S. is 335 million kilowatts; by 1980 we shall need a capacity close to 700 million kilowatts and, by 2000, an estimated 1.5 billion. Unless this energy is available, the opportunities for jobs, for creative growth of individuals and the society, and for improving the quality of life will be diminished.

There is no doubt that an important fraction of the world's future energy requirements will be furnished by fossil fuels—coal, oil, and natural gas. However, before long we shall have to depend in large measure on the atom for our energy sources. We now know that atomic energy can be released by the fusion of nuclei of light atoms and also by the fission of nuclei of heavy elements. In both processes the mass of the products is less than the mass of the reactants. The mass "lost" appears as energy. We know how to carry out the fission process in a controlled manner in a nuclear reactor; the energy released appears as electrical or mechanical power. A diagram of a nuclear reactor and a discussion of the fission and fusion processes appear in Chapter 23. As an illustration of the fuel efficiency of nuclear power sources, it is estimated that 1.5 billion kilowatts of power needed for the year 2000 will necessitate burning 10 million tons of coal per day, whereas the same number of nuclear power plants would consume 3 tons of nuclear fuel per day.

A major problem with the use of nuclear power plants arises because there is considerably more waste heat (60 per cent more) discharged from a nuclear reactor than from a plant using fossil fuels. This waste heat must be discharged into the air or into natural waters and, if not controlled, may lead to thermal pollution and a potential hazard to life and the balance of nature.

Solutions to this problem are being developed. They involve: a) more precise control of the release of energy in the nuclear process; b) using the waste heat to desalinate seawater by evaporation, for irrigation, to improve the efficiency of sewage treatment, or for sea farming; c) less harmful ways of discharging the heat, such as creating artificial lakes and designing cooling towers that would allow more gradual dissipation of the heat.

While danger from radiation emitted by the nuclear fuel has not been eliminated completely, this is not now and need not become a serious problem. Some aspects of radiation safety are given in Chapter 23.

More efficient methods of transmitting electricity from nuclear power plants to use centers are being developed. The use of cryogenic superconducting cables (certain metals, at very low temperatures, conduct electricity with much less resistance than do ordinary metals) will enable us to send ultrahigh voltage electricity great distances through underground cables. A single 8-inch diameter carrier may be able to carry economically the electricity needed for a city of several million over distances as much as a thousand miles. Sometime in the future, power might be transmitted across oceans by satellite, the energy being converted to a laser beam that would be reflected from the satellite to other continents. *However, all of this depends upon a better knowledge of the properties of matter.*

The need for cheap and plentiful power is emphasized by still another critical problem: the availability of adequate supplies of fresh water. In the year 1900, consumption of fresh water in the U.S. was 40 billion gallons per day. Today it is nearly 400 billion gallons per day, and by the year 2000 it could reach 1000 billion gallons per day. Despite this, the best estimates of the available and tappable stream flow in the U.S. are 600 billion gallons per day. It would thus appear that to meet our fresh water needs during the next quarter century, we must make provision for desalinating large quantities of seawater. Nuclear power can make this economical.

Research now well underway suggests that controlled thermonuclear fusion will enable us to use the heavy hydrogen isotopes in the ocean waters as a virtually limitless source of fuel—enough to last for millions of years. The roots of the research which has made nuclear power possible and which may provide solutions to these problems lie in the knowledge and in the kind of scientific thought outlined in this chapter.

Obviously, the needs and aspirations of man go far beyond the availability and management of physical resources and power needs. However, without excellence in the management of these resources and without inspired creativity in the development of new and more efficient methods of using them, all citizens may be denied opportunities and even basic rights. Decisions on the management of physical resources are in the hands of the citizenry, upon whose willingness to inform themselves properly and to act with wisdom the fate of all depends.

SUMMARY   From the evidence cited we conclude that there are at least three fundamental particles of which matter is composed: the *electron*, a negatively charged particle of mass 0.0005486 amu; the *proton*, a positively charged particle, equal in magnitude of charge but opposite in sign to that of the electron, and of mass 1.0073 amu; and the *neutron*, an electrically neutral particle of mass 1.0087 amu.

The heavier particles—protons and neutrons—are found in the nucleus, a small dense region at the center of the atom. Around this central positive nucleus the electrons are distributed; the manner of this arrangement will be our early major concern. The atom is electrically neutral and thus contains the same number of protons as electrons.

In the lighter elements there are roughly equal numbers of protons and neutrons in the nucleus. For example, helium has two neutrons and two protons; lithium has three protons, and either three or four neutrons in its stable isotopes; beryllium has four protons and five neutrons; boron has five protons and five or six neutrons; carbon has six protons and six or seven neutrons; nitrogen has seven protons and seven or eight neutrons; oxygen has eight protons and eight, nine, or 10 neutrons in its stable isotopes, and so forth. In the heavier elements there is an increased proportion of neutrons over protons; thus, in the uranium atom there are 92 protons and 142, 143, or 146 neutrons in the isotopes found in nature. The sum of the number of protons and the number of neutrons in the atom gives the mass number and the approximate mass of the isotope.*

Exact atomic masses and the relative abundances of isotopes may be determined in the mass spectrograph or in the mass spectrometer. The weighted value of the atomic masses is known as the *atomic weight*.

IMPORTANT TERMS

Electron	**Positive rays**	**Atomic nucleus**
cathode ray	mass spectrograph	radioactivity
cathode-ray tube	mass spectrometer	alpha and beta particles
charge-to-mass ratio	protons	gamma rays
oil-drop experiment	atomic masses	metal foil experiments
charge on electron	isotopes	X rays
mass of electron	atomic weights	atomic numbers
		subatomic particles

QUESTIONS AND PROBLEMS

1. Give experimental evidence indicating that: (a) the nucleus is made up of several components; (b) the atom is porous; (c) the nucleus is positively charged; (d) the nucleus occupies a small volume; (e) the electrons are outside the nucleus; (f) many elements have isotopes.

2. Summarize the information attainable from experiments with cathode-ray tubes. Classify this information as qualitative or quantitative.

3. Describe the motion of an electron beam moving transversely: (a) between two parallel plates, one charged negatively, the other charged positively; (b) between the pole pieces of a magnet.

4. In the oil-drop experiment, state whether similar results could have been obtained had: (a) light instead of X rays been used to ionize the air; (b) helium instead of air been used in the chamber; (c) a magnetic field been used instead of the electrostatic field to deflect the charged drop. Why or why not?

5. The atomic weight of carbon is listed as 12.01115 amu. However, carbon at 12 amu exactly is the standard by which comparative atom weights are measured. Explain.

6. Only one isotope of the element fluorine exists in nature. Why is its atomic weight not an even number?

7. Radon-228, actinium-228, and thorium-228 are different, yet have the same mass numbers. How do they differ?

8. For each of the following particles give the sign of the charge, the magnitude of the charge, and the mass: (a) proton, (b) electron, (c) lithium atom, (d) chloride ion, (e) magnesium ion, (f) alpha particle, (g) magnesium nucleus, (h) potassium nucleus.

9. An atom has 16 neutrons in its nucleus. By the addition of two extranuclear electrons, the ions of this element have a double negative charge

---

*Actually, the sum of the weights of the protons and neutrons in a nucleus is always greater than the measured mass of the nucleus, as indicated for helium in Chapter 24; this difference in mass is called the *mass defect* and is related to the binding of the nuclear particles in the atom. This is discussed in Chapter 24.

and have then the same number of extranuclear electrons as there are in a neutral atom of argon. (a) What is the symbol of this element? (b) How many protons are in the nucleus of the negative ion?

10. What is the composition of the nucleus of each of the following atoms: (a) titanium, $^{48}_{22}$Ti; (b) arsenic, $^{75}_{33}$As; (c) indium, $^{115}_{49}$In; (d) cesium, $^{133}_{55}$Cs?

11. What is the composition of the nuclei of the isotopes of (a) potassium (element 19) having masses of 39 and 41 amu, respectively; (b) strontium (element 38) having masses of 86, 87, and 88 amu, respectively?

12. How many protons, neutrons, and electrons are there in each of the following?
    (a) $^{79}_{34}$Se (b) $^{80}_{35}$Br$^-$ (c) $^{40}_{20}$Ca^{+2} (d) $^{11}_{5}$B (e) $^{133}_{55}$Cs (f) $^{238}_{92}$U

13. An atom is transformed into an ion with a +4 charge. In the ion there are 88 extranuclear electrons and 146 neutrons. (a) What is the symbol of the atom? (b) An isotope of the element has a mass number four units smaller. How many neutrons are in this isotope nucleus? (c) Another atom has the same mass number as the isotope in (b) above, and it has 143 neutrons in its nucleus. What is the symbol of this atom?

14. If sulfur were burned in ordinary air, how many physically different molecules of SO_2 would be possible? (Refer to table of natural isotopes.)

15. Consult the table of natural isotopes in Appendix Table E.4. How many physically different molecules of CH_2Cl_2 probably exist in a sample of that compound?

16. Assuming that the kinetic energy of electrons in a cathode-ray beam is given by $V \times e$, where V is the accelerating voltage of the cathode tube and e is the charge on the electron, calculate the kinetic energy of an electron in a cathode tube operating at 4000 volts. Express the answer in joules and in calories.

17. (a) A container of balls of identical weight is weighed 12 times, each time with a different number of balls in it. The weights of the contents follow: 150 g, 315 g, 90 g, 135 g, 30 g, 285 g, 105 g, 240 g, 425 g, 195 g, 15 g, 60 g. What is the maximum possible weight for one ball? (b) Follow your reasoning in the problem above. If, in an oil-drop experiment (similar to the Millikan experiment), the charges on an oil drop were found to be 1.44×10^{-18} coul, 9.6×10^{-19} coul, 1.76×10^{-18} coul, 6.4×10^{-19} coul, and 1.28×10^{-18} coul, what is the maximum possible charge of the electron? (c) What is the charge on each drop in terms of the fundamental unit?

18. The isotopes of chlorine found in nature have atomic masses 37 amu and 35 amu and occur in relative abundance of 24% and 76%, respectively. What would be the average atomic weight of a sample of such a mixture?

19. Given the following data, calculate the weighted average atomic weight of copper.

Isotope Mass (amu)	% of Isotope in Nature
62.930	69.09
64.928	30.91

20. An element consists of three isotopes as shown in the table:

Isotope	%Occurrence	Mass (amu)
1	92.0	28.0
2	5.0	29.0
3	3.0	30.0

Compute its atomic weight.

21. Calculate the atomic weight of each of the following elements from the data provided:

Isotope	Mass (amu)	Abundance (%)
^{6}Li	6.0151	7.40
^{7}Li	7.0160	92.60
^{28}Si	27.977	92.3
^{28}Si	28.976	4.6
^{30}Si	29.974	3.1

22. Values for the ratio of charge to mass for two isotopes of hydrogen obtained from the mass spectrometer are 4.7×10^4 and 3.1×10^4 coul/g, respectively. Recalling that the charge on the electron is 1.60×10^{-19} coul, calculate the mass in grams of each of these isotopes. What are their masses in amu?

23. Using the atomic weights given on the inside back cover, calculate the weight in grams of each of the atoms of the following monoisotopic elements: bismuth, gold, thulium, praseodymium, cesium, and iodine.

24. There are 80 electrons in an atom of mercury. The mass of an electron is 5.5×10^{-4} amu and the mass of a mercury atom is 200 amu. What percent of the total mass of the mercury atom is due to the mass of the electrons in it?

25. Two isotopes of masses 4.20×10^{-23} g/atom and 4.27×10^{-23} g/atom make up a naturally occurring element. The isotopes occur in the ratio of 3 to 4, respectively. What is the atomic weight of this element in amu?

26. It has been determined by the mass spectrograph that the ratio of the mass of the hydrogen atom to that of the electron is 1.84×10^3. (a) What value would you ascribe to the atomic weight of the electron from these considerations? (b) What is its mass in amu?

27. A 47.16 g sample of a certain element occupies 6.00 cc and is composed of 5.08×10^{23} atoms. (a) Find the mass (in grams) of one atom of this element. (b) Find the mass (in amu) of one atom of the element. (c) Find the volume occupied by one atom of the element.

28. Aluminum has an atomic weight of 26.9815 amu. Its density is 2.699 g/cc. (a) What is the average volume occupied by a single atom of aluminum? (b) If an aluminum atom is spherical and has the volume computed in (a), what is its diameter?

29. The radius described by a ^{12}C ion, 12.000, singly charged, in the mass spectrograph is 2.7 A. (a) What is the radius of the ^{13}C ion, 13.003, singly charged? (b) What is the radius of the ^{12}C ion, 12.000, doubly charged?

1. A stream of alpha particles is projected at a thin strip of gold foil, 4.5×10^{-4} mm thick. A detecting device counts 87 particles which have been deflected into the region between $93°$ and $180°$ relative to the initial path in a given period. When the experiment is repeated in the same manner except that the thickness of the foil is increased to 5.5×10^{-4} mm, 112 particles are counted which have been deflected into this region. (a) Why should the increase in the thickness of the "wall" cause more particles to rebound? How does the gold "wall" being struck by alpha particles differ from a brick wall being struck by a rubber ball? (b) What would you expect to happen when one of the alpha particles collides with an electron in a gold atom?

2. In a cathode-ray tube, a beam of electrons (the cathode ray) is passed between two metal plates which have been connected to a voltage source to create an electric field. In passing through the field, the electrons are deflected. The deflection is measured on a calibrated fluorescent screen upon which the electrons impinge. The deflection caused by the electric field may be counteracted by the imposition of a magnetic field that is capable of restoring the beam to its original position. The forces F_E (due to the electrical field) and F_M (due to the magnetic field) which act upon the electron, are described as follows:

$$F_E = eL$$

where
  F_E is in dynes

when

  e is the charge on the electron measured in electrostatic units of charge (esu)
  L is the field strength measured in electrostatic units of field strength (statvolts/cm)

$$F_M = \frac{mv^2}{r} = evH$$

where
  F_M is in dynes

when

  e is the charge on the electron measured in electromagnetic units of charge (emu)
  v is the velocity measured in cm/sec
  H is the field strength measured in gauss
  m is the mass of the electron in grams
  r is the radius of the circular path described by the electron

(a) Suppose that a cathode ray is deflected by an electrical force of 10^3 volts/cm (the field strength) and that the beam is returned to its original position by the imposition of a magnetic field of 10^3 gauss. It is obvious that the two forces acting on the electron, F_E and F_M, must be equal for this to be true.
Given the following definition of units, *calculate the velocity of the electrons.*
3.00×10^{10} esu $= 1$ emu

1 emu = 10 coul
1 joule = 1 volt-coul = 10^7 dyne-cm
1 statvolt = 300 volts

(b) If the electron beam has a velocity of 4.22×10^8 cm/sec and undergoes a deflection of radius 2.11 cm when subjected to a magnetic field strength of 11.4 gauss, what would be the e/m of the electron?

3. When an alpha particle "collides head-on" with a nucleus, it is deflected 180°—directly back along the path on which it came—or is captured by the nucleus. In such an instance, the kinetic energy must become zero at some instant. At this instant the initial kinetic energy of the particle becomes its potential energy and

$$1/2\ mv^2 = \frac{ZeZ'e}{d}$$

where

m is the mass of the alpha particle
v its initial velocity
Z its atomic number
d the closest approach the alpha particle may make to the nucleus
e is the electronic charge
Z' is the atomic number of the target atom

(a) Calculate the closest approach distance, d, to an ^{197}Au nucleus, if the alpha particle's initial velocity is 2.0×10^9 cm/sec and the electronic charge, e, is 4.8×10^{-10} esu.

$$1\ esu^2 = 1\ dyne\text{-}cm^2$$

Note: If the nearest approach is 1×10^{-12} cm, the alpha particle will probably be captured.

4. Do atoms exist? Explain your answer.

REFERENCES

Aston, F. W., "Isotopes and Atomic Weights," in W. C. Dampier and Margaret Dampier, (eds.), *Readings in the Literature of Science*, Harper, New York, 1959.
Brecia, F., "Analogy for Forces and Quantum Field Theory," *J. Chem. Educ.*, **47**, 642 (1970).
Garrett, A. B., *The Flash of Genius*, Van Nostrand Reinhold Co., Princeton, 1963.
Kay, W. A., "Recollections of Rutherford" in D. E. Gershenson and D. A. Greenberg, *The Natural Philosopher*, Vol. 1., Xerox College Publishing, Lexington, Mass., 1963.
Morrow, B. A., "On the Discovery of the Electron," *J. Chem. Educ.*, **46**, 584 (1969).
Roller, D., and Roller, D. H. D., "The Development of the Concept of Electrical Charge," in *Harvard Case Histories in Experimental Science*, Case 8, Harvard University Press, Cambridge, 1954.

MOLES; SYMBOLS, FORMULAS, AND EQUATIONS

When you can measure what you are speaking about and express it in numbers, you know something about it. . . .

Lord Kelvin (1824–1907)

Some of the fundamental concepts chemists use as they pursue their objective of understanding the universe were introduced in Chapter 1. A few of these are:

1. Matter is made of atoms.
2. The masses of atoms can be determined.
3. Symbols are used to designate elements.
4. Formulas are used to designate compounds.
5. Equations are used to indicate and to give information about chemical changes.

The chemical symbol and combinations of symbols were devised by chemists not only to refer to elements and compounds, but also to communicate information about those elements and compounds, as well as atoms and molecules. The chemical symbol system of notation is simple and convenient. It uses abbreviations (symbols) to designate elements or atoms and a combination of symbols to represent compounds or molecules. Such is the language of chemistry.

This notation is powerful in that symbols and formulas can be used not only to form equations to indicate chemical changes (reactions); they also lend themselves to many interpretations concerning the amounts of elements and compounds, or the numbers of atoms and molecules involved in these reactions. The symbols and formulas thus make it possible to apply mathematics, another important communication device, to elements and compounds, and to chemical changes.

One of the earliest problems that had to be resolved was how to determine and designate weights of atoms. This was done by a simple, ingenious method of relative atomic weights (see Chapter 2). Now the absolute weights of atoms can be determined very accurately with the mass spectrometer (see Chapter 2).

In measuring amounts of materials for chemical reactions, the number of atoms or molecules present is important. However, no convenient, large-scale atom-counting device has been developed. Weighing the sample is usually the most feasible way to measure the quantity of material present. So the question arises: Can we find a unit connecting the weight in grams of a sample of an element (or compound) and the number of atoms (or molecules) present?

The units agreed upon for this purpose are: the *gram-atomic weight* of an element, defined as the number of grams of an element numerically equal to the number of atomic mass units in its atomic weight (for example, 16 g for oxygen), and the gram-molecular weight of compounds, defined as the number of grams of the compound numerically equivalent to the number of atomic mass units in its formula weight (for example, 44 g for CO_2).

The importance of these definitions is this: *All gram-atomic weights contain the same number of atoms.* Thus, there are equal numbers of atoms in 16.0 g of oxygen, in 12.0 g of carbon, and in 55.8 g of iron. *All gram-molecular weights contain the same number of molecules.* Thus 44 g of CO_2, 18 g of H_2O, and 180 g of $C_6H_{12}O_6$ all contain the same number of molecules.

Now we ask: "How many atoms of an element are present in a gram-atomic weight of that element?" We can find the answer to this question by dividing the gram-atomic weights of several elements by the measured mass of one atom of those elements. This is shown in Table 3.1.

From the table, we see that the gram atomic weight is the weight of 6.02×10^{23} atoms; this is called a gram-mole or simply a mole of an element. The number of atoms in a mole is called the Avogadro number.

Chemists now use the word mole to refer to the Avogadro number of particles of any chemical species. Thus, 1 g-at. wt (4 g) of helium is often referred to as a mole of helium; 2 g of helium constitutes one-half mole, while 10 g of helium is 2.5 moles. A mole of water is 6.02×10^{23} water molecules (which weighs 18 g); a mole of chloride ions is 6.02×10^{23} ions and weighs 35.5 g (Figure 3.1).

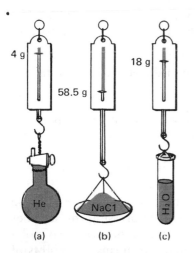

Figure 3.1 Weights (in grams) of one mole of several substances. (a) Helium, 6.02×10^{23} atoms; (b) sodium chloride, 6.02×10^{23} chloride ions (35.5 g) plus 6.02×10^{23} sodium ions (23 g); (c) water, 6.02×10^{23} molecules.

CHEMICAL SYMBOLS

Symbols have been used as a chemical shorthand since the advent of alchemy. Modern symbols are either derived from or are abbreviations of the accepted names of elements. The symbol of an element is used to represent or signify any or all of the following:

1. The name of that element.
2. The presence of that element in a compound. The formula HCl indicates that this substance is composed of the elements hydrogen and chlorine.

Table 3.1 Calculations of the Avogadro Number

For hydrogen:	$\dfrac{1.008 \text{ g}}{1 \text{ g-at. wt.}} \times \dfrac{1 \text{ atom}}{1.674 \times 10^{-24} \text{ g}} =$	6.02×10^{23} atoms in a g-at. wt. of hydrogen
For helium:	$\dfrac{4.003 \text{ g}}{1 \text{ g-at. wt.}} \times \dfrac{1 \text{ atom}}{6.65 \times 10^{-24} \text{ g}} =$	6.02×10^{23} atoms in a g-at. wt. of helium
For oxygen:	$\dfrac{16.00 \text{ g}}{1 \text{ g-at. wt.}} \times \dfrac{1 \text{ atom}}{26.6 \times 10^{-24} \text{ g}} =$	6.02×10^{23} atoms in a g-at. wt. of oxygen

3. *One atom* of the element or a weight of the element equivalent to the weight of one atom. The symbol F may indicate one atom of fluorine, 19 amu of fluorine, or 31.55×10^{-24} g of fluorine.
4. *One mole* of the element—that is, 6.02×10^{23} atoms of the element.
5. A *symbol weight* of the element in any weight units; thus it may represent one gram-atomic weight, one pound-atomic weight, one ton-atomic weight, or a weight in any units numerically equivalent to the atomic weight in atomic mass units. The symbol Al may represent 26.98 amu of aluminum, 26.98 g of aluminum, 26.98 lb of aluminum, or 26.98 tons of aluminum, and so on.

It is usually obvious from the context just which of the above interpretations of the symbol is intended.

CHEMICAL FORMULAS

The formula of a substance is a symbolic representation of its composition. Several kinds of formulas are commonly used. The choice of a particular kind of formula depends upon the ideas the chemist wishes to convey.

All types of formulas illustrate quantitatively the fundamental characteristic of pure substances—that they have a definite composition. Thus the formula NO_2 states that the composition of this substance is fixed at the ratio of two atoms of oxygen for each atom of nitrogen. Expressing this another way, we might say that for every mole of nitrogen atoms, there are two moles of oxygen atoms present in this compound. Stated in terms of weights of materials, there are 14 g of nitrogen for every 32 g of oxygen or, in terms of per cent by weight:

$$\text{Per cent nitrogen present} = \frac{14}{32 + 14} \times 100 = 30\%$$

$$\text{Per cent oxygen present} = \frac{32}{32 + 14} \times 100 = 70\%$$

Three types of formulas are of particular interest at this point:

1. The *empirical* (or simplest) *formula*, which gives the symbols for the elements present and indicates the simplest whole-number ratio of the respective atoms in the substance under consideration.
Examples: CH, $HgCl$, NO_2, P_2O_5, and CO_2.
2. The *molecular formula*, which gives the actual number and kinds of atoms in one molecule of the substance under consideration.
Examples: C_2H_2, C_6H_6, Hg_2Cl_2, N_2O_4, P_4O_{10}, and CO_2.
3. *Structural formulas,* which give the relative arrangements of atoms in space and may indicate the types of chemical bonds connecting the atoms and, in some cases, the shape of the molecule.
Examples (the wedge-type line indicates that the attached atom lies in front of the plane of the paper):

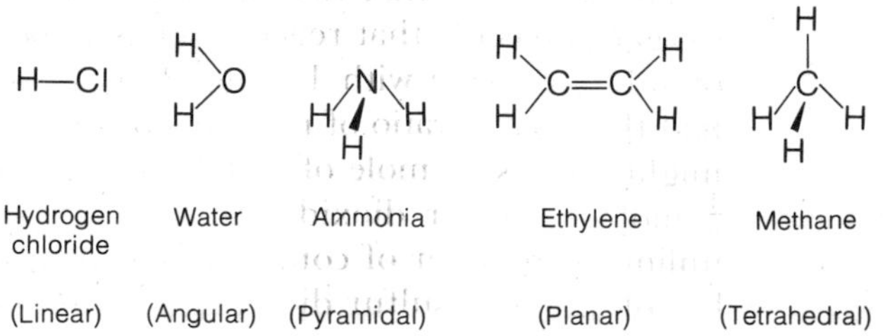

Hydrogen chloride	Water	Ammonia	Ethylene	Methane
(Linear)	(Angular)	(Pyramidal)	(Planar)	(Tetrahedral)

Before discussing the methods used to determine each type of formula, it might be desirable to look at the conventions chemists have adopted for measuring bulk quantities of substances represented by formulas.

Moles of Molecules. Just as it is convenient to use the mole or gram-atomic weight of an element as the standard measure for atoms in bulk, so it has become useful to use the gram-molecular weight (or gram-formula weight) as a standard measure for molecules in bulk. The molecular weight of a substance is obtained from the molecular formula by adding the weights of all atoms present in the molecule. The molecular weight of nitrogen dioxide is 46 amu (resulting from contributions of 14 amu for the nitrogen atom and 2×16 or 32 amu for the two oxygen atoms). The gram-molecular weights of carbon dioxide, sulfur dioxide, and hydrogen chloride are 44 g, 64 g, and 36.5 g, respectively.

The number of molecules in one gram-molecular weight of a substance may be calculated by dividing the gram-molecular weight by the average weight of one mole of the substance. Thus, for nitrogen dioxide (NO_2),

$$\frac{46 \text{ g}}{1 \text{ g-mol. wt.}} \times \frac{1 \text{ molecule}}{7.64 \times 10^{-23} \text{ g}} = 6.02 \times 10^{23} \text{ molecules in 1 g-mol. wt.}$$

Since the gram-molecular weight of a substance contains the Avogadro number of molecules, it has become the practice to refer to this quantity as a mole of the material under consideration. Thus, 46 g of nitrogen dioxide is referred to as a mole of nitrogen dioxide, 23 g is $\frac{1}{2}$ mole, and 230 g is 5 moles. In 11 g of carbon dioxide ($\frac{11}{44}$ or $\frac{1}{4}$ g-mol. wt. or mole), there are $\frac{1}{4} \times 6.02 \times 10^{23}$ molecules; in 7.3 g of hydrogen chloride ($\frac{7.3}{36.5}$ or $\frac{1}{5}$ g-mol. wt. or mole), there are $\frac{1}{5} \times 6.02 \times 10^{23}$ hydrogen chloride molecules; in 6 g of glucose, $C_6H_{12}O_6$ ($\frac{6}{180}$ or $\frac{1}{30}$ mole), there are $\frac{1}{30} \times 6.02 \times 10^{23}$ glucose molecules.

One apparent anomaly arises in considering the mole concept in connection with elements which normally appear as molecules; the familiar O_2 molecule is an example. When a chemist speaks about a mole of oxygen, he may mean 6.02×10^{23} atoms of oxygen (16 g) or he may mean 6.02×10^{23} molecules of oxygen (32 g). Since the stable form of oxygen at room conditions is the molecular form, O_2, the expression "a mole of oxygen" usually means 6.02×10^{23} molecules of oxygen. To avoid confusion, however, most chemists will refer to a mole of oxygen atoms or to a mole of oxygen molecules, as the case may be.

Use of Mole in Calculations When molecular species are involved in chemical reactions, the mole concept makes it easy to choose the correct number of moles for reaction. Consider, for example, the equation

$$2SO_2 + O_2 \longrightarrow 2SO_3$$

This indicates that two molecules of sulfur dioxide are needed for every oxygen molecule that reacts, and an experimenter might use 2 moles (128 g) of sulfur dioxide with 1 mole (32 g) of oxygen molecules and be certain he had the correct ratio of numbers of molecules for the reaction. However, he might also use 1 mole of sulfur dioxide with $\frac{1}{2}$ mole of oxygen molecules, or $\frac{1}{5}$ mole of sulfur dioxide and $\frac{1}{10}$ mole of oxygen molecules, or any of an unlimited number of combinations, provided only that the ratio of the number of moles of sulfur dioxide to the number of moles of oxygen is 2:1.

If 2 g of sulfur dioxide is converted to sulfur trioxide, what weight of oxygen is consumed and what weight of sulfur trioxide is formed?

2 g of sulfur dioxide constitutes $\frac{2}{64}$ or $\frac{1}{32}$ mole. Since the ratio of moles of SO_2 to moles of O_2 is 2:1, $\frac{1}{64}$ mole O_2 is consumed. The weight of $\frac{1}{64}$ mole of O_2 is $\frac{1}{64} \times 32$ g O_2/mole or $\frac{1}{2}$ g O_2. The weight of sulfur trioxide formed is 2.5 g, consisting of a contribution of 2 g from the SO_2 and 0.5 g from the molecular oxygen.

Moles of Simplest Formula Units

Chapter 5 shows that many compounds, such as sodium chloride, are composed of ions rather than of simple molecules. Other compounds, such as silicon dioxide, are made of *macro*molecules, molecules that are very large, and often as large as the crystals of the substance. It is evident that ionic and macromolecular compounds cannot be adequately represented by molecular formulas. Instead, the simplest formula usually is used to represent these substances. What, then, shall be the standard measure for those substances in bulk? Chemists have agreed to use the gram-formula weight as the standard; they have defined this as the mole of the ionic or macromolecular compound. It is the weight in grams of the substance numerically equivalent to the weight of the atoms in the simplest formula expressed in atomic mass units. Thus, in sodium chloride, NaCl, the formula weight is 58.5 amu, resulting from contributions of 23.0 amu from the sodium ion and 35.5 amu from the chloride ion. One gram-formula weight or 1 mole of sodium chloride has a mass of 58.5 g and may be regarded as containing 6.02×10^{23} of the simplest formula units—6.02×10^{23} units, each composed of one sodium ion and one chloride ion. A gram-formula weight, or 1 mole, of calcium fluoride, CaF_2, has a mass of 78 g and may be regarded as containing 6.02×10^{23} units, each composed of one calcium ion and two fluoride ions. A mole of silicon dioxide is 60.1 g, and may be regarded as containing 6.02×10^{23} units, each composed of one silicon atom and two oxygen atoms present in a macromolecular structure.

Determination of Formulas

All chemical formulas must, of course, agree with experimental data. The kinds of data necessary to determine the simplest formula and the molecular formula are weight relationships, number relationships, or both.

Simplest Formulas. The experiments required to determine the simplest formula must tell us the simplest whole number ratio of the respective atoms in the substance under consideration. But these experiments are most often performed on samples large enough to be weighed—that is, on bulk quantities of the substance. If we could determine the simplest ratio of moles of the respective elements in the substance, we would then know the simplest ratio of atoms.

The simplest ratio of moles of the respective elements may be obtained from the number of grams of these elements in a given sample of the compound. For example, a certain sample of silver oxide was found to contain 1.7271 g of silver and 0.1281 g of oxygen.

PROBLEM

$$\text{Number of moles of silver present in the sample} = 1.7271 \text{ g of silver} \times \frac{1 \text{ mole silver}}{107.9 \text{ g silver}}$$

$$= 0.016 \text{ mole of silver}$$

Solution | Number of moles of oxygen present in the sample $= 0.1281$ g of oxygen $\times \dfrac{1 \text{ mole oxygen atoms}}{16 \text{ g oxygen}}$

$$= 0.008 \text{ mole of oxygen atoms}$$

The ratio of moles of silver atoms to moles of oxygen atoms is $0.016/0.008$, or 2:1. Thus, the simplest formula is Ag_2O.

PROBLEM The compound methyl formate is found to contain 40.00% carbon, 6.71% hydrogen, and 53.28% oxygen. Calculate the simplest formula.

Solution In 100 g of methyl formate, there are 40.00 g of carbon, 6.71 g of hydrogen, and 53.28 g of oxygen.

Moles of carbon in 100 g of methyl formate $= 40.00$ g carbon $\times \dfrac{1.00 \text{ mole carbon atoms}}{12.01 \text{ g carbon}}$

$$= 3.33 \text{ moles carbon}$$

Moles of hydrogen in 100 g of methyl formate $= 6.71$ g hydrogen $\times \dfrac{1.00 \text{ mole hydrogen atoms}}{1.008 \text{ g hydrogen}}$

$$= 6.66 \text{ moles hydrogen atoms}$$

Moles of oxygen in 100 g of methyl formate $= 53.28$ g oxygen $\times \dfrac{1.00 \text{ mole oxygen atoms}}{16.00 \text{ g oxygen}}$

$$= 3.33 \text{ moles oxygen atoms}$$

The ratio of moles of C:H:O is $3.33/6.66/3.33$ or 1:2:1, and the simplest formula becomes CH_2O.

Figure 3.2 Comparison of the gram-molecular volume of a gas at standard conditions with a regulation-size basketball.

Molecular Formulas The molecular formula indicates the number and kind of atoms in one molecule of the substance under consideration. It may be the simplest formula or a multiple of the simplest formula. Once the simplest formula has been obtained, the molecular formula may be determined if the molecular weight is known. To do this, several methods are available:

1. Using the mass spectrometer. In this instrument, the weights of molecules, as well as of atoms, may be obtained with high precision.

2. Measuring vapor density. A mole of gas at 0°C and 760 torr (standard conditions of temperature and pressure) occupies approximately 22.4 liters. The gram-molecular weight of gases and volatile liquids may be determined by measuring the density of the gas or vaporized liquid and calculating the weight of 22.4 liters of the gas or vapor at standard temperature and pressure (Chapter 10).

3. Determining the depression of the freezing point or elevation of the boiling point of a solvent caused by the presence of a solute. One mole of a molecular solute such as sugar dissolved in 1000 g of water will lower the freezing point of water 1.86°C. By dissolving a weighed sample of a molecular solute in a known weight of water and measuring the freezing point of the solution, it is possible to calculate the weight of the solute which, when dissolved in 1000 g of water, will lower the freezing point 1.86°C. This is the weight of one mole of the solute.

Since the methods of determining molecular weights will be presented later (Chapters 10 and 12), the following example will suffice to illustrate the determination of the molecular formula from the simplest formula.

PROBLEM

The simplest formula for methyl formate has been found to be CH_2O (see problem, page 45). The molecular weight of this compound has been found to be 60 amu by the freezing-point depression method. What is the molecular formula?

Solution

The molecular weight must be a multiple of the weight of the simplest formula unit, CH_2O. Since the weight of CH_2O is 30 amu, it is evident that the molecular weight is twice this, which corresponds to the molecular formula $C_2H_4O_2$.

Structural Formulas. One modern method used to determine the arrangement of the atoms within a molecule or an ion is to study the pattern given by X rays directed toward a sample of the substance under consideration. Microwave spectroscopy also is used for this purpose (Chapter 10).

CHEMICAL EQUATIONS

A chemical equation is simply a sentence written with chemical symbols and formulas; it represents facts which are already known from experiment; it indicates with symbols and formulas the rearrangement of atoms, ions, and molecules that occurs during a chemical reaction; it is a record, in chemical notation, of what chemical species and what relative amounts of them enter into the reaction and what species and what relative amounts are formed. Consider, for example, the reaction of hydrogen with oxygen to form water (Figure 3.3):

In words:

2 molecules of hydrogen react with 1 molecule of oxygen to form 2 molecules of water.

In equation form:

$$2H_2(g) + O_2(g) \longrightarrow 2H_2O(l)$$

Note that the reactants (the materials consumed) are placed on the left-hand side and the products (the materials produced) are placed on the right-hand side of the chemical equation. The letters (*s*), (*l*), or (*g*) are sometimes used to indicate solid, liquid, or gaseous phase. The numbers placed before the formulas are called coefficients. The arrow indicates the direction of predominant change.

Figure 3.3 Reactants and products in the formation of water from hydrogen and oxygen.

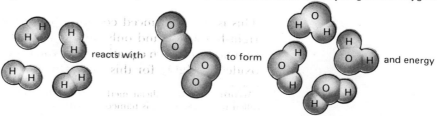

reacts with   to form   and energy

It must be emphasized that an equation is a statement of *facts learned by experiment*. In other words, we cannot attempt, by equation writing, to find out what the products of a chemical reaction are, or what their formulas are. But having discovered these products in actual experiments, we can record the facts in an equation.

Modern chemists usually use *net equations*—equations that specify only the chemical species consumed and the species formed in the chemical reaction. Occasionally they use equations to show a series of intermediate compounds that often are formed in the process of the reaction, or an *over-all equation* to show all the reactants and products in which they are interested.

In words:

A solution of silver nitrate (containing Ag^+ and NO_3^-) is poured into a solution of sodium chloride (containing Na^+ and Cl^-) and a white *precipitate* of silver chloride (AgCl) appears.

In equation form:

Over-all equation: $AgNO_3(aq) + NaCl(aq) \longrightarrow AgCl(s) + NaNO_3(aq)$

Net equation: $Ag^+(aq) + Cl^-(aq) \longrightarrow AgCl(s)$

The net equation indicates that the product of the reaction, silver chloride, was formed by the reaction of silver ions with chloride ions. No statement is made about the presence or the fate of the sodium and nitrate ions present in the reaction mixture. Since these ions remain unchanged during the reaction, they are not included in the net equation. An abbreviation (*aq*) often is used to indicate that the species is in water (aqueous) solution.

Writing and Balancing Equations

When the chemical reaction is known, the *first* step in writing and balancing an equation for that reaction is to write the correct formulas for the reactants and products. This involves symbols and *subscripts* only.

All atoms entering into the reaction must be accounted for in the products of that reaction; this is a requirement of the law of conservation of mass. The *second* step is to balance the equation with respect to atoms. This merely involves making the numbers of each kind of atom the same on both sides of the equation by changing the *coefficients* only, but without changing the formulas of the species reacting or the species formed. Atoms may be bonded in the products to atoms different from those to which they were bonded in the reactants. They may also exist as a different ionic species in the products than in the reactants; but their number must remain constant.

PROBLEM 1   Write the equation for the reaction of mercuric oxide decomposition.

Solution   The expression for the chemical fact that mercuric oxide decomposes to give mercury and oxygen is first written as follows:

$$HgO \longrightarrow Hg + O_2$$

This is not a balanced equation because there are two oxygen atoms on the right-hand side and only one on the left-hand side. To balance the expression (and make an equation of it) you cannot change the formula for mercuric oxide* to HgO_2, for this would be contrary to the results of the experiments

*According to the official method of nomenclature, the oxide of mercury (HgO), commonly called mercuric oxide, is named mercury(II) oxide.

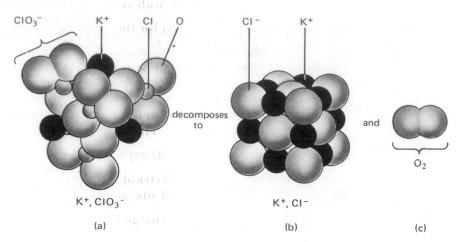

CIO₃⁻   K⁺   Cl   O   Cl⁻   K⁺

decomposes
to   and

O₂

K⁺, ClO₃⁻   K⁺, Cl⁻

(a)   (b)   (c)

Figure 3.4 An intimate look at the reaction $2KClO_3 \longrightarrow 2KCl + 3O_2$. (a) The probable arrangement of K^+ and ClO_3^- ions in a fragment of a potassium chlorate crystal. Note the ClO_3^- clusters. (b) The probable arrangement of K^+ and Cl^- ions in a fragment of a potassium chloride crystal. (c) An oxygen molecule.

used to determine the formula. However, by doubling the amount of mercuric oxide in the expression, we obtain the equation:

$$2HgO \longrightarrow 2Hg + O_2$$

Note that in doubling the amount of mercuric oxide, we have doubled the number of mercury atoms on the left in the expression. This means we must double the number of mercury atoms on the right. This equation indicates that two formula units of mercuric oxide decompose to give two atoms of mercury and one oxygen molecule.

PROBLEM 2   Write the equation for the reaction of potassium chlorate decomposition (Figure 3.4).

Solution   The expression for the decomposition of potassium chlorate to give potassium chloride and oxygen is initially written as follows:

$$KClO_3 \longrightarrow KCl + O_2$$

The next step required to make this into an equation is to note that the three oxygen atoms on the left might be regarded as being transformed into one and one-half oxygen molecules on the right. The equation then becomes:

$$KClO_3 \longrightarrow KCl + 1\tfrac{1}{2}O_2$$

Since many students find it difficult to think of fractions of molecules, both sides of the equation can be multiplied by 2, giving the equally correct and perhaps more common equation for the decomposition of potassium chlorate:

$$2KClO_3 \longrightarrow 2KCl + 3O_2$$

Note that the difficulty of fraction of molecules does not arise if the equation is interpreted in terms of moles; it is clear to say that 1 mole of $KClO_3$ decomposes to produce 1 mole of KCl and $1\tfrac{1}{2}$ moles of oxygen.

To be correct, an equation must be balanced with respect to electrical charges as well as with respect to mass, as shown in the following problem.

PROBLEM 3   Write the equation for the reaction of a piece of metallic zinc with a solution of iron trichloride.*

Solution   The equation might be represented by:

$$Zn(s) + Fe^{+3}(aq) \longrightarrow Fe(s) + Zn^{+2}(aq)$$

While this expression is balanced with respect to mass, it is not balanced with respect to charge. The correct equation is:

$$3Zn(s) + 2Fe^{+3}(aq) \longrightarrow 3Zn^{+2}(aq) + 2Fe(s)$$

The number of electrical charges on each side of the equation is +6 and the masses of iron and zinc are the same on each side of the equation.

Very often the charge is balanced in the process of balancing mass (Problem 4).

PROBLEM 4   Write the equation for the reaction of the dichromate ion with hydroxide ions in aqueous solution, which gives rise to the chromate ion and water.

Solution   First write:

$$Cr_2O_7^{-2}(aq) + OH^-(aq) \longrightarrow CrO_4^{-2}(aq) + H_2O$$

In balancing the masses, you obtain:

$$Cr_2O_7^{-2}(aq) + 2OH^-(aq) \longrightarrow 2CrO_4^{-2}(aq) + H_2O$$

This equation is also balanced with respect to charge.

Facility in balancing equations comes largely from practice. Most simple equations can be balanced by inspection, which usually involves working back and forth from one side of the expression to the other, changing the coefficients preceding the formulas until the expression is balanced with respect to both mass and charge. A more systematic but longer method of balancing equations is presented in Appendix C. That method is useful for complex reactions.

Additional Information Shown in Chemical Equations

Chemical equations may be written to show not only the reactants and products but also some conditions of the experiment:

1. Chemists sometimes indicate on the arrow the optimum conditions or the use of a catalyst for the reaction.

$$2KClO_3 \xrightarrow[200°C]{MnO_2} 2KCl + 3O_2$$

The information on the arrow states that the reaction takes place in the presence of manganese dioxide, which speeds up the reaction (it is thus a *catalyst*), and that the temperature at which the reaction occurs is 200°C.

2. The heat liberated or absorbed may be indicated in the equation as:

$$2H_2(g) + O_2(g) \longrightarrow 2H_2O(g) + energy\ (116\ kcal)$$

This equation must now be read: When 2 moles of hydrogen gas react with 1 mole of oxygen gas to give 2 moles of water vapor, 116 kilocalories (kcal) of heat are liberated (Figure 3.3).

---

*According to the official rules of nomenclature this chloride of iron ($FeCl_3$) may be named iron trichloride, ferric chloride, or iron(III) chloride.

**Some Limitations
of Many Equations**

Most equations do not indicate a number of important facts about a reaction. These include:

1. The equation gives no information about the path by which the reactants were converted to products. The equation $2H_2 + O_2 \longrightarrow 2H_2O$ does not imply that two molecules of hydrogen collide with a molecule of oxygen whereupon two molecules of water are formed. We now know that the actual path of this reaction involves a series of processes.

2. The equation gives no information about the reaction rate or even whether the reaction will occur in a finite time. The reaction of hydrogen with oxygen will not take place at an appreciable rate at room temperature unless set off by a spark or a catalyst (Chapter 14); then the reaction probably will proceed with explosive violence.

3. The equation gives no information regarding whether a reaction is complete or, if incomplete, of the extent of the reaction. Chemists have good reason to believe that reactants never completely disappear in a reaction mixture. In many reactions the concentrations of reactants left when the reaction has "run its course" may be extremely small, and the reaction is said to have proceeded to completion. In many other reactions, the concentrations of reactants are appreciable when the reaction has "run its course."

Reactions that do not go to completion are usually represented by equations using a double arrow, indicating that both forward and reverse reactions occur and a state of equilibrium develops:

$$N_2 + 3H_2 \rightleftharpoons 2NH_3$$

**Predicting Products
in Chemical Reactions**

While it is true that the experimental facts must be known before an equation can be written, it is the function of chemical theory to provide a basis for predicting reliably the products of chemical reactions. If the theory were highly developed or if the behavior of chemical substances were known well enough, it should be possible to predict the products for any reaction. It would then be unnecessary to remember the results of experiments to write equations. But the theory of chemistry is far from this goal at the present time. For this reason, the student will find many reactions whose products he can readily predict from his knowledge of theory, and it is one of the aims of this book to increase that knowledge. There will remain many other reactions, however, whose products he (and every chemist) must simply remember from having carried out the reactions or having read accounts of others who have done so.

Chemical Calculations

In discussing the balancing of equations, emphasis was placed on the numbers of atoms, molecules, or ions involved in the reaction. However, the equations can equally well represent changes not of atoms alone but also of matter in bulk. To make the transition from the molecular scale to bulk scale, you need only to recall that symbols or formulas may represent one chemical species or a mole (or a pound-mole or a ton-mole) of this species. This way, an equation might simultaneously represent both a molecular-size view and a macro-size view of the reaction. For example, the equation

$$4Li + O_2 \longrightarrow 2Li_2O$$

may be regarded as simultaneously stating both of the following:

1. If 4 atoms of lithium reacts with oxygen, 1 molecule of oxygen will be consumed and 2 formula units of lithium oxide will be formed.

2. If 4 moles of lithium atoms react with oxygen, 1 mole of oxygen molecules will be consumed and 2 moles of lithium oxide will be formed.

Since moles may be converted to weight, the equation may be regarded as stating the weight relationships in the reaction. Thus, if 27.8 g (or 4 moles) of lithium reacts with oxygen, 32.0 g (or 1 mole) of oxygen will be consumed and 59.8 g (or 2 moles) of lithium oxide will be formed. Of course, if 6.94 g (1 mole) of lithium were used, only 8.00 g ($\frac{1}{4}$ mole) of oxygen would be consumed and 14.94 g ($\frac{1}{2}$ mole) of lithium oxide would be formed.

The following problems illustrate weight relationships in chemical equations:

PROBLEM 1   What weight of carbon dioxide is produced when 640 g of methane burns? How much oxygen will be consumed and how much water will be formed?

Solution   You first need a correctly written and balanced equation. It is

$$CH_4 + 2O_2 \longrightarrow CO_2 + 2H_2O$$

From the equation, the mole relation between the material given and that desired can be obtained if we assume the reaction goes to completion. In this case the relation is: for every mole of methane burned, 1 mole of carbon dioxide will be produced.

The problem states that 640 g or

$$\left(640 \text{ g} \times \frac{1 \text{ mole } CH_4}{16 \text{ g } CH_4}\right)$$

or 40 moles of methane are to be burned.

Evidently, then, 40 moles of carbon dioxide will be formed:

$$\left(40 \text{ moles} \times \frac{44 \text{ g } CO_2}{1 \text{ mole } CO_2}\right)$$

or 1760 g of carbon dioxide are formed.

To find the weight of oxygen required, the relation between methane (the material given) and oxygen (that desired) is obtained from the equation. It is: for every mole of methane burned, 2 moles of oxygen are consumed.

Since 40 moles of methane are to be burned, evidently 80 moles, or

$$\left(80 \text{ moles} \times \frac{32 \text{ g } O_2}{1 \text{ mole } O_2}\right)$$

or 2560 g of oxygen will be consumed.

To find the weight of water formed, the equation gives the relation: for every mole of methane burned, 2 moles of water form. Since 40 moles of methane are to be burned, evidently 80 moles or

$$\left(80 \text{ moles} \times \frac{18 \text{ g } H_2O}{1 \text{ mole } H_2O}\right)$$

or 1440 g of water will be formed (Figure 3.5).

PROBLEM 2   How many grams of iron can be produced when 6.54 g of zinc are placed in a saturated solution of iron trichloride?

Figure 3.5 Interpretation of a chemical equation in terms of moles and grams of reactants and products.

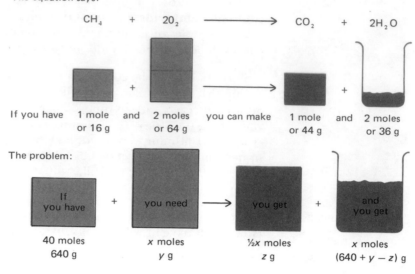

The equation says:

$$CH_4 \quad + \quad 2O_2 \quad \longrightarrow \quad CO_2 \quad + \quad 2H_2O$$

If you have   1 mole   and   2 moles   you can make   1 mole   and   2 moles
or 16 g   or 64 g   or 44 g   or 36 g

The problem:

If you have + you need → you get + and you get

40 moles   x moles   ½x moles   x moles
640 g   y g   z g   (640 + y − z) g

Solution   The balanced equation for the reaction is:

$$3Zn + 2Fe^{+3} \longrightarrow 3Zn^{+2} + 2Fe$$

Assuming the reaction goes to completion, we need first to find the relation between zinc and iron from the equation. It is for each 3 moles of zinc which react, 2 moles of iron will be formed. With this information the problem can be solved in a manner analogous to that used in Problem 1. The problem states that 6.54 g of zinc are to be consumed. This is

$$\frac{6.54 \text{ g}}{65.4 \text{ g/mole Zn}}$$

or 0.100 mole Zn.

The equation states that for every 3 moles of zinc consumed, 2 moles of iron will be formed.

The number of moles of iron that can be formed from 0.100 mole of zinc is

$$\text{Moles of iron} = 0.100 \text{ mole of zinc} \times \frac{2 \text{ moles of iron}}{3 \text{ moles of zinc}} = 0.067$$

The number of grams of iron formed then is:

$$\text{Grams of iron} = 0.067 \text{ mole of iron} \times \frac{55.8 \text{ g of iron}}{1 \text{ mole of iron}} = 3.74$$

These steps can be shortened to one relation:

$$\text{Grams of iron formed} = \frac{\text{grams of zinc consumed}}{\text{grams of zinc per mole}} \times \frac{\text{moles of iron}}{\text{moles of zinc from eq.}}$$

$$\times \frac{\text{grams of iron}}{\text{moles of iron}}$$

$$\text{Grams of iron formed} = \left(6.54 \text{ g of zinc} \times \frac{1 \text{ mole of zinc}}{65.4 \text{ g of zinc}}\right)$$

$$\times \left(\frac{2 \text{ moles of iron}}{3 \text{ moles of zinc from eq.}}\right) \times \frac{55.9 \text{ g of iron}}{1 \text{ mole of iron}}$$

Grams of iron formed $= 3.74$

Note the three steps: (a) conversion of grams zinc to moles zinc; (b) use of the relation in the equation which tells how many moles of iron can be produced from a given number of moles of zinc; (c) conversion of the moles of iron to grams of iron. These appear in order in the arithmetical equation and in such a way that all units except the grams of iron may be canceled in the equation.

PROBLEM 3 How many grams of sulfur trioxide can be prepared from 24 g of sulfur dioxide and 8 g of oxygen?

Solution The equation is

$$2SO_2 + O_2 \longrightarrow 2SO_3$$

The equation indicates that the ratio of sulfur dioxide to oxygen is 2 moles to 1 mole. Since the weights of both reactants are given, it is desirable to check to see if these weights are in the proper ratio for the reaction or if one reactant is in short supply.

The number of moles of SO_2 available is

$$24 \text{ g } SO_2 \times \frac{1 \text{ mole } SO_2}{64 \text{ g } SO_2} = \frac{3}{8} \text{ mole } SO_2$$

The number of moles of O_2 available is.

$$8 \text{ g } O_2 \times \frac{1 \text{ mole } O_2}{32 \text{ g } O_2} = \frac{1}{4} \text{ mole } O_2$$

For complete reaction, $\frac{1}{4}$ mole of O_2 would require:

$$\frac{1}{4} \text{ mole } O_2 \times \frac{2 \text{ moles } SO_2}{1 \text{ mole } O_2} = \frac{1}{2} \text{ mole of } SO_2$$

But only $\frac{3}{8}$ mole of SO_2 is available. Here sulfur dioxide is in short supply; when this has reacted, oxygen will be left over.

The problem now becomes: How much SO_3 can be prepared from $\frac{3}{8}$ mole SO_2 and excess oxygen? From the equation, we see that $\frac{3}{8}$ mole of SO_3 can be made from $\frac{3}{8}$ mole of SO_2.

$\therefore$ The mass of SO_3 prepared will be $\frac{3}{8}$ mole $SO_3 \times \frac{80 \text{ g } SO_3}{1 \text{ mole } SO_3} = 30 \text{ g } SO_3$

PROBLEM 4 As an example of the calculation of the quantities present in an incomplete reaction, consider the following: An experimenter mixes 1.0 mole of nitrogen and 3.0 moles of hydrogen in a reactor under such conditions that, at equilibrium, 0.9 mole of nitrogen remains unreacted. Calculate the quantities of hydrogen and ammonia present.

The chemical equation is:

$$N_2 + 3H_2 \longrightarrow 2NH_3$$

Moles of nitrogen reacted = 1.0 mole − 0.9 mole = 0.1
$\quad\quad\quad\quad\quad\quad\quad\quad\quad$ at start $\quad$ at end

Moles of hydrogen reacted = 0.1 mole $N_2 \times \dfrac{3 \text{ moles } H_2}{1 \text{ mole } N_2} = 0.3$

The ratio 3 moles H_2 to 1 mole N_2 is obtained from the chemical equation.

Moles of hydrogen left unreacted = 3.0 moles − 0.3 mole = 2.7
$\quad\quad\quad\quad\quad\quad\quad\quad\quad\quad\quad$ at start $\quad\quad$ reacted

Moles of ammonia formed = 0.1 mole $N_2 \times \dfrac{2 \text{ moles } NH_3}{1 \text{ mole } N_2} = 0.2$

Thus, the final mixture obtained from the reaction of 1.0 mole of nitrogen and 3.0 moles of hydrogen under the particular conditions of incomplete reaction used in the experiment contains 0.9 mole N_2, 2.7 moles H_2, and 0.2 mole NH_3.

In 1811, the Italian chemist Amedeo Avogadro stated a principle, verified often since then, that equal volumes of different gases at the same temperature and pressure contain the same number of molecules. This has important consequences in reactions of gases, as illustrated in Figure 3.6 and in the following problem.

$$2H_2(g) + O_2(g) \longrightarrow 2H_2O(g)$$

PROBLEM 5 You want to carry out the reaction:

$$2C_2H_6(g) + 7O_2(g) \longrightarrow 4CO_2(g) + 6H_2O(g)$$

A flask containing 3 liters of ethane under a pressure of 1 atmosphere and at 25°C is available, and it is desired to supply exactly the correct amount of oxygen to oxidize the ethane. How much oxygen is required?

Solution According to Avogadro's hypothesis, 3 liters of oxygen at 25°C and 1 atm pressure contain the same number of oxygen molecules as there are ethane molecules in the 3-liter flask of ethane at the same temperature and pressure. The equation shows, however, that $3\frac{1}{2}$ times as many oxygen as ethane molecules are needed, so we need $3 \times 3\frac{1}{2} = 10\frac{1}{2}$ liters of oxygen measured at 1 atm and 25°C to do the job.

Figure 3.6 Relative volumes of hydrogen, oxygen, and water vapor observed in the reaction of hydrogen and oxygen to form gaseous water.

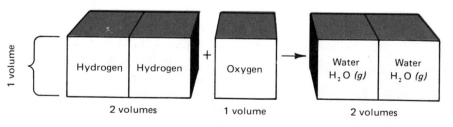

Avogadro's Law and the Properties of Gases

Avogadro's Law may be stated in symbols as

$$V = KN$$

N is the number of molecules, V is the volume of the gas at a specified temperature and pressure, and K is a proportionality constant. It is evident that doubling N will double the volume; this use of the equation was made in solving Problem 5. The volume of a gas is, in addition, related to the pressure and the absolute temperature (Chapter 10) by:

$$V = K'\frac{T}{P}$$

so that

$$V = KK'\frac{NT}{P}$$

If the right-hand side of this equation is multiplied and divided by Avogadro's number, we have

$$V = KK'N_o\frac{\frac{N}{N_o}T}{P}$$

N/N_o is the number of moles of gas present, and the product of the three constant quantities K, K' and N_o is, of course, also a constant. Denoting this product by R and solving for $N/N_o = n$, we have

$$n = \frac{PV}{RT}$$

From the experimental fact that 1 mole of gas occupies 22.4 liters at standard conditions (1 atmosphere pressure and 273K), the value of R is calculated to be 0.08206 liter atm mole^{-1}K^{-1} by direct substitution in the above equation.

The equation above is known as the general gas law and is discussed more fully in Chapter 10. Its use in determining molecular weights and gas volumes is illustrated in the following problems.

PROBLEM 6 A 2.2-g quantity of a gaseous compound is found to occupy 400 cc at 300K and 0.92 atm pressure. What is the molecular weight of the gas?

Solution The number of moles present is found to be

$$n = \frac{PV}{nT} = \frac{0.92 \text{ atm} \times 0.400 \text{ liter}}{0.00206 \text{ liter atm mole}^{-1} \text{ K}^{-1} \times 300\text{K}} = 0.0149 \text{ mole}$$

Since 0.0149 moles weighs 2.2 g., 1 mole must weigh

$$\frac{2.2 \text{ g}}{0.0149 \text{ moles}} = 148 \text{ grams per mole}$$

The molecular weight is 148.

PROBLEM 7 Solution of a problem similar to Problem 1 has shown that 748 g CO_2 is produced in a chemical reaction. What volume will this occupy at 1 atm and 27°C? The Kelvin temperature is equal to (273 + temperature in degrees Celsius); thus the absolute temperature is $273 + 27 = 300$K.

Solution   The number of moles of carbon dioxide is

$$748 \text{ g } CO_2 \times \frac{1 \text{ mole } CO_2}{44 \text{ g } CO_2} = 17 \text{ moles}$$

Solving the general gas law equation for the volume and substituting,

$$V = \frac{nRT}{P} = \frac{17 \times 0.08206 \times 300}{1} = 41.9 \text{ liters}$$

The Importance of Measurement Standards.   An almost unbelievable portion of what we do and what we use in our daily lives depends upon accurate measurement standards.   We expect things we use to work with precision—our cars, our watches, our TV sets, even our institutions and industry.   Time schedules of airlines and commuter trains, radio and television programs, the regimen of automated business run by unseen computer networks—all have assumed a certain dominance over our life style.   Such precision requires reliable, accurate, and universally accepted standards for measuring whatever it is that controls operation of the object or system under consideration.   The unit of length used by the manufacturer of spark plugs must be identical to that used by the engineer who designs an automobile engine; and the clock in the communications center in Rome or in the Apollo spacecraft must be synchronized with that in New York or in Houston to about a millionth of a second; the mercury levels in foods must be several orders of magnitude below that demonstrated to cause symptoms in laboratory animals.

Today's technology requires standards for the accurate measurement of chemical levels, color, light and sound intensity, electric current, X ray and nuclear radiation dosages and many other phenomena in addition to the four master measures length, mass, time, and temperature.   The assembly of a TV set, a refrigerator, or an automobile depends on dozens of such measurements.   Construction and maintenance of an airplane require a far greater number of standards if the aircraft is to operate successfully over the wide range of atmospheric and mechanical conditions it will be exposed to.

Defining basic units and standards has proven to be a difficult and continuing task.   The goal is to provide a unit that can be reproduced anywhere with high fidelity by anyone with access to certain technical facilities.   The early unit of length was a particular fraction of the earth's dimension—specifically, the one 1/10-millionth of the length of the quadrant of earth's meridian passing through Barcelona and Dunkirk.   The early unit of time was specified in terms of the relative positions of the earth and the sun, specifically 1/86,400th of the mean solar day.   But both these definitions proved inadequate; neither could provide the accuracy needed for the modern world.   For example, the redefinition of the second became necessary when it was found that the rotation of the earth is too erratic to provide accuracy much beyond one part in 10^8.   The irregularities apparently are caused by magnetic fields, earthquakes, winds, and tides.   Dependence on atomic clocks enables us to measure time to one part in 10^{12}.

Among the most difficult to establish and control are standards of measurement associated with dosages of drugs, nutrients and fluids in medical practice, and with the quality control of numerous consumer items, including water, foods, medicines, detergents, insecticides, herbicides, and gasoline.

While we pride ourselves that our food and drinking water are the safest, cleanest, and the most abundant and varied available anywhere, we recognize that this has been possible only by imposing rigid standards of sanitation and testing, enforced by a demanding public. However, improvements in the quality of food and drinking water require not only constant upgrading of sanitation and testing standards, but also developing improved purification and preservation procedures, including removing certain chemicals and adding others. It is important to realize that food and water in the amounts needed for our population could not be as safe and healthful as they are without adding chemicals and without careful control over what chemicals and how much of them are present.

Food manufacture and preparation is now largely a matter of factory production controlled by chemical compounds. "Food improvers" have been developed for virtually every property of natural food. Among these are preservatives, mold inhibitors, antisprouting agents, fumigants, germicides, antioxidants, emulsifiers and stabilizers, shortening extenders, humectants, antistaling agents, artificial flavors, flavor enhancers, tenderizers, artificial colors, firming agents, plasticizers, ion-exchange agents, vitamin and mineral enrichment chemicals, leavening agents, bleaching agents, acidifying ingredients, neutralizers, enzymes, films and waxing compounds, artificial sweeteners, plastic wrappers, treated papers, antifungicidal wrappers, lubricants, antifoaming agents, propellants, baking pan glazes, and others. Existing laws recognize the need for additives in food but require that these chemicals be pretested in advance of use to assure safety. A food chemical for use by man cannot be used in amounts in excess of 1/100th of the maximum amount demonstrated to be without harm to experimental animals.

Comparable controls have been developed for chemicals applied to crops, such as pesticides, soil conditioners, seed-treating compounds, fertilizers, fungicides, rodenticides, soil fumigants, growth and maturation regulators, preharvest defoliants, antibiotics for plant disease control, and many others. Under the law, raw agricultural commodities and manufactured foods including canned meat and fish cannot be marketed if they bear a residue of a pesticide chemical in excess of safe tolerance limits.

Water quality control is subject to similar restrictions, as we shall see in Chapter 9.

Continued public concern and vigilance over food and water quality standards are essential. Therefore, an understanding and appreciation of the concepts of chemical quantification and its measurement, such as those given in this chapter, would appear to be vital to the welfare of anyone who would participate actively and intelligently in the life of the modern world.

SUMMARY   In this chapter, the relation was established between the mass of a sample of matter and the number of atoms, molecules, or ions it contains. In addition, both the qualitative and quantitative meanings of chemical symbols, formulas, and equations were developed. The concept of the mole (6.02×10^{23} particles) was introduced as a useful unit for chemical arithmetic and in comparisons between chemical species.

Atomic weights
relative atomic weight
gram-atomic weight
gram-molecular weight
gram-formula weight

Chemical formulas
empirical formula
molecular formula
structural formula
chemical symbol

Chemical equations
net equation
over-all equation
balanced equation

Gases
Avogadro's Law
General Gas Law

Moles
gram-mole
pound-mole
ton-mole
Avogadro number

QUESTIONS AND PROBLEMS

1. Distinguish between: (a) atomic weight and gram-atomic weight, (b) mass of one atom and atomic weight, (c) mole and molecule, (d) gram-molecular weight and gram-formula weight, (e) empirical formula and molecular formula, (f) gram-mole and pound-mole, (g) over-all equation and net equation.

2. In 1811 chemists spoke of "Avogadro's hypothesis," but today they speak of "Avogadro's law." How can this change in terms be justified?

3. How many atoms are present in each of the following: (a) 9 g of aluminum, (b) 64 g of oxygen, (c) 10 g of calcium, (d) 10.7 g of silver, (e) 11 g of carbon dioxide, (f) 126 g of hydrogen nitrate (HNO_3)?

4. How many molecules are present in each of the following: (a) 8.0 g of oxygen, (b) 7.6 g of fluorine, (c) 64 g of methane (CH_4), (d) 1.52 g of carbon disulfide (CS_2), (e) 126 g of hydrogen nitrate (HNO_3)?

5. What is the weight in grams of 1.2×10^{24} atoms of each of the following: (a) helium, (b) lithium, (c) nitrogen, (d) neon, (e) iron, (f) lead?

6. How many moles of atoms and how many moles of molecules are present in 100-g samples of each of the following: (a) Cl_2, (b) I_2, (c) S_8, (d) P_4, (e) C_2H_6, (f) $SiCl_4$?

7. What is the weight in grams of (a) 0.2 mole of $KClO_3$, (b) 12 moles of sucrose ($C_{12}H_{22}O_{11}$), (c) 5 moles of Na_2HPO_4, (d) 3 moles of dichromate ions ($Cr_2O_7^{-2}$), (e) 1×10^{-5} mole of NH_4^+ ions?

8. How many sodium ions are present in each of the following: (a) 2 moles of Na_3PO_4, (b) 5.8 g of NaCl, (c) a mixture containing 14.2 g of Na_2SO_4 and 2.9 g of NaCl?

9. Using the data provided, calculate empirical formulas for the compounds indicated: (a) an oxide of nitrogen, a sample of which contains 6.35 g of nitrogen and 3.65 g of oxygen; (b) an oxide of copper, 1.0000 gram of which contains 0.7989 g of copper; (c) an oxide of carbon that contains 42.85 per cent carbon; (d) a compound of potassium, chlorine, and oxygen containing K = 31.97%, O = 39.34%; (e) a compound of hydrogen, carbon, and nitrogen containing: H = 3.70%, C = 44.44%, N = 51.85%.

10. It is found that 15.5 g of phosphorus combine with 12.0 g of oxygen to form a compound whose molecular weight is 220. What is the molecular formula of the compound?

11. Chloroform is a colorless liquid having a molecular weight of approximately 110. Analysis shows that it contains 10.05% carbon, 0.84% hydrogen, and 89.10% chlorine. Calculate the molecular formula for this compound.

12. This question refers to the equation $2NO_2 + 7H_2 \longrightarrow 2NH_3 + 4H_2O$ which, for purposes of this problem, we shall assume proceeds to completion:
    (a) How many moles of NH_3 can be produced from 1 mole of NO_2 and excess H_2? How many grams of NH_3 is this?
    (b) How many moles of NH_3 will be produced from 1 mole of H_2 and excess NO_2? What weight of water is produced?
    (c) How many moles of NH_3 and of water will be produced from 3 moles of NO_2 and 12 moles of H_2?
    (d) How many grams of NO_2 and H_2 must be taken to produce 604 g of NH_3?

13. (a) Write a balanced equation for the reaction of methanol (CH_3OH) with oxygen to give carbon dioxide and water.
    (b) Calculate the number of grams, the number of moles, and the number of molecules of carbon dioxide produced when 9.6 g of methanol is oxidized with excess oxygen.

14. When 1.5 g of a gaseous compound of carbon and hydrogen were burned, there were formed 4.4 g of carbon dioxide and 2.7 g of water. If the molecular weight of the compound of carbon and hydrogen is 30, write a balanced chemical equation for the burning process.

15. How many grams of hydrogen will be liberated when 5.0 g of zinc are added to a solution containing 60 g of H_2SO_4?

16. How many moles and how many grams of sodium carbonate are needed to prepare 220 g of carbon dioxide by the action of hydrochloric acid on sodium carbonate: $Na_2CO_3 + 2HCl \longrightarrow 2NaCl + H_2O + CO_2$?

17. Native sulfur ore containing some rock impurities is used to make sulfuric acid. The reactions are:

$$S + O_2 \longrightarrow SO_2$$

$$2SO_2 + O_2 \longrightarrow 2SO_3$$

$$SO_3 + H_2O \longrightarrow H_2SO_4$$

From 10.0 g of the ore, 29.4 g of sulfuric acid were obtained. What is the per cent of sulfur in the ore?

18. Calculate the number of particles in a pound-mole.

19. Given the following percentages, calculate the empirical formulas of the compounds:
    (a) Fe $= 53.73\%$,   S $= 46.27\%$
    (b) C $= 63.1\%$,   H $= 11.92\%$,   F $= 24.97\%$
    (c) C $= 47.37\%$,   H $= 10.59\%$,   O $= 42.04\%$
    (d) B $= 31.3\%$,   O $= 68.7\%$

20. How many moles are there in:
    (a) 66 g of propane, C_3H_8
    (b) 0.22 g of C_3H_8

(c) 3.01×10^{23} molecules of methane, CH_4

(d) the amount of NH_3 that contains 6 g of hydrogen?

21. Analysis of a sample of a compound reveals that it contains 4.50 g-atoms of oxygen and 1.80 g-atoms of phosphorus.
    (a) What per cent by weight of each element is present in the compound?
    (b) What is the weight of the sample?
    (c) Calculate the empirical formula of the compound.
    (d) Calculate the molecular formula if the molecular weight of the compound is 284.

22. The equation

$$CaCO_3(s) \longrightarrow CaO(s) + CO_2(g)$$

illustrates the thermal decomposition of calcium carbonate in which carbon dioxide is driven off. According to this reaction, how many grams of $CaCO_3$ must be decomposed to yield
    (a) 3 moles of CO_2
    (b) 4.0 g of CO_2
    (c) 1.0 mole of CaO

23. The atomic mass of zinc is 65.4 amu.
    (a) What is the mass of one atom of Zn (expressed in grams)?
    (b) How many atoms are there in 13.1 g of Zn?
    (c) How many gram-atoms are there in 13.1 g of Zn?
    (d) If Zn reacts with sulfur to form ZnS, how many grams of sulfur are needed to react with 13.1 g Zn to form ZnS?
    (e) How many grams of ZnS can be formed from 13.1 g Zn and 3.5 g S?

24. Consider the following equation:

$$NO_2(g) + H_2(g) \longrightarrow NH_3(g) + H_2O(g)$$

    (a) Is it balanced? If not, balance it.
    (b) How many grams of water would be formed by the above reaction from 9.03×10^{23} molecules of hydrogen?

SPECIAL PROBLEMS

1. (a) How many grams of carbon dioxide and water can be obtained when 200 g of heptane (C_7H_{16}) are burned in the presence of 30 moles of oxygen gas? (b) Which starting material remains unreacted? How much of it?

2. Current market prices for zinc and iron are 12 and 5 cents per pound, respectively. Which metal would be the cheaper to use in the preparation of hydrogen by the reaction $M + H_2SO_4 \longrightarrow MSO_4 + H_2$?

3. If n atoms of element A weigh 25.6 g and $3n$ atoms of element B, having an atomic weight of 28, weigh 67.2 g, what is the atomic weight of element A?

4. The burning of magnesium is represented by the following equation:

$$2Mg(s) + O_2(g) \longrightarrow 2MgO(s)$$

Suppose 2.0 g of magnesium are burned in 2.0 g of oxygen.
    (a) Which reactant is excessive?
    (b) By how much?
    (c) How many grams of the product are formed?

5. A sample of iron oxide was heated and treated with a stream of hydrogen gas, converting it completely to metallic iron. The original sample weighed 3.50 g and the resultant iron metal weighed 2.45 g. What is the empirical formula of the original compound?

6. Equal weights of magnesium and iodine react to form MgI_2 until all the iodine is used up. What percentage of the original magnesium weight is left unreacted?

7. A sample of NaCl and $CaCl_2$ combined, weighing 2.11 g, reacted with carbonate ion, CO_3^{-2}, to precipitate all the calcium as $CaCO_3$. The $CaCO_3$ precipitate was then converted to CaO by heating. If the final weight of the CaO was 0.480 g, what fraction by weight of the original mixture was $CaCl_2$?

8. A 7.00-g sample containing a mixture of $CaCO_3$ and $NaHCO_3$ is heated, causing the two compounds to decompose according to the following equations:

$$2NaHCO_3(s) \longrightarrow Na_2CO_3(s) + H_2O(g) + CO_2(g)$$

$$CaCO_3(s) \longrightarrow CaO(s) + CO_2(g)$$

If, after the decomposition is complete, there remain 4.00 g of solid residue, how much of the original sample was $NaHCO_3$?

9. Aspirin may be synthesized from benzene. A summary of the chemistry involved follows:

Reaction	yield (%)
C_6H_6(benzene) $+ Cl_2 \longrightarrow C_6H_5Cl + HCl$	80
$C_6H_5Cl + NaOH \longrightarrow C_6H_5OH + NaCl$	90
$C_6H_5OH + CO_2 \longrightarrow C_6H_4(OH)CO_2H$	70
$C_6H_4(OH)CO_2H + CH_3COCl \longrightarrow$	
$C_6H_4(O_2CCH_3)CO_2H$(aspirin) $+ HCl$	90

Considering the per cent yield in each step, if synthesis began with a half ton of benzene, how much aspirin (in pounds) would be produced?

10. The following equation

$$C_7H_{16} + 11O_2 \longrightarrow 7CO_2 + 8H_2O$$

represents the burning of a component of gasoline (heptane). According to this equation, how many pound-moles of oxygen will be needed to oxidize 1 pound-mole of heptane. What are the weights of the heptane and oxygen in pounds and in grams?

REFERENCES   AVOGADRO, A., "Essay on the Manner of Determining the Relative Masses of Elementary Molecules and the Proportions in Which They Enter into These Compounds," in W. C. Dampier and Margaret Dampier, (eds.), *Readings in the Literature of Science*, Harper, New York, 1959.

KIEFFER, W. F., *The Mole Concept in Chemistry*, Van Nostrand Reinhold Co., New York, 1962.

NASH, L., *Stoichiometry*, Addison-Wesley, Reading, Mass., 1966.

THE ELECTRON CONFIGURATION OF ATOMS

Where Are the Electrons in the Atom?

There is convincing evidence from experiments with alpha-particle bombardment that the heavy fundamental particles, the protons and the neutrons, are concentrated in a very small nucleus in the center of the atom (Chapter 2). But since the diameter of the atom is about 10,000 times that of the nucleus, we must assume that the electrons are arranged in some fashion in the relatively vast space between the nucleus and the outer sphere of influence which characterizes the volume of the atom.

The Role of Electrons in Determining the Effective Size of the Atom

Since the atoms of which all matter is made are mostly space, you may wonder why one object does not easily penetrate another, or how we can hammer a nail and not have the nail pass right through the hammer head, To explain this, we must assume that the outer surfaces of all atoms contain electrons, and the close approach of one object to another requires the electrons of one object to come near to those of the other. The approach continues until the strong repulsive force of the similarly charged electrons becomes dominant. By virtue of these electronic repulsions, an object appears to be solid (even though we now know it is very porous) when struck with another piece of matter.

Only the neutral particles are free from these electrostatic repulsive forces; neutrons can thus pass through matter in a fashion which reflects its true openness.

It is our plan to consider here some of the experimental evidence and theoretical considerations that point to the detailed arrangements of the electrons in atoms. Through a knowledge of this arrangement we shall find a valuable key to the chemical reactions of atoms, their combining capacity, and the spatial arrangement of atoms in molecules.

A THEORY FOR THE ARRANGMENT OF ELECTRONS IN ATOMS

Following his bombardment studies, Rutherford postulated that moving electrons, attracted by the nucleus of the atom, may be circling it in orbits just as the planets of the solar system move about the sun. This idea was appealing but it raised the important question of what keeps the electrons in their orbits; the theory of electrodynamics insists that moving electrons would spiral into the nucleus and the atom would collapse. In answering this question, physicists opened a Pandora's box of serious problems for classical physics. The first clues to the answer were found in studies of the light emitted from excited atoms.

Light Emission by Atoms

Early in the history of chemistry, it was observed that when elements were heated in a flame (and, later, when they were excited in an electric discharge tube) colors of light characteristic of the material used were seen. For example, discharge tubes containing neon gas at low pressure (as in a neon sign) glow red, sodium vapor emits yellow light, and mercury provides a greenish glow. Color is associated with the wave properties of light. The wavelengths for the various colors of the visible spectrum are given in Table 4.1. Examination of the light from a single element such as hydrogen in a spectroscope (Figure 4.1) shows that not one color (or wavelength) of light is emitted by hydrogen, but that several different wavelengths are present in the visible region and many more in the ultraviolet and infrared regions.

In a spectroscope, the image of the entrance slit is focused on the receiving screen, and the different wavelengths in the visible region appear as colored "spectral lines." Each element gives a unique pattern of spectral lines. Each line has a distinct wavelength. The wavelengths of the emission spectra in the visible region are given for several elements in Figure 4.2.

To understand how a study of atomic spectra led to an improved theory for the arrangement of electrons in atoms, it will be necessary to review briefly the nature of light and its origin.

Table 4.1 Wavelengths Associated with Various Colors of Visible Light

Color	Wavelength Range (Angstroms)
Red	7500 to 6100
Orange	6100 to 5900
Yellow	5900 to 5700
Green	5700 to 5000
Blue	5000 to 4500
Violet	4500 to 4000

Figure 4.1 Diagram of the principle of operation of a spectroscope. The prism separates the light into its component wavelengths.

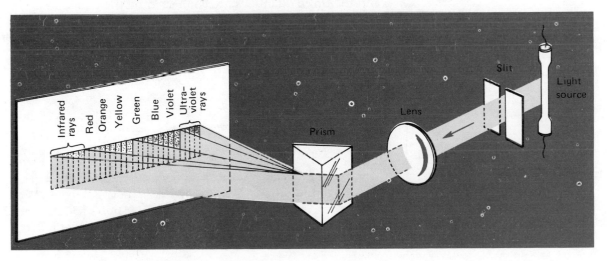

Figure 4.2 Emission of spectra of several elements (wavelengths in Angstroms).

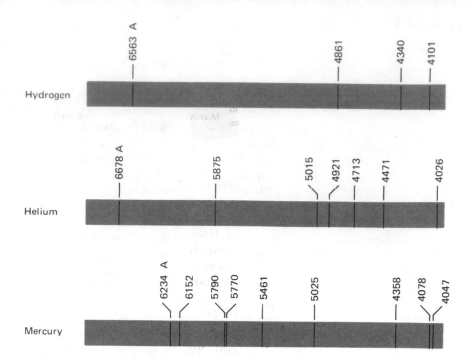

Hydrogen

6563 A 4861 4340 4101

Helium

6678 A 5875 5015 4921 4713 4471 4026

Mercury

6234 A 6152 5790 5770 5461 5025 4358 4078 4047

The Nature of Light   We know that light is one form of energy; we can feel the warming effect of the sun's rays on our bodies; we can concentrate the rays of the sun with a lens and heat a piece of paper until it catches fire. Light energy is a form of electromagnetic radiation similar to radio waves, gamma rays, etc. All of these forms of radiation exhibit certain properties, such as refraction, diffraction, interference, that may be explained in terms of a wavelike character of the radiation. In terms of the wave theory of light, light energy is propagated through space by means of wave motion of electric and magnetic fields, as represented diagrammatically in Figure 4.3. The distance between repeating units on the wave is called the wavelength of the radiation, symbolized by λ (lambda); the number of times per second that the amplitude at some fixed point completes a cycle from maximum to zero to negative maximum to zero to maximum again is the frequency, ν (nu). All types of electromagnetic radiation in free space travel with the same velocity $c = 3 \times 10^{10}$ cm/sec, and the three quantities are connected by the equation $c = \lambda\nu$.

Figure 4.3 Light is propagated through space by waves, which have both an electric and a magnetic component.

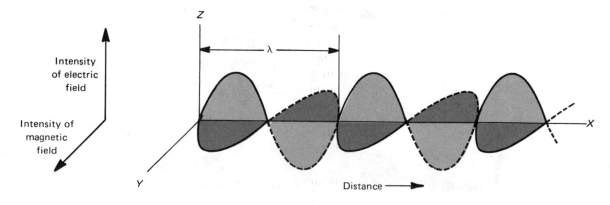

Z

λ

Intensity of electric field

Intensity of magnetic field

Y

X

Distance

Figure 4.4 Range of electromagnetic radiations.

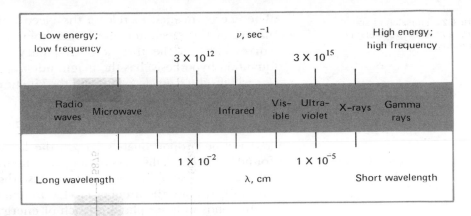

Figure 4.4 illustrates the range of electromagnetic radiations from the very long wavelength (low-frequency) radio waves to the very short wavelength (high-frequency) gamma rays. In the visible region, red light has the longest wavelength and violet light the shortest.

Light Comes in Bundles

Quantum Theory. Although the wave theory is useful in the description of many of the properties of light, it does not explain certain observations involving light, such as the variation of intensity with wavelength in the radiation emitted by hot objects (black-body radiation) and the ejection of electrons when light strikes metal surfaces (photoelectric effect).

Hot objects emit radiation, and the wavelength of the predominant radiation decreases as the temperature increases. Thus we see a piece of iron in a forge change from the "red hot" of longer wavelengths to "white hot" as its temperature becomes higher and the intensity grows in the shorter wavelength radiation. This change in the distribution of intensity among different wavelengths as the temperature of the emitting surface changes could not be properly explained by physicists and remained a puzzle until Max Planck, in 1901, postulated that light carried its energy in definite amounts or "quanta." He assumed that the energy of a light wave had to be an integral number of quanta, where the energy of each quantum is given by the equation

$$E = h\nu$$

where E is the energy of one quantum, ν is the frequency of the light wave, h, known as Planck's constant, has a value of 6.62×10^{-27} erg sec. When this value is used for h, the energy of the quantum will be in ergs.*

Studies of the photoelectric effect show that electrons are not ejected from a metal unless the frequency of the light used is greater than a threshold value ν_0 characteristic of the metal, and the kinetic energy of the electrons ejected at frequencies above the threshold value is proportional to the frequency. Albert Einstein explained these observations in 1905 by applying Planck's quantum theory to the phenomenon, applying the term photon to the quantum in this connection. If light comes in quanta, then the absorption of a quantum and the energy of the ejected electron must obey the law of conservation of energy

$$\begin{bmatrix} \text{Energy of incident} \\ \text{quantum, } h\nu \end{bmatrix} = \begin{bmatrix} \text{Energy needed to release} \\ \text{electron from metal} \end{bmatrix} + \begin{bmatrix} \text{Kinetic energy of} \\ \text{ejected electron} \end{bmatrix}$$

---

*An erg is the energy equivalent to 10^{-7} joule (Appendix A).

The energy needed to release the electron from the metal is fixed by the nature of the metal and the forces holding the electron in the surface. It corresponds to the threshold frequency required and, according to the quantum hypothesis, has the magnitude $h\nu_0$. Thus, light of energy less than $h\nu_0$ will produce no electrons, and the kinetic energy of those elections ejected for $\nu > \nu_0$ will be given by the equation

$$KE = h\nu - h\nu_0$$

and will be proportionately larger, the larger the value of frequency ν, as found experimentally. Increasing the intensity of light of fixed frequency (above the threshold ν_0) does not increase the energy of an ejected electron, but does increase the number ejected in unit time, since the greater intensity corresponds to more photons (each of energy $h\nu$) striking a unit area of surface in unit time. This experimental fact fits the further assumption of Einstein that a single molecule or atom absorbs or emits only a single quantum of light at one time.

The success of the quantum theory in explaining two important but previously puzzling phenomena of physics established it as a useful and powerful tool.

Since the frequency of the light wave is related to its wavelength by the equation $\nu = c/\lambda$, where c is the velocity of light and λ is the wavelength, the energy of a quantum of light is related to the wavelength by the equation

$$E = h\frac{c}{\lambda}$$

This equation indicates that the shorter the wavelength of light, the greater the energy of its quanta. This means, for example, that a quantum of red light carries less energy than a quantum of green light.

In accepting the quantum theory, scientists were forced to regard light as having both particle and wave properties, and photons were said to be "particle-waves." Although this view of light at first seemed cumbersome, it successfully explained a great variety of experimental facts which defied explanation in terms of the older wave theory of light. One group of these experimental facts concerns the relationship between the wavelengths of the spectral lines and the internal structure of atoms.

The Spectrum of the Hydrogen Atom

The visible emission spectrum of hydrogen atoms consists of a characteristic group of lines spaced between the violet and red regions of the visible spectrum (Figures 4.2 and 4.5). In terms of the quantum theory, each of these lines is the result of millions of photons of a single wavelength striking the eye or the photographic plate of the spectrometer. Moreover, since an atom may emit only one photon at a time, it is evident that many atoms are giving off identical photons simultaneously as the spectrum is being observed. From this it follows that a hydrogen atom may lose only discrete quantities of energy—specifically, those quantities corresponding to the energies of the photons in its spectral lines (and no other energy quantities). The reasonable explanation for this observation is that the internal energy content of atoms changes by definite steps, and not continuously. The successive internal energies which an atom may possess are called its *energy levels*. Emission or absorption of a photon occurs when an electron moves from one energy level to another. The energy of the photon is the difference between the energies

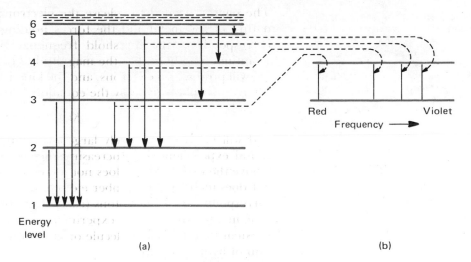

Figure 4.5 Origin of the visible spectrum of hydrogen.

of the electron in the energy levels involved in the transition; the wavelength of the photon is related to its energy by the Planck equation,

$$E_{\text{photon}} = E_{\text{level }X} - E_{\text{level }Y} = h\nu = h\frac{c}{\lambda}$$

The spectrum of hydrogen then results from electron transitions between energy levels, as shown in Figure 4.5. In this figure, the horizontal lines represent the internal energy of the hydrogen atom when the electron is in the successive levels, and the arrows (a) represent the transitions between energy levels resulting in the liberation of photons which appear as lines in the spectrum (b).

The suggestion of transitions between energy levels within atoms as the source of spectral lines was a great contribution made by Niels Bohr in 1913. Bohr went further and showed the connection between Rutherford's planetary model of the atom and the energy level concept. Considering the simplest atom, hydrogen, he suggested that

1. The electron can exist only in certain orbits.
2. The energy levels of the atom arise because the electron possesses a fixed but different energy in each of the successive orbits.
3. Photons are emitted or absorbed when an electron moves from one orbit to another.

The Rutherford-Bohr theory predicted satisfactorily the energy levels, and thus the wavelengths of the spectral lines for the hydrogen atom—not only those observed previously but many which were to be discovered later. It

---

BOHR'S REASONING   Bohr postulated that the radius of an orbit was fixed by the requirement that the force of attraction between a nucleus of charge Ze and an electron of charge $-e$, as given by Coulomb's law,

$$F = \frac{-(Ze)e}{r^2} \qquad (1)$$

may be alternatively written as the centripetal force of the electron moving in the orbit

$$F = -\frac{mv^2}{r} \qquad (2)$$

In these equations, r is the distance between the two charges and is therefore the radius of the

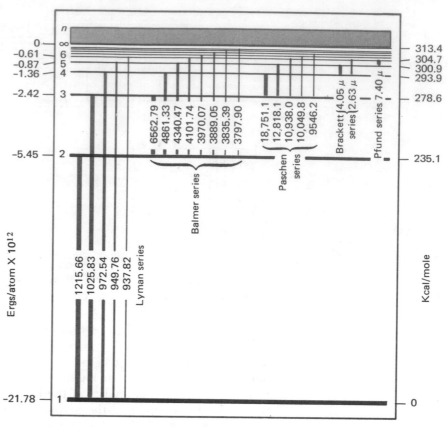

Figure 4.6(b) Energy levels in the hydrogen atom assigned on the basis of the emission spectrum of hydrogen. Wavelengths of spectral lines are given in Angstroms unless otherwise noted. The energy scale at the left gives the energy in ergs/atom, negative from zero for complete separation of the nucleus and the electron ($n = \infty$). The scale at right gives the energy in kcal/mole of atoms above the ground level with ($n = 1$) taken as zero.

orbit, m is the mass of the electron, and v its velocity; both forces are negative because they act to decrease r. Bohr's postulate equates these two forces

$$-\frac{Ze^2}{r^2} = -\frac{mv^2}{r} \tag{3}$$

We know that the angular momentum mvr of a system has to be constant, and Bohr required the constant value of the angular momentum to be equal to $h/2\pi$ for the first orbit, $2h/2\pi$ for the second orbit, $3h/3\pi$ for the third, and so on. Generalized, this becomes:

$$mvr = n\frac{h}{2\pi} \qquad n = 1, 2, 3, \ldots \tag{4}$$

Solving the preceding equation for v, substituting for v^2, and canceling terms

$$Ze^2 = \frac{n^2 h^2}{4\pi^2 mr} \tag{5}$$

or

$$r = \frac{n^2 h^2}{4\pi^2 mZe^2} \tag{6}$$

For hydrogen, where $Z = 1$, the orbits have radii corresponding to

$$\frac{n^2}{4\pi^2 me^2}, \quad \frac{4h^2}{4\pi^2 me^2}, \quad \frac{9h^2}{4\pi^2 me^2}, \quad \text{etc., for } n = 1, 2, 3, \ldots$$

The total energy for the nth orbit is given by the sum of the kinetic energy $\frac{1}{2}mv^2$ and the potential energy $-Ze^2/r$. The potential energy is negative because the zero of potential energy is taken for infinite separation difference, and the energy stored in the system at distance r acquires a negative sign (see Figure 6.1). Thus

represented an important step in the development of the theory of atomic structure.

Subshells

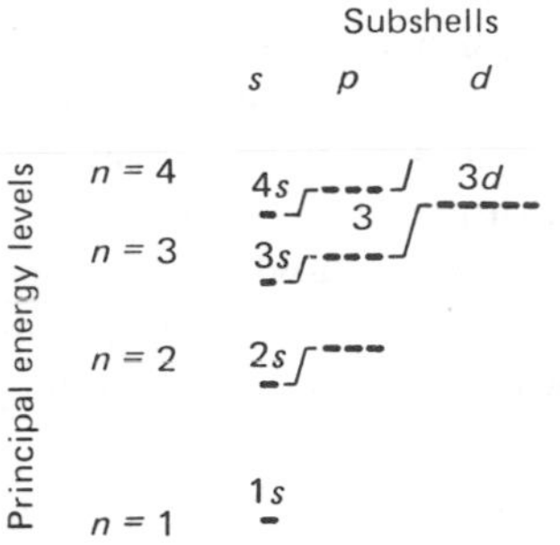

Advocates of the Rutherford-Bohr theory encountered a series of failures when they used it to predict the spectra of elements other than hydrogen. Moreover, as research in spectra continued, additional spectral lines were observed. The spectra of sodium and potassium atoms, for example, though somewhat similar to the spectrum of hydrogen, require for their interpretation the introduction of more energy levels. These were fitted into the model by assuming that the main orbits or principal energy levels, n, are composed of a series of *subshells*, which are given the symbols s, p, d, and f, as shown in Figure 4.7. *The first principal energy level (shell) contains only one subshell, an s level; the second principal energy level contains two subshells, called s and p levels; the third shell contains three subshells, an s, p, and d, and so on.* But even this modification of the Rutherford-Bohr model could not account for all the observations of the spectroscopists.

Figure 4.7 Subshell energy levels for elements 1 through 20.

Orbitals

When an atom is excited in a magnetic field, some of the spectral lines are found to separate or split and appear as several lines. To explain such line-splitting, it was assumed that a subshell, under the influence of the external field, may separate into more than one energy level. The different mathematical descriptions of the atom in these energy levels are called orbitals.

In the absence of interaction with other atoms or with external electrical or magnetic fields, all of the orbitals of a given group have the same energy. Interaction with or application of a field causes the orbitals to appear in several new energy levels; the appearance of these new levels permits new energy transitions and hence additional lines in the spectrum. Studies of

$$E = \frac{1}{2}mv^2 - \frac{Ze^2}{r} \tag{7}$$

But by equation (3),

$$\frac{1}{2}mv^2 = \frac{1}{2}\frac{Ze^2}{r} \tag{8}$$

and

$$E = \frac{1}{2}\frac{Ze^2}{r} - \frac{Ze^2}{r} = -\frac{1}{2}\frac{Ze^2}{r} \tag{9}$$

Substituting for r from (6) and setting $Z = 1$, we obtain the energy levels for the hydrogen atom:

$$E = -\frac{1}{2}\frac{e^2}{\dfrac{n^2h^2}{4\pi^2me^2}} = -\frac{2\pi^2me^4}{h^2} \times \frac{1}{n^2}, \; n = 1, 2, 3, \ldots$$

The energy will be given in ergs per atom if m is in grams and e is in electrostatic units. It is evident that the limit of E is zero for n very large, and that E becomes negatively larger as n becomes smaller, as seen on the left-hand scale of Figure 4.6. If the energy zero is shifted to assign zero energy for $n = 1$, then positive values of energy appear for larger values of n, as given by the right-hand scale. Here, the energy is given as kilocalories per mole of atoms.

From its relationship to the quantum of energy emitted when an electron jumps from one orbit to another, the integer n, which appears in the equation above, is known as a *quantum number*.

this line-splitting led to the conclusion: *The s subshell has just one orbital; the p subshell has three orbitals; the d subshell has five orbitals; and the f subshell has seven.*

Spinning Electrons

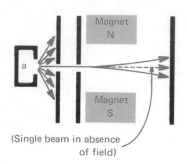

(Single beam in absence of field)

Figure 4.8 Diagram illustrating the Stern-Gerlach experiment. Vaporized metal atoms such as those of silver or sodium emerge from a heated oven (*A*) and move toward a screen with a hole in it. A beam of atoms moves through the hole and into the magnetic field where the beam is split into two equivalent beams. Physicists interpreted this to mean that the unpaired electron in each of these atoms behaves just like a spinning top that can spin in either of two directions, corresponding, perhaps, to a clockwise and a counterclockwise spin.

Very careful measurements of the spectra of certain atoms with uneven numbers of electrons, such as sodium and potassium, showed that some of the lines were doublets—that is, two lines of nearly, but not exactly, equal wavelengths. To explain the existence of these doublets, it was assumed that electrons behave like spinning tops, some spinning in one direction and others in the opposite direction. The spinning electrons would act like small magnets, and the interaction of these magnets with the electrons moving in orbits would produce the effect of additional energy levels. This assumption was verified by experimental evidence from a beam of sodium atoms in a magnetic field. Using an apparatus such as that described in Figure 4.8, Otto Stern and W. Gerlach (1922) observed that the beam was divided by the magnetic field, at high temperatures of the oven (A), into two beams, indicating that there are two, and only two, different kinds of magnets in the beam. This was interpreted to mean that five of the electrons in the sodium atom (atomic number 11), spinning in one direction, just canceled the magnetic affects of five others spinning in the opposite direction, but that the 11th electron could—and did—spin in either direction (in half the atoms in one direction and in half in the other direction). The applied magnetic field sorts the atoms into those two different kinds.

The 10 electrons of the sodium which "neutralize" each others' spins are said to be "paired" or to have "paired spins," while the 11th is said to be unpaired. The presence of one or more unpaired electrons in an atom always gives the atom magnetic properties and, as we shall see, has important consequences for the chemical behavior of the atom.

Quantum Numbers to Describe the Energy of Electrons in Atoms

The atomic model derived to fit the above observations is founded on the idea that the internal energy content of atoms changes only in steps. As indicated earlier, the successive internal energies which an atom may possess are called its energy levels. Each energy level is described by a set of numbers called quantum numbers. These are:

Table 4.2 Development of Electronic Energy Level Patterns for Atoms Showing How These Patterns Developed as a Result of Spectroscopic Studies, Starting with the Original Bohr Theory

Original Bohr Theory	*Modification I*	*Modification II*	*Modification III*
Spectrum of hydrogen atom explained by principal energy levels designated by integral values of n.	Additional lines in spectra of hydrogen-like atoms require introduction of sub-shell energy levels designated s, p, d, and f.	Effect of magnetic fields on spectra requires introduction of orbital energy levels represented by the magnetic quantum number, m.	Electron spin requires two additional quantum numbers, $+\frac{1}{2}$, $-\frac{1}{2}$ or ↑ and ↓; two spinning electrons may occupy each orbital.
$n = 3$	$3d$ $3p$ $3s$	Five d orbitals Three p orbitals One s orbital	⇅ ⇅ ⇅ ⇅ ⇅ $3d$ ⇅ ⇅ ⇅ $3p$ ⇅ $3s$
$n = 2$	$2p$ $2s$	Three p orbitals One s orbital	⇅ ⇅ ⇅ $2p$ ⇅ $2s$
$n = 1$	$1s$	One s orbital	⇅ $1s$

Table 4.3 Relationships Among Quantum Numbers

n	1	2				3								
l	0 (s)	0 (s)	1 (p)			0 (s)	1 (p)			2 (d)				
m	0	0	-1	0	$+1$	0	-1	0	$+1$	-2	-1	0	$+1$	$+2$
Symbol for orbital*	$1s$	$2s$	$\binom{2p_x}{2p_y}$	p_z	$\binom{2p_x}{2p_y}$	$3s$	$\binom{3p_x}{3p_y}$	$3p_z$	$\binom{3p_x}{3p_y}$	$\binom{3d_{xy}}{3d_{x^2-y^2}}$	$\binom{3d_{xz}}{3d_{yz}}$	$3d_{z^2}$	$\binom{3d_{xz}}{3d_{yz}}$	$\binom{3d_{xy}}{3d_{x^2-y^2}}$

*For a number of reasons it is not possible to associate a specific magnetic quantum number with a particular p or d orbital. As indicated in the table, the -1 quantum number, for example, may be associated with either the $2p_x$ or $2p_y$ orbitals, or with the $3d_{xz}$ or $3d_{yz}$ orbitals. The orbitals np_x and np_y do not correspond directly to the wave functions $\psi_{n,1,+1}$ and $\psi_{n,1,-1}$ but are formed, for mathematical convenience, from the sum and differences of those two. Thus $np_x = 1/\sqrt{2}\,(\psi_{n,1,+1} + \psi_{n,1,-1})$ and $np_y = -\sqrt{-1}/\sqrt{2}\,(\psi_{n,1,+1} - \psi_{n,1,-1})$. Similarly, four of the d orbitals are formed from the pairs $\psi_{n,2,\pm1}$ and $\psi_{n,2,\pm2}$. These mathematical manipulations still produce only three p orbitals and five d orbitals, corresponding to the number of allowed values of m given in the third row of the table.

1. The principal (or shell) energy levels, represented by the principal quantum number n.
2. Within each shell there is a series of subshell energy levels known as s, p, d, or f and represented by the subshell quantum number l.
3. Within each subshell there is a series of orbital energy levels represented by the magnetic quantum number m.
4. Within each orbital the electron may have either of two possible spin orientations represented by the spin quantum number s.

The particular energy level occupied by each electron in an atom can be specified in terms of these four quantum numbers. The four quantum numbers identify the energy level of the electron just as the postal zip code identifies geographical locations. The relations among quantum numbers are given in Table 4.3.

THE WAVE NATURE OF ELECTRONS (THE ELECTRON CLOUD MODEL OF THE ATOM)

Wave Nature of Electrons

The postulates of the quantum theory endowed light, in its description as photons, with a particle nature, although most of the phenomena dealing with light previously had been described in terms of waves. It was natural that some scientists should consider the converse and suggest that electrons, which since Thomson's discovery had been presumed to be particles, might be described in terms of waves. This was first seriously proposed in 1924 by Louis de Broglie, who suggested that with each particle there must be associated a wave of wavelength given by $\lambda = h/mv$, where m is the mass, v the

---

USE OF MODELS

One of the main contributions of Bohr was to devise a model for the atom and demonstrate by the aid of the model that he could account for spectral lines in terms of energy levels within the atom.

However, models, like analogies, are scarcely ever perfect. Even today the chemist is uncertain what the atom looks like, and he does not yet have an atomic model that satisfies him. But he is thoroughly convinced that there are various energy levels in atoms and he uses his imperfect models to explain the appearance of the variety of spectral lines that result from electron transitions between various energy states. To do this he uses two quite different models and different types of calculations. We have called these models the *modified Bohr model* and the *electron-cloud model*.

A startling result of work with these two models is that they give almost identical answers in the calculation of the energy levels of atoms in spite of their radically different nature and the entirely different type of mathematics used to make the calculations.

velocity of the particle, and h Planck's constant. Experimental proof of the wave nature of electrons came in 1926, when C. J. Davisson and L. H. Germer showed that a beam of electrons striking a nickel crystal gave a diffraction pattern, just as if a beam of X rays had been allowed to strike the crystal (Chapter 13), and quite different from the pattern to be expected when a beam of particles strikes a target.

Particle-Wave Probability

This discovery put the electron and the photon in the same class, for both behave as particles and waves. Moreover, from their studies of photons, physicists had learned two important facts:

1. The particle-waves of small mass cannot both be located and their velocity determined precisely at the same time, as we are accustomed to do with larger bodies. For example, we can know both the position and the speed of a ship at sea, a football in flight, or a satellite. However, Werner Heisenberg concluded that it is impossible for us to know both the precise position and the exact velocity of a photon or an electron at the same time. If we know the velocity of an electron quite accurately, we shall then know its energy accurately, but we cannot know its position precisely. Since energy levels are especially important in our view of the atom, we shall seek to determine the energy of the electron in the atom, but in so doing we cannot know the precise position of the electron.

2. When the position of a particle-wave cannot be determined precisely, the mathematics of wave behavior can be used to describe the probability of the particle-wave being in a certain place.

For these reasons, modern descriptions of the atom include precise statements about the energies of electrons and refer always to the probability of the electron's being in a certain region (Figure 4.9).

The Wave Function

The probability or chance of finding a particle-wave (electron) in a certain region in space is related to the intensity of the wave in that region times the size of the region. If we could calculate the intensity (charge density) of the electron wave in the neighborhood of a nucleus, we could outline the region in which the electron is most likely to be found.

The behavior of a wave can be described in mathematical terms by the wave function denoted by the symbol ψ; the intensity of the electron wave is related to the square of the wave function, ψ^2. The square of the function, ψ^2, at any point is related to the probability of finding the electron at that point. The magnitude of ψ can be calculated for electrons in the neighbor-

---

THE MEANING OF
THE WAVE FUNCTION

Because of its particle-wave nature, an electron in the field of an atomic nucleus may be thought of as a wave. One way to describe wave behavior is in terms of the disturbance the wave creates at various points in space. In the case of a water wave, for example, this disturbance is measured by the height of the wave; for electromagnetic radiation, it is the strength of the electrical or magnetic field associated with the wave.

Since the disturbance changes from point to point along the path of propagation of the wave, it often is convenient to express the over-all picture of this disturbance in mathematical terms. The mathematical quantity called the wave function ψ is the mathematical description of the wave in terms of the disturbance it creates at different points in space.

The mathematical form of the wave function for an electron in an atom is obtained by solving the Schrödinger wave equation. But realistic solutions to the wave equation are possible only

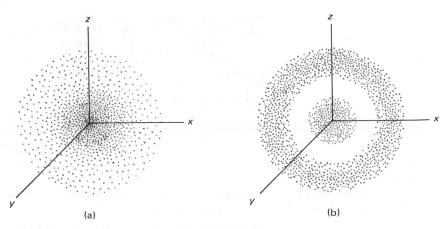

Figure 4.9 Probability representations of electrons in atoms. Both plots are for *s* electrons. For these, the probability of finding the electron at a particular point depends only on the *distance* from the nucleus and not on the *direction* ("north," "east," "south," "west") from the nucleus; hence the regions of high probability are spherical in shape. The drawings represent *sections*, that is, slices through the centers of these spheres. Imagine the page to be cross-ruled into tiny squares, and each dot to represent some constant fraction, perhaps 1 ten-millionth, of the total probability of finding the electron *somewhere* on the page. If the probability of finding the electron in a particular square is 5 ten-millionths, there will be five dots in that square; if the probability is 1 ten-millionth, there will be one dot in that square; if it is $\frac{1}{5}$ of one ten-millionth, there will be one dot for five squares, and so on. The electron thus has a high probability of being found where the dots are close together, and a lesser chance of being found where the dots are scattered. The 1s electron in hydrogen is represented in (a). This electron can be found anywhere from near the nucleus to a great distance away, but it is most likely to be found about 0.5×10^{-8} cm from the nucleus. This is because although the dots are closest together near the nucleus, the number of dots on a circular ring 1 square wide is greater a short distance from the nucleus than the number in a similar ring of smaller radius. A section for a *2s* electron in hydrogen, drawn on a much smaller scale than for (a), is represented in (b). Here, there is a region about 1×10^{-8} cm from the nucleus where the probability of finding the electron is zero and no dots appear there.

hood of nuclei and other electrons by solving a complex mathematical equation developed by Erwin Schrödinger in 1925 (see Appendix D). To obtain a valid solution, the potential energy of interaction between nuclei and electrons and between electrons and electrons must be known as a function of the positions of the particles. It is found that a solution to the Schrödinger

if the electron (and the atom) has certain energy values—that is, the system may exist only in certain energy states.

The solutions found give ψ in terms of several variables, one of which is the distance from the nucleus r, but the mathematical dependence on the variables is different for each energy state. Once these solutions have been found, it is possible to calculate ψ for each energy state at any point outside the nucleus. This value will be the amplitude of the wave at that point.

In studies of the wave nature of electromagnetic radiation, the intensity of the radiation often is more important and useful than the amplitude of the wave. In light, intensity is related to the square of the amplitude. It is this which is related to the probability of finding a photon at a given point. Similarly, with the electron, ψ^2 is considered a measure of the intensity (or charge density) of the electron wave. It measures the probability of finding the electron in an infinitesimally small region of space situated at the point for which ψ^2 has been evaluated.

equation is possible, under conditions which correspond to physical reality, only for particular values of the total energy of the system of particles. The values of the energy for which a solution is possible depend, in turn, upon the values of three integers, which again are known as quantum numbers—in fact, the same quantum numbers as used in the modified Bohr theory.

In effect, then, the mathematical treatment of the electron as a charged particle-wave in the field of the nucleus produces a model of the atom as a system of energy levels. To characterize the energy state (or level) of any electron, it is necessary to specify its magnetic quantum number m, subshell quantum number l, and principal quantum number n.

Electron Spin;
The Pauli Exclusion Principle

The interpretation of spectra on the basis of the Bohr model of the atom required a fourth quantum number (called the spin quantum number) to explain spectra more completely. In the wave theory, the concept of electron spin must also be introduced, and finds some theoretical confirmation when the Schrödinger equation is solved, taking into account the demands of relativity theory. Thus the existence of four quantum numbers receives a reasonable explanation.

Careful study has shown that *no two electrons in an atom can have the same four quantum numbers*. This statement, first presented in 1925, is known as the Pauli exclusion principle

All of this adds up to an overwhelming confirmation of the atom's energy-level pattern developed by spectroscopists. In addition, it provides a reliable mathematical tool for extending the concept of energy levels to areas of chemistry and physics widely removed from spectroscopy.

Relationships Among
the Quantum Numbers

The three quantum numbers obtained from solution of Schrödinger's equation are found to be related to each other, and they correspond directly to those discovered by the spectroscopists. The principal quantum number, n, can have any integral value other than zero. For any chosen value of n, a subshell quantum number, l, sometimes called the azimuthal quantum number, can have integral values from 0 to a maximum of $n - 1$. For any chosen values of n and l, the magnetic quantum number, m, can have positive or negative integral values from $-l$ to $+l$, including zero. Each combination of the three quantum numbers specifies an energy level, so that we may say that an electron is in the energy level characterized by the values $n = 2$, $l = 1$, $m = +1$, for example, or by the values $n = 2$, $l = 0$, $m = 0$, and so on. The relationships among the quantum numbers for $n = 1$, 2, and 3 are shown in the first three rows of Table 4.3.

Electron Clouds: Regions
of Highest Probability

Let us now look at some energy levels in an atom and discover, if we can, the regions in space where electrons in each of these states spend most of their time. Each wave function obtained from the solution to the Schrödinger equation and designated by integral values of the three quantum numbers n, l, and m has associated with it the value of the total energy specified for that solution. This function, $\psi_{n,l,m}$, is often called an orbital. The value of ψ can be calculated for any point in space, and the square of this value specifies the probability per unit volume of finding an electron (having the energy characteristic of that orbital) at that point. Thus ψ^2 for a point P, multiplied by a volume element dV that includes that point, measures the probability of finding the electron within that volume element.

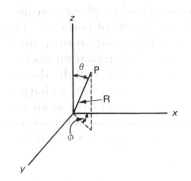

Figure 4.10 The relationship of the spherical coordinates R, θ, ϕ, to the Cartesian coordinates x, y, z.

The complete orbital, $\psi_{n,l,m}$, can be written as the product of two functions, $\psi(R)$ and $\psi(\theta, \phi)$. The *radial part* of the wave function, $\psi(R)$, depends only upon the distance R from the nucleus. The *angular part* of the wave function, $\psi(\theta, \phi)$ depends upon the angles θ and ϕ (corresponding to latitude and longitude); its value thus depends upon the *direction* of the point P from the nucleus (Figure 4.11). The angular wave function is of particular value in discussing directional effects in chemical bonding, and pictorial representations of the probability of finding an electron in a volume element located in a particular direction from the nucleus are frequently used. These representations commonly are boundary surfaces drawn for that angular region of space which encloses a particular fraction—for example, 90%—of the charge distribution, as worked out from the locus of probability points. Such boundary surfaces are often referred to as *orbital shapes.**

Figure 4.12 is a sketch of some orbital shapes. It is important to note that the shapes of all *s* orbitals (constructed from angular wave functions with the *l* quantum number 0) are spherical, and that all *p* orbitals (from angular wave functions with $l = 1$) produce egg-shaped lobes, and that lobes from *p* orbitals of the same principal quantum number are mutually perpendicular to each other.

Since these diagrams represent the distribution of electron charge, the distributions which they represent are often spoken of as the *electron clouds*. The term comes from the fact that we imagine the charge on the electron to be smeared out to cover the regions indicated, but with fuzzy and indistinct edges, like a cloud. Each cloud or orbital can be occupied by one or two electrons; if by two, they must have opposite spins, according to the Pauli principle. Thus the 14 orbitals listed in the last row of Table 4.2 can accommodate 28 electrons.

---

*The term orbital shape is also used for the surface for which $\psi(\theta, \phi)$, as contrasted to $[\psi(\theta, \phi)]^2$, has a fixed value.

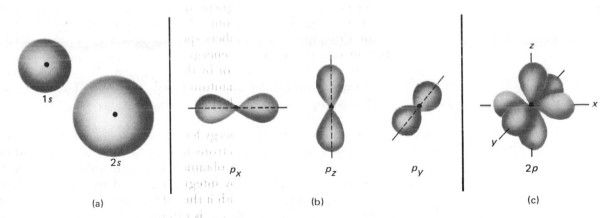

(a)

(b)

(c)

Figure 4.11 Boundary surfaces for the angular portion of various orbitals. (a) The orbital shapes for *s* orbitals are spherically symmetrical. (b) The shapes for *p* orbitals are distorted ellipsoids with two lobes that join at the nucleus; for p_x orbitals, these extend along the *x* axis, for p_y orbitals along the *y* axis, and for p_z orbitals along the *z* axis of an arbitrarily chosen system of Cartesian coordinates. (c) The shapes of the three perpendicular 2*p* orbitals as they might be imagined to appear in the atom.

Figure 4.12 Order of increasing energy in the ground states of light elements. The orbitals corresponding to the same values of *l* are included together since, in general, these have the same energy in the absence of an electrostatic or magnetic field.

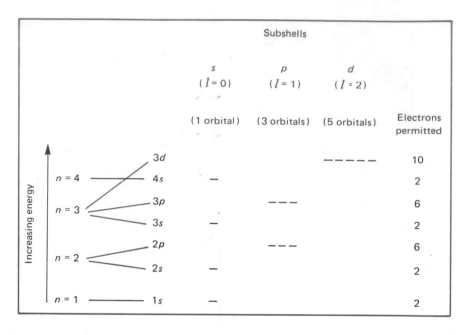

ELECTRON CONFIGURATION OF ATOMS

With the information now available, a model of the electron arrangement in the atoms of the elements can be described.

Levels of Lowest Energy Fill First

When an atom is not heated in a flame or excited in a discharge tube, it adopts the state of lowest energy, or the "ground state." In such a state the electrons of the atom are in the lowest energy levels allowed to them by the Pauli exclusion principle. For atoms with more than two electrons, however, it is not possible to put all the electrons into the energy state corresponding to $n = 1$, since this is forbidden by the Pauli principle. Therefore, the normal state of the lithium atom (three electrons) will have two electrons in the state $n = 1$, and one in the state $n = 2$. If we examine each element in turn, allowing the nuclear charge to increase by one unit from one element to the next, we can determine the arrangement of electrons in the energy levels by assigning them always to the lowest levels permitted by the exclusion principle. Examination of the spectra of successive elements permits us to identify the energy levels of each experimentally. The order of increasing energy as the quantum number of the energy level changes is given for the elements of low atomic weight in Figure 4.13. This order is different for elements of higher atomic number as shown in Figure 4.14.

The right-hand column in Figure 4.13 gives the total number of electrons possible in each designated subshell energy level (l), counted by allowing integral m values from $-l$ to $+l$ and values of the spin quantum number of $+\frac{1}{2}$ and $-\frac{1}{2}$ for each m value. The symbols (s, p, d, f) give the conventional designation as energy levels, listing the principal quantum number $n = 1$, 2, 3, . . ., and the letter designation for $l = 0, 1, 2, . . .$ (the s, p, d, f subshells).

Atoms of Elements from Hydrogen to Calcium

Figure 4.13 enables us, starting with the lightest element, hydrogen, to write down the electron configuration of the ground state for the atoms of each successive element. Hydrogen will have its single electron in the $1s$ level. There can be two electrons in the $1s$ level (spin $= +\frac{1}{2}, -\frac{1}{2}$), and helium adopts this configuration ($1s^2$). Lithium can have two of its three electrons in the $1s$

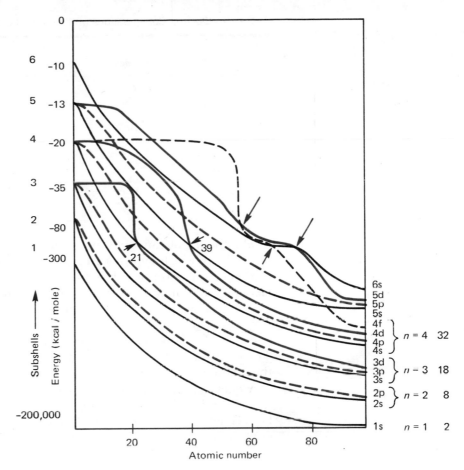

Figure 4.13 A diagrammatic sketch showing how the relative values of the energy levels of atoms change with atomic number. There is a general decrease in the energy corresponding to a particular value of the principal quantum number n as the nuclear charge increases and exercises a greater attraction for the electrons. There is also a shift in the relative values of the d and f orbitals as indicated, for example at element 21 where the $3d$ level drops below the $4s$ level. Similar changes occur in the other transition series. (Note that the energy scale at the left is not linear but grossly distorted toward large values at the bottom.)

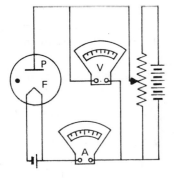

Figure 4.14 Diagram of apparatus for determining ionization potentials. Electrons from the filament F are accelerated toward the positively charged plate P and, in transit, some of them collide with the atoms of the gas in the tube. If the electrons have sufficient energy, they will ionize the gaseous atoms by knocking off electrons. By increasing the voltage difference between the plate and the filament, the energy of the bombarding electrons may be increased at will so any gas can be ionized. Eventually all electrons find their way to the plate and contribute to the current passing through the ammeter (A). As voltage (V) is increased systematically, there appear sudden increases in current as the bombarding electrons acquire sufficient energy to ionize the gas, thereby producing many more electrons. The voltage at these points is an ionization potential.

Table 4.4 Electron Configurations of the First 10 Elements

Element	Atomic Number	Configuration
H	1	$1s^1$
He	2	$1s^2$
Li	3	$1s^2\,2s^1$
Be	4	$1s^2\,2s^2$
B	5	$1s^2\,2s^2\,2p^1$
C	6	$1s^2\,2s^2\,2p^2$
N	7	$1s^2\,2s^2\,2p^3$
O	8	$1s^2\,2s^2\,2p^4$
F	9	$1s^2\,2s^2\,2p^5$
Ne	10	$1s^2\,2s^2\,2p^6$

level without violating the Pauli principle, but the third must go into the next lowest level, which Figure 4.13 shows to be $2s$. Again a total of two electrons can enter this level, which is filled at beryllium (four electrons). The fifth electron of boron must now go to the next lowest level, the $2p$ level. Since for this level m may have values -1, 0, $+1$ and the spin may have values $+\frac{1}{2}$ and $-\frac{1}{2}$, a total of six electrons may appear here, and as additional electrons are added in the atoms from carbon (6) to neon (10), these six places are filled. The electron configurations in the ground states of the first 10 elements are thus those given in Table 4.4.

A question arises as to whether the two $2p$ electrons of carbon are in the same p orbital, with $s = +\frac{1}{2}$ and $s = -\frac{1}{2}$, or in different p orbitals, with s for both equal to $+\frac{1}{2}$ (or both $-\frac{1}{2}$). Two electrons can get farther away from each other if one is in each of two orbitals than if they are both in the same orbital. Since electrons repel each other, the former is the state of lower energy, and the atom adopts the former configuration, with the two spin quantum numbers the same. Similarly, the three $2p$ electrons in nitrogen occupy three different p orbitals and have the same spin quantum numbers. In oxygen, however, the fourth p electron must pair with an electron of opposite spin quantum number in one of the p orbitals. A detailed description of the electron configurations of the elements carbon through neon might then be given as in Table 4.5, where the p orbitals along the x and y axes (p_x and p_y) have arbitrarily been designated as the ones which the $2p$ electrons of carbon occupy. It is a general rule (Hund's rule) that electrons in different orbitals of the same subshell maintain identical spin quantum numbers wherever possible.

Electron Configurations of Elements from Sodium to Potassium

Returning now to electron configurations of the elements beyond neon, we note that the 11th electron to be added to form the sodium atom cannot appear in the $2p$ level, since this would require two electrons to have the same four quantum numbers; and it must go to the $3s$ level. The electron config-

Table 4.5 Electron Configurations of Atoms of the Elements Carbon Through Neon

C	$1s^2\,2s^2\,2p_x^{\,1}\,2p_y^{\,1}$
N	$1s^2\,2s^2\,2p_x^{\,1}\,2p_y^{\,1}\,2p_z^{\,1}$
O	$1s^2\,2s^2\,2p_x^{\,2}\,2p_y^{\,1}\,2p_z^{\,1}$
F	$1s^2\,2s^2\,2p_x^{\,2}\,2p_y^{\,2}\,2p_z^{\,1}$
Ne	$1s^2\,2s^2\,2p_x^{\,2}\,2p_y^{\,2}\,2p_z^{\,2}$

Table 4.6 Electron Configurations of Atoms of the Elements Sodium Through Argon

Element	Atomic Number	Configuration
Na	11	$1s^2\ 2s^2\ 2p^6\ 3s^1$
Mg	12	$1s^2\ 2s^2\ 2p^6\ 3s^2$
Al	13	$1s^2\ 2s^2\ 2p^6\ 3s^2\ 3p^1$
Si	14	$1s^2\ 2s^2\ 2p^6\ 3s^2\ 3p^2$
P	15	$1s^2\ 2s^2\ 2p^6\ 3s^2\ 3p^3$
S	16	$1s^2\ 2s^2\ 2p^6\ 3s^2\ 3p^4$
Cl	17	$1s^2\ 2s^2\ 2p^6\ 3s^2\ 3p^5$
Ar	18	$1s^2\ 2s^2\ 2p^6\ 3s^2\ 3p^6$

uration of the sodium atom is thus $1s^2\ 2s^2\ 2p^6\ 3s^1$, so that sodium is like lithium in having its electron of highest energy in an s level. This is related to the chemical similarity of sodium and lithium. Beyond sodium, the $3s$ and $3p$ orbitals fill as before, as shown in Table 4.6.

With argon (atomic number 18), the $3p$ orbitals are filled. Hence, the next element, potassium (atomic number 19), has one electron in the energy level lying immediately above the $3p$ level (Figure 4.13). This is the $4s$ orbital and the configuration for potassium is:

$$_{19}K \quad 1s^2\ 2s^2\ 2p^6\ 3s^2\ 3p^6\ 4s^1$$

Calcium (atomic number 20) has two $4s$ electrons, thereby filling this orbital:

$$_{20}Ca \quad 1s^2\ 2s^2\ 2p^6\ 3s^2\ 3p^6\ 4s^2$$

Atoms of Elements Beyond Calcium

With element 21 (scandium), a new phenomenon appears as electrons are added to the atom (see Figure 4.14). If we visualize electrons being added one by one around the nucleus of scandium, the first 18 electrons will fill the $1s$, $2s$, $2p$, $3s$, and $3p$ levels in that order, but the 19th electron will go into one of the $3d$ orbitals rather than the $4s$. This happens because, when the nuclear charge is 21 and there are 19 electrons, the $3d$ level is below the $4s$ level, as shown by the spectrum of Sc^{+2}. The electron configuration of this ion is:

$$_{21}Sc^{+2} \quad 1s^2\ 2s^2\ 2p^6\ 3s^2\ 3p^6\ 3d^1\ 4s^0$$

This one electron in the $3d$ level now alters the relative positions of the $3d$ and the $4s$ levels so that the last two electrons, the 20th and 21st, go into the $4s$ level rather than into any of the other $3d$ orbitals, producing the neutral atom:

$$_{21}Sc \quad 1s^2\ 2s^2\ 2p^6\ 3s^2\ 3p^6\ 3d^1\ 4s^2$$

Thus, upon ionization by removal of electrons, the two $4s$ electrons are removed first, then the $3d$ electron comes off.

In the case of element 22 (titanium), the 19th and 20th electrons go into $3d$ orbitals:

$$_{22}Ti^{+2} \quad 1s^2\ 2s^2\ 2p^6\ 3s^2\ 3p^6\ 3d^2\ 4s^0$$

This causes the energy of additional electrons in the $4s$ level to be lower than they would be in other $3d$ levels, and hence the last two electrons, the 21st and 22nd, go into the $4s$ level:

$$_{22}Ti \quad 1s^2\ 2s^2\ 2p^6\ 3s^2\ 3p^6\ 3d^2\ 4s^2$$

This type of shifting of the relative positions of the d and s orbitals is found in each of the series of elements 21–29; 39–47; 57, and 72–79, which are called *transition elements* (Figure 4.14). A similar shifting of d and f orbitals is found in the series 58–71, called the *lanthanide series*, and in 89–103, called the *actinide series*. In the atoms of elements in these series, inner f electron orbitals as well as inner d orbitals fill before any more electrons are added in levels beyond the outer s level, as shown in Table 4.6. In each of these elements, however, the most loosely bound electrons are the outer s electrons, as a result of the rearrangement of the relative values of the orbital energies as the nuclear charge increases.

There is a special stability associated with half-filled and filled subshells. Thus the six outer electrons of $_{24}Cr$ have the configuration $3d^5 4s^1$ (two half-filled subshells) instead of the regular pattern $3d^4 4s^2$, and $_{29}Cu$ is $3d^{10} 4s^1$.

Periodic Relations Among the Elements

The Rutherford-Bohr model of the atom, with its modifications developed by the work of a generation of spectroscopists, and the more comprehensive and unified electron-cloud model, are in remarkably good agreement concerning the energy-level pattern for electrons in atoms. There are, however, two other sources which provide additional confirming evidence for this pattern. These are the periodic classification of elements and the table of ionization potentials.

OTHER EVIDENCE FOR ELECTRON CONFIGURATIONS

In 1868, Dmitri Mendeleev in Russia observed that the physical and chemical properties of the elements appeared to be periodic functions of their atomic weights, and suggested a classification of the elements based on this periodicity. (By a *periodic function* we mean that some quantity changes in a cyclic fashion, returning to a comparable value after a definite period, while another quantity changes in a uniform fashion. A familiar example is given by the periodic changes in the phases of the moon, which occur while the time changes in uniform fashion during the year.) Mendeleev's classification was known as a periodic table of the elements. H. G. J. Moseley later (1912) showed that the periodicity was a function of the atomic number rather than of atomic weight, and from this emerged the periodic law: *The physical and chemical properties of the elements are periodic functions of their atomic numbers.* The periodic arrangement is shown in the periodic table, which appears in Figure 1.6 and inside the front cover; the elements in each column (family) have similar properties.

This chart is a dramatic reflection of the pattern of electron configuration in atoms. In the first horizontal row there are two elements corresponding to the places available for electrons in the $1s$ orbitals, often known, from their position in the Bohr model, as electrons in the first shell. In the second row there are eight elements arranged in groups of two and six respectively. The group of two corresponds to the number of places available in the $2s$ orbitals; the group of six corresponds to the number of places in the $2p$ orbitals. These constitute the eight "electrons in the second shell," divided into two subshells of two and six. In the third row there are again eight elements in groups of two and six, corresponding to the $3s$ and $3p$ orbitals. The fourth row contains 18 elements in three groups of two, ten, and six elements, matching the $4s$, $3d$, and $4p$ orbitals. In potassium and calcium, confirming the spectral data, the $4s$ orbitals are filled before the $3d$ orbitals. The remainder of the table completes the pattern for the arrangement of electrons in atoms.

Table 4.7 Electron Configurations of Atoms of the Elements

Atomic Number (Z)	Element	1 s	2 s	2 p	3 s	3 p	3 d	4 s	4 p	4 d	4 f	5 s	5 p	5 d
1	H	1												
2	He	2												
3	Li	2	1											
4	Be	2	2											
5	B	2	2	1										
6	C	2	2	2										
7	N	2	2	3										
8	O	2	2	4										
9	F	2	2	5										
10	Ne	2	2	6										
11	Na	2	2	6	1									
12	Mg	2	2	6	2									
13	Al	2	2	6	2	1								
14	Si	2	2	6	2	2								
15	P	2	2	6	2	3								
16	S	2	2	6	2	4								
17	Cl	2	2	6	2	5								
18	Ar	2	2	6	2	6								
19	K	2	2	6	2	6		1						
20	Ca	2	2	6	2	6		2						
21	Sc	2	2	6	2	6	1	2						
22	Ti	2	2	6	2	6	2	2						
23	V	2	2	6	2	6	3	2						
24	Cr	2	2	6	2	6	5	1						
25	Mn	2	2	6	2	6	5	2						
26	Fe	2	2	6	2	6	6	2						
27	Co	2	2	6	2	6	7	2						
28	Ni	2	2	6	2	6	8	2						
29	Cu	2	2	6	2	6	10	1						
30	Zn	2	2	6	2	6	10	2						
31	Ga	2	2	6	2	6	10	2	1					
32	Ge	2	2	6	2	6	10	2	2					
33	As	2	2	6	2	6	10	2	3					
34	Se	2	2	6	2	6	10	2	4					
35	Br	2	2	6	2	6	10	2	5					
36	Kr	2	2	6	2	6	10	2	6					
37	Rb	2	2	6	2	6	10	2	6			1		
38	Sr	2	2	6	2	6	10	2	6			2		
39	Y	2	2	6	2	6	10	2	6	1		2		
40	Zr	2	2	6	2	6	10	2	6	2		2		
41	Nb	2	2	6	2	6	10	2	6	4		1		
42	Mo	2	2	6	2	6	10	2	6	5		1		
43	Tc	2	2	6	2	6	10	2	6	5		2		
44	Ru	2	2	6	2	6	10	2	6	7		1		
45	Rh	2	2	6	2	6	10	2	6	8		1		

Table 4.7   *(Continued)*

		MAIN SHELLS AND SUBSHELLS																	
		1	2		3			4				5				6				7
Atomic Number (Z)	Element	s	s	p	s	p	d	s	p	d	f	s	p	d	f	s	p	d	f	s
46	Pd	2	2	6	2	6	10	2	6	10										
47	Ag	2	2	6	2	6	10	2	6	10		1								
48	Cd	2	2	6	2	6	10	2	6	10		2								
49	In	2	2	6	2	6	10	2	6	10		2	1							
50	Sn	2	2	6	2	6	10	2	6	10		2	2							
51	Sb	2	2	6	2	6	10	2	6	10		2	3							
52	Te	2	2	6	2	6	10	2	6	10		2	4							
53	I	2	2	6	2	6	10	2	6	10		2	5							
54	Xe	2	2	6	2	6	10	2	6	10		2	6							
55	Cs	2	2	6	2	6	10	2	6	10		2	6			1				
56	Ba	2	2	6	2	6	10	2	6	10		2	6			2				
57	La	2	2	6	2	6	10	2	6	10		2	6	1		2				
58	Ce	2	2	6	2	6	10	2	6	10	2	2	6			2				
59	Pr	2	2	6	2	6	10	2	6	10	3	2	6			2				
60	Nd	2	2	6	2	6	10	2	6	10	4	2	6			2				
61	Pm	2	2	6	2	6	10	2	6	10	5	2	6			2				
62	Sm	2	2	6	2	6	10	2	6	10	6	2	6			2				
63	Eu	2	2	6	2	6	10	2	6	10	7	2	6			2				
64	Gd	2	2	6	2	6	10	2	6	10	7	2	6	1		2				
65	Tb	2	2	6	2	6	10	2	6	10	9	2	6			2				
66	Dy	2	2	6	2	6	10	2	6	10	10	2	6			2				
67	Ho	2	2	6	2	6	10	2	6	10	11	2	6			2				
68	Er	2	2	6	2	6	10	2	6	10	12	2	6			2				
69	Tm	2	2	6	2	6	10	2	6	10	13	2	6			2				
70	Yb	2	2	6	2	6	10	2	6	10	14	2	6			2				
71	Lu	2	2	6	2	6	10	2	6	10	14	2	6	1		2				
72	Hf	2	2	6	2	6	10	2	6	10	14	2	6	2		2				
73	Ta	2	2	6	2	6	10	2	6	10	14	2	6	3		2				
74	W	2	2	6	2	6	10	2	6	10	14	2	6	4		2				
75	Re	2	2	6	2	6	10	2	6	10	14	2	6	5		2				
76	Os	2	2	6	2	6	10	2	6	10	14	2	6	6		2				
77	Ir	2	2	6	2	6	10	2	6	10	14	2	6	7		2				
78	Pt	2	2	6	2	6	10	2	6	10	14	2	6	9		1				
79	Au	2	2	6	2	6	10	2	6	10	14	2	6	10		1				
80	Hg	2	2	6	2	6	10	2	6	10	14	2	6	10		2				
81	Tl	2	2	6	2	6	10	2	6	10	14	2	6	10		2	1			
82	Pb	2	2	6	2	6	10	2	6	10	14	2	6	10		2	2			
83	Bi	2	2	6	2	6	10	2	6	10	14	2	6	10		2	3			
84	Po	2	2	6	2	6	10	2	6	10	14	2	6	10		2	4			
85	At	2	2	6	2	6	10	2	6	10	14	2	6	10		2	5			
86	Rn	2	2	6	2	6	10	2	6	10	14	2	6	10		2	6			
87	Fr	2	2	6	2	6	10	2	6	10	14	2	6	10		2	6			1
88	Ra	2	2	6	2	6	10	2	6	10	14	2	6	10		2	6			2
89	Ac	2	2	6	2	6	10	2	6	10	14	2	6	10		2	6	1		2
90	Th	2	2	6	2	6	10	2	6	10	14	2	6	10		2	6	2		2
91	Pa	2	2	6	2	6	10	2	6	10	14	2	6	10	2	2	6	1		2
92	U	2	2	6	2	6	10	2	6	10	14	2	6	10	3	2	6	1		2

Table 4.7   (Continued)

Atomic Number (Z)	Element	MAIN SHELLS AND SUBSHELLS																		
		1	2		3			4				5				6				7
		s	s	p	s	p	d	s	p	d	f	s	p	d	f	s	p	d	f	s
93	Np	2	2	6	2	6	10	2	6	10	14	2	6	10	4	2	6	1		2
94	Pu	2	2	6	2	6	10	2	6	10	14	2	6	10	5	2	6	1		2
95	Am	2	2	6	2	6	10	2	6	10	14	2	6	10	7	2	6			2
96	Cm	2	2	6	2	6	10	2	6	10	14	2	6	10	7	2	6	1		2
97	Bk	2	2	6	2	6	10	2	6	10	14	2	6	10	8	2	6	1		2
98	Cf	2	2	6	2	6	10	2	6	10	14	2	6	10	9	2	6	1		2
99	Es	2	2	6	2	6	10	2	6	10	14	2	6	10	10	2	6	1		2
100	Fm	2	2	6	2	6	10	2	6	10	14	2	6	10	11	2	6	1		2
101	Md	2	2	6	2	6	10	2	6	10	14	2	6	10	12	2	6	1		2
102	No	2	2	6	2	6	10	2	6	10	14	2	6	10	14	2	6			2
103	Lr	2	2	6	2	6	10	2	6	10	14	2	6	10	14	2	6	1		2
104																				
105	Hn																			

The Table of Ionization Energies

A second source of supporting data for the validity of our present theory of atomic structure is obtained from the table of ionization energies of atoms (Table 4.8).

The energy necessary to remove an electron from an atom or an ion in the gaseous state is known as the *ionization energy*. Figure 4.5 shows one type of apparatus that can be used to measure this quantity. Since in this apparatus the quantity actually measured is the potential in volts necessary to pull electrons from the atoms, the ionization energy is the product of the charge on the electron and the *ionization potential*, and is commonly expressed in electron volts. The ionization energy is often loosely (but improperly) called the ionization potential, since the numerical values of the two are the same when the former is expressed in electron volts and the latter in volts.

When the first electron is removed from the neutral atom,

$$M(g) + \text{energy } (E_1) \longrightarrow M^+(g) + e^-$$

the minimum energy required, E_1, is called the *first* ionization energy. The minimum energy, E_2, to remove the second electron from the singly ionized atom,

$$M^+(g) + \text{energy } (E_2) \longrightarrow M^{+2}(g) + e^-$$

is termed the *second ionization energy*. The third ionization energy refers to the minimum energy to remove the third electron from the doubly charged atom:

$$M^{+2}(g) + \text{energy } (E_3) \longrightarrow M^{+3}(g) + e^-$$

Table 4.8 summarizes the ionization energies for the atoms of the first 20 elements. Some interesting variations in the numerical values of these energies are evident:

(a) Note that it is always easier to remove the first electron from an atom than to remove the second, because removal of the second leaves double the attractive positive charge, compared to removal of the first electron. The third electron is more difficult to remove than the second for a similar reason.

(b) This increase in going from the first ionization energy to the second and from the second to the third is not regular, however. Compare the increase from the removal of the first electron to the removal of the second, for beryllium, with the corresponding increase for lithium, for example, and note the big jump between the second and third ionization energies for beryllium.

(c) With boron, a discontinuity in the ionization energy data occurs between the third and fourth ionization energy; with carbon, a marked jump occurs between the fourth and fifth ionization energy; with nitrogen, a jump occurs between the fifth and the sixth, and so on.

(d) Sodium, like lithium, shows a discontinuity between the first and second ionization energies; magnesium, like beryllium, shows one between the second and third; aluminum, like boron, shows one between the third and the fourth, and so on.

(e) The pattern for sodium and magnesium is repeated for potassium and calcium.

These discontinuities are marked, in Table 4.7, by the stepped lines, which show again the build-up of the electron configurations in terms of the main shells. Comparison of the data for sodium through argon, for example, with the configurations given in Table 4.5 shows that it is relatively easier to remove the first, second, third, and so forth, electrons from the $3s$ and $3p$ levels than to break into the $2s$ and $2p$ levels. Therefore, when the first electron has been removed from sodium, a large increase in energy is needed to remove a second, which must come from the $2p$ level. It is not too difficult to remove two electrons from magnesium, since these are in the $3s$ level, but a great increase is again necessary to break into the shell of quantum number 2. The distribution of electrons in levels of different energies can be similarly verified for other atoms in the table.

Table 4.8 Ionization Energies (in kcal/mole)

Element	First	Second	Third	Fourth	Fifth	Sixth	Seventh	Eighth
H	313							
He	567	1254						
Li	124	1743	2822					
Be	215	420	3547					
B	191	579	874	5978	7843			
C	260	562	1103	1468	9037	11240		
N	325	683	1097	1784	2256	11660	15300	
O	314	809	1271	1781	2624	3170	16960	
F	402	806	1444	2009	2633	3606	4244	
Ne	495							
Na	118	1090						
Mg	176	346	1847					
Al	138	434	656	2766				
Si	188	377	772	1040	3838			
P	253	453	695	1184	1499			
S	239	539	808	1090	1670	2020		
Cl	299	549	920	1255	1563	2232	2622	
Ar	362							
K	100	733						
Ca	141	274	1180					

Table 4.9 Ionization Energies of the Elements Related to Position in the Periodic Table

X ← Symbol
00 ← Ionization energy (kcal/mole)

H_{313}																	He_{567}
Li_{124}	Be_{215}											B_{191}	C_{260}	N_{336}	O_{314}	F_{402}	Ne_{497}
Na_{118}	Mg_{176}											Al_{138}	Si_{188}	P_{254}	S_{239}	Cl_{300}	Ar_{363}
K_{100}	Ca_{141}	Sc_{151}	Ti_{158}	V_{156}	Cr_{156}	Mn_{171}	Fe_{182}	Co_{181}	Ni_{176}	Cu_{178}	Zn_{216}	Ga_{138}	Ge_{187}	As_{231}	Se_{225}	Br_{273}	Kr_{323}
Rb_{96}	Sr_{131}	Y_{152}	Zr_{160}	Nb_{156}	Mo_{166}	Te_{167}	Ru_{173}	Rb_{176}	Pd_{192}	Ag_{175}	Cd_{207}	In_{133}	Sn_{169}	Sb_{199}	Te_{208}	I_{241}	Xe_{280}
Cs_{90}	Ba_{120}	La_{129}	Hf_{127}	Ta_{138}	W_{184}	Re_{182}	Os_{201}	Ir_{212}	Pt_{207}	Au_{213}	Hg_{241}	Ti_{141}	Pb_{171}	Bi_{185}			Rn_{248}

The energy levels of the subshells also appear in the ionization-energy data. For example, the first ionization energy of boron is less than that of beryllium because the $2p$ orbital of boron holds the electron less tightly than the two electrons of beryllium are held in the $2s$ orbital.

Similarly, Table 4.7 shows the stability of the half-filled shells. According to Hund's rule, the three $2p$ electrons in nitrogen are each in a different orbital, whereas the fourth $2p$ electron in oxygen must be paired. The decrease in the first ionization energy in going from nitrogen to oxygen indicates that it is easier to remove an electron from the pair than it is to break into the stable configuration of three unpaired electrons in the half-filled subshell of nitrogen.

Ionization Energies and the Periodic Table

The stepped curve of Table 4.8 clearly shows the separation of the elements into periods such as H-He, Li-Ne, and Na-Ar. The separation into families within the periods is also shown—elements on corresponding "steps" of the curves belong together. This is also shown in Table 4.9 which records the *first* ionization energies in the periodic arrangement. These energies generally increase in going from left to right across the table, because of the increase in nuclear charge. The higher the nuclear charge, the greater the attraction of the nucleus for the electron being removed. The ionization energies generally decrease in going from top to bottom in a family because the growing number of inner shells shield the electron from the nuclear charge.

How We See Nonluminous Objects. The interaction of light with matter discussed in this chapter provides a basis for appreciating the mechanism by which we are able to perceive objects and color. Most of the things we see do not emit light of their own; they are visible because they re-emit part of the light that impinges upon them from a primary source such as the sun or an electric lamp. We recognize the shape, the degree of opacity, and the color of objects by the light they reflect or transmit to the collectors in our eyes. When white light strikes the surface of some material, it is either scattered instantly without change in wavelength or absorbed by the atoms or molecules present, to be released (all or in part) later or to be transformed into heat energy.

Absorption of light by atoms and molecules should be understandable on the basis of the earlier discussions in this chapter. Electrons in atoms (and molecules) exist in definite energy states. An atom can absorb only those photons that carry energy identical to the energy difference between two of these states. The spacing of energy levels is unique for each kind of atom or molecule, which means that the wavelengths of the photons absorbed by atoms or molecules of a given substance will differ from those absorbed by atoms or molecules of other substances. If a beam of white light is allowed to pass through an assembly of identical atoms or molecules in the gaseous state, these particles will selectively absorb photons of certain wavelengths (certain colors) and the photons not absorbed will pass through the material. An observer viewing the transmitted light will no longer see white light but white light without the color(s) absorbed. This registers as the complementary color to that absorbed. Examples of this are the gases Cl_2, NO_2, and I_2, which appear greenish-yellow, reddish-brown, and purple to the eye because the molecules of these substances absorb photons corresponding to the colors blue-violet, blue-green, and yellow-green, respectively.

Color in most liquids and solids also is the result of selective absorption of certain wavelengths of light. The colors of flowers, grass, and leaves, of the dyes present in our fabrics and photographic emulsions, of the pigments used in paints and inks, and of the color additives to our drinks and other foods are directly traceable to the electron energy-level spacing in the atoms and molecules present in these materials. Chlorophyll, the green pigment of plants, absorbs heavily in the blue-violet (4000–4500 A) and also in the red (6450–6800 A), transmitting most of the light from 4800 to 6400 A, which corresponds to some blue, green, yellow, orange, and red. When mixed, this combination appears as green.

The blue of the sky, however, cannot be explained by the absorption of orange or red photons by the nitrogen and oxygen molecules that are the major components of the atmosphere. Neither of these molecules absorbs light in the visible region of the spectrum. An explanation for this that also helps explain the phenomenon of light reflection by all objects is that a photon striking any object may interact with the electrons in that object without being absorbed, causing the electrons to vibrate. This vibration is related to the changing amplitudes of the photon wave (Figure 4.3). We can call this a nonabsorptive or a nonresonance interaction.

While this interaction usually is weak—meaning that only a relatively small fraction of the photons striking the object may experience it—the result is that photons interacting this way are scattered from the interacting center in all directions. This is the basis for reflected light, without which we could see only primary light sources. In gases, the scattering is greatest for the short-wavelength photons of violet and blue light. Therefore, in looking into the sky but not directly at the sun, we are in effect viewing the sun's rays from the side; we see coming at us the blue photons that have been scattered from the direct sunlight by interaction with nitrogen and oxygen molecules in the air. Looking directly at the sun, we see white light with some orange and red. Sunsets are richer in these colors. Again, this is the result of greater scattering of the shorter wavelengths, for in these cases the sun's rays come directly to us with small amounts of the blue and violet having been removed by scattering.

In liquids and solids, reflection or nonresonance interaction and scattering is more pronounced than in gases, but the preference for scattering the short wavelengths is reduced. Substances such as ice, glass, sodium chloride, and sucrose (table sugar) do not absorb in the visible region. We therefore expect that these substances will be transparent or white since the light transmitted to us from them will be the normal mixture of wavelengths. One interesting aspect of all these materials is that while they are transparent as large crystals, all of them appear white when crushed to powders or in cakes or pellets made by compacting the powders.

An explanation for this is that the large crystals with their orderly arrangement of atoms, molecules, or ions extending for great distances (compared with the photon wavelength) form an ideal medium for propagating light waves. While some light is reflected from the surface, much of it moves through the crystal, being scattered in a systematic manner from the regularly spaced particles along the way. The eye receives small amounts of light from many points in the crystal and this reflected light coming from the orderly arranged centers allows us to "see through" the object.

Even in liquids such as water, the arrangement of molecules is orderly enough to permit propagation and reflection from many points in the liquid. In the powder form of transparent crystals one must imagine that microscopic crystals are pressed against one another so the orderly arrangement of building blocks is not contiguous from one tiny crystal to another. The result is that light cannot be propagated effectively over any distance in the powder; more of it is reflected to the eye from each of the haphazardly arranged tiny crystals, thus giving the appearance of a white opaque material. Paper and snow are examples of compacted fibers or powders. Clouds are examples of compacted but uncoalesced droplets of liquid.

Metals are among the best reflecting materials known because some of the electrons in metals are more free to interact with light waves than are electrons in other materials. This greater photon-electron interaction results in unusually high reflectivity from the metal surface.

SUMMARY   This chapter summarizes the accumulated evidence and thinking of scientists as related to electron configurations of atoms. It emphasizes how many sources, both experimental and theoretical and both qualitative and quantitative, have been used in developing a consistent pattern. It also shows how a successful, simple theory evolved, and how that theory was modified and adjusted to embrace more complicated atoms. It develops the current view of the electron not as a ball revolving around the nucleus but as a particle-wave with a specified energy having a certain probability of being in a given region in the space outside the nucleus.

IMPORTANT TERMS

Nature of light
wavelength
frequency
energy
quantum theory
photon
Planck equation

Spectrum

Spectroscopy

Energy levels in atoms
Rutherford-Bohr theory
shells, subshells, orbitals
electron spin
Pauli exclusion principle

Wave nature of electrons
particle-wave duality
particle-wave probability
wave function

quantum numbers
uncertainty principle

Electron-cloud model
orbital shapes
s, p, d, f

Electron configurations
periodic relations among
    elements
ionization energies

1. Summarize the quantum theory of light.

2. Account for the presence of *discrete lines* in atomic spectra.

3. Cite experimental evidence for the existence of energy levels in atoms.

4. Distinguish between emission and absorption spectra of elements.

5. Explain in terms of an energy-level picture why elements 1 (hydrogen) and 3 (lithium) are highly reactive, whereas element 2 (helium) is inert.

6. What is the maximum number of electrons that can be accommodated in (a) all energy levels with $n = 4$, (b) all $4f$ orbitals, (c) all orbitals with $n = 5$, $l = 4$?

7. (a) List all the elements that have an s orbital containing only one electron. (b) List all the elements that have a single electron in a p subshell. (c) List all the elements that have only three electrons in d orbitals. (d) List all the elements that have only five electrons in f subshells.

8. Write electron configurations for all the elements (a) in Group V of the periodic table; (b) in the first transition series ($_{21}$Sc–$_{30}$Zn).

9. (a) Which group in the periodic table contains elements with the highest ionization energies? (b) Which period contains elements with the highest ionization energies?

10. The quantum numbers of all the electrons in the outer shell of each of the atoms (a), (b), and (c) are given below; each inner shell is filled. Identify the three atoms.

(a)

n	l	m	s
2	0	0	$-\frac{1}{2}$
2	0	0	$+\frac{1}{2}$

(b)

n	l	m	s
2	0	0	$-\frac{1}{2}$
2	0	0	$+\frac{1}{2}$
2	1	-1	$-\frac{1}{2}$

(c)

n	l	m	s
3	0	0	$-\frac{1}{2}$
3	0	0	$+\frac{1}{2}$
3	1	-1	$-\frac{1}{2}$
3	1	0	$-\frac{1}{2}$
3	1	$+1$	$-\frac{1}{2}$

11. Offer an explanation for the fact that the spectrum of atomic hydrogen consists of a series of lines even though the hydrogen atom has only one electron.

12. Draw a picture of what you think the electron cloud of the outermost electrons in each of these atoms looks like: (a) helium, (b) lithium, (c) boron, (d) nitrogen, (e) neon.

13. Compare and contrast the models of the atom as viewed by Dalton, Rutherford, Bohr, and modern quantum mechanics.

14. The cadmium and magnesium atoms both have configurations ending in ns^2. Why are they not placed in the same family of the periodic table?

15. List the electron configurations of the following atoms and ions and indicate which of them contain the same number of electrons: Mn, S^{-2}, Ar, Fe, Br$^-$, Kr, Co^{+2}.

16. Prepare a table for each of the following categories showing the electron

configurations of all the elements whose final electrons conform to the particular pattern indicated:

(a) ns^1

(b) $ns^2 np^6$

(c) $(n-1)d^2 ns^2$

(d) $(n-1)d^5 ns^1$

17. What experimentally controlled factors might give rise to an increase or a decrease in (a) the number of lines in the hydrogen spectrum; (b) in the intensity of the lines in the hydrogen spectrum?

18. What evidence is there for the idea that electrons have wave properties?

19. Show that you know what is meant by each of the following: (a) wave function, (b) electron cloud, (c) ψ^2, (d) probability distribution of an electron in an atom.

20. When heated in a Bunsen burner flame, calcium can emit light resulting from an electron transition of 3.49×10^{-12} erg. Under the same conditions, barium light may be emitted as the result of a 3.62×10^{-12} erg transition, and strontium from a 4.33×10^{-12} erg transition. (a) Calculate the corresponding wavelengths, and (b) refer to Table 4.1 to identify the appropriate colors.

21. The ionization energy of potassium is 100 kcal/mole. What must be the minimum frequency of a photon that can effect this ionization?

SPECIAL PROBLEMS

1. The following are the characteristic X-ray wavelengths for some metals: Rb, 0.93 A; Mg, 9.87 A; Cr, 2.29 A; Ca, 3.35 A. (a) Calculate the energy-level differences in these atoms indicated by the emission of the X rays. (b) What orbitals might be involved in these transitions?

2. In the copper atom, when an electron makes the transition from the $2p$ to the $1s$ orbital, radiation of wavelength 1.54 A is emitted. (a) Calculate the energy drop in this transition. (b) Calculate the energy difference between the hydrogen $1s$ and $2p$ orbitals.

3. Refer to Table 4.1. The sodium flame test emits radiation from 5896 A to 5890 A, the lithium flame test does from 6708 A to 6104 A, and the potassium flame emits light having a frequency of $7.41 \times 10^{14} \sec^{-1}$. (a) Identify the color of these emissions, and (b) calculate the energy differences to which they correspond. (c) Compare the sizes of the energy transitions involved.

4. Examine the wavelengths of the Lyman, Balmer, and Paschen series given in Figure 4.6. (a) Do any of these appear in the visible spectrum? (b) If so, what colors do they represent? (c) Identify the others as to their inclusion in the ultraviolet or infrared portion of the spectrum.

5. Using the hydrogen atom as the model, (a) calculate the energy when the principal quantum number of its electron is 8, and (b) calculate the work done in kcal/mole required to ionize a hydrogen atom whose electron was originally in the $n = 2$ level. (c) An "excited" electron cascades from $n = 6$ to $n = 3$, then from $n = 3$ to the ground state in the hydrogen atom. How many photons are emitted in these transitions and what is the energy of each? (d) The frequency? (e) The wavelength?

6. The hydrogen spectrum can be represented by the equation:

$$\nu = 3.289 \times 10^{15} \left(\frac{1}{n_1^2} - \frac{1}{n_2^2} \right)$$

where ν is the frequency of the line per sec, 3.289×10^{15} sec^{-1} is the Rydberg constant (for hydrogen), and n_1, n_2 are dimensionless whole numbers representing energy levels ($n_2 > n_1$). Calculate the energy, in ergs per atom, of the line in the visible spectrum of hydrogen corresponding to the electron energy transition $n = 3 \longrightarrow n = 2$.

7. According to Bohr, the energy of each orbit in the hydrogen atom is given by the equation

$$E = -\frac{2\pi^2 m e^4}{h^2 n^2}$$

where m, e, h, and n are, respectively, the mass of the electron, charge on the electron, velocity of light (cm/sec), Planck's constant, and the principal quantum number. (a) Using this equation, calculate the energy of each of the first two orbits for hydrogen—that is, when $n = 1$ and $n = 2$ —and calculate the difference in energy between these states. (b) Calculate also the energy of each of the first two orbits in the helium ion He$^+$. To do this multiply the energy for each orbit in the hydrogen atom by Z^2, where Z is the nuclear charge on helium. Compare the energies of the first and second orbits in He$^+$ with those in hydrogen. (c) If the electron of He$^+$ could take on any energy value, what kind of an emission spectrum would you predict for the helium ion?

8. Solving Bohr's quantum condition for the radius, r, of an orbit, we get

$$r = \frac{nh}{2\pi m v}$$

where h is Planck's constant, and m and v are mass and velocity of the electron, respectively. (a) Calculate the radii of the Bohr orbits at energy levels $n = 2$ and $n = 3$. (b) Given the Bohr radius for the $n = 1$ orbit as 0.530 A, and using your answers from (a), calculate the energy of a hydrogen electron in each of the three orbits: $n = 1$, $n = 2$, $n = 3$, from the equation:

$$E = \frac{-Ze^2}{2r}$$

where Z is the atomic number and e is the electronic charge $= 4.80 \times 10^{-10}$ esu (1 esu $= 1$ g$^{1/2}$ cm$^{3/2}$/sec). (c) Using your results from (b), calculate the frequencies and wavelengths of photons emitted as a result of all possible electron transitions involving these energy levels (that is, $n = 3 \rightarrow n = 1$, $n = 3 \rightarrow n = 2$, $n = 2 \rightarrow n = 1$).

9. Several of the transition elements do not progress in the regular fashion through the $(n\text{-}1)d$ energy levels. Instead, there are apparent variations in these elements such that they have more $(n\text{-}1)d$ electrons than predicted by the regular progression. Tabulate the elements that show these variations, comparing their configurations as might be expected according to the regular progression with their actual configuration variations.

10. Making the assumption that the Bohr equation holds for the sodium atom (it is only valid for hydrogen!), (a) calculate the atomic radius and the ionization energy of sodium using $n = 3$ and the effective nuclear charge as 2.2. (b) What is the wavelength of the emitted photon when this electron drops from $n = 5$?

REFERENCES   CROMER, D. T., "Stereo Plots of Hydrogen-Like Electron Densities," *J. Chem. Educ.*, **45**, 626 (1968).

GAMOW, G., *The Atom and Its Nucleus*, Prentice-Hall, Englewood Cliffs, N. J., 1961.

GARRETT, A. B., *The Flash of Genius*, Van Nostrand Reinhold, Princeton, 1962.

KLEIN, M. J., "Einstein's First Paper on Quanta," in Gershenson, D. E. and Greenberg, D. A. (eds.), *The Natural Philosopher*, Vol. 2, Xerox College Publishing, Lexington, Mass., 1963.

LINNETT, J. W. and BORDASS, W. T., "A New Way of Presenting Atomic Orbitals," *J. Chem. Educ.*, **47**, 672 (1970).

SCHRÖDINGER, E., "What Is Matter?" *Sci. Amer. Reprint* 241, Freeman, San Francisco, 1963.

WHAT HOLDS ATOMS TO EACH OTHER?

While persistent experimenters and theoreticians were painstakingly working out the detailed structure of the atom, other groups of scientists were trying to find out how atoms are held to each other in the myriads of substances found in the world. In the course of their examination of this problem, they were led to recognize three types of chemical bonds: covalent, electrovalent, and metallic.

Nature of Combining Forces of Atoms

It is well known that a particle bearing an electric charge exerts a force, known as an electrostatic force, on other charged particles in its vicinity. Since atoms are made of negatively and positively charged particles, we might expect to find that the combining forces involved in chemical combinations are electrostatic, and that they arise from the opposite electrical charges on nuclei and electrons. It is true that both nuclei and electrons show magnetic characteristics also, which we describe in terms of their spins, but the evidence is that with rare exceptions magnetic forces have little to do with holding atoms to each other. We shall find, then, that *the forces between atoms are those between charged particles, attractive between charges of opposite sign, repulsive between charges of like sign, and obeying Coulomb's law*. Coulomb's law states that the electrostatic force between two charged bodies is proportional to the product of the charges and inversely proportional to the square of the distance between them. Oppositely charged bodies tend to get as close together as possible, whereas like-charged bodies tend to get as far away from each other as possible. The force of attraction or repulsion decreases sharply as the charged bodies are separated. In mathematical form, Coulomb's law may be stated

$$F = \text{a constant} \times \frac{e_1 e_2}{r^2}$$

where F is the force acting between bodies bearing charges e_1 and e_2 and at a distance r from one another.

If we try to classify or group different chemical substances on the basis of their properties, we seem at first to get only baffling confusion. Patient examination in a search for gross similarities, however, shows that three general types of substances emerge. Some substances are soft and have low melting points; others are hard and have high melting points; and still others have a metallic luster not generally shown by either of the other two kinds. Further observations indicate that the known substances are not equally divided among these three classes; there appear to be many more materials in the first class. Furthermore, there is no definite line separating one class from the other; rather, the classes merge gradually from one to the other, so that some substances cannot definitely be assigned to any specific group.

The differences in properties have been explained as the result of differences in how the electrostatic forces act in holding the atoms to each other, and we speak of atoms held to each other by *covalent bonding, electrovalent bonding,* and *metallic binding.* By an extension of the concept, we speak of *covalent substances, electrovalent substances,* and *metallic substances.*

Properties of Covalent, Electrovalent, and Metallic Substances

Some of the distinguishing characteristics of these three classes of substances are listed in Figure 5.1 and Table 5.1. The covalent substances, in addition to being soft and low melting, are nonconductors of electricity, are usually insoluble in water, and are liquids only within a short range of temperature. Electrovalent substances are hard and brittle, have high melting points and a long liquid range, are nonconductors except when melted, and undergo chemical reactions when an electric current passes through them. Metallic substances have the metallic properties all of us recognize, such as metallic luster and good conduction of electricity and heat. There are exceptions to these characteristics, however. While covalent substances as a class tend to be insoluble in water, some (such as sugar and alcohol) are water soluble. While metals as a class are solids of high melting point, one (mercury) is a liquid,

Figure 5.1 Classification of substances by bond type. (a) Covalent substances: soft, low-melting, nonconductors of electricity, short liquid range. Examples: Dry ice, wax, mothballs. (b) Ionic substances: hard and brittle, high-melting, long liquid range. Examples: salts (NaCl, CaSO$_4$). (c) Metallic substances: metallic luster, good conductors of heat and electricity. Examples: copper, iron, sodium.

CO$_2$

Cl$^-$
Na$^+$

Atom

Dry ice   Salt   Metal

(a)   (b)   (c)

Table 5.1 Properties of Substances Compared

Properties	Small-Molecule Covalent Substance	Network Covalent Substance	Electrovalent Substance	Metallic Substance
Hardness	Soft	Hard	Hard	Soft or hard
State at room temperature	Gas, liquid, or solid	Solid	Solid	Solid
Melting point	Within 200°C above or below room temperature	Very high	Well above room temperature	Room temperature and above
Boiling point	Short liquid range (100°C)	Very high	Long liquid range	Long liquid range (1500°C)
Solubility in water	Usually insoluble	Insoluble	Often soluble	Insoluble
Solubility in oils	Usually soluble	Insoluble	Insoluble	Insoluble
Electrical conductivity	Nonconducting	Nonconducting	Conducts when melted or in solution	Conducts in solid and in melt
Chemical reaction on conduction	Nonconducting	Nonconducting	Chemical reaction occurs when current passes	No chemical reaction on passage of current
Heat conductivity	Insulator	Insulator	Insulator	Conductor
Opacity	Transparent	Transparent	Transparent or translucent	Opaque
Reflectivity	——	——	——	Metallic luster
Typical example	Methane, CH_4	Quartz $(SiO_2)_x$, diamond, C_x	Sodium chloride, NaCl	Aluminum, Al

and several others melt a little above room temperature. Nevertheless, there is enough divergence among the three groups to show that the separation into these three classes is a useful classification.

There is a fourth class of substances, small in number, that includes such extremely hard and high-melting-point materials as diamond and silicon carbide (carborundum). These substances consist of huge *networks* of atoms held together by covalent bonds. The result is that the entire crystal may be a single structure.

Let us now seek the explanation for the different properties of these classes.

Relation of Structural Units to Properties

To explain such properties as hardness, boiling point, and electrical conductivity of the three classes of substances, we must make some hypotheses about the structural units and binding in each.

We might expect hardness and rigidity to be associated with an extended structure, in which all particles are tied to other particles by inflexible linkages, as in the steel skeleton of a skyscraper. Softness and fluidity could arise from unlinked particles, as in a collection of glass beads in a dish, which may be easily penetrated by an inserted pencil and will pour like a liquid. Electrical conductivity requires charged particles, and these must be able to move when other electric forces are applied. Charged particles that are unable to move could also produce inflexibility. If the electrical forces could be satisfied, for example, by only one particular arrangement of particles of opposite charge, any stress that tended to change this arrangement would be resisted if it tended to bring particles of like charge closer together. Inflexibility in such a solid at low temperatures might change to conductivity in the liquid melt at higher temperatures if the forces resisting change in the solid were

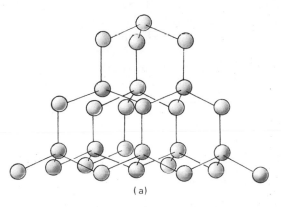

(a)

Figure 5.2 Rigidity is associated with extended structures in which the atoms are connected in three-dimensional networks. (a) Three-dimensional network of carbon atoms in diamond held to each other with covalent bonds. (b) An analogy is the steel network structure of a large building.

(b)

overcome in the liquid so that the charged particles would be unlinked and free to move. The unlinking of particles is carried to an extreme in gases, where the particles are widely separated from each other; we thus interpret boiling as the result of adding energy to separate the particles of a liquid. More energy is needed for this separation (higher boiling point) when the attracting forces holding the particles together in the liquid are large.

These concepts can give reasonable explanations of the properties of the three classes. Small-molecule covalent compounds are of the unlinked-particle type (softness); they are made of small neutral molecules held to each other only by weak forces of attraction (low boiling points). Electrovalent compounds are of the second inflexible type mentioned; they are made of

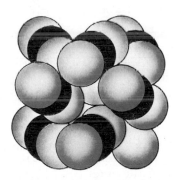

Figure 5.3 CO_2 crystal. Softness and flexibility are associated with weakly linked particles. Covalent substances are made of molecules held to each other with relatively weak binding forces. Note the CO_2 molecules in the structure.

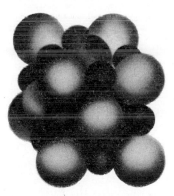

Figure 5.4 NaCl crystal. Hardness and high melting point are associated with strongly linked particles. Ionic substances are made of ions held to each other with strong electrostatic forces between the charged particles. Note that Na^+ ions and Cl^- ions are in the lattice, but no NaCl molecules appear.

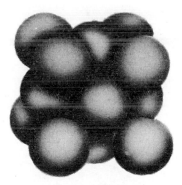

Figure 5.5 Metal crystal. Electrical conductivity is associated with extended structures containing easy-to-move electrons. Atoms of a pure metal are all the same size and usually arranged so the largest number are packed into the smallest space.

charged particles (ions), both positive and negative, held firmly in a tight arrangement in the solid, but freed to move in the liquid state or when dissolved, to become carriers of an electric current (conductivity in the melt and in solution). Metals have free current carriers even in the solid state; we guess that these may be free electrons, since the passage of the current does not produce any chemical change in the metal.

What Property Determines How Atoms Combine?

The structural units suggested above in explanation of the observed properties all have the same fundamental units—atoms. They exist in one case as combined atoms in molecules, in another as ions, and in the third case as atoms. We must look then for some characteristics of atoms that permit them (a) to combine to form the molecules characteristic of covalent compounds; (b) to form the ions characteristic of electrovalent compounds; or (c) to form metallic solids with free electrons. Ordinarily we do not find atoms of the same element appearing in more than two of the three types of substances.

To find the answer, we must look for differences in the structure and properties of atoms—that is, in the *number* of electrons contained in the atom, the *energy levels* occupied in the atom's normal energy state, and the *ionization energy*—as a guide to the preferred types of behavior. Perhaps it would be helpful first to examine the number ratios in which atoms of one element combine with another, particularly among the elements of lower atomic number.

Atomic Ratios in Compounds

Consider first the formulas of some compounds of hydrogen with one other element (Table 5.2). Remember that each formula has been determined by careful examination and represents the natural combining ratio of the atoms. Note that in these the number of hydrogen atoms combined with other atoms is 1, 2, 3, or 4. When we observe that the hydrogen atom has a single $1s$ electron, we guess that the atom ratios arise from the fact that hydrogen has *one* electron while the other atoms have 1, 2, 3, or 4 electrons to use in bonding. This is confirmed by examination of lithium and sodium compounds. Like hydrogen, lithium and sodium have one electron in the outermost shell and form compounds in which there is not more than one atom of the other element per atom of lithium or sodium: LiH, LiCl, Li_2O, NaCl, Na_2O.

When we move on to helium, with its two electrons, we find that it does not combine chemically with other elements. However, other atoms with two electrons in the outer shell (beryllium, magnesium, calcium) do form compounds, and often in the ratio of 1:2, as in CaH_2, $BeCl_2$, and MgF_2. Except for helium, the two outer electrons, then, seem to be associated with a combining ratio of 2. However, the fact that the two electrons of helium do not lead to chemical combination agrees with the fact that the two $1s$ electrons in lithium and in beryllium apparently do not play any role in compound formation either, since only the outer electron in lithium (or the outer two in beryllium) seems to be involved. This is what you might expect from examination of the ionization energies (Chapter 4), from which it is evident that the $1s$ electrons in lithium and beryllium are held much more tightly than the $2s$ electron or electrons of those elements. This difference in the energies of $2s$ and $1s$ electrons appears to limit compound formation only to the less tightly bound electrons.

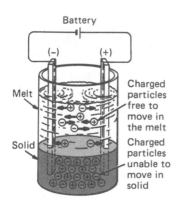

Figure 5.6 Diagram of a partially melted solid (ionic) in which two electrodes are inserted. The ionic solid is a nonconductor but, when melted, becomes a good conductor of electricity because of the freedom of the ions to move toward the electrodes.

Table 5.2 Atom Ratios in Several Substances

Compound	Ratio: $\dfrac{\textit{Atoms of Second Element}}{\textit{Atoms of Hydrogen}}$
H_2	1:1
HCl	1:1
H_2O	1:2
NH_3	1:3
LiH	1:1
CH_4	1:4
H_2O_2	1:1
N_2H_4	1:2

The outer or valence electrons of sodium, magnesium, and calcium do not overlie a shell of two electrons, but a shell of eight. Examination of the ionization energies again shows that large amounts of energy must be expended to break into the lower shell while relatively small amounts are needed to remove the outer electrons. Hence, compound formation is again limited to the less tightly bound electrons.

Further confirmation of the stability of the underlying shell of eight electrons is obtained by the observation that neon and argon, each with eight electrons in the outer shell, are inert, like helium, and normally form no compounds.

Figure 5.7 (a) Helium atoms with two 1s electrons are unreactive. (b) Hydrogen atoms with one electron form H_2 molecules.

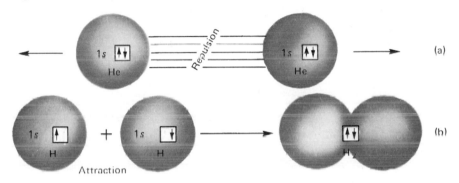

Figure 5.8 Eight electrons in a shell often confer stability. This configuration may arise as a result of ionization processes, the atoms (a) acquiring or (b) losing electrons to present an outer shell of eight electrons. This configuration also may result from a sharing of pairs of electrons between two atoms, as illustrated in Figure 5.9.

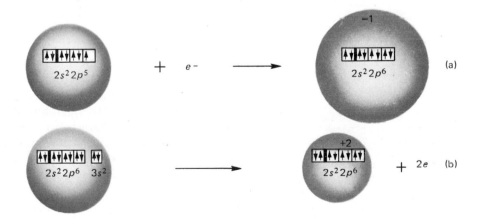

Figure 5.9 Atoms use outer-shell
electrons in combining with other
atoms. Here, a pair of electrons is
shared in the outer shells of both
atoms, as in F + F ⟶ F_2.

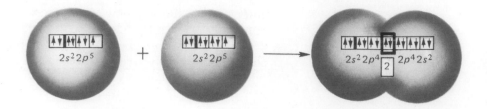

**Stability and Noble
Gas Configuration**

From the above evidence we might conclude that the number of atoms of a
given element that one atom of another element can combine with depends
upon the number of electrons in the outermost (valence) shell, and that a
shell of two or eight represents a very stable configuration. The second part
of this conclusion is correct. But the first part might lead us to predict that a
chlorine atom, with seven electrons, might combine with seven other atoms.
No such compounds of chlorine are known. In many of its common compounds, chlorine combines with one other atom, as in HCl, LiCl, and NaCl.
If we note that hydrogen, lithium, and sodium are elements that have one
usable electron per atom, then we suggest that a chlorine atom adds one
electron to its seven to make eight in the outer shell to arrive at the peculiar
stability of the eight-in-the-shell configuration of argon. This point of view
is confirmed on examination of the formulas H_2O, NH_3, and CH_4.

A chlorine atom with seven electrons	needs one from a hydrogen atom	to form an outer shell of eight.
$:\ddot{C}l\cdot$	$\cdot H$	$:\ddot{C}l:H$
An oxygen atom with six electrons	needs two from two hydrogen atoms	to form an outer shell of eight.
$:\ddot{O}\cdot$	$\cdot H$ $\cdot H$	$:\ddot{O}:H$ H
A nitrogen atom with five electrons	needs three from three hydrogen atoms	to form an outer shell of eight.
$\cdot\ddot{N}\cdot$	$\cdot H$ $\cdot H$ $\cdot H$	$H:\ddot{N}:H$ $\ddot{H}$
A carbon atom with four electrons	needs four from four hydrogen atoms	to form an outer shell of eight.
$\cdot\dot{C}\cdot$	$\cdot H$ $\quad$ $\cdot H$ $\cdot H$ $\quad$ $\cdot H$	H $H:\ddot{C}:H$ $\ddot{H}$

Take the atoms of chlorine and hydrogen in HCl, for example. If we permit each of the atoms held together by the pair of electrons to have a share in
both of the electrons in the pair, then chlorine not only has eight electrons,
but hydrogen has two. Hydrogen has thus acquired the stable configuration,
two-in-the-outer-shell, characteristic of helium. This stable two-electron
configuration for hydrogen is also present in each of the other compounds.

From consideration of these compounds, then, we might conclude that
one type of chemical combination arises from the tendency of atoms to acquire, by sharing valence electrons in pairs, the electron configuration of a
noble gas, with two electrons (for light elements) or eight electrons in the

outer shell. This is an empirical generalization, based simply on a consideration of the atom ratios in a few chemical compounds. It remains to be seen if the generalization covers all compounds and if there is a reasonable explanation for combinations occurring in this manner.

We must immediately conclude that it is doubtful if all compounds are formed in this manner. If they are, then their properties might be expected to be similar, whereas we have found three classes, indicating three different kinds of bonding. Some compounds in both the covalent and electrovalent classes satisfy the numerical values of two and eight valence electrons, so these numbers must be common to both these classes. Thus, of the compounds listed on page 96, LiH, LiCl, Na_2O, for example, have electrovalent properties, while HCl, H_2O, NH_3, and others have covalent properties. Nevertheless, we may examine the significance of the presence of a pair of electrons shared by two atoms in a molecule.

Electron-Dot or Lewis Formulas

Since atoms appear to become more stable by pairing electrons and acquiring either eight or two outer-shell electrons, it is oftentimes (but not always) possible to predict the number and type of covalent bonds which might be formed between several atoms in a molecule. To do this we need only to write the valence-shell electron configuration of each of the uncombined atoms, pair the electrons, and form as many covalent bonds as needed to give stable outer-shell configurations. For example, hypochlorous acid, HOCl, should contain two covalent bonds, one between the oxygen and hydrogen atoms, and one between the oxygen and chlorine atoms. These arise when a hydrogen atom, H·, and a chlorine atom, ·Cl̈:, share one electron each with an oxygen atom, ·Ö·, to give H:Ö:Cl̈:, a structure in which all electrons are paired and all atoms have acquired stable valence-shell electron configurations. Similarly, formaldehyde, H_2CO, may be written

$$\text{H:C::Ö:}$$
$$\text{H}$$

Here, all electrons are paired and all atoms have stable valence electron configurations when the oxygen atom is bound by two pairs (doubly bonded) and the hydrogen atoms are each bound by one pair (singly bonded) to the carbon atom. Formulas that include dots to represent valence electrons are known as Lewis formulas, in recognition of the work of the American chemist G. N. Lewis, who early conceived the idea of electron pairing and valence-shell stability as an explanation for atom combination.

Other Lewis formulas are:

$$\text{:Ö:}$$
$$\text{H:Ö:Cl̈:Ö:} \qquad \text{:Cl̈:P:Cl̈:} \qquad \text{H:Ö:N::Ö:}$$
$$\text{:Ö:} \qquad \text{:Cl̈:} \qquad \text{:Ö:}$$

Perchloric acid   Phosphorus trichloride   Nitric acid

Covalent Bonds

Let us make the assumption that chemical combination occurs when an electron on one atom pairs off with an electron on another atom, which has the opposite spin. The electron clouds of the two electrons can then occupy the same space without violating the Pauli exclusion principle and can act as a binding force of negative electricity to hold the positively charged nuclei together. According to the picture of electron orbitals developed in Chapter

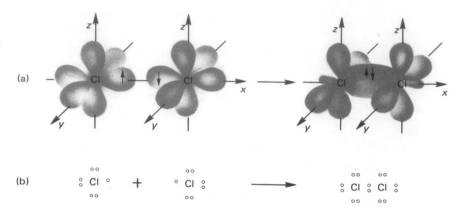

4, combination by pairing of unpaired electrons would then consist in an overlapping of an electron cloud from each of two atoms.

Figure 5.10 represents such a process for *p* electrons of two chlorine atoms; the shape of the cloud changes as a result of combination. The corresponding electron-dot picture also is shown in Figure 5.10. Calculation shows that a combination of two electrons of opposite spin, one from each of two atoms, does form a configuration of lower energy than the energy of the separated atoms, and hence is a stable state. The bond thus formed is called a covalent bond. Covalent bonds are most often formed between the atoms of the nonmetallic elements.

Experimental confirmation of the pairing of electron spins in compound formation is found from magnetic measurements. Unpaired electrons give a type of magnetism called paramagnetism. Many atoms are paramagnetic because they have unpaired electrons. When these electrons pair off with electrons of opposite spin, paramagnetism is no longer observed, but the compound shows a type of magnetism called diamagnetism. Covalent compounds are consistently diamagnetic (with few exceptions), showing that their electrons are paired (see Figure 23.7).

The significance of the two or eight valence-shell configuration in stable compounds is simply that these numbers represent the maximum number of electrons that can be held in low-energy states in the lighter elements. Combination takes place until all of the unpaired electrons in the atoms have paired with electrons from other atoms, and these maxima are reached. The nitrogen atom, for example, has three unpaired electrons, and will form bonds with other atoms until these three are paired. If each of the other atoms brings one unpaired electron, as hydrogen does, the formula of the compound will be in the atom ratio of 1:3, as NH_3. In combination with it-self, however, a nitrogen atom will pair its three unpaired electrons with the three on another nitrogen atom, to share three *pairs* between them, as :N⋮⋮⋮N:. If the electrons of the three pairs are credited to both atoms, then each atom again has eight electrons in the outer shell. The nitrogen atoms, sharing three pairs of electrons, are said to have a triple bond between them.

Exceptions to the Rule of Eight

Compound formation by pairing of unpaired electrons does not always produce eight electrons in the valence shell, however. Consider, for example, boron compounds. In the boron atom there are three electrons in the valence shell. In the configurations in Table 4.3, two of these are shown paired in

Figure 5.11 Illustration of the principle that combination takes place until all unpaired electrons are paired. (a) This atom has two unpaired electrons; (b) each of these atoms has one unpaired electron; (c) all electrons are paired.

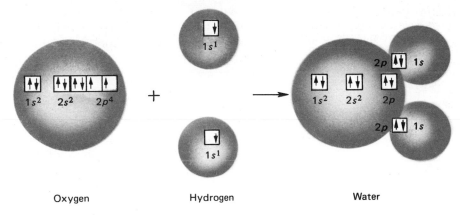

Oxygen   Hydrogen   Water

the s orbital and one unpaired in the p orbital. You might suppose then that boron should combine with one chlorine atom to give a compound like this: $\ddot{B}{:}\ddot{\underset{..}{Cl}}{:}$; all the electrons would be paired in such a compound, but only chlorine has eight in the outer shell. This compound has not been found. Apparently, under the forces of other atoms, the s electrons of the boron atom unpair, one of them being promoted to a p orbital, so that the atom behaves as if it had three unpaired electrons, and forms the compound BCl_3:

$$:\overset{..}{\underset{..}{Cl}}:\overset{..}{B}:\overset{..}{\underset{..}{Cl}}:$$
$$:\overset{..}{\underset{..}{Cl}}:$$

Covalent bond

$$\overset{oo}{\underset{oo}{O}} \overset{xx}{\underset{xx}{Cl}} \overset{oo}{\underset{oo}{O}} \;\; -$$
$$\overset{oo}{\underset{oo}{O}}$$

Coordinate covalent bonds

Figure 5.12 The chlorate ion structure illustrating that two of the oxygen atoms are bonded by a coordinate covalent bond. Here, the seven valence electrons on the chlorine atom are indicated by x; the valence electrons on oxygen are indicated by o for convenience. An additional electron □ contributing the negative charge also is shown. There is, of course, no difference between electrons, and one cannot distinguish those originally belonging to oxygen from those originally belonging to chlorine.

Note, however, that the boron atom has less than eight electrons in the outer shell; nevertheless, it reaches a moderately stable state with all its electrons paired.

In other compounds, more than eight electrons may be used in binding. In gaseous PCl_5, for example, the phosphorus atom has 10 electrons in the valence shell. Again, the two $3s$ electrons, present in the phosphorus atom, $1s^2\,2s^2\,2p^6\,3s^2\,3p_x{}^1\,3p_y{}^1\,3p_z{}^1$, unpair in the presence of chlorine atoms, and the five unpaired electrons pair with one electron on each of five chlorine atoms. Presumably the phosphorus atom uses one of the $3d$ orbitals to accommodate the fifth pair of electrons.

The Coordinate Covalent Bond

Although boron compounds in which the boron atom has six electrons in the outer shell are stable, boron forms other compounds in which there are eight electrons in the outer shell. Since the boron atom has only three unpaired electrons, such compounds can only be formed if some other atom presents boron with the two electrons needed for the fourth pair. This occurs with such atoms as the nitrogen of ammonia, which forms compounds with boron trifluoride according to the following scheme:

$$\begin{array}{ccc}
H & :\overset{..}{\underset{..}{F}}: & H:\overset{..}{\underset{..}{F}}: \\
H:\overset{..}{\underset{..}{N}}: + \overset{}{B}:\overset{..}{\underset{..}{F}}: \longrightarrow & H:\overset{..}{\underset{..}{N}}:\overset{}{B}:\overset{..}{\underset{..}{F}}: \\
H & :\overset{..}{\underset{..}{F}}: & H:\overset{..}{\underset{..}{F}}:
\end{array}$$

The bond shown between the nitrogen atom and the boron atom, in which both electrons of the shared pair come originally from one of the bonded atoms, is called a coordinate covalent bond. In general, a coordinate covalent bond is a single bond similar in properties to the type of covalent bond in which one of the electrons of the bond came from each of the bonded atoms. It differs from the usual type only in the ancestry of its electrons. Coordinate covalent bonds are common, especially in ions containing several atoms such as sulfate ion, SO_4^{-2}, chlorate ion, ClO_3^-, and phosphate ion, PO_4^{-3}. In chlorate ion, for example, two of the oxygen atoms are linked to the chlorine atom by such a bond. (Figure 5.12.)

Polar and Nonpolar Covalent Bonds

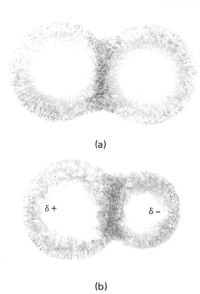

(a)

(b)

Figure 5.13 Nonpolar and polar covalent bonds. (a) Br_2—a nonpolar bond; the electron cloud is symmetrical around both nuclei. Examples: H_2, Cl_2, F_2. (b) BrF—a polar covalent bond; the electron cloud is unsymmetrical and more dense (dark shading) around one nucleus. Examples: HCl, ICl.

What would happen if the electrons in the shared pair were not shared equally between the two atoms held together by the bond—that is, if one of the atoms had more than its share of the negative charge of the electron pair? This would mean that there would be more negative charge on one end of the bond; since the molecule as a whole is electrically neutral, this would leave the other end positive, for the nuclear charge on the atom of that end would not be balanced by the number of electrons left to it. This would produce a bond which had two poles, a positive pole and a negative pole; it would be called a *polar covalent bond*. Such a bond is frequently represented by writing the signs δ^+ (delta plus) and δ^- (delta minus) over the charged atoms of the polar bond, thus $\overset{\delta+ \; \delta-}{HCl}$. Molecules having such polar covalent bonds are usually called dipoles (two poles) or dipolar molecules.

Molecular polarity is measured experimentally in terms of a quantity called the dipole moment. The dipole moment is the product of the magnitude of the charge at one end of the dipole and the distance between the two charges. The dipole moment of hydrogen chloride, for example, is 1.03×10^{-18} electrostatic unit-centimeter (esu-cm). Since the distance between the hydrogen and the chlorine nuclei is 1.27×10^{-8} cm, and the charge on the electron is 4.80×10^{-10} esu,* this datum tells us that, on the average, the chlorine atom, having the higher electron affinity, has an excess of negative charge on it, having attracted the electron cloud more strongly than the hydrogen atom did.

As a consequence of its dipolar character, a molecule having a dipole moment will tend to orient itself in the electric field between the plates of a condenser, so that the positive end is toward the negative plate and the negative end is toward the positive plate (Figure 5.14). It is this property that permits us to measure the dipole moment, since the effect of the oriented dipoles is to shield the one plate from the opposite charge on the other, thus changing, in a measurable way, the electrical characteristics of the condenser.

Most covalent bonds have polar character. Perhaps the only nonpolar bonds are those between identical atoms, as in H_2, Cl_2, and I_2. But carbon-hydrogen bonds in most organic compounds appear to be nonpolar.

Electrovalent Binding

The discussion relative to Table 5.2 has suggested that atoms combine with each other in such number ratios as to produce, among the elements in the first few rows of the periodic table at least, units in which each atom has eight (or two) electrons in its outer shell. The implications of this for covalent compounds, in which this stable configuration is reached by sharing electrons between two atoms, has been discussed on the preceding pages. The sharing process leaves each combined atom essentially neutral electrically. If, however, the ionization energies of two atoms about to combine are very different, it may be cheaper, in terms of energy, for the atoms to "buy" this stable con-

*One electrostatic unit = 3.336×10^{-10} coul.

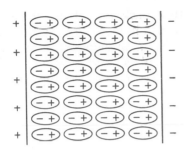

Figure 5.14 Orientations of dipoles in an electric field.

figuration by transferring an "extra" electron (over the eight-in-the-outer-shell) to an atom that needs only one more to complete its outer shell to eight—even though they both must pay for this by becoming charged electrically. The resulting particles (ions) would then be oppositely charged and would produce the hardness and rigidity associated with the properties of electrovalent compounds.

The transfer of the "extra" electron may be illustrated by the following electron-dot picture of the formation of sodium chloride from sodium and chlorine atoms:

$$\text{Na·} + \text{.Cl̈:} \longrightarrow \text{Na}^+ + \text{:Cl̈:}^-$$

The sodium ion would be positive by one unit of electronic charge, since its charge of 11 protons in the nucleus is not balanced by the 10 electrons outside ($1s^2$, $2s^2$, $2p^6$); the chlorine ion would be negative by one unit of electronic charge (17 protons, with 18 electrons). Again, both atoms (ions) have no unpaired electrons and, indeed, both have eight electrons in the outer shell. In this case, however, the electrons are not shared. Rather, the sodium atom loses control of its valence electron, which becomes the property of the chlorine atom, and the pair of ions Na^+, Cl^- results. The "bond" so formed is that resulting from the attraction of two oppositely charged particles for each other, and it is called an electrovalent or ionic bond *Ionic bonds are most often formed between metals and nonmetals.*

There is a great difference, however, between the covalent bond in hydrogen chloride and the ionic bond in sodium chloride. In a covalent bond, the two atoms are directly joined to each other; they belong to each other and the molecule exists as a unit in the crystal at low temperatures. By contrast, in sodium chloride, a chloride ion has no greater tendency to be paired with a particular sodium ion than with any other sodium ion at the same distance. As a consequence, when groups of ions are close together as in solid sodium chloride, the ions arrange themselves in such a way that every chloride ion is surrounded by as many sodium ions as can conveniently crowd around it without getting in each other's way (namely, six). Each sodium ion is similarly surrounded by six chloride ions to give the three-dimensional array shown in Figure 5.4. This contrasts with HCl, which forms crystals of *molecules* of HCl similar to those of CO_2 in Figure 5.3. The array of oppositely charged ions extends almost indefinitely in all three directions. Within the crystal there is *no* structural unit that we can point to and call a "sodium chloride molecule, NaCl." On the contrary, each chloride ion belongs to the six sodium ions that are its nearest neighbors, and, to a lesser extent, to the other sodium ions farther away; each sodium ion belongs similarly to many chloride ions. For this reason, even though we sometimes speak of an ionic bond, we cannot speak of any real bond between two specific ions. Rather, we must speak of

Figure 5.15 Transfer of an electron from the outer shell of one atom to the outer shell of the other, as in Na + Cl → NaCl.

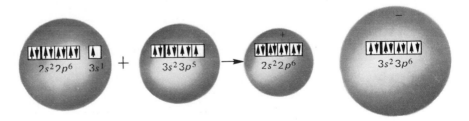

electrovalent binding as the type of force that holds ionic crystals together. The formula NaCl thus does not represent a molecule of sodium chloride, but represents the simplest ratio of Na^+ and Cl^- ions in the compound.

For atoms differing from sodium in the number of electrons outside the noble gas shell of eight, and from chlorine in the number of electrons short of a noble gas shell, we would expect a different number of electrons to be lost or gained, if ions are formed. This is actually the case, and we observe ionic ratios such as in $Ca^{+2}2Cl^-$, in which a calcium atom has released two electrons to chlorine atoms (one electron going to each of two atoms) and in $Mg^{+2}O^{-2}$, in which the two electrons released by magnesium have gone to fill the second shell of a single oxygen atom from six to eight. Here again, the crystal structure may be pictured as an infinite three-dimensional array of alternating positive and negative ions. The same situation occurs in other ionic compounds.

It must not be supposed that all simple ions have eight electrons in the outer shell. One common ion that does not is the zinc ion, Zn^{+2}, which has the electron configuration $1s^2\ 2s^2\ 2p^6\ 3s^2\ 3p^6\ 3d^{10}$, with 18 electrons in the outer shell. There are many others with more than eight among the transition elements. In the first 20 elements, however, the eight-in-the-outer-shell configuration for ions is commonly observed.

Many ions are charged particles containing several atoms that are themselves held together by covalent or coordinate covalent bonds. Thus, ions such as NH_4^+, ClO_3^-, SO_4^{-2}, and others are well known. In the solid state, these ions are present in crystal lattices or networks associated with ions of opposite charge in a fashion similar to that for simple ions, such as the sodium and chloride ions.

Criterion for Bond Type
Electronegativity

Which pairs of elements will combine to form electrovalent compounds, polar covalent compounds, or nonpolar covalent compounds? The type of compound formed will evidently be determined by the relative attraction of the two nuclei for electrons, as modified and screened by the inner shells of electrons. This modification and screening lead us to consider only the electrons in the outermost shell, the *valence electrons*, which are the ones concerned in chemical binding.

Let us consider the possibility of combination of two atoms A and B, each of which has a single unpaired electron in its valence shell. If electrovalent binding occurs between A and B, one of the atoms must have a greater ability than the other to attract the unpaired electron. If A and B form a nonpolar covalent bond, the two atoms must have nearly the same ability to attract electrons. Polar covalent bonds should arise when the two atoms have different electron-attracting abilities but not so widely different that an ionic bond is formed. If some measure of this electron-attracting power were available, the type of bond formed between any two atoms could be predicted.

A number of ways of estimating the relative electron-attracting power of atoms have been devised, and the values are recorded in what is known as a scale of *electronegativities* of the elements. Electronegativity might be defined as a measure of the attraction exerted by the atom upon electrons in its valence shell; this attraction determines the type of bond the atom is capable of forming.

The type of binding formed—whether covalent, polar covalent, or electro-

Table 5.3 Electronegativity Values of the Elements*

Metals												Nonmetals				
Li 1.0	Be 1.5											B 2.0	C 2.5	N 3.0	O 3.5	F 4.0
Na 0.9	Mg 1.2											Al 1.5	Si 1.8	P 2.1	S 2.5	Cl 3.0
K 0.8	Ca 1.0	Sc 1.3	Ti 1.5	V 1.6	Cr 1.6	Mn 1.5	Fe 1.8	Co 1.8	Ni 1.8	Cu 1.9	Zn 1.6	Ga 1.8	Ge 1.8	As 2.0	Se 2.4	Br 2.8
Rb 0.8	Sr 1.0	Y 1.2	Zr 1.4	Nb 1.6	Mo 1.8	Tc 1.9	Ru 2.2	Rh 2.2	Pd 2.2	Ag 1.9	Cd 1.7	In 1.7	Sn 1.8	Sb 1.9	Te 2.1	I 2.5
Cs 0.7	Ba 0.9	(rare earths) 1.1–1.2	Hf 1.3	Ta 1.5	W 1.7	Re 1.9	Os 2.2	Ir 2.2	Pt 2.2	Au 2.4	Hg 1.9	Tl 1.8	Pb 1.8	Bi 1.9	Po 2.0	At 2.2
Fr 0.7	Ra 0.9	(actinides) 1.1–1.7														

*Values given by Pauling for elements in their common oxidation states.

valent—thus depends upon the *electronegativity* of the atoms. You might suppose, from its relative nature, that each compound would have to be considered separately. Fortunately, however, there is sufficient regularity in how atoms of a particular element interact with atoms of other elements to permit us to assign numerical values of electronegativity to those elements for which necessary data have been obtained. These numerical values express semiquantitatively the tendency of the atoms to appropriate electrons available to take part in bonding. In this electronegativity scale, the noble gases are assigned the value 0, and fluorine—the element the atoms of which have the greatest attraction for electrons—the value 4.0. The other elements then fall between these limits, as shown in Table 5.3.

The values are such that electrovalent compounds between two elements are represented by differences in electronegativity *values* for the two elements greater than approximately 1.7. In general, covalent compounds have difference values less than 1.7, and the nearer the difference is to zero, the more nearly nonpolar is the covalent bond formed. For example, sodium chloride, an electrovalent compound, has an electronegativity difference 3.0 (for chlorine) − 0.9 (for sodium) = 2.1. This is greater than 1.7, and in accord with the electrovalent character of sodium chloride. Hydrogen chloride, a polar covalent compound, has an electronegativity difference 3.0 (for chlorine) − 2.1 (for hydrogen) = 0.9. This is less than 1.7, and in accord with the polar covalent character of hydrogen chloride. A chlorine molecule has a nonpolar covalent bond, as recorded by the difference 3.0 − 3.0 = 0 for the electronegativity values.

Trends in Electronegativity Examination of the electronegativity values in Table 5.3 shows that the trends in electronegativity are those to be expected from a consideration of the nuclear charge, electron configuration, and size of the atoms. Small atoms will attract electrons more strongly than large atoms, and there is a general trend toward smaller electronegativity from top to bottom of each column in the non-transition elements. For atoms of about the same size, the attraction for electrons should increase with increasing nuclear charge. Thus there is a general trend toward larger electronegativity from left to right in the rows of the periodic table.

Exceptions to the above generalizations are more likely among very small atoms, very large atoms, or those in higher oxidation states. For example, in both beryllium fluoride, BeF_2 (electronegativity difference 2.0), and silicon tetrafluoride, SiF_4 (electronegativity difference 2.2), the bonding is known to be at least partially covalent. Silicon tetrafluoride has a melting point of $-90°C$. In both cases, the covalent bonding is a result of several factors, including the fact that both cations, should they form, would be very small and would have extremely high charge densities. In Be^{+2} a charge of $+2$ would be distributed over an ion having a radius of 0.38 A; in Si^{+4} a charge of $+4$ would be distributed over an ion having a radius of 0.41 A. Such high charge densities would result in strong attraction for and polarization of the electron clouds of the fluoride ions, thereby giving the bonds their covalent character.

Comparison of the Properties of Covalent and Electrovalent Compounds

Covalent Compounds Made of Small Molecules. Earlier we made the hypothesis that the typical covalent compounds of the first column of Table 5.1 were made of small neutral particles (molecules), and that only weak forces of attraction existed between these molecules. The assumption of weak forces between molecules explains the low boiling points and small liquid ranges, since these properties mean that it is easy to separate the molecules from each other, moving them from the close distances of solids and liquids to the large distances characteristic of gases. One reason for this is that nearly all of the charges—positive and negative—present in the atoms are satisfied in a single molecule. The electric forces acting outside the molecule, which could attract nuclei and electrons of atoms in other molecules, are very weak. Separation of the molecules from each other thus involves only these weak *van der Waals forces of attraction of the electrons and nuclei of atoms in one molecule for the nuclei and electrons of atoms in another molecule.*

The presence of polar bonds and dipoles in a molecule often increases the tendency of molecules to stick to each other. The positive end of one molecule attracts the negative end of another. The positive end of the second attracts the negative end of another, and so on. The result is to increase the boiling point of a compound containing polar molecules, compared to that of similar compounds whose molecules are nonpolar, and to make it more soluble in water (which is also a polar compound).

Covalent Compounds Made of Large Molecules. A crystal of diamond or silicon dioxide may be one giant molecule of network type with all atoms bonded with covalent bonds (Figure 5.16).

No weak van der Walls forces are involved in holding this structure together as in the collection of small covalent molecules; rather, all of the bonds are the strong covalent type. The rigidity of this structure can be destroyed only by breaking some of these bonds to make smaller fragments. High temperatures are therefore necessary in order to melt or vaporize such compounds. Many types of large-molecule covalent compounds exist and will be studied later; some of these are starch, cellulose, and proteins (Chapter 28).

Figure 5.16 Macromolecules. These are three-dimensional structures: (I) a quartz crystal or (II) a diamond is a three-dimensional network of atoms, and is known as a macromolecule. (a) Structure of quartz. (b) Structure of diamond—small unit of large network structure showing more magnified structure.

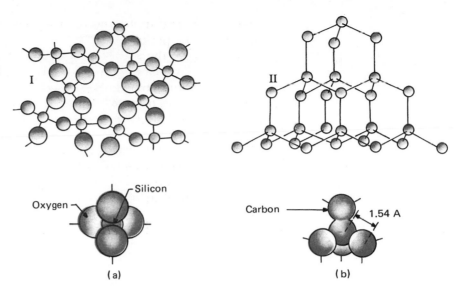

Electrovalent Compounds. The properties of electrovalent compounds are those listed in Table 5.1. The rigid pattern of charged ions in the crystal, tightly interlocked by the electrostatic forces between the alternating particles of opposite charge (Figure 5.4), gives to these compounds properties quite different from those of covalent compounds. In covalent compounds of small molecules, the molecules are only weakly bound to each other, and it is easy to separate one from another. In electrovalent compounds, however, the forces are equally strong between all particles in the crystal, and separation of one unit from another is a matter of extreme difficulty. Hence the melting points are high, and the crystals are hard. The attractive forces between charged ions persist in the melted material; it is difficult to separate the ions by boiling, and the liquid range is large. The ions are free to move in the melt, however. Released from the rigid pattern of the crystal, they can, by their motion, carry electric charge across the melt which, therefore, conducts electricity.

Metallic Binding What holds the atoms of metals to each other? Certainly these forces must be different from those in covalent or electrovalent compounds. They produce not only the hard metals such as chromium, vanadium, and tungsten, but also the malleable metals such as copper and silver.

Perhaps the most unusual characteristic of a solid metal and the property that distinguishes a metal from other substances is the ease with which it conducts electricity. We believe that this ease of conductivity arises, in turn, from the ease with which electrons can move through metals; a current in a wire is the result of the passage of electrons along the wire. No chemical change accompanies this passage. Thus, the movement of electrons does not result from ionization processes. The electrons are fed in at one end of the wire from a dynamo or battery and, in the completed electrical circuit, return to the dynamo or battery at the other end of the wire.

There must then be electrons in the metal of the wire that are not attached to specific atoms, that can be exchanged for other electrons from the dynamo, and that are free to travel through the wire. Upon investigation, we find that

Table 5.4 Electron Configurations of Metals Among Elements 1 to 20

Element	Electron Configuration
Li	$1s^2\,2s^1$
Be	$1s^2\,2s^2$
Na	$1s^2\,2s^2\,2p^6\,3s^1$
Mg	$1s^2\,2s^2\,2p^6\,3s^2$
Al	$1s^2\,2s^2\,2p^6\,3s^2\,3p^1$
K	$1s^2\,2s^2\,2p^6\,3s^2\,3p^6\,4s^1$
Ca	$1s^2\,2s^2\,2p^6\,3s^2\,3p^6\,4s^2$

this freedom of motion which electrons have in metals also accounts for the high heat conductivity and for the luster. Heat conductivity, like electrical conductivity, is unique to metals. For instance, you can hold a wooden stick in a flame without discomfort; a metallic rod, however, quickly transmits the heat to your hand.

Electron Structure of Metal Atoms

In reconciling this freedom of electrons in metals with the requirement that electrons be used in binding atoms together, we must note that the atoms of the metallic elements have only a few electrons in the outermost shells. Among the first 20 elements, for example, the metals include only those listed in Table 5.4.

In a pure metal, all atoms are alike. Having so few electrons, it is not possible for metal atoms to combine with each other to produce the eight-in-the-outer-shell configuration of a noble gas. The nearest approach to eight could be made by aluminum, but not even this element can produce more than six electrons for every two atoms, either by sharing or by transfer.

Electrons Shared by Several Metal Atoms: Community Sharing

An approach to the stability of the eight-in-the-outer-shell configuration can be made, however, if we assume that electrons can be shared among more than two atoms. This apparently occurs in metals, and the sharing is really on a grand scale. Each atom in a metal crystal contributes all its valence electrons to the community, retaining only a small share in them but acquiring, at the same time, a small share in the electrons of all the other atoms. Since even a small bit of metal may contain as many as 10^{20} valence electrons, a small share may still add up to a considerable number; their bonding effect is thus significant.

Energy Levels in Metals

This grand sharing, however, must still operate within the restrictions of the Pauli exclusion principle. There must therefore be energy levels within the metal crystal, as there are in isolated atoms, and no more than two electrons

Figure 5.17 Electron cloud distortion in various types of bonds. (a) Very little distortion: ionic bond. (b) Distortion caused by interaction: covalent bond. (c) Distortion caused by interaction among many atoms: metallic bond.

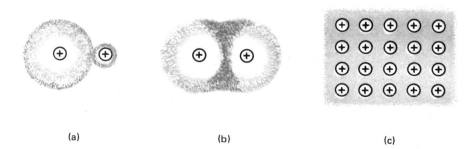

(a)   (b)   (c)

may be in any one of those energy levels. A difference between the metal crystal and the isolated atom, however, is that the energy difference between one level and another in the metal is very small. It is easy for an electron to move from one level to another, provided that the level into which it is moving is not filled with its quota of two electrons. Further, each of these energy levels belongs to the crystal as a whole; the electron cloud that occupies it extends through the crystal as a whole, and is a part of the valence shell of each atom in the crystal.

On the average, each atom has a sufficient share in the many electron clouds to satisfy its valence requirements. These electron clouds are at the same time satisfying the requirements of all the other atoms in the crystal, and the attraction of all the atoms for the super-cloud of electron clouds gives rise to metallic binding.

Location of Valence Electrons in Different Types of Bonding

1. *Covalent Binding: Localized Electrons and Directional Bonds.* A sharp distinction among the three types of binding is apparent when we examine the distribution of negative charge about the atomic nuclei. In a covalent molecule the negative charge of the two electrons of the bonding pair is *localized* between the two nuclei held together by the covalent *bond*. If such localized spots of negative charge are on the same atom, they will tend to repel each other. Consequently, covalent bonds from the same atom are *directional*. This is implied in the spatial arrangement of the *p*-electron clouds on isolated atoms.

2. *Electrovalent Binding: Localized Electrons and Nondirectional Bonds.* In an electrovalent substance, negative charge is still *localized* on the negative ions, but the forces between the ions are *nondirectional*. The ions behave like charged spheres, positive and negative, and the attractions and repulsions depend only upon the distances between the charges and not upon the direction from one charge to another. In a crystal of an electrovalent substance, therefore, the arrangement of ions depends upon two general factors; how the ions pack together in such a ratio that the solid as a whole is neutral, and the need to keep positive and negative ions as far away as possible from other ions of like charge but as close as possible to ions of opposite charge. (In some cases, the relative size also is a factor.) No one ion has any particular attachment to any other specified ion, however, and if we add enough energy to melt the crystal, the ions are free to move about, exchanging partners promiscuously under the influence of the heat motion.

3. *Metallic Binding: Nonlocalized Electrons and Nondirectional Bonds.* Metallic binding differs from both of the other types by having the negative charge *nonlocalized*, that is, spread throughout the crystal. This situation has been likened to a diffuse "sea" of negative charge bathing the positive units left behind when the valence electrons leave to become a part of the sea. The electrostatic forces between these positive units are again *nondirectional*, and the atoms dispose themselves in a regular array determined by the way in which they can be packed to be as far apart from every other atom as possible, yet closely grouped to share in the negative charge that surrounds them. Since, in a pure metal, all atoms are alike, the structural grouping is almost always that which would be reached by packing spheres together with as many points of contact between the spheres as possible (Chapter 13).

Perhaps one of the properties of the three types of solids easiest to interpret in terms of binding is the difference in how they behave under stress. A metal is characterized by the fact that it can be forged into varied shapes, or drawn into wire. The ease of forging means that the units of the structures move easily with respect to each other; the ease of drawing means that the units maintain their cohesion even while changing position. This is easily understood on the basis of the concept of metallic binding just developed.

In a pure metal, all the positive units are alike, and moving a group of them with respect to the others makes little change in the energy of the whole. Contrast this with the behavior of an electrovalent compound if we should try to change the position of one of the ions. Motion of, say, a positive ion, in whatever direction, must bring it closer to another positive ion. Such motion is opposed by the repulsion of the second ion, and it turns out to be energetically easier for the crystal to break apart than to suffer deformation. In a crystal of a nonpolar covalent substance, the forces holding the molecules together are so small that molecules may move over each other with relative ease. These substances cannot be drawn, however, because application of even a small force will pull the crystal apart.

Materials and the Quality of Life. In essence, this chapter has given us an overview of the nature of the materials found on earth or extractable from it. We know that many human ills and much poverty have been eliminated or at least minimized by learning to use these materials to provide better food, clothing, and shelter for all. However, as populations expand and the quality of life increases, there is a serious question as to whether our supplies of raw materials are sufficient to meet our needs.

To support one American citizen, an estimated 25 tons or 2300 kilograms of materials of all kinds must be extracted from the earth and processed each year. The increasing need for metals, stone, and concrete to be used for tools, shelter, transportation, and communication is a reality of the contemporary world. Current consumption of steel in the U.S. is about 640 kilograms per person per year. Actually, we use about 180 kilograms per person of common salt and phosphate rock, 82 kilograms of lime, and 250, 43, and 35 kilograms per person of clay, gypsum, and sulfur, respectively. Estimates are that for every person in this country, there is now in use about 150 kilograms each of lead and copper, just over 100 kilograms of aluminum, just under 100 kilograms of zinc, and roughly 20 kilograms of tin.

The 18 most prosperous nations, with a total population of about 680 million, consume steel at rates varying between 300 and 700 kilograms per capita annually. For example, Japan and the Soviet Union consume over 515 and over 415 kilograms per capita per year, respectively. The 13 developing countries, with a population totaling about 1.4 billion, use between 10 and 25 kilograms per person per year. India and Brazil use about 13 and 47 kilograms of steel per person per year, respectively. During the decade 1957-1967, the worldwide *rate of increase* in per capita steel consumption was 44 per cent, compared to the U.S. increase of 12 per cent. This country's demand for metal is expected to increase 4.5 times between 1970 and 2000.

Since the inorganic raw materials needed to supply these needs are the products of extremely slow processes in the lithosphere and are therefore not self-renewing (as are most of the organic materials of the biosphere), enormous strains are being placed on the earth's resources and on man's ability and ingenuity to extract the needed amounts.

Rough estimates of the lifetimes of metal reserves (ores containing amounts of the metals extractable economically by existing technology), based on the assumption that use of metals will continue to rise with population growth and increasing per capita requirements, indicate that the known U.S. reserves of platinum, silver, gold, aluminum, tin, lead, tungsten, nickel, chromium, and manganese will be all but exhausted by 1980. And by 2000, only iron and molybdenum among the commonly used metals will still be available from traditional U.S. sources. World supplies of these metals will last only a few decades longer than our own reserves.

Despite this, the prospects for sustaining and even improving the level of technological development in this country are good, provided we take steps now to deal with the rapidly developing problems. The same is true for the rest of the world, although the magnitude of the task faced by the developing nations is such that without a massive infusion of production facilities and know-how, it will take the better part of a century for them to reach production levels sufficient to feed, clothe, and shelter their populations adequately.

To maintain adequate supplies of raw materials, we will have to recycle waste materials more effectively. We will also have to develop adequate substitutes for metals, such as plastics and materials made from ordinary rock. The potential in recycling metals can be illustrated with steel. On a per capita annual basis in the U.S., about 350 kilograms of steel becomes obsolete or lost, most of it by corrosion. Of this, about 40% is recovered and recycled (as junked automobiles or other worn-out iron or steel products). About 210 kilograms per person is lost every year.

Obviously, improved efficiency of recycling is desirable for all solid wastes. Each year, the average American throws away almost 300 cans, 150 bottles, and about 150 kilograms of paper. Properly recycled, our solid wastes could provide raw materials for the steel, glass, aluminum, and plastics industries. Such recycling, together with extracting needed metals from ordinary rock, when combined with the plastics and nonmetallic inorganic substitutes for metals now being developed by chemists and other materials scientists, could keep us for millennia in the style of living to which we have become accustomed. However, prudent action by an informed citizenry is needed to ensure this.

The ideas on the structure of materials given in this chapter and developed in subsequent parts of this text are the basis from which scientists and engineers must work to develop adequate substitutes for metals, more efficient processes for recycling obsolete materials, and for extracting metals from low-grade ores. They also form the basis from which an enlightened citizen can make intelligent judgments regarding our mineral resources and the inorganic materials he uses in his life.

SUMMARY   About 100 different elements exist in the universe—meaning that only about 100 different kinds of atoms exist (excluding isotopes)—but the universe contains several millions of compounds! Why so many compounds?

Furthermore, why the difference in the properties of an element in a compound, as compared to the properties of a free element? For example: The element chlorine is a noxious, irritating gas, and sodium is a soft, very reactive metal in air and water. But table salt, composed of these two elements, is a stable substance, used on our food and essential for life processes. The element carbon in pure form is usually a black, insoluble solid found as graphite or in an impure form as soot, coal, or charcoal, but it represents a large percentage of the compound called sugar, a white, soluble solid also used in our foods. How do we account for these differences in properties as we go from the free element to compounds of the element?

And finally, a metal is made of atoms bound together to give a substance usually with a high melting point, but the mass of metal has the property of being a good conductor of an electric current whereas many other bonded atoms do not have this property. Why?

Such are the questions that the chemist is often asked. He finds his main answers in two sources: the electron structure of atoms (Chapter 4) and the type of bonds between atoms.

We have examined in this chapter the nature of the forces binding atoms together in the various types of known substances. We have found that atoms combine by sharing pairs of electrons (covalent bonds), by transfer of electrons (electrostatic binding), or by community sharing of electrons (metallic binding). We have seen that the combining power of an atom is related to the number of electrons in its valence shell, and that light atoms combine until they acquire a stable configuration of eight or, in some cases, two outer-shell electrons. The concept of electronegativity was introduced and was used as a criterion for predicting the kind of binding to be expected between atoms. Finally, the properties of substances were related to the kind of binding present.

IMPORTANT TERMS

Combining forces of atoms	**Atom ratios in compounds**
Types of binding	**Stability and noble gas configuration**
covalent	**Lewis formulas**
electrovalent	**Coordinate covalent bonds**
metallic	**Polar and nonpolar covalent bonds**
Types of substances	**Dipole moment**
covalent	**Electronegativity**
electrovalent	**Localized and nonlocalized binding**
metallic	**Paramagnetic**
network structures	**Diamagnetic**

QUESTIONS AND PROBLEMS

1. Using the Lewis electron-dot method, represent the following atoms: (a) hydrogen, (b) fluorine, (c) sulfur, (d) phosphorus, (e) boron, (f) silicon, (g) Group IA elements, (h) Group IIA elements, (i) the noble gases.

2. Using electron-dot structures write formulas for the following: (a) Cl_2, (b) HF, (c) ClBr, (d) NF_3, (e) CCl_4, (f) CO_2, (g) NO_2, (h) Li_3N, (i) $CaCl_2$, (j) $KClO_3$, (k) H_2SO_4, (l) SF_6, (m) $XeOF_4$.

3. Write electron-dot formulas for each of the following: (a) HNO_3, (b) N_2O_5, (c) Cl_2O_7, (d) HCN, (e) HNO_2, (f) H_2NNH_2, (g) CS_2.

4. Comment on each of the following statements: (a) The greater the difference in electronegativities of two atoms, the stronger is the bond between them. (b) It is possible to state with assurance that chemical bonds between atoms are either electrovalent or covalent. (c) Covalent bonds are more likely to occur between atoms of nearly the same electronegativity than between atoms of widely different electronegativities.

5. Classify each of the following as covalent, electrovalent, or metallic on the basis of the properties indicated: (a) This is a white solid that melts at 772°C to give a transparent liquid that conducts electricity. (b) This is a silver-white solid that melts at 98°C to give a silvery liquid. Both the solid and liquid are good heat conductors. (c) This is a yellow solid melting at 113°C to give a clear yellow liquid. Both liquid and solid are poor conductors. (d) This is a dark metallic-appearing solid that sublimes readily, giving a purple vapor. It is a poor conductor of heat and electricity.

6. Which of the substances, magnesium fluoride or sulfur hexafluoride, is expected to have: (a) the lower melting point; (b) the better heat conductivity; (c) the better ductility; (d) a tendency to be brittle; (e) the higher melting point.

7. Offer an explanation for the trend in dipole moments μ among the following: $CH_4, \mu = 0$; $CO, \mu = 0.12$; $HI, \mu = 0.38$; $HCl, \mu = 1.03$ Debye units 1.0×10^{-18} esu-cm.

8. Element A is a good conductor of electricity; element B is a nonconductor. A and B are both solids. Predict which element probably (a) is the most malleable; (b) is the better heat conductor; (c) has the greater number of valence electrons; (d) has the higher ionization energy; (e) is brittle.

9. The bonds in diamond and in the hydrogen molecule are completely nonpolar. Despite this similarity, hydrogen is a gas at ordinary temperatures and diamond has a very high boiling point. Reconcile this disparity.

10. Compare the two hydrides, SiH_4 and LiH. The melting point of the latter is much higher. Explain this based on the factors which contribute to varying physical properties.

11. The following compounds are ionic. Write electronic structures for them. (a) BaO_2, barium peroxide; (b) CaC_2, calcium carbide; (c) $LiBH_4$, lithium borohydride.

12. Saturation of valence occurs when the atoms involved in formation of a compound exhibit the limits of their valence toward one another. With this in mind, indicate how the following compounds show saturation of valence: (a) KCl, (b) $MgCl_2$, (c) PCl_5, (d) HBr.

13. Consider the atoms A, B, C, D, and E occurring in the *same family* such that their valence electrons are in energy levels 2, 3, 4, 5, and 6, respectively. They each possess 6 valence electrons. Predict the answers to the following: (a) Will the bond between A and B be primarily ionic or covalent? (b) Between B and E? (c) Will any polarity be exhibited in the interfamily molecules possible here? (d) Write the electronic formula for a compound between any two members. (e) Between which two atoms would the electronegativity values be more diverse? (f) Assuming the electronegativity values in (e) to be 3.5 and 2.0, could any possible interfamily molecule of this group be ionic in nature?

14. Atoms A, B, C, D, and E occur in the *same period*. They have one, two, four, five, and seven valence electrons, respectively. Predict: (a) The formulas for the compounds between A and E, B and E, and C and E. (b) Which electronegativity value is greater, that of B or D. (c) Whether the compound between B and E is ionic or covalent. (d) The electronic formula

for the compound between *D* and *E*. (e) Which of these atoms might form a diatomic molecule. (f) Which of the five has the highest ionization potential. (g) Which has the highest electron affinity.

15. Predict the bond type for the following compounds: (a) CO, (b) $MgCl_2$, (c) $TeBr_2$, (d) S_8, (e) Na_2S, (f) ICl, (g) PCl_3.

16. Using electron-dot formulas, write equations representing the following reactions: (a) A potassium atom and a chlorine atom forming potassium chloride. (b) A magnesium atom and an oxygen atom forming magnesium oxide. (c) The formation of the nitrogen molecule from two nitrogen atoms. (d) The formation of the water molecule from two hydrogen atoms and an oxygen atom.

17. Silicon carbide, SiC, is marketed under the general name Carborundum. The compound contains no multiple bonds. (a) Can you suggest a structure for it, observing the restrictions of the covalency and octet rules? (b) Predict a melting-point range for Carborundum based on that structure and any other properties that seem evident.

18. The radii of positive ions are always smaller than those of their respective neutral atoms, whereas the radii of negative ions are always larger than their respective neutral atoms. Explain.

19. Compare and give the reason for similarities and differences in the physical properties of the following compounds: (a) $MgCl_2$ and $BeCl_2$; (b) SiO_2 and SiF_4; (c) KH and HF.

20. Compare PF_3 and SiF_4. The former is a polar compound. Since silicon and fluorine are not greatly separated in the periodic chart, you would expect the same characteristic for SiF_4. However, SiF_4 is a nonpolar substance. Why is this so?

21. How does coordinate covalency occur? Each of the following molecules has one or more coordinate bonds. Diagram the formation of each. (a) SO_4^{-2}, (b) PO_4^{-3}, (c) ClO_3^-, (d) H_3O^+, (e) $C_3H_5NH_2BF_3$.

22. The following molecules contain some double and triple bonds. Write an electronic formula for each observing the octet rule. (a) N_2O_4, (b) H_2CO_3, (c) C_3H_4, (d) HCHO, (e) C_2N_2, (f) HCN, (g) C_2Cl_4.

SPECIAL PROBLEMS
1. Based on the criteria put forth in this chapter and with reference to Table 5.1, predict the properties of the following substances: (a) CsF, (b) $SnCl_4$, (c) crystalline As_4, (d) solid HF. (e) Check your answers by looking up these compounds in a handbook.

2. The structures of the crystal lattices of potassium chloride, sodium fluoride, and magnesium oxide are all alike. (a) Assuming that the interionic distance is the same for each compound, calculate the densities of the crystals. (b) Compare the melting points of the three compounds—KCl (790°C), NaF (992°C), and MgO (2800°C). Explain the difference.

3. In the HF molecule, the nuclei are separated by a distance of 0.917 A. The dipole moment is 1.91×10^{-18} esu-cm. (a) What is the charge at one end of the molecule? (b) If the compound were completely ionic, the ions would bear the full electronic charge of 4.8×10^{-10} esu. Calculate the par-

tial ionic nature of the molecule. (c) Compare the H—Cl bond with the H—F bond as to polarity. Explain your answer.

4. Offer an explanation for the fact that the elements in Group V of the periodic table—nitrogen, phosphorus, arsenic, antimony, and bismuth—change in properties from distinctly covalent substances to distinctly metallic substances in proceeding from nitrogen to bismuth.

5. Several elements, including tin and antimony, exist in either a metallic or a nonmetallic form, depending on the temperature. Which form would you expect to be stable at lower temperatures? Why?

6. Mixtures of many molten metals solidify on cooling to form solid solutions. Is this consistent or inconsistent with our ideas on metallic binding?

7. Predict the relative values for the electrical conductivity at room temperature and at 1000°C of copper, melting point (m.p.) 1083°C; sodium chloride, m.p. 808°C; and naphthalene, C_8H_{10}, m.p. 80°C. Discuss the basis for your predictions.

REFERENCES FLANAGIN, D., *Materials, A Scientific American Book,* Freeman, San Francisco, 1967.
GEHMAN, W. G., "Standard Ionic Crystal Structures," *J. Chem. Educ.,* **40**, 61 (1963).
PAULING, L., *The Nature of the Chemical Bond,* third edition, Cornell University Press, Ithaca, 1960.
VERHOEK, F. H., "What Is a Metal?" *Chemistry,* **37**, 6 (1964).
VON HIPPEL, A. R., "Molecular Designing of Materials," *Science,* **138**, 91 (1962).

MOLECULAR STRUCTURE AND
INTERMOLECULAR FORCES OF ATTRACTION

Studies of covalent, electrovalent, and metallic binding have led to several especially interesting and useful concepts of attractive forces between atoms. These concepts have enabled chemists to account for the properties of many known substances and to devise methods to synthesize many previously undiscovered materials having properties that could be predicted prior to their syntheses.

Because of its great power in correlating and predicting properties of matter, the theory of chemical bonding will be developed in more detail in this chapter, with particular emphasis on the nature of the covalent bond and the shapes of molecules.

Molecules are here defined* as *particles consisting of two or more atoms held together by covalent bonds.* These species may be neutral, such as CH_4, C_2H_5OH, H_2, H_2O; or they may be charged, such as sulfate ion, SO_4^{-2}, where the four oxygen atoms are viewed as being bonded covalently to the sulfur atom, and ammonium ion, NH_4^+, where four hydrogen atoms are covalently bonded to the nitrogen atom.

Nature of the Covalent Bond

The idea that a pair of electrons shared by two atoms constitutes a covalent bond was discussed in Chapter 5. The strong interaction between atoms that occurs when orbitals containing single electrons from two atoms overlap results in an attractive or binding force between the atoms if the electrons

---

*The term *molecule* was originally used by chemists in connection with the kinetic-molecular theory of gases, in which context it was defined as the smallest independently existing particle in the gas. In recent years many chemists have found it convenient to use the newer definition given above. This newer definition is somewhat more limited than the older one used in the kinetic theory, but more definitive.

have opposite spins. When the chemical bond forms, energy is released, and we say that the bonded atoms have a lower potential energy than the unbonded (free) atoms. This is illustrated in Figure 6.1.

Many scientists believe that this binding force arises because the electron charge becomes highly concentrated in the region between the atomic nuclei with the result that the positively charged nuclei are attracted toward the negative electronic charge between them (Figure 6.2). It is reasonable to expect that the magnitude of the force between the electronic charge and the two nuclei will depend upon the charges on the nuclei, the distances between the nuclei, and on the other atoms or groups of atoms that are bonded to the atoms in question. Chemists have measured the energy needed to break a covalent bond, the covalent bond energy (Table 6.1), and have determined the distance between nuclei held by a covalent bond, the covalent bond distance.

Bond Energies and Bond Distances

Some covalent bond energies and covalent bond distances are given in Table 6.1 and Figure 6.3. Note the very high bond energies for bonds having bond distance less than one Angstrom (A). Higher bond energies are generally associated with small atoms and relatively large differences in electronegativity between the bonded atoms. Compare these covalent bond energies with the energy binding water molecules to each other in liquid water (9700

Table 6.1 Some Bond Energies and Bond Distances

Bond	Bond Energy (kcal/mole of bonds)	Bond Distance (A)
H—H	103	0.75
H—O (in water)	110	0.96
H—C (in hydrocarbons)	98	1.09
H—N (in ammonia)	93	1.01
I—I	35	2.66
F—F	36	1.42
H—F	135	0.92
H—I	71	1.62
N≡N	225 (for total of 3 bonds)	1.10
N=N	100 (for total of 2 bonds)	1.23
N—N	38	1.47
C—C	82	1.54
C=C	145 (for total of 2 bonds)	1.35

NOTE: The bond energy is the average energy required, per mole of bonds, to break all the bonds of the type indicated, in gaseous molecules.

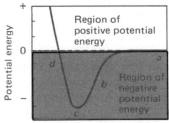

Distance between atoms

Figure 6.1 A plot of potential energy of a pair of atoms which form a diatomic molecule as a function of distance between atoms. At point a the atoms are widely separated and are not interacting. We say that the potential energy in this region is zero. At point b the bonding orbitals of the atoms are overlapping somewhat, the bond starts to form, and the potential energy drops. But the atoms have not yet reached the equilibrium bond distance. At c the bond has formed. Overlapping is at a maximum, potential energy is at its lowest, and distance between atoms is the equilibrium bond distance. Starting from the internuclear distance c, energy must be added to reach any other point on the curve. At d the atomic separation is less than the bond distance, and the potential energy rises because of repulsion between the two nuclei. At all points of negative potential energy on the curve, the net attractive force of the two atoms for one another is greater than the repulsive forces acting to separate the atoms.

Figure 6.2 Illustrating the concentration of electronic charge between two nuclei, which gives rise to the attractive force in the covalent bond.

cal/mole), and with the ionic bond energy (181,000 cal/mole) necessary to break all bonding in the ionic solid sodium chloride.

Chemists often make use of bond energies in estimating energy changes in reactions. For example, if gaseous water is to be formed from hydrogen and oxygen, the net energy change must be the difference between (1) the energy needed to break the H—H bonds in two molecules of hydrogen and the O—O bond in oxygen and (2) the energy released in forming four O—H bonds in the two molecules of water formed. It is as if we imagined the reaction to go in steps, first dissociation of reactants into atoms, then combination of these atoms to form the product molecules. On a molar basis this becomes

$$2(\text{H—H}) + (\text{O—O}) \xrightarrow[\substack{2 \times 103 \\ +1 \times 117}]{\text{add energy}} \begin{vmatrix} 4\text{H} \\ 2\text{O} \end{vmatrix} \xrightarrow[4 \times 110]{\substack{\text{energy} \\ \text{released}}} 2(\text{H—O}_{\text{H}})$$

and the net energy released in forming 2 moles of water from its elements is $4 \times 110 - (2 \times 103 + 1 \times 117) = 117$ kcal. This topic will be discussed further in Chapter 18.

MOLECULAR STRUCTURE

Orientation of Covalent Bonds in Space: Stereochemistry

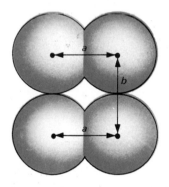

Figure 6.3 Two types of internuclear distances in covalent substances: a, covalent bond distance and b, van der Waals distance. In molecules composed of similar atoms, the covalent radius is $\frac{1}{2}a$ and the van der Waals radius is $\frac{1}{2}b$.

That molecules have definite shapes was deduced by organic chemists about the middle of the 19th century from studies of the number and kinds of products obtained in a variety of chemical reactions. More recently, such techniques as molecular spectroscopy, X-ray and electron diffraction, and electron microscopy have confirmed this deduction and have provided precise information regarding molecular shape. The water molecule, for example, is angular or bent; the angle

$$\underset{\text{H} \qquad \text{H}}{\text{O}}$$

is 104.5°. Table 6.2 gives the shapes of some simple molecules.

As the knowledge of molecular shapes developed, chemists interested in chemical binding attempted to extend the theory of the covalent bond to account for the reported shapes and bond angles. They realized that molecular shape must result from orientation of the covalent bonds in definite directions in space. They reasoned that the shape was controlled by a "central" atom bonded to two or more other atoms. Their reasoning and method of predicting can be illustrated in the following specific cases:

Case I. Hydrogen Sulfide, H_2S—A central atom combined with two other atoms. From molecular spectroscopy studies, it is known that the atoms in this molecule do not lie in a straight line but that the molecule is angular having an H—S—H bond angle of 92°. Why should this molecule be angular? And is there reason to expect a 92° angle between the two covalent bonds on the sulfur atom?

To answer these questions we might well ask still another: What electron clouds (orbitals) on the sulfur atom are used to form the two covalent bonds with hydrogen? Examination of the electron configuration of an unreacted sulfur atom reveals that there are two $3p$ orbitals that are only half-filled, i.e., that contain only one electron each. These must be the electron clouds

Table 6.2 Shapes of Some Simple Molecules

Molecule	*Shape*	*Bond Angles*
Water, H_2O	O H H	104.5°
Fluorine oxide, F_2O	O F F	101.5°
Hydrogen sulfide, H_2S	S H H	92.2°
Hydrogen selenide, H_2Se	Se H H	91°
Hydrogen telluride, H_2Te	Te H H	89.5°
Ammonia, NH_3	N H H H	106.75°
Phosphine, PH_3	P H H H	91.6°
Stibine, SbH_3	Sb H H H	91.5°
Nitrogen trifluoride, NF_3	N F F F	102.2°
Methane, CH_4	H C H H H	109.5°

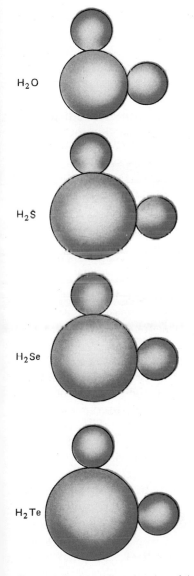

H_2O

H_2S

H_2Se

H_2Te

Figure 6.4 Angular molecules of the hydrides of oxygen-family elements.

used in forming the two covalent bonds with the two hydrogen atoms. Moreover, we believe that p orbitals are oriented at approximately right angles to one another in space (see Chapter 4). Hence it seems reasonable to expect that the two covalent bonds directed from a sulfur atom are oriented at approximately right angles to one another, as shown in Figure 6.4.

If the molecular shape is caused by the space orientation of the bonding orbitals on the central atom, we would predict that water, hydrogen selenide, H_2Se; and hydrogen telluride, H_2Te, should be angular molecules. The "central" atom in each of these compounds—oxygen, selenium, and tellurium —has two half-filled valence shell p orbitals oriented at right angles to each other and available for bonding. The data of Table 6.2 confirm the prediction that the molecules are angular. However, the bond angle in the water molecule is somewhat larger than that in the other H_2X molecules listed. No doubt this is related to the oxygen atom's being smaller than the sulfur, selenium, and tellurium atoms, so there should be more repulsion among the

NH₃

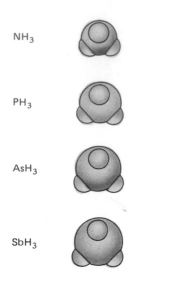

PH₃

AsH₃

SbH₃

Figure 6.5 Pyramidal shaped molecules of nitrogen-family hydrides viewed from the base of the pyramid.

CH₄

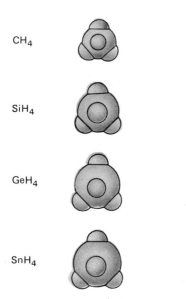

SiH₄

GeH₄

SnH₄

Figure 6.6 Tetrahedral shaped molecules of the hydrides of carbon-family elements.

electron pairs in the two O—H bonds than there is between those of the X—H bonds of the other atoms in the group. This repulsion could result in an increase in the bond angle.

The fact that all of the H_2X molecules listed above are angular constitutes significant support for the idea that the *space orientation of covalent bonds depends upon the orbitals used in bond formation.*

Case II. Ammonia, NH_3—A central atom combined with three other atoms. Further support for this idea of orientation of bonds comes from an examination of the shapes of other simple molecules. A typical example is the ammonia molecule, NH_3. Here, there are three covalent bonds directed from the nitrogen atom to the hydrogen atoms. The orbitals used by the nitrogen atoms must be the three half-filled $2p$ orbitals. Since these orbitals are oriented at right angles, the shape of the ammonia molecule is expected to be pyramidal with the nitrogen at the apex and the three hydrogen atoms forming the corners of the base of the pyramid, as shown in Figure 6.5. Experimental evidence from several sources shows that this is correct; the ammonia molecule is pyramidal in accordance with the concept of space orientation of covalent bonds. Here, as in the case of water, the bond angles are somewhat greater than 90°, undoubtedly because of the small size of the nitrogen atom. (However, see page 123.)

The shapes of many simple molecules can be accurately predicted by simply writing the electron-dot formula for the molecule, identifying the orbitals used in forming the covalent bonds, and recalling how those orbitals are oriented in space. Thus the molecules Cl_2O, F_2O, SCl_2, $SeCl_2$, and $TeBr_2$, in which oxygen, sulfur, selenium, and tellurium use two p orbitals, are all angular like water; the molecules NCl_3, PH_3, AsH_3, and PCl_3 are all pyramidal like ammonia because the "central atom uses three p orbitals to form bonds.

Not all molecules containing three atoms bonded to a "central" atom are pyramidal. Boron trifluoride, BF_3, is planar; the boron atom in this molecule uses s and p orbitals. This will be discussed in the following sections.

Case III. Methane, CH_4—A central atom combined with four other atoms; hybridization. An important extension of the concept of space orientation of covalent bonds is necessary to account for the shape of molecules in which a central atom is bonded to *four* other atoms. Examples include CH_4, $SiCl_4$, SO_4^{-2}, and NH_4^+. In all of these examples, the central atom is found by experiment to be at the center of a tetrahedron with the four surrounding atoms located at the corners of the tetrahedron, as shown in Figure 6.6. The central atom apparently uses one s and three p orbitals to form the four covalent bonds in all of these species. The s orbital is spherically shaped, and there is no preferred direction for a bond formed from it. However, we would guess that a group attached to it would avoid the other groups attached to the p orbitals, so that a molecule such as NH_4^+ would correspond to the ammonia pyramid with a fourth hydrogen attached at the top. Since the bonding orbitals are of two kinds, we would expect the energy of the bond using the s orbital to be different from those using the three p orbitals, and the fourth hydrogen to be closer to the nitrogen than the other three. This appears *not* to be the case; all four bonds have the same energy,

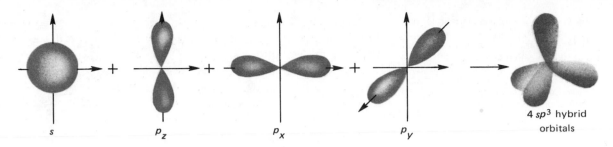

s p_z p_x p_y 4 sp^3 hybrid orbitals

Figure 6.7 Orbital shapes for s and p orbitals and for the four sp^3 hybrid orbitals.

are of equal length, and are oriented at equal angles in space so that the four hydrogen atoms lie at the corners of a regular tetrahedron!

To account for equal energies and equal lengths of the four bonds, it has been postulated that when one s and the three p orbitals are used to form single covalent bonds from the same atom, an *orbital mixing,* called *hybridization,* occurs, such that the four atomic orbitals are converted to four new or hybrid orbitals. These new hybrid orbitals are called sp^3 *hybrids* to imply that each new orbital is a mixture of portions from one s and three p orbitals (Figure 6.7.) The four hybrid orbitals are thought of as being oriented from the central atom so the four pairs of valence electrons assume the most symmetrical space arrangement possible—that is, toward the corners of a regular tetrahedron. When covalent bonds are formed with these orbitals, the molecular shape becomes the characteristic tetrahedron such as that found in methane, CH_4, or in sulfate ion, SO_4^{-2}. In cases such as these, where four identical atoms are bound, all four bonds have the same length and the same energy. In most molecular species involving a central atom surrounded by four other atoms, the structure is tetrahedral; a few square planar structures also are known, such as $PtCl_4^{-2}$, $Ni(CN)_4^{-2}$.

Hybridization, which is treated in bonding theory as a mathematical mixing of the wave functions of the s and p orbitals (or of s, p and d orbitals), can be used to explain the space orientation of bonds in a great many cases in which the "central" atom uses more than one kind of orbital in forming bonds, as when s and p orbitals or s, p, and d orbitals are used.

Case IV. Other Types of Hybridization. The covalent bonds in many boron compounds are believed to involve hybrid orbitals from the boron atom. Since boron atoms contain three valence electrons and often form compounds in which three covalent bonds are directed from boron, it seems reasonable to imagine that one s and two p orbitals on the boron atoms are involved in bond formation. To do this, an electron must be "promoted" from the s to an empty p orbital. This gives the atom an electron configuration in which single electrons are present in the s and in each of two of the p orbitals. This promotion requires energy, but much more energy is released when the bonds form.

We now imagine that as the bonds form, the s and the two p orbitals hybridize, forming three new sp^2 hybrid orbitals. The net result of promotion plus hybridization and bonding is a system more stable than the original configuration. Since boron compounds of the type BX_3 are known to be planar

Figure 6.8 Imagined steps in formation of bonds in BF$_3$. (a) Promotion of electron from 2s to p orbital in B atom. (b) Hybridization of s and p orbitals on B accompanies the formation of the B—F bonds. (c) A p orbital on each F atom overlaps with an sp² hybrid orbital on the B atom.

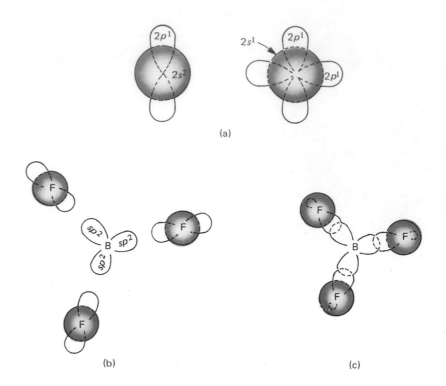

(a)

(b)

(c)

with angles of approximately 120° between the atoms bonded to boron, the indication is that *sp*² hybrid orbitals are oriented in the same plane with angles of 120° between them, as shown in Figure 6.8. This again is the most symmetrical possible arrangement of the three pairs of valence electrons. Certain compounds of carbon probably have *sp*² hybrid orbitals. Examples of this type are discussed later in the section on multiple bonding.

Compounds of beryllium of the type BeX$_2$ are linear and undoubtedly involve *sp* hybrid orbitals from the beryllium atoms. Here, the beryllium atom must "promote" an electron from the *s* to a *p* orbital, thus creating two half-filled orbitals that might hybridize to give *sp* hybrid orbitals. Other examples of *sp* hybrids are known. In all cases these hybrids appear to be oriented at 180° from one another, as shown in Figure 6.9, consistent with the symmetry expected with two electron pairs.

Other types of hybridization are recognized (Figure 6.9) and will be introduced when needed. For the present, it is important to remember that electron pairs in the valence shell tend to orient symmetrically in space. Thus the two *sp* hybrid orbitals are oriented at 180° from one another, the three *sp*² hybrid orbitals are oriented in one plane with angles of 120° between them, and the four *sp*³ hybrid orbitals are tetrahedrally oriented.

A Symmetry Criterion for Molecular Shape

For molecules containing a central atom surrounded by two or more other atoms, the molecular shape often can be predicted by determining the number of electron pairs in the valence shell of the central atom and then imagining that the orbitals containing these electron pairs arrange themselves in the most symmetrical fashion possible around the central atom. The atoms

Figure 6.9 Shapes of hybrid orbitals: (a) *sp* linear; (b) *sp²* trigonal planar; (c) *sp³* tetrahedral; (d) *dsp²* planar; (e) *dsp³* trigonal bipyramid; and (f) *d²sp³* octahedral.

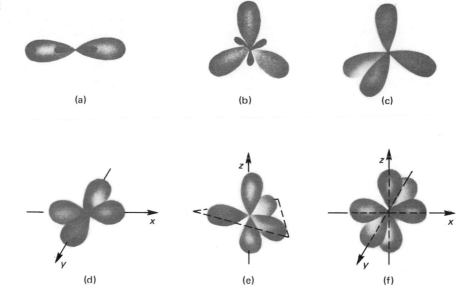

(a)   (b)   (c)

(d)   (e)   (f)

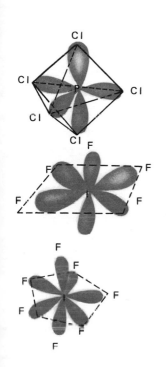

Figure 6.10 Illustrating the trigonal bipyramidal PCl₅, the square pyramidal IF₅, and the pentagonal bipyramidal IF₇ structures.

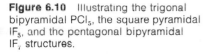

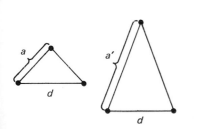

Figure 6.11 Showing that the distance *d* which determines repulsion can be the same at a small angle as at a large angle if *a'* is greater than *a*.

bonded to the central atom would of course be bonded to these symmetrically arranged orbitals. Thus, in SF_6, for example, the sulfur atom has six pairs of electrons in its valence shell. These, when arranged symmetrically, will be directed toward the corners of an octahedron—four pairs pointing toward a square lying in a plane around the S atom, one pair directed above and one pair directed below this plane. The result is the octahedral structure

Similarly, PCl_5, in which the phosphorus atom has five electron pairs in its valence shell, has a trigonal bipyramid shape—three electron pairs are oriented around the "equator" of the phosphorus atom, one pair is directed above this plane, and one pair directed below it. In IF_5, we find a central atom with six pairs of valence electrons. Once again the structure is that predicted on symmetry grounds—namely, the five fluorine atoms form at five of the six corners of an octahedron. In IF_7, the structure is that of a pentagonal bipyramid

The non-bonded, lone pairs of electrons on the central atom play a role similar to that of the pairs to which atoms are bonded in determining the symmetry of the arrangement of orbitals. Since no nucleus of a bonded atom is present to attract the charge of the lone pair, however, the lone pairs tend to occupy more space than bond pairs, and the repulsion between lone pairs (if there is more than one) tends to distort the symmetry. Distortion also occurs because of repulsion between lone pairs and bond pairs, but to a lesser extent. Repulsion increases in the order bond pair-bond pair < bond pair-lone pair < lone pair-lone pair. This distortion of the symmetry accounts for the fact that the bond angles in ammonia and water, both of which have four pairs of electrons in the valence shell of the central atom, are less than the tetrahedral angle 109° expected on the basis of symmetry. The lone pair on the nitrogen of the ammonia molecule repels the three N—H bond pairs more than these repel each other, and the N—H bonds are pushed together to the angle 107°. This effect is still more marked in the water molecule, where the two lone pairs on the oxygen atom repel each other strongly, pushing the bond pairs of the O—H bond closer, and make the H—O—H angle 105°.

The bond pair-bond pair repulsion depends upon the nature of the bonded atoms. The bond angles in F_2O and Cl_2S are smaller than in H_2O because, in the larger F and S atoms, the bonds are longer. This means that for these molecules the repulsive forces among electron pairs can be as large at a smaller angle as they are at the larger angle for H_2O.

Multiple Bonding

It was pointed out in Chapter 5 that double or triple covalent bonds may be formed between two atoms under certain conditions. In a double bond, four electrons are shared; in a triple bond, six electrons are shared. From what has been said about the orientation of covalent bonds in space, it is apparent that the formation of double or triple bonds requires distortion of the bond angles to get appreciable overlapping (Chapter 5) of two or three pairs of orbitals between nuclei.

Much thought has been given to considerations of how orbitals in double and triple bonds overlap in forming the bond. It is now generally believed that one of the bonds in a multiple bond is formed by end-to-end overlap of the atomic orbitals and the remaining bonds are formed by sideways overlap of p orbitals, as shown in Figure 6.12.

A *covalent bond* formed by end-to-end overlap of orbitals is called a sigma (σ) bond and one formed by sideways overlap of p orbitals is called a pi (π) bond.

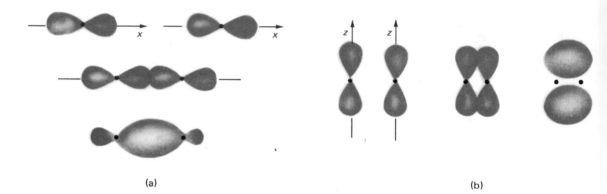

(a)   (b)

Figure 6.12 Electron clouds and sigma and pi bonds: (a) *sigma bond* formed by end to end overlap of two atomic orbitals; (b) *pi bond* formed by sideways overlap of two p atomic orbitals.

GUIDELINES FOR PREDICTING MOLECULAR SHAPE

The following guidelines for predicting shapes of simple molecules are not foolproof, but they are useful in most cases.

The correct electron-dot formula is an essential prerequisite for determining the shape of any molecule. To determine this formula:

1. Determine the total number of valence-shell electrons of all atoms, taking cognizance of the charge on the species if it has one:

Total number = sum of valence electrons considering the + or − charge.

2. Arrange the atoms symmetrically and place the total number of valence-shell electrons around them to obey the octet rule, if possible. Sometimes the atoms are not arranged symmetrically and, in some common compounds of phosphorus, sulfur, and the heavier halogens, the octet rule is not followed. In these cases, refer to the text for similar structures.

Once the correct electron dot formula is derived, the bonding and shape may be determined as follows:

3. Determine the total number of lone pairs and σ bond pairs around the central atom from the electron-dot formula:

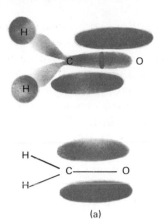

$C \equiv N^{-1}$ (2 π and 1 σ bonds)

(a)

(b)

Figure 6.12 illustrates the σ and π bonding in formaldehyde, H_2CO, and in cyanide ion, CN^-. These species contain double and triple bonds, respectively.

The bond distance in double and triple bonds is less than that of the corresponding single bond, and the bond energy for multiple bonds is greater than that for the corresponding single bond. This is illustrated in Table 6.1, where the N—N, N=N, and N≡N bond distances are given as 1.47, 1.23, and 1.10 A, respectively; the N—N bond energy is cited as 38,000 cal/mole, compared with 225,000 cal/mole for N≡N. While many compounds containing multiple bonds are more reactive than those with only single bonds, mainly because of the greater accessibility of the π bonds to attacking reagents, the nitrogen molecule, which contains a triple bond, is among the more stable molecules known.

Structures of Simple Molecules Containing Multiple Bonds

To illustrate the principles just discussed for predicting molecular shapes and types of bonding of several simple molecules, let us consider as examples molecules of formaldehyde, H_2CO, acetylene, C_2H_2, and propylene, C_3H_6.

Formaldehyde. The electron-dot formula for formaldehyde is

$$\text{H}$$
$$\overset{..}{\text{H} \cdot \text{C} \cdot \cdot \overset{..}{\text{O}} \cdot}$$

Total pairs = sum of (σ bond pairs + non-bonded electron pairs) around the central atom.

4. Use the total pairs to predict hybridization and "overall" shape, assuming that these pairs will orient in space as symmetrically as possible.

5. In describing the molecular shape, relate the spatial positions of the atoms (not the lone pairs) to a geometric figure.

Example: Predict the shape of NO_2^-

1. $5 + 2(6) + 1 = 18$ total electrons
2. give $\overset{..}{:O}:\overset{..}{N}::\overset{..}{O}:^-$ as the electron-dot formula
3. total pairs = $2 + 1 = 3$
4. sp^2; trigonal planar orientation
5. ∴ shape is angular

$$\overset{..}{N}:$$
$$:\overset{..}{O}: \quad :\overset{..}{O}:$$

H—C≡C—H

Figure 6.14 Bonding in acetylene.

The carbon atom is bonded to three other atoms with a double bond between the carbon and oxygen atoms. The molecule is known to be planar.

The planar structure and the fact that three atoms are bonded to the carbon atom suggest that the carbon atom is using three sp^2 hybrid orbitals to form three σ bonds, one with each of the hydrogen atoms and one with the oxygen atom. The fourth orbital on the carbon (presumably an unhybridized p orbital) is used to form the π bond with oxygen. The oxygen atom uses one of its p orbitals to form the σ bond and a second p orbital to form the π bond with carbon. Each hydrogen atom uses its s orbital to form a σ bond with an sp^2 hybrid orbital on carbon.

Acetylene. The electron-dot formula for acetylene is

$$H:C:::C:H$$

Each carbon atom is bonded to two other atoms. A triple bond connects the carbon atoms. The molecule is linear.

The linear structure and the fact that each carbon atom is bonded to two other atoms suggest that sp hybrid orbitals are being used in the σ bonds connecting the carbon atoms to one another and to the hydrogen atoms. The two π bonds arise from unhybridized p orbitals on both carbon atoms. The hydrogen atoms use s orbitals to form the σ bonds with carbon. (See Figure 6.14.)

An Important Generalization

The structures above are those that would be predicted on the basis of symmetrical space orientation of σ bonds with π bonds forming perpendicular to the σ bonds wherever possible. This is generally true, and is further illustrated by propylene, which has the electron-dot formula

$$\begin{array}{ccc} & H & H \\ H:C: & :\ddot{C}: & \ddot{C}:H \\ & \ddot{H} & \ddot{H} \end{array}$$

Two of the carbon atoms (the first two starting from the left) are bonded to three other atoms; the third carbon atom is bonded to four other atoms. Each of the first two carbon atoms has three σ and one π bond. The three σ bonds are expected to orient symmetrically in space to give a trigonal planar symmetry. The π bond forms between the two carbon atoms but perpendicular to the trigonal planes formed by the σ bonds. The result is that the two carbon atoms and the four atoms bonded to them lie in the same plane, restricted from rotating by the π bond. The third carbon atom has four σ bonds, three to hydrogen atoms and one to a carbon atom. These four bonds orient tetrahedrally, as expected on symmetry grounds.

The overall shape of this molecule can be described as a planar unit attached to a tetrahedral unit, as illustrated in Figure 6.15.

Resonance: Delocalized π Bonding

Many covalent molecules containing multiple bonds cannot be presented adequately by an electron-dot formula. An example is the sulfur dioxide molecule, which is sometimes represented by the formula

$$\begin{array}{c} \ddot{S}::\ddot{O}: \\ :: \\ :\ddot{O}: \end{array}$$

Figure 6.15 Shape and bonding in propylene.

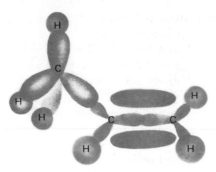

This formula implies that the molecule contains one oxygen atom singly bonded to sulfur and a second oxygen atom doubly bonded to sulfur. If this were the case, the bond lengths would be different, since single bonds are longer than double bonds. Reliable measurement of the sulfur-oxygen bond distances reveals that both bonds are the same length; this length is intermediate between the single- and double-bond sulfur-oxygen bond distances.

This situation can be accounted for by recognizing that two equivalent electron-dot formulas can be written for sulfur dioxide, namely

$$:\ddot{O}::S \qquad \text{and} \qquad :\ddot{O}:\ddot{S}$$
$$:\ddot{O}: \qquad \qquad \qquad \ddot{O}:$$

$$\text{I} \qquad \qquad \qquad \text{II}$$

The difference between the structures is the location of the π bond between sulfur and oxygen. In formula I, the π bond is between the sulfur atom and the oxygen atom on its left; in formula II, the π bond is between the sulfur atom and the oxygen atom below it. If you imagine that the true structure of this molecule is neither that represented by formula I nor formula II but is some structure intermediate between the two, the observation of equal bond lengths and bond energies can be accounted for. This situation whereby the actual structure of a molecule is intermediate between two (or more) structures differing only in the positions of the electrons is known as   .

Perhaps the simplest view of bonding in structures exhibiting resonance is to imagine that the electrons in the π bonds are not localized between two

---

CONDITIONS FOR RESONANCE

Resonance is said to exist only when an alteration of one electronic formula to give another is possible without a change in the positions of the atomic nuclei. To decide whether resonance exists in a structure, the following criteria may be applied:

1. Can two or more acceptable Lewis-type electronic formulas differing only in the positions of paired electrons be written for the structure?

2. Do the several electronic formulas contain the same number of paired electrons (usually no unpaired electrons)?

3. Do the several electronic formulas contain the same number of shared electrons?
Example—$:\ddot{C}l\cdot\ddot{C}l:$ and $:\ddot{C}l::\ddot{C}l$ cannot be considered equivalent structures because they contain different numbers of shared electrons. Sometimes resonance hybrids with different numbers of shared electrons are written, but these are usually considered less important than those having the same number of shared electrons.

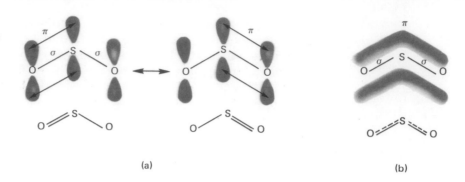

Figure 6.16 Delocalization of pi bonds in SO₂: (a) localized structures; (b) delocalized structure.

(a)

(b)

atoms but are able to interact with more than two atoms.* This interaction is known as *delocalization* and arises because of overlapping of the orbitals used to form π bonds between two atoms with orbitals from adjacent atoms; it is illustrated in Figure 6.16.

Examples of other molecules that we describe in terms of the resonance concept are:

1. *Sulfur trioxide.* The actual structure is believed to be intermediate among

and   and

2. *Dinitrogen oxide.* This molecule, known to have the structure N—N—O, exhibits resonance. Its structure is intermediate between

and

I   II

3. *Benzene.* In this molecule, three π bonds are involved in resonance. The true structure is intermediate among the following structures and some more complex structures.

and

I   II

In benzene, the delocalization of electrons in the π bonds is extensive so that the electron clouds representing them are fused into a ring covering all six carbon atoms, as shown in Figure 6.17.

Molecules exhibiting resonance are much more stable than we would predict them to be if their true structure were that represented by any one

---

*The true structure is not an equilibrium mixture of the two or of a system oscillating between the two, but a unique one that our graphic art cannot represent by the dot formula or by the bond formula.

of the resonance forms. For example, if the structure of benzene was represented by one of the formulas above, it would require $3 \times 145 + 3 \times 82 = 681$ kilocalories to break all the carbon-to-carbon bonds in a mole of benzene. Actually it requires more than 700 kilocalories to break the 6 moles of carbon-to-carbon bonds, indicating the greater stability of the delocalized structure. It is also observed that the multiple bonds in benzene are much less reactive than in ethylene, $CH_2{=}CH_2$, in which the π bonds are not delocalized.

Molecular Orbitals

The theoretical treatment of covalent bonding forces involves the postulate that to have bonding between two atoms, the atomic orbitals and nuclei of the atoms in question must interact so that electrons from both atoms are attracted simultaneously to both nuclei. In mathematical terms this means that as the bond forms, the wave functions for the orbitals on the two atoms must overlap.

Two theories relative to the nature of the orbitals present in molecules have been profitably developed. The first, known as the valence bond theory, starts with atoms and their atomic orbitals. It assumes that when a covalent bond forms, a pair of atomic orbitals overlap in such a way that each electron in the bond is attracted to both nuclei. In this theory, the atomic orbitals retain their identities in the covalent bond. The second theory, known as the molecular orbital theory, assumes the initial existence of *molecules*. These molecules consist of nuclei located at distances similar to those in the final molecule, and electrons subject to the influence of the nuclei. The arrangement and energies of the electrons are then described in terms of molecular orbitals. In molecular orbitals, the electrons are interacting with more than one nucleus.

Each molecular orbital has a characteristic energy. Electrons occupy these orbitals in accordance with the same rules as those used to build atoms in Chapter 4, that is, the electrons fill the lowest energy levels first. Orbitals of the same energy will be occupied by one electron each before a second electron is added to any one of the equal-energy orbitals. The Pauli exclusion principle prohibits the appearance of more than two electrons in any one orbital. The number of orbitals in the molecule formed from two atoms must be the same as the number of orbitals originally present in the valence shells of the two atoms, and the number of electrons distributed among the molecular orbitals must be the same as the number originally present in the two atoms.

Figure 6.17 Bonding and structure in benzene: (a) localized structures and formulas; (b) delocalized structure and formula.

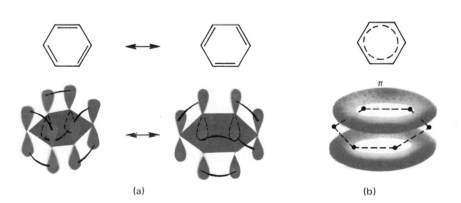

(a)   (b)

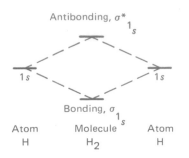

Antibonding, σ^*_{1s}

$1s$ $1s$

Bonding, σ_{1s}

Atom	Molecule	Atom
H	H_2	H

Figure 6.18 Atomic and molecular energy levels in H atoms and in the H_2 molecule.

A complete determination of the molecular orbitals would require the solution of the Schrödinger equation for the system of two (or more) nuclei and their electrons. This has never been done because of formidable mathematical obstacles. A mathematically more tractable approximate procedure is to postulate that the molecular orbitals are linear combinations of atomic orbitals. This is known as the LCAO treatment. For example, if ψ_A and ψ_B are the interacting atomic-orbital wave functions for atoms A and B, a linear combination of ψ_A and ψ_B would produce two molecular orbitals—one by an addition of the two wave functions and a second by subtraction of the two wave functions. Since these molecular orbitals might not be the result of equal contributions from the two wave functions, weighting factors for each wave function are included. Thus, a mathematical expression for the two molecular orbitals arising which ψ_A and ψ_B interact is

$$\psi_{MO_I} = a\psi_A + b\psi_B$$

$$\psi_{MO_{II}} = c\psi_A - d\psi_B$$

The constants a, b, c, and d are weighting factors representing the fractions of ψ_A and ψ_B which must be used to give the best representations of the molecular orbitals ψ_{MO_I} and $\psi_{MO_{II}}$. If A and B are atoms of the same element, a, b, c, and d are equal in magnitude.

In terms of electron density, MO_I above gives a high concentration of electron charge between nuclei and is called a bonding molecular orbital, while MO_{II} gives a very low concentration of electron charge between nuclei and is known as an antibonding molecular orbital. The bonding molecular orbital lies lower in energy than the antibonding molecular orbital.

Both bonding and antibonding molecular orbitals may be either σ or π orbitals. As noted earlier, σ orbitals are cylindrical around an axis connecting the nuclei; π orbitals are not cylindrical around this axis.

Molecular Orbitals in Hydrogen-Like Systems

If two atoms with $1s$ electrons combine, two molecular orbitals might be formed, as shown in Figure 6.18. The σ_{1s} or bonding molecular orbital lies lower in energy than the σ_{1s}^* antibonding molecular orbital. Each of these molecular orbitals can hold two electrons and, in the hydrogen molecule, the two electrons reside in the orbital of lowest energy—the bonding orbital, σ_{1s}; thus a stable molecule is formed. If, however, an attempt is made to form a diatomic helium molecule, He_2, four electrons must be fitted into the molecular orbitals. Two electrons are expected to go to σ_{1s} and two to the antibonding orbital of high energy, σ_{1s}^*. Filling the antibonding orbital nullifies the effect of the filled bonding orbital. The net result is that no bond is formed and therefore He_2 does not exist. The hydrogen molecule H_2^+ should have its one electron in a bonding molecular orbital and thus have some stability. This species is known to exist; its binding energy is 61 kcal/mole—a relatively high bond energy.

Second-Period Diatomic Molecules

In a more complex molecule such as O_2, bonding involves the p orbitals of the atom as well as s orbitals; p orbitals are mutually perpendicular to each other. It is convenient to consider the x axis as the axis through the centers of the two oxygen atoms about to be joined, so that we designate the p orbital along that axis as p_x, while the orbitals p_y and p_z are perpendicular to that axis.*

---

*Some authors prefer to choose the z axis as the axis through the atom centers.

Figure 6.19 Coordinates chosen for p
orbitals of the oxygen molecule.

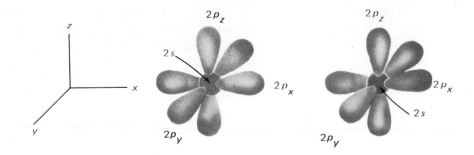

Thus the orbitals in the oxygen atoms about to form a molecule can be considered to be oriented as in Figure 6.19, which also shows the spherical $2s$
orbitals. (The *filled* $1s$ orbitals are buried in the core or kernel of the oxygen
atoms, and probably play only a minor role in bonding or antibonding.)
As the atoms are brought together, the s orbitals can overlap, forming σ_s and
σ_s^* molecular orbitals; the p_x orbitals also overlap, forming another set of σ
molecular orbitals of different energy, σ_{p_x} and $\sigma_{p_x}^*$. The p_y and p_z orbitals
however, overlap only "side-to-side" (Figure 6.10), and form a quite different
set of molecular orbitals known the π_{p_y} and π_{p_z} and $\pi_{p_y}^*$ and $\pi_{p_z}^*$. The
energies of the molecular orbitals and their relation to the atomic orbitals
from which they came are shown in Figure 6.20.

The molecular orbitals π_{2p_y} and π_{2p_z} have lower energies than the atomic
orbitals p_y and p_z from which they came. The atomic orbitals p_y and p_z have
the same energies. They are physically indistinguishable since the only
identification they have is that the y and z axes are perpendicular to the
central axis of the molecule and to each other; the labels y and z are purely
arbitrary. The π_{2p_y} and π_{2p_z} molecular orbitals into which the p_y and p_z

Figure 6.20 A molecular orbital
diagram for a diatomic molecule
similar to oxygen.

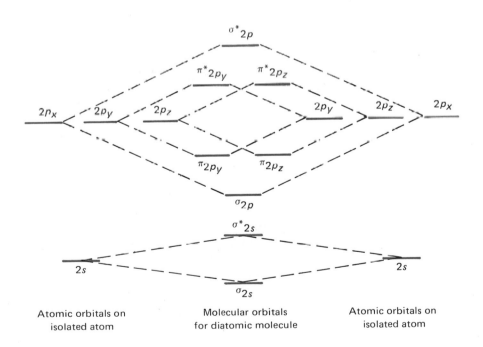

Figure 6.21 Molecular orbital diagrams for homonuclear diatomic molecules containing second period elements.

$\sigma^*_{2p_x}$ $\sigma^*_{2p_x}$ $\sigma^*_{2p_x}$

$\pi^*_{2p_y}$ $\pi^*_{2p_z}$ $\pi^*_{2p_y}$ $\pi^*_{2p_z}$ $\pi^*_{2p_y}$ $\pi^*_{2p_z}$

π_{2p_y} π_{2p_z} π_{2p_y} π_{2p_z} π_{2p_y} π_{2p_z}

σ_{2p_x} σ_{2p_x} σ_{2p_x}

σ^*_{2s} σ^*_{2s} σ^*_{2s}

σ_{2s} σ_{2s} σ_{2s}

For O_2 For N_2 For F_2

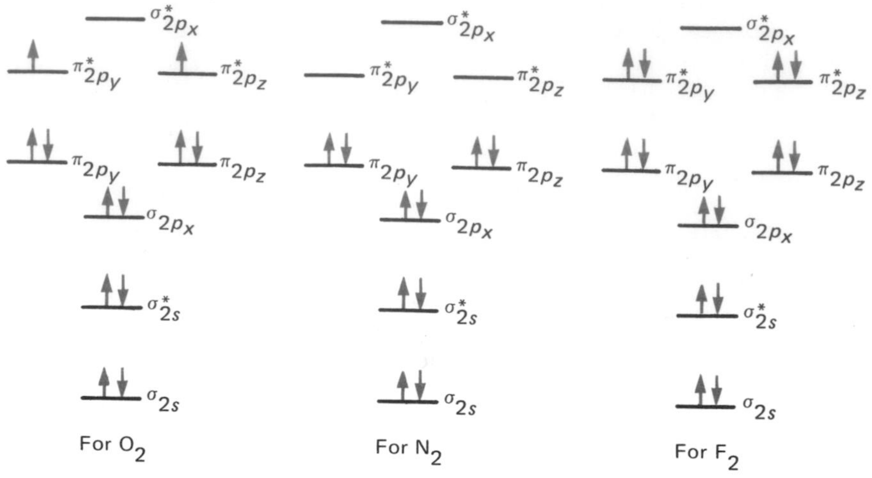

atomic orbitals coalesce also are of equal energy. Similarly, the high-energy antibonding orbitals $\pi_{2p_y}*$ and $\pi_{2p_z}*$ have the same energy.

The 12 valence electrons of the two oxygen atoms must now be placed into the molecular orbitals in the center of Figure 6.19. They fill the lowest energy levels first, so two each enter the σ_{2s}, $\sigma_{2s}*$, σ_{2p_x}, π_{2p_y}, and π_{2p_z}. This accounts for 10 of the 12. The two remaining electrons enter the $\pi_{2p_y}*$ and $\pi_{2p_z}*$ orbitals, remaining unpaired in spins in accordance with Hund's rule (page 78). The increase in energy in filling two electrons into the antibonding $\sigma_{2s}*$ orbital almost exactly offsets the decrease in energy when two electrons sink into the bonding σ_{2s} orbital, and the net bonding effect for those two orbitals is very small. Similarly, the energy requirements to place two electrons in the $\pi_{2p}*$ orbitals nullifies the bonding effect of one of the pairs in the π_{2p} orbitals. The molecule thus contains one clear σ bond (the σ_{2p}) and a net of one π bond to give the oxygen molecule what amounts to double-bond (one σ bond plus one π bond) character. The electron configuration of oxygen can be sketched as shown in Figure 6.21, where the arrows represent electrons with parallel (↑ ↑), or antiparallel (↑ ↓), spins. The formation of the oxygen molecule can be written as

$O[1s^2\, 2s^2\, 2p^4] + O[1s^2\, 2s^2\, 2p^4]$
$$\longrightarrow O_2[(1s^2)(1s^2)(\sigma_{2s})^2(\sigma_{2s}*)^2(\sigma_{2p_x})^2(\pi_{2p_y})^2(\pi_{2p_z})^2(\pi_{2p_y}*)^1(\pi_{2p_z}*)^1$$

where identification of the $1s$ electrons of the atom cores with their individual atoms is indicated by retaining for them the notations of atomic orbitals.

The diagram of Figure 6.20 with altered energies appropriate to the specific molecule under consideration is used for other diatomic molecules formed from second-period elements. For example, in N_2, 10 valence electrons must be added to the diagram of Figure 6.20. For F_2, 14 valence electrons must be added. This is illustrated in Figure 6.21, from which it is apparent that in both N_2 and F_2 there are more electrons in bonding molecular orbitals than in antibonding molecular orbitals.

A convenient measure of the net bonding in molecular orbital theory is the *bond order*, defined as one-half the difference between the number of electrons in bonding molecular orbitals and the number of electrons in antibonding molecular orbitals. For several diatomic cases this is:

Bond order

For O_2	$\frac{1}{2}(8-4)=2$
For N_2	$\frac{1}{2}(8-2)=3$
For F_2	$\frac{1}{2}(8-6)=1$
For Ne_2	$\frac{1}{2}(8-8)=0$ (unstable)

INTERMOLECULAR FORCES OF ATTRACTION

One problem that intrigued physical scientists for many years and which still remains only partially solved is: Why do the noble gases form liquids? The atoms of these gases are unreactive; they have virtually no tendency to combine with other atoms, except with fluorine (Chapter 9). Yet they must possess some attractive forces that keep them together in the liquid state.

The same question can be asked of many molecules. Why, for example, is carbon tetrachloride a liquid? Why do not carbon tetrachloride molecules, all the atoms of which have acquired stable electron configurations, remain separated as a gas, refusing to condense into clusters characteristic of the liquid state? The only reasonable answer is that there must exist *forces of attraction between stable molecules*. What then is the nature of these forces, and how do they differ from forces responsible for chemical bonds?

The most widely accepted current theories recognize three types of forces that cause attraction between molecules: *van der Waals forces, dipolar forces,* and *hydrogen bonding.*

Van der Waals Forces

These are very short-range attractive forces believed to be present in all atoms, molecules, and ions. They are very weak forces compared with bonding forces. They are important, for example, in keeping carbon tetrachloride a liquid at room temperature, and in making possible liquefaction of the noble gases. Unlike valence forces, which seem to become satisfied or "saturated" when an appropriate number of electrons is shared or transferred, *van der Waals forces* in a given species can attract as many molecules as can surround the species.

It is suspected that these forces arise from a mutual polarization or distortion of the outermost electron clouds in the species interacting. For example, if two argon atoms were to come very close to one another, the exterior electron clouds might be distorted because of electrostatic repulsion of the electrons of the two atoms. This distortion would result in electrically lopsided atoms—atoms with the nucleus displaced toward one side, leaving a high concentration of negative charge on the opposite side. However, such an arrangement is unstable and would result in an oscillation of the electron cloud. The atoms would thus become oscillating dipoles and attract other atoms that are similarly oscillating. This way noble gas atoms and stable molecules may generate forces of attraction great enough to hold molecules together in the liquid and solid states.

Ordinarily, molecules and atoms with easily distorted or polarized electron clouds exhibit stronger van der Waals forces than those with stiff, hard-to-distort structures. For example, van der Waals forces are stronger in iodine, I_2, than in fluorine, F_2, presumably because the outermost electrons in iodine are much farther from the nucleus than those in fluorine. The nuclei in iodine molecules thus have less control over the electrons and the molecules are more polarizable than are fluorine molecules.

One would expect van der Waals forces to increase as the number of electrons and nuclei is in each molecule increases. This, in a general way, is

Table 6.3 Variation of the Boiling Point of the Noble Gases with Number of Electrons

Element	Nuclear Charge	Number of Electrons	Boiling Point (°C)
He	2	2	−269
Ne	10	10	−246
Ar	18	18	−186
Kr	36	36	−153
Xe	54	54	−107
Rn	86	86	−62

the case, as may be seen from the data in Table 6.3. That table contains the boiling points of the simplest types of "molecules," the atoms of the noble gas elements. Each of these has a single nucleus, but the charge on the nucleus, and thus the number of electrons around the nucleus, differ from one to another. The increase in van der Waals forces between molecules as the number of nuclei and electrons increases is also shown in Table 6.4, which gives the boiling point increases for a group of normal hydrocarbons. When the number of nuclei and electrons in a molecule becomes very large, the van der Waals forces also become very large.

Dipolar Forces   It was observed in Chapter 5 that many covalent bonds have a polar character. The presence of dipoles in molecules results in intermolecular attraction. The positive end of the dipole in one molecule is attracted to the negative end of the dipole in an adjacent molecule. The positive end of the second attracts the negative end of another, and so on. These attractions increase as the dipole moment increases but, in some cases, it is less important than

Table 6.4 Variation of the Boiling Point of Normal Hydrocarbons with Number of Nuclei and Electrons

Compound	Nuclei	Electrons	Boiling Point (°C)
CH_4	5	10	−162
C_2H_6	8	18	−88
C_3H_8	11	26	−42
C_4H_{10}	14	34	0
C_5H_{12}	17	40	+36
C_6H_{14}	20	46	+69

Table 6.5 Estimated Values for Intermolecular Attractive Energies in Simple Molecular Solids (in kcal/mole)

Molecule	Dipolar Energy	Van der Waals Energy	Total Attractive Energy
Argon	0.00	2.03	2.03
Carbon monoxide, CO	1×10^{-4}	2.09	2.09
Hydrogen iodide, HI	6×10^{-3}	6.18	6.21
Hydrogen bromide, HBr	0.164	5.24	5.52
Hydrogen chloride, HCl	0.79	4.02	5.05
Ammonia, NH_3	3.18	3.52	7.07
Water, H_2O	8.69	2.15	11.30

Figure 6.22 Hydrogen bonding in water. (a) Note that the unshared electrons in one molecule are interacting with a hydrogen atom in another molecule. (b) Hydrogen bonding gives rise to an "open" structure in ice, thereby accounting for the decrease in the density of the water system on freezing.

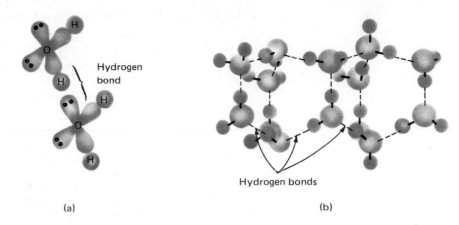

(a)

(b)

Hydrogen bonds

van der Waals attraction. Table 6.5 gives some estimated values for the dipolar attractive energies and for the van der Waals energies in some simple molecules. The total attractive energy includes a minor contribution from interaction between permanent dipoles and dipoles induced on neighboring molecules.

The Hydrogen Bond

Certain molecules that contain hydrogen atoms bonded to other atoms having high electron-attracting power, such as fluorine, chlorine, oxygen, and nitrogen, exhibit stronger forces of attraction than might be expected from their van der Waals forces or measured dipole moments. The additional attractive force in these cases is called a "hydrogen bond." It is believed that this bond arises because the hydrogen atom covalently bonded in one molecule is highly polarized and attracted by the electronegative atom in a neighboring molecule. In hydrogen fluoride, for example, the situation might be visualized as

$$H\!:\!\ddot{F}\!:\!-\!-\!-\!H\!:\!\ddot{F}\!:$$
Hydrogen bond

Hydrogen bonding is also found in molecules such as H_2O, NH_3, and HCN (Figure 6.22). The boiling points, melting points, and heats of vaporization of compounds exhibiting hydrogen bonding are considerably higher than for comparable compounds in which such bonding is unlikely. This is illustrated in Figure 6.23, where the boiling point of the hydrides of the elements of the nitrogen, oxygen, and halogen families are plotted against the period number. In all three families, anomalous boiling points of the substances exhibiting hydrogen bonding are observed when compared to boiling points of the other members of the series. For example, in the oxygen family, the boiling points of hydrogen telluride, H_2Te; hydrogen selenide, H_2Se; and hydrogen sulfide, H_2S, are $-1.8°C$, $-41.5°C$, and $-60.8°C$, respectively. If this trend were to continue, the boiling point of water would be about $-75°C$ rather than the observed value of $100°C$.

The fact that water expands on freezing is attributed to the formation of hydrogen bonds. This will be discussed in detail in Chapter 14.

Although hydrogen bonds are weaker than other types of chemical bonds, they may be responsible for the shape and stability of certain chemical structures. In protein molecules, which are huge structures, it is possible to have several thousand or more hydrogen bonds in one molecule. This number of

Figure 6.23 Boiling points of nonmetal-family hydrides. The period number is the number of the period in the periodic table to which the central atom belongs.

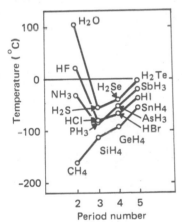

hydrogen bonds gives great stability to the protein structure, and biochemists believe that the behavior of certain proteins is directly related to the shape of the protein which, in turn, is related to the hydrogen bonding present. This will be developed further in Chapter 29.

In summary, van der Waals forces, dipole forces, and hydrogen bonds give rise to the principal forces of attraction between molecules. The van der Waals forces are especially important because they are present in all chemical species.

Odor and Molecular Geometry. There are three chemical senses: taste, smell, and the common chemical sense. Of these, the sense of smell is by far the most sensitive, being about 10^4 times as sensitive as the other two.

The common chemical sense is one of chemical irritation, such as that caused by ammonia, acid fumes, or lachrymators. Receptors for chemical irritants are in mucous membrane. While the common chemical sense is distinct from smell, taste, or even pain, and is transmitted to the brain by nerves different from those for the other senses, a chemical irritant can stimulate olfactory receptors giving, in addition to the irritant sensation, one of odor as well.

Studies of taste have not yet produced a simple theory relating taste and molecular properties. The sense organs for taste—taste buds—are located primarily on the tongue. However, the tongue surface is not uniformly sensitive. The middle surface is insensitive to taste, but sensitivity for sweet occurs mainly at the tip, bitter at the back, sour on the sides, and salt around the edges.

The olfactory apparatus in man consists of two patches of yellow tissue, of about 1 square inch, one on each side of the nose at the top of the nasal chamber. Olfactory tissue contains two types of nerve fibers, the endings of which receive and detect "scented" molecules.

Molecular geometry appears to be the key to the sense of smell. Seven primary odors have been identified; each of these is detected by an appropriately shaped receptor at the olfactory nerve endings. Most odors are composed of several of these primary scents combined in various proportions. Table 6.6 lists the seven primary odors, illustrates the nature of the receptor sites, and describes the molecular geometry appropriate to each receptor.

To have an odor at all, a substance must be volatile and be somewhat soluble in both water and fatty substances. Volatility, of course, enables the molecules to reach the receptor sites. Water solubility is needed so the molecules will not be repelled by the water film covering the receptor ends. Fat solubility enables the substance to penetrate the fat layer on the surface of the nerve endings.

As Table 6.6 shows, five of the seven primary odors are associated with a molecular geometry and a molecular size. Thus the disk, characteristic of a musky odor, is too large to fit into the spherical receptor reserved for camphoraceous molecules. Interestingly, the less pleasing pungent and putrid odors are not associated with molecular shape but with the presence of what can be called electron-rich or electron-poor reaction centers in the molecules.

The stereochemical theory of odor was advanced in the early 1960's. In the short time it has been available, it has proved to be a most valuable intellectual tool for analyzing in fine detail many of the complex flavors we encounter daily. From it have come, and will continue to come, new and

Table 6.6 Primary Odors and Corresponding Olfactory Receptors and Molecular Geometry

Primary Odor	Familiar Example	Receptor Site*	Molecular Geometry
Camphoraceous	Moth balls		Roughly spherical Diameter ~ 7 A
Ethereal	Dry-cleaning fluids		Rod-shaped
Floral	Roses		Disk with tail
Pepperminty	Mint candy		Wedge-shaped with H-bond receptor
Musky	Damp basement		Disk-shaped Diameter ~ 10 A
Pungent	Vinegar	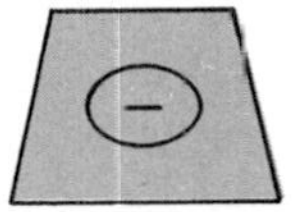	Strong affinity for electrons
Putrid	Rotten eggs	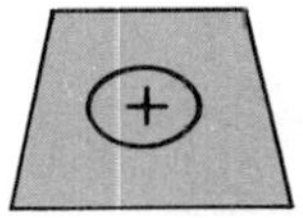	Strong affinity for + center or empty orbital

*Source. Figures adapted from Amoore, J. E., Johnston, Jr., J. W., and Rubin, M., *Sci. Amer.*, **210**, no. 2, 42–49 (1964).

perhaps more exciting fragrances. It also should enable us to rid ourselves of some obnoxious odors and possibly discourage some pesky insects whose habits are controllable in large degree by certain odors, notably those of sex attractants.

SUMMARY

Some important aspects of molecular structure have been presented in this chapter. The nature of the covalent bond, its strength and length, the orientation of the bonds in space, hybridization of atomic orbitals, make-up of multiple bonds, the concept of resonance, and an introduction to molecular orbital theory have been developed. In addition, the forces of attraction between stable molecules—van der Waals forces, polar covalent bonds, and hydrogen bonds—have been identified.

Perhaps as a result of this elementary introduction, it is clear that chemists have some well-developed ideas about molecules, their shapes, the forces that are responsible for these shapes, and the forces that cause liquids and solids to form.

IMPORTANT TERMS

Nature of the covalent bond
orbital overlap
directional character
localization of electrons
potential-energy diagram
bond energy
bond length

Stereochemistry
space symmetry
lone pair
bond pair
hybridization
sp, sp^2, sp^3
dsp^3, d^2sp^3

Multiple bonds
resonance

Molecular orbitals
σ and π bonds

Intermolecular attractive forces
van der Waals
dipolar
hydrogen-bonding

QUESTIONS AND PROBLEMS

1. Why do atoms combine?

2. (a) Describe the changes that occur in the atomic orbitals of two hydrogen atoms as they combine in the formation of a covalent bond. (b) Contrast this with the corresponding changes that occur when two fluorine atoms combine.

3. Write the molecular formulas and structural formulas using dashes to represent electron pairs for the following compounds: (a) carbon tetrachloride, (b) sodium trisulfide, (c) carbon disulfide, (d) hydrogen peroxide.

4. Explain the following experimental observations: (a) The Br—Br bond is stronger than the I—I bond. (b) The H—F bond energy is greater than the H—Cl bond energy. (c) The C≡O bond energy is higher than the bond energy for C—O, C≡S, and C—S. (d) The strongest bonds formed by carbon are with fluorine. This also is true of boron and fluorine; however, it is not true of nitrogen and fluorine or oxygen and fluorine. Explain.

5. Draw structural formulas indicating the stereochemistry in each of the following: (a) $BeCl_2$, (b) H_2Se, (c) PCl_3, (d) PCl_5, (e) NH_4^+, (f) $SiCl_4$, (g) PO_4^{-3}, (h) ClF_4^-, (i) H_2NOH, (j) H_2NNH_2.

6. Predict the geometric structures for the following species and cite the theoretical base which led to your decisions: (a) $SeCl_4$, (b) BF_4^-, (c) SF_6, (d) SiF_6^{-2}, (e) $AgCl_2^-$, (f) ICl_2^+.

7. Write electron-dot diagrams and predict the shapes of the following: (a) H_3O^+, ClO_3^-, NH_3, SO_3^{-2}; (b) BO_3^{-3}, SO_3, BF_3; (c) BF_4^-, ClO_4^-, SO_4^{-2}, NH_4^+, $POCl_3$; (d) NH_3, NH_4^+, NH_2^-.

8. Discuss the nature of the bonding in each of the following. Indicate the number of σ bonds, the stereochemistry, and the type of hybridization expected. (a) F_2O, (b) $HOCl$, (c) BCl_2^+, (d) AsF_5, (e) SO_4^{-2}, (f) NF_3, (g) H_3BO_3, (h) CH_3OH, (i) IF_7.

9. (a) Show by electron-dot diagrams how the following reactions take place: (1) $K^+ + OH^- + H_2SO_4 \longrightarrow K^+ + HSO_4^{-2} + H_2O$. (2) $NH_3 + HCl \longrightarrow NH_4^+ + Cl^-$. (3) $NH_4^+OH^- \longrightarrow NH_3 + H_2O$. (4) $Na^+ + CO_3^{-2} + H_2O \longrightarrow Na^+ + OH^- + H_2CO_3$. (b) Predict the geometry of all polyatomic species in (a).

10. A compound formed between PCl_5 and BCl_3 appears to be ionic, probably composed of PCl_4^+ and BCl_4^-. Are such ions reasonable in terms of the theory of bonding? What is their stereochemistry?

11. A compound formed between PCl_5 and ICl probably is composed of the ions PCl_4^+ and ICl_2^-. What is the expected stereochemistry of ICl_2^-?

12. On the basis of electron-pair repulsions, predict the geometry for the following compounds. (1) AsH_3, (2) SCl_2, (3) $SeCl_3^+$, (4) IF_5, (5) Cl_2O.

13. Indicate the shapes of the following molecules, using the concept of electron pair repulsion: (a) XeF_4, (b) XeF_6, (c) AsF_6^-, (d) $TeCl_5^-$, (e) SF_4, (f) SiF_4.

14. One of the allotropic forms of phosphorus exists as the P_4 molecule. The geometric configuration of the molecule is a regular tetrahedron with each atom at a corner thereof and all bonds equivalent. (a) Diagram the molecule, including its electron-dot structure. (b) Indicate bond angles for P_4 and compare the molecule to PH_3 (Table 6.2). (c) Would you have predicted tetrahedral bonds for P_4?

15. Ordinary double bonds such as the one in ethylene, $H_2C{=}CH_2$, do not allow rotation about the axis of the bond. Explain this on the basis of σ and π bond theory. The acetylene molecule has a triple bond, $HC{\equiv}CH$. Would you expect rotation about such a bond?

16. The nitrite ion, NO_2^-, is angular. The nitrogen atom is centrally located and the oxygen atoms are equidistant from the nitrogen atom. The oxygen atoms are equally negative. Diagram the possible resonance forms for the nitrite ion.

17. The atoms in the nitrogen tetroxide molecule are oriented in the same plane. The general arrangement of the atoms is:

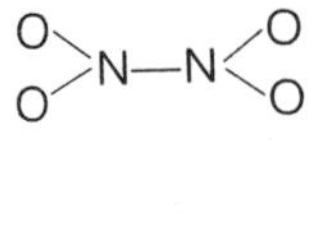

All four oxygen atoms are equivalent. (a) Write resonance forms to explain this structure. (b) What type of orbital hybridization is involved?

18. Discuss the nature of the bonding in each of the following. Indicate the number of σ and π bonds, the stereochemistry, and type of hybridization expected. (a) CO_2, (b) $\begin{matrix} Cl \\ \\ Cl \end{matrix}\!\!>\!\!C\!\!=\!\!O$, (c) $H\!-\!C\!\!\underset{O\!-\!H}{\overset{O}{<}}$, (d) HNO_3, (e) (e) H_2CO_3, (f) HCN, (g) $HONO_2$.

19. Write electron-dot formulas for, and discuss the bonding and stereochemistry, in the following: (a) O_2^{-2}, (b) NO_2^+, (c) BH_4^-, (d) S_3^{-2}, (e) OF^+.

20. Discuss the resonance possibilities in each of the following: (a) NO_3^-, (b) NO_2^-, (c) HCO_2^-, (d) CO, (e) N_3^-.

21. Discuss as many aspects as you can of the structure and bonding in each of the following: (a) $CaCO_3$, (b) Na_2C_2 (sodium acetylide), (c) $H_2N\!-\!\overset{\parallel}{\underset{O}{C}}\!-\!NH_2$, (d) $H_2C\!\!=\!\!C\!\!=\!\!CH_2$.

22. It is said that the van der Waals forces between I_2 molecules are stronger than those between Cl_2 molecules, and that the van der Waals forces between N_2 molecules are stronger than those between O_2 molecules. Offer an explanation for each case.

23. Arrange the following in order of increasing boiling point and justify your assignments: CH_3F, CH_3Cl, CH_4Br, CH_3I.

24. (a) Construct a molecular orbital diagram for nitric oxide, NO, similar to that in Figure 6.19. (b) How many bonds are represented in your diagram? (c) Consider the additional species, NO^+ and NO^-. How many bonds would each have? (d) Is any of these species paramagnetic?

25. (a) Indicate the molecular orbital designation, as was done for oxygen on Page 131, for the following: (1) HeH^+, (2) He_2^+, (3) Cl_2, (4) Be_2, (5) Ar_2. (b) Predict whether the existence of each species is likely or unlikely. Explain.

SPECIAL PROBLEMS

1. The compound $XeOF_4$ is believed to have a square pyramid structure with the four fluorine atoms surrounding the xenon atom in the base plane; the oxygen atom lies above or below this plane bonded to the xenon atom. This structure is said to contain five σ bonds and one π bond. Suggest a rationale for this structure in terms of the orbitals available and suggest other possible structures for this substance.

2. There is a limit to the proximity of approach of two molecules. This limit is governed by van der Waals attractive forces and forces of electronic repulsion. The values in the following tabulation of van der Waals radii represent one-half of this distance. The distance of closest approach is twice these radii for like species and the sum of radii for unlike.

Van der Waals' Radii (Angstroms)

H 1.2				
	C 2.0	N 1.5	O 1.4	F 1.35
		P 1.9	S 1.85	Cl 1.80
		As 2.0	Se 2.00	Br 1.95
		Sb 2.2	Te 2.20	I 2.15

(a) Using these data, calculate the closeness of approach for: (1) two molecules of Br_2; (2) two molecules of I_2; (3) a molecule of Br_2 and a molecule of I_2; (4) two molecules of HCl oriented three possible ways: HCl—HCl, HCl—ClH, and ClH—HCl. (b) How would hydrogen bonding effect the HCl—HCl approach? (c) Referring to the last column, what correlation is there between size of the atom and proximity of approach for homodiatomic molecules?

3. Explain: (a) Ethyl alcohol, C_2H_5OH, boils at 80°C. Although its molecular weight is higher than that of water, its boiling point is lower. (b) When water and ethyl alcohol are mixed, the volume decreases. (c) Salts of the anion HCl_2^- exist.

4. Explain why: (a) Hydrogen gas exists as diatomic molecules, H_2, but the He_2 molecule has never been found. (b) The Li_2 molecule can form under unusual circumstances. (c) The B—F bond distance in the BF_4^- ion is longer than in boron trifluoride, BF_3.

REFERENCES

COMPANION, A. L., *Chemical Bonding,* McGraw-Hill, New York, 1964.

DOUGLAS, B. E., and McDANIEL, D. H., *Concepts and Models of Inorganic Chemistry,* Xerox College Publishing, Lexington, Mass., 1965.

SEBERA, D. K., *Electronic Structure and Chemical Bonding,* Xerox College Publishing, Lexington, Mass., 1965.

GILLESPIE, R. J., "The Electron-Pair Repulsion Model for Molecular Geometry," *J. Chem. Educ.,* **47**, 18 (1970).

PETERSON, Q. R., "Some Reflections on the Use and Abuse of Molecular Models," *J. Chem. Educ.,* **47**, 24 (1970).

RODERICK, W. R., "Current Ideas on the Chemical Basis of Olfaction, *J. Chem. Educ.,* **43**, 510 (1966).

PART TWO

STRUCTURE AND REACTIVITY

PRINCIPLES OF REACTIONS; PATTERNS OF REACTIVITY; THE CHEMISTRY OF HYDROGEN

The preceding chapters have been planned to introduce two fundamental and underlying themes of chemical thought. The first focuses on the basic laws of chemical combination and the constructs and concepts developed by chemists to use and extend the knowledge and understanding these laws provide. Symbols, formulas, equations, atomic weights, formula weights, moles, and classifications such as chemical and physical changes are facets of this theme.

The second theme is that of structure, and includes the structure of atoms, molecules, and ions and the nature of the forces that give rise to these structures. The important implication in understanding structural relations in chemistry is that they lead to greater insight and predictability relative to the behavior of chemicals and chemical systems.

Both themes are essential to understanding still a third and possibly the most pervasive in chemical science, that of reactivity—the tendency to undergo chemical change and the manner in which substances undergo such change. The principles of reactivity can be appreciated most readily by studying and comparing actual behavior patterns of substances such as those of important elements and families of elements. In this and in the two chapters that follow, we summarize the chemistry of the simplest element, hydrogen, and the chemistry of families of representative metals and non-metals, giving particular emphasis to principles and patterns that can help provide the basis for viable models to correlate and predict reactivity.

Before beginning our study of reactivity, it will be helpful to introduce some common features of chemical reactions and to describe some approaches for correlating reactivity based on these features. Because hydrogen is the simplest element, its chemistry will provide especially useful illustrations. Our introduction to chemical equilibrium, acid-base chemistry, and oxidation-reduction will be brief and sufficient only to enable us to study the descriptive chemistry which follows. Later we shall discuss these concepts in considerable detail.

Chemical Equilibria

Experiments have shown that may reactions never reach completion. In such reactions the starting materials (reactants) never completely disappear to form products, no matter how long the reaction is allowed to continue. At the start of the reaction the reactants are present in relatively large amounts. As time proceeds, the amounts of reactants decrease. Eventually, this decrease ceases and the reaction appears to stop, even though appreciable amounts of reactants may still be present. When this occurs, the reaction is most likely in a state of *chemical equilibrium.*

Equilibrium in the Reaction
$H_2 + I_2 \longrightarrow 2HI$

The principle underlying chemical equilibrium is that many chemical reactions are reversible. The reaction of hydrogen with iodine

$$H_2 + I_2 \rightleftharpoons 2HI$$

illustrates this reversibility and how chemical equilibrium is attained.

Starting with a mixture of hydrogen and iodine, the reaction proceeds so that hydrogen iodide is formed. As the amount of hydrogen iodide increases, the molecules of this substance react with one another to form hydrogen and iodine. Ultimately, a condition is reached in which the speed or rate of formation of HI from H_2 and I_2 is exactly equal to the rate of decomposition of HI to form H_2 and I_2 again. When these rates become equal, the reaction appears to stop and is said to be in a state of chemical equilibrium, or simply in equilibrium. Reversibility is commonly indicated in equations by two arrows, one pointing to the right and the other to the left.

Table 7.1 consists of data taken from an experiment in which 0.0100 mole of hydrogen and 0.0100 mole of iodine vapor were placed in a 1-liter flask at 410°C. These data show the changes in concentration and in the reaction rate for the reversible formation and decomposition of hydrogen iodide. The various columns in the table show how the rate of the forward reaction decreases and that of the reverse reaction increases as the reaction proceeds, together with the changes in concentration of hydrogen and hydrogen iodide causing these changes in rate. These data are plotted in Figure 7.1.

Both Figure 7.1 and Table 7.1 show that the rates of the two opposing

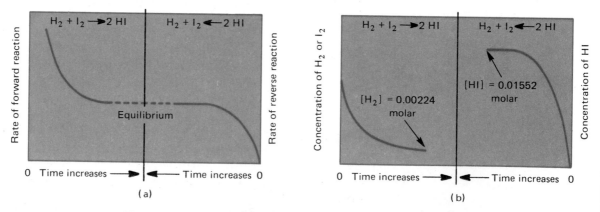

Figure 7.1 The approach to equilibrium in the formation and decomposition of hydrogen iodide: $H_2 + I_2 \rightleftharpoons 2HI$. (a) Graph of changes in reaction rates. (b) Graph of changes in concentrations.

Table 7.1 Rates of the Reactions Indicated by $H_2 + I_2 \rightleftharpoons 2HI$ at 410°C

Time (min)	Concentration of Hydrogen and of Iodine (molar) × 10^{+4}*	Rate of Forward Reaction (mole HI formed/liter/min) × 10^{+6}*	Rate of Reverse Reaction (mole HI disappearing/liter/min) × 10^{+6}*	Concentration of Hydrogen Iodide (molar) × 10^{+4}*
0	100	150	0	0
34	80	96	0.5	40
90	60	54	2.0	80
220	40	24	4.5	120
470	30	13.5	6.1	140
960	23	7.9	7.3	154
1400	22.5	7.6	7.4	155
Very long	22.4	7.5	7.5	155

*All experimental values have been multiplied by 10^x to make for easier reading of the table. The value of x is chosen to eliminate all exponentials from table values. The actual value of the first entry in the second column is 100×10^{-4}.

reactions gradually approach each other and become equal at equilibrium. When equilibrium has been reached, the concentrations of hydrogen and iodine have been reduced to 0.00224 mole/liter and the concentration of hydrogen iodide is 0.0155 mole/liter. Thus, 22.4 per cent of the original gases remain when equilibrium has been established.

The theory of chemical equilibria is of great importance in indicating the extend of reaction and will be discussed in considerable detail in Chapters 19, 20, and 21.

Indication of Extent of Reaction In this chapter we need to consider only qualitative statements describing the position (or point) of equilibrium of a reaction. For this purpose we define the position of equilibrium as the measure of the relative amounts of reaction products (in the equation as written) and reactants. If, when equilibrium has been reached, more than half of the reactants have been converted to products, we say that the position of equilibrium lies to the right. On the other hand, if the equilibrium mixture contains more of the reactants than the products, we say that the position lies to the left. Alternatively, we may say that the reaction goes, or proceeds, to a great or slight extent, meaning that the concentration of products at equilibrium is high, or low, respectively. For example, in the hydrogen iodide reaction (Table 7.1), since, at 410°C, 77.6 per cent of the reactants have been changed to products when equilibrium is established, we say that the position of equilibrium lies to the right at this temperature.

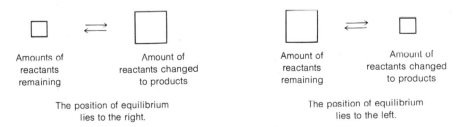

The position of equilibrium is determined by the particular reaction in question, the temperature, and the relative concentrations of two or more starting materials. The laws of equilibrium are so well known that it is possible to give an exact numerical statement about the extent of reaction (the position of equilibrium) if information about these three factors is available.

EQUILIBRIA INVOLVING ACIDS AND BASES

The terms *acid* and *base* have been used by chemists for several centuries. The definitions in current use are interrelated, but each focuses attention on a different aspect of the properties of acids or bases. The choice of definition often depends upon the information sought or the goal desired in its use. For example, one definition is limited to reactions of hydrogen compounds in water solutions (the Arrhenius definition), another to hydrogen compounds in any solvent (the Brønsted definition), and another emphasizes the making and breaking of coordinate covalent bonds in acid-base reactions (the Lewis definition). In this section we shall emphasize definitions of acids and bases that are of use in considering water solutions.

Accordingly, an acid may be defined as a hydrogen-containing substance which, when dissolved in water, forms a solution that tastes sour, turns litmus red, reacts with many metals, and conducts electricity well. A base may be defined as a substance which, when dissolved in water, forms a solution that tastes brackish or bitter, turns litmus blue, conducts electricity well, and reacts with an acid to destroy the acid properties. These definitions are based on experimental observations, and any substance suspected of being an acid or a base can be so classified if it meets the experimental criteria specified in the definition.

However, it sometimes is more useful to use the definitions which emphasize the chemical makeup of acids and bases so that you can predict whether a given substance will act as an acid or a base when placed in water. Such definitions were advanced in 1923 by J. N. Brønsted and M. Lowry, who defined an acid as a proton donor, and a base as a proton acceptor. In effect, they said that an acid-base reaction involved the *transfer of a proton* from an acid to a base.

Perhaps the two different definitions can be interrelated if we consider first some acids and then some bases in the light of both viewpoints.

Acids

The gaseous compound, hydrogen chloride, is an acid. This means, among other things, that it is a proton donor and it dissolves in water to produce a solution that conducts electricity. Since electrical conductivity in water means that ions are present, it is reasonable to write the equation for the reaction of the acid with water

$$HCl + H_2O \rightleftharpoons H_3O^+ + Cl^-$$

Here, the hydrogen chloride molecule donates a proton to a water molecule giving the *hydronium ion*, H_3O^+, and a chloride ion. Other acids should donate protons to water in a similar manner. For example, equations for the reactions of nitric, sulfuric, acetic, and hydrocyanic acids with water may be written:

$$HNO_3 + H_2O \rightleftharpoons H_3O^+ + NO_3^-$$

$$H_2SO_4 + H_2O \rightleftharpoons H_3O^+ + HSO_4^-$$

$$HSO_4^- + H_2O \rightleftharpoons H_3O^+ + SO_4^{-2}$$

$$CH_3CO_2H + H_2O \rightleftharpoons H_3O^+ + CH_3CO_2^-$$

$$HCN + H_2O \rightleftharpoons H_3O^+ + CN^-$$

Note that in each case the hydronium ion* is a product of the reaction. Evidently, sour taste, reaction with litmus, and reactivity with metals are characteristics of the hydronium ion.

Returning to the definition of an acid, we may note that any compound that reacts with water to give hydronium ions can meet the experimental criteria needed to be designated an acid, but to produce hydronium ions, the compound must have donated protons to water molecules. This, then, is the interrelation between the Arrhenius and the Brønsted-Lowry definitions of an acid.

All reactions of acids with water are examples of acid-base reactions in which water acts as the base or proton acceptor. All such reactions are reversible in that the hydronium ion formed in the reaction may donate a proton to the anion, giving the starting acid and water. For example:

$$\text{Forward reaction} \quad HCN + H_2O \longrightarrow H_3O^+ + CN^-$$

$$\text{Reverse reaction} \quad H_3O^+ + CN^- \longrightarrow H_2O + HCN$$

The strength of an acid is measured by the extent of the over-all reaction of the acid with water. For strong acids the over-all reaction goes far to the right—in fact, virtually to completion. For weak acids the over-all reaction goes only slightly to the right before the rates become equal.

Before discussing equilibrium in acid-base reactions, we shall attempt to interrelate the definitions of a base.

Bases   A number of compounds—ammonia, sodium hydroxide, and calcium oxide— produce solutions having a bitter taste, the ability to destroy acids, and other properties characteristic of bases. Examination shows that each of these solutions contains an appreciable concentration of hydroxide ions, OH^-, and leads to the Arrhenius definition of a base as a substance that produces hydroxide ions in solution. Often, as in the case of sodium hydroxide, these hydroxide ions are present in the original base and are simply dissolved in

---

*The actual structure of the positive ion produced when an acid reacts with water may be more complex than H_3O^+; the species $H_9O_4^+$ has been suggested. This structure may be regarded as a hydrated hydronium ion (Chapter 8). For lack of more precise information, we shall use the hydronium ion to represent the acid cation.

THEORIES OF ACIDS COMPARED   *The Arrhenius Theory.* An acid is a compound that produces H_3O^+ ions in water solution. Examples: HCl, HNO_3, and H_2SO_4. A base is a compound that produces OH^- ions in solution. Examples: NaOH, $Ba(OH)_2$, NH_3, Na_2O, and BaO.

The Brønsted-Lowry Theory. An acid is a proton donor. Examples: HCl, HNO_3, H_2SO_4, H_2O, and NH_4^+. A base is a proton acceptor. Examples: NH_3, H_2O, $CH_3CO_2^-$, NH_2^-, OH^-, and O^{-2}. In accepting the proton, the base must provide a lone pair of electrons. Therefore, species with lone pairs are potential Brønsted bases.

The Lewis Theory. An acid is an electron-pair acceptor. Examples: H_3O^+, $AlCl_3$, and BF_3. A base is an electron-pair donor. Examples: H_2O, NH_3, OH^-, NH_2^-, and CH_3OH. To accept an electron pair, the acid must provide an "empty orbital"; to donate an electron pair, a base must provide a lone pair of electrons. Neutralization is the formation of a coordinate covalent bond between an acid and a base.

$$H_3N: \ + \ AlCl_3 \longrightarrow H_3N:AlCl_3$$

Lewis base   Lewis acid   Coordinate
                                covalent bond

the water. In other cases, the hydroxide ions are formed by reaction with water as shown in the equations

$$NH_3 + H_2O \longrightarrow NH_4^+ + OH^-$$

$$Ca^{+2}O^{-2} + H_2O \longrightarrow Ca^{+2} + 2OH^-$$

In each of these equations there has been a transfer of a proton from the water molecule to another entity, and this leads to the Brønsted definition of a base as a proton acceptor. Thus, ammonia and oxide ion, O^{-2} (proton acceptors), are bases according to the Brønsted theory. The hydroxide ion itself is a base, since it can accept a proton, as illustrated by the equation

$$H_2O^+ + OH^- \rightleftharpoons 2H_2O$$

where, by accepting a proton, the hydroxide ion becomes a water molecule.

In the Brønsted theory, bases may be neutral (NH_3), negative (OH^-), or positive [$Al(H_2O)_5OH^{+2}$],

$$H_2O + Al(H_2O)_5OH^{+2} \longrightarrow Al(H_2O)_6^{+3} + OH^-$$

Arrhenius theory recognizes only neutral bases (NaOH, NH_3, CaO). Both theories include neutral and negative acids (H_2SO_4, HSO_4^-), but only the Brønsted theory includes positively charged acids (NH_4^+):

$$NH_4^+ + H_2O \rightleftharpoons H_3O^+ + NH_3$$

We shall use the Brønsted theory generally in what follows:

Reactions of bases with water to give hydroxide ions are acid-base reactions in which water acts as an acid.

$$H_2O + NH_3 \longrightarrow NH_4^+ + OH^-$$

Water also acts as a base in some reactions.

$$HCl + H_2O \longrightarrow H_3O^+ + Cl^-$$

It can use an unshared pair of electrons on the oxygen atom to form a co-ordinate covalent bond with a proton.

Equilibrium in Acid-Base Reactions

Let us now examine a typical acid-base reaction, the reaction of acetic acid, CH_3CO_2H, with water:

$$CH_3CO_2H + H_2O \rightleftharpoons H_3O^+ + CH_3CO_2^-$$

In the forward reaction the acetic acid molecule donates a proton to the water molecule, which acts as the base. In accepting the proton, the water molecule becomes an acid (it now has an extra proton), and the acetic acid molecule, having lost a proton, now becomes the acetate ion, $CH_3CO_2^-$, a base. Thus, in the forward reaction, one acid and one base are destroyed but a new acid and a new base are formed. The new acid and new base are called the *conjugate acid* and the *conjugate base* of the original base and acid. Thus hydronium ion is the conjugate acid of the base water, and acetate ion is the conjugate base of acetic acid. Note that the conjugate acid or base has one more or one less proton than the parent base or acid. Some acids and their conjugate bases are given in Table 7.2.

Table 7.2 Some Acids and Their Conjugate Bases Arranged in Order of Acid and Base Strength

	Acid	Conjugate Base	
Strong acids	$HClO_4$	ClO_4^-	Weak bases
	HNO_3	NO_3^-	
	HCl	Cl^-	
	H_2SO_4	HSO_4^-	
	H_3O^+	H_2O	
	HSO_4^-	SO_4^{-2}	
Increasing acid strength	H_3PO_4	$H_2PO_4^-$	Increasing base strength
	CH_3CO_2H	$CH_3CO_2^-$	
	$Al(H_2O)_6^{+3}$	$Al(H_2O)_5(OH)^{+2}$	
	H_2CO_3	HCO_3^-	
	H_2S	HS^-	
	NH_4^+	NH_3	
	HCN	CN^-	
	HCO_3^-	CO_3^{-2}	
	HS^-	S^{-2}	
Weak acids	H_2O	OH^-	Strong bases
	NH_3	NH_2^-	

Let us now consider the reverse reaction in the acetic acid-water system. Here the hydronium ion donates a proton to the acetate ion to give one molecule of acetic acid and one of water. As a result of the reverse process, chemical equilibrium will be established and the forward reaction will appear to stop. At equilibrium, two acids (CH_3CO_2H and H_3O^+) and two bases (H_2O and $CH_3CO_2^-$) will be present in the reaction mixture. In this particular case, experiment shows that the forward reaction will proceed only slightly to the right before equilibrium is established, so that at equilibrium the relative amounts of hydronium ion and acetate ion will be much less than those of acetic acid and water.

Applying the reasoning of the preceding two paragraphs to all acid-base reactions, we can state the principle, *the equilibrium mixture in any acid-base reaction always contains at least two acids and two bases.* This is illustrated by the following equations:

$$Acid + base \rightleftharpoons conjugate\ acid + conjugate\ base$$

$$HCl + H_2O \rightleftharpoons H_3O^+ + Cl^-$$

$$HNO_3 + H_2O \rightleftharpoons H_3O^+ + NO_3^-$$

$$H_2SO_4 + H_2O \rightleftharpoons H_3O^+ + HSO_4^-$$

$$HSO_4^- + H_2O \rightleftharpoons H_3O^+ + SO_4^{-2}$$

$$H_3O^+ + OH^- \rightleftharpoons H_2O + H_2O$$

$$H_2O + NH_3 \rightleftharpoons NH_4^+ + OH^-$$

$$CH_3CO_2H + NH_3 \rightleftharpoons NH_4^+ + CH_3CO_2^-$$

$$HCN + OH^- \rightleftharpoons H_2O + CN^-$$

One of the initial questions in acid-base reactions is: How far will the over-all reaction go as written before equilibrium is established? We define the strength of an acid in terms of the position of equilibrium in its reaction with water. If the position of equilibrium lies far to the right, we say that the acid is a strong acid. If the position of equilibrium lies to the left, we say that the acid is a weak acid. Similarly, the strength of a base is measured by the position of equilibrium in its reaction with water. Thus, ammonia in the equation third from the bottom (above) is classified as a weak base because, at equilibrium, a relatively small fraction of the ammonia originally added is present as ammonium ion.

What now is the behavior on reaction of acids and bases with each other? A strong acid and a strong base react almost completely, whereas a weak acid and a weak base react to only a slight extent before the rate of the reverse reaction becomes equal to the rate of the forward reaction and equilibrium is established. If one knows the relative strengths of the acids or bases involved in the equilibrium, it is possible to make a qualitative prediction of the extent of the reaction. The general principle is: *At equilibrium, the weaker acid and the weaker base predominate in concentration.*

Table 7.2 is a list of some common acids arranged in order of decreasing acid strength. These acid strengths were determined by experiment, using a method to be described in Chapter 19. The acid strength is a measure of the tendency of the acid to lose a proton to a water molecule.

Opposite each acid in Table 7.2 is its conjugate base. The order of base strengths for these bases is exactly opposite that of the corresponding acids. A strong acid has a strong tendency to lose a proton but, once the proton is lost, the conjugate base that remains has only a slight tendency to accept a proton. Thus, a strong acid produces a weak conjugate base. Conversely, a weak acid which loses its proton reluctantly forms a strong conjugate base—one that has a great affinity for protons.

The tendency of an acid to lose a proton to water is a measure of the strength of an acid and can be expressed by a value called the ionization constant of the acid, K_{ion}. This constant is a measure of the extent to which the forward reaction of the acid with water proceeds before equilibrium is established. It is also a measure of the relative amounts of products and reactants in the reaction mixture. The form of the equilibrium constant is known to be (Chapter 19)

$$K_{ion} = \frac{[H_3O^+]_{eq} \times [X^-]_{eq}}{[HX]_{eq}}$$

the symbols $[H_3O^+]_{eq}$, and $[HX]_{eq}$ represent the equilibrium concentrations in moles per liter of the hydronium ion, H_3O^+, the anion, X^-, and the acid, HX, respectively.

High values for K_{ion} mean that the over-all reaction proceeds far to the right before equilibrium is established. This is the case with strong acids such as hydrochloric acid, nitric acid, and perchloric acid. In these cases the over-all reactions proceed virtually to completion and K_{ion} becomes infinity because $[HX]_{eq}$ approaches zero.

Low values for K_{ion} mean that the over-all reaction proceeds only slightly

Table 7.3 Ionization Constants of Some Acids in Water

Name	Formula	Ionization Constant (25°C)
Acetic	CH_3CO_2H	1.8×10^{-5}
Aluminum ion	$Al(H_2O_6)^{+3}$	1.1×10^{-5}
Ammonium ion	NH_4^+	5.7×10^{-10}
Boric	H_3BO_3	6.0×10^{-10}
Hydrochloric	HCl	$\sim 10^7$
Hydrocyanic	HCN	2.0×10^{-9}
Hydrofluoric	HF	6.7×10^{-4}
Hydrogen sulfide	H_2S	1.0×10^{-7}
Hypochlorous	$HClO$	3.0×10^{-8}
Iodic	HIO_3	1.6×10^{-1}
Nitric	HNO_3	Very large
Nitrous	HNO_2	5.1×10^{-4}
Perchloric	$HClO_4$	Very large
Phosphoric	H_3PO_4	5.9×10^{-3}
	$H_2PO_4^-$	6.2×10^{-8}
	HPO_4^{-2}	4.5×10^{-13}
Sulfuric	H_2SO_4	Very large
	HSO_4^-	1.2×10^{-2}

to the right before equilibrium is established. This is the case with weak acids such as acetic, where K_{ion} is 1.8×10^{-5} at 25°C. Table 7.3 gives values for the ionization constants of some weak acids.

RELATION BETWEEN THE MAGNITUDES OF K_{ion} AND THE STRENGTH OF AN ACID

Assume that three acids, HA, HB, and HC, each of a different acid strength, are dissolved in water and the acid-base reaction allowed to come to equilibrium. The situation may be described as follows. In the diagrams the boxes represent relative amounts of material present at equilibrium.

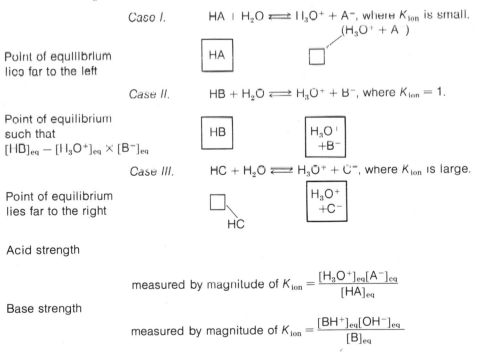

Case I. $HA + H_2O \rightleftharpoons H_3O^+ + A^-$, where K_{ion} is small.
$(H_3O^+ + A^-)$

Point of equilibrium lies far to the left

Case II. $HB + H_2O \rightleftharpoons H_3O^+ + B^-$, where $K_{ion} = 1$.

Point of equilibrium such that $[HB]_{eq} = [H_3O^+]_{eq} \times [B^-]_{eq}$

Case III. $HC + H_2O \rightleftharpoons H_3O^+ + C^-$, where K_{ion} is large.

Point of equilibrium lies far to the right

Acid strength

measured by magnitude of $K_{ion} = \dfrac{[H_3O^+]_{eq}[A^-]_{eq}}{[HA]_{eq}}$

Base strength

measured by magnitude of $K_{ion} = \dfrac{[BH^+]_{eq}[OH^-]_{eq}}{[B]_{eq}}$

We shall discuss K_{ion} and other equilibrium constants in detail in Chapter 19. In the following chapters we shall use the value of K_{ion} as an indication of the strength of acids and bases.

Other Theories of Acids and Bases

Thus far we have limited our discussion of acids and bases to a consideration of water solutions. At other points in the text, the concept of acids and bases will be expanded to include other solvents and to systems in which proton transfer does not occur. We shall find that the reasoning used in this chapter will serve as the basis for consideration of acids and bases in nonaqueous media.

OXIDATION-REDUCTION REACTIONS

Competing in importance with acid-base reactions is the large class of chemical changes known as oxidation-reduction reactions. The important characteristic of these reactions is that certain elements involved undergo a change in oxidation state during the reaction.

Consider this reaction:

$$2Cu + O_2 \longrightarrow 2CuO \qquad (1)$$

Long usage has called a reaction of a substance with oxygen an oxidation reaction. The product of the reaction, copper(II) oxide, is an electrovalent compound, and contains the ions Cu^{+2} and O^{-2}. Oxidation of the copper has thus consisted of a removal of two electrons from a copper atom to form the positive ion Cu^{+2}:

$$Cu \longrightarrow Cu^{+2} + 2e^-$$

Now consider this reaction:

$$Cu + Cl_2 \longrightarrow CuCl_2$$

Again the product is an electrovalent compound, containing the ions Cu^{+2} and Cl^-. The change in the copper has again been a removal of two electrons:

$$Cu \longrightarrow Cu^{+2} + 2e^-$$

Therefore, if we are to call the reaction of copper with oxygen an oxidation, we also must refer to the reaction of copper with chlorine as an oxidation—the same thing happens to the copper in both cases.

By extension, any reaction in which an atom loses electrons to form a positive ion is an oxidation; by further extension, any reaction involving a

---

SUMMARY OF REACTIONS OF BRØNSTED ACIDS AND BASES

In water solutions, acids may be defined as proton donors and bases as proton acceptors. An acid-base reaction involves transfer of a proton from an acid to a base. The products in such reactions, also acids and bases, are called conjugate acids or bases. The equilibrium in an acid-base reaction involves two acids and two bases as represented in the relation

$$\underset{\text{Acid}}{HA} + \underset{\text{Base}}{B} \rightleftharpoons \underset{\substack{\text{Conjugate} \\ \text{acid}}}{HB^+} + \underset{\substack{\text{Conjugate} \\ \text{base}}}{A^-}$$

The distribution of material at equilibrium always lies in favor of the weaker acid and base. Strong acids have large values of K_{ion}; weak acids have small values.

loss of electrons is an oxidation. Thus all the following are examples of oxidation processes:

$$Cu \longrightarrow Cu^{+2} + 2e^-$$

$$Fe^{+2} \longrightarrow Fe^{+3} + e^-$$

$$2Cl^- \longrightarrow Cl_2 + 2e^-$$

Consider this reaction:

$$CuO + H_2 \longrightarrow Cu + H_2O$$

As with oxidation reactions, long usage has called this a reduction reaction. The effect on the copper has been to restore the two electrons to the copper ion of copper oxide to form a copper atom:

$$Cu^{+2} + 2e^- \longrightarrow Cu$$

By extensions similar to those above, all processes in which electrons are added to chemical species are described as reductions:

$$Cu^{+2} + 2e^- \longrightarrow Cu$$

$$Fe^{+3} + e^- \longrightarrow Fe^{+2}$$

$$Cl_2 + 2e^- \longrightarrow 2Cl^-$$

$$O_2 + 4e^- \longrightarrow 2O^{-2}$$

The last two equations show that oxidation of copper by oxygen or by chlorine, characterized by a loss of electrons from the copper atom, is accompanied by a reduction of oxygen or chlorine, characterized as a gain of electrons by those atoms. We see, then, that oxidation and reduction are complementary processes and that oxidation does not occur without reduction, nor reduction without oxidation. The reactions of copper with oxygen or with chlorine are seen to be oxidation-reduction reactions (the hyphen is important), which involve an electron transfer from copper atoms to oxygen or chlorine atoms.

$$Cu\text{:} + \text{:}\ddot{O} \longrightarrow Cu^{+2} + \text{:}\ddot{O}\text{:}^{-2}$$

$$Cu\text{:} + \text{:}\ddot{C}l\cdot + \text{:}\ddot{C}l\cdot \longrightarrow Cu^{+2} + \text{:}\ddot{C}l\text{:}^- + \text{:}\ddot{C}l\text{:}^-$$

How can we reconcile this point of view with the reduction of copper oxide by hydrogen, where the product of the reaction is H_2O, a covalent compound and not an ion or atom? The electrons gained in the reduction of the copper ion,

$$Cu^{+2} + 2e^- \longrightarrow Cu$$

must come from somewhere. If we write

$$O^{-2} + H_2 \longrightarrow H_2O + 2e^-$$

we have an equation that balances electrically and provides the electrons needed. But what has been oxidized? The oxygen had eight electrons in its valence shell as the oxide ion; it still has eight in the water molecule.

$$\text{:}\ddot{O}\text{:}^{-2} \text{ has become } H \text{:}\ddot{O}\text{:}$$
$$H$$

If there is no change in the oxygen, we must conclude that the hydrogen has been oxidized. If the water molecule was ionic, this conclusion would have been an obvious one. We would have written

$$H_2 \longrightarrow 2H^+ + 2e^-$$

The hydrogen in water would have appeared as a hydrogen ion, and the oxygen would have remained as an oxide ion, unchanged in the reaction.

The Oxidation State

The above concepts of oxidation and reduction, which apply exactly to oxidation-reduction reactions involving electrovalent compounds, have proven to be so useful in discussing chemical reactions that chemists have adopted an arbitrary system for treating oxidation-reduction processes which, for this purpose, treats all compounds as if they were electrovalent even though they are not. This is done by introducing the concept of *oxidation state*. This concept assigns to each atom in a compound a number (positive or negative) the same as its electrovalence if it is an ion, or determined by a set of arbitrary rules when the atom is held by covalent bonds. The effect of these rules is to define the oxidation state of an element in a compound as the charge which an atom of that element would have if it existed as an ion in the substance in question. The rules are:

   1. *The oxidation state of a free element (that is, one not combined with another element) is zero.*
   2. *Hydrogen has an oxidation state of* I *and oxygen has an oxidation state of* −II when they are present in most compounds. Exceptions to this are that hydrogen has an oxidation state of −I in hydrides of active metals, and oxygen has an oxidation state of −I in peroxides, and of II in its compound with fluorine.
   3. *The algebraic sum* (the sum taking account of positive and negative signs) *of the oxidation states of all atoms in a neutral molecule must be zero.* If the substance is an ion rather than a molecule, the algebraic sum of the oxidation states of the atoms in the ion must equal the charge on the ion.

For simple ions, the oxidation state is equal to the charge on the ion: the electrovalence. For compounds containing covalent bonds, the oxidation state of an element must be determined by applying rules such as those above.

The common oxidation states of several elements are given in Table 7.4. You can see from the table that some elements have more than one oxidation state. When it is necessary to specify the oxidation state, the symbol of an element and its oxidation state are written together, thus: Fe(II), Cr(VI), and N(V).

EXAMPLES OF THE
CALCULATION OF
OXIDATION STATE

PROBLEM 1

Calculate the oxidation state of sulfur in sulfur dioxide (SO_2).

Solution

If the oxidation state of oxygen is −II, the total oxidation state of two oxygen atoms must be −4.* Hence, the oxidation state of sulfur in sulfur dioxide must be IV so that Rule 3 may be satisfied.

---

*Current rules of nomenclature call for Roman numerals to designate oxidation states. In numerical problems, however, it is often more convenient to use Arabic numbers.

Table 7.4 Oxidation States of Some Common Elements

Element	Type of Compound	Example	Oxidation State of Element
Aluminum	All	$AlCl_3$	III
Bromine	Bromides	NaBr	−I
Calcium	All	$CaSO_4$	II
Chlorine	Chlorides	NaCl	−I
Chlorine	Chlorates	$KClO_3$	V
Chromium	Chromic	$CrCl_3$	III
Chromium	Chromates	K_2CrO_4	VI
Copper	Cupric	$CuSO_4$	II
Iron	Ferrous	$FeSO_4$	II
Iron	Ferric	$FeCl_3$	III
Manganese	Manganous	$MnSO_4$	II
Manganese	Permanganates	$KMnO_4$	VII
Nitrogen	Nitrides	AlN	−III
Nitrogen	Nitrates	$NaNO_3$	V
Phosphorus	Phosphates	$Ca_3(PO_4)_2$	V
Potassium	All	KCl	I
Sodium	All	$NaNO_3$	I
Sulfur	Sulfides	CuS	−II
Sulfur	Sulfites	Na_2SO_3	IV
Sulfur	Sulfates	$BaSO_4$	VI

$$2O = 2 \times (-2) = -4$$

$$S = \underline{+4 \text{ (IV)}}$$
$$0$$

PROBLEM 2 Calculate the oxidation state of sulfur in hydrogen sulfate (H_2SO_4).

Solution

The total oxidation state for 2H is	$2H = 2 \times (+1) = +2$
The total oxidation state for 4O is	$4O = 4 \times (-2) = -8$
Therefore, the oxidation state of S in H_2SO_4 is	$S = \underline{+6 \text{ (VI)}}$
	0

PROBLEM 3 Calculate the oxidation state of chlorine in potassium chlorate ($KClO_3$).

Solution

The electrovalence of potassium is +1; hence, the oxidation state of 1K is	$K = +1$
The total oxidation state for 3O is	$3O = 3 \times (-2) = -6$
Therefore, the oxidation state of Cl in $KClO_3$ is	$Cl = \underline{+5 \text{ (V)}}$
	0

PROBLEM 4 Calculate the oxidation state of manganese in the permanganate ion (MnO_4^-).

Solution

The total oxidation state for 4O is	$4O = 4 \times (-2) = -8$
Therefore, the oxidation state of Mn must be	$Mn = \underline{+7 \text{ (VII)}}$
in order that the charge on the ion may be −1 (Rule 3).	-1

Oxidation State and Oxidation and Reduction

In terms of the oxidation state, oxidation is defined as an increase in oxidation state and reduction as a decrease in oxidation state (Figure 7.2).

Oxidizing Agents and Reducing Agents

An oxidizing agent is a reagent or chemical species that can oxidize another species. In oxidizing another species, it is itself reduced. In the reaction $Cu + Cl_2 \longrightarrow CuCl_2$, chlorine, Cl_2, is the oxidizing agent (it is reduced to

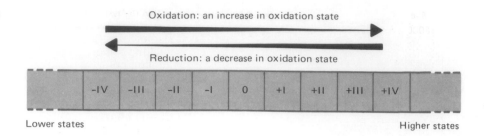

Cl⁻). Similarly, a reducing agent is a reagent or chemical species that reduces another species. In reducing the other species, it is itself oxidized. In reaction (1), copper is the reducing agent (it is oxidized to Cu^{+2}).

Equilibrium in Oxidation-Reduction Reactions

Oxidation-reduction reactions are reversible. When a strong oxidizing agent and a strong reducing agent are allowed to react, the over-all reaction will approach completion. When a weak oxidizing agent and a weak reducing agent are allowed to react, the over-all reaction will proceed only slightly to the right before the forward and reverse rates become equal and equilibrium is established. In many ways, oxidation-reduction equilibria are analogous to acid-base equilibria.

Later it will be shown that a measure of the strength of an oxidizing or reducing agent is a value called the oxidation potential, $E°$ (see Chapters 8 and 21).

Balancing Oxidation-Reduction Equations

Balancing more complex oxidation-reduction equations requires more guidelines than have been provided here. Appendix C gives some procedures for balancing these equations.

THE CHEMISTRY OF HYDROGEN

Hydrogen is unique among all the elements. It has the lightest and also the simplest atom—one containing only one electron. It has the lowest density of any element or compound. It is the element from which other elements are synthesized at the high temperatures of the sun and the other stars. It combines with nearly every other active element. Because of its intermediate electronegativity, it forms compounds of a variety of properties; for example: (a) compounds with highly electronegative elements, such as fluorine and oxygen, in which the hydrogen atoms acquire a positive character; (b) compounds with elements having low electronegativity, such as calcium and aluminum, in which the hydrogen atoms acquire a negative character; and (c) compounds in which the bonds are essentially nonpolar, such as those with carbon.

SOME FACTS ABOUT HYDROGEN

Occurrence

Hydrogen is by far the most abundant element in the universe—about 92 per cent of the atoms of the universe are hydrogen atoms. Any free hydrogen that was originally in the earth's atmosphere when the earth was formed has no doubt escaped to interstellar space because of the high velocity of the gaseous hydrogen molecules, which are moving at an average velocity of

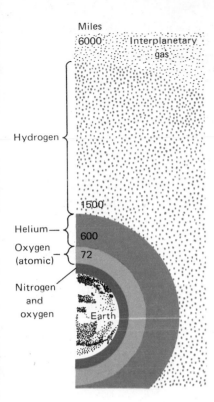

Figure 7.3 Principal components of the atmosphere up to 6000 miles above the earth.

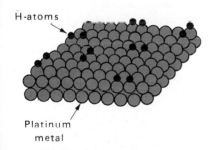

Figure 7.5 Hydrogen molecules adsorb on surfaces of some metals as atoms because of the catalytic action of the metals, which weakens or breaks the bonds between hydrogen atoms.

Table 7.5 Some Properties of Hydrogen

Atomic weight	1.008 amu
Melting point	14.1 K*
Boiling point	20.4 K
Density (S.T.P.)	0.0899 g/liter
Atomic diameter	1 A
Covalent bond length	0.75 A
H—H bond strength	103 kcal/mole

*K is read "kelvin" and is defined as 273.15 + °C.

over 4000 miles per hour at ordinary temperatures.* On the earth hydrogen occurs in large amounts combined in substances such as water, petroleum, and plant and animal tissue. In terms of the number of atoms, it is the third most abundant element in the materials of the earth's crust†, the oceans, and the atmosphere. Water, which is combined hydrogen and oxygen, contains 11.2 per cent hydrogen by weight.

Hydrogen is a gas at ordinary temperatures. It is odorless, tasteless, and colorless, and it has a very low solubility in water. In some metals it dissolves readily, usually as atoms rather than as diatomic molecules. Since hydrogen atoms are more reactive than hydrogen molecules, the metals serve as catalysts for reactions of hydrogen with other substances.

Because of their electron configuration (a single electron in the $1s$ shell), hydrogen atoms are much too reactive to exist in the free state under ordinary conditions, especially if there are other atoms or groups of atoms to

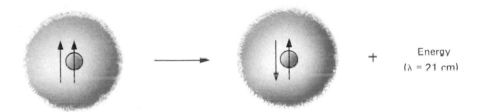

Figure 7.4 Before 1965, only two chemical species were identified in interstellar space—hydrogen atoms and hydroxyl radicals. Deuterium and helium atoms were detected in 1966. The hydrogen atoms are identified by the energy emitted when atoms having parallel electron and nuclear spins undergo a transition during which the spins become antiparallel. The hydroxyl radicals are identified by the energy absorbed by the several types of rotational changes this radical is known to undergo.

combine with. Therefore, we find atomic hydrogen only in interstellar space. Hydrogen molecules, H_2, exist over a wide temperature range; even at 4000°C, about 40 per cent of the particles in a volume of hydrogen gas are H_2 molecules.

The hydrogen molecule is nonpolar. This helps explain the low boiling point (20.4 K) and freezing point (14.1 K), since hydrogen molecules are but slightly polarizable and will have only very weak van der Waals forces between them.

---

*This is about one-sixth the escape velocity of a particle from the earth's gravitational field—25,000 miles per hour.
†The earth's crust may be defined as the surface layer of the earth, which has a thickness of about 10 miles.

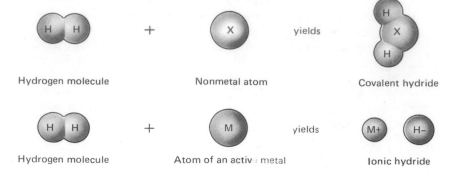

Figure 7.6 Hydrogen reacts with nonmetals to form covalent hydrides; it forms ionic hydrides with active metals.

Hydrogen molecule + Nonmetal atom yields Covalent hydride

Hydrogen molecule + Atom of an active metal yields Ionic hydride

Uses Hydrogen is one of the most important industrial gases. Large amounts are used for the synthesis of ammonia, hydrogenation of vegetable oils to produce butter and lard substitutes, hydrogenation of hydrocarbons in the production of gasoline, hydrogenation of carbon monoxide in the production of methyl alcohol, as a reducing atmosphere for metallurgical operations, as a coolant for large generators, and as one of the most powerful rocket fuels. In the future, large amounts may also be used in the hydrogenation of coal and carbon monoxide to produce lubricating oil, gasoline, and other products.

Hydrogen Isotopes Of the three isotopes of hydrogen, the mass 1 isotope is by far the most abundant, making up about 99.98% of naturally occurring hydrogen. Most of the other 0.02% is deuterium, D or ^2_1H, with only 10^{-15}% tritium, T or ^3_1H.

Deuterium The isotope of mass 2 is obtained by electrolysis of water. The O—H bond is slightly weaker than the O—D bond. Hence, on electrolysis of water, the O—H bonds break in preference to the O—D bonds and the D_2O concentrates in the water mixture being electrolyzed. The D_2O can then be electrolyzed to produce D_2. High-purity deuterium is now available, its chief use being as a tracer in studying the mechanism of chemical reactions. Deuterium has a boiling point of 23.6 K and a freezing point of 18.6 K.

Water made of deuterium rather than hydrogen has the following properties:

Boiling point	101.4°C
Freezing point	3.8°C
Density at 25°	1.108 g/cc
Temperature of maximum density	11.2°C

In general, the properties of isotopes of a given element are very similar. However, for very light elements, several of the properties that depend on bond strength, mass, and reaction rates show wide variation. These are indicated in the differences in some of the properties of D_2O and H_2O. Several other differences are:

Bond energy	O—H = 110 kcal/mole	O—D = 121 kcal/mole	
Bond energy	H—H = 103 kcal/mole	D—D = 105 kcal/mole	
Relative rates of reaction	$H_2 + Cl_2$ is 13 times as fast as $D_2 + Cl_2$ under comparable conditions.		

These differences, now called *isotope effects*, are appreciable only where the difference in mass of the isotopes is a significant fraction of the mass of the isotope. Hence, isotope effects are observed mainly with the isotopes of the light elements.

Tritium

This isotope of mass 3 is present in nature as the result of the nuclear reaction* of neutrons (*n*) in the upper atmosphere with nitrogen nuclei:

$$_0^1n + \,_7^{14}N \longrightarrow \,_1^3H + \,_6^{12}C$$

Tritium is radioactive. Half of an original sample of tritium will decompose in 12 years, forming helium (see Chapter 23). The main use of tritium is as a radioactive tracer. Tritium shows an even greater isotope effect than deuterium.

Stable Chemical States (Oxidation States)

Hydrogen exhibits three stable chemical states called +I, 0, and −I oxidation states.

An example of hydrogen in the −I oxidation state is sodium hydride, NaH. Here, the hydrogen atom has accepted an electron from a sodium atom, thereby forming the hydride ion, H^-. An example of hydrogen in the +I oxidation state is hydrogen chloride, HCl. Here, a covalent bond exists between hydrogen and chlorine atoms but, because the chlorine atom has a higher electronegativity than the hydrogen atom, the oxidation state of the hydrogen atom is arbitrarily assigned a value of +I, suggesting that the hydrogen atom donates a good portion of its electron to the chlorine atom. For hydrogen the stable species in the 0 oxidation state is the hydrogen molecule, H_2.

The fact that hydrogen forms three stable oxidation states is related to the structure of the hydrogen atom. This atom may complete its valence shell by adding one electron or it may donate (at least in part) its one valence electron.

Reactions of Hydrogen Atoms

About 103 kcal of energy are necessary to break all the covalent bonds in a mole of hydrogen molecules, thereby forming hydrogen atoms. Hydrogen atoms are also formed from hydrogen molecules on the surfaces of some catalysts, and by passing electricity through gaseous hydrogen. The hydrogen atom is a very vigorous reactant since it has a single unpaired electron which can pair off readily with the single electron of another atom or group of atoms to form a new bond. For example,

$$H\cdot + H\cdot \longrightarrow H:H + energy \quad (103 \text{ kcal/mole})$$

$$2H\cdot + \cdot \ddot{O}: \longrightarrow H:\ddot{O}: + energy \quad (220 \text{ kcal/mole})$$
$$\phantom{2H\cdot + \cdot \ddot{O}: \longrightarrow H:}H$$

Hydrogen atoms may react with molecules such as those of chlorine to form hydrogen chloride and chlorine atoms as follows:

$$H\cdot + Cl_2 \longrightarrow HCl + \cdot \ddot{\underset{\cdot\cdot}{Cl}}:$$

The heat generated in the atomic hydrogen torch results from the recombination of hydrogen atoms to form hydrogen molecules.

Hydrogen atoms, in reactions with a few metals such as sodium and lithium, will take on another $1s$ electron and form the negative ion, H^-.

---

*Nuclear equations are written as chemical equations, but, in addition, the mass number and atomic number of each atom are indicated with superscripts and subscripts, respectively.

Reactions of Hydrogen Nuclei

The nucleus of hydrogen isotope 1 is a proton that has a very high affinity for an electron and, therefore, forms hydrogen atoms readily. About 310 kcal of energy are necessary to pull all the electrons away from all the hydrogen atoms in a mole of hydrogen atoms, thereby leaving hydrogen nuclei (Table 4.7).

$$H \cdot \longrightarrow H^+ + e^-$$

A high concentration of hydrogen nuclei is never formed except at the high temperatures of the sun and stars or of the electric arc.

At about 10,000,000°C, the velocities of protons are so high that the repulsion of the positive charges is overcome and the protons come so close together on collision that the extremely strong but very short-range forces of the atomic nuclei become effective, causing the nuclei to fuse together to form helium nuclei and release a tremendous amount of energy. This is one of the main energy-producing reactions of the universe; it is also the first step in the formation of the elements.

$$4{}_1^1H + 2\,{}_{-1}^0e \longrightarrow {}_2^4He + 6.5 \times 10^{11} \text{ cal/mole}$$

This is about 1,000,000 times more energy than is released when hydrogen burns to form water:

$$2H_2 + O_2 \longrightarrow 2H_2O + 1.4 \times 10^5 \text{ cal/2 moles}$$

Reactions of Hydrogen Molecules

Most of the reactions of hydrogen molecules are those in which the covalent bond between hydrogen atoms is broken and another one is formed with another atom. For example:

The combustion of hydrogen	$2H_2 + O_2 \longrightarrow 2H_2O$
The synthesis of ammonia	$3H_2 + N_2 \longrightarrow 2NH_3$
The reaction with halogens	$H_2 + Cl_2 \longrightarrow 2HCl$
The hydrogenation of unsaturated organic compounds	$H_2 + C_2H_4 \longrightarrow C_2H_6$

In reactions of hydrogen with active metals such as lithium, sodium, potassium, and calcium, the covalent bond in the hydrogen molecule is broken, but the new bonds formed are predominantly ionic in character. For example:

$$2Li + H_2 \longrightarrow 2LiH$$

$$Ca + H_2 \longrightarrow CaH_2$$

As stated earlier, hydrogen forms compounds with nearly all active elements. However, it has not been possible (as yet) to bring about the direct reaction of hydrogen with all active elements. Hydrogen will react directly with all elements of low electronegativity (1.2 or less) and with those of high electronegativity (2.3 or higher). To synthesize hydrogen compounds of elements of intermediate electronegativities, it is usually necessary to use reactions of the type illustrated for the preparation of aluminum hydride, AlH_3, and silane, SiH_4.

$$AlCl_3 + 3LiH \longrightarrow AlH_3 + 3LiCl$$

$$Mg_2Si + 4H_2O \longrightarrow 2Mg(OH)_2 + SiH_4$$

Preparation of Hydrogen: Application of Oxidation-Reduction Principles

Since molecular hydrogen does not occur in abundance in nature, it must be prepared from compounds of hydrogen. Nearly all such compounds contain hydrogen in the I oxidation state, so the problem is one of reducing the hydrogen from the I to the zero oxidation state. A reducing agent is needed and it remains only to select reducing agents strong enough for this purpose. Three types of reducing agents are used to reduce the hydrogen in water or in acids: very active or moderately active metals, heated carbon, and electrons at a cathode.

Hydrogen from Water or Acids and Metals

Metals are reducing agents because they are oxidizable—they can be converted from zero to positive oxidation states as they lose electrons in chemical reactions. While being oxidized in a reaction, metals make it possible for another element to be reduced. The most active metals are those which are oxidized most readily and are, therefore, the strongest reducing agents among the metals.

Water and strong acids such as hydrochloric and sulfuric are common laboratory sources of hydrogen. In water solutions of acids, the species that reacts with the metal is the hydronium ion, H_3O^+; in water itself, it may be the water molecule that reacts with the metal. The following equations for reactions of metals with water and hydronium ions indicate that the hydronium ion appears to be a stronger oxidizing agent than water. As a result, the very active metals will react with both water and acids to produce hydrogen, moderately active metals will react only with acids, and less active metals will react with neither water nor acids to produce hydrogen.

1. Active metals + water
$$2Li + 2H_2O \longrightarrow 2LiOH + H_2$$
$$2Na + 2H_2O \longrightarrow 2NaOH + H_2$$

Reactions of strong reducing agents (active metals) with a moderately weak oxidizing agent (water) proceed readily.

2. Moderately active metals + steam
$$Mg + H_2O \longrightarrow MgO + H_2$$
$$Ca + H_2O \longrightarrow CaO + H_2$$

Reactions of moderately strong reducing agents with a moderately weak oxidizing agent proceed at elevated temperatures.

2. Moderately active metals + acids
(H_3O^+ from HCl or dilute H_2SO_4)
$$Mg + 2H_3O^+ \longrightarrow Mg^{+2} + H_2 + 2H_2O$$
$$Zn + 2H_3O^+ \longrightarrow Zn^{+2} + H_2 + 2H_2O$$

Reactions of moderately strong reducing agents with a stronger oxidizing agent (H_3O^+) proceed readily.

4. Less active metals + acids
(HCl or dilute H_2SO_4)
$$Cu + H_3O^+ \longrightarrow \text{no reaction}$$
$$Ag + H_3O^+ \longrightarrow \text{no reaction}$$

Reactions of weak reducing agents with the oxidizing agent H_3O^+ are unsuccessful.

Hydrogen from Reaction of Water with Hot Carbon. An important industrial process, called the Bosch process, for the production of hydrogen is the reaction of steam with hot carbon. Two reactions may occur:

$$H_2O + C \xrightarrow{hot} CO + H_2$$

$$CO + H_2O \xrightarrow{hot} CO_2 + H_2$$

Hydrogen by Electrolysis of Water. Another industrial method for the preparation of hydrogen is the electrolysis of water. The electrolysis is done in an alkaline or acidic solution.

In acid solution:

At cathode $2H_3O^+ + 2e^- \longrightarrow H_2 + 2H_2O$

At anode $6H_2O \longrightarrow O_2 + 4H_3O^+ + 4e^-$

In alkaline solution:

At cathode $2H_2O + 2e^- \longrightarrow H_2 + 2OH^-$

At anode $4OH^- \longrightarrow O_2 + 2H_2O + 4e^-$

Chemical Processes in the Biosphere. We often fail to appreciate just how much chemical change is taking place around us and within us every second. And we seldom contemplate just how much we depend upon chemical reactions for our very existence. However, we may be unwise if we simply take for granted those processes on which life itself depends.

The biosphere, that part of the earth in which life exists, contains a mixture of compounds composed mostly of carbon, oxygen, nitrogen, and hydrogen in a continuous state of creation, transformation, and change. Facile chemical change is the central feature of the biosphere, for it is this that sustains and controls life. The mechanisms used are grand cycles involving assimilation and production of energy and a few reactions of chemical elements. The chemical processes in these cycles are a myriad of oxidation-reduction reactions involving mainly the four elements named above.

Energy from the sun is absorbed by plants which utilize it to turn carbon dioxide and water into organic molecules and oxygen. The organic molecules serve as both the fuel and the medium for maintaining life and growth. The oxygen is, at the same time, the supporter and the potential destroyer of life.

Certain microorganisms in soil react with nitrogen of the air, converting it to oxidation states in which it can be used by plants to incorporate nitrogen atoms into the organic molecules produced by photosynthesis, thereby synthesizing the building blocks of proteins, the compounds that control the chemical processes in living cells.

Plants and animals die with accompanying decomposition of the organic materials to carbon dioxide, water, and simple nitrogen compounds such as ammonia or urea, NH_2—CO—NH_2. These compounds are returned to the soil to be absorbed and used by a new generation of plants and animals, or to be decomposed into elemental nitrogen and returned to the atmosphere. The carbon dioxide and water are returned to the air or the soil where they can again serve as chemical reactants in the photosynthetic processes.

The existence on the surface of the earth of substantial quantities of water, of a gravitational field strong enough to retain an atmosphere, of a continuous and ample energy supply from an external source, and of massive areas of contact among the water, air, and land provide an environment in which the chemical cycles can operate efficiently and facilely.

While all of this suggests a steady state or equilibrium condition in which nothing new emerges once the equilibrium has been established, we know that the biosphere has undergone some profound changes during the 4.5 billion years of the earth's existence. Indications are that the early atmosphere contained no oxygen but much hydrogen with some methane and ammonia. About 2 billion years ago, marine organisms emerged that utilized the sun's energy to make simple organic compounds from carbon dioxide and water, releasing oxygen to the atmosphere. Gradually an oxygen atmosphere built up, screening out much of the destructive ultraviolet radiation from the sun. New species evolved that obtained more energy and more viability by more efficient respiration in the oxygen-rich air, and the present chemical cycles were established.

Evolution of advanced forms of animal life—those possessing tissue and organs—probably would not have been possible without the high levels of energy release that characterize the oxidation of carbohydrates found in plants and animals. To illustrate this, higher animals regularly oxidize glucose to carbon dioxide and water, a process that liberates 686 kcal/mole of glucose oxidized. Fermentation of glucose, an oxidative process used at times by certain lower animals, releases only 50 kcal/mole of glucose used. An animal as complex as man requires the ready availability of unusually large reservoirs of energy for hour-by-hour existence. Carbohydrates in the blood or stored in the liver or muscles are the principal fuel of the human engine.

However, nature has provided in oxygen the ambivalence to control the characteristics and, to some extent, the lifetimes of organisms. Free oxygen can react readily with many of the organic compounds of living cells, liberating more energy than the cells can handle, thereby destroying or changing them. For this reason, those organisms that conduct their oxidations anaerobically—mainly by removing hydrogen from foodstuffs rather than by direct reaction with oxygen—have the greatest chance of survival in the earth's biosphere. Even these organisms must constantly repair damage done by reactions of oxygen with cellular materials.

We have seen from this discussion that the energy we use to move, to think, and to exist comes from the sun. It was captured and stored in the carbohydrates formed in the photosynthetic processes which, in turn, were used to make proteins and other organic materials needed for health and growth. Without the large quantities of food that are supplied continuously as part of the chemical processes of the biosphere, and without the subtle destruction wrought by oxygen even as it supports life, the quality and viability of life cycles of living things would be diminished.

SUMMARY   The purpose of this chapter is threefold: to introduce certain principles related to chemical reactions—for example, chemical equilibria and the fundamentals of both acid-base and oxidation-reduction reactions; to summarize the chemistry of hydrogen, the simplest element; and to use these principles to discuss the reactions of hydrogen and some of its compounds.

Many of the ideas associated with chemical reactions presented here will be used constantly throughout the remainder of the text. Concepts and terms such as extent of reaction, position of equilibrium, acid, base, ionization constant, oxidation, reduction, oxidizing agent, and reducing agent are essential parts of a basic understanding of the elementary principles of chemical reactions.

Acids
Arrhenius theory of acids
Brønsted theory of acids
conjugate acid
conjugate base
Lewis theory of acids
proton acceptor
proton donor
strong acid
weak acid
hydronium ion

Equilibrium
chemical equilibrium
equilibrium constant
position of equilibrium

Escape velocity

Electronegativity

Reaction
rate of reaction
reversible reaction
electrolysis

Oxidation-Reduction
oxidation state
oxidizing agent
reducing agent

Radioactive tracer

Hydrogen
tritium
deuterium
hydride

QUESTIONS AND PROBLEMS

1. What is meant by the expression "chemical equilibrium"?

2. Cite evidence to indicate that chemical reactions are reversible.

3. What is meant by the rate of a reaction? Is the rate normally dependent or independent of the concentrations of the reacting substances? Explain. (See Table 7.1.)

4. If, at equilibrium, the rate of the forward reaction equals the rate of the reverse reaction, does this mean that half of the reactants have reacted? Explain. What determines the position of equilibrium?

5. Compare the three theories of acids, illustrating each by chemical equations.

6. Give examples of several weak and strong acids and of several weak and strong bases.

7. What species and what relative amounts of them are present in a dilute HCl solution?

8. Write the equations for the reactions of each of the following bases with water and with one other acid. (a) $CH_3CO_2^-$, (b) NH_3, (c) NH_2^-, (d) H_2O, (e) CN^-, (f) Cl^-, (g) SO_4^{-2}.

9. Write the formula for the conjugate base of each of the following: (a) NH_4^+, (b) H_2S, (c) H_2SO_4, (d) $H_2PO_4^-$, (e) NH_3, (f) HSO_4^-.

10. Write the formula for the conjugate acids of each of the following: (a) SO_4^{-2}, (b) S^{-2}, (c) $H_2PO_4^-$, (d) HSO_4^-, (e) NH_3.

11. List the following in order of decreasing H_3O^+ concentration: (a) 0.1 M HCl, (b) 0.1 M $HC_2H_3O_2$, (c) 0.1 M HNO_2, (d) 0.1 M H_2SO_4.

12. Calculate the oxidation state of the underlined element in each of the formulas: (a) $H_2\underline{S}O_4$, (b) $H\underline{N}O_3$, (c) $\underline{N}H_3$, (d) $\underline{C}Cl_4$, (e) $NaH_2\underline{P}O_4$, (f) $H_2\underline{S}O_3$, (g) $K\underline{Cl}O_3$, (h) $\underline{C}H_2Cl_2$.

13. Identify the oxidizing agent and reducing agent in each of the following:
(a) $Fe + Cl_2 \longrightarrow FeCl_3$
(b) $MnO_2 + NaCl + H_2SO_4 \longrightarrow MnSO_4 + NaHSO_4 + H_2O + Cl_2$
(c) $KMnO_4 + H_2S + H_2SO_4 \longrightarrow K_2SO_4 + MnSO_4 + H_2O + S$
(d) $KOH + Cl_2 \longrightarrow KClO_3 + KCl + H_2O$
(e) $K_2Cr_2O_7 + HCl \longrightarrow HCl + CrCl_3 + H_2O + Cl_2$

14. Does it appear that there is a relation between the electronegativity of an

element and its strength as an oxidizing or reducing agent? Justify your answer (See page 000 and Table 5.3).

15. Indicate the location and relative abundance of hydrogen in the free state and in the combined state.

16. Of what importance is hydrogen in the formation of plant and animal tissue? In the formation of the chemical elements?

17. Describe several physical properties of hydrogen, deuterium, and tritium.

18. Show by chemical equations: (a) preparation of hydrogen from water by electrolysis; (b) preparation of hydrogen from water by active metals; (c) preparation of hydrogen from acids; (d) reduction property of hydrogen in reaction with an oxide; (e) formation of atomic hydrogen from molecular hydrogen; (f) neutralization of an acid with a base.

19. Name several commercial uses of hydrogen.

20. Calculate the mass loss when four moles of protons and two moles of electrons combine to form one mole of helium ions with an evolution of 6.5×10^{11} cal. If 1 g of coal gives 7000 cal when it burns, calculate how many tons of coal must be burned to give the amount of heat that would be evolved when 1 g of matter is converted to energy.

21. How many grams of sodium hydroxide are required to neutralize 50 g of perchloric acid?

22. How many grams of sulfuric acid in dilute solution are required to oxidize 25 g of iron from Fe^0 to Fe^{+2}?

23. In the reaction

$$H_2 + I_2 \rightleftharpoons 2HI$$

at a given temperature, equilibrium is established when the following concentrations are present:

$$H_2 = 0.1 \text{ mole/liter; } I_2 = 0.1 \text{ mole/liter; } HI = 1.8 \text{ moles/liter}$$

If the equilibrium constant expression is given by

$$K = \frac{[HI]^2_{eq}}{[H_2]_{eq} \times [I_2]_{eq}}$$

calculate the value for K at this temperature. Does this value of K reflect the position of equilibrium? Why or why not?

24. If 1.0 mole of hydrogen and 1.0 mole of iodine are added to a 5.0-liter container at a given temperature and the system is allowed to come to equilibrium, and if at this time 0.10 mole of hydrogen iodide has been formed, calculate the value of the equilibrium constant at that temperature.

SPECIAL PROBLEMS 1. In terms of the velocity required for gas molecules to escape from the earth's atmosphere, and knowing that, on the average, oxygen molecules travel at about one-fourth the speed of hydrogen molecules, account for the relative abundance of hydrogen and oxygen in the atmosphere. Is there an additional factor that may account for the larger amount of oxygen?

2. Compare the physical properties of ordinary water with heavy water. From this comparison, can you see any relationship of physical properties to mass?

3. Discuss and correlate the changes in bond type and properties among the compounds RbH, SrH_2, InH_3, SbH_3, TeH_2, and HI.

4. Calculate the equilibrium constant for the hypothetical reaction

$$W(g) + 2X(g) + 3Y(g) \rightleftharpoons 4Z(g)$$

where

$$K = \frac{[Z]^4_{eq}}{[W]_{eq} \times [X]^2_{eq} \times [Y]^3_{eq}}$$

if, at equilibrium, a 1.5 liter bottle contains 2.0 moles of W, 1.5 moles of X, 1.0 moles of Y, and 0.5 moles of Z. (b) At the same temperature and volume, how many moles of W will be in equilibrium with 1.0 mole of X, 1.5 moles of Y, and 2.0 moles of Z?

5. In each of these reactions, the equilibrium position lies to the right.

$$H_2C_2O_4 + H_2PO_4^- \rightleftharpoons HC_2O_4^- + H_3PO_4$$

$$H_3PO_4 + F^- \rightleftharpoons HF + H_2PO_4^-$$

(a) Make a list of all the species that act like Brønsted acids in order of decreasing strength. (b) Do the same for the Brønsted bases.

6. Interpret the following reactions in terms of the Lewis theory:

(a) $BF_3 + F^- \longrightarrow BF_4^-$

(b) $S + SO_3^{-2} \longrightarrow S_2O_3^{-2}$

(c) $AlCl_3 + Cl^- \longrightarrow AlCl_4^-$

(d) $Ag^+ + 2NH_3 \longrightarrow Ag(NH_3)_2^+$

7. A 0.01 M solution of NaA is more basic than a 0.01 M solution of NaB. Which acid is the stronger, HA or HB? Explain. (Hint: The important reaction here is $X^- + H_2O \rightleftharpoons HX + OH^-$.)

8. (a) Which solution is more basic, 0.1 M $NaC_2H_3O_2$ or 0.1 M $NaNO_2$? ($HC_2H_3O_2$ is a weaker acid than HNO_2) Support your answer with theoretical considerations. (b) Can you devise a simple experiment to show that $HC_2H_3O_2$ is a weaker acid than HNO_2?

REFERENCES   CANAGARATNA, S. G., and SELVARATAM, M., "Analogies Between Chemical and Mechanical Equilibria," *J. Chem. Educ.*, **47**, 759 (1970).

GOLDISH, D. M., "Component Concentrations in Solutions of Weak Acids," *J. Chem. Educ.*, **47**, 65 (1970).

GOODSTEIN, M. P., "An Interpretation of Oxidation-Reduction," *J. Chem. Educ.*, **47**, 452 (1970).

ROBERTS, M. S., "Hydrogen in the Galaxies," *Sci. Amer.*, **208**, no. 6, 94 (1963).

ROBINSON, B. J., "Hydroxyl Radicals in Space," *Sci. Amer.*, **213**, no. 1, 26 (1965).

VAN DER WERF, C. A., *Acids, Bases and the Chemistry of the Covalent Bond*, Van Nostrand Reinhold, New York, 1961.

THE ALKALI AND ALKALINE EARTH ELEMENTS: TWO FAMILIES OF REPRESENTATIVE METALS

This chapter is the first of many that come later in this book in which the properties of specific elements and their compounds are discussed. The authors' aim in all these chapters on descriptive chemistry is to correlate bulk properties with a few tabulated properties of the atoms to show you, the student, that the observed chemical behavior of the substances is a consequence of, or develops predictably from, fundamental properties. Thus the properties of the elements of Group IA and Group IIA are found to be those that would be predicted from a knowledge of their electron configurations, ionization energies, atom or ion size, and crystal structure. The interplay among these factors is complex but gives rise to certain trends in properties within the families. Chemists have found it useful to look for these trends and, having found them, to explain them in terms of atomic and crystal structure. These explanations are helpful correlative tools for organizing and remembering the properties of the elements and predicting possible new uses of these metals.

In subsequent chapters as well, the student is urged to search for the relationships between the properties of elements and compounds and the fundamental nature of their atoms, and to train himself to predict properties as a logical consequence of the nature of the atoms. Only thus can the massive amount of chemical information become a unified whole and not a hodgepodge of unrelated facts.

COMPARISON OF THE TWO FAMILIES

The six alkali metal elements—lithium, sodium, potassium, rubidium, cesium, and francium—have a single valence electron beyond a well-shielded nucleus. The alkaline earth elements—beryllium, magnesium, calcium, strontium, barium, and radium—found in the adjacent group in the periodic table have two valence electrons beyond a well-shielded nucleus. In compounds, the elements of these families show only a single oxidation state, I and II, respectively, and therefore the chemistry of their reactions is relatively simple. Their low ionization energies and shortage of valence electrons make for easy loss of electrons to form positive ions (Figure 5.17), and provide the conditions favorable to community sharing characteristic of metallic binding. All these elements are thus metals; as a matter of fact, these two families (Figure 8.1) contain the most reactive of the metals.

Properties of Atoms

The electron configurations assigned the atoms of Group IA and Group IIA elements (Table 8.1) imply that:

1. The elements should readily lose their valence electrons, thereby forming positive ions having stable noble gas configurations.
2. The ease of formation of these ions should increase as the amount of shielding due to intervening shells of electrons increases (as the atomic weight increases) because the nucleus will exert less attraction on valence electrons.
3. The atomic and ionic radii should increase within each family as the atomic weight increases because of the greater number of electron shells that accompany an increase in atomic weight.

Each of these implications can be verified by experimental data.

Ready Loss of Valence Electrons

A comparison of the ionization energies recorded in Table 8.2 with those in Tables 4.7 and 4.8 show that the energy needed to remove an electron from an atom of one of these elements is lower than that required for almost all other elements. It also is evident that the alkali metal elements lose only one electron easily; removal of the second following the first requires a much higher energy. Consequently, alkali metals will have an oxidation state of I, only, in compounds.

A similar comparison of first, second, and third ionization energies for alkaline earth metals shows that these atoms may lose two electrons easily, but not a third, confirming, on the basis of ionization energies, the experimentally observed single oxidation state, II, for alkaline earth metals in compounds.

The higher first ionization energies for an alkaline earth metal, compared to an alkali metal having the same configuration of internal electron shells, is an expected consequence of the higher nuclear charge on the alkaline earth metal, producing a greater attraction for the valence electron. (Compare values for Li, Be; for Na, Mg; etc.)

$_3$Li$_{6.939}$	$_4$Be$_{9.0122}$
$_{11}$Na$_{22.9898}$	$_{12}$Mg$_{24.312}$
$_{19}$K$_{39.102}$	$_{20}$Ca$_{40.08}$
$_{37}$Rb$_{85.47}$	$_{37}$Sr$_{87.62}$
$_{55}$Cs$_{132.905}$	$_{56}$Ba$_{137.34}$
$_{87}$Fr$_{(223)}$	$_{88}$Ra$_{(226)}$

Figure 8.1 Elements of Groups IA and IIA.

Table 8.1 Electron Configurations of the Atoms of the Alkali and Alkaline Earth Elements

Alkali Metals		Alkaline Earth Metals	
Li $\boxed{1s^2}\, 2s^1$		Be $\boxed{}\, 2s^2$	
Na $\boxed{1s^2\, 2s^2\, 2p^6}\, 3s^1$		Mg $\boxed{}\, 3s^2$	
K $\boxed{1s^2\, 2s^2\, 2p^6\, 3s^2\, 3p^6}\, 4s^1$		Ca $\boxed{}\, 4s^2$	
Rb $\boxed{1s^2\, 2s^2\, 2p^6\, 3s^2\, 3p^6\, 3d^{10}\, 4s^2\, 4p^6}\, 5s^1$		Sr $\boxed{}\, 5s^2$	
Cs $\boxed{1s^2\, 2s^2\, 2p^6\, 3s^2\, 3p^6\, 3d^{10}\, 4s^2\, 4p^6\, 4d^{10}\, 5s^2\, 5p^6}\, 6s^1$		Ba $\boxed{}\, 6s^2$	
Fr $\boxed{1s^2\, 2s^2\, 2p^6\, 3s^2\, 3p^6\, 3d^{10}\, 4s^2\, 4p^6\, 4d^{10}\, 4f^{14}\, 5s^2\, 5p^6\, 5d^{10}\, 6s^2\, 6p^6}\, 7s^1$	Ra $\boxed{}\, 7s^2$	

NOTE: Boxed sections indicate filled orbitals; the boxed section in each alkaline earth element contains the same number of orbitals and electrons as the comparable element among the alkali metals.

Table 8.2 Ionization Energies for Alkali and Alkaline Earth Elements (kcal/mole of atoms)

Alkali Element	1st Electron	2nd Electron	Alkaline Earth Element	1st Electron	2nd Electron	3rd Electron
Li	124	1743	Be	215	420	3547
Na	118	1090	Mg	176	346	1847
K	100	733	Ca	141	274	1180
Rb	96	634	Sr	131	254	
Cs	90	548	Ba	120	200	

Table 8.2 also shows that the ionization energies decrease from the elements of lower to those of higher atomic weight in each family as predicted in the implication above concerning the ease of ion formation.

Atomic and Ionic Radii

The values of the radii of the atoms and of the corresponding ions of Group IA and Group IIA elements estimated from crystal structure studies (Chapters 5 and 13) given in Figure 8.2 verify the prediction that the atomic radius increases as the atomic weight increases within each family. Thus, the atomic radii for the Group IA atoms increase steadily from 1.55 Å for lithium to 2.67 Å for cesium. Similar trends exist for Group IIA atoms and for the ions of both families.

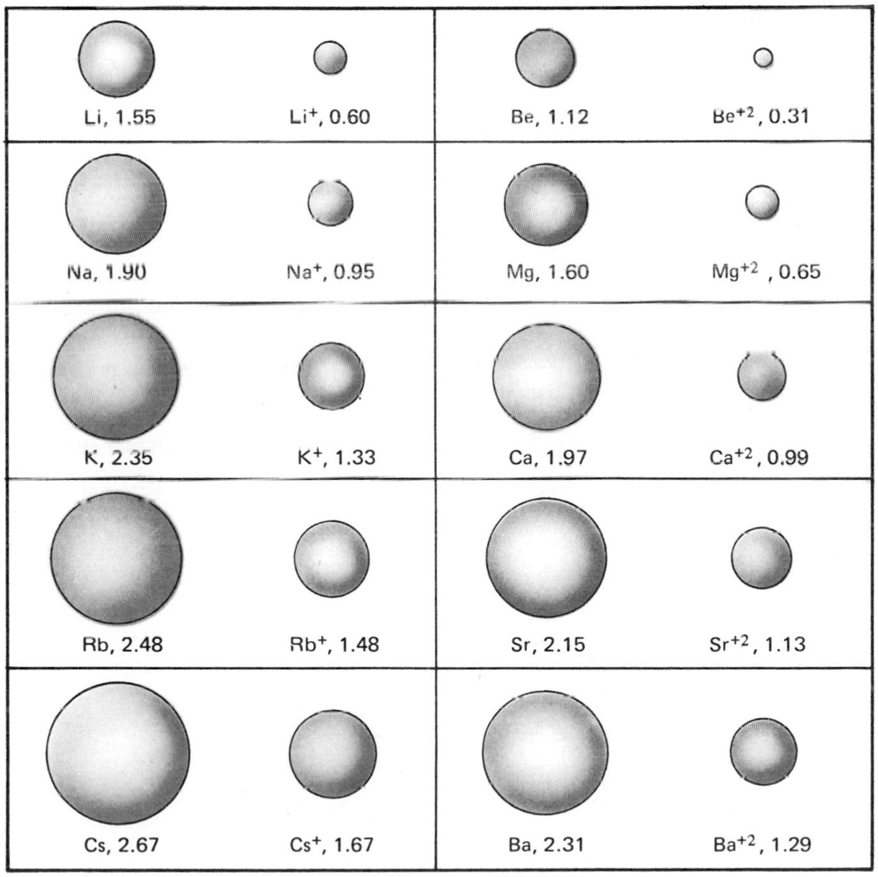

Figure 8.2 Relative sizes (in angstroms) of atoms and ions of Group IA and Group IIA elements.

Li, 1.55   Li$^+$, 0.60   Be, 1.12   Be^{+2}, 0.31

Na, 1.90   Na$^+$, 0.95   Mg, 1.60   Mg^{+2}, 0.65

K, 2.35   K$^+$, 1.33   Ca, 1.97   Ca^{+2}, 0.99

Rb, 2.48   Rb$^+$, 1.48   Sr, 2.15   Sr^{+2}, 1.13

Cs, 2.67   Cs$^+$, 1.67   Ba, 2.31   Ba^{+2}, 1.29

Three additional comparisons pertaining to atomic and ionic radii are important:

(a) Taken as a group, the atoms and ions of these elements are the largest atoms and ions among the metals. For example, the lithium atom is larger than the gold atom (1.55 A compared with 1.44 A) even though gold weighs more than 28 times as much as lithium, and the sodium atom is larger than the lead atom (1.90 A compared with 1.75 A).

(b) In all cases the *ions are smaller than the corresponding atoms*, as might be anticipated since the valence shell of the atom is vacated when the ion is formed. The actual decrease is quite large. For example, the beryllium atom, Be, shrinks from 1.12 A to 0.31 A in the ion Be^{+2}, and the sodium atom, Na, drops from 1.90 A to 0.95 A in the ion Na^+.

(c) The ions of the Group IIA elements are smaller than the corresponding Group IA ions when the electron configurations of the two corresponding *ions* are identical. The Group IIA ion is always the smaller of the pair because it has a higher nuclear charge than the Group IA ion. This higher charge causes the electron cloud to be drawn closer to the nucleus. In sodium ion, for example, 10 electrons are attracted to a nucleus having a nuclear charge of +11; in magnesium ion, 10 electrons are attracted to a nucleus of nuclear charge +12. The radius of the sodium ion, Na^+, is 0.95 A; the radius of the magnesium ion, Mg^{+2}, is 0.65 A.

Physical Properties

Let us examine some physical properties of the Group IA and Group IIA elements given in Tables 8.3, 8.4, and 8.5 and compare the magnitudes of these properties with those of typical nonmetals, other metals, and other members of Group IA and Group IIA.

Density

Although the densest elements are metals, the Group IA and Group IIA metals are among the least dense of the elements. These data are shown in Table 8.3. Furthermore, these data indicate that the densities of the Group IIA elements are significantly greater than those of the corresponding members of Group IA.

The especially low densities of these elements has been explained in terms of the relatively large sizes of their atoms and the packing pattern of their atoms in the metallic crystal. The relatively large size of the Group IA and Group IIA atoms limits the number of atomic nuclei that can be packed in a given volume of the metal, thereby lowering the density. The greater

Table 8.3 Comparison of Densities of Some Common Elements (room temperature, g/cc)

Element	Density	Element	Density
Alkali Metals		*Some Heavy Metals*	
Lithium	0.53	Iron	7.86
Sodium	0.97	Mercury	13.6
Rubidium	1.53	Gold	19.3
Cesium	1.90		
Alkaline Earth Metals		*Some Nonmetals*	
Beryllium	1.8	Phosphorus (violet)	2.35
Magnesium	1.7	Sulfur (rhombic)	2.07
Calcium	1.6	Silicon	2.4
Barium	3.5		

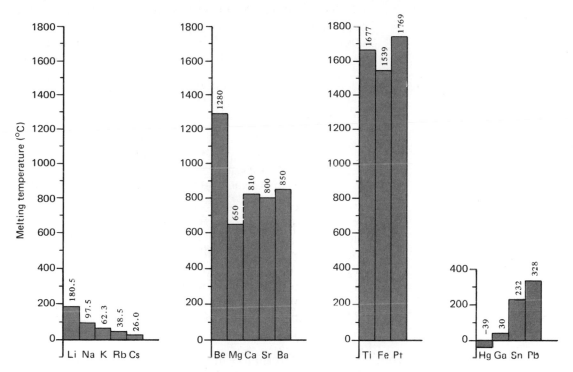

Figure 8.3 Comparison of melting points of Group IA and Group IIA metals. The melting points of several other metals are listed for comparison.

densities of Group IIA elements compared with those of Group IA can be attributed to the significantly smaller sizes of the Group IIA atoms compared with those of Group IA. Also important is the fact that the Group IA atoms are arranged or packed in the crystal in a much more open structure or less efficient manner than are those of Group IIA (see Chapter 13).

The gradual increase in density in moving down each family is a result of the fact that in moving from one member to the member below it in a family, the increase in nuclear mass is greater than the increase in atomic radius. Thus, in passing from potassium to rubidium, the mass changes from 39 to 85 amu, an increase of about 120%, while the atomic radius changes from 2.35 to 2.48 Å, an increase of about 6% (the volume per atom changes by about 20%). This results in a greater mass per unit volume—that is, in a greater density.

Melting Point The Group IA elements have low melting points which decrease systematically from 180.5°C for lithium to 28.6°C for cesium; the melting points for the Group IIA elements are much higher (Figure 8.3).

Low melting points are usually associated with weak forces of attraction among the building blocks of the crystal. For these families of elements, the metallic bonding is apparently much weaker in the Group IA metals than in those of Group IIA. Could this be because Group IA metals contribute one electron per atom to the bonds between atoms, while Group IIA elements contribute two electrons per atom?

Is the systematic decrease in melting point in proceeding down the two families due to a weakening of the metallic bond in the heavier elements brought about because the atomic nuclei have less control of the bonding electrons as the atoms increase in size? These questions are examples of problems encountered by research workers who study the metallic solid state.

Boiling Point

There is a very wide liquid range between the boiling points of these elements and their melting points; this is a general characteristic of metallic substances. For example:

	Melting Point	Boiling Point
Lithium	180°C	1326°C
Potassium	63°C	757°C
Calcium	810°C	1492°C

This wide liquid range suggests that the forces holding the atoms in the liquid are quite strong. It has consequently been suggested that when a metal melts, only a small fraction of the metal bonds is broken and that considerable metallic bonding exists in the liquid state. When the metal boils, all of the metallic bonds must be broken. This occurs at considerably higher temperatures than those required for melting.

The boiling points of the alkaline earth metals are much higher than those of the alkali metals, again pointing toward stronger metallic bonds among the Group IIA elements.

Heats of Fusion and Vaporization

The heat of fusion is the *energy needed to melt a solid at its melting point*; the heat of vaporization is the *energy needed to vaporize a liquid at its boiling point*. In essence, the heat of fusion is a measure of the energy needed to collapse the crystal; the heat of vaporization is a measure of the energy needed to separate the building blocks of the liquid.

For these two families of elements, the heats of fusion are very small, decreasing from top to bottom in each family (Table 8.4). However, the

Table 8.4 Comparison of Heats of Fusion and Vaporization of Several Elements and Compounds (in kcal/mole)

Substance	Heat of Fusion	Heat of Vaporization
Metals		
Alkali		
Lithium	0.72	32.2
Sodium	0.62	21.3
Potassium	0.55	18.5
Rubidium	0.52	16.5
Cesium	0.50	15.6
Alkaline earth		
Beryllium	2.8	70.4
Magnesium	2.14	30.8
Calcium	2.1	35.8
Strontium	2.2	33.2
Barium	1.8	36.1
Other Substances		
Water	1.45	9.7
Iodine	2.97	6.1
Sodium chloride	6.8	40.1

Table 8.5 Electrical and Thermal Conductivities of Group IA Metals

Metal	Electrical Conductivity (microohm-cm^{-1} at 0° to 20°C)	Thermal Conductivity (cal/cm/sec/deg at 25°C)
Li	0.117	0.17
Na	0.238	0.32
K	0.163	0.23
For Comparison		
Cu	0.598	0.99
Ag	0.628	1.01

heats of vaporization of the elements in these two families are especially high by comparison with the heats of fusion.

The low values for the heats of fusion of the alkali and alkaline earth metals emphasize again that only a small fraction of the metallic bonds need be broken to collapse the crystal. The very high values for the heats of vaporization indicate that much energy is needed to break all metallic bonds and to separate the atoms. It is interesting to note that the energy needed to break a large fraction of the metallic bonds and to separate the atoms of these elements is about the same as that required to vaporize the ions from melted ionic compounds, such as sodium chloride, and considerably higher than that needed to separate water or iodine molecules.

Thermal and Electrical Conductivity

Among the elements, only silver, copper, gold, and aluminum are better conductors of heat and electricity than are the members of these families. Table 8.5 shows that the trends in thermal conductivity parallel those of electrical conductivity in Group IA and in copper and silver. This suggests that the factors responsible for electrical conductivity also are responsible for thermal conductivity. Since electrical conductivity is related to the relative electron mobility, it is reasonable to conclude that heat also is conducted in metals by moving electrons.

Chemical Properties

Group IA and Group IIA elements are the most reactive metals known. They react readily with oxygen of the air, with halogens, and with sulfur. All except beryllium react with water—some react spectacularly. In this section we shall:

1. Summarize the reactivity pattern of these elements, emphasizing some specific reactions and some trends in reactivity within each series, drawing comparisons between the two series.
2. Present a quantitative measure of the reactivity of the metals in water solutions.
3. Attempt to relate energy changes to chemical reactivity for representative members of the families.
4. Discuss two important reactions of the ions of these metals.

Stable Oxidation States

The alkali elements exhibit only the 0 and I oxidation states in normal reactions, while the alkaline earth elements show only the 0 and II oxidation states. These are precisely those anticipated from families of elements having one or two valence electrons.

Group I A
M, M+
Group II A
M, M+2

Figure 8.4 Stable oxidation states of the alkali and alkaline earth metals.

These elements are very reactive metals because the valence electrons are not tightly bound to the atom. This is illustrated by an examination of the ionization energies of the elements. In general, the reactivity increases with increasing atomic weight in each series in accord with opposite trends of the ionization energies (Table 8.2), indicating that the *large atoms lose electrons more readily than* the *small ones.* In several cases, the Group IIA elements have especially high second ionization energies. In these cases, the energy needed to remove the second electron from an atom must be supplied by some additional process if the divalent ion is to form. When the metal reacts with water, this additional reaction is that caused by the attraction of the metal ion for the dipolar water molecules. The reaction is highly exothermic for small divalent ions so that elements such as magnesium and calcium form divalent ions in water in spite of the relatively high energies needed to remove their second electrons. When the metal reacts in the absence of a polar solvent, the additional reaction often is the formation of the ionic crystal lattice, which is a highly exothermic process.

Some Typical Reactions

The following equations represent some typical reactions of Group IA and Group IIA metals.

Group IA

Group IIA

With halogens (X = any halogen)
$$2M + X_2 \longrightarrow 2MX \text{ (halide)} \qquad M + X_2 \longrightarrow MX_2 \text{ (halide)}$$

With sulfur
$$16M + S_8 \longrightarrow 8M_2S \text{ (sulfide)} \qquad 8M + S_8 \longrightarrow 8MS \text{ (sulfide)}$$

With hydrogen
$$2M + H_2 \longrightarrow 2MH \text{ (hydride)} \qquad M + H_2 \longrightarrow MH_2 \text{ (hydride)}$$
$$\text{(except Be and Mg)}$$

With nitrogen
$$6M + N_2 \xrightarrow{\text{spark}} 2M_3N \text{ (nitride)} \qquad 3M + N_2 \longrightarrow M_3N_2 \text{ (nitride)}$$

In nearly all of these reactions, an increase in reactivity with increasing atomic number is observed within each family.

The alkali metals give some unexpected and unusual oxides in reactions with excess oxygen. These reactions are illustrated by the equations

$$4Li + O_2 \longrightarrow 2Li_2O \text{ (lithium oxide)}$$

$$6Na + 2O_2 \longrightarrow 2Na_2O + Na_2O_2 \text{ (sodium oxide and sodium peroxide)}$$

$$3K + 2O_2 \longrightarrow K_2O_2 + KO_2 \text{ (potassium peroxide and potassium superoxide)}$$

$$Rb + O_2 \longrightarrow RbO_2 \text{ (rubidium superoxide)}$$

$$Cs + O_2 \longrightarrow CsO_2 \text{ (cesium superoxide)}$$

Only lithium gives the expected oxide, while sodium gives both the oxide and the peroxide. Potassium gives both a peroxide and a superoxide, while rubidium and cesium give superoxides.

The peroxides contain the O_2^{-2} ion and the superoxides contain the O_2^{-1} ion. Both of these species are strong oxidizing agents. When lithium reacts with air, the nitride as well as the oxide is formed.

Among the alkaline earth elements, beryllium, magnesium, and calcium

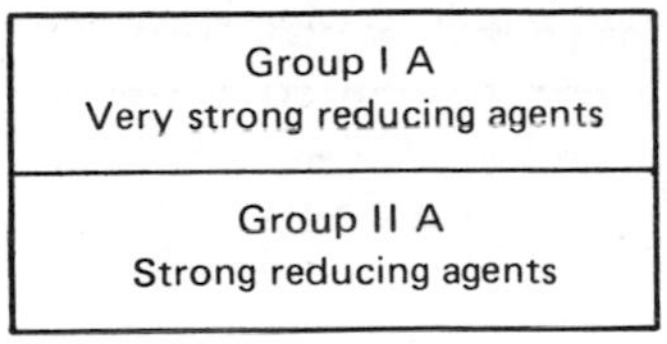

Group I A
Very strong reducing agents
Group II A
Strong reducing agents

Figure 8.5 Strength of Group IA and Group IIA metals as reducing agents.

yield monoxides on treatment with oxygen, strontium produces both a peroxide and a superoxide, barium gives a peroxide.

All of the alkali metals react with cold water according to the equation

$$2M + 2H_2O \longrightarrow 2M^+ + 2OH^- + H_2$$

giving strongly basic solutions. Lithium is somewhat slower in reacting than the others. This presumably is because the heat of reaction is sufficient to melt the lower melting elements but is not sufficient to melt lithium. The liquid metals react considerably more vigorously than the solid metals.

The alkaline earth elements are somewhat more selective in their reaction with water. Beryllium will not react with water even at high temperatures; magnesium reacts readily with boiling water; calcium, strontium, and barium react vigorously even with cold water in accordance with the equation

$$M + 2H_2O \longrightarrow M^{+2} + 2OH^- + H_2$$

Metals of both families react with aqueous solutions of acids according to the equations:

$$2M + 2H_3O^+ \longrightarrow 2M^+ + H_2 + 2H_2O$$

$$M + 2H_3O^+ \longrightarrow M^{+2} + H_2 + 2H_2O$$

Strength as Reducing Agents
In all of their reactions, these metals act as reducing agents; they are oxidized during the process. Since they are the most reactive metallic elements, they are also excellent reducing agents. A quantitative measure of their strength as reducing agents in water solutions is given by their own tendency to be oxidized. This tendency is measured by the oxidation potential obtained in experiments with electrochemical cells.

The standard oxidation potential $E°$ indicates in volts the reducing strength of the substance in a 1 M* solution compared with the reducing strength of hydrogen taken as a standard.

Values for the standard oxidation potentials for the Group IA and Group IIA elements are given in Table 8.6. These values show that the Group IA elements are stronger reducing agents than Group IIA elements, as anticipated from a comparison of their ionization potentials.

Two Common Reactions of Ions
The ions of these elements undergo a number of reactions, two important ones being hydration (or solvation) and reduction.

Table 8.6 Standard Oxidation Potentials for Group IA and Group IIA Metals

Group IA Elements	$E°$ Volts	Group IIA Elements	$E°$ Volts
$Li \rightleftharpoons Li^+ + 1e^-$	3.045	$Be \rightleftharpoons Be^{+2} + 2e^-$	1.85
$Na \rightleftharpoons Na^+ + 1e^-$	2.71	$Mg \rightleftharpoons Mg^{+2} + 2e^-$	2.37
$K \rightleftharpoons K^+ + 1e^-$	2.924	$Ca \rightleftharpoons Ca^{+2} + 2e^-$	2.76
$Rb \rightleftharpoons Rb^+ + 1e^-$	2.925	$Sr \rightleftharpoons Sr^{+2} + 2e^-$	2.89
$Cs \rightleftharpoons Cs^+ + 1e^-$	2.923	$Ba \rightleftharpoons Ba^{+2} + 2e^-$	2.90

*See Chapter 21 for a more precise statement concerning these concentrations.

Table 8.7 Hydration Energies of Group IA and Group IIA Ions (in kcal/mole)

Li^+	Na^+	K^+	Rb^+	Cs^+	Be^{+2}	Mg^{+2}	Ca^{+2}	Sr^{+2}	Ba^{+2}
-123	-97	-77	-70	-63	-587	-460	-395	-355	-305

NOTE: The negative sign indicates energy is evolved.

Hydration Reactions

These involve the reaction of the ion with water. The attractive force here is the interaction of the positive charge on the ion with the negative end of the water dipole. Considerable heat is liberated in this process, as shown in Table 8.7. These data show that the hydration energies increase as the ions become smaller and as their charge increases. In the attraction of the ion for the water dipole, small, highly charged ions exert very strong attraction for water molecules.

The high oxidation potential for lithium (Table 8.6), greater than that of cesium, is at first glance surprising, since one might expect that this value, which measures the tendency of the ionization reaction of the metal

---

THE OXIDATION POTENTIAL, $E°$

One of the most useful methods of indicating the strength of an atom, molecule, or ion as an oxidizing or a reducing agent *in solution* is by means of the quantity, $E°$, which indicates the ease of removal of electrons from the molecule, atom, or ion *in aqueous solution*. $E°$ is called the standard oxidation potential. The more positive the $E°$ value, the easier the atom or ion can be oxidized, and hence the better reducing agent it is.

The value of $E°$ for a metal is obtained as follows:

A voltaic cell is constructed containing hydrogen gas adsorbed on platinum (an inert electrode) in contact with a hydrochloric acid solution having a concentration approximately one molar* as one electrode, and a metal in contact with a solution of one of its salts as the other electrode; the voltage is measured when no current is flowing.

Consider such a cell in which the metal electrode is a strip of zinc immersed in a solution containing zinc ions. When the cell is connected, it is observed that electrons flow from the zinc to the hydrogen electrode. The cell voltage at 25°C is 0.76 volt when the H_3O^+ and Zn^{+2} concentrations are each very nearly 1 mole/liter.*

Since electrons flow from zinc to hydrogen in this cell, the zinc electrode reaction may be written

$$Zn \longrightarrow Zn^{+2} + 2e^-$$

and the reaction at the hydrogen electrode is

$$2e^- + 2H_3O^+ \longrightarrow H_2 + 2H_2O$$

If the cell voltage is regarded as a sum of contributions from the two electrode systems,

$$E_{cell} = E_{H_2\ electrode} + E_{Zn\ electrode}$$

we can obtain relative values for the electrode voltages by assigning a value to one electrode system and relating all other electrode systems to this one.

In practice, the hydrogen electrode (at 25°C, 1 atm H_2 pressure, 1 M HCl) is assigned a potential known as the standard oxidation potential, $E°$, of 0. The standard oxidation potential, $E°$, of any electrode system can now be evaluated by measuring the voltage of a cell containing the desired electrode (at 25°C and having ionic concentrations of 1 mole/liter) and the standard hydrogen electrode. The zinc-hydrogen cell described above illustrates this method. The cell voltage is 0.76 volt, but since the $E°$ value for the hydrogen electrode is 0, the $E°$ value for the zinc electrode becomes 0.76 volt.

Cells containing hydrogen electrodes and electrodes from other metals, and set up the same way, are used to obtain $E°$ values for these metals.

The value of $E°$ can be used as a measure of the tendency of the oxidation reaction of the metal to go as written. If the data are obtained for a series of elements, we can then list these elements in the order of their relative reducing power in solution. Such a list is called the *electrochemical series*.

$$Me(s) \longrightarrow Me^+ (aq) + e^-$$

to take place in the presence of hydronium ion in water solution, would be related to the energy of the similar reaction in the gas phase,

$$Me(g) \longrightarrow Me^+(g) + e^-$$

as measured by the ionization energy. The ionization energy for lithium, 124 kcal/mole (Table 8.2), is 34 kcal higher than that for cesium, 90 kcal/mole, yet in water solution lithium ionizes more readily than cesium. The explanation for this apparent anomaly becomes obvious if we imagine the reaction to take place in steps, and write the overall reaction in solution as the sum of three processes

	Energy change (kcal/mole)	
	Li	Cs
$Me(s) \longrightarrow Me(g)$	+33	+16
$Me(g) \longrightarrow Me^+(g) + e^-$	+124	+90
$Me^+(g) \xrightarrow{H_2O} Me^+(aq)$	−123	−63
$Me(s) \xrightarrow{H_2O} Me^+(aq) + e^-$	+34	+43

The energy of the overall reaction must be the sum of the energies of each separate step, by the law of conservation of energy. The sums for lithium and cesium are compared at the right of the equations. It is evident that the larger energy requirement for ionization of lithium, as compared to cesium, is more than compensated for by the larger hydration energy of this small ion. The effect of the large hydration energy is such as to make lithium in water solution a better reducing agent than cesium in spite of the higher ionization energy of lithium. The other Group IA ions also react with water but release less energy (see Table 8.7).

The equation for the hydration of ions may be written

$$M^+ + xH_2O \longrightarrow M(H_2O)_x{}^+$$

The values of x are not precisely known but are believed to be 4 for lithium and beryllium, 8 for barium, and 6 for all other members of the two families.

Water tightly bonded to the ion often becomes a part of the framework of crystals. This is the case in hydrated salts such as $Na_2SO_4 \cdot 10H_2O$. Usually, hydrates are observed in salts containing small or highly charged cations, such as lithium, sodium, beryllium, magnesium, or calcium salts.

In many cases, the bonding forces of the metal ion are so strong that the

Figure 8.6 Electrolysis of molten sodium chloride. In the molten state, sodium chloride is composed of the ions of Na and Cl in thermal motion. Such a liquid is an excellent conductor of an electric current. The positively charged ions will move to the negatively charged electrode and pick up electrons, as shown by the half-reaction $Na^+ + e^- \longrightarrow Na$, to become electrically neutral atoms. Hence, the Na^+ ions are reduced.

At the positive electrode, the chloride ions, Cl^-, give up electrons and become chlorine molecules, Cl_2. Hence, the Cl^- ions are oxidized.

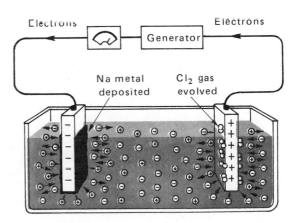

H—O bond in the water molecule is weakened and the hydrated ion undergoes an acid-base reaction in which the ion acts as an acid and the water solvent acts as the base. This accounts for the observed acidity of solutions of beryllium salts:

$$Be(H_2O)_4{}^{+2} + H_2O \rightleftharpoons Be(H_2O)_3OH^+ + H_3O^+$$

Aluminum, copper, zinc, and all small multivalent cations act in a similar manner.

Reduction Reactions

Reduction of the ions in the alkali and alkaline earths involves converting (reducing) the ion to the metal atom. Strong reducing agents are needed for this reaction because the ions do not accept electrons readily; they are thus extremely weak oxidizing agents. This is anticipated, for if the metal atoms lose electrons very readily, their ions will accept electrons only with some difficulty.

The most widely used process for reducing these ions is reaction at the cathode of an electrolysis cell (Figure 8.6). The method used is to electrolyze the fused salt. Thus, the reaction

$$2MX \xrightarrow[\text{fused}]{\text{electrolysis}} 2M + X_2$$

is a general reaction for reduction of the metal ion and for the preparation of the metals.

Sodium is prepared in this way from fused sodium chloride; to do this, sodium carbonate or calcium chloride is added to lower the melting temperature of the salt. Most of the magnesium produced today is obtained from seawater, from which it is precipitated as the hydroxide. The hydroxide is dissolved in hydrochloric acid solution and magnesium is recovered by electrolysis of this solution. All of the other metals of these families are prepared by electrolysis.

Occasionally, an active metal is used to reduce the ion, as in the processes

$$3BaO + 2Al \longrightarrow Al_2O_3 + 3Ba$$

$$KCl + Na \longrightarrow NaCl + K$$

Both of these processes take place at very high temperatures in the absence of oxygen. Their success is due to the relatively low boiling points of barium and potassium, which evaporate from the reaction mixtures.

Occurrence

Calcium, sodium, potassium, and magnesium are among the most abundant elements in the earth's surface, ranking fifth, sixth, seventh, and eighth, respectively, in order of abundance (Table 8.8). The remaining members

Table 8.8 Distribution of Alkali and Alkaline Earth Elements in Nature

Element	Per Cent of Earth's Crust	Element	Per Cent of Earth's Crust
Lithium	4×10^{-5}	Beryllium	1×10^{-5}
Sodium	2.65	Magnesium	1.94
Potassium	2.40	Calcium	3.39
Rubidium	1×10^{-6}	Strontium	2×10^{-4}
Cesium	1×10^{-7}	Barium	5×10^{-2}
Francium		Radium	1×10^{-12}

of the two groups are present in relatively small amounts, with francium being almost nonexistent.

Because of their reactivity, all elements in both groups occur in nature in the combined state. Simple compounds of the alkali metals have high solubilities in water. Consequently, the largest sources of these elements are the oceans and salt seas, or salt beds well protected from contact with ground water. They also occur in a variety of clays which are not soluble in water. In general, compounds of the alkaline earth elements are much less water-soluble than are those of the alkali elements. As a result, some important sources of the Group IIA elements are insoluble deposits of compounds of alkaline earth metals such as the carbonates, sulfates, phosphates and some silicates.

Uses of the Metals

Sodium is the only alkali metal having wide commercial use, being used for the manufacture of soap, organic chemicals, dyes, tetraethyl lead, sodium peroxide, sodium cyanide, and a variety of other products. Liquid sodium has been used as a coolant in nuclear reactors and in the centers of exhaust valves in aircraft engines because it is a rapid and effective heat conductor. The metal is also used as a reducing agent in the preparation of other metals from their chlorides.

Cesium and rubidium are used to a limited extent in photoelectric cells, because these metals emit electrons when visible light falls on them.

Beryllium is used in small quantities in alloys to harden the base metal. Certain alloys of beryllium in copper have a higher tensile strength than steel, make excellent springs, and are sometimes used in electric motors to reduce sparking.

Magnesium alloys have lightness and high tensile strength and, consequently, are used extensively in the construction of airplanes and portable tools, where excessive weight would be a disadvantage. In addition, magnesium is used in incendiaries—signals and flares—because the metal emits an intense white light when it burns.

SOME IMPORTANT COMPOUNDS OF GROUP IA AND GROUP IIA ELEMENTS

Compounds of Sodium

Sodium carbonate and sodium hydroxide are among the most important industrial chemicals. The raw material for these is usually sodium chloride, since it is abundant and inexpensive.

Preparation of Sodium Carbonate

In the U.S., most of the sodium carbonate is made from sodium chloride by a process devised by the Belgian chemical engineer Ernest Solvay, and known as the Solvay process. In this process, ammonia and carbon dioxide are passed into a concentrated solution of sodium chloride. This causes the formation of sparingly soluble sodium hydrogen carbonate which precipitates, is removed by filtration, and changed to sodium carbonate by heating. The ammonia is recovered by heating the ammonium chloride with calcium hydroxide.

$$NH_3 + CO_2 + NaCl + H_2O \longrightarrow NaHCO_3 + NH_4Cl$$

$$2NaHCO_3 \xrightarrow{\text{heat}} Na_2CO_3 + CO_2 + H_2O$$

$$2NH_4Cl + Ca(OH)_2 \longrightarrow 2NH_3 + 2H_2O + CaCl_2$$

Some sodium carbonate also is obtained from natural deposits of the carbonate or bicarbonate.

Preparation of Sodium Hydroxide

The Electrolysis of NaCl Solution. If a direct electric current is passed through an aqueous solution of sodium chloride, using platinum electrodes, the products of the reaction are found to be hydrogen, chlorine, and sodium hydroxide. Evidently water molecules decompose at the cathode, giving hydrogen and leaving hydroxide ions in solution.

$$4H_2O + 4e^- \longrightarrow 2H_2 + 4OH^-$$

At the other electrode (anode) the chloride ions will give up their electrons readily. Hence chlorine gas is formed at this electrode. Thus, as the chloride ions are gradually converted to chlorine and are removed from solution, and as water molecules decompose, giving hydrogen and hydroxide ions, the two ions that predominate are sodium and hydroxide ions. This is the industrial process for making much of these three substances.

$$2NaCl + 2H_2O \xrightarrow{\text{electrolysis}} 2NaOH + H_2 + Cl_2$$

If a mercury cathode (negative electrode) instead of a platinum electrode is used in the electrolysis of a sodium chloride solution, sodium rather than hydrogen is produced. This is caused by the formation of a sodium-mercury amalgam.

Properties of Sodium Hydroxide

This compound is a white, crystalline, brittle solid. It is often called caustic soda. It is a corrosive substance and has a strong disintegrating action upon both animal and vegetable tissue. When exposed to air, it absorbs both moisture and carbon dioxide and is changed into sodium carbonate. It is very soluble in water, and a great deal of heat is liberated in the process of solution due to the hydration of the ions. Its solution has a soapy feel and a strong cleansing action.

Production and Uses of Sodium Hydroxide

About 8,800,000 tons of sodium hydroxide are produced annually in the U.S. (Table 8.11), more than half of it by the electrolytic process.

In general, the methods of preparing compounds of the other alkali metals are similar because of their similar properties. The compounds of sodium have considerably wider use because they are more abundant and far less expensive than those of other metals.

HISTORICAL

One of the earliest practical applications of electrochemistry was in the production of the alkali metals by Sir Humphry Davy in 1807. He had tried first to decompose saturated solutions of the alkali metals, but he succeeded only in decomposing water. He reported, "The presence of water appearing thus to prevent any decomposition, I used potash in igneous fusion" (October 6, 1807). To his surprise, on electrolysis of fused alkali he noticed a bright light and a flame at the negative electrode. When he reversed the current, he observed this flame again at the negative electrode. After several experiments he identified the metal which was being formed at the electrode and which was igniting in the air to give the flame and bright light as a new element. He named it potassium because he had obtained it from potash. A few days after isolation of potassium he prepared another new element, sodium, by the same electrolytic process. This work soon led to the discovery of the alkaline earth metals by a similar electrochemical method. Beryllium was discovered twenty years later by Wöhler, and radium by Madame Curie in 1898.

Figure 8.7 Electrolysis of aqueous sodium chloride. More complex cells are used in industry to prevent mixing of the chlorine and sodium hydroxide. Cells using diaphragms for this purpose are common. Another cell important in industry uses mercury as an amalgam—no hydrogen is released at this electrode.

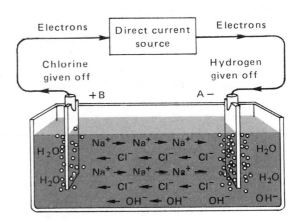

Table 8.11 Approximate Yearly Production of Sodium Compounds

Compound	Tons	Compound	Tons
Chloride	19,000,000	Bicarbonate	5,000,000
Carbonate	8,969,000	Phosphate	1,191,000
Hydroxide	8,800,000	Chromate	133,872
Silicate	719,000	Metal	125,566
Sulfate	1,350,000		

Calcium Compounds   The most abundant calcium compound in nature is calcium carbonate.

Calcium Carbonate, $CaCO_3$.   Calcium carbonate occurs in nature in two crystalline forms, represented by the minerals *calcite* and *aragonite* (Figure 8.8). *Marble* is made up of minute, snow-white calcite crystals. In the impure form, calcium carbonate occurs as *limestone*. It is used in agriculture to "sweeten" (make less acid) soil, in the building industry as a structural

Figure 8.8 Steps in the preparation of important sodium compounds starting with sodium chloride.

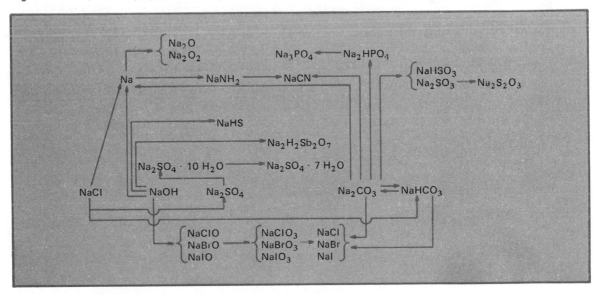

stone, and in the steel industry as a flux in blast furnaces. On heating, calcium carbonate is converted to the oxide, lime (Figure 8.9).

THE PERIODIC TABLE AND TRENDS OBSERVED IN GROUP IA AND GROUP IIA

In the early pages of this chapter, attention was called to the trends in ionization energy and atomic size within the alkali metal family (Group IA) and within the alkaline earth family, and between the Group IA and Group IIA elements of the same configuration of the inner-shell electrons (Table 8.2 and Figure 8.2). A reasonable question might be whether these trends continue in other families in the periodic table.

Atomic Size and Ionization Energy

A comparison of the data in Table 8.2 and Figure 8.2 shows that the ionization energy increases as the atomic size decreases. This is true both for changes within either family, from high atomic weight to low atomic weight, and for "horizontal" changes between the Group IA and IIA elements of the same inner-shell configuration (such as Li-Be and Na-Mg). *This trend is general throughout the periodic table;* other conditions being comparable, a small atom has a higher ionization energy than a large one, as shown in Figure 8.10. This is understandable, since the smaller the atom, the closer the electrons in the valence shell must be to the nucleus and the more tightly they are held. The evidence for the trends listed in Figure 8.10 can be obtained by comparing the sizes given in Figure 8.11 with the ionization energies of Figure 4.8. In general, the atomic size decreases from left to right in a period and increases from top to bottom in a group. The ionization energies show the opposite trend.

Two Effects: Increased Charge on Nucleus Versus Increased Number of Electrons. A question now arises: How is it that the sizes of the atoms increase from top to bottom in most columns in the periodic table, in spite of the increased nuclear charge, which would be expected to exert a more constricting effect on the electrons in its neighborhood? The answer must

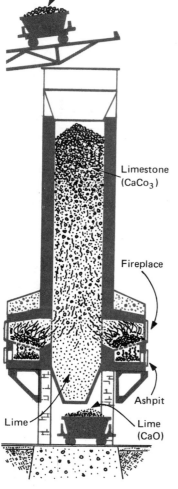

Figure 8.9 A vertical section of a limekiln.

Figure 8.10 Relation of atom radius to ionization energy.

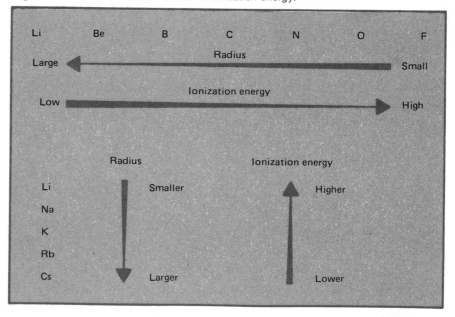

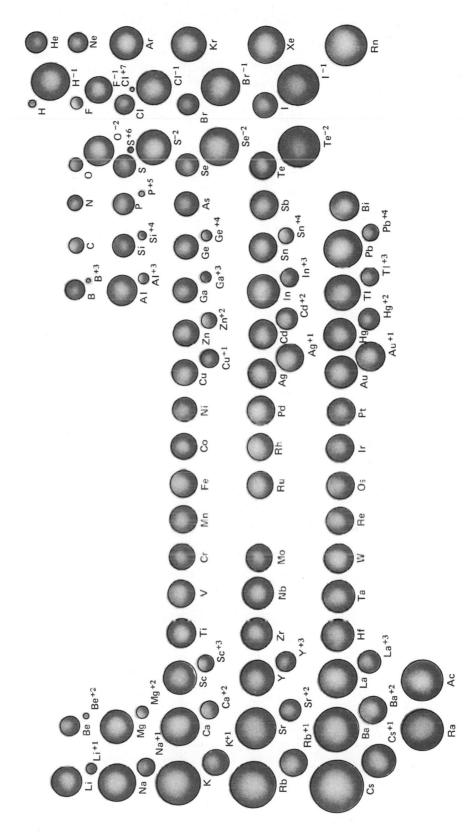

Figure 8.11 Periodic table showing relative sizes of atoms and ions. The atoms have been assumed to be spherical, and the circles have the relative radii of these spheres. The scale is 0.2 cm = 1 Å [Adapted from Campbell, *J. Chem. Educ.*, **23** 525 (1946).]

lie in the effectiveness of electrons near the nucleus in screening the nuclear charge from exerting its full effect on the more distant electrons. This is understandable on the basis of the Bohr theory, from which we can easily imagine that the electrons in the inner orbits also act to shield the electrons in the outer orbits from the full effect of the positive charge on the nucleus. The calculations on the basis of wave mechanics lead to a similar conclusion. For each model, the addition of electrons in the next higher principal quantum number level causes a notable expansion in the size of the atom, as shown in the increases from fluorine (atomic number 9) to sodium (atomic number 11), from chlorine (atomic number 17) to potassium (atomic number 19).

When there is no change in the principal quantum number of the added electrons, the expected decrease in atomic size with increasing nuclear charge is observed, as in the first two rows of the periodic table. In the third row, however, there are exceptions to this prediction; note, for example, the increase in atom size in the Ni-Cu-Zn series.

Group Number and Chemical Similarities

The elements in each vertical column or family in the periodic table have similar valence electron configurations. In the case of the transition elements, this similarity sometimes extends to the outermost and next inner shell of electrons.

This similarity of electron configurations gives to each group chemical properties that set it apart from the other groups. We have seen, for example, that all the atoms of Group IA form monovalent positive ions such as Na^+ by loss of the single outer electrons; those of Group IIA form divalent ions (Ca^{+2}) by loss of two electrons. We shall see in the next chapter that the elements of Group VIIA form monovalent negative ions (Cl^-) by gaining a single electron which, added to the seven originally present, gives to the ion a noble gas configuration.

Oxidation State

Just as the oxidation states of elements of Groups IA and IIA are I and II in compounds, so, the group number in the periodic table generally records the maximum positive oxidation state. The adjective "maximum" appears in the last sentence because the elements other than those in Groups IA, IIA, and IIIA have more than one oxidation state. For those elements showing a negative oxidation state, its value is 8 minus the group number. For example, sulfur, a Group VI element, has a maximum oxidation state of VI and a minimum oxidation state of −II. There are several exceptions to those generalizations among the transition elements, and even among the representative elements (the A groups), chlorine, for example, forms compounds in the I, III, IV, and V states as well as in the expected −I and VII states (Figure 8.12). The chemistry of chlorine and the other members of Group VIIA is discussed in the next chapter.

Metals and Nonmetals

The elements of Groups IA and IIA have been characterized as *metals*. This is because they have only a few valence electrons, which become delocalized, and are conduction electrons in the pure solid elements. They also are elements with relatively low electronegativity, and ionization energies generally lower than those of *nonmetals*, with which they are contrasted. Nonmetals include those elements those atoms have many valence electrons and form covalent compounds or electrovalent compounds in which the non-

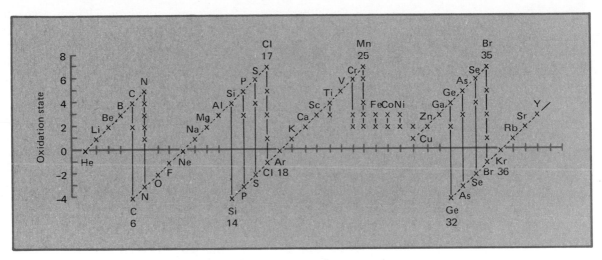

Figure 8.12 Periodicity in some common oxidation states of the lighter elements.

metal is found in the negative ion; they are elements with relatively high electronegativity. The heavy stairstep line from boron to astatine in the periodic table appearing on the inside front cover of this book separates the two groups; it is evident that there are many more metallic elements than nonmetallic elements.

It must not be supposed, however, that there is an abrupt change from metallic to nonmetallic in crossing this line. As we have seen, the ionization potential generally increases across a row in the periodic table, and decreases down a column. The first of these effects means that the metals at the left-hand end of a row of the table *gradually* change toward more metallic characteristics as you examine the elements from top to bottom. As you might expect, the properties of the intermediate elements tend to be intermediate between the properties of metals and nonmetals. Thus, arsenic and antimony show, although weakly, some properties both of metals and nonmetals.

Simple Ions in Human Health. Four of the ions discussed in this chapter —sodium, potassium, calcium, and magnesium—have so profound an influence on the processes that keep our bodies alive and healthy that no discussion of the chemistry of these ions is complete without mention of their vital biochemical roles.

Sodium ion is in many ways the work-horse of body chemistry. It is the cation in largest supply in all extracellular fluids such as plasma and the interstitial fluids—those fluids that surround the cells and provide the medium for transporting food and nourishment to them. Sodium ions are responsible in several ways for the transport of materials into and out of the cells. Thus concentration of sodium ions in the extracellular fluid regulates the flow of water across the cell membrane, thereby maintaining an optimum water level in cells. Sickness can result if the cellular water level becomes too high (edema) or too low (Addison's disease). Sodium ion is also involved in the mechanism by which foods and cellular building blocks (glucose and amino acids) and essential inorganic ions such as potassium are transported into cells. This transport takes place by what amounts to a pumping action by which sodium ions are pumped out of the cell and potassium ions, glucose, and amino acids pumped in.

Sodium ion also appears in the extracellular fluid as sodium bicarbonate. This salt acts with carbonic acid, H_2CO_3, and carbon dioxide as a buffer—a chemical system that maintains a specific acid-base balance in the fluid thereby permitting certain reactions to occur but inhibiting others. Sodium bicarbonate also plays an important role in the transport of carbon dioxide from the cells to the lungs.

The body's source of sodium ion is food, most often as sodium chloride. However, many foods contain such small amounts of salt that additional quantities must be ingested to meet minimum requirements. Body salt is maintained at an approximately constant level by the action of the kidneys. If we eat more salt than we need, the excess is excreted in the urine. A small amount of salt is lost by sweating. This can become a problem if the individual is exposed to intensely hot, dry air.

Inside the cell, potassium ion is the work-horse. It is the most abundant cation present. Its concentration is remarkably constant at between 100 and 150 millimoles/liter. These relatively high concentrations of potassium ion are required for maximum efficiency in the synthesis of proteins, the compounds that control the chemical reactions of the cell. Potassium ion also is required in the glycolysis portion of the oxidation of glucose, the processes from which energy to do the work of the cell is obtained. It is said that potassium ion exerts a relaxing effect on heart muscle during heart beats, and that specific concentrations of sodium and potassium ions on each side of nerve and muscle cell membranes are required for maximum sensitivity of nerves to stimuli and for the transmission of nerve impulses.

Calcium ion is needed in the formation of bones and teeth; it is important in the clotting of blood, in the formation of milk during lactation, and in the proper rhythm of heartbeat. There is much evidence to indicate that a flow of calcium ions in response to a nerve signal gives rise to muscle contraction and that withdrawal of these ions results in muscle relaxation.

The loss of potassium and, to a lesser extent, calcium ions in severe diarrhea may cause the concentration of cellular K^+ and Ca^{+2} to drop. Failure to replace these ions may bring about disturbances in the rhythm of the heartbeat and muscle weakness.

Magnesium ion is needed at all stages of protein synthesis. The ribosomes —the sites of protein synthesis in the cell—are stable and functional only in the presence of relatively high concentrations of magnesium ion. Several steps in the oxidation of glucose and in the transfer of energy at the cell membrane require the presence of magnesium ion. Membranes that are particularly sensitive to this Mg^{+2}-catalyzed energy transfer are those of excitable cells such as brain, nerve, and muscle. Thus, the transmission of impulses along the nerves requires the presence of magnesium ion. However, Mg^{+2} is a depressant. Injection of its salts into the blood stream produces anesthesia. This treatment has been used to reduce convulsions.

Because magnesium ions are not readily absorbed through the intestinal wall, and because they have a tendency to attract water molecules (because of their small size and large charge), they are used to pull water into the intestines, thereby acting as purgatives.

Any significant loss of fluid from the body gives rise to dehydration. Dehydration causes thirst, reduced urine output, fever, dry mucous membrane, and coma. Not only must water be replaced, but electrolytes also must be added to restore the acid-base balance—the sodium and potassium balance outside and inside the cell—and to reestablish the transport of essential materials across the cell membrane.

SUMMARY   In this chapter we have examined the chemistry of the most reactive metals. We have seen that not only their reactivity but also their physical properties can be explained in terms of the electron configurations and ionization energies of their atoms, the sizes of their atoms and ions, and the crystal structure and binding in the metals. We have evidence that the connection between properties and atomic structure described for the Group IA and Group IIA elements has a general application to other elements in the periodic table. We have introduced a quantitative measure of reactivity—the standard oxidation potential.

Once again we have applied the principles of oxidation-reduction, acid-base chemistry, and ion-dipole interactions to explain and to help us predict the behavior of these elements and their compounds. But we also have learned that not all properties are predictable in terms of our simple theory. For example, the reactions of the metals with oxygen give, in many cases, unexpected peroxides and superoxides.

IMPORTANT TERMS

Metal
alkali metals
alkaline earth metals
nonmetal
strength as reducing agent

Electrochemistry
anode; cathode
electrical conductivity
electrolysis
hydrogen electrode
ionization energy
oxidation potential

Energy changes
heat of fusion
heat of vaporization
thermal conductivity
hydration energy

Compounds of metals
peroxide
superoxide
mineral
Solvay process

QUESTIONS AND PROBLEMS

1. Why is the second electron much more difficult to remove from the alkali metal atoms than the first electron?

2. How do you account for the fact that the ionic radii of the alkali and alkaline earth metals increase as the atomic weights increase?

3. What factors are involved in causing the densities of the alkali metals to increase as the masses and atomic radii increase?

4. Tabulate the ion sizes and the hydration energies for the ions of the alkali metals and of the alkaline earth metals. Do you find a correlation between the two (a) within each family? (b) between the families? Why, or why not, should such correlations be expected?

5. Select a typical metal from Group IA and another from Group IIA. Compare them with respect to (a) ionic size, (b) ionization energy (first and second), (c) melting point, (d) hardness, (e) metallic luster, (f) oxidation potential.

6. The clean surfaces of most metals have a very similar luster. In terms of the absorption and emission of visible light, how would you account for this?

7. The range between the melting point and the boiling point is much greater for metals than for water, ammonia, and other liquids. What does this suggest with reference to metallic bonds? Explain.

8. Describe the nature of the bond between atoms in metals which makes it possible to explain high electrical and thermal conductivity.

9. How do you account for the fact that a small atom such as lithium has almost the same $E°$ value as a large atom such as cesium?

10. In terms of $E°$ values, which of the following in each pair of elements is the best reducing agent? (a) Li and Cs, (b) K and Mg, (c) Ba and Na, (d) Li and Ba.

11. Predict the following properties of francium and radium from the data on the other members of those two families of elements: (a) ion size, (b) electron configuration, (c) ionization energy, (d) density, (e) oxidation state, (f) strength as reducing agent.

12. Lithium is harder than cesium but softer than beryllium. Explain.

13. All of the Group IA metals form crystals of the same crystal structure, and hence the solids may be fairly compared with each other. (a) Using the data of Figure 8.2, calculate the volume of a mole of each of the alkali metals relative to that of a mole of potassium atoms taken as unity. (b) From the results of (a), calculate the densities of the Group IA elements relative to the density of potassium as unity, and compare your result with a similar calculation of relative densities from the data of Table 8.3.

14. Hydrated positive ions act as stronger acids the smaller the radius of the ion and the greater its charge. (a) Cite an example of such an ion from this chapter. (b) Suggest other hydrated ions that might be expected to be acidic.

15. Considering oxidation potential, ionization energy, electronegativity, and the like, which element is the most metallic metal? Which element is the most nonmetallic nonmetal? Ignoring the lanthanides, actinides, and the noble gases, where do these two elements lie in the periodic table with respect to the stair-step line which separates metals from nonmetals?

16. Compare the reactions that occur when (a) fused sodium chloride is electrolyzed and (b) a solution of sodium chloride is electrolyzed. Explain why any difference in the two processes occurs.

17. Write balanced equations for each of the following: (a) reaction of lithium with hydrogen; (b) exposure of potassium to air; (c) reaction of sodium (i) with water (ii) with air; (d) formation of Na_2O.

18. Calcium hydride is an electrovalent compound. What ions are present in the solid? What products would you predict for its reaction with water? Write the equation.

19. (a) What reactions take place when oxides of the alkaline earth metals react with water? (b) Which oxide of an alkaline earth metal may react both with acids and with bases? Write equations for the reactions.

20. Why should the alkali metals not be handled with bare hands? Suggest reactions which would be harmful to the skin.

21. Do all the alkaline earth metals form peroxides in reaction with oxygen? Compare with Group IA elements in this respect.

22. Illustrate by equations the general methods for preparing Group IA metals.

23. Give a general method for preparing barium, strontium, and calcium.

24. Given the formulas of the following anions, write correct formulas for the sodium and calcium salts of nitric, sulfuric, and phosphoric acids.

Nitrate ion, NO_3^- Phosphate ion, PO_4^-

Sulfate ion, SO_4^{-2} Dihydrogen phosphate ion, $H_2PO_4^-$

Hydrogen sulfate ion, HSO_4^- Monohydrogen phosphate ion, HPO_4^{-2}

25. Calculate the number of grams of sodium hydroxide formed when excess water reacts with sodium to form 10 g of hydrogen.

26. How many grams of chlorine will be produced in an electrolysis of melted sodium chloride that produces 10 g of sodium?

27. (a) Calculate the per cent of sodium in Na_2CO_3. (b) Calculate the per cent of chlorine in *carnallite*, $KCl \cdot MgCl_2 \cdot 6H_2O$, one of the potassium minerals.

28. Write the equation for the neutralization of calcium hydroxide with sulfuric acid. Calculate the weight of calcium hydroxide required to neutralize 500 g of sulfuric acid.

29. Calculate the percentage of sodium and of calcium in each of the compounds of Problem 24.

SPECIAL PROBLEMS

1. In going from the metallic state to the hydrated ion, a mole of atoms of an alkali metal can be imagined to pass through the transitions represented by the diagram

$$A(s) \xrightarrow{\text{ I }} A(l) \xrightarrow{\text{ II }} A(g) \xrightarrow{\text{ III }} A^+(g) \xrightarrow{\text{ IV }} A^+ \cdot nH_2O$$

In the transition labeled I, the metal melts. The energy required for this step is the heat of fusion. (a) What energies are involved in transitions II, III, and IV? (b) How do you expect these energies to vary from element to element in Group IA? (c) Suggest reasonable explanations for the variations observed.

2. Below are tabulated the energies involved when metals A, B, and C become hydrated ions, in kcal/mole.

	Element		
	A	B	C
Energy of fusion	1	2	3
Energy of vaporization	50	70	90
Energy of ionization	120	110	100
Energy of hydration of ion	70	−90	−110

(a) Using the table, list these elements in order of increasing energy required for the formation of hydrated ions from the metals. (b) Using the values recorded in Tables 8.2, 8.4, and 8.7, calculate the energy change in the formation of a mole of hydrated alkali metal ions from a mole of solid metal.

3. Describe a process, together with any equilibria involved, for the following preparations: (a) Na_2CO_3 from $NaCl$, (b) $NaCl$ from Na_2CO_3, and (c) $NaOH$ from $NaCl$.

4. The carbon dioxide needed in the Solvay process is commonly prepared by heating limestone, and the resulting lime is changed to calcium hydroxide for use in the recovery of the ammonia. (a) Show that the net result of such a Solvay process is to change calcium carbonate and sodium chloride into sodium carbonate and calcium chloride. (b) Calculate the weight of sodium chloride and of limestone needed to prepare 2 tons of sodium carbonate.

5. Suppose you wished to neutralize a quantity of hydrochloric acid containing a ton of hydrogen chloride. If sodium hydroxide cost twice as much per pound as sodium carbonate, which neutralizing agent would be cheaper for this purpose, and by how much?

REFERENCES   FERN, W. O., "Potassium," *Sci. Amer.*, **181**, no. 2, 16 (1949).
GILMAN, H., and EISCH, J. J., "Lithium," *Sci. Amer.*, **208**, no. 1, 88 (1963).
HEILBRUNN, L. V., "Calcium and Life," *Sci. Amer.*, **184**, no. 6, 60 (1951).
HOYLE, G., "How Is Muscle Turned On and Off?" *Sci. Amer.*, **222**, no. 4, 84 (1970).
LEDDY, J. J., "Salt—A Pillar of the Chemical Industry," *J. Chem. Educ.*, **47**, 396 (1970).
SCHUBERT, J., "Beryllium and Berylliosis," *Sci. Amer.*, **199**, no. 2, 27 (1958).
WILLIAMS, L. P., "Humphry Davy," *Sci. Amer.*, **202**, no. 6, 106 (1960).

THE HALOGEN FAMILY: A FAMILY OF NONMETALS

In the preceding chapter, the chemistry of two families of metals was discussed. In this chapter we shall study a group of the most reactive nonmetals, the halogen family. The members of this family are fluorine, chlorine, bromine, iodine, and astatine. So little is known about astatine that we shall not discuss its properties here. The halogen family constitutes Group VIIA of the periodic table and contains some of the most highly electronegative elements. It is our plan here to point out those properties characteristic of the halogens and their compounds, to illustrate the trends in these properties in proceeding from one halogen to another, and to relate these characteristic properties and their trends to atomic structure, atomic or ionic size, and electronegativity.

A BROAD OVERVIEW OF THE HALOGEN FAMILY

The halogens differ from the metals of Groups IA and IIA in three especially important properties:

1. The atoms of the halogen elements need only one more electron to give them a stable noble gas configuration. The metals of Group IA and Group IIA have but one or two valence electrons beyond a noble gas configuration. Table 9.1 gives the electron configurations of the halogen atoms.

2. The electronegativity of the halogens is the highest of any family of elements and much higher than that of the Group I and Group II metals. Compare, for example, the electronegativity of lithium (1.00) and strontium (1.00) with that of fluorine (4.0) and iodine (2.5).

3. Except for fluorine, the halogens exhibit a variety of oxidation states; the Group IA and Group IIA elements show only two oxidation states (zero and one other). Chlorine, for example, forms stable species in the $-I$, 0, I, III, IV, V, and VII oxidation states. Examples of halogen species in the various oxidation states are given in Table 9.2.

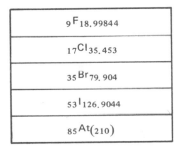

Figure 9.1 The halogen elements.

Table 9.1 Electron Configurations of the Atoms of the Halogen Elements

Element	Electron Configuration
Fluorine	$\boxed{1s^2}\ 2s^2\ 2p^5$
Chlorine	$\boxed{1s^2\ 2s^2\ 2p^6}\ 3s^2\ 3p^5$
Bromine	$\boxed{1s^2\ 2s^2\ 2p^6\ 3s^2\ 3p^6\ 3d^{10}}\ 4s^2\ 4p^5$
Iodine	$\boxed{1s^2\ 2s^2\ 2p^6\ 3s^2\ 3p^6\ 3d^{10}\ 4s^2\ 4p^5\ 4d^{10}}\ 5s^2\ 5p^5$

Table 9.2 Some Halogen Species in Various Oxidation States

Oxidation State	Fluorine	Chlorine	Bromine	Iodine	Names
VII		$HClO_4$		$HIO_4,\ H_5IO_6$	Perhalic acids
V		$HClO_3$	$HBrO_3$	HIO_3	Halic acids
III		$HClO_2$			Chlorous acid
I		$HClO$	$HBrO$	HIO	Hypohalous acids
0	F_2	Cl_2	Br_2	I_2	Halogens
−I	F^-	Cl^-	Br^-	I^-	Halide ions

Bonding

Because of their high electronegativity the halogen elements usually form ionic bonds when combined with elements having low electronegativity (the metals). In such compounds the halogens will exist as negative ions, as in sodium chloride or potassium iodide. In combination with hydrogen or other nonmetals—elements with moderate to high electronegativity—the halogens form covalent bonds—for example, hydrogen chloride, HCl, dichlorine oxide, Cl_2O, and nitrogen trifluoride, NF_3. This change in bond type with its accompanying change in properties is illustrated in Table 9.3, in which the melting points of the fluorides and chlorides of the elements of the third period of the periodic table are recorded. Most compounds of the halogens with metals have the high melting points characteristic of ionic bonding; compounds of the halogens with nonmetals have the low melting points characteristic of covalent bonding.

In general, the stability of the covalent compounds of the halogens increases as the differences in electronegativity between the halogen and other nonmetals increases. The most stable halogen species appears to be the halide ion, X^-. This species not only possesses a noble gas configuration, but, in it, none of the electrons is shared with other elements.

IMPORTANT PROPERTIES OF THE HALOGEN FAMILY

1. Their atoms have high electronegativity.
2. They exhibit multiple oxidation states (−I to VII).
3. Bonding varies from covalent, X_2, to ionic, NaX.
4. Many are strong oxidizing agents.
5. They form a variety of oxyacids, HXO_n.

Compound	NaF	MgF_2	AlF_3	SiF_4	PF_3	S_2F_2	ClF
Melting point (°C)	995	1263	1290	−90.3	−151.3	−120.5	−155.6
Compound	NaCl	$MgCl_2$	$AlCl_3$	$SiCl_4$	PCl_3	S_2Cl_2	Cl_2
Melting point (°C)	808	714	192	−68	−92	−80	−102.4

Positive Oxidation States

The positive oxidation states exhibited by the halogens may be thought of as arising when the halogen atom uses one or more of its valence electrons to form covalent bonds with elements of higher electronegativity, as in dichlorine oxide, Cl_2O. Here, the chlorine has a I oxidation state,

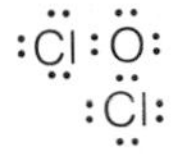

$$:\ddot{C}l:\ddot{O}:$$
$$:\ddot{C}l:$$

Dichlorine oxide

and the oxygen has a −II oxidation state. Other examples are perchloric acid, $HClO_4$, and bromic acid, $HBrO_3$, as illustrated below. In these compounds, the chlorine is using seven valence electrons in bond formation and has an oxidation state of VII; the bromine atom is using five valence electrons in bond formation and has an oxidation state of V.

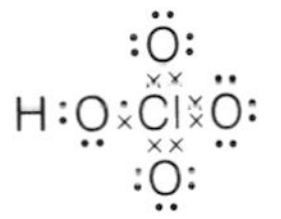

$$:\ddot{O}:$$
$$H:\ddot{O}\overset{\times\times}{\times}\ddot{C}l\overset{\times}{\times}\ddot{O}:$$
$$:\ddot{O}:$$

Perchloric acid

$$H:\ddot{O}\overset{\times\times}{}\ddot{B}r\overset{\times\times}{}\ddot{O}:$$
$$:\ddot{O}:$$

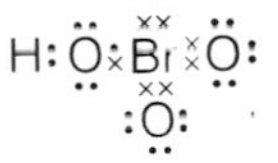

Bromic acid

The highest stable oxidation state of the halogens is the VII state, in which all seven valence electrons in the halogen atom are used in bond formation. Perchloric acid is an example of a halogen compound containing chlorine in its highest oxidation state. The V, III, and I states also exist and represent cases in which the halogen has one, two, or three fewer covalent bonds than in perchloric acid.

Oxidizing Agents

Some of the most powerful oxidizing agents are halogens or their compounds. Many other halogen species are good-to-strong oxidizing agents. This is explained as follows: As a rule, a chemical species tends to react to attain a more stable state. Among the halogens, the halide ion, X^- (the −I oxidation state), has great stability. For this reason, halogens and their compounds in all but the −I state have a strong tendency to change to the −I state in reactions. Under these circumstances the halogen species acts as an oxidizing agent (it is reduced). For example, potassium chlorate, $KClO_3$, is a strong oxidizing agent. Here, chlorine is in the V oxidation state. The resultant chlorine-containing product of most reactions in which potassium *chlorate* acts as an oxidizing agent is potassium *chloride*; in this, chlorine is in its very stable −I state. Thus the halogen ions in all but the −I state are good-to-strong oxidizing agents.

Acids

The halogens combine with hydrogen or with hydrogen and oxygen to form numerous acids. All four acids of the general formula HX are known and all but hydrofluoric acid, $(HF)_x$, are strong acids in water. In the positive oxidation states, chlorine, bromine, and iodine form a number of oxyacids.

These compounds have the formula HXO_n, where n may vary from one to four, as shown in Table 9.2. The strength of these oxyacids increases with increase in electronegativity of the halogen for a fixed value of n, and it increases as n increases for a given halogen. A detailed examination of the reasons for these changes in acid strength will be presented later.

Occurrence
Because of their high reactivity, the halogen elements are not found in nature in the free state. All four members are found in the form of the very stable halide ions in seawater.

Most of the fluorine is found in the minerals cryolite, Na_3AlF_6, and fluorspar, CaF_2. The principal sources of chlorine and bromine are seawater, brine wells, and mineral deposits. Sodium and chloride ions make up about 2.5 per cent of seawater. The most important source of iodine is the sodium iodate found in the huge nitrate deposits of northern Chile.

It is estimated that fluorine occurs to the extent of 0.1 per cent in the earth's crust, and that chlorine, bromine, and iodine make up 0.2, 0.001, and 0.001 per cent of the earth's crust, respectively.

Uses
Large quantities of fluorine, available commercially, are used in the preparation of fluorine-containing hydrocarbons (called fluorocarbons). These include Freons, used as refrigerants and as the inert pressurizing ingredient in cans of hair spray, insecticide, and the like, and plastics such as Teflon, an excellent insulator and chemically inert material suitable for containers for corrosive chemicals. Hydrogen fluoride is an important intermediate in a number of industrial processes including the manufacture of high-octane gasolines.

Chlorine is one of the most important industrial chemicals. It is used as a bleaching agent, especially for wood pulp and paper and for cotton cloth. Compounds of chlorine such as hypochlorites are used as bleaching agents and disinfectants.

Strong oxidizing agents such as the chlorates are used in explosives and rocket propellants. Other uses of chlorine include the manufacture of important organic compounds used as drugs and dyes.

The demand for bromine has fluctuated greatly in recent years. Potassium and sodium bromides find use as medicinal agents, and silver bromide is used in photography. Bromine is used in the preparation of ethylene dibromide, which has been an important constituent of "ethyl gasoline." It is also used in the manufacture of certain dyes and other organic compounds.

The chief use of iodine is in the manufacture of potassium and sodium iodides and of certain organic drugs and dyes. Silver iodide (AgI) is an important compound of iodine used in photography. The ordinary tincture of iodine, used largely as an antiseptic, is a solution of iodine and potassium iodide in alcohol; a colloidal dispersion of the element is also used for the same purpose. Iodine is also a valuable reagent in the laboratory.

PROPERTIES OF THE
HALOGEN ELEMENTS
The halogen elements occur as the diatomic molecules F_2, Cl_2, Br_2, and I_2. The single covalent bond between the atoms enables each atom to obtain a stable electron configuration, as illustrated below.

$$\ddot{\underset{\cdot\cdot}{:F}} \! : \! \ddot{\underset{\cdot\cdot}{F}} : \qquad :\!\overset{\cdot\cdot}{\underset{\cdot\cdot}{Cl}}\!:\!\overset{\cdot\cdot}{\underset{\cdot\cdot}{Cl}}\!: \qquad :\!\overset{\cdot\cdot}{\underset{\cdot\cdot}{Br}}\!:\!\overset{\cdot\cdot}{\underset{\cdot\cdot}{Br}}\!: \qquad :\!\overset{\cdot\cdot}{\underset{\cdot\cdot}{I}}\!:\!\overset{\cdot\cdot}{\underset{\cdot\cdot}{I}}\!:$$

Table 9.4 Some Physical Properties of the Halogen Elements

Element	Density (g/cc)	Melting Point (°C)	Boiling Point (°C)	Heat of Fusion (kcal/mole)	Heat of Vaporiza-tion (kcal/mole)	Color of Gas
Fluorine	1.108	−219.62	−187.9	0.061	0.782	Pale yellow
Chlorine	1.57	−102.4	−34.0	0.77	2.439	Yellow-green
Bromine	3.14	−7.2	58.2	1.26	3.59	Brown-red
Iodine	4.942	113.6	184.2	1.87	4.99	Violet

These elements are very reactive oxidizing agents; fluorine is the most reactive, followed by chlorine, bromine, and iodine in that order.

Physical Properties

Some physical properties of the halogen elements are given in Table 9.4. The melting and boiling points and the heats of fusion and vaporization are very low, suggesting that the forces of attraction between X_2 molecules are not strong. But all four of these properties rise systematically in proceeding from fluorine to iodine (Table 9.4), indicating that the attractive forces increase regularly as the atomic number increases from fluorine to iodine (as illustrated in Figure 9.2). This increase has been attributed to an increase in the polarizability (Chapter 6) of the atoms as the atomic diameter increases. As the valence electrons get farther away from the nucleus in the larger atoms, they are less tightly bound by the nuclear charge and come more readily under the influence of the nuclei of neighboring molecules—thus we say they are more polarizable.

This increase in attractive forces presumably is the reason iodine is a solid and bromine is a liquid, but both fluorine and chlorine are gases at room temperature and pressure.

The color of the halogen element is the result of the absorption of certain wavelengths of visible light. The energy absorbed excites valence electrons to higher energy levels. The energy difference between the excited state and the ground state changes in the same order as do the ionization potentials of the atoms: F > Cl > Br > I (Table 9.5). This means that short-wavelength photons are required to excite fluorine, but progressively longer-wavelength photons are required to excite chlorine, bromine, and iodine. As a consequence, fluorine absorbs short-wavelength violet radiation and appears yellow while iodine absorbs longer-wavelength yellow and green light and appears violet. This is again a result of the weaker attraction of the nuclei for valence electrons in the larger atoms, reflected both in the lower ionization energies of the atoms and in the colors of their molecules.

Properties of Atoms

Table 9.5 lists the electronegativities, atomic and ionic radii, and X—X bond energies of the halogens. As stated earlier these elements have a very high electronegativity—*fluorine has the highest electronegativity of all the elements.* In proceeding down the family, the electronegativity decreases. However, even iodine has a relatively high electronegativity, for only oxygen, nitrogen, and the other halogens have higher values.

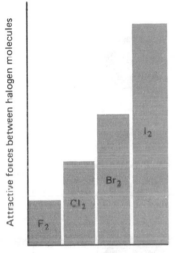

Attractive forces between halogen molecules

F_2 Cl_2 Br_2 I_2

Figure 9.2 Attractive forces between halogen molecules.

Table 9.5 Some Properties of the Halogen Atoms, Molecules, and Ions

Element	Electro-negativity of Atom	Ionization Energy (kcal/mole)	Atomic Radius (A)	Ionic Radius (A)	X—X Bond Energy (kcal/mole)
Fluorine	4.0	402	0.72	1.36	36
Chlorine	3.0	300	0.99	1.81	57
Bromine	2.8	273	1.14	1.95	45
Iodine	2.5	241	1.33	2.16	35

Both the atomic and ionic radii of the halogen atoms and the halide ions increase regularly from fluorine to iodine, as illustrated in Figure 9.3. The atomic radii are determined from the X—X bond distance in halogen molecules while the ionic radii are determined from X-ray studies of crystals containing halide ions. The halide *ion is invariably larger than the corresponding atom.* This is due in part to a greater repulsion among the valence electrons in the ion compared with that in the atom, and in part to the fact that atomic radii are one-half the X—X bond distance, which actually is shorter than one-half the atomic diameter (Figure 9.3).

Figure 9.3 Relative sizes of halogen atoms and halide ions. Note the differences among: (a) the van der Waals radii; (b) the covalent radii; (c) and the ionic radii for these elements. (See Figure 6.3 for comparison of covalent radii and van der Waals radii.)

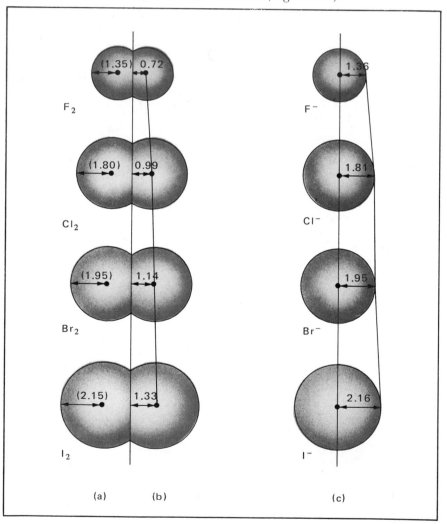

The *energy needed to dissociate* the halogen *molecules into atoms* decreases regularly from chlorine to iodine, indicating that the covalent bond in iodine is weaker than that in chlorine, Cl_2, or in bromine, Br_2. This is the order of the decrease in the ability of the various atoms to attract electrons.

Fluorine has a much lower dissociation energy than chlorine, which is somewhat surprising in view of its higher electronegativity. No completely satisfactory explanation for this observation has been reported. However, it has been suggested that the F—F bond is weakened because nonbonding electrons in the small, very stiff and nonpolarizable fluorine atom repel the bonding electrons to the point of weakening the bond considerably.

The X—X bond energies are important factors in the reactions of halogen molecules because energy is required to break these bonds if the reactions are to proceed.

Chemical Properties

In a great many of their reactions the halogens act as oxidizing agents, being reduced to the −I oxidation state in the process. The chemical properties of these elements may be summarized with four types of reactions:

Reaction with Metals. Fluorine and chlorine react directly with all metals to give metal halides; bromine and iodine are somewhat less reactive but react directly with all but the noble metals, such as gold and platinum.

Reaction with Nonmetals. Fluorine and chlorine react directly with all nonmetals except nitrogen, oxygen, and most of the inert gases. Bromine and iodine are even more selective in reactions with nonmetals. Although the halogens do not react directly with certain nonmetals, compounds of the halogens with all nonmetals except certain noble gases can be prepared by indirect methods.

A good example of the reaction of halogens with nonmetals is their reaction with hydrogen:

$$H_2 + X_2 \rightleftharpoons 2HX$$

This reaction occurs with all halogens, but the position of equilibrium at 298 K lies farthest to the right with fluorine and is displaced successively more to the left in proceeding from chlorine to iodine. The position of equilibrium in these reactions parallels the trend of the H—X bond energies, which is HF $\gg$ HCl $\gg$ HBr $\gg$ HI.

Reactions with Water. Halogen elements react with water two ways. In the first type,

$$2X_2 + 6H_2O \rightleftharpoons 4H_3O^+ + 4X^- + O_2 \qquad (1)$$

the halogen displaces oxygen from the water. In the second type,

$$X_2 + 2H_2O \rightleftharpoons H_3O^+ + X^- + HOX \qquad (2)$$

the halogen is converted to both the −I and the I oxidation states. Such a reaction may be called a hydrolysis reaction, since in principle both the halogen and the water molecules are *split apart*, giving initially the HX and XOH species as illustrated:

$$
\begin{array}{ccc}
\begin{matrix} X \\ | \\ X \end{matrix} + \begin{matrix} O\text{—}H \\ | \\ H \end{matrix}
& \longrightarrow &
\left[\begin{matrix} X \cdot\cdot \; O\text{—}H \\ : \quad\quad : \\ X \cdot\cdot \; H \end{matrix} \right]
& \longrightarrow &
\begin{matrix} X\text{—}OH \\ \\ X\text{—}H \end{matrix}
\end{array}
$$

The HX species, a strong acid, ionizes almost completely in water, giving hydronium ions and X^- ions (except in the case of HF).

The first reaction—displacement of oxygen—proceeds very rapidly with fluorine and much less rapidly with the other halogens. The second reaction is not observed with fluorine because the first process is so rapid and complete. However, chlorine, bromine, and iodine all hydrolyze to some extent in water.

Interestingly, all halogens react with hydrogen sulfide, H_2S, to displace sulfur in a reaction similar to reaction (1) above.

$$8X_2 + 8H_2S \longrightarrow 16HX + S_8 \tag{3}$$

This behavior with respect to hydrogen sulfide suggests that the ready occurrence of reaction (1) requires that the electronegativity of the halogen be greater than that of oxygen. All of the halogens are more electronegative than sulfur and all react readily according to Equation (3).

Reactions with Other Reducing Agents. The halogen elements will react with (are reduced by) strong and moderately strong reducing agents such as sulfides, sulfites, arsine (AsH_3), many metal ions such as tin(II) ion (Sn^{+2}), iron(II) (Fe^{+2}), and vanadyl (VO^{+2}) ions, and a great many organic compounds.

Formation of Polyhalide Anions. Anions such as I_3^-, IBr_2^-, and ICl_2^- are formed in solution and are stable in salts with large cations.

Oxidation Potentials The values of the standard oxidation potentials, $E°$, for the halide ions are given in Table 9.6. The values are a measure of the tendency for the reaction

$$X^- \longrightarrow \tfrac{1}{2}X_2 + e^- \tag{4}$$

to occur in water at 25°C. The negative values indicate that the reactions proceed only with difficulty and that the reaction involving fluoride is the most difficult—has the largest negative value. Conversely, the reaction with iodide, though difficult, proceeds more readily than the others. In essence this means that in water solution fluoride ions lose electrons only with great difficulty while iodide ions do so with less difficulty.

Equation (4) represents an oxidation (the halogen changes from the −I to the 0 oxidation state) so the $E°$ values are measures of the relative strengths of X^- ions as reducing agents. Evidently, iodide ion can be oxidized most easily of the group and must therefore be the strongest reducing agent. By the same reasoning, fluoride ion is the most difficult to oxidize so it must be the weakest reducing agent of this group of poor reducing agents.

The reverse of the reaction represented by Equation (4)—that is,

$$e^- + \tfrac{1}{2}X_2 \longrightarrow X^- \tag{5}$$

Table 9.6 Standard Oxidation Potentials for Halide Ions

Conversion	$E°$ (volts)
$I^- \rightleftharpoons \tfrac{1}{2}I_2 + e^-$	−0.53
$Br^- \rightleftharpoons \tfrac{1}{2}Br_2 + e^-$	−1.07
$Cl^- \rightleftharpoons \tfrac{1}{2}Cl_2 + e^-$	−1.36
$F^- \rightleftharpoons \tfrac{1}{2}F_2 + e^-$	−2.87

is a reduction, and the $E°$ values with their signs reversed are measures of the ease with which these reactions occur. In this case fluorine has the largest and iodine the smallest tendency to be reduced, with all reactions being strongly favored. Evidently, then, fluorine is the strongest oxidizing agent and iodine is the weakest in this group of strong oxidizing agents. This trend is in accord with electronegativity considerations, which indicate that fluorine should attract electrons more strongly than does iodine.

PREPARATION OF THE HALOGEN ELEMENTS

In most cases the halogen elements are prepared by oxidizing the halide ions, which are the largest natural source of the halogens. Compounds of iodine in the V or VII oxidation state occur in nature; these compounds must be reduced to produce the free element (the 0 oxidation state species).

Use of Oxidizing Agents

If the naturally occurring halide ions are used to prepare the halogen elements, an oxidizing agent must be used to oxidize the halide ion from its −I to the 0 oxidation state. The general reaction then becomes

$$X^- + \text{oxidizing agent} \rightleftharpoons \tfrac{1}{2}X_2 + \text{reduced form of oxidizing agent} \qquad (6)$$

Since the X_2 species is itself an oxidizing agent, the oxidizing agent used should be at least as strong as X_2 if the reaction is to proceed very far to the right before equilibrium is established.* A number of oxidizing agents or systems have been used, but the following three are used most often.

Electrolysis (Oxidation by an Anode Reaction). The fused halide salts of all the halogens can be electrolyzed to give the halogen element and the metal of the salt used (Figure 8.6). For example, the electrolysis of fused sodium chloride gives chlorine at the anode and sodium at the cathode. The over-all reaction is

$$NaCl \xrightarrow[\text{electrolyzed}]{\text{fused}} Na + \tfrac{1}{2}Cl_2$$

At the anode the reaction is

$$Cl^- \longrightarrow \tfrac{1}{2}Cl_2 + e^-$$

This method can be used for all the halogens and is the only method suitable for the preparation of fluorine. For fluorine preparation, cells of Monel metal or copper must be used because of the great reactivity of this halogen.

Chlorine, bromine, and iodine can be also prepared by electrolyzing water solutions of their halide salts. However, in fluoride solutions, water is decomposed at the anode more easily than fluoride ion is oxidized; therefore, oxygen is produced rather than fluorine gas. Dilute solutions containing chloride ions give both oxygen and chlorine when electrolyzed, suggesting that the ease with which water is decomposed and the ease of discharge of chloride ion are about the same.

Manganese Dioxide. Acid solutions containing chloride, bromide, or iodide ions give the corresponding halogens upon reaction with a strong

---

* Under certain conditions a weaker oxidizing agent may be used. For example, if one of the products is removed from the reaction mixture as fast as it is formed, the reverse reaction in Equation (6) is prevented from taking place. Meanwhile, the forward reaction will proceed at its normal pace. The result will be a continuous production of product that can be removed rapidly. Thus a slightly weaker oxidizing agent may produce amounts of product in a non-equilibrium situation that it could not produce under equilibrium conditions.

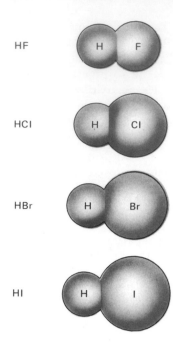

HF

HCl

HBr

HI

Figure 9.4 Relative sizes of HX molecules.

oxidizing agent such as manganese dioxide. The reaction may be represented:

$$2X^- + 4H_3O^+ + MnO_2 \rightleftharpoons Mn^{+2} + X_2 + 6H_2O$$

This method is commonly used to prepare small quantities of the halogen elements in the laboratory.

Manganese dioxide, a weaker oxidizing agent than fluorine, is stronger than bromine or iodine and just slightly weaker than chlorine. The oxidation of chloride ion is successful because the gaseous chlorine escapes from the reaction mixture so that equilibrium is never established.

Manganese, in the course of this reaction, is reduced from the IV to the II oxidation state.

A Stronger Oxidizing Halogen. The oxidizing strength of the halogen elements decreases in the order $F_2 > Cl_2 > Br_2 > I_2$, as indicated earlier. Conversely, the reducing strength of halide ions is in the opposite order, increasing regularly from fluoride to iodide ion. This means that while fluorine attracts electrons strongly, iodide ion loses electrons relatively easily.

These facts can be used to select methods of preparing the weaker oxidizing halogens from stronger ones. For example, in the reaction

$$\tfrac{1}{2}Cl_2 + I^- \rightleftharpoons \tfrac{1}{2}I_2 + Cl^-$$

the position of equilibrium lies far to the right because chlorine is a stronger oxidizing agent than iodine, and iodide ion is a stronger reducing agent than chloride ion. As indicated in Chapter 7, the equilibrium lies in favor of the weaker oxidizing and reducing agents.

This situation may be generalized as follows: A stronger oxidizing halogen can be used to prepare a weaker oxidizing halogen from its anion. In practice, chlorine water is used to produce bromine and iodine from bromides and iodides, respectively. Such reactions are not limited to halogens but apply equally well to other nonmetals and their ions.

SOME COMPOUNDS OF THE HALOGENS

Interhalogens

Compounds formed between two different halogens are called interhalogens. The best known of these is iodine monochloride, ICl, which is used as an indicator and in the synthesis of other iodine compounds. Examples of some interhalogen compounds which have been made and characterized are given in Table 9.7.

The existence of compounds of the type XY_n, where n is three or more, is probably due to the fact that larger halogen atoms may use d orbitals as well as s and p orbitals in bonding. Interhalogens are usually prepared by the reaction of two halogen elements with each other. The structures of IF_5 and

Table 9.7 Some Interhalogen Compounds and Ions

ClF	BrCl	BrF	IBr	IF	ICl_2^-
				ICl	
ClF_3		BrF_3		I_2Cl_6	ClF_4^-
		BrF_5		IF_5	IF_6^-
				IF_7	

Table 9.8 Some Properties of the Hydrogen Halides

| | Compound | | | |
Properties	HF	HCl	HBr	HI
Bond length (A)	0.917	1.275	1.410	1.62
Dipole moment ($D*$)	1.91	1.03	0.80	0.42
Bond energy (kcal/mole)	135	102	87	71
Melting point (°C)	−83.07	−114.19	−86.86	−50.79
Boiling point (°C)	19.9	−85.03	−66.72	−35.35
Heat of fusion (kcal/mole)	1.094	0.4750	0.5751	0.6863
Heat of vaporization (kcal/mole)	7.24	3.86	4.210	4.724
Approximate ionization constant (K_{ion} at 25°C)	10^{-4}	10^7	10^9	10^{11}

*One Debye unit (D) is 1×10^{-18} esu-cm.

IF_7, which have 12 and 14 electrons, respectively, around the central atom, are given in Figure 6.10.

The Hydrogen Halides and Halide Ions

Some properties of the hydrogen halides are listed in Table 9.8. From the relative magnitudes of bond length, dipole moment, and bond energy, it is apparent that in passing from hydrogen fluoride to hydrogen iodide the H—X bond distance increases regularly as the size of the halogen atom increases, polarity of the H—X bond decreases regularly as the halogen atom decreases in electronegativity, and the H—X bond energy decreases, indicating that hydrogen iodide can be dissociated into atoms more easily than can hydrogen fluoride. The relative sizes of the four HX molecules are shown in Figure 9.4.

Physical Properties

The relatively low values of the melting and boiling points and the heats of fusion and vaporization (Table 9.8) indicate that the forces of attraction between molecules are relatively weak except in hydrogen fluoride, in which hydrogen bonding significantly increases the boiling point, melting point, and the heats of fusion and vaporization. Hydrogen bonding is also responsible for the excellent solvent properties of liquid hydrogen fluoride, which acts like an assembly of very large aggregates of hydrogen fluoride molecules.

Interestingly, the melting and boiling points of the other three HX acids increase in the order HCl < HBr < HI (Figure 6.23). This suggests that the attractive forces between molecules also increase in this order. Since the polarity of the molecules is greatest in HCl and least in HI, evidently van der Waals forces are more important here than are the polar forces.

Chemical Properties

As reducing agents, the HX compounds and their anions, X^-, are relatively poor, with fluoride ion being the weakest and iodide ion the strongest, as previously indicated. An example of the reducing strength of hydrogen iodide is the fact that solutions of this compound will react slowly with oxygen from the air to give iodine, which is indicated by the brown color that gradually forms in hydrogen iodide solutions.

As acids, the HX compounds ionize in water according to the reaction

$$HX + H_2O \rightleftharpoons H_3O^+ + X^- \tag{7}$$

Water solutions of HX compounds are called *hydrohalic acids*. This equilibrium lies far to the right; in fact, the reaction is essentially complete for all

HX compounds except hydrogen fluoride. For this reason, hydrochloric, hydrobromic, and hydroiodic acids are considered strong acids in water.

Approximate values for the ionization constants of the hydrohalic acids are given in Table 9.8. The order of acid strengths is HF ≪ HCl < HBr < HI. The explanation for this order is complex, but in essence it is related to the concept that the small fluoride ion has a greater attraction for protons than do the increasingly larger chloride, bromide, and iodide ions. In the ionization process, the proton must be separated from the newly formed halide ion. And this is most difficult in the case of fluoride ion.

OXYACIDS

The common oxyacids of the halogens are listed in Table 9.2. The generalized names of the acids and their ions are given in Table 9.9.

Nomenclature

The nomenclature pattern of Table 9.9 can be applied to any halogen oxyacid by substituting the appropriate stem *chlor*, *brom*, or *iod*, for the more general stem *hal* used in the table. The two periodic acids listed in Table 9.2 are known as *meta*periodic and *para*periodic acid, respectively.

Structure

The *electronic structures* and *shapes* of four oxyanions are illustrated in Figure 9.5. The tetrahedral shape of a perhalate ion, XO_4^-, arises because the halogen atom uses sp^3 hybrid orbitals in forming the covalent bonds with the oxygen atoms. The angular shape of the halite ion is the result of the halogen atom using two of its p orbitals to form the bonds with oxygen. Like water this structure also may contain sp^3 hybrid orbitals.

The structure of *para*periodic acid is

This structure presumably arises because of the large size of the iodine atom, its relatively low electronegativity, and the fact that d orbitals on the iodine atom can be used to form bonds. Salts, such as $Na_2H_3IO_6$, containing this structure can be obtained from water solutions of sodium *meta*periodate.

Chemical Properties

Two important chemical properties of the halogen oxyacids are their abilities to act as oxidizing agents and as acids.

These compounds are good-to-strong oxidizing agents; they are easily reduced to lower oxidation states, often giving the halogen in the −I state. In general, the chloro oxyacids are stronger oxidizing agents than the cor-

Table 9.9 Names of Halogen Oxyacids and Oxyacid Anions

Formula of Acid	Name	Formula of Ion	Name
HXO_4	Perhalic acid	XO_4^-	Perhalate ion
HXO_3	Halic acid	XO_3^-	Halate ion
HXO_2	Halous acid	XO_2^-	Halite ion
HXO	Hypohalous acid	XO^-	Hypohalite ion

Figure 9.5 Oxyanions of the halogens.

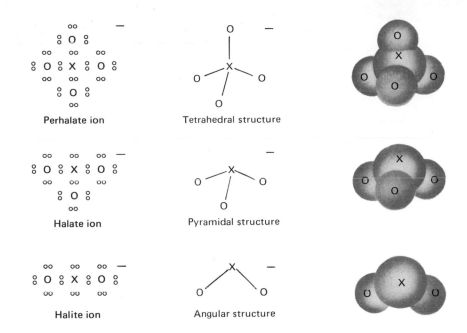

Perhalate ion   Tetrahedral structure

Halate ion   Pyramidal structure

Halite ion   Angular structure

responding bromo and iodo compounds. This is consistent with electronegativity considerations since chlorine, the most electronegative of the three halogens (chlorine, bromine, and iodine), has the most stable −I oxidation state and hence has the strongest tendency to reach that state.

Among the chloro oxyacids, hypochlorous acid is the strongest and perchloric acid the weakest oxidizing agent, suggesting that one important point here is that perchlorate ion must break four bonds while hypochlorite ion must break only one bond to reach the −I oxidation state.

These oxidizing properties are used in many ways. For example, the hypochlorite ion is used to prepare other oxyanions such as chromate ion (CrO_4^{-2}), selenate ion (SeO_4^{-2}), periodate ion (IO_4^{-}), and ferrate ion (FeO_4^{-2}) from compounds of the respective elements in their lower oxidation states. Hypochlorites and chlorites are excellent for bleaching cotton or other cellulose-type fibers or cloth. They are also used as disinfectants and antiseptics. Mixtures of chlorates with sulfur, sugar, or other reducing agents are powerful explosives. All three halate ions are used in quantitative analysis procedures, in which the halate ion acts as an oxidizing agent. For example, iodate is used to determine the amount of antimony in a sample or to estimate the concentration of chloride in a solution. Salts of perchloric acid are used as the oxidants in solid rocket fuels.

Acid Strengths   The halogen oxyacids vary in acid strength from very strong to extremely weak acids, as illustrated by their ionization constants (Table 9.10). While the data are incomplete, two trends are apparent: The first is that *the acid strength increases as the number of oxygen atoms in the molecule increases, provided the halogen atom remains the same.* For example, in the series HClO, $HClO_2$, $HClO_3$, $HClO_4$, the acid strength increases regularly from a very low to a high value. A similar trend is observed among the iodoxy acids.

Table 9.10 Acid Strengths of Halogen Oxyacids

Acid	Ionization Constant (at 25°C)	Acid	Ionization Constant (at 25°C)	Acid	Ionization Constant (at 25°C)
$HClO_4$	Very large			H_5IO_6	5×10^{-4}
$HClO_3$	Large	$HBrO_3$	Large	HIO_3	2×10^{-1}
$HClO_2$	1×10^{-2}				
$HClO$	3.8×10^{-8}	$HBrO$	2×10^{-9}	HIO	1×10^{-11}

The second trend is that *the acid strength increases as the electronegativity of the halogen atom increases, provided the number of oxygen atoms remains constant.* For example, the acid strengths decrease systematically in the series HOCl > HOBr > HOI. A similar trend appears among the halic acids.

Both of these trends may be explained in terms of the electronegativity or the apparent electronegativity of the halogen atoms in the compounds in question.

As the number of oxygen atoms in the molecule of the oxyacid increases, the halogen atom is sharing more and more of its valence electrons. In effect, this increases the electronegativity of the halogen atom which then attracts electrons from the hydrogen-oxygen bond, thereby increasing the ease of proton removal. These effects also tend to stabilize the anion relative to the acid. In the hypohalous acids, a highly electronegative halogen atom attracts electrons from the oxygen atom to which it is bonded. In effect, this increases the electronegativity of that oxygen atom (since its nucleus now has less control of the electrons) so that it attracts electrons from the oxygen-hydrogen bond. This then increases the ease of proton removal. The magnitude of these effects depends upon the electronegativity of the halogen atom. For example, in hypochlorous acid, there is probably a net attraction of electrons toward the chlorine and away from the hydrogen-oxygen bond. In hypo-iodous acid, this attraction is less because the electronegativity of iodine is less than that of chlorine. The proton is removed more easily in the chloro acid because the bonding electrons are pulled away from the proton and toward the oxygen to a greater extent here than in the iodo acid. Again, these electron-withdrawing effects tend to stabilize the anion relative to the acid.

Both of these trends appear to be general for the oxyacids of the nonmetals. Thus the acid strengths of the oxyacids of the Group V and Group VI elements increase with the number of oxygen atoms in the molecule and with increasing electronegativity of the central atom, just as for the halogen oxyacids of Group VII.

Preparation Preparation of the oxyacids is often difficult, but their salts can be prepared by the reaction of the halogen elements with bases or by electrolysis of solutions of halides.

The Reaction of Halogen Elements with a Base. The position of equilibrium in the general reaction

$$X_2 + 2H_2O \rightleftharpoons HOX + H_3O^+ + X^-$$

lies to the left for chlorine, bromine, and iodine, as indicated in reaction (2) (see p. 199). However, in the presence of cold sodium or potassium hydroxide, the acids are converted to their salts and the position of equilibrium is displaced sharply to the right. This reaction may be written as two steps, or simply as

$$X_2 + 2OH^- \longrightarrow X^- + OX^- + H_2O$$

If hot base is used, the hypohalite ion undergoes a disproportionation reaction of the type

$$3XO^- \longrightarrow 2X^- + XO_3^-$$

This is a method for preparing the halate salts. For example, potassium chlorate may be prepared by the action of chlorine on hot aqueous potassium hydroxide. When the resulting solution is evaporated, the potassium chlorate, being much less soluble than potassium chloride, separates first. It can be obtained in the pure state by repeated crystallization.

This method of making halates is not economical because, in the over-all reaction of halogen to halate,

$$3X_2 + 6OH^- \longrightarrow 5X^- + XO_3^- + 3H_2O$$

five moles of halide are produced for every mole of halate formed.

Electrolysis of Halides. More economical methods for preparing halates and hypohalites involve the electrolysis of halides in basic solution. By controlling the temperature and other experimental variables, it is possible to oxidize halide ions to any stable higher oxidation state. Thus chlorine, hypochlorites, chlorites, chlorates, and perchlorates can be prepared electrolytically. Electrolytic methods are not used to prepare iodates, periodates, or hypoiodites since the ready oxidizability of iodides make use of chemical oxidizing agents more suitable. An example is the oxidation of iodide ion to iodate by hypochlorites:

$$I^- + 3HOCl + 3H_2O \longrightarrow IO_3^- + 3Cl^- + 3H_3O^+$$

Chemical Solutions to Some Everyday Health Problems. The four common halogens are present, either in elemental form or in compounds, in so many health-related chemicals that a brief summary of the use of chemicals in some everyday health problems of the individual might be appropriate at this point.

The use of elementary chlorine in water purification and as a general disinfectant in swimming pools is well known, as is the use as an antiseptic of elemental iodine dissolved in ethyl alcohol. Minute quantities (0.7 to 1 part per million) of sodium fluoride in drinking water are employed to reduce dental caries in children. Less well known are the uses of bromine compounds, including sodium bromide, as sedatives, hypnotics, anticonvulsants; of chlorine compounds, including chloroform ($CHCl_3$), as antitussive agents (against coughs), analgesics, muscle relaxants, sedatives, amebicides, diuretics, and urinary antiseptics.

Between 800 and 1000 different drugs are in rather extensive use in the U.S. and the new ones appear at a rate of about 30 to 40 per year. Generally, the newer drugs are more powerful and much more effective than those used a decade or two ago. Hence they must be used with greater care, often only under a doctor's care.

Drugs for the relief of pain (analgesics), as sedatives, laxatives, antiseptics, emollients, and nasal and throat sprays are used by hundreds or thousands of people with little or no harmful side effects. Alkalizers, cold tablets, "pep" pills and tranquilizers also are widely used, but the incidence of undesirable side effects with many of these preparations is enough to cause the enlightened individual to approach their use with caution.

Reactions to drugs differ with different people and in the same person at different times and under different conditions. Moreover, use of a drug may serve only to cover up temporarily a much more serious situation. For example, an "acid stomach" usually can be palliated by taking an alkalizer such as aluminum hydroxide. Unfortunately, the biochemical processes of the stomach are such that if too much acid is neutralized, more acid will be produced. If the overacidity is caused by an ulcer or a disfunctioning gallbladder, no amount of alkalizer can solve the problem.

Headaches and other minor aches and pains are an occasional inconvenience to many. Often these pains are caused by swelling and minor inflammation of tissue, resulting in pressure on peripheral nerves and accompanied by an increase in body temperature. This in turn irritates nerve endings. Aspirin or acetylsalicylic acid (Chapter 28) is the most widely used analgesic for this condition. It functions by reducing the swelling and inflammation of tissues and by reducing fever.

In some preparations, aspirin is accompanied by caffeine (Chapter 28), which serves as a diuretic, stimulating the secretion of urine, thereby helping remove undesirable chemicals from the body.

Sneezing, a running or blocked nose, and running eyes are the familiar symptoms of a cold or a nasal allergy such as hay fever. Nasal allergies often are characterized by a clear, watery excretion from the nose, whereas sinus infections are indicated by thick yellow or green secretions. Nasal allergy, sinus infections, and asthma (allergy of the bronchial tubes) require treatment by an experienced physician.

The allergic symptoms are caused by the release of a chemical, histamine, in the system. Histamine causes dilation of capillaries (small blood vessels) and the smooth muscles in the nasal and bronchial passages. It also can dilate cerebral blood vessels, causing severe headaches. The antihistamines can provide temporary relief to a blocked or running nose by preventing the action of histamine, thus "drying up" the nasal passages. Unfortunately, antihistamines are not effective when large amounts of histamine are liberated, and they do not get at the basic difficulty—the release of histamine by a distressed system.

Bacterial infections of superficial tissues, such as might accompany any break in the skin, are combated medically by antiseptics. Deep-seated or generalized infections of the human system usually are treated by chemotherapeutic agents, such as antibiotics, which are administered by mouth or by injection. A very large number of chemical compounds possess the property of killing bacteria, but many of these also exhibit properties that profoundly affect or prohibit their use for this purpose. The very property that destroys bacterial cells also can bring harm to the body's cells and tissue. An antiseptic is most valuable when there is a wide margin between its bactericidal and its toxic concentrations.

Chlorine and iodine are extremely effective skin antiseptics and may be used in high dilution in the absence of blood or serum. Ethyl alcohol also is an excellent skin antiseptic when diluted with water to a concentration of

70 per cent. Complex salts of mercury, copper, silver, and zinc have been used as antiseptics, but most are slow acting and many of the mercury salts are toxic. Soaps and detergents possess a bactericidal as well as a cleansing action and are of considerable value in preventing superficial skin infections.

SUMMARY
In this chapter, we have discussed some of the important chemistry of the halogen elements, a series of nonmetals having high electronegativity and forming both ionic and covalent compounds in a variety of oxidation states. This family is outstanding for its numerous strong oxidizing agents and its many acids of varying strength.

The structure, properties, and reactivity of the elements and compounds belonging to this family have been related to the principles of atomic and molecular structure established earlier in the text. In addition, the ideas of acid-base equilibria and oxidation-reduction presented earlier have been applied and extended in introducing the important reactions of the family.

IMPORTANT TERMS

Halogens
covalent and ionic radii
electronegativity
hydrohalic acid
interhalogen
oxyhalogen
oxyacids
multiple oxidation states

Halogen structures
sp^3 hybridization
square pyramid
pentagonal bipyramid

Dipole moment

Dissociation energy

Ionization constant

QUESTIONS AND PROBLEMS

1. What physical and chemical properties of the halogens characterize them as nonmetals?

2. (a) Give the electronic configurations of atoms of each of the halogen elements, listing them in order of increasing atomic size. (b) How does the stability of each diatomic molecule in this group vary? (c) F_2 is a gas, I_2 is a solid. Explain this difference in physical state on the basis of van der Waals forces. (d) Account for the high reactivity of this family of elements.

3. Compare the elements in the halogen family with respect to: (a) oxidizing power, (b) ionization energies, (c) electronegativity, (d) ease of preparation from the corresponding halides. Interpret similarities and trends in terms of electronic configurations, sizes, and so forth, and try to explain any apparent contradictions to what you might predict on these bases.

4. Cite evidence to support the statement that fluorine is the most active nonmetal.

5. How do you account for the following: (a) a chloride ion (Cl^-) has a greater stability than a chlorine atom; (b) the ionic radii of the halide ions increase with the atomic weight; and (c) hydrogen chloride is a covalently bonded compound, but in water solution it forms a strong acid?

6. What is the oxidation state of the Group VII element in each of the following compounds? (a) $NaBrO_3$, (b) HCl, (c) $HClO$, (d) $HClO_2$, (e) $NaBrO_2$, (f) $NaIO_4$, (g) $Ca(IO_4)_2$, (h) HIO, (i) KBr, (j) Cl_2, (k) CCl_4.

Write the formulas for the following compounds and indicate the oxidation state of the halogen present: (a) potassium bromate, (b) hypoiodous acid, (c) sodium hypobromite, (d) zinc chlorite, (e) zinc chlorate, (f) copper(II) chlorite, (g) aluminum periodate, (h) chlorous acid, (i) hydrobomic acid.

7. List the oxidation states of chlorine. In which ones will chlorine be an oxidizing agent? In which is it a reducing agent? Under what circumstances will this be true?

8. Write the electron-dot formula for each of the following species: (a) CI_4, (b) ClO^-, (c) H_5IO_6, (d) CH_3Cl, (e) $HClO_2$.

9. List some commercial uses of fluorine. What property of fluorine makes it more desirable than chlorine for these purposes?

10. How can you account for the relatively low values of the boiling point, heat of fusion, and heat of vaporization of the halogens?

11. Give equations for three methods of preparation of chlorine, bromine, and iodine. What methods of preparation of these elements cannot be used for the preparation of fluorine? Why not?

12. In terms of the van der Waals forces of attraction between the molecules, explain the trend in volatility in the halogens.

13. (a) What are the two usual ways by which Cl^-, Br^-, and I^- may be oxidized to Cl_2, Br_2 and I_2? (b) Can F^- be chemically oxidized to F_2? Explain. (c) How is F_2 usually prepared?

14. (a) Name the following: $NaIO_3$, $Ca(OCl)_2$, aqueous solution of HBr, KIO, $KClO_4$. (b) Write the formulas of hypobromous acid, sodium hypochlorite, barium periodate, hydroiodic acid, and potassium chlorate.

15. Indicate the structures of the following oxychloride ions: ClO^-, ClO_2^-, ClO_3^-, ClO_4^-.

16. Predict the shape of the BrF_4^- ion. Where would you expect to find the two pairs of nonbonding electrons? Why?

17. How do you account for the facts that: (a) hydrofluoric acid is a weak acid but hydrochloric acid is a strong acid; (b) the polarity of hydrogen halides decreases as the electronegativity of the halogen decreases; and

18. Describe a method of preparation of an acid of chlorine in the $-I$, I, III, and V oxidation states.

19. (a) Compare the electrode potentials for the reduction of each halogen element to its respective halide ion. (b) Show why these data are consistent with the decreasing oxidizing abilities of the group.

20. What is the distinction among covalent radius, ionic radius, and van der Waals radius of halogen atoms?

21. Calculate the weights of the five halogens required to oxidize 10 g of hydrogen from the 0 to the I oxidation state.

22. Calculate the weight of calcium that can be oxidized with 10 g of fluorine. (See Chapter 3.)

23. Show by equations and oxidation numbers how oxygen can be oxidized by fluorine. Indicate the oxidation state changes.

24. The melting points and boiling points of H_2O, NH_3, and HF are much higher than might be expected from comparison of the hydrogen compounds in subsequent periods of the same families. This is attributed largely to hydrogen bonding. Explain why hydrogen bonding is less important in the third period representatives of these groups—that is, in H_2S, PH_3, and HCl.

25. In the series of oxyacids of chlorine, including $HClO$, $HClO_2$, $HClO_3$, and $HClO_4$, acid strength increases as the oxidation state of the chlorine increases. Explain.

SPECIAL PROBLEMS

1. Fill in the blanks with the appropriate halogen or halogen compound.
   (a) _____ etches glass.
   (b) _____ is a yellow gas.
   (c) _____ is a reddish-brown liquid.
   (d) _____ sublimes.
   (e) _____ was the first to be discovered.
   (f) _____ is the most abundant halogen.
   (g) _____ is a very dark solid.
   (h) _____ is the most reactive halogen.
   (i) _____ may occur free in nature.

2. Using equations to illustrate, show the preparation of each of the following: (a) bromine from seawater, (b) fluorine from calcium fluoride, (c) hydrogen fluoride from calcium fluoride, (d) hydrogen iodide from sodium iodide.

3. What are the products of the following electrolysis processes? Use equations to illustrate the reaction in each case. The electrolyte is (a) fused NaCl, (b) water solution of NaCl.

4. Only a few elements are liquid at or near room temperature. Bromine and mercury are among these. Mercury finds frequent application in thermometers and manometers. (a) Discuss the possible utility of bromine as a substitute liquid in these instruments. (b) Do the same for water.

5. Compare the following species: SF_6, BrF_5, BrF_4^-. Which of these would be polar and which would be nonpolar?

6. Below are two statements dealing with the comparative ionization energy of fluorine and iodine. Read them and decide which is better stated. Explain your choice.
   (a) "The ionization energy of iodine is much lower than that of fluorine; from this we conclude that the electrons in the valence shell of iodine are farther from the nucleus, hence, less tightly bound."
   (b) "Because the valence electrons of iodine are farther from the nucleus than those of fluorine, its ionization energy is lower."

7. Using periodic table considerations, explain the following physical properties of solid iodine: dark purple crystals that are lustrous, having some electrical conductivity that increases with temperature.

8. *Special Project* (will require a search for appropriate data) Predict whether

the following reactions are possible. Where applicable, complete and balance.

(a) $Cl_2 + KBr \longrightarrow$

(b) $I_2 + NaCl \longrightarrow$

(c) $Br_2 + NaI \longrightarrow$

(d) $Cl_2 + AgNO_3 \longrightarrow$

(e) $Cl^- + AgNO_3 \longrightarrow$

(f) $CaF_2 + H_2SO_4 + heat \longrightarrow$

(g) $Fe + Br_2 \longrightarrow$

(h) $FeCl_2 + Cl_2 \longrightarrow$

(i) $NaCl + CaBr_2 \longrightarrow$

(j) $F_2 + H_2 \longrightarrow$

(k) $Br_2 + H_2S \longrightarrow$

(l) $Ca + Br_2 \longrightarrow$

(m) $I_2 + Au \longrightarrow$

(n) $Pt + Br_2 \longrightarrow$

(o) $F_2 + H_2O \longrightarrow$

(p) $F_2 + I_2 \longrightarrow$

(q) $Cl_2 + KF \longrightarrow$

REFERENCES

DOUGLAS, B. E., and McDANIEL, D. H., *Concepts and Models of Inorganic Chemistry,* Xerox College Publishing, Lexington, Mass., 1965.

JOHNSON, R. C., *Introductory Descriptive Chemistry,* Benjamin, New York, 1966.

PHILLIPS, C. S. G., and WILLIAMS, R. J. P., *Inorganic Chemistry, Vol. 1, Principles and Non-Metals,* Oxford, New York, 1965.

REMY, H., *Treatise on Inorganic Chemistry,* Vol. 1 and 2, Elsevier, New York, 1956.

SNYDER, A. E., "Desalting Water by Freezing," *Sci. Amer.,* **207**, no. 6, 41 (1962).

WELLS, A. F., *Structural Inorganic Chemistry,* 3rd edition, Oxford, New York, 1962.

WILLIAMS, L. P., "Humphry Davy," *Sci. Amer.,* **202**, no. 6, 106 (1960).

PART THREE

ENERGY AND THE
STATES OF MATTER

 **ENERGY AND MOLECULES OF GASES**

Common air, when reduced to half its wonted extent, obtained near about twice as forcible a spring as it had before; so this thus comprest air being further thrust into half this narrow room, obtained thereby a spring about as strong as that it last had, and consequently four times as strong as that of the common air.

Robert Boyle (1627–1691)

The search for an understanding of the nature of the gaseous state of matter began early in the history of science, and still continues. The early work is a classic example of the deductive method of reasoning that led from the laws or habits of gases to a theory of the gaseous state—the kinetic-molecular theory. In its original form, the kinetic-molecular theory explained the behavior of gases in terms of the kinetic energy of motion of minute spheres without internal structure. The preceding chapters of this book have shown, however, that molecules are made up of atoms held together by bonding electrons, and have structures characteristic of the nature and arrangement of these atoms. Hence this chapter will later discuss the kind of mental models the scientist has made to interpret the nature of the internal make-up of molecules that makes it possible for them to hold energy.

THE GAS LAWS The springiness of gases, their compressibility and diffusibility, aroused the curiosity of the scientists of the 17th century. They sought to know the nature of this thing or substance that man could not see but which had such curious properties. The scientists' studies on gases led to general laws or statements of the habits of all types of gases, and to a theory of the nature of the gaseous state. The gas laws are known by the names of their discoverers: Boyle, Charles, Gay-Lussac, Dalton, and Graham.

Boyle's Law While studying the effect of pressure on the springiness and compressibility of air, Robert Boyle, in 1662, obtained data somewhat like that in Table 10.1 when he measured the change in volume with the change in pressure while keeping the amount of air and the temperature constant. The data of Table 10.1 are plotted in Figure 10.1. Comparison of the data in columns 1 and 2 shows that the volume decreases as the pressure increases.

Data similar to those of Table 10.1 can be obtained for any kind of gas;

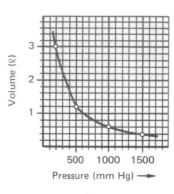

Figure 10.1 Graphic representation of the data in Table 10.1—effect on the volume of change in the external pressure of the gas at a constant temperature.

Table 10.1 Change in Volume with Change in Pressure for a Given Amount of Gas at a Constant Temperature

Pressure (mm Hg)	*Volume* (liters)	$P \times V$
200	3.00	600
500	1.20	600
1000	0.600	600
1500	0.400	600

they show that the qualitative effect of a decrease in volume when the pressure is increased is common to all gases.

A further study of the data in column 3 of Table 10.1 reveals that the product of the pressure times volume is a constant value:

$$PV = \text{constant value} = K \qquad \text{[amount and temperature constant]} \qquad (1)$$

or, for a given quantity of gas at two different pressures, P_1 and P_2:

$$P_1V_1 = P_2V_2, \qquad \text{or} \qquad \frac{V_1}{V_2} = \frac{P_2}{P_1} \qquad (2)$$

The results above may be summarized in the *quantitative* statement known as Boyle's law: *The volume which a gaseous substance occupies is inversely proportional to the pressure under which it is measured, provided that the temperature and the amount of gas are held constant.* The significance of the phrase "inversely proportional" may be better understood by dividing both sides of the experimental Equation (1) by P to obtain the equation in the following form:

$$V = K \times \frac{1}{P} \qquad \text{[amount and temperature constant]} \qquad (3)$$

This shows in symbols that the volume is proportional to the reciprocal of the pressure, that is, *inversely* proportional to the pressure, and that K is the proportionality constant. The value of K, however, depends upon the amount and kind of gas, upon the temperature, and, of course, upon the units used for P and V.

Methods of Measuring Gas Pressures: The Mercury Barometer

In the experiments recorded in Table 10.1, the pressure is given in millimeters of mercury. This unit of measurement implies that the pressure-measuring instrument was adapted from a mercury barometer. Such an instrument may be constructed by filling a glass tube (about 80 cm long and closed at one end) with mercury, then inverting it and thrusting the open end into a well of mercury. The mercury column falls until the weight of the mercury in the tube is exactly equal to the weight of a column of air of a cross section equal to that of the mercury column and extending from the surface of the mercury in the well to the top of the atmosphere. As the weight of the air changes, owing to changes in the atmospheric pressure, the level of the mercury in the tube moves up or down. The height of the mercury column, read off on a meter stick placed alongside, thus serves as a measure of the atmospheric pressure (see Figure 10.2).

Above the mercury in the tube is a vacuum called the Torricellian vacuum in honor of the Italian, Evangelista Torricelli, who, while a young student assistant of Galileo, made the first barometer in very much this way about 1643.

From the measurement of the height of the mercury column the gas can support with its pressure, we obtain the term *millimeters of mercury,* or *torr* (after Torricelli), as the pressure unit. Other units are also used frequently. The average barometer reading at sea level is 760 mm Hg (760 torr), and the pressure may be expressed in multiples or fractions of an *atmosphere* (abbreviated *atm*). A pressure, it may be remembered, is a *force per unit area*, and this is expressed as *dynes per square centimeter* or, in the common English units of pressure: *pounds per square inch* (abbreviated *psi*).

Experiments show that all gases expand when the temperature is raised (if the pressure is kept constant). To make this *qualitative* statement quantitative, experiments must be performed to show how much the volume of a given quantity of gas changes for a *measured* change in temperature at constant pressure. The results of four such experiments are given in Table 10.2.

By trying various mathematical operations with these data, we find that a graph of the volume against the temperature gives, for each case, a straight line, as shown in Figure 10.3. The slopes and intercepts of these lines differ according to the amount of gas used (*A* compared to *C*), the (constant) pressure under which the measurements are carried out (compare *A* with *B*), and the kind of gas used (*C* and *D*). The curious fact emerges, however, that if the curves are extended (extrapolated) to lower and lower temperatures,

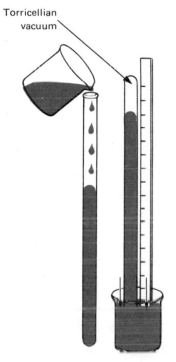

Torricellian vacuum

Figure 10.2 A simple barometer made by filling a glass tube with mercury and inverting it in a well of mercury.

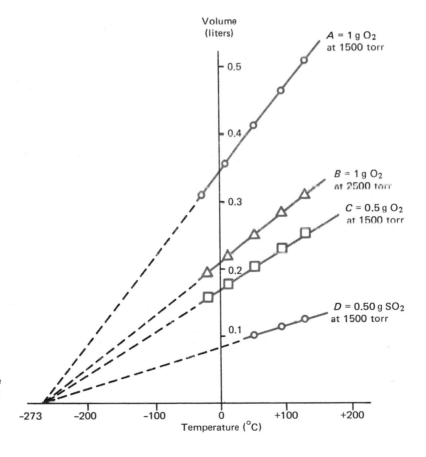

Figure 10.3 Graphic representation of the data in Table 10.2—effect on the volume of varying the temperature of a gas at constant pressure. The curves, when extrapolated, all cross the temperature axis at the same point.

Table 10.2 Change in Volume of Gases with Change in Temperature, for Constant Mass and Constant Pressure

Temperature (°C)	A Volume (liters) of 1g O_2 at 1500 torr	B Volume (liters) of 1 g O_2 at 2500 torr	C Volume (liters) of 0.5 g O_2 at 1500 torr	D Volume (liters) of 0.5 g SO_2 at 1500 torr
−33	0.312	0.187	0.156	
+7	0.364	0.218	0.182	
+47	0.416	0.250	0.208	0.104
+87	0.468	0.281	0.234	0.117
+127	0.520	0.312	0.260	0.130

all four meet at the same point, which lies on the temperature axis at −273°C (more precisely, −273.15°C). This means that if we could lower the temperature of a gas to −273°C, the gas would occupy zero volume, regardless of the kind of gas, the weight of it present, or the pressure under which it was measured. Of course, such a contraction to zero volume could never occur, because before such a low temperature was reached the gas would have changed to a liquid or a solid.

Absolute Temperature

Nevertheless, we should take advantage of such a uniformity of behavior of all gases at temperatures where they are still gases, and we do this by changing the zero of the temperature scale to −273°C. On this new *absolute temperature scale,* zero of temperature corresponds to zero of volume, and the *volume* becomes *directly proportional to the absolute temperature, provided that the quantity of gas and its pressure remains constant.* This italicized sentence is a statement of Charles' law; expressed in symbols, it is

$$V = K'T \quad \text{[amount and pressure constant]} \tag{4}$$

where T is an absolute temperature and K' is a proportionality constant. This is an equation for a straight line of zero intercept, as shown in Figure 10.4 for the data of Table 10.2. The value of the proportionality constant K' again depends upon other conditions, specified in the concluding phrase in the statement of the law: "provided that the quantity of gas and its pressure remains constant."

Two absolute temperature scales are in use. The more common one, known as the *Kelvin* scale, maintains the size of the degree the same as the size of the Celsius degree, but changes the zero point by 273, so that 0K = −273°C; 273K = 0°C; 373K = 100°C; etc. These values are given in the upper row of figures on the horizontal axis of Figure 10.4. The other scale is known as the *Rankine* scale, built on the Fahrenheit scale by shifting the zero point by 460 (precisely, 459.67). Thus 0°R = −460°F, 460°R = 0°F, 492°R = 32°F, 32°F = 0°C, and, of course, 0°R = 0K. The lower row of figures in Figure 10.4 gives the Rankine temperatures. The Rankine scale is frequently used by engineers.

A useful form of Equation (4) for the same quantity of gas at two different temperatures, but at the same pressure, is

$$\frac{V_1}{V_2} = \frac{T_1}{T_2} \tag{5}$$

Figure 10.4 The gas volumes of Table 10.2 plotted as a function of absolute temperature.

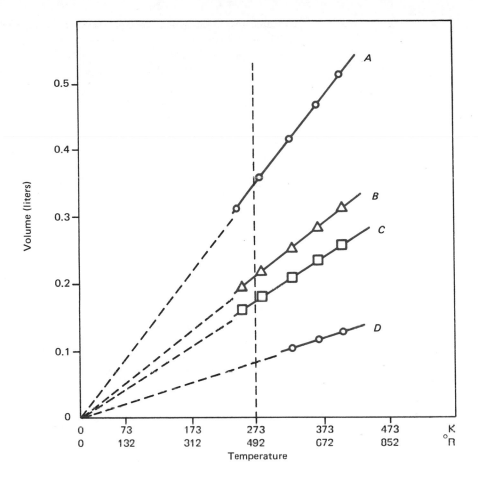

Volume (liters)

0.5

0.4

0.3

0.2

0.1

0

A

B

C

D

Temperature

| 0 | 73 | 173 | 273 | 373 | 473 | K |
| 0 | 132 | 312 | 492 | 672 | 852 | °F |

Boyle's law and Charles' law describe the dependence of volume upon the pressure and temperature for constant quantity of gas. What, now, is the dependence of volume upon the quantity, for fixed pressure and temperature?

The data necessary are already available in Table 10.2. If we choose the temperature 47°C and the pressure 1500 torr, we find the three sets of values in the first and second columns of Table 10.3. Comparison of the first two values in these two columns shows that the volume is directly proportional to the weight of oxygen present, since the ratio of volumes 0.416/0.208 (2:1) is the same as the ratio of weights, 1.00/0.50 (2:1). This is reflected in the third column, which shows that the density of oxygen (the weight per unit volume) is constant for the fixed temperature and pressure, 47°C and 1500 torr. The density of sulfur dioxide in these units is different from that of oxygen (third row).

However, if we use the data in the fourth and fifth columns to compute the density in moles per liter, we see that oxygen and sulfur dioxide have the same value. Since a mole represents a definite number of molecules, the implication is that the volume is determined by the number of molecules present, as if each molecule of a gas required a certain amount of space no matter what its weight or chemical nature. This was in fact suggested in 1811

Table 10.3 Volume and Density as a Function of Quantity of Gas at 47°C and 1500 torr

Quantity	Volume (liters)	Density (g/liter)	Molecular Weight	Moles	Density (moles/liter)
1.00 g O_2	0.416	2.404	32	0.0312	0.075
0.50 g O_2	0.208	2.404	32	0.0156	0.075
0.50 g SO_2	0.104	4.808	64	0.0078	0.075

by Amedeo Avogadro in his statement, often known as Avogadro's law, that *equal volumes of all gases, measured at the same temperature and pressure, contain the same number of molecules.*

The relationship suggested by the data for oxygen and sulfur dioxide is found to be general, and we write for the density in moles per liter:

$$\frac{n}{V} = \text{constant, or } V = K''n \quad \text{[constant temperature and pressure]} \tag{6}$$

where n is the number of moles present. Equation (6), like the two preceding equations, Equations (3) and (4), holds for all gases, with the value of K'' depending upon the values of the chosen temperature and pressure.

The General Gas Law   The three variables considered above—namely, pressure, temperature, and number of moles—are independent of each other. The experimenter can change the volume and any one of the three without affecting the values of the other two. Equations (3), (4), and (6) show that the volume is proportional to $1/P$, to T, and to n. It is an axiom of algebra that a quantity that is proportional to each of several independent variables is proportional to their product, and we write

$$V = R \times n \times T \times \frac{1}{P} \tag{7}$$

using the conventional symbol, R, for the over-all proportionality constant. Rearrangement gives

$$PV = nRT \tag{8}$$

We say that a perfect or ideal gas is one that *obeys this equation.*

Equation (8) connects all the experimental variables affecting the expansion or compression of a perfect gas. By rearrangement to the form

$$\frac{PV}{nT} = R \tag{9}$$

the product of pressure times volume divided by the product of moles times absolute temperature has a constant value, R, for a perfect gas.

Actually, under ordinary laboratory conditions, most *real* gases obey Equation (8) closely enough so that it can be used for all purposes in which high accuracy is not required. However, Equation (8) cannot be used for accurate calculations of gas volumes at high pressure and low temperature for reasons that will be discussed later. But for all gases, the value PV/nT approaches the constant value of R at low pressures and high temperatures.

Derivation of the Gas Laws
from the General Equation
$PV = nRT$

Since the general equation or law was obtained from Equations (3), (4), and (6), we must be able to derive or retrieve those equations from it.

Boyle's Law. If we keep n and T constant, then the equation $PV = nRT$ becomes $PV = K$ or $V = K \times 1/P$, which is Equation (3). This can be converted to the form

$$\frac{V_1}{V_2} = \frac{P_2}{P_1}$$

which is Equation (2).

PROBLEM Calculate the volume occupied by a certain sample of gas at 760 torr if at 700 torr and at the same temperature its volume is 1000 cc.

Solution Since n and T are fixed ("a certain sample"; "at the same temperature"), the conditions of Equation (2) are satisfied. Tabulation gives

$$P_1 = 700 \text{ torr} \qquad P_2 = 760 \text{ torr}$$

$$V_1 = 1000 \text{ cc} \qquad V_2 \text{ is to be found}$$

Substituting and rearranging,

$$V_2 = \frac{700 \text{ torr} \times 1000 \text{ cc}}{760 \text{ torr}} = 923 \text{ cc}$$

Charles' Law. If we keep n and P constant then equation $PV = nRT$ reduces to $V = K'T$, which is Equation (4) and which can be converted to the form

$$\frac{V_1}{V_2} = \frac{T_1}{T_2}$$

which is Equation (5).

PROBLEM Calculate the volume occupied by a certain sample of gas at 25°C if, at the same pressure and at 0°C, its volume is 1.14 liters.

Solution Since n and P are fixed ("a certain sample"; "at the same pressure") Equation (5) applies. Tabulation gives

$$V_1 = 1.14 \text{ liters} \qquad V_2 \text{ to be found}$$

$$T_1 = 0 + 273 = 273\text{K} \qquad T_2 = 25 + 273 = 298\text{K}$$

$$\frac{V_1 \times T_2}{T_1} = \frac{1.14 \times 298}{273} = V_2 = 1.24 \text{ liters}$$

Gay-Lussac's Law. If we keep n and V constant, then equation $PV = nRT$ reduces to $P = K''T$, which is a law derived by Gay-Lussac.

Volume as a Function of Both Pressure and Temperature at Constant Number of Moles. In the laboratory we often wish, for comparison with data of other investigators, to convert the volume of a gas measured at our laboratory conditions of pressure P_1 and temperature T_1 to the volume it would occupy at some other pressure P_2 and temperature T_2. Here the number of moles n is a constant and equation (7) becomes

$$PV = K'''T \qquad \text{or} \qquad \frac{PV}{T} = K'''$$

Thus, for values of V_1 at P_1 and T_1 as well as V_2 at P_2 and T_2,

$$\frac{P_1 V_1}{T_1} = \frac{P_2 V_2}{T_2} \tag{10}$$

Standard Conditions

To compare gas volumes, a standard set of pressure and temperature conditions have been chosen as *760 torr pressure and 0°C*. Any convenient set could have been chosen, but these are agreed upon and called standard conditions.

PROBLEM

Calculate the volume at standard conditions which will be occupied by a gas which measured 635 cc at 721 torr and 27°C.

Solution

Equation (10) applies, since no change in the number of moles is indicated. We have

$P_1 = 721$ torr $\qquad\qquad$ $P_2 = 760$ torr (standard pressure)

$T_1 = 27 + 273 = 300$K $\qquad$ $T_2 = 0 + 273 = 273$K (standard temperature)

$V_1 = 635$ cc $\qquad\qquad$ V_2 is to be found

Solving Equation (10) for V_2 and substituting,

$$V_2 = \frac{P_1 V_1 T_2}{P_2 T_1} = \frac{721 \times 635 \times 273}{760 \times 300} = 549 \text{ cc}$$

Answers to problems such as this should be checked to see that they are at least qualitatively correct. In this case both the pressure increase and the temperature decrease require a diminution of volume, by Boyle's law and Charles' law. Since $549 < 635$, the answer is of the correct magnitude.

Numerical Values of R

In all the solved problems above, the conditions were such that the constant R canceled. To use the general Equation (7), the numerical value of R must be known. To obtain this, we measure the volume V occupied by a known number of moles at a measured low pressure (P) and high temperature (T), where the gas behaves more nearly as a perfect gas, and substitute these values in Equation (9), $PV/nT = R$.

For $n = 1.00$ mole, $P = 1.00$ atm, $T = 273$K, we find $V = 22.4$ liters and

$$\frac{PV}{nT} = R = \frac{1 \text{ atm} \times 22.4 \text{ liters}}{1 \text{ mole} \times 273\text{K}} = 0.0821 \text{ liter-atm/K/mole}$$

The value of R and its units will depend upon the units used for P, V, n, and T (Table 12.4). In this chapter

$$R = 0.0821 \text{ liter-atm/K/mole}$$

will be used most often. Later we shall find that

$$R = 1.987 \text{ cal/K/mole}$$

will be used often.*

---

*R in English Units. Although a mole has been defined as an Avogadro number of particles (molecules), it is also shown that the weight of a mole of molecules is equal to its molecular weight in grams. This has prompted engineers to define a mole (usually called a pound-mole) as equal to the molecular weight in pounds. Such a pound-mole contains 2.73×10^{26} molecules and, of course, a pound-mole of a gas at a given temperature and pressure occupies a larger volume than a gram-mole. If the other variables in Equation (8) are also expressed in English units—pounds per square inch, cubic feet, degrees Rankine—the value of R is found to be 10.71 psi cubic feet per degree Rankine per pound-mole.

Table 10.4 Values and Units of R

Pressure	Volume	Temperature	Quantity	Value of R
Atmospheres	Liters	Degrees Kelvin	Gram-mole	0.0821 liter atm/deg/mole
Torr	Cubic cm	Degrees Kelvin	Gram-mole	62,400 torr cc/deg/mole
Dynes per sq cm	Cubic cm	Degrees Kelvin	Gram-mole	8.31×10^7 ergs/deg/mole
Dynes per sq cm	Cubic cm	Degrees Kelvin	Gram-mole	1.987 cal/deg/mole
Pounds per sq in.	Cubic ft	Degrees Rankine	Pound-mole*	10.71 psi ft^3/deg R/lb-mole

Calculations from the Perfect Gas Equation When R Is Known

The perfect gas law may be used to calculate a value for any one of the four variables if the other three are known.

PROBLEM   Calculate the pressure that will be exerted by 0.200 mole of a perfect gas in a 50.0-liter flask at 25°C.

Solution   Solving the perfect gas law for pressure, we obtain

$$P = \frac{nRT}{V} = \frac{(0.200 \text{ mole}) \times \left(0.0821 \frac{\text{atm} \times \text{liters}}{\text{moles} \times \text{K}}\right) \times (298\text{K})}{(50.0 \text{ liters})}$$

$$= 0.0979 \text{ atm}$$

The perfect gas law may also be used to calculate the gram-molecular weight of a gas. The perfect gas law equation must be revised as follows to calculate this value.

The number of moles is equal to the mass of gas, g, divided by the gram molecular weight of the gas, M, or $n = g/M$, so we may substitute g/M for n in the perfect gas law and obtain the expression

$$PV = \frac{g}{M}RT \qquad \text{or} \qquad M = \frac{gRT}{PV} \qquad (11)$$

This equation says that if we know the pressure, temperature, volume, and mass of a perfect gas we can calculate its gram-molecular weight.

PROBLEM   0.533 g of a perfect gas occupies a volume of 250 cc at 740 torr pressure and 25.0°C. Calculate its gram-molecular weight.

Solution

$$M = \frac{gRT}{PV} = \frac{(0.533 \text{ g}) \times \left(0.0821 \frac{\text{atm} \times \text{liters}}{\text{moles} \times \text{K}}\right) \times (298\text{K})}{\left(\frac{740}{760} \text{ atm}\right) \times (0.250 \text{ liter})} = 33.2 \text{ g/mole}$$

Dalton's Law and Graham's Law

In addition to the generalization obtained from experiments on the pressure-volume-temperature-amount relationship of gases, the results of two other sets of experiments, formalized as Dalton's law and Graham's law, throw light on the properties of gases.

Dalton's Law. As a result of a large number of experiments which he conducted with mixtures of gases, John Dalton found that, *in a mixture of* gases, the total pressure is equal to the sum of the pressures each gas would exert if it alone were present in the volume occupied by the mixture. Thus if p_1 and p_2 are the partial pressures that gases 1 and 2 would exert if they were alone, the total pressure is

$$P = p_1 + p_2 \qquad (12)$$

A frequent use of Dalton's law is to calculate the pressure p_2 of a gas when P and p_1 are known, as in the example below.

PROBLEM If oxygen is collected over water at 25°C and a total pressure of 740 torr, what is the pressure due to the oxygen (that is, its partial pressure) (Figure 10.5)?

Solution $P = 740$ torr (if the water levels inside and outside the test tube are equal)

$p_{water} = 24$ torr (This is the pressure of water vapor over liquid water at 25°C. See Appendix Table E.3.)

$p_{oxygen} = P - p_{water} = 740 - 24 = 716$ torr

The partial pressure of each individual gas is determined by the number of moles of it present; if n_1 and n_2 moles of each is present, then

$$P = \frac{n_1 RT}{V} + \frac{n_2 RT}{V} \qquad (13)$$

where V is the total volume of the mixture.

Graham's Law. The Scot, Thomas Graham, carried out a series of studies on the rate at which gases would pass through the fine holes in unglazed pottery. He found that the *volumes* of different gases, *measured at the same pressure and temperature*, which *passed through in a given time depended inversely upon the square root of the densities of the gases.* If the volume passing through in unit time is defined as the rate of diffusion, we may express this result as

$$\frac{\text{Rate gas}_1}{\text{Rate gas}_2} = \sqrt{\frac{\text{density gas}_2}{\text{density gas}_1}}$$

The general gas law shows that the density is directly proportional to the molecular weight at constant pressure and constant temperature [density $= g/V = PM/RT$, from Equation (11)], so that Graham's law may also be written

$$\frac{\text{Rate gas}_1}{\text{Rate gas}_2} = \sqrt{\frac{\frac{P}{RT}M_2}{\frac{P}{RT}M_1}} = \sqrt{\frac{M_2}{M_1}} \qquad (14)$$

where M_1 and M_2 are the molecular weights of gases 1 and 2, respectively. Note that the lighter gas diffuses more rapidly than the heavier.

PROBLEM Calculate how much more rapidly hydrogen will diffuse than oxygen if both gases are at the same temperature and pressure.

Solution Substituting molecular weights in Equation (14),

$$\frac{\text{Rate H}_2}{\text{Rate O}_2} = \sqrt{\frac{32}{2}} = \frac{4}{1}$$

Hydrogen diffuses four times more rapidly than oxygen.

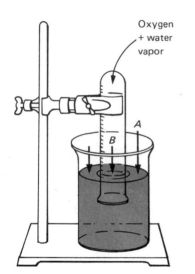

Oxygen + water vapor

A

B

Figure 10.5 Measuring the volume of a gas collected over a liquid. When the liquid levels are the same inside and outside the container, the pressure of the atmosphere (at A) is exactly balanced by the pressure inside the vessel (at B).

The Meaning of Laws in Science

The five laws just considered are merely general statements in regard to the conduct of gases as determined by experiment. Like all other scientific laws, they offer no explanation of the facts they state, nor do they place any restriction upon nature that compels obedience, as the laws of a country bind society. They are simply concise statements of what might be called the habits of nature as observed in experiment. But that such habits of nature exist, enabling us to make definite statements about how a gas will behave under specified conditions in an experiment we have not yet made, suggests that the universe is a place of order, and leads scientists to feel that a diligent search will uncover other laws describing other habits of nature.

THE KINETIC-MOLECULAR THEORY
Forming a Theory

Now that we have found it possible to describe in concise form (laws) the conduct of gases under varying conditions of temperature and pressure, many questions arise in our minds. Why do all gases expand and contract in the same way, regardless of their other widely differing properties? Why does heating a gas cause it to expand? How does a gas exert pressure?

To answer these questions, we begin by a process of imagining. We imagine that the similar conduct of all gases is probably due to some simple mechanical structure they all share, and we try to form a mental picture of this structure. *The process of constructing a mental picture of this kind is called forming a theory;* the structure which we imagine is often called a *model*. After constructing a theory and a model that answers all our questions, we then try to make the theory and the model serve as the basis for the prediction of new effects, and try to find new facts and laws that will subject the theory and model to severe and critical test. One of the most penetrating and far-reaching of all our theories, and one of the most fertile in prediction and forecast, is the famous *kinetic-molecular theory of gases.* So much evidence has now been collected in support of this theory that scientists no longer doubt that it is a truthful picture of things as they really are. To illustrate how a theory might evolve, let us take several important properties all gases have in common and imagine, if we can, the kind of internal structure a gas must have to display these properties. For convenience, we can prepare three columns: one for the observation, another to summarize the thinking in connection with it, and a third for a summarizing statement or postulate of the evolving theory.

A continuous fabric resists stretching.

Observation	Thinking	Postulate
Gases are infinitely *expansible.*	For this to be possible the internal structure of a gas must be discontinuous—composed of small pieces. If it were continuous, the fabric could be stretched so far and no farther, and there would be a limit to the expansibility.	1. All gases are made up of minute particles. Let us call them molecules; in a few cases these particles are atoms.
Gases move from place to place in *diffusion.*	If the particles are in motion, they can move from place to place. Since gases move in all directions, the motion of the particles must be in all directions.	2. The molecules are in constant chaotic motion.

225 ENERGY AND MOLECULES OF GASES

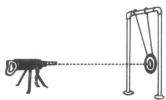

Individual particles may be separated easily.

Observation	Thinking	Postulate
Gases are highly *compressible*.	There must be a tremendous amount of empty space available for compressing the gas.	3. The molecules are very small compared to the distances between them.
The gas exerts a *pressure*, that is, a force, on the walls of the container, or on a piston in a cylinder containing the gas.	This would be easy to explain if the gas were continuous, since the gas, like a compressed spring, would push back on the piston.	
	It could be explained on the basis of moving particles if the particles bombard the piston, exerting a force somewhat like that produced by a stream of machine-gun bullets striking a target.	4. Pressure results from the bombardment of the walls of the container by the moving molecules.
The pressure remains *constant over long periods of time*, if the temperature, volume, and amount of gas remain constant.	The average speed and average kinetic energy of the particles must remain constant under these conditions, even though collisions between particles must be occurring constantly.	5. No energy is lost by the molecules as a result of collisions; that is, all collisions between molecules and molecules and between molecules and the walls are elastic.
At constant volume and constant amount of gas, the pressure is directly proportional to the absolute temperature (Gay-Lussac's law—a quantitative statement).	For the pressure to increase, either the frequency of collisions with the wall must increase, or else the force of each collision must increase, or both. Since it is unreasonable to think that the mass of the molecule changes with change in temperature, we guess that the speed of motion changes. However, if the speed were proportional to the temperature, both the frequency of collision and the force of each collision would change, and the pressure would increase as the square of the temperature. To make the pressure proportional to the first power of the temperature, the speed must be proportional to the square root of the temperature.	6. The kinetic energy, $\frac{1}{2}mv_2$, is proportional to the absolute temperature.

A stream of bullets directed at a suspended target pushes the target back.

A stream of molecules directed at a piston tends to push the piston back. An opposing force on the piston required to keep the piston from moving, when divided by the area of the piston, measures the pressure of the gas.

The postulates given above constitute a formal statement of the kinetic-molecular theory. Although these postulates were developed much more extensively and convincingly than we have done here, nevertheless the treatment given above is illustrative of a scientist's approach to model building or theory making.

Molecular Motion

The implications of the kinetic-molecular theory are many and exciting. Not only does it postulate the existence of tiny particles, each of which must have a definite mass, but it also postulates that the particles are in constant motion, having velocities that increase as the temperature increases.

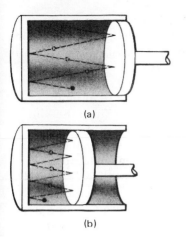

Figure 10.6 Illustrating the motion of gas molecules in a container.

The ceaseless random motion, or thermal agitation, of the molecules is really *heat* itself, or, rather, the *kinetic energy of this random motion of the molecules is heat.* A great many effects of various sorts find their explanation in terms of this heat motion. For instance, the tendency of a gas to spread, or *diffuse*, is attributed to the moving molecules darting here and there. The molecules travel at high velocity, like bullets, in straight lines except when they strike the walls of the containing vessel or bump into one another and bounce off like billiard balls. The pressure of a gas against the containing walls is the result of a continuous bombardment of the walls by the moving molecules, just as the flying fists of a fighter in the prize ring push his opponent back against the ropes. The magnitude of the pressure depends upon the rapidity of the bombardment—the number of collisions with the walls occurring in unit time—and the force with which the molecules strike. This concept of the pressure of a gas as the result of a bombardment of the walls by the moving molecules makes it possible to explain the observed pressure-volume-temperature relationships of gases as recorded in the gas laws.

Boyle's Law

Consider an idealized case in which there is only a single gas molecule in a container; let it be moving in a horizontal direction so that it bounces back and forth from one wall to the other [Figure 10.6(a)]. It will be moving at a certain speed, so it will take a definite time to get back and forth. Suppose that the size of the container and the speed are such that gas molecule makes 4000 collisions with the right-hand wall in a certain time interval—say, one second—and in so doing produces a certain pressure. Now suppose that the volume of the vessel is decreased to one half by moving the wall on the right closer to the wall on the left [Figure 10.6(b)]. In the same time interval as before, the molecule will make 8000 round trips instead of 4000, since it has only half as far to go. Therefore, it will make twice as many collisions with the right-hand wall as before, and the pressure, which depends upon the number of collisions in a given time, will be twice as great. This is Boyle's law: that halving the volume has doubled the pressure, or *the volume is inversely proportional to the pressure.*

Gay-Lussac's Law and Charles' Law

According to the sixth postulate of the kinetic-molecular theory, the kinetic energy is proportional to the absolute temperature. The kinetic energy of any moving particle is given by the expression $\frac{1}{2}mv^2$, where m is the mass and v the speed. Since experiment shows that mass is not altered by a change in temperature, any change in the kinetic energy of a molecule that is produced by a change in temperature must come about by a change in the speed with which the molecule is moving. If we return to the situation pictured in Figure 10.6 (a) and raise the temperature of the gas, the molecule at the higher temperature will make the transit from wall to wall in a shorter time than before. Furthermore, since the molecule is moving faster, it will hit harder. As a result of these two effects (an increased frequency of impact and a greater impact), the pressure on the walls is increased at the higher temperature. This is Gay-Lussac's law: that *the pressure of a gas is proportional to the absolute temperature, at constant volume.* If the pressure is to be kept constant, then to compensate for the increased speed of the molecules, the walls must be drawn farther apart so they will not be hit as often. This is Charles' law: that *if the pressure is kept constant, the volume is proportional to the absolute temperature.*

Dalton's Law Since a gas volume is made up largely of free space, it follows that in a mixture of gases, the molecules of each type of gas are acting independently of the molecules of any other type. Thus, *the partial pressure exerted by a single gas is that which it would exert if it alone occupied the total volume.*

Graham's Law The kinetic-molecular theory postulates that all molecules are in motion and that the kinetic energy increases as temperature increases. Since the kinetic energy depends upon v^2, the square of the molecular speed, the molecular speed must increase also. The question immediately arises: Do all the molecules in a sample of a certain gas have the same speed at a given temperature? Surely not, for collisions among molecules will result in some molecules increasing in speed while others decrease in speed. As a consequence, the molecules of a gas sample will have a variety of speeds centered around some *average value.* It is the corresponding *average kinetic energy* to which the postulate of the kinetic-molecular theory refers. James Clerk Maxwell proved from the postulates of the theory that at any one temperature, the average kinetic energy of the molecules of all gases is the same; specifically, for any two gases at the same temperature, the average kinetic energy of the molecules of one gas is equal to the average kinetic energy of the molecules of the other. This is true regardless of how large or small the masses of the two different molecules may be:

$$\tfrac{1}{2}m_1{v_1}^2 = \tfrac{1}{2}m_2{v_2}^2$$

If m_1 is small, v_1 must be relatively large; if m_2 is large, v_2 is relatively small. Solving for v_1 and v_2, we obtain

$$\frac{{v_1}^2}{{v_2}^2} = \frac{m_2}{m_1} \qquad \text{or} \qquad \frac{v_1}{v_2} = \frac{\sqrt{m_2}}{\sqrt{m_1}}$$

Since the molecules continually collide with one another, they do not travel far between collisions. Nevertheless, it is reasonable to suppose that the rate at which the molecules work forward in a particular direction depends upon the molecular speeds. Accepting this, the rate of diffusion should be proportional to the speeds, and we obtain Graham's law: *The rate of diffusion is inversely proportional to the square root of the masses of the molecules, or the molecular weights.*

Distribution of Molecular Speeds We have seen above that the molecules in a sample of gas at a given temperature have a *distribution of molecular speeds* about an average value. It is possible to measure the speeds (and kinetic energies) of gas molecules and to estimate the number of molecules having a given speed. Figure 10.7 is a sketch of the apparatus for determining molecular speeds. The results of such measurements are summarized in the graph of Figure 10.8, in which the number of molecules having a given speed is plotted against speed. This figure shows

---

THE KINETIC-MOLECULAR THEORY AND PROPERTIES OF MOLECULES

The kinetic-molecular theory states or implies that molecules have the following properties:

mass — m
speed — v
momentum — mv
kinetic energy — $\tfrac{1}{2}mv^2$

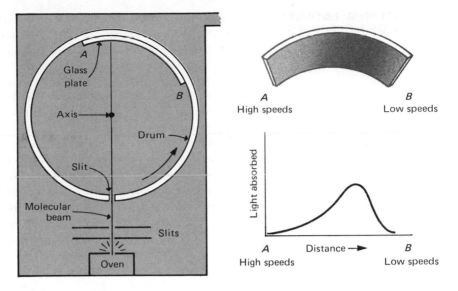

Figure 10.7 Determining the distribution of molecular speeds. Atoms of a metal escape from an oven into a vacuum. Those not going in the right direction are blocked off by a series of slits, and the molecular beam enters the slit of a rotating drum each time the opening of the drum passes the beam. The atoms impinge upon a glass plate *AB* opposite the opening. If a slowly moving atom enters the drum, the drum will have turned a long way before the atom meets the glass, and it will land near *B* on the plate. A fast-moving atom will travel across the drum very quickly and will land near *A* before the drum has turned much. After an experiment the plate is examined in a bright light. The light absorbed at different positions on the plate measures the relative numbers of metal atoms deposited at these positions. A plot of light absorbed against distance from *A* to *B* is obtained, which is the reverse of that given in Figure 10.8.

that most molecules have speeds near the average value, but a few have considerably higher-than-average and few have considerably lower-than average speeds.

Figure 10.8 also shows the distribution of speeds for the gas at a higher temperature. Increasing the temperature causes a general shift of the speed distribution toward one with more molecules having high speeds. Moreover, at the higher temperature, the distribution is more random—there are fewer molecules having speeds near the average value and a more uniform distribution of speeds among the molecules.

The Scientific Method

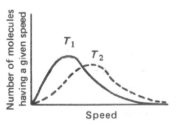

Figure 10.8 Distribution of speeds in a gas. At the higher temperature, T_2, the maximum is lower and the curve is broader than at T_1.

What we have been saying in the last few pages is a fine illustration of the application of the scientific method. During the past two or three centuries, measurements of various sorts were made in connection with gases; thus, *facts* were gathered, and this knowledge was organized and formulated in *laws*. Then a *theory*, the kinetic-molecular theory, was invented to explain the laws, and to give a plausible reason for the various ways in which gases behave. The theory was successful from the beginning, and was fortunate in having such masters as Rumford, Joule, Clausius, Maxwell, Kelvin, Boltzmann, Jeans, and Langmuir to nurse it along and develop it. The theory suggested further pioneering experimental work, and was itself gradually improved and strengthened until now it gives us a convincing and deeply satisfying insight into the nature of material-particle behavior.

The kinetic-molecular theory has assumed that gases are made up of molecules having no internal structure. But molecules are made up of atoms which, in turn, have nuclei and electrons. The presence of positive nuclei and negative electrons produces deviations from the general gas law because they generate attractive forces between the molecules; this feature of real gases has been discussed in Chapter 5 and will appear again in Chapter 11. The internal structure also permits molecules to absorb and hold energy internally in addition to the kinetic energy of motion (translational energy) of molecules as a whole, which is the essence of the kinetic-molecular theory. A close study indicates that important aspects of the energy of a molecule can be described in terms of four components or modes of motion: a component due to the translational motion of the molecule, a component due to the vibrations of the atoms with respect to each other in the molecule, a component due to the rotation of the molecule around its center of mass, and a component related to the electron arrangement in the molecule. Moreover, molecules absorb only discrete quantities of energy in increasing their vibrational, rotational, and electronic energies.

Absorption Spectrum

If the internal energy of the molecules is changed by absorption of light, then the light quantum absorbed must raise the energy by an amount corresponding to the separations of the electronic, vibrational, or rotational levels. Since there are many energy levels, quanta of several different magnitudes may be absorbed, producing an absorption *spectrum*. Examination of the spectra of molecules reveals that the spacing between the energy levels in molecules is greatest between electronic levels and least between rotational levels.

Table 10.5 Energy Required and Wavelengths of Photons Needed for Electronic, Vibrational, and Rotational Transitions

Type of Transition	*Energy Required* (kcal/mole)	*Wavelength of Photon Needed* (A)
Electronic Excitation		
In general	10–300	1000–8000 Ultraviolet—visible
Electrons in several types of bonds:		
C—C		Below 1600
C≡C		1625
C≡C		1775
C═O		1875
Vibrational Excitation		
In general	3–8	10,000–500,000 Visble—near infrared
Atoms in several types of molecules:		
O—H		28,000–32,000
C═C		60,000–62,000
C═O		54,000–61,000
Rotational Excitation		
In general	1	5,000,000–3×10^9 Near infrared—microwave

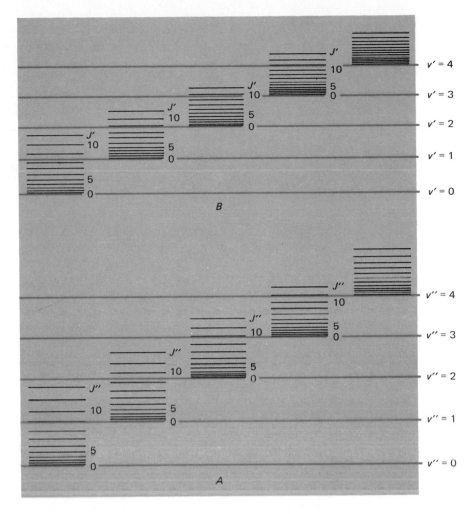

Figure 10.9 Schematic diagram of energy levels in a molecule. *A* and *B* represent different levels of electronic energies; *v'* and *v"* represent different levels of vibrational energy in the two electronic states. *J'* and *J"* represent rotational levels. Transitions from one of the *B* levels to one of the *A* levels will produce a spectral line in the ultraviolet; transition within the same electronic state from a higher vibrational level to a lower level will produce a line in the infrared.

For example, for the carbon monoxide molecule, CO, the energy level separation is of the order of 10^{-12} ergs for electronic levels, 10^{-13} ergs for vibrational levels, and 10^{-16} ergs for rotational energy level separation.

The data in Table 10.5 give the energy required and the wavelength of photons needed for various electronic energy transitions, vibrational energy transitions, and rotational energy transitions. The relative separations between the levels are illustrated in Figure 10.9.

The data in Table 10.5 tell us that many molecules may be raised to higher rotational energy levels by absorbing photons in the far infrared or microwave portion of the spectrum (because microwave photons have energies of about 10^{-16} erg, corresponding to the energy difference between rotational levels). If infrared photons (energy $\sim 10^{-13}$ erg) are absorbed, many molecules may be raised to higher vibrational energy levels. But in being elevated to a higher vibrational level, the molecule also may change its rotational energy level. This would mean that the molecule would change simultaneously in its vibrational amplitude and its rotational velocity.

When molecules absorb visible or ultraviolet light, electrons within them are raised to higher energy levels. Since this visible or ultraviolet light has high enough energy to cause electronic transitions, it can also change the vibrational frequency of the atoms and the rotational velocity of the molecule.

Types of Rotational and Vibrational Motion

It is of interest to picture the motions of the molecule as it absorbs energy. In the simple diatomic molecule CO, the rotational motion is about axes through the center of mass of the molecule perpendicular to the line joining the carbon and oxygen atoms. There are two such axes—in the plane of the paper and perpendicular to the plane of the paper. The vibrational motion is a stretching and compressing of the bond between the two atoms.

For more complicated molecules, both bending and stretching vibrations are possible. Certain of these are known as the fundamental modes of vibration, and all actual vibrations may be described as mixtures of these fundamental modes. The fundamental modes for the linear triatomic molecule, carbon dioxide, are shown in Figure 10.10.

Nonlinear molecules, such as H_2O or NH_3, have similar bending and stretching vibrations and three mutually perpendicular axes about which rotation can occur.

Figure 10.10 Rotational and vibrational motions of carbon dioxide. (a) Translation of center of gravity; (b) rotation about two perpendicular axes; (c) vibration in four normal modes.

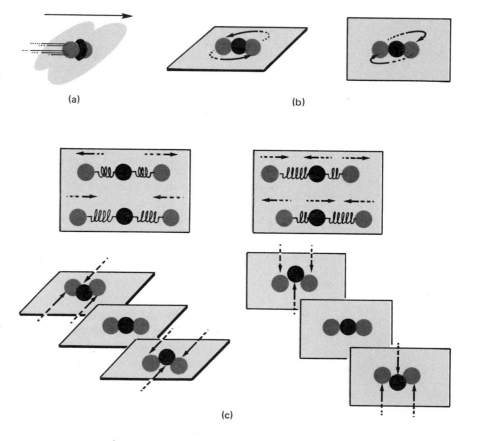

(a)

(b)

(c)

Figure 10.11 Bond angles and bond distances determined from microwave spectra.

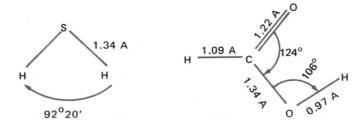

Spectra and Molecular Structure

By observing the spectrum of a molecule, scientists learn much about molecular structure. For example, from the microwave spectra of simple molecules, it is possible to learn not only the allowed rotational energies but often the molecular geometry, including the bond distances and the bond angles. This is illustrated in Figure 10.11. From the visible and ultraviolet spectrum of a molecule, information concerning the types of bonds present in the molecule sometimes is obtained. For example, the double bonds of the carbonyl group

$$\diagdown C=O$$

absorb ultraviolet light at wavelengths close to 1875A. The energy corresponding to this wavelength, $hc/1875 \times 10^{-8}$ erg, is just sufficient to promote an electron in the π bond to a higher energy level. Absorption at this wavelength, therefore, is a "trademark" of compounds containing the C=O structure.

Infrared spectra of molecules give data on vibrational energy levels and equally important information on the kinds and strengths of the bonds present. In fact, infrared spectra provide chemists with one of their most useful tools for studying the structure of molecules and for identifying substances, as illustrated in the following paragraph.

Infrared Spectrophotometry

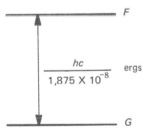

Figure 10.12 Ground state, G, and first excited electronic state of π bond in

$$\diagdown C=O$$

Infrared spectrophotometers are instruments in which a beam of infrared radiation of known wavelength and intensity is allowed to fall on a cell containing the sample being studied, and the intensity of the radiation after passing through the cell is measured. The difference between the incident and the transmitted intensity represents the light absorbed, and this magnitude is displayed upon a recorder. The wavelength of the incident beam is now slowly varied, so that the per cent of light absorbed is automatically traced out by the recorder as a function of wavelength.

Such a recorder graph is shown in Figure 10.13. The valleys correspond to wavelengths at which the molecule shows absorption. The wavelength of the valleys (transposed to represent $h\nu$) gives the energy-level difference between the vibrational-rotational levels in the molecules. Each valley is associated with the presence in the molecule of a certain structure or bond. Molecules containing the O—H bond absorb photons having wavelengths in the 28,000–32,000 A range; those containing the C=C structure absorb in the 60,000–62,000 A wavelength range; and those containing the C=O group absorb in the 54,000–61,000 A region. The chemist soon learns to know in general what absorption position in the infrared region to expect from a certain group of atoms in a given environment and can then use the infrared spectrum to determine the presence or absence of various groups of atoms in a given sample.

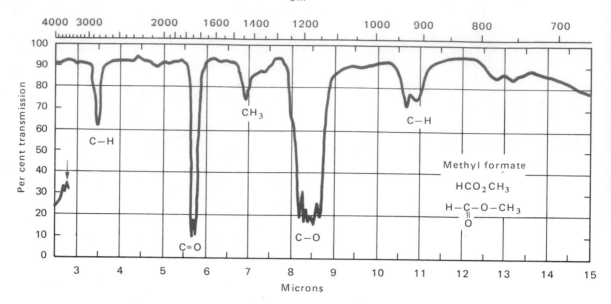

Figure 10.13 The infrared absorption spectrum of methyl formate, HCO_2CH_3.

Bond-Breaking Energies

We have described how molecules can absorb energy and move faster or undergo transitions to higher energy states of electronic energies, vibrational energies, or rotational energies. The next question to be asked is: Can a molecule absorb enough energy to break the bonds or to ionize one or more atoms? We are now prepared to answer this question from data given in Table 6.1 on bond strengths. Since the energy of a photon is related to its wavelength by $E = hc/\lambda$ (Chapter 4), we find these values for the energy associated with visible light:

	λ *Wavelength* (A)	*Photon Energy* (ergs)	*Energy per Mole of Photons* (kcal)
Violet light	λ 4000	49.7×10^{-13}	71.5
Red light	λ 7000	28.4×10^{-13}	40.9

Thus, if a molecule has energy levels that can and do absorb radiation having greater energy than the bond strength, we expect the bond to break. Iodine molecules, for example, have bond strengths of 36,000 cal/mole and are broken by radiation of $\lambda = 7000$ A. But a molecule must absorb energy before it can be used to break bonds, and a sufficient amount of that energy must be concentrated in the vibration frequency of the particular bond to cause the bond to break.

What do X rays and γ rays do to molecules? The answer is that their wavelengths are so short, and therefore their energies so high, that both ionization and bond breaking occurs when they are absorbed. This is the main cause of radiation damage of biological tissue by X rays and γ rays.

HEAT CAPACITY The change of energy in a molecule from one quantum state to another by absorption of radiation is specific; each quantum absorbed, $h\nu$, changes the energy of a single molecule from a lower to a higher energy state. The energy of a group of molecules can also be changed by adding heat to the group. For monatomic gases, the heat added goes only into increasing the kinetic energy of the molecules. But for more complex molecules, the energy distributes itself among translational, rotational, and vibrational energies. (Electronic excitation may also occur but, in view of the large quantum jumps, this is not significant except for high temperatures.) Since the rotational and vibrational energies represent additional ways in which complex molecules can hold energy, compared to the translational way (only) for a monatomic gas, it takes more heat to raise the temperature of a given number of complex molecules by a fixed amount than it does for the same number of monatomic molecules. In other words, the complex molecules have a greater capacity for energy (heat). To be quantitative, we speak of the molar heat capacity, defined as the *amount of heat necessary to raise the temperature of one mole of substance by 1°C.* For a monatomic gas, the molar heat capacity at constant volume is $\frac{3}{2}R$; for diatomic gases, it is approximately $\frac{5}{2}R$ at low temperatures and increases toward $\frac{7}{2}R$ at high temperatures as a result of the increased energy of molecular vibration.

PLASMA At several thousand degrees, gas molecules are traveling with such high energies than on collision they drive off electrons. Electrons may also be driven off in other processes involving the absorption of a great deal of energy. The term *plasma* is used for a high concentration of particles that have lost one or more electrons.

At very high temperatures, 10,000,000°C and higher, the plasma particles are nearly bare nuclei and have so much translational energy that, on collision, their nuclei approach within 10^{-12} cm of each other. Then the powerful nuclear attractive forces take effect and cause nuclear fusion processes to occur. Much of the material in the sun and stars is thought to be in the plasma state, and the nuclear reactions there produce elements of low atomic weight. The energy released in these reactions accounts for the stars' brilliance.

Atmospheric Stability and Air Quality. Each day, the average male human requires roughly 30 pounds of air, 2.75 pounds of food, and 4.50 pounds of water. A man can live about five weeks without food, about five days without water, and about five minutes without air. The senses of sight, smell, and hearing also are affected by air.

Air, which, as we know, is made up of four important components—oxygen (20.9% of dry air), nitrogen (78.1%), water (in variable amounts depending on atmospheric conditions), and carbon dioxide (0.032%)—is available to us from our atmosphere, an approximately 10-mile-thick gaseous envelope that surrounds the earth and fades out gradually at higher altitudes. Most planets and a few satellites of planets in the solar system have atmospheres. The existence and stability of an atmosphere depends primarily on the planet's gravitational field strength and, to a lesser extent, on its temperature.

To escape from this gravitational field, a gaseous molecule of the atmosphere must acquire the escape velocity. The earth's escape velocity is just under 7 miles/sec; the moon's escape velocity is about 1.5 miles/sec. The molecules present in air at ordinary temperatures move with average speeds of less than one half mile per second. We know, however, that there is a distribution of molecular velocities such that some molecules have greater and some less than the average velocity, and that should even a fraction of the molecules of the atmosphere acquire the escape velocity, the entire atmosphere would eventually escape.

Presumably this is what has happened on the moon, which has no atmosphere. The moon's escape velocity is so close to the average velocity of the molecules at its surface temperature that the faster-moving molecules in any atmosphere that did exist on the moon escaped. The remaining molecules, in attaining thermal equilibrium, reestablished the original distribution of molecular speeds, with the result that more molecules acquired the escape velocity. This continued until the atmosphere was virtually exhausted.

On the earth, however, the escape velocity is so much greater than the average velocity of the gaseous molecules in the atmosphere that the chances of escape, especially in the cold upper atmosphere, are remote. Consequently, there is little danger that the earth will lose its valuable air blanket.

A far greater danger to the stability of the atmosphere, in terms of its capacity to support life, comes from air pollution. Air pollution is defined as the presence in the outdoor atmosphere of one or more contaminants or combinations thereof, in such quantities and of such duration that they are or may tend to be injurious to human, animal, or plant life, or to property, or that they may interfere with the comfortable enjoyment of life or property or with the conduct of business.

Atmospheric contaminants are being poured into the air from domestic and industrial sources at rates that stagger the imagination. A 1966 U.S. Public Health Service study revealed that 72 million tons of carbon monoxide, 26 million tons of sulfur oxides, 13 million tons of nitrogen oxides, 19 million tons of hydrocarbons, and 12 million tons of particulate matter are added to the air over the continental U.S. each year. Motor vehicles add the largest fraction of the carbon monoxide (66 million tons), nitrogen oxides (6 million tons), and hydrocarbons (12 million tons). Power plants and industry adds the largest fraction of sulfur oxides (21 million tons) and particulate matter (9 million tons). Regulations that became effective in 1970 have reduced these quantities, but the situation calls for continued improvement.

All of these pollutants can be considered imminent or potential health hazards. Carbon monoxide is a known poison. It can produce a variety of symptoms ranging from mild discomfort through headache, visual difficulty, ataxia, paralysis, coma, and death. It combines with hemoglobin approximately 210 times more readily than oxygen. Estimates are that a person breathing air containing 100 parts per million of carbon monoxide for a period of time will reduce the oxygen carrying capacity of his blood by about 15 per cent. As the level of carbon monoxide rises above 300 parts per million, the person can literally smother from the lack of oxygen available to his tissue cells. In heavy traffic, the concentration of carbon monoxide in the air frequently exceeds 100 parts per million by a wide margin.

The sulfur oxides, most of which enter the air from the combustion of sulfur-containing fuels (especially coal), are highly irritating gases. One part per million of sulfur dioxide can produce physiological distress in man, most often as watery eyes and running nose. Sulfur dioxide also hinders the movement of air in and out of the lungs. Sulfur trioxide is more of an irritant than sulfur dioxide, a property that is enhanced by its reaction with water to give highly corrosive sulfuric acid.

The oxides of nitrogen (principally NO and NO_2), formed at the high temperatures of the operating automobile engine, are highly reactive molecules. In the presence of sunlight, they react with oxygen to produce ozone, O_3. Even in one-part-per-million concentration, ozone can cause bronchial irritation. In addition, ozone reacts with many hydrocarbon pollutants to produce substances known as secondary pollutants, many of which may be harmful to plant and animal life even at low concentrations.

Carbon dioxide pollution also constitutes a potential health hazard. As more and more fuel is burned, producing ever increasing amounts of carbon dioxide, and as millions of square miles of carbon-dioxide-absorbing vegetation are replaced by cities and highways, the atmospheric concentration of carbon dioxide will increase. This increase can result in an increase in the earth's temperature. It has been postulated that such a temperature increase could cause violent air circulation and much more violent storms than are currently seen.

Means are available to maintain a safe and stable air supply, but money, personal restraint, and commitment, as well as good sense, are needed to assure success.

SUMMARY   A study of gases reveals that all gases obey certain laws of nature that have become known as gas laws.

A theory, known as the kinetic-molecular theory, has been devised to explain the gas laws and to give a concept of the gaseous state of matter.

A general equation can be derived to make calculations of the pressure, volume, and temperature changes for n moles of a gas:

$$PV = nRT$$

The kinetic energy of a gas, $\frac{1}{2}mv^2$, is proportional to the absolute temperature.

In addition to translational energy, gas molecules can have rotational, vibrational, and electronic energy.

Rotational, vibrational, and electronic energies are not continuous but are quantized. The separations of the energy levels are indicated by the frequencies of lines in the band spectra (molecular spectra).

Interpretation of molecular spectra gives useful information about the identity and structure of molecules.

A collection of high-energy, charged particles is called plasma. At very high temperature (and therefore at high translational energies), fusion reactions of the light elements occur.

Gas laws and theory
Boyle's law
Charles' law
general gas law
Gay-Lussac's law
Dalton's law
Graham's law
kinetic-molecular theory

Spectra
infrared spectra
band spectrum
excitation spectrum
spectrophotometer

Energy
kinetic energy
electronic energy
vibrational energy
rotational energy
heat capacity

Properties of gases
volume
pressure
partial pressure
expansible
compressible
diffusion
ideal gas or perfect gas
elastic collision
barometer
atmosphere pressure
standard conditions
R (gas constant)
pound-mole
distribution of molecular speeds
plasma

QUESTIONS AND PROBLEMS

1. How does the kinetic-molecular theory account for the experimental observations that gases (a) are infinitely expansible, (b) are highly compressible, (c) diffuse readily, (d) exert a pressure, (e) exert a pressure that is constant at constant temperature, constant volume, and constant quantity, and (f) exert a pressure proportional to the absolute temperature at constant volume and constant quantity?

2. What experimental facts lead to the choice of $-273.15°C$ as the zero of temperature on the Kelvin scale?

3. On a single set of axes, draw curves showing the distribution of molecular speeds for three temperatures, T_1, T_2, and T_3, such that $T_3 > T_2 > T_1$. Label the axes.

4. What kinds of energy are possible in a monatomic gas? A diatomic gas?

5. What are the relative magnitudes of the quantum jumps in the several kinds of energy a polyatomic gas molecule may have? How do the relative wavelengths of the light absorbed vary for the different kinds of energy?

6. Predict the molar heat capacity for the following gases: (a) argon, (b) krypton, (c) iodine, (d) carbon dioxide, (e) carbon tetrachloride.

7. Calculate the volume at 1000 torr and 28°C of a quantity of gas that occupied 250 cc at 700 torr and 28°C.

8. Calculate the volume occupied at 87°C and 950 torr by a quantity of gas that occupied 20 liters at 27°C and 570 torr.

9. A sample of a gas fills a 200-cc container at 1.0 atm. of pressure. If temperature is held contant, what volume will the gas occupy at (a) 50 torr, (b) 2.00 atm., (c) 1.00×10^{-2} torr?

10. A sample of an ideal gas occupies 21.1 liters at 16°C and 800 torr. What will its volume be at 20°C and 500 torr?

11. One hundred cubic centimeters of oxygen are collected over water at a temperature of 25.0°C and a pressure of 760 torr. Calculate the volume the dry gas will occupy under standard conditions. (The vapor pressure of water at 25.0°C is 23.8 torr.)

12. The gauge pressure of an automobile tire is 29.0 lb/in² at 22°C. The tire's capacity is 50.0 liters. What would the gauge pressure be after a long trip during which the volume expands to 50.4 liters and the temperature of the enclosed air reaches 40°C? The pressure of the atmosphere remains constant at 14.7 lb/in².

13. Using modern vacuum techniques, it is possible to evacuate a system to the very low value of 10^{-10} torr. Calculate the volume occupied by 3.00×10^5 molecules of an ideal gas at 20°C and this pressure.

14. Sixteen grams of sulfur dioxide, SO_2, were measured at 20°C and 740 torr pressure. What was the volume found?

15. Calculate the molecular weight of the gas for which 12 lb occupied 35 ft³ at 40°F and 12 lb/in³ pressure.

16. The density of a gas is 3.54 g/l at 25°C and 1520 torr. (a) Find the density of the gas (g/l) at standard temperature and pressure. (b) Find the molecular weight of the gas.

17. Find the density (g/l) of argon at 100°C and 80 torr. Assume ideality.

18. Find the molecular weight and molecular formula of mercuric chloride from the following experimental data:

Weight of mercuric chloride (vaporized)	1.48 g
Temperature	410°C
Volume of flask	500 cc
Pressure of vapor	459 torr

19. What is the molecular weight of a certain gas if a 585-cc sample of it collected at 92°C and 730 torr weighs 1.50 g?

20. A flask that can withstand an internal pressure of 2500 torr, but no more, is filled with a gas at 21°C and 758 torr, then heated. At what temperature will it burst?

21. A flask of volume 1.2 liters is filled with carbon dioxide at room temperature to a pressure of 650 torr. A second flask, of volume 900 cc, is filled, also at room temperature, with nitrogen to a pressure of 800 torr. A stopcock connecting the two volumes is then opened and the gases allowed to mix at room temperature. What is the partial pressure of each gas in the final mixture, and what is the total pressure of the mixture?

22. Calculate the number of molecules present in a quantity of gas that occupies 26,880 cc at 546°C and 380 torr.

23. In a mercury barometer, the space above the column of liquid mercury contains only mercury vapor. At 26°C, the vapor pressure of mercury is 0.001691 torr. Calculate the concentration of mercury in atoms/cc under these conditions, assuming mercury to be an ideal monatomic gas.

24. At a given temperature, a gas diffuses at a rate of 34% of that of hydrogen at the same temperature. Find the molecular weight of the gas.

25. A certain gas diffuses through an apparatus in 10.0 minutes. The same volume of oxygen diffuses through the apparatus in 12.6 minutes. Calculate the molecular weight of the gas in question.

SPECIAL PROBLEMS

1. Prove that Avogadro's law follows logically from the general gas law.

2. How does the kinetic-molecular theory account for the quantitative information (a) that the volume of a given mass of gas is proportional to the absolute temperature at constant pressure; (b) that the pressure of a given mass of gas is inversely proportional to the volume at constant temperature; (c) that the partial pressure of a gas does not depend upon the pressure of other gases in the same container; (d) that the ratio of the rates of diffusion of two gases is inversely proportional to the ratio of the square roots of their molecular weights; (e) that the pressure of a gas, at constant volume and constant temperature, is proportional to the mass of gas present?

3. A sample of neon was collected by bubbling it through water. The volume of gas collected was 125 cc at 25°C, and the gas pressure inside the bottle was 740 torr. (a) Calculate the volume the dry neon would occupy at standard pressure and temperature. (b) Calculate the weight of neon in the 125-cc sample.

4. An iron tank of 3.0-ft³ capacity is filled with a mixture of nitrogen and oxygen. All the oxygen eventually reacts with the iron to form a solid iron oxide of negligible volume. The original pressure was 760 torr and the final pressure is 600 torr. (a) Find the volume of nitrogen remaining. (b) Calculate the partial pressures of the two gases initially and finally.

5. Calculate the atomic weight of the hypothetical element "statine" from the following data on the volumes of its gaseous compounds and their analyses:

Compound	Volume (liters) of 5 g	Temperature (°C)	Pressure (torr)	Per Cent "Statine"
With hydrogen	5.2	117	741	93.7
With oxygen	2.4	27	361	55.2
With oxygen	3.4	47	380	38.1
With carbon	11.1	247	95	76.7

6. A certain compound composed of the elements carbon, hydrogen, and nitrogen is a gas. A given amount of the compound is combined with exactly the right amount of oxygen needed for its combustion, the combustion products being carbon dioxide, water vapor, and nitrogen gas. The mixture before burning totaled 300 cc. After burning, 133 cc of carbon dioxide, 200 cc of water vapor, and 66 cc of nitrogen were produced at the same temperature and pressure. (a) How many cc of oxygen was needed for the combustion? (b) Find the molecular formula of the compound.

7. Titanium reacts with hot hydrochloric acid to liberate hydrogen. At 100°C and 720 torr, 2.42 l of hydrogen are produced by the action of 2.40 g of titanium. Calculate (a) the number of moles of hydrogen liberated; (b) the number of moles of titanium used. (c) Write the balanced net equation for the reaction.

8. Ten liters of ammonia and 10 l of oxygen are mixed and a reaction occurs according to the following equation:

$$4NH_3(g) + 5O_2(g) \longrightarrow 4NO(g) + 6H_2O(g)$$

At the end of the reaction, how many liters of each substance are present?

9. An organic compound composed only of carbon, hydrogen, and oxygen is a gas of density 0.420 g/l at 673 K and 200 torr. An 11.0 mg sample of it contains as much hydrogen as is contained in 9.0 mg of water, and as much carbon as is contained in 12.4 cc of carbon dioxide at 23°C and 740 torr. Find the molecular formula of the compound.

10. An ideal gas occupied an unknown volume at 600 torr. A sample of the gas was withdrawn and found to occupy 3.04 cc at a pressure of 1 atm. The pressure of the gas remaining in the original container was measured at 550 torr after the withdrawal of the sample. What was the volume of the original container? (Assume all temperatures remain constant.)

REFERENCES   BARROW, G. M., *The Structure of Molecules; An Introduction to Molecular Spectroscopy*, W. A. Benjamin, Menlo Park, Calif., 1963.

CAMPBELL, R. J., "Gas Energy Transfer Processes," *J. Chem. Educ.*, **45**, 156 (1968).

FITZGEREL, R. K., and VERHOEK, F. H., "The Law of Dulong and Petit," *J. Chem. Educ.*, **37**, 545–549 (1960).

HILDEBRAND, J. H., *An Introduction to Molecular Kinetic Theory*. Van Nostrand Reinhold, New York, 1963.

MASON, E. A., and EVANS, R. B., III, "Gases in Motion, Simple Demonstrations of Graham's Laws," *J. Chem. Educ.*, **46**, 358 (1969).

WOLFENDEN, J. H., "The Noble Gases and the Periodic Table," *J. Chem. Educ.*, **46**, 569 (1969).

Attractive Forces; Three States of Aggregation

The gas laws hold true for gases at low pressures and at ordinary and high temperatures. Thus under these conditions gases behave as ideal gases. But at high pressures deviations from the gas laws appear. Why do these deviations occur at high pressures rather than at low pressures? If gases behave ideally when the molecules are far apart (large volume), why should they deviate from ideal behavior when the molecules are close to each other? What is wrong with the model postulated in the kinetic-molecular theory? Could it be that tiny but significant attractive forces exist between gas particles? Could these eventually cause a gas to liquefy?

Deviations from Perfect-Gas Behavior

To correct the model, let us first examine how the behavior of a real gas differs from that predicted by the general gas law, $PV = nRT$. Let us measure corresponding values of P, V, and T for 1 mole of nitrogen, multiply $P \times V$ and divide by $R \times T$ to form the *function PV/RT*, known as the compressibility factor, and plot it against the pressure, as in Figure 11.1.

Since we chose 1 mole, PV/RT would always be 1 for an ideal gas, as indicated by the horizontal dotted line in the figure. The actual behavior does not give a horizontal straight line; at high pressures and lower temperatures, the curves deviate considerably from the ideal.

If we consider the curve for $-50°C$, it is seen that the nature of the deviation from the perfect-gas law is of two kinds: In one, there is a pressure region in which the compressibility factor of the gas falls below the perfect-gas curve; in this region the gas is more compressible than the perfect gas. In the second deviation, at higher pressures the compressibility factor changes

---

* "Thus far we have always considered molecules to be material points, and thus have introduced a simplification in the model which is at once in contradiction with the actual phenomena. Even a molecule in the simplest form, consisting of a single atom, must have a certain extension; the various forces exerted upon it by the other molecules can thus not be considered to act upon a single point." (The quotation is from van der Waals' doctoral dissertation "Over de Continuiteit van den Gas- en Vloeistof-Toestand," presented at Leiden, Holland, June 14, 1873.)

Figure 11.1 Compressibility factor of nitrogen as a function of pressure at three temperatures.

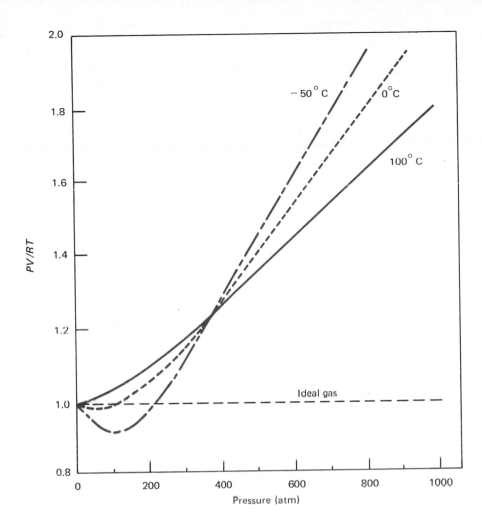

and gradually rises above the perfect-gas curve; in this region the gas is less compressible than the perfect gas. These deviations become less pronounced at higher temperatures, as shown by the curves for 0°C and 100°C, which are coming closer to the horizontal line for the ideal gas. If we can find an explanation for the two kinds of deviations, then we may have a deeper understanding of gas behavior than that provided by the kinetic-molecular theory.

Causes of the Deviations: Attractive Forces and Volume of Molecules

If the gas is more compressible than a perfect gas, one imagines that something in addition to the applied pressure is causing the volume to decrease. Could this be the result of attractive forces between molecules? If they are present, such forces would have the effect of slightly decreasing the volume. The presence of such forces would also cause the pressure of the real gas to be less than that of an ideal gas having no intermolecular attractive forces under the same conditions. As a hypothesis, we might suppose that intermolecular attractive forces are responsible for the observation that real gases are sometimes more compressible than perfect gases.

But what causes a gas to be less compressible than a perfect gas? This occurs at very high pressures and small volumes. Is it possible that, under these conditions, the "compressible space" in the gas has decreased to the point that a further increase in pressure cannot bring about a corresponding decrease in volume? This would happen if there were insufficient empty or free space between molecules. Another way to look at this is to imagine that at high pressures, the gas molecules themselves are occupying an appreciable fraction of the total volume of the gas, whereas at low pressures, the gas molecules are occupying only a very small fraction of the total volume. At low pressures, there is ample compressible space; at high pressures, compressible space is limited and the gas becomes less compressible.

Modification of the $PV = nRT$ Equation

If we assume that deviations from perfect-gas behavior arise from intermolecular attractive forces and from a decrease in compressible space brought about by the volume occupied by the molecules themselves, then perhaps we can modify the perfect-gas-law equation to account for these deviations. The modified equation should be more reliable than the perfect-gas law for predicting compressibility behavior of real gases. Several such equations have been devised; one such equation was developed by the Dutch chemist Johannes van der Waals and bears his name. The form of van der Waals equation is based upon the following new considerations about real gases:

1. The pressure of the real gas will be less than that of a perfect gas because of the intermolecular attractive forces. These require a pressure correction term, an^2/V^2, to be added to the real-gas pressure, where a is a constant, n, the number of moles, and V, the volume. Hence, the pressure term to be substituted in the perfect gas equation is

$$\left(P + \frac{an^2}{V^2}\right)$$

2. The volume of the real gas is made up of the free space and the volume the molecules appear to occupy (sometimes called the effective volume). If b is the effective volume occupied by one mole of molecules, the compressible volume of n moles of a real gas is

$$(V - nb)$$

The corrected equation then is

$$\left(P + \frac{an^2}{V^2}\right)(V - nb) = nRT$$

This is the van der Waals equation. The constants a and b can be obtained for each gas from experiments similar to those summarized in Figure 11.1.

When Does a Real Gas Fit the PV = nRT Equation? Using the van der Waals equation as a guide, we may say that whenever the correction terms an^2/V^2 and nb become small, compared to P and V, respectively, the gas will behave like a perfect gas. Both of these terms will become unimportant if the volume becomes very large while the number of moles of gas remains small. Of course, the volume becomes large at very low pressures and at high temperatures, so it is under these conditions that real gases most closely approach perfect-gas behavior.

The Van der Waals Constant a and Values of Boiling Points

We have already used the concept of intermolecular attraction to explain the fact that gases condense to liquids, and we have used boiling point as a measure of the strength of the attractive forces (Chapter 6). We now see

Table 11.1 Boiling Points and van der Waals Constants

Gas	Boiling Point (°C)	$a \left(\dfrac{\text{liters}^2 \text{ atm}}{\text{mole}^2}\right)$	$b \left(\dfrac{\text{liters}}{\text{mole}}\right)$	Gas	Boiling Point (°C)	$a \left(\dfrac{\text{liters}^2 \text{ atm}}{\text{mole}^2}\right)$	$b \left(\dfrac{\text{liters}}{\text{mole}}\right)$
He	−269	0.0341	0.0237	CH_4	−162	2.253	0.0428
Ne	−246	0.2107	0.0171	C_2H_6	−88	5.489	0.0683
Ar	−186	1.345	0.0322	C_3H_8	−42	8.664	0.0845
Kr	−153	2.318	0.0398	C_4H_{10}	0	14.47	0.1226
Xe	−107	4.194	0.0511	C_5H_{12}	+36	19.01	0.1460

that van der Waals' *a* is also a reliable measure of the strength of these forces; hence, boiling point and *a* should change in a parallel manner. Table 11.1 shows that they do; the higher the boiling point, the larger the value of *a*.

Further Evidence of Attractive Forces

Surface Tension. The molecules in the interior of a liquid are surrounded by other molecules, but those in the surface have no liquid molecules on top of them. The attractive forces on a surface molecule are thus different from those on an interior molecule, and give rise to a pull of the molecules toward the interior of the liquid, which tends to make the surface as small as possible for a given amount of liquid. The attractive force is called *surface tension*. (See Figure 11.2.)

Cooling by Expansion. All gases but hydrogen and helium cool when, at room temperature, they are allowed to expand through a valve into a space at lower pressure. Hydrogen and helium cool also when expanded at temperatures well below room temperature. This cooling effect is called the *Joule-Thomson effect*. The loss in energy that results in the cooling is due to the energy used to overcome the attractive forces in separating the molecules from each other.

Figure 11.2 Surface tension, making the surface as small as possible for a given volume, pulls liquids into spherical drops.

Factors Involved in Liquefaction of Gases

One requirement to liquefy a gas is to cool the gas (reduce the velocity of the particles) to a sufficiently low temperature so the attractive forces can hold them to each other. The *temperature above which it is not possible to liquefy a gas* is called the critical temperature. A second requirement to liquefy a gas is to increase the pressure sufficiently to push the molecules close enough together so the attractive forces can hold the particles to each other. The *pressure required to liquefy a gas at the critical temperature* is called the critical pressure. Values of the critical temperature and critical pressure of several gases are given in Table 11.2.

All gases can be liquefied. Helium has the lowest boiling point of any gas; hydrogen is next. The Joule-Thomson effect is made use of in liquefying

Table 11.2 Critical Temperature and Pressure of Several Gases

Gas	Critical Temperature (°C)	Critical Pressure (atm)
He	−268	2.3
H_2	−240	12.8
O_2	−119	49.7
N_2	−147	33.5
H_2O	374	218
CO_2	31	73

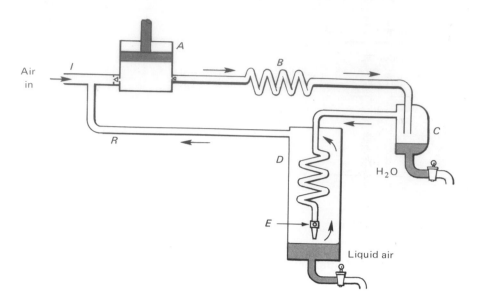

Figure 11.3 Liquefaction of air. Entering (*I*) and recirculated (*R*) air is compressed in *A* and passes through the cooling coil *B* to remove heat produced in the compression. Water vapor present condenses and is removed at *C*. The dry gas then passes through a coil in *D* arranged so that the gas, cooled by expansion through the throttling valve *E*, circulates back over the coil and cools the incoming compressed gas. On expansion through *E*, part of the gas liquifies and the remainder is recirculated (*R*).

gases. After expansion through a throttling valve, the cooled gas is allowed to flow back over the compressed gas, which becomes so cold that, after the still further cooling produced by additional expansion, it will liquefy (Figure 11.3).

Liquids and Gases

Liquids differ from gases in that they do not distribute themselves uniformly throughout the whole volume of a containing vessel, but retain their own volumes. Because of their mobility, however, liquids do take the shape of the portion of the containing vessel they occupy. Further, liquids contract only slightly even upon application of great pressure—a result that finds an interpretation in terms of our picture of the molecules of a liquid as being already crowded together with little free space between molecules.

Evaporation. When a liquid such as water is placed in an open vessel, it gradually passes into the air in the form of gas or vapor, and the process is called *evaporation*. In a confined space, as in a closed bottle (Figure 11.4), evaporation proceeds until the air above the liquid contains a definite percentage of gaseous water and then apparently ceases; the air is said to be *saturated* with water vapor. But the evaporation process does not cease— it simply establishes an equilibrium with the condensation process. We can understand this if we consider the question from a kinetic point of view. The molecules of a liquid are in random heat motion, just as they are in a gas; some of them have speeds greater than the average and others have speeds less than the average (Figure 10.8). The more rapidly moving molecules will escape from the surface of the liquid, breaking free from the attraction of their neighbor molecules. They will then move about in the air space above

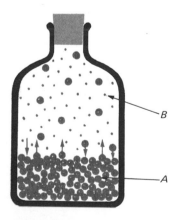

Figure 11.4 Diagram of a liquid (large black dots, *A*) evaporating into air (small dots, *B*) in a closed bottle.

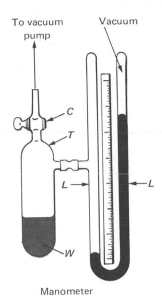

Figure 11.5 Measurement of vapor pressure with a manometer. A liquid, such as water, W, is placed in the tube, T, and the stopcock C, is opened to a vacuum pump. If air and water vapor are pumped out of the system on the left-hand side of the U-tube, the mercury will be at the same level (L, L) in both arms of the manometer. Then, if the stopcock is closed, the water evaporates and establishes an equilibrium vapor pressure at the temperature of the tube T. The difference in the level of the mercury columns indicates the vapor pressure of the liquid.

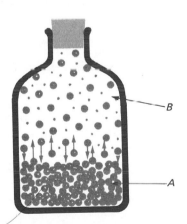

Figure 11.6 Diagram of the same liquid as in Figure 11.4, but at a higher temperature.

as gas molecules and, as more and more of them accumulate there, many will strike the surface of the liquid and return to the liquid, as represented in Figure 11.4. Eventually, enough molecules of water vapor will accumulate in the air space so that *the number returning to the liquid in one second will be equal to the number escaping from the liquid into the vapor in the same time*, and a state of **equilibrium** (Chapter 7) will be reached. Although molecules continue to escape, there is no further net loss of molecules from the surface. The equilibrium between the tendencies to escape into the gas space (evaporation) and to return to the liquid (condensation) may be represented in the following manner:

Liquid ⇌ vapor

Vapor Pressure. The *pressure of the saturated vapor in equilibrium with the liquid* is the **vapor pressure of the liquid**. It is commonly *expressed in millimeters of mercury* or *torr*, and it is a measure of the tendency of the liquid to evaporate. Experimentally, the value of the vapor pressure of a liquid at any temperature may be measured in many different ways. A simple procedure is illustrated in Figure 11.5. Liquids differ greatly among themselves in the magnitude of their vapor pressures at a given temperature. Those with high vapor pressure at room temperature are said to be *volatile*. Such liquids include ether, alcohol, and benzene.

Effect of Temperature upon Vapor Pressure. Measurements show that vapor pressure increases with rising temperature; in Figure 11.7, the experimental values of the vapor pressures of four liquids are plotted against temperature. How can this effect be explained?

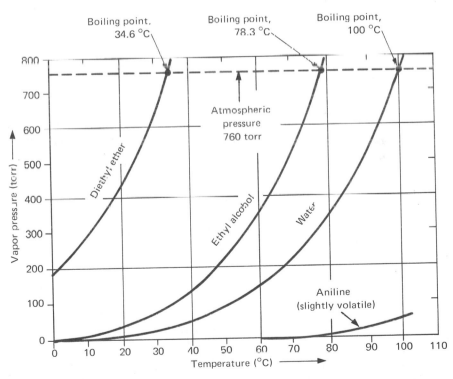

Figure 11.7 Vapor pressure curves showing boiling points of several liquids.

Suppose we have a liquid in equilibrium with its vapor in an enclosed space. If the temperature is raised, the average kinetic energy of the molecules in both the vapor and the liquid is increased. But the small fraction of the liquid molecules that have a kinetic energy far above the average, the "hot" molecules that are able to escape through the surface into the vapor, is markedly increased. Consequently, the rate of evaporation becomes greater than the rate of condensation, and the concentration of molecules in the vapor phase increases. As this increase takes place, however, the rate of condensation will begin to increase and will shortly overtake the evaporation rate and establish a new equilibrium:

$$\text{Liquid} \rightleftharpoons \text{vapor}$$

The concentration of vapor molecules and the vapor pressure will be higher than at the lower temperature (compare Figures 11.4 and 11.6).

Boiling Point

During the heating of a liquid at ordinary temperatures, a portion of the energy given to it goes to raise its temperature, and a portion to change it into a vapor at its surface. When the pressure of the vapor arising from the liquid just exceeds the opposing atmospheric pressure, all the heat energy goes to change the liquid into vapor (in freeing the molecules from the attraction of their neighbors) and into mechanical work in pushing back the atmosphere. The temperature then remains constant, notwithstanding the fact that heat is being applied to the liquid. This temperature is called the boiling point under the pressure of the experiment. The boiling point may be defined as *the temperature at which the vapor pressure of the liquid just exceeds the opposing pressure of the atmosphere.* The boiling point at standard atmospheric pressure is often called the *normal* boiling point. By suitably altering the pressure on a liquid, it may be caused to boil at temperatures higher or lower than its normal boiling point. Thus, at high altitudes, the boiling point of water is less than 100°C (Figure 11.8).

Heat of Vaporization and Condensation

Since only the fastest-moving ("hottest") molecules in a liquid are able to escape through its surface, the liquid becomes colder when it loses these molecules. Heat must be supplied to maintain the evaporation. The *quantity of heat absorbed in changing 1 g of a liquid into 1 g of vapor at constant temperature* is called the heat of vaporization. Conversely, the condensation of a gas or vapor to a liquid liberates a quantity of heat exactly equal to the heat of vaporization. The heat of vaporization of water at its boiling point is exceptionally large, amounting to 539 cal/g or (18×539) cal/mole. Molar heats of vaporization for several liquids are given in Table 11.3.

Table 11.3 Molar Heats of Vaporization at the Boiling Point

	cal/mole		cal/mole
Ammonia	556	Mercury	14,166
Argon	150	Methane	2320
Ethane	3810	Methyl alcohol	8410
Ethyl alcohol	7344	Sulfur dioxide	607
Helium	24	Water	9700

Figure 11.8 Boiling point of water at various elevations above sea level.

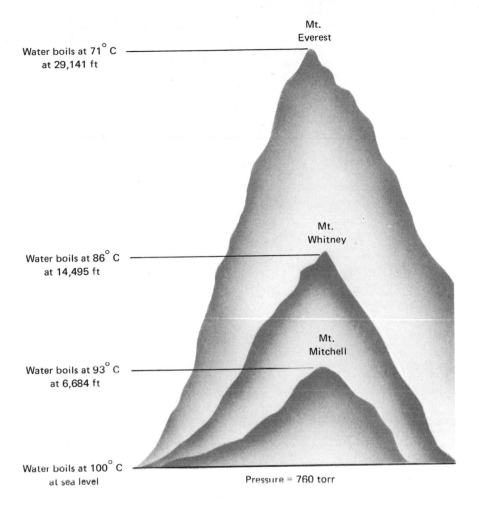

Mt. Everest

Water boils at 71° C at 29,141 ft

Mt. Whitney

Water boils at 86° C at 14,495 ft

Mt. Mitchell

Water boils at 93° C at 6,684 ft

Water boils at 100° C at sea level

Pressure = 760 torr

Changing a Liquid to a Solid

As a liquid is cooled, the translational motion of the molecules becomes less and less. At a certain temperature, the translational motion is reduced to a point where the attractive forces between the particles of the liquid overcome the kinetic energies of the particles, and the particles take positions in a pattern or latticework of particles that we call a crystalline solid. The temperature at which this change occurs for a pure liquid is called the *freezing point* or the *melting point*; the temperature remains constant until all of the liquid has crystallized. The particles in the solid retain a vibrational motion about the fixed lattice points of the solid.

Liquids composed of large molecules become so viscous near the freezing point that the big molecules cannot orient themselves in a regular crystal pattern. Such substances are called *amorphous* solids. Glasses, glazes, tar, glue, and gum are examples of such materials.

Freezing Point and Melting Point

Sometimes when a crystallizable liquid is cooled, it does not begin to crystallize at its freezing point. Indeed, liquid water has been cooled far below 0°C without freezing. A liquid below its freezing point is said to be *undercooled*. If a crystal of the substance once forms, or if a "seed crystal" is dropped into the undercooled liquid, solidification begins at once, heat is given out in the process, and the temperature rises to the true freezing point and remains there as long as both liquid and solid are present. The freezing point is defined as *the temperature at which both solid and liquid will remain in contact with each other without change of temperature.* The more viscous a liquid is at its freezing point, the more readily undercooling takes place; and with very viscous liquids, crystallization may never occur, as in the glasses.

Conversely, when a crystalline solid is slowly heated, its temperature steadily rises to the melting point (unless the solid should undergo decomposition) and remains constant at that temperature (if pure) until the melting is complete, then rises again. While the solid and liquid are in contact with each other, in equilibrium, the addition of heat energy will cause the solid to change to liquid (at the melting point), and the removal of heat will result in the conversion of the liquid to solid (at the freezing point):

$$\text{Solid} \rightleftharpoons \text{liquid}$$

But there will be no change in temperature as long as both phases are present.

Heat of Fusion

The *heat absorbed in converting 1 g of a solid at its melting point into a liquid at the same temperature* is called the heat of fusion of the substance. For ice, this amounts to about 80 cal/g or 1440 cal/mole. The heat of fusion represents the difference in energy between the orderly, rigid arrangement of molecules in a solid and the relative disorder and mobility of molecules in a liquid.

Vapor Pressure of Solids; Sublimation

The odor in a room where pieces of such solids as camphor and naphthalene (moth balls) are present is evidence that the particles of solids are still in motion and can escape as vapor, even at ordinary temperatures. Here again, an equilibrium exists between the solid and the vapor:

$$\text{Solid} \rightleftharpoons \text{vapor}$$

As the temperature is raised, the pressure of the vapor increases. If, before the melting point is reached, the vapor pressure of the solid increases to the point where it just exceeds the pressure of the atmosphere, the solid cannot be heated to a higher temperature in an open vessel, just as a liquid cannot be heated (at the pressure of the atmosphere) above its boiling point. To melt such a solid, it is necessary to heat it in a closed vessel. This behavior is illustrated by solid carbon dioxide; its vapor pressure reaches 760 torr at $-78.5°C$, but it does not melt until $-56.6°C$. At this latter point, its vapor pressure is more than 5 atm. When the vapors from such solids are cooled, they pass directly back into the solid form. The *process of converting a solid into a vapor and condensing the vapor to a solid again without going through the liquid state* is called sublimation, and the solid is said to *sublime* on heating. The corresponding process with liquids is called *distillation.* Solids having a sufficient vapor pressure are often separated from nonvolatile impurities by sublimation.

Diagram Illustrating Equilibria Among Solid, Liquid, and Vapor

In the preceding pages we have seen that a number of different types of equilibrium exist in a system containing a substance in different physical states. Thus, with water we have an equilibrium between the vapor and the liquid, between the vapor and the solid, and between the liquid and the solid. We can show the general character of these three equilibria in a very compact form in a diagram (Figure 11.9), in which we represent temperature along the horizontal axis and pressure along the vertical axis. The curve *OA* is a plot of the experimentally measured vapor pressures of the liquid over a range of temperatures—for example, the vapor pressure is P_1 at temperature t_1, and P_2 at t_2. This curve ends at a point, *A*, which is the critical temperature, above which the substance can no longer exist in the liquid state. *OB* is the vapor-pressure curve of the solid, usually called the sublimation curve. The point *O*, where these two curves intersect, is the freezing point of the liquid under its own vapor pressure; at that point, the vapor pressures of solid and liquid are equal. Since the two curves intersect at this point, the solid, liquid, and vapor can coexist at this temperature and pressure. But they can do so at no other point, so long as no air or dissolved matter is present. If no solid makes its appearance at the freezing point, the vapor pressure of the liquid will be represented by the extension of the curve *AO* toward *S*. It will be seen that the vapor pressure of such an undercooled liquid is greater than that of the solid at the same temperature, as at *t*. This indicates that the undercooled liquid is in a more unstable condition. Very few (if any) solids can be heated above the melting point without melting; so the curve *BO* can rarely be prolonged beyond the melting point, as at *X*.

Figure 11.9 Diagram illustrating the equilibria among solid, liquid, and gaseous states.

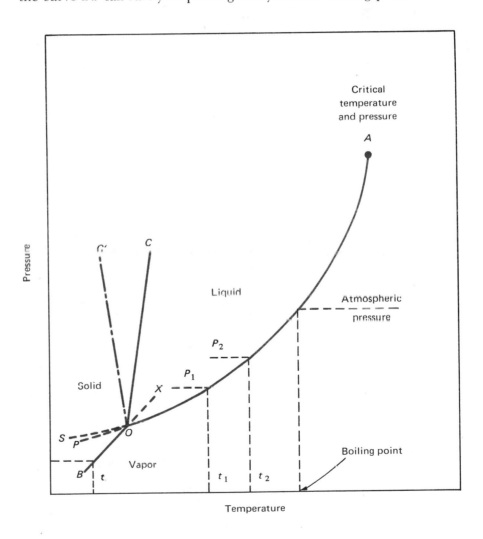

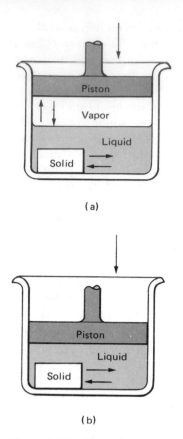

(a)

(b)

Figure 11.10 Illustrating the Le Chatelier principle in equilibria among states.

If, when the solid, liquid, and vapor are all in equilibrium at the point O (often called the *triple point*), you apply a steady outside pressure (as with a piston) greater than the pressure of the vapor [Figure 11.10(a)], the vapor will all condense, and only solid and liquid will remain [Figure 11.10(b)]. Then, if the applied pressure is increased more and more, the result is that the freezing point, or melting point, is raised for all known substances (with several exceptions). This effect is indicated by the line OC (Figure 11.9), which represents the equilibrium between melting solid and freezing liquid.

Fogs, Clouds, and Precipitation. The condensation of water from cooled air is an everyday occurrence that has profound implications for every inhabitant of the earth.

Dew forms when the ground surface is cooled by rapid radiation of heat at night; this lowers the air temperature near the ground. When the temperature of the air is lowered so the vapor pressure of the water it carries is greater than the vapor pressure of pure water at that temperature (above 100 per cent humidity), tiny water droplets form and deposit on cool surfaces such as blades of grass. Fogs are produced when large masses of air are cooled, often by contact with the colder sea or earth, so that the tiny water droplets formed remain suspended in the air, reducing visibility and sometimes creating safety or health hazards.

When air cools by expansion on rising, clouds may form as the air temperature falls below the saturation point of the water vapor present. Here again, the tiny droplets of water remain suspended in the air unless there is very rapid cooling of the large air mass of which they are a part. Such rapid cooling leads to growth of the tiny particles of liquid or solid water, producing rain, snow, or hail. Rapid cooling on a scale sufficient to cause precipitation most often takes place when the air mass rises rapidly, thereby expanding and cooling.

Low lying clouds (below 3000 feet) and fogs contain liquid droplets at temperatures above about −30°C. Cirrus clouds, which usually occur at high altitudes (16,000 to 45,000 feet), and all clouds at very low temperatures contain tiny ice crystals. The liquid droplets in clouds are nearly spherical, ranging in diameter between 2 and 50 microns. The concentration of drops varies between 20 and 500 per cubic centimeter of air. The opacity of a cloud depends mainly on the concentration of drops; the higher the concentration, the more dense the cloud. A cloud such as a wet sea fog may contain a half gram of water per cubic centimeter.

Just what prevents the tiny water droplets from falling to the ground has been a puzzle, especially since liquid water is roughly 800 times more dense than air. The best explanation for this so far advanced is that the droplets do indeed fall toward the ground, as predicted. However, under cloud-creating conditions, they are buoyed up by rising air currents or are continually replaced by new droplets formed by the condensation of water vapor in the air.

Studies reveal that fogs can form at humidities less than 100 per cent. The droplets form on tiny dust particles in the air. Where dust concentration is high, fogs can form at humidities as low as 65 per cent. Dust particles with high affinity for water will cause droplets to grow and fogs to become thicker. In this way, fogs in highly industrialized areas may become more dense, dirtier, and may persist longer than in areas of relatively clean air.

In recent years, fogs have become more and more a problem in and around cities. Dense fogs at airports and on superhighways are dangers we have all experienced. Fog in combination with air pollution can be lethal.

A dense fog settling over a city already blanketed by polluted air can bring about a strong inversion—warm air lying over cold air. This will prevent removal of polluted air by vertical movement. If winds are light, the pollutants will not be removed by horizontal movement. Should business and industry continue as usual, tons of smoke, dust, and fumes will be added to the trapped air and fog. The fog will become darker, visibility will be reduced, and the air will become more toxic. In a situation similar to this in London in December, 1952, at least 4000 people were killed and tens of thousands suffered permanent injury, mainly in the form of respiratory disorders.

Efforts to ameliorate problems created by fogs have resulted in the testing of procedures not only for dissipating them or inhibiting their formation, but also for creating them in circumstances where they might be helpful, as in rural areas where crops can be protected from frost by artificial fogs.

One of the most promising approaches to fog dispersal is "seeding the fog" with particles of a very cold substance, such as liquid propane droplets or solid carbon dioxide particles. Water droplets in the fog condense onto these cold centers, resulting in precipitation of part of the water droplets and evaporation of others (into the partially dried air). At present, this method is applicable only for supercooled fogs, which amount to only about 5 per cent of all U.S. fogs.

An extension of the "seeding" concept is the addition of salt crystals to the fog. A similar method using potassium iodide crystals has been used to produce rain from nonexpanding clouds. Spreading a chemical film over lakes and swamps near airports or highways to reduce water evaporation and the humidity has also been used to inhibit fogs in limited areas.

Better understanding of the chemical, physical, and electrical properties of fogs and clouds are needed to help us control these important masses of matter which have so much effect on the lives of all.

SUMMARY   The deviations of real gases from ideal behavior can be accounted for in terms of attractive forces between molecules and the volume occupied by the molecules themselves. As the temperature of a gas is lowered, the average kinetic energy of the molecules decreases but the attractive forces remain approximately constant. At some temperature for each gas, the attractive forces should overcome the molecular momentum and the molecules should cluster together forming the liquid state.

Continuing to use this model, we predict that liquids should exhibit properties reflecting an interplay between the attractive forces and the kinetic energy of the molecules. The equilibrium between a liquid and its vapor is a dramatic corroboration of this prediction. Other properties of liquids— boiling point, surface tension, viscosity, and heat of vaporization—are all generally explicable in terms of this interplay.

Even the equilibrium between a liquid and a solid may be viewed as a process in which some molecules acquire sufficient energy to overcome the attractive forces in the solid and pass to the liquid, while an equal number of liquid molecules lose enough kinetic energy so that they are captured and held in the solid.

Temperature is of special importance in nearly all physical properties of gases, liquids, or solids, because a change in temperature brings about a corresponding change in kinetic energy. Diagrams showing the equilibrium among the solid, liquid, and gaseous states of a single substance over a wide range of temperatures and pressures are useful in obtaining a view of the regions of stability of the various phases.

IMPORTANT TERMS

Gas
compressibility factor
van der Waals equation
critical temperature
critical pressure
Joule-Thomson effect
manometer

Liquids
surface tension
evaporation
vapor pressure
volatility
boiling point
heat of vaporization
freezing point
distillation

Solids
heat of fusion
amorphous
sublimation

Equilibrium
triple point

QUESTIONS AND PROBLEMS

1. How does a real gas differ from an ideal gas?

2. What two features of real gases are taken into consideration in introducing the van der Waals constants a and b into the equation for ideal gases?

3. Why do most gases cool when they expand at room temperature?

4. Why are liquids much less compressible than gases?

5. Does the vapor pressure of a liquid depend upon the amount of liquid surface exposed to the vapor? On the basis of the equilibrium picture, explain why or why not.

6. What density changes might be expected as a substance goes from the solid to the liquid to the gaseous state? Explain.

7. A pure solid substance is heated at a constant rate. At first the temperature rises regularly, then it stays constant for a measurable time, and then rises again. Why?

8. Will a tightly covered saucepan half filled with water boil at 100°C? Explain.

9. How much heat must be applied to 500 g of solid xenon to melt it at its melting point? The molar heat of fusion for xenon is 549 cal/mole.

10. How many calories are required to raise the temperature of 30 g of ice at −15°C to steam at 120°C? (The specific heat of ice and steam is approximately 0.5 cal/g°C.)

11. A sample of water is cooled to −8°C without freezing. When a small piece of ice is dropped into the sample, freezing occurs. If the two described phenomena are observed experimentally, can we conclude from these observations that the freezing point of water is −8°C? Explain.

12. How many kilograms of ice could be melted at 0°C by the heat liberated from the condensation of 1.5 kg of steam at 100°C?

13. What is the boiling point of water if the gas pressure on the liquid surface is kept at 300 torr? At 600 torr?

14. The normal boiling points of ammonia and sulfur dioxide are −33.3°C and −10°C, respectively. Which of the two substances is expected to have the higher (a) critical temperature, (b) melting point, (c) vapor pressure at −45°, (d) molar heats of fusion and vaporization, (e) intermolecular attractive forces?

15. Explain how, if at all, each of the following affects the vapor pressure of a liquid: (a) surface area of the liquid; (b) volume of the liquid; (c) temperature of the liquid; (d) presence of other gases in the vapor.

16. What factors determine how high a liquid will rise in a vertical pipe attached to a suction pump? What effects, if any, would be manifested by the following factors: (a) density of liquid, (b) vapor pressure, (c) altitude of system?

17. Sketch curves showing the compressibility factor as a function of pressure at constant temperature for a typical gas for two temperatures, $T_1 > T_2$. How do you explain the fact that the curve for the higher temperature T_1 is closer to the curve for an ideal gas than the curve for the lower temperature?

18. The vapor pressure of water at 22°C is 20 torr. If the partial pressure of water vapor in the laboratory at a time when the temperature is 22°C is 14 torr, what is the relative humidity in the laboratory?

19. Calculate the weight of water vapor in the air of a room 20 × 12 × 8 ft if the humidity is 65% at 70°F.

20. A certain quantity of water at 80°C was cooled to 30°C by the addition of 50 g of ice at 0°C. To what volume of water was the ice added? (Density of water = 1 g/cc.)

21. (a) Compute the pressure exerted by 0.5 mole of ammonia at 25°C when confined to a volume of 0.5 l, assuming ideality. (b) Repeat the calculation using the van der Waals equation (van der Waals constants for ammonia: $a = 4.17$ l² atm/mole², $b = 0.0371$ l/mole).

SPECIAL PROBLEMS

1. Do heats of vaporization remain constant over a temperature range or do they, in general, change with temperature?

2. Show, by algebraic operation, that the van der Waals equation for 1 mole gives for the compressibility factor

$$\frac{PV}{RT} = \frac{1}{1 - \frac{b}{V}} - \frac{a}{RTV}$$

Using this equation, show that the compressibility factor becomes unity when the volume becomes very large.

3. The curve OC' (Figure 11.9) has a slope corresponding to the ice ⇌ water equilibrium at not-too-high pressures. It is claimed that ice skating is possible because liquid water acts as a lubricant under the blades of the skates. Discuss the relationship between these two facts.

4. A 300-g sample of ice initially at −5°C is converted to steam at 100°C. The heat energy is supplied at the rate of 120 cal/min. (a) Calculate the time necessary for the conversion. (b) Plot the conversion showing the temperature of the sample as a function of time.

5. Given the following data concerning a given substance, calculate the amount of heat necessary to convert 25 g of the solid at the melting point to vapor at the boiling point.

Melting point	20°C	Heat of vaporization	180 cal/g
Boiling point	150°C	Heat of fusion	30 cal/g
		Specific heat (liq)	0.6 cal/g°C

6. According to Charles' law, when the absolute temperature of a gas is doubled, the volume doubles. However, this is not true for hydrogen fluoride. In the case of this gas, the volume may exhibit a more than twofold increase. Explain.

7. Two grams of water are confined in a 1-liter sealed container at a temperature of 25°C. Vapor pressure of water at that temperature is 23.7 torr after equilibrium is established. (a) How much water is left in liquid form? (b) If the remaining liquid water were all electrolyzed to hydrogen and oxygen, what would be the partial pressure of each gas in the container?

8. On expansion at room temperature, carbon dioxide gas cools, hydrogen gas becomes warmer, but helium gas shows very little change in temperature. Explain.

9. Calculate the compressibility factor and the pressure exerted by 1 mole of butane when it is confined at 300°C in the volumes (a) 20, (b) 2.0, (c) 0.40, (d) 0.30, (e) 0.20 liters, using the van der Waals equation. (f) Plot the compressibility factor against the pressure for butane at 300°C. (g) How would the curve be altered if the temperature was 300 K? (h) Sketch as a dotted line the 300 K curves on your graph. (i) If butane were an ideal gas at 300°C, what would be the pressure of 1 mole confined in 2 liters? (j) Mark this point on your graph.

10. It is claimed that the vapor pressure of a liquid increases because the rate of evaporation increases as the temperature is raised. How does an increase in temperature affect the rate of condensation? Suppose a system could be devised to prevent evaporation and allow condensation only, and the system were arranged to feed in additional vapor molecules as fast as they condensed, thus keeping the number of vapor molecules constant. Would the rate of condensation change if the temperature increased? In light of your answer, discuss the vapor-pressure equilibrium.

11. Using the equation $\log \dfrac{P_2}{P_1} = \dfrac{\Delta H_v}{2.3R}\left[\dfrac{1}{T_1} - \dfrac{1}{T_2}\right]$ (where P_2 and P_1 are the vapor pressures of the liquid at temperatures T_2 and T_1, respectively; ΔH_v is the molar heat of vaporization; and R is the gas constant, 1.987 cal/mole-deg) calculate the vapor pressures of ammonia, ethane, and ethyl alcohol at 30° below their boiling points, using the data from Table 11.3 and the boiling points: −33°C for ammonia, −103.8°C for ethane, and 78°C for ethyl alcohol.

12. Calculate the heat of vaporization and the boiling point of carbon tetrachloride from the following vapor pressure data:

P(torr)	56.0	91.0	143.0	215.8
T(°C)	10	20	30	40

REFERENCES   BARTON, A. F. M., "Internal Pressure, a Fundamental Liquid Property," *J. Chem. Educ.*, **48**, 156 (1971).

BERNAL, J. D., "The Structure of Liquids," *Sci. Amer.*, **203**, no. 2, 124 (1960).

DREISBACH, D., *Liquids and Solutions*, Houghton Mifflin, Boston, 1966.

FRENKEL, J., *Kinetic Theory of Liquids*, Dover, New York, 1955.

HILDEBRAND, J. H., *An Introduction to Molecular Kinetic Theory*, Van Nostrand Reinhold, New York, 1963.

MYERS, J. N., "Fog," *Sci. Amer.*, **219**, 75 (1968).

PUPEZIN, J., JANCSO, G., and VAN HOOK, W. A., "The Vapor Pressure of Water: A Good Reference System," *J. Chem. Educ.*, **48**, 114 (1971).

ROWLINSON, J. S., *Liquids and Liquid Mixtures*, Butterworth, London, 1959.

SOLUTIONS—HOMOGENEOUS MIXTURES

Almost all the chemical processes which occur in nature, whether in animal or vegetable organisms, or in the non-living surface of the earth, . . . take place between substances in solution.

Wilhelm Ostwald (1853–1932)

Why— Do sugar and salt dissolve in water, but not in benzene?
Does sand not dissolve in water?
Does alcohol dissolve in water, but gasoline does not?
Do large volumes of ammonia dissolve in water, but only tiny volumes of oxygen?
Does sodium chloride dissolve in water but barium sulfate does not?
Do water solutions of sodium chloride and hydrogen chloride conduct an electric current?

THE PROCESS OF DISSOLVING   In the search for an answer to these questions, let us use the information we have gathered about attractive forces between particles of solids, liquids,

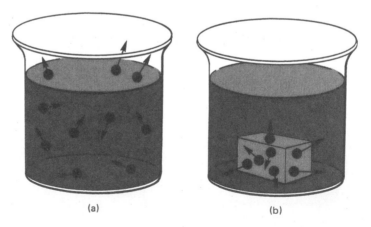

Figure 12.1 Both evaporation (a) and dissolution (b) involve the escape of particles at the surface.

(a)   (b)

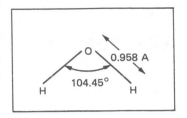

Figure 12.2 Water molecule.

and gases. Let us compare the process of the passage of a substance into solution with the process of evaporation (Figure 12.1). In each case the faster-moving particles escape from the surface. But the dissolution process differs from evaporation in one respect. In evaporation, the vapor space does not attract the particles of the evaporating substance, whereas the particles of dissolving substance, the solute, are attracted by the molecules of a solvent. These attractive forces may be van der Waals forces of various strengths, such as those that cause deviations from the gas laws (Chapters 6 and 11), or they may be the very strong forces which produce chemical combination. If the attractive forces of the solvent for the solute are strong enough to overcome the attractive forces that hold the particles of the solute to each other, the substance will dissolve.

As a rule, substances with a very high melting point, whose particles are bound very rigidly together, do not dissolve appreciably in a solvent. Examples are silica, the heavy-metal silicates, barium sulfate, calcium carbonate, and the heavy-metal sulfides. The ions of some compounds, even though they are bound very tightly together, have such a strong attraction for water molecules that they become jacketed, or *solvated*, by water molecules; therefore, they dissolve. Examples of such compounds are sodium chloride, calcium chloride, potassium nitrate, and copper sulfate.

Oil and water proverbially do not mix, or dissolve in each other. Water does not dissolve in oil because water molecules attract one another so strongly that they will not respond to the much feebler attraction of the oil molecules for them; nor will the water molecules, because of this strong attraction, separate from one another far enough to allow the oil molecules to penetrate among them.

Water as a Solvent

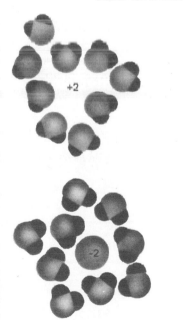

Figure 12.3 Hydration of ions.

Probably the most abundant compound on the earth is water. Water is a good solvent for many substances; it also forms hydrates with many atoms, molecules, and ions. The reason for the good solvent properties of water can be understood when we recall that water is an angular, polar covalent molecule; as a result, it is a dipole (Figure 12.2). Hence we find ion-dipole attractive forces causing ions to be *surrounded by water molecules* (Figure 12.3); this process is called hydration or solvation. Neutral atoms and molecules that are polar, such as sucrose, may also become hydrated by dipole-dipole forces. The forces that cause hydration or solvation seem to be the main cause of solution of soluble compounds.

Bonding between a water molecule and an ion can become essentially covalent if the ion has empty orbitals to which the water molecule can donate an unshared pair of electrons. Examples of hydrated ions in which bonding of this type occurs are $Co(H_2O)_6^{+3}$, $Zn(H_2O)_6^{+2}$, and $Cu(H_2O)_6^{+2}$.

When a solvent sheath is formed around an ion, the next step usually is the dissolving of the ion. The solvent sheath keeps the + and − ions separated and insulates them from each other. Water has a high dielectric constant—that is, it has good insulating properties.

If, however, the electrostatic attraction between a pair of ions is high, as it is with aluminum ions (Al^{+3}) and phosphate ions (PO_4^{-3}), the solvent sheath will be squeezed out, and the ions will come together and precipitate from solution.

Polar Solvents	Water is one of the most highly polar solvents known, thus it is one of the best solvents for ions. Methyl alcohol has polar properties somewhat similar to water, as do ethyl alcohol and other alcohols, and many other molecules such as acetone, CH_3COCH_3; acetic acid, CH_3CO_2H; ammonia, NH_3; and hydrogen chloride, HCl.

Closely associated with polar properties are the acid-base properties of many solutes and solvents as these are defined in terms of Lewis theory—electron-pair donors and acceptors. For example, water, alcohol, ether, and ammonia have electron pairs that can be donated. These electron pairs become important factors in increasing the solubility of substances by coordinate covalent bond formation. For example, ammonia dissolves in water in large amounts probably because of the reaction

Acid-Base Solvents

$$NH_3 + H_2O \rightleftharpoons NH_4^+ + OH^- \tag{1}$$

Hydrogen chloride is very soluble in water largely because of the reaction

$$HCl + H_2O \rightleftharpoons H_3O^+ + Cl^- \tag{2}$$

Energy of Hydration

A measure of the strength of the bonds formed when ions hydrate is obtained from calculations of the *heat energy released on hydrate formation*. Table 12.1 gives the hydration energies for some common ions. One of the highest energies of hydration is observed when sulfuric acid dissolves in water. This is largely due to the very high hydration energy of the proton (259 kcal/mole). So much energy is released in the dilution of concentrated sulfuric acid that it is dangerous to add water to the concentrated acid because the first small amount of water may be suddenly turned to steam.

Heat of Solution

The heat of solution is the *sum of the energy changes that occur when a substance dissolves*. Energy is released when the solute particles become solvated or form new bonds; energy is absorbed in demolishing the structure of solids, in separating molecules in the crystal from each other, or in ionizing molecules such as acetic acid. Whether the over-all solution process results in heating or cooling depends upon the relative amounts of heat absorbed and evolved in each step. Dissolving sulfuric acid gives off a large quantity of heat; sodium chloride shows only a very small temperature change, but ammonium nitrate cools on dissolving.

Constituents and Varieties of Solutions

We think of one constituent of a solution (usually the more abundant one) as the medium in which the other (less abundant) constituent is dissolved; we call the medium the *solvent*, the dissolved constituent the *solute*.

Table 12.1 Hydration Energies of Some Common Ions

Ion	Hydration Energy (kcal/mole)	Ion	Hydration Energy (kcal/mole)
Li^+	121.3	Zn^{+2}	485
Na^+	95.2	Al^{+3}	1,110
K^+	75.0	Fe^{+3}	1,042
Ag^+	111.9	OH^-	87
Mg^{+2}	456	F^-	122
Ca^{+2}	377	Cl^-	89.8

SOLUTIONS OF SOLIDS IN LIQUIDS

A solid dissolved in a liquid is by far the most familiar type of solution; an example is a solution of sodium chloride in water. It is sometimes said that zinc "dissolves" in hydrochloric acid. In that case, however, the solution is preceded by a chemical reaction whereby the zinc is converted into zinc chloride, and it is this compound that is obtained when the solution is evaporated. Solutions of solids in liquids, such as we are now considering, are those in which evaporation will leave the solute in its original chemical condition.

Equilibrium in Saturated Solutions

When a lump of sugar is placed in a small beaker and covered with water (Figure 12.4), it gradually passes into solution—that is, particles leave it due to attraction to the solvent molecules and wander through the solvent. If there is enough sugar, and if a long enough time elapses, the concentration of the sugar in the solution reaches a definite limiting value, and we say that the solution is saturated. The solution action, however, does not cease when the saturation point is reached. Molecules of the sugar continue to leave the lump and pass into the solution, while other molecules of sugar, previously dissolved, return to the lump from the solution. The rate of these two processes is exactly equal at saturation, so that the number of molecules of solute leaving the solid and entering the solution in unit time is equal to the number of solute molecules leaving the solution and crystallizing out on the solid in the same unit of time. This is, then, another example of an equilibrium process (Chapter 7). A saturated solution may be defined as one in which *the dissolved solute is in equilibrium with the undissolved solute.*

Figure 12.4 Representation of equilibrium in a saturated solution.

$$\text{Undissolved solute} + \text{solvent} \rightleftharpoons \text{dissolved solute}$$

The *solubility* of a solute is the *amount present in a given quantity of solvent* when the solution is *saturated.*

We can demonstrate that the molecules or ions of a solute are still dissolving and crystallizing after saturation is reached by making a saturated solution, using "tagged" atoms of a radioactive substance, and then dropping in a crystal of the same material that does not contain any radioactive atoms. We soon find that the nonradioactive crystal has acquired some radioactive atoms from the solution, having traded some of its ordinary molecules for some dissolved molecules containing radioactive atoms.

Factors That Affect the Rate of Solution and the Solubility

The size of the solid solute particles, the rate of stirring, and the temperature affect the rate of solution of a given solute in a given solvent. But the particle size and stirring rate do not affect the solubility; they simply control the rate at which the solubility equilibrium is attained. A change in temperature changes the solubility of a substance (Figure 12.5). Most solid solutes are more soluble as the temperature is increased, but a few, including calcium hydroxide, $Ca(OH)_2$, decrease in solubility as the temperature increases.

Supersaturated Solutions, a Nonequilibrium Condition

Since the solution of most solids increases with increasing temperature, we can readily obtain a saturated solution by approximately saturating a solution at a high temperature and then reducing the temperature to the desired point, taking care to have some of the solid present all the time. The excess of solute will crystallize out as the temperature is lowered, and almost at once the solution will come to saturation equilibrium at the lower temperature.

Figure 12.5 Change of solubility with change in temperature.

Figure 12.6 Preparation of a 1-molar solution. One mole of solute is added to water. After the solute has dissolved, more water is added (stirring to keep the composition uniform) to bring the final volume to the 1-liter mark.

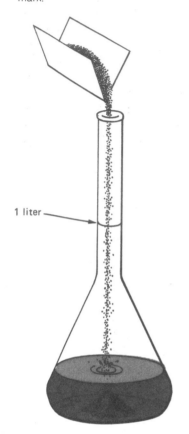

1 liter

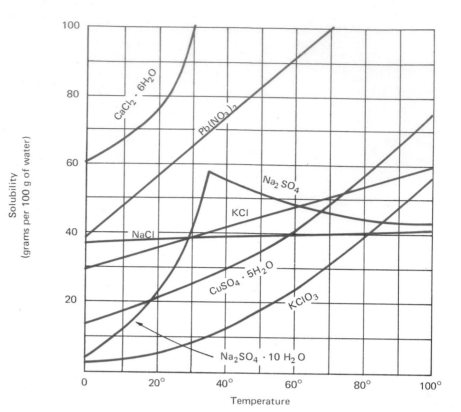

If we neglect to have some of the solid present while a concentrated solution cools, it may happen that the excess of solute will fail to crystallize. The solution will then contain *more than the normal saturation quantity of the solute* and is said to be **supersaturated** Supersaturation occurs especially in the case of very soluble solids, with salts containing much water of hydration—for example, with sodium sulfate ($Na_2SO_4 \cdot 10H_2O$) and sodium thiosulfate ($Na_2S_2O_3 \cdot 5H_2O$). The introduction of even the smallest fragment of the solid solute into a supersaturated solution will bring about crystallization of the excess of solute, and thus restore the equilibrium condition appropriate to the lowered temperature.

Concentration

The term **concentration** denotes *the quantity of solute dissolved in a given quantity of solven or of solution.* The concentration of a solution may be expressed in a number of ways. Sometimes it is stated on a percentage basis. More often it is desirable to state the number of moles which a given volume of the solution contains. If, for example, a given solution contains *one mole of solute for each liter of solution,* we say this is *a 1-molar (1M) solution;* similarly, a 0.5M solution contains one-half mole of solute for each liter of solution, and a 6M solution contains six moles of solute per liter of solution (Figure 12.6). Other examples using weights of solutes are: a 1M solution of sodium hydroxide contains 40.00 g NaOH per liter of solution; 63.02 g (one mole) of nitric acid dissolved in 2 liters of solution is a 0.5M solution. By dipping up or pouring out the proper number of milliliters of a solution of known molar concentration, you can measure out a known fraction of a mole of the solute substance.

Because molar solutions depend on the volume of solution (and this is temperature dependent), it is often convenient to use a concentration unit known as molality. Molality is defined as the *number of moles of solute in each kilogram (1000 g) of solvent.* Thus, for example, a 2-molal (2*m.*) solution contains two moles of solute for each kilogram of solvent, and a 0.1*m.* aqueous sucrose solution contains 0.1 mole (34.3 g) of sucrose in 1000 g of water.

A third concentration unit is known as the mole fraction. It is defined as *the number of moles of solute divided by the number of moles of solvent* plus the number of moles of solute—that is, by *the total number of moles.* Thus, a solution having a mole fraction of 0.1 contains one mole of solute for every nine moles of solvent, or

$$\frac{1 \text{ mole solute}}{9 \text{ moles solvent} + 1 \text{ mole solute}} = 0.1$$

Properties of a Solvent Modified by a Solute

The properties which a solvent (such as water) possesses when pure are often greatly modified when a solid is dissolved in it. Among these properties are vapor pressure, boiling point, and freezing point. In the present treatment we shall deal only with covalent solutes, which do not furnish ions; later, solutions containing ions will be considered.

Lowering of the Vapor Pressure. A solution of a solid in a liquid has a lower vapor pressure than that of the liquid itself. This can be easily shown, with a manometer, for solutions of sugar in water (Figure 12.7). No completely satisfying explanation for the lowering is known. Perhaps the sugar molecules act as a "screen" and interfere with the escape of solvent molecules and so lower the vapor pressure.

After many years of careful experimental work, the French scientist François Raoult found that (a) *in dilute solutions, the lowering of vapor pressure at a given temperature is directly proportional to the molal concentration of the solution;* and (b) *the percentage decrease in vapor pressure is the same for all solutes, regardless of their chemical nature, provided that they do not have an appreciable vapor pressure of their own and do not dissociate into ions.* In other words, at a given temperature, the lowering of the vapor pressure of a given solvent depends on the *number* and not the *kind* of solute molecules in a given quantity of the solvent. Raoult's law may be written

$$\Delta vp = k_p m.$$

where Δvp is the lowering of the vapor pressure, k_p is a proportionality constant, and m is the molality.

Elevation of the Boiling Point. Since the boiling point of a liquid is the temperature at which its vapor pressure just exceeds the opposing pressure of the atmosphere, it is evident that any condition that lowers the vapor pressure will raise the boiling point, since the liquid will have to be heated to a higher temperature to regain its original vapor pressure. This is shown schematically in Figure 12.8, where the curve showing the change of vapor pressure as a function of temperature for a pure liquid (as in Figure 11.9) is compared with the corresponding curve for a solution of a definite concentration. In the course of his experiments, Raoult found that the *elevation of the boiling point,* like the lowering of the vapor pressure, *is proportional to*

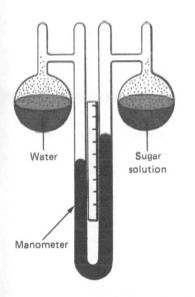

Water   Sugar
         solution

Manometer

Figure 12.7 Vapor-pressure lowering. The air is pumped out of the apparatus. The difference in levels of the mercury columns gives the difference in pressure between the water and the solution.

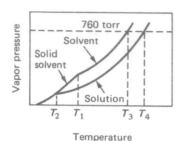

Figure 12.8 Comparison of the change of vapor pressure with temperature for solution and solvent. Comparison of T_4 with T_3 and of T_2 with T_1 shows that the boiling point of the solution is raised and the freezing point lowered, compared to the corresponding values for the pure solvent.

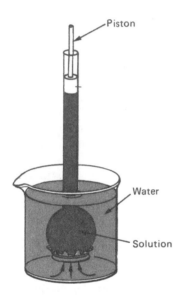

Figure 12.9 Illustrating the meaning of osmotic pressure. Reverse osmosis, applying a pressure in excess of the osmotic pressure to the solution and thereby causing water to move from the solution into the pure solvent, is one suggested procedure for desalinating seawater.

the molal concentration of the solution. In mathematical terms, $\Delta T_b = k_b m$, where ΔT_b is the rise of the boiling point, m is the molality, and k_b is a proportionality constant.

It follows that molar quantites of any substances not electrovalent and not having a vapor pressure of their own, when dissolved in a definite weight of a solvent, produce the same rise in the boiling point of the solution. The boiling-point elevation produced by 1 mole of solute added to 1000 g of water (a molal solution) is 0.52°C. Since the boiling-point elevation is proportional to the concentration, the boiling point of a 0.50-molal solution will be $0.50 \times 0.52 = 0.26°C$ higher than that of water. The rise in boiling point is different for different solvents.

Lowering of the Solvent's Freezing Point. When an unsaturated solution is cooled sufficiently, it does not freeze as a solution, but crystals of the pure solvent form. Since at the freezing point the vapor pressure of the solid solvent is equal to that of the liquid, and since the vapor pressure of the solution is less than that of the pure solvent, it is evident, as shown in Figure 12.8, that the freezing point of the solution is below that of the pure solvent. It is to lower the freezing point of water that we add various compounds (usually alcohol or ethylene glycol) to the water in an automobile radiator in cold weather.

Raoult found that the *freezing-point depression is proportional to the molal concentration of the solution,* and is the same for covalent solutes in the same solvent, as shown for water in Table 12.2. Thus a mole of solute dissolved in 1000 g of water lowers the freezing point 1.86°C (called the freezing-point constant, k_f, for water).

Every solvent has its own characteristic freezing-point constant; for acetic acid it is 3.9°C; for benzene, 5.1°C; for naphthalene, 6.8°C. The equation relating the freezing-point depression to the molality is

$$\Delta T_f = k_f m$$

where ΔT_f is the freezing-point depression, m is the molality, and k_f is the freezing-point constant.

If the concentration of the solution is such that the solution is saturated at the freezing point, then a lowering of the temperature will cause both solid solute and solid solvent to crystallize. The mixture of solids formed is called the *eutectic*, and the temperature at which the eutectic forms is called the *eutectic temperature*: it is the lowest freezing point of the solution (Table 12.3).

Table 12.2 Lowering of the Freezing Point of Water

Solute	Formula	Lowering Produced in 1000 g of Water by 1 Mole of Solute (°C)
Methyl alcohol (methanol)	CH_3OH	1.86
Ethyl alcohol	C_2H_5OH	1.83
Dextrose	$C_6H_{12}O_6$	1.90
Glycerin	$C_3H_5(OH)_3$	1.92
Urea	$CO(NH_2)_2$	1.86

Table 12.3 Eutectic Temperature and Composition of Ice with Various Compounds

Salt	Eutectic Temp. (°C)	Eutectic Composition (moles compound/mole water)
Potassium nitrate	−2.8	0.022
Ammonium nitrate	−16.7	0.168
Sodium chloride	−21.3	0.092
Calcium chloride	−49.8	0.070
Hydrogen chloride	−86.0	0.163

Molecular Weight Determination

Since the vapor-pressure lowering, boiling-point elevation, and freezing-point depression are each proportional to the molal concentration of a solution, that is, to the number of solute molecules in a solution, it is possible to determine the molecular weight of a dissolved compound from measurements of one of those properties—for example, the freezing-point depression.

PROBLEM

On a sensitive thermometer, the freezing point of water was found to be exactly 0°C, while a solution of 0.775 g of ethylene glycol, dissolved in 25.0 g of water, froze at −0.93°C. Calculate the molecular weight of ethylene glycol.

Solution

Since the freezing-point depression is proportional to the molal concentration, the molal concentration is evidently

$$m = \frac{\Delta T_f}{k_f} = \frac{0.93}{1.86} = 0.5 \; m$$

Thus the number of moles of solute present in the 25.0 g of water taken is

Number of moles = molality × kg water = 0.5 × 0.0250 = 0.0125

This tells us that the 0.775 g of ethylene glycol solute contains 0.0125 mole so the molecular weight (the number of grams per mole) is

$$\text{Molecular weight} = \frac{\text{grams}}{\text{mole}} = \frac{0.775 \text{ g}}{0.0125 \text{ mole}} = 62.0$$

Osmosis

Closely related to the lowering of the vapor pressure, to the elevation of the boiling point, and to the depression of the freezing point is the phenomenon of *osmosis*. This is the passage of a solvent from a dilute solution (or from a pure solvent) through a membrane into a more concentrated solution. This is an important process in plant and animal life.

The membrane must be semipermeable—that is, such that it will allow molecules of solvent to pass through readily, but will not allow molecules of solute to pass. A piece of parchment paper or nonwaterproof cellophane is semipermeable in this sense, and may be used to demonstrate osmosis. The membrane is fastened tightly over the bell of a thistle tube (Figure 12.9). A sugar solution is placed in the vessel so constructed, which is then immersed in pure water. Osmosis starts, and the volume of the sugar solution slowly increases as the water enters through the membrane.

Osmotic Pressure

If we could fit the tube in Figure 13.9 with a piston, as shown, and with it apply enough pressure on the solution, osmosis would stop. In fact, by pushing hard enough on the piston, water could be squeezed out of the solution, through the membrane, and back into the outer vessel.

The pressure that, when applied to a solution, will just prevent the entrance of solvent into it through a semipermeable membrane is the osmotic pressure of the solution. The pressure need not be applied with a piston. The hydrostatic pressure of a long vertical column of the solution may serve the same purpose.

If the solution on each side of the membrane were of the same concentration, no pressure would develop. Solutions having the same osmotic pressure are called *isotonic solutions*.

The osmotic pressure of a solution is approximately equal to the gas pressure the solute would exert if it were a gas occupying the same volume as the solvent at the same temperature. For instance, 1 mole of a gas, in 22.4 liters at 0°C, exerts a pressure of 1 atm. At 20°C the pressure is $293/273 \times 1$ atm $= 1.07$ atm. Similarly, 1 mole of sugar dissolved in 22.4 liters of water, at 20°C, has an osmotic pressure of about 1.07 atm.

OTHER TYPES OF SOLUTIONS

Other types of solutions very often encountered are solutions of liquids in liquids; solutions of gases in gases, liquids, and solids; and solutions of solids in solids.

Solutions of Liquids in Liquids

Two liquids may conduct themselves toward each other in either of two ways: They may each reach a definite limit of saturation with the other, or they may be freely soluble (or miscible) in all proportions. Several properties of liquid solutions are of importance.

1. *Vapor pressure.* Before mixing, each liquid has its own characteristic vapor pressure at the temperature in question. After mixing, it is found that each liquid has diminished the vapor pressure of the other, depending upon the relative amounts of the two liquids. The vapor pressure of the solution is never as great as the sum of the two original pressures; it may be greater or less than that of either liquid taken separately, or it may have an intermediate value.

2. *Boiling point.* On heating a solution of one liquid in another, the total vapor pressure increases. And when it just exceeds the opposing pressure of the atmosphere, the solution boils. From what has been said in regard to the vapor pressure of solutions, it will be seen that the boiling point of a solution may be lower or higher than that of either constituent. Usually it has an intermediate value [Figure 12.10(a)].

3. *Fractional distillation.* When distilling a solution, *the component having the greater vapor pressure will in general pass away from the solution more rapidly than the one of lower vapor pressure* (higher boiling point). If the vapors are condensed and the resulting liquid collected in successive portions by changing the receiver at intervals, the first portions will be richer in the more volatile constituent, and the higher-boiling liquid will be largely obtained in the later portions. By repeating the process with each portion obtained in the first operation, the two liquids may in time be separated from each other. Such a process is called *fractional distillation*.

4. *Constant-boiling solutions.* Not all liquid mixtures show the simple boiling temperature characteristic illustrated in Figure 12.10(a). More complex types are shown and described in Figure 12.10(b) and (c). A common example of a constant-boiling solution with a maximum boiling point [Figure 12.10(b)] is a solution of approximately 95 per cent ethyl alcohol and 5 per cent water by volume.

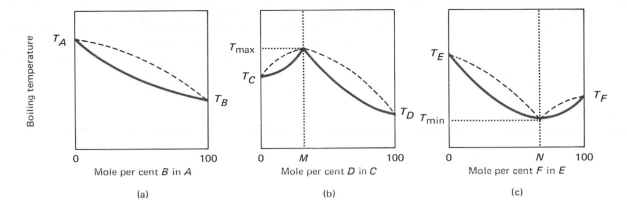

Figure 12.10 Boiling point as a function of composition for three types of liquid pairs boiling under constant pressure. In (a), fractional distillation will separate the liquids into the two components. Graphs (b) and (c) show maximum and minimum boiling points for liquid pairs C, D and E, F. Both graphs may be separated into two parts (vertical dotted lines)—each of which looks like (a). Fractional distillation will separate only into the extremes of the two parts—for example, into C and M (b) or M and D, but not into C and D. At each temperature, the composition of the vapor (dashed curves) is richer in the lower boiling material except for the constant-boiling mixtures where liquid and vapor have the same composition. Hence, these constant-boiling mixtures boil unchanged, and no separation of the two components occurs. At a different pressure, the position of the curves will be different, just as, for example, T_A or T_B differ at different pressures.

Solutions of Gases in Gases Dalton's law of partial pressures applies to such systems.

Solutions of Gases in Liquids The solubility of a gas in a liquid depends upon the temperature and pressure and may be expressed in a variety of ways; the most prevalent usage is to state the number of volumes of gas dissolved in 100 volumes of the solvent.

With respect to solubility, gases fall roughly into two classes: those of rather small solubility, such as oxygen, nitrogen, and hydrogen; and those of much larger solubility, such as ammonia, hydrogen chloride, and sulfur dioxide. In the case of the latter group there is always a change in the volume of the solvent when a large volume of gas is absorbed; the volume may increase markedly, as for solutions of ammonia in water. The slightly soluble gases do not cause any appreciable volume change in the solvent.

Factors Affecting the Solubility of Gases The main factors affecting the solubility of gases are the specific properties of the solute and solvent (Table 12.4), the pressure, and the temperature.

The high solubility of ammonia and sulfur dioxide in water is that an acid-base reaction occurs, represented by the equations:

$$NH_3 + H_2O \rightleftharpoons NH_4^+ + OH^- \tag{1}$$

$$SO_2 + 2H_2O \rightleftharpoons H_3O^+ + HSO_3^- \tag{2}$$

Other very soluble gases undergo similar reactions with water, forming ions.

Gases become *more* soluble in liquids, as the pressure is increased. The change in solubility with change in pressure follows no standard rules for the more soluble gases. For the slightly soluble gases, however, such as oxygen or hydrogen, *the weight of gas dissolved by a definite quantity of a given solvent is directly proportional to the pressure, provided the temperature remains constant*

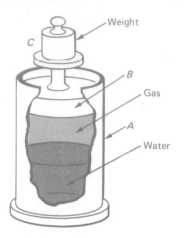

Figure 12.11 A gas standing over water in the cylinder, *A*, dissolves under the pressure of a weight, *C*, which is placed on the piston, *B*.

Table 12.4 Solubility of Gases in 100 ml of Water

| Gas | Amount Dissolved at 0°C and under 760 torr *Pressure* | |
	(g)	*(cc)*
Hydrogen	0.000191	2.14
Nitrogen	0.0029	2.33
Oxygen	0.0070	4.96
Carbon dioxide	0.339	171.3
Hydrogen sulfide	0.67	437
Sulfur dioxide	23.35	7,979
Hydrogen chloride	82.97	50,600
Ammonia	100.12	129,890

(Figure 12.11). This statement is known as Henry's law, after the English chemist William Henry who discovered the relationship in the early years of the 19th century.

When a mixture of two or more gases of low solubility is maintained over a liquid, *each dissolves independently of the others and in accordance with its own partial pressure.* This is a necessary consequence of Dalton's law.

With a rise in temperature, gases become less soluble and, except in special cases in which constant-boiling solutions (page 266) are formed, all of a dissolved gas can be driven out by boiling the solution. Gases do not dissolve appreciably in ice.

Solutions of Gases in Solids

Although not commonly recognized, solutions of gases in some solids appear to exist. The helium gas found in some uranium ores is probably very nearly in atomic dispersion since it is formed one atom at a time. Other cases that appear to be solutions of gases in solids may be due to adsorption over a large active surface, and some cases may be caused by occlusion in imperfections or voids in the crystal lattice. Some metals, notably palladium, dissolve large quantities of hydrogen; this metal also adsorbs hydrogen. Some large complex organic molecules form cages that occlude atoms and small molecules; such substances are called clathrate compounds.

Solutions of Solids in Solids

Solutions of solids in solids are discussed in Chapter 13.

IONIZATION

Irregularities in the Effects Produced by Solutes in Liquid Solvents

Experiments show that compounds can be divided into two classes on the basis of the effects they produce on the properties of the solvent liquid: (a) those that give the normal molecular effects on freezing point, vapor pressure, boiling point, and osmotic pressure discussed earlier in this chapter; and (b) those that produce excessive effects on these properties. This distinction may be seen by considering the lowering of the freezing point (Table 12.5). The first three substances are examples of a very large group of compounds for which the freezing point depression is (1.86°C) × (molality). The *freezing point depressions* of the second and third groups are *greater than (1.86°C) × (molality)* and are known as abnormal freezing-point depressions.

The Theory of Ionization

In his experiments on electricity, Michael Faraday (1791–1867) had observed that while pure water was a very poor conductor of electricity, solutions of certain substances made it a good conductor. He called these *substances, whose water solutions conduct an electric current,* electrolytes Svante Arrhenius noted in 1883 that Faraday's electrolytes included all the compounds whose solutions showed abnormal effects in freezing-point depression and boiling-point elevation. Arrhenius knew from Raoult's law that the freezing-point lowering was proportional to the number and not the kind of dissolved molecules; if each solute molecule separated on dissolving into two or more particles, and if each had the same effect on the freezing-point depression as that of the whole molecule, then the observed abnormal effect would appear. If, in turn, these fragments of molecules were electrically charged, and by motion through the solution would transport charge from one electrode to the other, their presence would also account for the conductive properties of the solution.

The assumption that substances dissociate into charged particles rather than neutral atoms was also necessary to explain chemical behavior. For example, a compound such as sodium chloride could not separate into sodium and chlorine atoms in water solution, since sodium reacts with water to produce hydrogen, and no hydrogen is evolved when sodium chloride dissolves in water. If it separated into the charged ions Na^+ and Cl^-, however, these would not be expected to show the properties of the uncharged atoms.

Ions from Some Covalent Molecules

Solutions containing ions are sometimes formed when covalent molecules are dissolved in water. It has been mentioned earlier than when ammonia dissolves, reaction takes place with the formation of ions:

$$NH_3 + H_2O \rightleftharpoons NH_4^+ + OH^- \tag{1}$$

Further, when the covalent gas hydrogen chloride dissolves in water, hydronium and chloride ions are formed as a result of reaction with the water. In the case of hydrogen chloride, this reaction is nearly complete, so that the solution contains practically no hydrogen chloride molecules; with ammonia or acetic acid, the reaction is incomplete and most of the solute remains as molecules rather than as ions.

Table 12.5 Freezing-Point Lowering Produced by Several Kinds of Solutes

Substance	Formula	Lowering of Freezing Point (°C) Produced by Dissolving 1 Mole in 1000 g of Water
Dextrose	$C_6H_{12}O_6$	1.90
Alcohol	C_2H_5OH	1.83
Urea	$CO(NH_2)_2$	1.86
Acetic acid	CH_3CO_2H	1.93
Ammonia	NH_3	1.93
Sodium hydroxide	$NaOH$	3.44
Sodium chloride	$NaCl$	3.37
Sodium nitrate	$NaNO_3$	3.02
Hydrogen sulfate	H_2SO_4	4.04
Calcium nitrate	$Ca(NO_3)_2$	4.59
Magnesium sulfate	$MgSO_4$	2.02

The data in Table 12.5 show that the freezing-point depression produced by the electrolytes is not as great as that to be expected for complete separation of the ions. For sodium chloride, for example, the freezing-point depression for the molal solution is not quite twice that for a molal solution of a nonelectrolyte, as would be expected if two ions were formed for each molecule. Similarly, the freezing-point depression for calcium nitrate is not that to be expected for three ions from each molecule, and for magnesium sulfate the effect is only slightly greater than for a nonelectrolyte. The reason for this was interpreted by G. N. Lewis and by Peter Debye. They showed that the electrostatic forces between the ions are not obliterated completely in the solution, but are still sufficiently powerful *to pull the ions toward one another*. As a result of this interionic attraction, the ions are not completely freed from one another, and the partial effects recorded in the table are observed experimentally. Similar partial effects are observed for vapor-pressure lowering, boiling-point elevation, and osmotic pressure.

1. *Effect of charge.* As one would suppose, the electrostatic forces are greater the higher the charge on the ions. This explains the difference between sodium chloride, for example, and magnesium sulfate (Table 12.5). The doubly charged magnesium (Mg^{+2}) and sulfate (SO_4^{-2}) ions have a greater attraction for each other than Na^+ and Cl^-, and consequently a smaller effect on the depression of the freezing point.

2. *Effect of solvent.* The magnitude of the attractive forces between ions depends upon the solvent in which they are dissolved, and is determined by the insulating properties of the solvent as expressed by the dielectric constant. In water, with a high dielectric constant, the electrostatic forces are small, so that the ions are relatively free. Most other solvents have smaller dielectric constants than water, so that the electrostatic forces are greater, with consequent greater interionic attraction.

3. *Effect of dilution.* Besides the dependence of the electrostatic forces on the charge of the ions and on the solvent, they also depend upon the distance between the ions, being smaller the farther apart the ions are.

Strong and Weak Electrolytes

In the data of Table 12.5, it is seen that the freezing-point depression for the second and third groups of compounds is greater than the average value of 1.86 for the covalent compounds dextrose, alcohol, and urea. But the freezing-point depression for acetic acid and ammonia is much less than for sodium hydroxide and the other compounds of the third group. The difference in freezing-point depression of these two groups must be due to the difference in the number of ions present in the 1-molal solutions examined. For ammonia and acetic acid, ions are formed from covalent molecules in reversible reactions, illustrated by Equations (1) and (2); (see also Chapter 7).

$$CH_3CO_2H + H_2O \rightleftharpoons H_3O^+ + CH_3CO_2^- \tag{3}$$

At ordinary concentrations, the ionization is not very great, but it increases with dilution, owing to the lessened chance that the ions have of recombining the farther apart they are.

Electrolytes similar to sodium hydroxide and sodium chloride in their freezing-point depressions are called *strong electrolytes*; this class includes most salts and strong acids and bases such as hydrochloric acid and sodium hydroxide. Those of the acetic acid type are called *weak electrolytes*; ammonium hydroxide (ammonia) belongs to this class. The difference between the two classes is that strong electrolytes *have only ions present in solution,* held together to a greater or lesser extent by interionic attractive forces determined by the charge, the solvent, and the dilution. Weak electrolytes, on the other hand, *exist in solution predominantly as covalent molecules,* only a few of which are ionized at any one time. The difference between strong and weak electrolytes is experimentally more evident from a consideration of the conductivity of solutions containing electrolytes.

Electrolytic Conduction

If two oppositely charged electrodes are placed in a solution containing ions, the positive ions will be drawn toward the negative electrode, and the negative ions will be drawn toward the positive electrode. This movement of charged particles through the solution results in a transfer of electric charge; we recognize this as an electric current, and say that the solution conducts the electric current. This type of *conduction, which results from the motion of ions through the solution,* is called *ionic* or electrolytic conduction to distinguish it from electronic or metallic conduction, which *occurs when a current of electrons flows through a wire* (Figure 12.12).

The amount of charge transferred across the solution in unit time will depend upon the number of ions present in the solution and the speed with which they move. In the case of strong electrolytes, the dissolved substance is all in the form of ions and the conductivity is high; in the case of weak electrolytes, mostly covalent molecules are present in the solution, with comparatively few ions, and the conductivity is low.

Electrolytic conduction can also occur if an ionic crystal is melted. The ions are released from the rigid-patterned arrangement of the crystal (Figure 5.6) and are free to move. Thus if electrodes are placed in the melt, the ions will move toward the electrode of opposite charge, and the melt will conduct a current.

Speed of Ionic Reactions

Reactions involving only electrolytes in solution are usually much more rapid than most of those involving nonelectrolytes. For example, if solutions of sodium chloride and silver nitrate (both electrolytes) are mixed at room temperature, a precipitate of silver chloride forms immediately. By contrast, if silver nitrate solution is added to a nonelectrolyte containing chlorine, such as carbon tetrachloride, a precipitate forms only after heating for a long time.

Figure 12.12 Apparatus for testing conductivity of a solution. Equal volumes (to cover the electrodes to the same height in each case) of solutions of different substances having equal concentrations are placed in the bottle, and the intensity of the glow from the lamp is observed. In precise determinations of conductivity, electrical resistance of the solution is measured by special Wheatstone-bridge circuits using low-intensity alternating current.

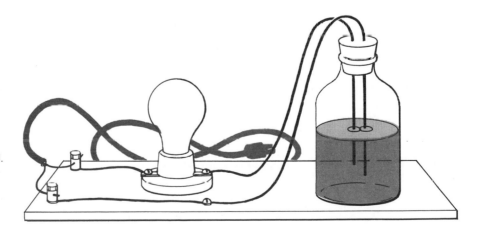

The reason for the behavior of the two electrolytes is that the formation of silver chloride consists solely in the combination of the silver ions and chloride ions already present in the solutions of silver nitrate and sodium chloride before mixing, so that the reaction is

$$Ag^+, NO_3^- + Na^+, Cl^- \longrightarrow AgCl(s) + Na^+, NO_3^- \qquad (3)$$

or simply,

$$Ag^+ + Cl^- \longrightarrow AgCl(s) \qquad (4)$$

No chloride ions are present in the solution of the nonelectrolyte carbon tetrachloride, and no reaction with silver ions occurs until hydrochloric acid has been formed by the slow chemical reaction of carbon tetrachloride with the water present.

Properties of Ionic Solutions Are Those of the Ions

It is evident also that all solutions containing the chloride ion will behave similarly toward the silver ions, since it makes no difference whether the chloride ion was originally paired with sodium, potassium, hydrogen, or any other ion. The essential reaction is still the simple combination of silver and chloride ions represented by Equation (4).

Just as all solutions containing chloride ions will show the property characteristic of that ion—forming a precipitate with silver ion—so the properties characteristic of other ions will be shown by all solutions containing those ions. The copper ion, for example, has a characteristic blue color. This blue of the copper ion is shown by all solutions of copper salts, unless the negative ion is also colored.

Concentration of Ionic Solutions

The customary concentration units—molarity, molality, and mole fraction—are used to express the concentration of ionic solutions; in addition, the term *normality* (N) is often used. For acids and bases, a normal solution of an acid contains, in 1 liter of solution, 1 mole of replaceable hydrogen ions. A normal solution of a base contains 1 mole of replaceable hydroxide ions. HCl has one replaceable H^+ ion, H_2SO_4 has two; sodium hydroxide has one replaceable OH^- ion, calcium hydroxide, $Ca(OH)_2$, has two.

In a normal solution of a strong acid such as hydrochloric acid, there is present a greater concentration of hydrogen ions than there is in a normal solution of a weak acid such as acetic acid. However, the concentration of potential hydrogen ions is the same in each solution. Consequently, the same amount of base is required to neutralize equal quantities of the two normal solutions. As rapidly as the hydrogen ions actually present in a solution of a weak acid enter into combination, other hydrogen ions are formed by the ionization of the weak acid.

Use of Normality

The use of normality to calculate the concentrations of solutions is demonstrated in the following problems. In many such calculations, the relationship

Volume of solution A × normality of solution A = volume of solution B × normality of solution B

or

$$V_A \times N_A = V_B \times N_B$$

is used.

Calculate the normality of a base 20 ml of which will neutralize 30 ml of 0.2 N acid.

Solution

$$V_A = 30 \text{ ml} \qquad V_B = 20 \text{ ml}$$

$$N_A = 0.2 \qquad N_B = \text{?}$$

$$30 \times 0.2 = 20 \times N_B$$

$$N_B = \frac{30 \times 0.2}{20} = 0.3 \; N$$

PROBLEM 2   How much barium hydroxide, $Ba(OH)_2 \cdot 8H_2O$, must be dissolved to make 500 ml of 0.1 N solution?

Solution   By definition, a 0.1 N solution contains 0.1 mole replaceable OH^- per liter

$$\frac{0.1 \text{ mole replaceable } OH^-}{1 \text{ liter}} \times \frac{1 \text{ mole } Ba(OH)_2 \cdot 8H_2O}{2 \text{ moles replaceable } OH} \times \frac{315.5 \text{ g } Ba(OH)_2 \cdot 8H_2O}{1 \text{ mole } Ba(OH_2 \cdot 8H_2O}$$

$$= 15.78 \text{ g } Ba(OH)_2 \cdot 8H_2O \text{ per liter}$$

For 500 ml, we need one-half of this, or 7.9 g.

Body Fluids: Precious Solutions.   It is said that living organisms, having evolved from the primordial sea, have effectively adapted to their salt-water environment and have even developed means of exploiting the unusual properties of water. One such property is the high specific heat of water, which is useful as a heat buffer in maintaining a relatively constant temperature in the organism even though the external temperature fluctuates. Another property is the high heat of vaporization of water, which is used to dissipate heat generated by the oxidation of foodstuffs, by other reactions, or during vigorous exercise. Thus the evaporation of water through the lungs and the skin in higher animals makes this dissipation of heat possible and helps assure a constancy of temperature in the body.

The excellent solvent properties of water are used to dissolve essential electrolytes and polar substances such as glucose and amino acids, and to assist in maintaining the proper acid-base balance. The relatively low viscosity of water is exploited in the circulatory system, which carries nutrient materials, enzymes, and regulatory substances such as hormones to the tissues and waste products to the excretory organs. Water also is used as a carrier to transport food and food products in the intestines. About two-thirds of the human body is water.

Water and electrolyte balance (Chapter 8) must be carefully maintained in the body. To some extent, both these balances are under hormonal control. The pituitary hormone, vasopressin, has an antidiuretic effect, functioning to retain water in cases of minor depletion. Under normal conditions, however, the water balance is maintained by an equality in the fluid uptake and excretion.

The body receives water from three sources: from ingested beverages, from ingested foods, and by oxidation of foodstuffs. An average size adult might drink 1 liter of water a day, acquire another 1.5 liters in the food he eats and perhaps 1/4 liter from oxidation of foods.

Excretion of water occurs from the lungs, skin, intestines, and kidneys. However, regulation of water balance is maintained only through that excreted by the kidneys. Water released through the lungs and skin serves only to control temperature, while that excreted by the intestines functions largely to maintain proper consistency of intestinal contents. The average-size adult may lose a liter of water a day from his lungs and skin, about 1.5 liters in urine, and about 1/4 liter in feces. Should he drink much more than usual and not excrete additional water in perspiration, he will void the excess as urine.

The daily intake and output of fluid is much less than the daily fluid turnover in the digestive system. In addition to the nearly 3 liters of "new" water daily, the body might develop about 1.5 liters of saliva, 2.5 liters of gastric juices, 3 liters of intestinal secretion, 0.75 liter of pancreatic juices, and 1/2 liter of bile.

Saliva, produced by glands under the tongue and in the inner surface of the mouth, is about 99.4 per cent water. It also contains urea, uric acid, and enzymes that hydrolyze starch.

Gastric juice is secreted by the stomach wall and is stimulated by the sight or odor of food, or by the presence of food in the mouth. It contains from 0.2 to 0.5 per cent hydrochloric acid and enzymes that hydrolyze proteins and fats. Intestinal secretion, pancreatic juices, and bile are juices that are secreted into the duodenum. All are basic. The first two contain sodium bicarbonate and a variety of enzymes. Bile contains no enzymes, yet it aids in the digestion of fats. Important among its components are the bile salts, which account for its basicity; the bile pigments, which account for its yellow-to-green color; and cholesterol (p. 640). The bile salts lower the surface tension of water and assist in emulsifying fats, thereby rendering them more readily attacked by the pancreatic juices. Bile salts also aid in the absorption of fatty acids through the walls of the intestines.

The internal fluids of the body include blood, lymph, tissue fluid, and intracellular fluid. Blood is the most active transport system and consists of a liquid medium, the plasma, and suspended cellular components. Lymph is a slow-moving fluid that occurs in body cavities and between cells. It also moves sluggishly through a system of vessels called the lymphatics. Tissue fluid surrounding the tissues is a gelatinous material through which fluid can readily pass.

Two-thirds of the blood volume is plasma; corpuscles make up the other third. Plasma contains the cations sodium (in by far the largest amount), potassium, calcium, and magnesium; the anions chloride (in largest amount), bicarbonate, hydrogen phosphate, sulfate, and carboxylate (p. 577). It also contains proteins. An average-size adult may have about 6 liters of blood.

Lymph is a colorless or yellowish fluid containing the same cations and anions as plasma but less protein. A primary purpose of the lymphatic system is to drain away from the tissues any excess protein that has escaped from the blood and, ultimately, to return it to the veins. The volume of lymph and tissue fluid in the average-size adult is 10.5 liters.

Intracellular fluid is the liquid found inside the cellular membrane. It is rich in potassium ion and hydrogen phosphate ion. The body of an average-size man contains about 35 liters of intracellular fluid.

SUMMARY The process of dissolving is seen as one in which the attractive forces of the solvent for the solute are strong enough to overcome the attractive forces that hold solute particles together. For polar solvents, such as water, the solvent molecules are dipoles that are attracted to ionic or polar solutes. Usually the solvent molecules surround the solute particle and tend to insulate it from other solute particles.

Convenient expressions of the concentrations of solutions are the molarity, M, defined as the number of moles of solute present in each liter of solution, and the molality, m, defined as the number of moles of solute dissolved in each kilogram of solvent. For acid-base titrations, the normality, N, is sometimes used.

The presence of a solute has a marked effect on the vapor pressure, boiling point, and freezing point of the solvent. The lowering of the vapor pressure and of the freezing point and the rise in the boiling point of the solvent caused by the presence of a nonvolatile, nonelectrolyte solute are proportional to the molality of the solution. Use is made of this fact to determine molecular weights of such solutes and to achieve temperatures below the freezing point of the solvent.

Solutions of certain solutes have unusually low vapor pressures and freezing points. The same solutions are good conductors of electricity. These observations are explained if the solutes are ionized in solution. Careful studies of solutions of electrolytes show that ions of opposite sign in solution still are attracted to one another. The effects of this interionic attraction are seen in the vapor-pressure lowering, the boiling-point elevation, the freezing-point depression, and in the conductivity of the solution.

If the solute is completely ionized, it is known as a strong electrolyte. Weak electrolytes are partially ionized in solution; the ions exist in equilibrium with nonionized species. Weak acids and bases are weak electrolytes in water solutions.

IMPORTANT TERMS

Concentration of solution
saturated solution
supersaturated solution
molar solution
molality
mole fractions
normality

Solubility
solute
solvent
heat of solution
solvation
hydrate
energy of hydration
Henry's law
eutectic
Raoult's laws

Dipole-dipole forces

Dielectric constant

Osmosis
Osmotic pressure

Fractional distillation
Constant-boiling solution

Electrolyte
abnormal freezing-point depression
interionic attraction
strong electrolyte
weak electrolyte
electrolytic conduction

QUESTIONS AND PROBLEMS

1. Compare evaporation and solution; melting and solution.

2. Carbon dioxide and sulfur dioxide are more soluble in water than are oxygen and nitrogen. Can you suggest a reason why this should be so?

3. Which of the following solutions will have the lower freezing point: (a) A solution of 100 g of urea in 600 g of water; (b) a solution of 100 g of alcohol (CH_3CH_2OH) in 460 g of water?

4. Calculate the freezing-point depression of a solution of 3.2 g of methyl alcohol (CH_3OH) dissolved in 50 g of water.

5. What weight of ammonium chloride is present in 600 g of a 2.0 molal solution of that salt?

6. Calculate the molar concentrations of the solutions prepared by dissolving the quantities of substances listed in the left-hand column and adding water until the final volume, given in the right-hand column, is reached.

Substance	Final Volume
60 g of urea, $CO(NH_2)_2$	500 ml
202 g of potassium nitrate, KNO_3	1600 ml
55.5 g of calcium chloride, $CaCl_2$	2.0 liters
21.9 g of calcium chloride, $CaCl_2 \cdot 6H_2O$	100 ml

7. Why does putting salt on an icy sidewalk in the winter cause the ice to melt? Does the sidewalk get warmer?

8. List the following in the order of decreasing freezing point: (a) water, (b) 1-molal urea solution, (c) 0.5-molal sugar solution, (d) 1-molal calcium chloride solution, (e) 0.5-molal sodium chloride solution, (f) 2-molal potassium chloride solution.

9. Solutions were prepared by dissolving 10.00 g of each of the following solutes in water, and adding water to the 100-ml mark in a volumetric flask. Find the molarity of each solution: (a) NaOH; (b) NaCl; (c) glucose, $C_6H_{12}O_6$; (d) K_2SO_4; (e) $KMnO_4$.

10. Calculate the number of grams of each of the following substances that must be dissolved in water to produce 250 ml of a solution that is 0.10 M in sodium ion. (a) Na_2CO_3, (b) NaCl, (c) $Na_2SO_4 \cdot 10 H_2O$, (d) Na_3PO_4.

11. To carry out a certain reaction, 4.2 millimoles of hydrogen chloride are needed. How many milliliters of 0.60 M solution of hydrochloric acid must be used?

12. How many grams of each of the following substances are contained in 55 ml of a 0.25 M solution of each substance? (a) $CaCl_2$, (b) $Ba(OH)_2$, (c) $Cr_2(SO_4)_3$, (d) $Zn(NO_3)_2$.

13. Make the conversions indicated for each of the following aqueous solutions:
    (a) For calcium chloride, $CaCl_2$, 2.5 M to mg/ml of solution.
    (b) For sodium carbonate, Na_2CO_3, 5.0 g/l of solution to molarity.
    (c) For disodium phosphate, Na_2HPO_4, 0.35 millimole/ml of solution to g/l of solution.
    (d) For barium chloride, BaCl2, 20% by weight solution to molality.
    (e) For ammonium sulfate, $(NH_4)_2SO_4$, 0.5 molal weight to % solution.

14. A 1.00-g sample of impure oxalic acid, $H_2C_2O_4$, requires 35.6 ml of a 0.150 N solution of NaOH to be neutralized. What is the per cent of oxalic acid in the sample?

15. The freezing point of a particular aqueous solution is −2.00°C. What is its boiling point?

16. A bottle of concentrated sulfuric acid is labeled as follows: "Assay 72%

H_2SO_4; Specific Gravity 1.63." (a) What is the molarity of the sulfuric acid in the bottle? (b) How would you prepare 250 ml of 2.0 M solution of sulfuric acid using the concentrated acid?

17. Seventy ml of 1.50 M potassium iodide and 30.0 ml of 1.25 M potassium iodide are combined. What is the molarity of the resultant potassium iodide solution?

18. A solution of 3.5 g of a covalent substance X dissolved in 80 g of water was found to raise the boiling point 0.302°C. What is the molecular weight of X?

19. Calculate the osmotic pressure developed by a solution of 4.4 g of dioxane ($C_4H_8O_2$) dissolved in 350 ml of water at 25°C.

20. The following pairs of solutions are mixed together. Which pairs would you expect to show an appreciable heat effect? Explain why or why not. (a) 1 M sodium chloride and 1 M potassium nitrate, (b) 1 M magnesium chloride and 1 M sodium nitrate, (c) 1 M sodium hydroxide and 1 M hydrochloric acid, (d) 1 M sodium sulfate and 1 M sulfuric acid.

21. Using data presented in Figure 12.5, calculate the number of moles of lead nitrate that would crystallize out if 500 g of a solution of lead nitrate $Pb(NO_3)_2$, saturated at 70°C, were cooled to 0°C.

22. Sodium thiosulfate dissolves with the absorption of heat. If a supersaturated solution of the salt is prepared and a small crystalline fragment of sodium thiosulfate is dropped into the solution, will you expect to observe a rise or fall in temperature as a result of the process initiated by the fragment? Explain the basis for your answer.

23. Calculate the number of milliliters of 0.500 N H_2SO_4 required to neutralize 100 ml of 0.300 N NaOH; of 100 ml of 0.0500 N $Ca(OH)_2$.

24. Calculate the number of grams of NaOH required to neutralize 0.500 gram of $H_2C_2O_4$ (which has two replaceable H^+ ions); 100 ml of 0.100 N $H_2C_2O_4$.

25. The freezing-point depression for a 1-molal solution of potassium nitrate is 2.6°C. What will happen if a 1 m KNO_3 solution is cooled from room temperature to −3°C?

26. Consider the hydration energies of the monovalent cations in Table 12.1. Can you suggest reasons why the order is $Li^+ > Ag^+ > Na^+ > K^+$? Will these same reasons explain the order $Mg^{+2} > Ca^{+2}$ and $F^- > Cl^-$?

SPECIAL PROBLEMS

1. What values of the ratios of the molal freezing-point depressions of salt solutions compared to those of sugar solutions would you expect to be approached at infinite dilution by the following salts: NaCl, $MgSO_4$, $AlCl_3$, $MgCl_2$, Li_2SO_4? If you were to plot a graph of this ratio against the concentration for each salt, how would you predict the curves to look in comparison to each other? Sketch six such curves on a graph.

2. Calculate the concentrations of potassium chloride, calcium chloride, potassium ion, calcium ion, and chloride ion in a solution prepared by dissolving 7.46 g of KCl and 5.55 g $CaCl_2$ in sufficient water to make 500 ml of final solution.

3. A certain salt has a solubility of 89 g/100 g H_2O at 10°C. At 70°C, the solubility increases to 172 g/100 g H_2O. A solution of the salt was pre-

pared at 70°C by dissolving as much salt as would dissolve in 500 ml of water. The solution was then cooled to 10°C. Describe in qualitative and in quantitative terms what is expected to happen.

4. In a 100-ml beaker, place 30 ml of water; in another, 30 ml of 0.10 molal sugar solution; and in a third, 30 ml of a 0.10-molal sodium chloride solution. Place a bell jar over all three so there is no loss of water vapor outside the bell jar. The temperature is 25°C. Allow the whole system to come to equilibrium. Will the amounts of liquid in each beaker remain the same? If not, estimate the relative amounts in each beaker.

5. Calculate the molecular weights for the nonelectrolyte solutes in the following. (a) The boiling point of water is raised 0.30°C when 4.8 g of the solute is dissolved in 160 g of water. (b) The freezing point of water is lowered 0.50°C when 6.0 g of solute is dissolved in 10 g of water. (c) The freezing point of benzene is lowered 0.53°C when 0.186 g of solute is dissolved in 10 g of benzene.

6. A weak acid is 15% ionized in a 0.100 m solution. What would be the expected freezing point?

7. An aqueous solution of a weak acid freezes at −0.500°C. (a) Express the concentration of the solution in solute particles/kg solvent. (b) If the molal concentration of the solution is 0.250 m, what is the per cent dissociation of the acid?

8. If 25.0 ml of 2.50M $MgCl_2$ is mixed with 75 ml of 3.40M $(NH_4)_2SO_4$, (a) what is the concentration of each substance in the final solution? (b) what is the concentration of each ion in the final solution?

9. At 0°C, 100 ml of water dissolves 4.96 ml of oxygen when the oxygen is at atmospheric pressure; for nitrogen, the corresponding figure is 2.33 ml. Using the approximation that air is $\frac{1}{5}$ oxygen and $\frac{4}{5}$ nitrogen, calculate the number of milliliters of each of the pure gases, measured at standard conditions, that are present in 100 ml of water saturated with air at atmospheric pressure at 0°C. Compare the concentration of oxygen in normal air with its concentration in water under these conditions, expressing each in moles per liter. Compare the ratio of oxygen to nitrogen in dissolved air with that in normal air.

10. Discuss what will happen in the following cases: (a) A solution of such concentration that it is not saturated at 20°C is prepared at 40°C and cooled to 20°C. (b) A solution of such concentration that it is saturated at 20°C is prepared at 40°C and cooled to 10°C. (c) A water solution of such concentration that it is not saturated at 0°C is prepared at 25°C and cooled to 0°C. (d) A 1-molal solution of urea in water is prepared at 25°C and cooled to −5°C.

REFERENCES CHAVE, K. E., "Chemical Reactions and the Composition of Sea Water," *J. Chem. Educ.*, **48**, 148 (1971).

DREISBACH, D., *Liquids and Solutions*, Houghton Mifflin, Boston, 1966.

HILDEBRAND, J. H., "A View of Aqueous Electrolytes Through a Watery Eye," *J. Chem. Educ.*, **48**, 224 (1971).

HOLDEN, A., and SINGER, P., "Crystals and Crystal Growing," Doubleday, New York, 1960.

THE METALLIC STATE: ARCHITECTURE OF SOLIDS

One of the major frontiers of research in modern physical science is the solid state. Such useful devices as the transistor, the laser or optical maser, and the solar battery have been developed as a result of scientific enquiries into the nature of solids. Although scientists have been interested in this topic for over a hundred years, modern crystal chemistry began during the early years of this century—about the time that Bohr and Rutherford were conducting their famous research into the structure of atoms. The electron energy-level theory of metals was devised even later; most of this research was done toward the middle of this century.

USE OF X RAYS TO DETERMINE CRYSTAL STRUCTURE

The development of X-ray diffraction, which led to a clear understanding of the internal structure of crystals, emerged from: (a) the idea, held by many 19-century crystallographers, that the external symmetry of crystals was the result of a symmetrical arrangement of the atoms or other particles within the crystal (Figure 13.1); (b) the prediction by the German scientist Max von Laue (1879–1960), dramatically confirmed by experiment, that because of the regularity in crystals, a beam of X rays would be scattered or diffracted from crystals giving a pattern that would indicate the arrangement of the particles in the crystal; and (c) development of an equation by the British scientists, Sir William Henry Bragg and his son, Sir William Lawrence Bragg, that made it possible to simplify the interpretation of the patterns.

---

*"If one agrees that some force or other acts, at the moment of crystallization, to cause the grouping being formed to favor a symmetric rather than an unsymmetric structure, it is clear that the grouping finally formed will belong to one of our seven classes . . . (Class 1: the cube, the body-centered cube, the face-centered cube; class 2: . . .). Examination of crystalline substances, natural or man-made, shows *a posteriori* that this is the case; further the geometrical classification of the groupings corresponds exactly to that which a patient and careful study has established for the different crystal systems." (The quotation is from *Journal de l'Ecole Polytechnique*, **33**, 127 (1850).

Figure 13.1 Photograph of crystals of sodium chloride, NaCl, and hydroquinone, $C_6H_6O_2$, taken through a microscope. The cubic NaCl crystals are the size found in ordinary table salt. The rod-like, hexagonal prisms of hydroquinone are approximately one cm long.

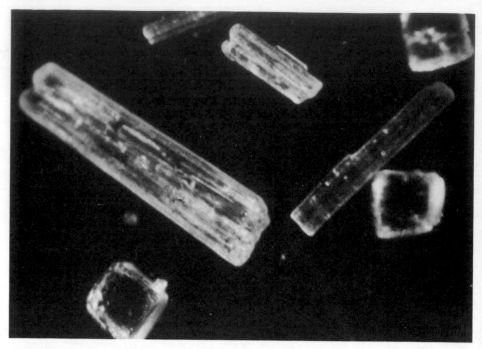

Subsequent investigations of the internal structure of crystals with beams of X rays have led to the following generalizations:

1. The macro crystal can be imagined to consist of stacks of repeating units or repeating groups. Each repeating unit is called the unit cell.

2. The unit cell may contain atoms, molecules, or ions arranged in systematic three-dimensional patterns in space.

3. The three-dimensional pattern also can be described in terms of arrays of identical points in space called space lattices. A space lattice can be constructed by replacing identical atoms, molecules, ions or similarly oriented groups in the crystal structure by dots. There are many different kinds of crystal structures and space lattices. For example, in the diamond structure, each carbon atom has four near neighbors; in sodium chloride, each ion has six near neighbors; in metallic zinc, each atom has 12 neighbors.

4. The arrangement of atoms, molecules, or ions in a crystal is related precisely to the external symmetry of the crystal. For example, there are cubic space lattices (three) which result in cubic crystals, tetragonal space lattices (two) which give tetragonal crystals, orthorhombic (four), rhombohedral (one), monoclinic (two), triclinic (one), and hexagonal (one) lattices which give crystals of the seven recognized crystal systems illustrated in Figure 13.2.

As an analogy, you might compare a crystal with a truckload of cases of canned goods. The unit cell can be imagined to be related to the case itself and to the way the cans are packed in the case; the external crystal symmetry to be related to the shape of the stack of cases. The shape of the stack is, of course, related to the shape of each case, and the shape of the case is related to the packing pattern of cans in the case.

Some examples of various types of crystals and their structures will be examined in this chapter with a view toward explaining the properties of solids. Perhaps before doing this, it will be useful to sketch an elementary picture of the principles underlying the interaction of X rays with crystals.

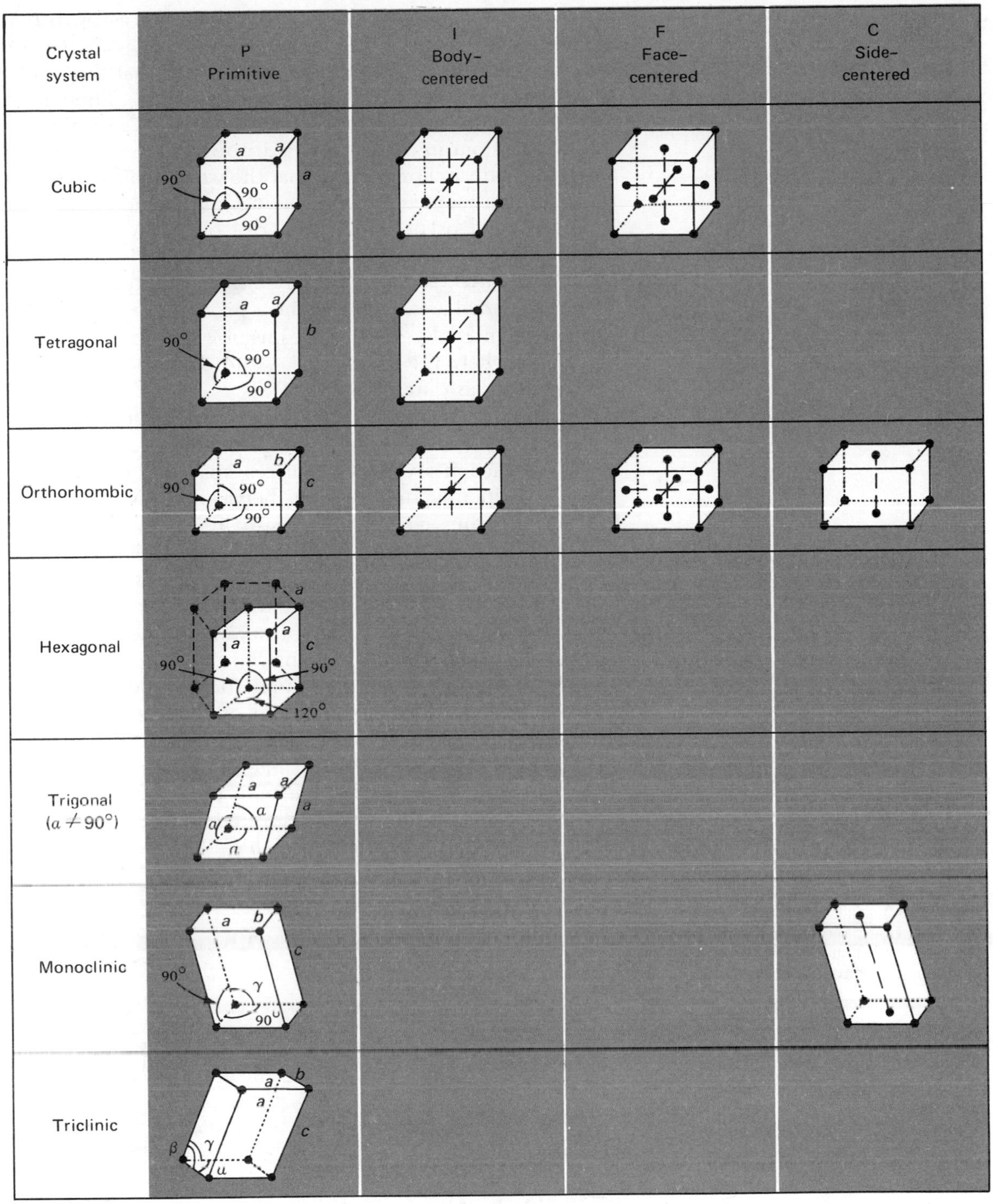

Figure 13.2 The seven crystal systems.

When X rays pass through a crystal, they interact with the electrons of the atoms or ions in the crystal in such a way that the atoms or ions act as if they were new sources of X radiation, and we say that the incident X-ray beam is scattered by the structural units in the crystal. The scattered beams move out from the structural units in all directions, as a wave on the surface of water moves out in circles from a dropped stone. The structural units are in a regular pattern, however, so that the waves from one unit overlap the waves from another, sometimes reinforcing, and sometimes destroying, the other. Reinforcement occurs when the crest of one wave coincides with the crest of another. This is shown in Figure 13.3(a), in which the semicircles represent advancing wave fronts from the sources *aa*, and two reinforced beams are shown as shaded areas. One of these proceeds in the straight-ahead direction, another at an angle. If these beams now fall upon a screen, you will observe light spots where the beams strike, and dark spots between. This phenomenon is known as diffraction.

The angular difference between the beams, for incident X rays of the same wavelength, depends upon the spacing of the scattering units. Figure 13.3(b) shows diffraction from two sources spaced farther apart than in Figure 13.3(a). Evidently the angle θ_b between the scattered beams is less than before. It is thus possible, by measuring the angles of diffraction, to determine the spacings (d) in the crystal, if the wavelength (λ) is known.

It should be noted that the formation of bright and dark areas on the screen results from the regularity of the arrangement of the scattering centers. If these were dispersed at random, no single line could be drawn tangent to the advancing wave fronts and, instead of a pattern of bright spots, you would observe only a diffuse general illumination of the screen. Hence the observation of bright spots on illumination of crystals with X rays, in the original experiments of Walter Friedrich and Paul C. M. Knipping suggested by Max von Laue (1913), is itself a confirmation of the regularity of the crystal lattice deduced earlier from the regularity of the external shape of crystals.

The possibility of distance measurement in crystals by X-ray diffraction was put to use in the Bragg method. This takes advantage of the fact that in the

Figure 13.3 Diffraction of waves (X rays). (a) Scattering from two point sources. (b) Scattering from two sources farther apart than in (a).

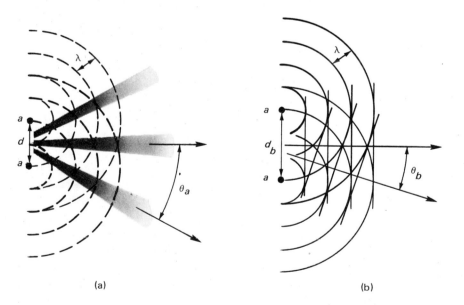

(a)

(b)

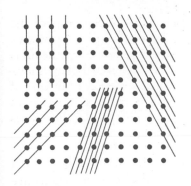

Figure 13.4 Planes in a regular arrangement of lattice points.

orderly arrangement of building blocks in a crystal, you can identify series of planes, as illustrated in Figure 13.4. The units in each plane act as the sources of Figure 13.3, and the incident X-ray beam will act as if it had been reflected from these evenly spaced planes. This too will give rise to reinforcement of the beam at certain angles and destruction at others, so that the spacings between planes can be determined. The condition for reinforcement is that the wavelength of the X ray, λ, and the distance, d, between planes are related to the angle of incidence (and "reflection") by

$$n\lambda = 2d \sin \theta$$

where n is an integer. Consequently, if you know λ for the X ray, and by experiment can determine the angle θ at which the intensity of the reflected beam is at a maximum, then you can calculate d, the only unknown quantity in the equation; d is the distance between the regularly spaced reflecting layers (Figure 13.5).

Electron Density Maps

Figure 13.5 "Reflection" of an X-ray beam. θ is the angle of deflection; d the distance between planes in the crystal.

Greater sophistication in the technique of X-ray diffraction has enabled crystallographers to learn from the direction of scattering something about the size and shape of the unit cell, and from the intensity of the scattering something about the distribution of electron density within the unit cell. To obtain information on the identity of the scattering units in complex crystals, one measures the intensity of the scattered radiation as a function of the scattering angle θ. For each scattering unit, the intensity depends upon the number of electrons in the unit, and the observed intensity is a sum of factors from all the scattering centers. By complicated and mathematically tedious processes, recently made easier by the employment of electronic computers, it is possible from the X-ray measurements to identify regions of high and low electron density within the crystal. These regions may be outlined on a contour map in which the regions of high and low density are represented as hills and valleys. Since the regions of high electron density represent the locations of atoms or groups of atoms, it is often possible to visualize, with the aid of the electron-density map, how the molecules are packed in the solid, or, as has been done in a few cases, to determine the particular arrangement of atoms within the molecules of the crystal. Figures 13.6 and 13.7 illustrate these two points. In Figure 13.6 the electron density map for naphthalene crystals is given, along with the structural formula for naphthalene molecules. The arrangement of these molecules in the crystal stands out clearly. In Figure 13.7 the electron-density map of the potassium benzyl penicillin crystal is given. Before this map was made, chemists did not know how the atoms were arranged in the penicillin molecule. The figure also shows the molecular structure assigned penicillin by crystallographers and since confirmed by several other methods.

METALS: THE ARRANGEMENT OF ATOMS IN METALS

Packing of Similar Spheres

In attempting to understand the structure of solids, it will be helpful to begin the study with metallic crystals, which are some of the simplest crystal systems. Metallic crystals are arrays of atoms. If the sample is a pure metal, all atoms will be of the same size and electron configuration. X-ray studies have shown that most metallic crystals can be regarded as a pile of similar spheres packed together so that as many spheres touch as possible. There are two "closest-

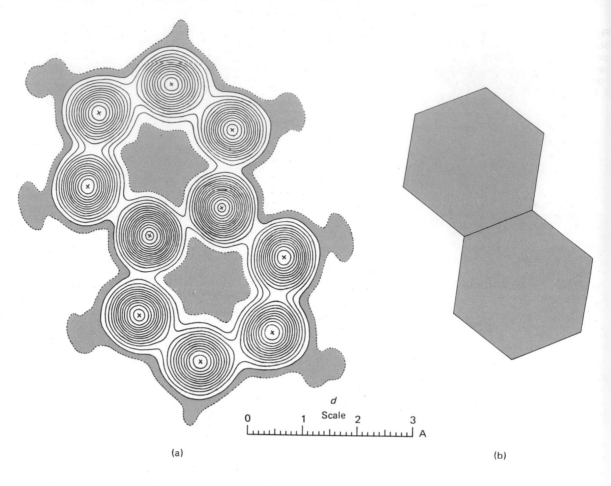

d
Scale
0   1   2   3
|.........|.........|.........|. A

(a)   (b)

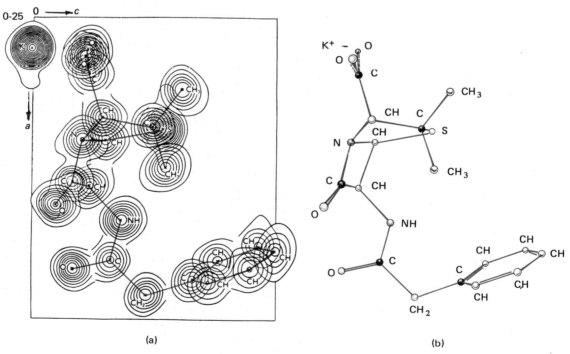

(a)   (b)

Figure 13.6 Naphthalene. (a) Electron density contours at intervals of 0.5 electron per cubic angstrom. (b) Conventional structural formula. [Part (a) from Robertson, J. M., *Acta Crystallogr.*, **2**, 241 (1949).

"packed" patterns for similar spheres, known as the hexagonal close-packed and the cubic close-packed structures. These are illustrated in Figure 13.8.

If a group of spheres (marbles, oranges, or baseballs) is placed on a level surface so that they are in contact with each other, an arrangement similar to that in Figure 13.9(a) is obtained. The second layer may be added by placing spheres in the depressions created by the groups of three spheres in the first layer. Only alternate depressions can be used since the spheres are too large to put into every depression. Imagine, then, that the second layer is formed by placing spheres in the depressions marked z in Figure 13.9(a).

When spheres are added to form the third layer, it is discovered that two types of depressions are available. One type of depression (x,x, . . .) is *directly above a sphere* in the first layer; the second type of depression (y,y, . . .) is *directly above a hole* in the first layer; see Figure 13.9(b). The two kinds of depressions alternate so that in forming the third layer, *all the spheres in that layer must be placed in only one kind of depression*—that is, all must be placed above spheres or all must be placed above holes. Herein lies the difference between hexagonal and cubic close-packing. If in the third layer (and in each successive third layer) spheres are placed over spheres in the first layer, hexagonal close-packing obtains; if spheres are placed over holes in the first layer, cubic close-packing obtains.

Another way to look at the differences between hexagonal and cubic close-packing is to note that in the hexagonal pattern, the layers appear to be in groups of two, whereas in the cubic pattern each layer appears to be related in the same way to the layers above and below.

The unit cells for both patterns are outlined in black on the stacks of spheres in Figure 13.8 and are represented more clearly in Figure 13.10. In the latter figure, the right angles between faces in the unit cell of the cubic form and the 120° and 60° angles between faces in the hexagonal cell are apparent. This figure also focuses attention on the number of atoms in the unit cell. Note that there are atoms (or fractions of atoms) at all corners and in the center of every face of the cubic cell. For this reason this close-packed structure is sometimes called face-centered cubic.

Figure 13.7 Potassium benzyl penicillin. (a) Electron density contours. This is a projection on a single one of the planes of the electron density contours taken for each atom on planes parallel to each other but passing through each atomic nucleus at different levels in the crystal. (b) A diagram showing the structural arrangement of the atoms in the benzyl penicillin anion. [Part (a) from Pitt, G. J., *Acta Crystallogr.*, **5**, 772 (1952).]

Another interesting aspect of the close-packed lattices is the fact that each point is surrounded by 12 close neighbors—three above, three below, and six around its equator. Each sphere in these structures is said to have a coordination number of 12. The coordination number is a useful concept for describing the packing in crystals. It will be used frequently in this chapter.

X-ray diffraction studies have shown that about 50 metallic elements crystallize in one or the other of these two close-packed structures. What causes a

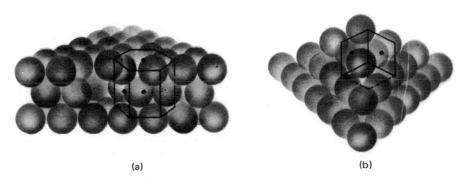

Figure 13.8 Close-packed structures: (a) hexagonal close-packing (one sphere in the second row has been removed to show the interior structure); (b) cubic close-packing.

(a)   (b)

Figure 13.9 Looking down on layers of spheres placed in contact on a level surface: (a) one layer of close-packed spheres; (b) two layers of close-packed spheres.

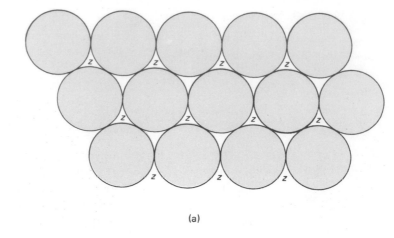

(a)

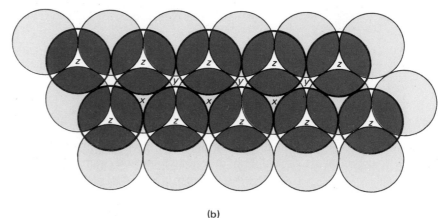

(b)

Figure 13.10 Unit cells of close-packed structures: (a) hexagonal close-packed unit cell; (b) cubic close-packed unit cell (face-centered cubic). (From the Chemical Bond Approach Project, *Chemical Systems*, McGraw-Hill, New York, 1964.)

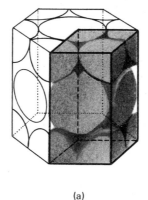

(a)

(b)

Figure 13.11 A body-centered unit cell. (From the Chemical Bond Approach Project, *Chemical Systems*, McGraw-Hill, New York, 1964.)

given metal to choose one or the other form is not known. Table 13.1 lists the packing patterns of the metallic elements. Some metals such as iron, cobalt, thallium, and lanthanum crystallize in more than one pattern; the different structures are stable under different temperature and pressure conditions.

About a dozen metals, including those in Group IA of the periodic table, crystallize at room temperature in what is known as a body-centered lattice,

Table 13.1 Crystal Structures of Metals

(Abbreviations: h, hexagonal; c, cubic close-packed; b, body-centered cubic; d, diamond; o, other)

Li,b	Be,h												
Na,b	Mg,h											Al,c	
K,b	Ca,c	Sc,o	Ti,h	V,c	Cr,c	Mn,o	Fe,b,c	Co,h,c	Ni,c	Cu,c	Zn,h	Ga,o	Ge,d
Rb,b	Sr,c		Zr,h			Tc,o	Ru,h	Rh,c	Pd,c	Ag,c	Cd,h	In,o	Sn,d
Cs,b	Ba,b	La,h	Hf,h	Ta,b	W,b	Re,o	Os,h	Ir,c	Pt,c	Au,c	Hg,o	Tl,h	Pb,c

illustrated in Figure 13.11. This is not a close-packed structure since the spheres at the corners do not touch each other. At lower temperatures, most of these metals revert to a close-packed structure. The coordination number in body-centered lattices is eight. The "open" packing results in a lower density for metals using the body-centered pattern.

Crystal Structure and Properties of Metals

Properties characteristic of metals include: metallic luster, high density, high melting point, malleability, ductility, high tensile strength, good heat and electrical conductivity, hardening, and thermionic emission. Not all metals possess all of these properties. For example, the melting points of mercury and the Group IA metals are very low; the densities of the Group IA metals are also low. The thermal conductivity of bismuth is only about 2 per cent that of silver; many of the transition metals are hard and brittle. Nevertheless, the metals as a class exhibit these properties. Presumably the properties are a result of the metallic bonding (p. 107) and the type of packing (p. 283) of the atoms in the crystal. Some of the properties of metals are discussed below.

Metal Luster. Metallic luster arises because the "free" electrons in the metal (Chapter 5) are able to absorb and quickly re-emit light photons of all frequencies. This makes for excellent reflecting surfaces.

Electrical Conductivity and Thermionic Emission. These properties are also caused by the "free" electrons in the metal. Electrical conductivity in solids is said to be a flow of electrons. Conductivity usually decreases with temperature because the vibrations of the atomic kernels in the crystal increase with temperature, thus providing an increased resistance to the flow of electrons. Thermionic emission is the emission of electrons from surfaces of heated metal cathodes placed in a vacuum. This phenomenon, which amounts to "boiling off" electrons from the hot metal surface, is the basis of an important part of the science of electronics.

Heat Conductivity. Heat conductivity is believed to result from electronic motion in metal crystals.

Malleability and Ductility. Malleability and ductility are related to deformation of the crystal lattice by an imposed stress. The stress is usually relieved by the slippage of adjacent crystal planes past one another, as illustrated in Figure 13.12. No fundamental change in the crystal structure occurs when this happens because the attractive forces holding the crystal together are unaltered. Slippage occurs most easily along close-packed planes, and is more likely to occur the greater the symmetry of the crystal. As a result, metals with cubic close-packed lattices are usually more malleable and ductile than those with body-centered cubic structures.

Slip planes

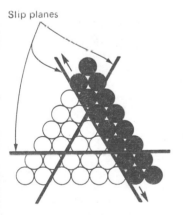

Figure 13.12 Slip planes in a metal crystal.

Work Hardening. In work hardening, the metal is made hard and brittle by bending or hammering. The hardening is believed to be caused in part by distortions of the crystal lattice that prevent adjacent planes from slipping past one another when the metal is put under stress.

Crystal Structure and Alloys

Most of the metals of commerce are alloys—mixtures of metal elements "tailor-made" for a given purpose. An enormous number of alloys is possible because a given alloy may contain more than two elements—perhaps up to five or more. Some common alloys and their properties are listed in Table 13.2.

The relation between the structures and properties of alloys has not been fully established. However, as a class, alloys are considerably harder than the pure metals. Presumably this hardness is due to the destruction of the slippage planes brought about by dispersing slightly larger or slightly smaller atoms throughout the crystal.

Several structurally distinct classes of alloys are known, and in certain cases it is possible to recognize a relation between properties and structure. Examples of some of these classes follow.

Solid Solutions

Solid solutions are *homogeneous solids whose composition can be varied over a wide range without loss of homogeneity.* The particles of the dissolved metals (the solute particles) are distributed at random among the particles of the solvent metal. Two important types of solid solutions are substitutional solutions and interstitial solutions.

Substitutional solutions arise when solute metal atoms are present at random in positions in the crystal lattice of the solvent metal normally occupied by atoms of the solvent metals (Figure 13.13). The existence of such solid solutions depends upon the relative sizes of the atoms of the two metals involved. Examination of a number of solid solutions reveals that in all cases, the size differences between atoms is 14 per cent or less. While solid solutions do not necessarily form if the atom sizes are within the limit, a size difference greater than 14 per cent is sufficient to prevent formation of solid solutions.

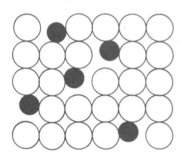

Figure 13.13 Substitutional solid solution (two-dimensional view).

Metal pairs that are within the 14 per cent range and form solid solutions often form alloys of more than one crystal structure. For example, pure copper has a face-centered cubic structure. Zinc dissolves in it without change in the crystal symmetry up to a ratio of about 40 atoms of zinc to 60 atoms of copper. Solutions containing a higher proportion of zinc take on a body-centered cubic structure, which persists until the atom ratio is about 60 atoms of zinc to 40 atoms of copper. Further addition of zinc produces other crystalline structures known as the λ and ε phases. Exactly similar changes in structure are noted for solutions of other metals in copper, though the atom ratios, at which the structure changes, depend upon the solute metal. Similar behavior is also observed for solvent metals other than copper.

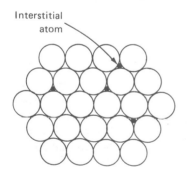

Figure 13.14 Interstitial solid solution (two-dimensional view).

Interstitial solutions arise when very small atoms such as those of carbon, boron, or nitrogen are dissolved in the transition metals. The atomic radii of the former elements are so small that they are accommodated in the holes within the close-packed crystal structure of the metal (Figure 13.14). Since there are so many holes in the structure, the solubility of these small atoms varies over wide limits. However, even very small amounts of interstitially substituted atoms may cause profound changes in the properties of the metal. In the case of carbon-iron alloys, as little as 0.5 per cent carbon causes suffi-

Table 13.2 Common Alloys and Some of Their Properties

Trade Name	Composition by Weight	Melting Point (°C)	Density (g/cc)	Use
Sterling silver	92.5% Ag, 7.5% Cu	920		Tableware
Monel	60% Ni, 33% Cu, 7% Fe	1360	8.9	Table tops
Plumber's solder	67% Pb, 33% Sn	275	9.4	
Type metal	82% Pb, 15% Sb, 3% Sn			Typesetting
Soft solder	50% Pb, 50% Sn	200–250	8.9	
Yellow brass	67% Cu, 33% Zn	940	8.4	
Babbitt metal	90% Sn, 7% Sb, 3% Cu	235		Bearings
Duralumin	95.5% Al, 4% Cu, 0.5% Mg			Structural purposes
Stainless steel	80.6% Fe, 0.4% C, 18% Cr, 1% Ni			Food processing and surgical utensils
Duriron	84.3% Fe, 0.85% C, 14.5% Si, 0.35% Mn	1265	7.0	Laboratory plumbing
Vanadium steel	98.9% Fe, 1% C, 0.1% V			Truck, auto, and train parts
Spring steel	98.6% Fe, 1% Cr, 0.4% C			Springs and saw blades

cient distortion of the close-packed structure to destroy most of the slippage planes, thereby creating the hard, high tensile structure characteristic of steel.

Extending the interstitial-solution concept, one would predict that boron, nitrogen, silicon, or beryllium could be substituted for carbon in steel. This has been done and, although carbon steels are by far the most versatile and readily made alloys of this class, boron steels are becoming more and more important. Some of the other steels also show promise.

Intermetallic Compounds

At least two types of intermetallic compounds are recognized. These may be called *superlattice compounds* and *saltlike compounds*. Like other compounds, these substances have a definite composition and a definite melting point or decomposition temperature. Some alloys containing compounds consist of a single compound; others are mixtures of compounds. Still others may be heterogeneous solids consisting of small crystals of a compound distributed throughout the metal, much like feldspar is distributed throughout granite.

An example of a superlattice compound is the gold-copper alloy containing 25 atom per cent gold. X-ray studies show that this alloy crystallizes in a face centered cubic structure in which the gold atoms occupy the corners of the cubes and the copper atoms occupy the centers of each face of the cube (Figure 13.15). The formula for this compound is Cu_3Au.

Examples of saltlike compounds of metals are $MgCu_2$, Mg_2Pb, Li_3Bi, and Mg_3Sb_2. In this class of compounds, the two metals involved have widely different electronegativities. The stabilities and melting points of these compounds generally decrease as the electronegativity difference between atoms decreases.

One example of the use of metal compounds to harden a metal element is the addition of about 4 per cent copper to aluminum. The product, known as *duralumin*, is used in the construction of large buildings. The hardness of this alloy is apparently produced by formation of the compound $CuAl_2$, which precipitates from the aluminum crystal. The small crystals of the compound are dispersed throughout the aluminum, thus distorting the crystal

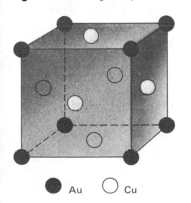

Figure 13.15 Cu_3Au superlattice.

● Au   ○ Cu

structure of the aluminum and providing a hard, high tensile structure. The tensile strength of duralumin is five times that of pure aluminum.

Addition of about 0.5 per cent magnesium to the copper aluminum alloy makes it possible to work annealed duralumin at room temperature. After a few days, precipitation of $CuAl_2$ occurs and the crystal becomes hard. This alloy has an advantage over steel because steel must be worked at high temperatures.

BAND THEORY OF METALS

Previous discussion has suggested that the valence electrons in metals are free to move, and that this freedom explains the high conductivity of metals. The amount of current carried increases with increase in the potential difference across the ends of the conductor so that, to make the explanation consistent, it must be assumed that the electrons move faster, permitting more charge to reach the end of the conductor in unit time. Continuing the argument, we note that this requires that the electrons have translational energies, and that these translational energies can increase continuously as the potential difference across the ends increases. Quantum theory predicts, however, that the energy of a system can only increase in quantum jumps from one energy level to another, and we know, from our experience with light absorption by atoms, that the energy separation between one electron level and another is quite large. How, then, can the electrons in metals change their energies so readily and by continuous increments?

The answer is given by the band theory of metals. Consider, as an example, sodium atoms each with its single $3s$ valence electron. When the atoms are far apart, the energy of the $3s$ electrons can be considered to be the same for all atoms. As the atoms approach each other, moving toward the final internuclear distances of the sodium crystal, these levels split, so that some $3s$ levels are higher and some lower than in the isolated atom. The situation is somewhat similar to that envisioned in the molecular orbital theory discussion (Figure 6.18) in which two identical levels, one on each atom, separate into bonding σ orbitals and antibonding σ^* orbitals as the nuclei of two atoms approach to form a diatomic molecule. Note that there still appear, in the diatomic molecule, as many levels (two, σ and σ^*) as there were levels (two) in the two isolated atoms. In a metal crystal, there are many more than two atoms. The number of energy levels is increased correspondingly, so that in a small bit of sodium weighing a milligram or so there will be as many as 10^{19} levels of different energy that can be traced back to the $3s$ levels of the same energy in the 10^{19} isolated atoms that have come together to form the crystal. The spacing between these levels is very small, so that very small increments of energy—corresponding to very small changes in potential difference, for example—can push the electron from one level to the next.

The group of closely spaced levels in the metal, derived from a single level in the isolated atom, is known as an energy band. Each energy level within the band can contain two electrons, but not more than two, and, if two, they must, according to the Pauli exclusion principle (Chapter 4), have opposite spins. As usual in filling energy levels, those of lowest energy fill first so that, in the sodium crystal, the single $3s$ electrons contributed by each atom are just enough to fill the lower half of the levels in the $3s$ band. This fact is important, because it means that there are many empty levels available in the band for one of the valence electrons to move to, by absorption of small amounts of energy; an electron cannot move into a level already filled with its quota of two.

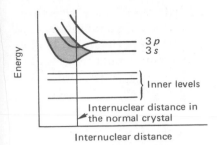

Figure 13.16 Energy bands in a metal.

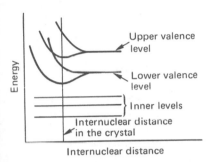

Figure 13.17 Energy bands in an insulator.

The 3s levels of sodium atoms produce one band in the crystal; the 3p levels produce other bands, and similarly for other energy levels within the atoms. The separation of the highest and the lowest levels in the bands and the separation between bands depend upon the nature of the elements in the crystal, the structure and directions in the crystal, and the internuclear distances. The situation for sodium is sketched diagrammatically in Figure 13.16. One imagines that the nuclei of all the atoms in the crystal, originally at large distances from each other, are brought closer and closer together. The highest and the lowest levels at each distance are marked by the heavy lines, and the closeness of spacing of the individual levels is suggested by the horizontal lines in the 3s band. At the distances shown, the 1s, 2s, and 2p levels have not separated into bands; the inner electrons occupying those levels are still under the control only of their own nuclei. To use an earlier wording, they have not been contributed to the community.

Insulators.   Note that, in sodium, there is considerable overlapping of the bands at the internuclear distance. This means that electrons occupying either s or p levels in this overlapping region may easily transfer from one band to another. For other substances, this is not necessarily the case. The corresponding figure for a typical insulator, for example, might look like Figure 13.17. Here, at the internuclear distance in the crystal, there is still an appreciable energy separation between the highest level of the lower band and the lowest level of the upper. Even though the levels of the upper band were empty and receptive to electrons from below, the energy requirement to lift electrons across the "forbidden" zone from the lower band to the higher band is so large that practically no electrons make the transition. If the substance has just enough valence electrons to fill all the levels of the lower band so that no easy transfer of electrons from filled to empty levels within that band is possible, the substance acts as an insulator, since no electrons can make the large jump from the lower band to the free spaces of the upper band.

Semiconductors.   In a few cases, the height of the forbidden zone, although large with respect to separation between levels within the band, is still not so large that some electrons, perhaps thermally excited, cannot move into the upper band. Once there, they are free to accept small amounts of energy and become conduction electrons. This is an explanation of the behavior of intrinsic semiconductors. As more electrons are thermally excited, they cross the forbidden zone and contribute to the conduction.

It is a general property of semiconductors that their conductivity increases with increase in temperature. This is in contrast to metallic conductors, for which the conductivity decreases with increase in temperature.

Doped Semiconductors.   The difference between metals, insulators, and semiconductors can be better illustrated by examining the energy bands at the equilibrium nuclear distances marked on Figures 13.16 and 13.17. The energy levels for the three cases discussed above could then be depicted as in Figure 13.18(a), (b), and (c).

It is possible to improve the conductivity of some intrinsic semiconductors by "doping" them with foreign materials. If, for example, there is added to a poor semiconductor such as silicon—in which each silicon atom is bonded to

Figure 13.18 Metals, insulators, and semiconductors.

Metal

Insulator

Semiconductor

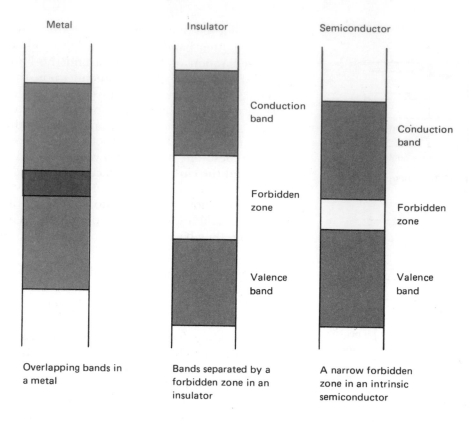

Conduction band

Forbidden zone

Valence band

Conduction band

Forbidden zone

Valence band

Overlapping bands in a metal

Bands separated by a forbidden zone in an insulator

A narrow forbidden zone in an intrinsic semiconductor

other silicon atoms by bonds formed from the four valence electrons on each atom—an amount of a Group III element such as gallium with only three electrons in its valence shell, the gallium atom replaces a silicon atom in the lattice. But it is able to furnish only three electrons instead of the usual four appropriate to that position. This creates a vacant electron position, or "hole." This "hole" is said to be positive relative to the mobile negatively charged electrons around it. By transfer of an electron from another place into it, this hole can move through the crystal, acting as a moving charge and conducting electricity. We have then a *p-type doped semiconductor* with positively charged current carriers [Figure 13.19(a)].

Alternatively, we may dope silicon with an element from Group V, such as arsenic. Now we have one electron more than the usual four, which can move from the "donor" level into the conduction band. There, it is free to move and to act as a negatively charged current carrier. We have than an *n-type doped semiconductor* [Figure 13.19(b)].

In both *p*- and *n*-type semiconductors, there are energy gaps that must be crossed, so the conductivity increases with increase in temperature. The energy needed to cross the gap can come from sources other than heat. In the Bell solar battery, for example, made from a "sandwich" of *n*-type and *p*-type semiconductors, light energy constitutes the driving force (Figure 13.20). Other combinations of *n*- and *p*-type semiconductors are used in transitors.

Figure 13.19 Donor and acceptor levels in semiconductors.

Acceptor

Donor

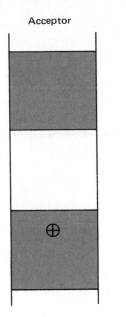

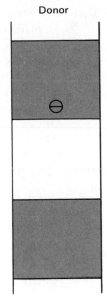

Semiconductor doped with impurity having a lower valence, leaving a conducting or positive hole

Semiconductor doped with impurity having a higher valence, electrons are transferred to the conduction band to give a negative current carrier

Figure 13.20 Diagram of a solar battery. Here a *p* type semiconductor is placed in contact with an *n*-type semiconductor such that the junction between the two is very close to the surface of the battery that is exposed to the sun. Photons from the sun excite electrons in the *p*-type layer. These electrons are drawn across the *p-n* junction creating a potential difference between the layers that increases as more light falls on the surface and releases more electrons. Terminals attached to the *p* and *n* layers, and connected to an external electrical circuit enable electrons to flow from the *n*-layer to the *p*-layer, thereby producing an electric current. (Figure adapted from "Introduction to the Utilization of Solar Energy," *University of California Engineering and Sciences Extension Series*, A. M. Zarem and Duane D. Erway, Editors. McGraw-Hill Book Company, Inc. New York, 1963. Figure 8.9, page 162, "The Direct Conversion of Solar Energy to Electrical Energy," D. M. Chapin.)

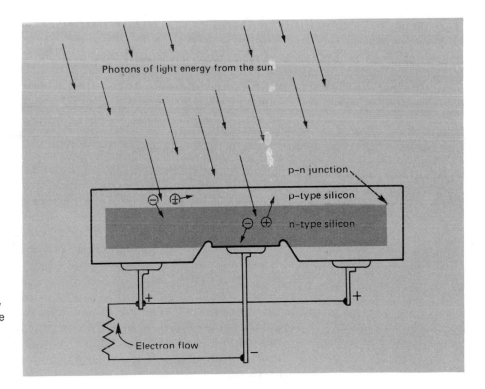

ARRANGEMENT OF IONS
IN CRYSTALS

It was pointed out in Chapter 5 that ionic compounds are solids at room temperature because of the strong attractive forces between oppositely charged ions in the crystal. X-ray studies have shown that the positive and negative ions in such crystals are arranged in alternating three-dimensional arrays, such as that shown in Figure 13.21. You might imagine that crystals grow large because the attractive forces between spherical ions extend uniformly in all directions, so that once a minute crystal starts to form, more and more ions will be attracted to it. Evidently, then, all ions in the interior of the crystal are surrounded by ions of opposite charge.

Packing of Dissimilar Spheres

The particles in ionic crystals differ from those in metals in two ways:

1. Two kinds of ions are present in ionic crystals.
2. The positive and negative ions in a given crystal nearly always differ in size whereas in pure metals the atoms are identical in size.

In ionic crystals, the first factor manifests itself in the alternating array of oppositely charged ions in the lattice; the second factor is responsible for the lattice symmetry, since the coordination number of the ions will depend upon the relative sizes of the two kinds of ions. For example, in a crystal such as that of lithium iodide, composed of small lithium ions and large iodide ions, each kind of ion has a coordination number of six. In cesium iodide, where the ions are closer to the same size, each kind of ion has a coordination number of eight. Since the coordination number determines the packing pattern of the ions, the crystal structure of lithium iodide is different from that of cesium iodide. Undoubtedly, there is a factor related to the relative sizes of the two kinds of ions that determines the coordination number of ions in any given case.

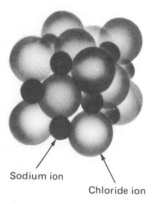

Sodium ion

Chloride ion

Figure 13.21 Crystal structure of sodium chloride.

Let us now proceed to a discussion of how the crystallographers sought this factor and, having found it, used it to predict the crystal structure of various ionic compounds. The important facets in the discovery of this factor may be summarized as follows.

1. Diffraction studies showed that most binary salts (compounds such as NaCl, containing equal numbers of positive and negative ions) occur in one of four structures. The coordination numbers of the ions in these structures are four (two forms), six, or eight, as illustrated in Figure 13.22.

2. Ionic radii for common ions were estimated from X-ray data. This can be done by determining the interionic distance by X-ray diffraction. Then, using the nuclear charge and an electronic screening factor for each ion, fractions of the interionic distance are calculated as the radii of the two ions.

3. A radius ratio r_+/r_-, defined as the ratio of the cation (positive ion) radius to that of the anion, was calculated for a large number of compounds. The reason for this calculation will become obvious.

4. Some simple relations between radius ratio and coordination number were immediately apparent. For example, it was observed that when the radius ratio was greater than 0.732, a coordination number of eight was usually observed; when the radius ratio was greater than 0.414 but less than 0.732, a coordination number of six was usually observed; when the radius ratio was greater than 0.225 but less than 0.414, a coordination number of four was often observed. Some examples of these relationships are given in Table 13.3.

Figure 13.22 Coordination numbers in ionic crystals: (a) body-centered cube, 38; (b) octahedron, 6; (c) tetrahedron, 4.

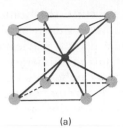 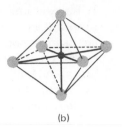 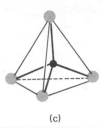

(a)　　　　　　　　　(b)　　　　　　　　　(c)

To explain this relation between radius ratio and coordination number, we should remember that attractive forces between ions of unlike charges are greatest when the oppositely charged ions are close together, and the repulsive forces between ions of like charges are greatest when the like-charged ions are close together. *Stable arrangements will therefore be those in which ions of unlike charges are as close together as possible, while ions of like charges are as far apart as possible.*

To understand the changes from one structure to another as the radius ratio changes, let us imagine anions of a particular size and consider the packing patterns that will appear, under the conditions of the preceding paragraph, as the size of the cation, around which the anions are grouped, decreases.

When the cation is about as big as the anion (radius ratio close to 1.0), we can easily arrange the anions at the corners of a cube with the cation at the center without having the anions touch. This arrangement is known as the cesium chloride structure (coordination number 8; see Figure 13.22). If now the cation is smaller so that the radius ratio is 0.732, we have the situation in which, with the cesium chloride structure, all nine ions of the unit cell are in contact. If the cation grows still smaller, the anions at the cube corners can no longer touch the cation at the center, and the attractive force between cation and anion can be increased by changing to a different structure in which they are closer together. In the sodium chloride structure (coordination number 6) of Figure 13.21, the cation will be at the center of an octahedron, as shown in Figure 13.22(b). In this structure, contact between cation and anion leaves the anions still separated, and the structure is more stable than the cesium chloride structure would be for the same radius ratio.

Table 13.3 Relation Between Coordination Number and Radius Ratio in Ionic Crystals

Coordination Number (Observed)	Salt	Radius Ratio	Salt	Radius Ratio
8	CsCl	0.93	TlCl	0.85
	CsBr	0.87	TlBr	0.78
	CsI	0.78		
6	LiF	0.44	NaBr	0.49
	NaF	0.70	KBr	0.68
	NaCl	0.52	NH_4Br	0.71
	KCl	0.73	NH_4I	0.65
4	ZnS	0.23	CuBr	0.26
	ZnO	0.31	HgS	0.33
	CuCl	0.28		

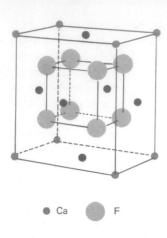

● Ca   ⬤ F

Figure 13.23 Calcium fluoride structure.

The sodium chloride structure reaches its limiting form when the radius ratio is 0.414. At this ratio, as shown in Figure 13.22(c), all seven ions are again in contact. Further decrease in cation size must again separate the anions from the cation, reducing the attractive force and making the structure unstable. Increased stability can be obtained if the pattern shifts to one of the two zinc sulfide structures (coordination number 4), in which the anions are arranged at the corners of a tetrahedron about the cation.

Contact of anions with the cation, for the tetrahedral structure with radius ratio 0.414, leaves the anions separated from each other, reducing the repulsive force between them and stabilizing the new structure. The limiting radius ratio for the tetrahedral structure is 0.225; at this value, all five ions would be in contact.

An alternative way of looking at the sodium chloride structure and at the two zinc sulfide structures is to imagine that, in sodium chloride, the chloride ions pack in a cubic close-packed pattern with the sodium ions occupying the holes surrounded by six chloride ions. In zinc sulfide, the sulfide ions pack in the cubic close-packed pattern in one structure (zinc blende) and in the hexagonal close-packed pattern in the second structure (wurtzite). In both structures, the zinc ions occupy alternate tetrahedral holes in the close-packed pattern—holes that are formed between groups of four sulfide ions.

Not all simple ionic structures can be predicted from the radius ratio of the ions, but it is usually true that for a given radius ratio, the coordination number may be less than predicted but rarely greater than predicted.

Finally, it can be shown that in binary compounds, the coordination number of both kinds of ions is the same so prediction or determination of this number for the smaller ion fixes it for the larger ion as well.

Structure of AB₂ Salts

In salts having the formula AB_2, such as calcium fluoride (CaF_2), the problem is one of efficiently packing two ions of one charge for each ion of the opposite charge while still maintaining the alternating arrangement of positive and negative ions. This can be accomplished if the coordination number of the ion in short supply is twice that of the second ion. Diffraction studies on calcium fluoride show that the coordination number of calcium ion is eight while that of fluoride ion is four. In magnesium fluoride, the coordination number of magnesium ion is six; that of fluoride ion is three. Here again, the coordination number is related to the radius ratio. In calcium fluoride, this is 0.73; in magnesium fluoride, it is 0.48. As in the case of binary compounds, the smaller radius ratio results in a lower coordination number.

The particular arrangement of ions to give different coordination numbers to the anion and cation is of interest. In calcium fluoride, this arrangement is illustrated in Figure 13.23. It might be described as a simple cube of fluoride ions placed just inside a face-centered cube of calcium ions. Each fluoride ion is then surrounded by three calcium ions from the faces and one calcium ion from a corner of the cube of calcium ions in a tetrahedral relationship. A calcium ion in the face of a unit cell is surrounded by four fluoride ions inside each of two adjacent unit cells, for a total of eight.

Structure of Complex Ionic Crystals

The structural ideas developed in this section are not limited to crystals with simple spherical ions but apply to all ionic crystals. However, the packing pattern becomes much more complex when polyatomic ions such as nitrate,

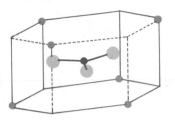

sulfate, carbonate, or phosphate occupy positions in the crystal lattice. For example, calcite ($CaCO_3$), which contains both calcium and carbonate ions, crystallizes in hexagonal units composed of the triangular-shaped carbonate ion surrounded by six calcium ions as shown in Figure 13.24.

Hydrates such as blue vitriol ($CuSO_4 \cdot 5H_2O$), are still more complex in that here the cation is surrounded by water molecules so that both ions may be nonspherical, thus making efficient packing more difficult.

Perhaps the most complex ionic crystals are those such as mica and asbestos, which contain fiberlike and layerlike anions held together by small cations placed at various positions between the fibers or layers (Figure 13.25).

ARRANGEMENT OF MOLECULES IN CRYSTALS

When molecules crystallize, they usually pack in the most efficient manner possible consistent with their shape, and with the most favorable orientation of their attractive forces. As a result, there are almost as many packing patterns for molecular crystals as there are varieties of molecules. Figure 13.26 shows four packing patterns with nitrogen, carbon dioxide, and water as examples.

Nitrogen The nitrogen molecule crystallizes in two allotropic forms—a hexagonal close-packed structure [Figure 13.26(a)] stable between 35K and the melting point, and a face-centered cubic structure [Figure 13.26(b)] stable below

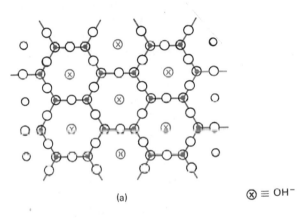

(a) $\otimes \equiv OH^-$

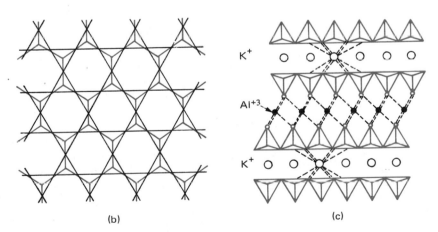

Figure 13.25 Sheets of SiO_4 tetrahedra as found in mica [muscovite, $KAl_3Si_3O_{10}(OH)_2$]: (a) structural representation showing positions of the OH groups; (b) diagrammatic representation; (c) sketch looking end-on at the sheets to show the packing pattern.

(b)   (c)

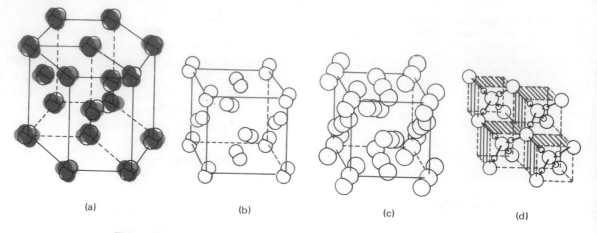

Figure 13.26 Crystalline forms of nitrogen, carbon dioxide, and water. (a) Hexagonal close-packed (high-temperature) allotropic form of solid nitrogen. (b) Face-centered cubic (low-temperature) form of solid nitrogen. (c) Structure of solid carbon dioxide. (d) Structure of solid water (ice). Four oxygen atoms are arranged tetrahedrally around a fifth and held together by hydrogen bonds.

35K. At each lattice point, there is one dumbbell-shaped molecule. In the high-temperature form, the distance between molecule centers is 4.07 A, and it appears that the molecules precess about these centers, presenting a nearly spherical shape to their neighbors. In the face-centered cubic structure, the distance between molecule centers is 3.99 A. But the molecules themselves appear to be fixed in the staggered arrangement shown, perhaps with some vibration about these positions. This is what might be expected. At the low temperatures where kinetic energies of vibration are small, the van der Waals forces between atoms on adjacent molecules are sufficient to hold the molecules in a preferred orientation. As the temperature rises, the vibrations become greater and greater and change to rotations, as the atoms lose their grip on the atoms of the neighboring molecules. The rotating molecule requires a little more space and the packing pattern changes. Still higher temperatures, of course, are needed to overcome the van der Waals forces holding the molecules together in the solid, and melting does not occur until 63.2K. In both allotropes, the distance between atoms is slightly greater than 1 A, as in gaseous nitrogen, emphasizing that the distance between molecules is considerably greater than the distance between atoms in the same molecule.

Carbon Dioxide These linear molecules pack so that oxygen atoms are oriented toward carbon atoms in adjacent molecules. The reason for this is that the oxygen atom is the negative end of a polar bond; the carbon atom is the positive end. Thus the packing is controlled by the most favorable orientation of the polar attracting forces in the molecule.

Water Hydrogen bonding is the important attractive and orienting force in ice. Here, each oxygen atom is surrounded by four hydrogen atoms. Two are covalently bonded to it. The other two, one from each of two neighboring molecules, are held by hydrogen bonds whose length is more than twice as great as that between the O—H in the water molecule. Each water molecule is now bonded to four other water molecules, giving a very strong structure with a very large number of hydrogen bonds.

Figure 13.27 (a) Diamond structure showing the tetrahedral bonding. In the right-hand drawing, the top front corner atom of the cube has been omitted for greater clarity. (b) The structure of the room-temperature form of quartz.

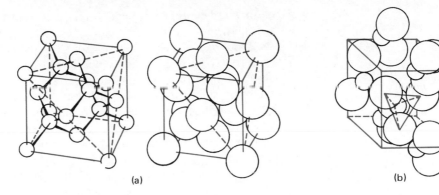

(a)   (b)

The strong orientation of hydrogen atoms toward oxygen atoms in adjacent molecules makes for inefficient packing in the crystal. This "open" structure is the reason ice floats on water—its density is less than that of liquid water near the freezing point.

Macromolecular Crystals

In macromolecular (often called network-type) crystals, the entire structure is tied together by covalent bonds. Examples include diamond, carborundum (silicon carbide), quartz (silicon dioxide), boron nitride, and a few other substances. In all of these crystals, the coordination number of the atoms is four. The resulting structures are characterized by great strength, high melting points, and extreme hardness. Figure 13.27 shows the tetrahedral network in diamond and the spiral arrangement in quartz.

Carbon and boron nitride may also crystallize in a pattern with a coordination number of three. This results in the layer or sheetlike crystals illustrated for graphite in Figure 13.28. Only van der Waals forces hold the sheets together, so when the crystal is placed under stress, these sheets will easily slip past one another. Three electrons per atom are used in the sp^2 hybrid bonds forming the hexagons, and the fourth is mobile, contributing an electrical conductivity in the direction parallel to the layers.

LATTICE ENERGY AND CRYSTAL STABILITY

Stability of Crystals

The stability of a crystal can be expressed in terms of its lattice energy. The lattice energy may be defined as *the energy needed* to disrupt the crystal and to *convert the building blocks into gaseous particles* (atoms, ions, molecules) at an infinite distance from each other. Methods of obtaining lattice energies are discussed in Chapter 18.

Applications of Solid State Models. Building on the type of knowledge and models of metallic crystals described in this chapter, metallurgical scientists and engineers have developed alloys to meet thousands of specific needs ranging from heat shields for reentering spacecraft to textured finished auto dash panels.

Examples of the application of basic principles of crystal chemistry in improving and modifying the properties of an alloy are provided by steels.

Steels are iron alloys containing carbon and usually other elements such as nickel, chromium, molybdenum, tungsten, and vanadium. One of the important features of iron in such alloys is that its crystal structure changes

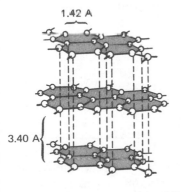

1.42 A

3.40 A

Figure 13.28 Structure of graphite.

from body-centered cubic at moderate temperatures to face-centered cubic at high temperatures.

Carbon atoms are small enough to fit into the tetrahedral holes in the face-centered structure, but they are too large to fit into the much smaller tetrahedral holes in the body-centered structure. (Compare the sizes of these two unit cells for identical atoms.) As a consequence, carbon dissolves in iron at high temperatures but precipitates out of the metal at low temperatures. However, the precipitate formed is that of the compound Fe_3C, known as cementite. This compound is hard but brittle, and its precipitation prevents the planes of atoms in the crystal from slipping over one another under stress as they normally do in pure iron. This, of course, hardens the steel—its formation is the important factor in tempering steel.

By controlling the amount of carbon added (usually between 0.1 and 1.7 per cent), and by controlling the heating or cooling rate of the steel, it is possible to produce a desired fineness of the precipitated particles. For example, slow cooling produces a steel that contains elongated plates of cementite dispersed throughout the iron. By contrast, quenching the hot steel in cold water and then reheating it to 300–400°C produces a dispersion of smaller spheroidal particles of cementite. Steel with cementite platelets is very strong, the alignment of the slip planes of iron atoms having been efficiently deterred. However, when such steel is stretched or bent, it can crack. The continuous stretching causes the metallic iron in the steel to tear away from the cementite platelets in much the same way that perforated paper tears under stress.

Steel containing the spheroidal particles of cementite is just as strong but much more tough and ductile. Under extreme stress, the spheroidal particles can move along with the metallic iron, thus reducing the amount of tearing of cementite from iron.

The effect of added metals such as nickel or chromium in altering the properties of steel often can be understood by considering the location of the alloying metal in the steel crystal structure. In general, these metals can be found in one or more of three possible phases: solid solution in iron; combined with carbon, replacing some or all of the iron in cementite; combined with nitrogen, sulfur, or oxygen as insoluble dispersions. Seldom does an alloying metal locate exclusively in one of these phases. However, its effect on the properties of steel usually can be attributed to its activity in a single phase. Nickel, chromium, and manganese, for example, are found primarily in solid solution in the iron, where they provide added strength to that phase. Manganese also serves as a scavenger for oxygen dissolved in the iron melt. Vanadium and molybdenum are found mainly in the cementite phase, and titanium is a scavenger of nonmetal atoms, forming precipitates of titanium sulfides, nitrides, or oxides.

The presence of nickel and chromium in the iron phase slows up the precipitation of cementite. This means that these alloy steels do not need to be cooled as rapidly to assure spherical dispersion of cementite. In fact, under certain conditions, nickel and chromium steels retain the face-centered cubic structure at room temperature. Such steels can be made much tougher and more ductile without sacrificing strength by a mechanical treatment known as strain hardening, a process of stretching the metal. Under these conditions, the iron phase in steels containing chromium, nickel, and other metals changes from the face-centered cubic to the body-centered tetragonal structure. In this new structure, the slip planes of atoms are not lined up as favorably as are those in the cubic structure. This provides increased toughness and ductility.

As increased knowledge of the internal structure of metals and alloys is developed, our ability to build safer, stronger, less expensive bridges, airplanes, autos, skyscrapers, pipelines, machine parts, and countless other metal structures multiplies. And with it, our opportunity to improve the lot of all is also increased.

IMPORTANT TERMS

Crystal structure
space lattice
Bragg equation
electron density map
hexagonal close-packed structure
cubic close-packed structure
unit cell
coordination number
octahedral structure
tetrahedral structure
molecular crystal
lattice energy
radius ratio

Metals
metallic luster
malleability
ductility
thermal conductivity
thermionic emission
work hardening

Alloys
substitutional solid solution
interstitial solid solution
intermetallic compounds
superlattice compounds
conductors
nonconductors
semiconductors
doped semiconductors
    (p- and n-types)

X rays
diffraction
interference

QUESTIONS AND PROBLEMS

1. Compare the number of atoms in the unit cell of (a) the simple cubic, (b) the face-centered cubic, (c) the body centered cubic structures.

2. Refer to Figure 13.21, which is a diagram of the sodium chloride structure. (a) Give the coordination number of sodium and of chloride ions. (b) Considering only chloride ions, what is the lattice structure? What is the lattice structure of sodium ions? (c) Draw the unit cell. How many ions of each element are situated in corners? On edges? On faces? In centers? (d) How many ions of each element are in the unit cell?

3. Refer to the diagram in Figure 13.22(c), which could be used to represent the cesium chloride structure. (a) Give the coordination number of cesium and of chloride ions. (b) Considering each element singly, what is the lattice structure of each? (c) How many atoms of each are in the unit cell?

4. A crystal contains atoms of element A and atoms of element B in a body-centered cubic arrangement. The A atoms occupy the centers and the B atoms occupy the corners. What is the simplest possible formula for such a substance?

5. The figure at left shows a face-centered cubic structure. Using the Pythagorean theorem where necessary, calculate the distance between planes through atomic centers (a) parallel to a plane through $ABCD$; (b) parallel to a plane through $CDEF$; (c) parallel to a plane through FBH.

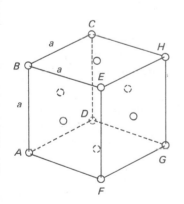

6. Consider the following figures, representing the visible faces of cubic forms.

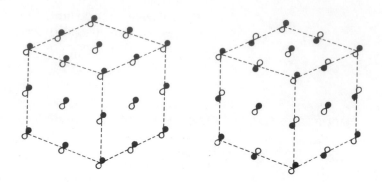

Do these have the same space lattice? The same "crystal structure"? Outline the unit cell in each figure. Make a sketch of the unit cell for each figure, showing the positions and orientations of the structural units on the visible faces.

7. How would you distinguish experimentally between a solid solution of two metals and a superlattice compound between the two metals?

8. Alloys, in general, are harder, lower melting, and poorer conductors of electricity than the pure metals of which they are composed. Why?

9. The combination of 1 mole of gaseous sodium atoms with 1 mole of gaseous chlorine atoms to form solid sodium chloride evolves 153 kcal of heat.

$$Na(g) + Cl(g) \rightleftharpoons NaCl(s) \text{ (heat evolved is 153 kcal)}$$

To ionize 1 mole of gaseous sodium atoms requires the addition of 118 kcal; ionization of 1 mole of chlorine atoms evolves 88 kcal. Calculate the lattice energy of sodium chloride.

10. What is the coordination number of Ca^{+2} in CaF_2? of F^-? (Figure 13.23.) How many moles of calcium fluoride are there in a unit cell of calcium fluoride? If the density of calcium fluoride is 3.18 g/cc, what is the length of the cube edge?

11. The metal nickel crystallizes in a face-centered cubic structure, as does sodium chloride. Its density is 8.90 g/cm³. Calculate (a) the radius of a nickel atom, (b) the length of the edge of a unit cell.

12. The potassium chloride crystal structure is that of a face-centered cube. The unit cell contains four potassium ions and four chloride ions. Given the side of a unit cell as 6.29 A, and the density of potassium chloride as 1.99 g/cm³, find the number of potassium ions and chloride ions in 1 g of the salt.

13. The sodium chloride structure is found to be the structure of many other salts. Magnesium oxide is one of these. The distance between magnesium and oxygen·atoms in the cell is 2.10 A. What is the density of magnesium oxide?

14. Lithium bromide crystallizes in the sodium chloride structure. The side length of the cell is 5.5013 A. Assume that the bromide ions are in contact as in Figure 13.21, and calculate the radius of the bromide ion.

15. (a) Calculate the length of a side of a cube that contains one mole of sodium chloride. The density is 2.165 g/cc. (b) Calculate the number of ions that lie along each edge of the cube if the distance between centers of adjacent sodium and chloride ions is 2.819 A. (c) Using this information, calculate Avogadro's number.

16. Argon crystallizes in a cubic system. The edge of the unit cell is 5.43 A and the density at 20K is 1.66 g/cc. From these considerations, what type of cubic structure does argon form?

17. Atomic volume is defined as the volume occupied by 1 mole of the solid substance. Calculate the atomic volume of an element whose atoms have a radius 1.4 A when the atoms are in the (a) cubic closest-packed, and (b) body-centered cubic structure. Give the length of an edge of the unit cell in each case.

18. A primitive cubic structure may be thought of as eight spheres situated at the right corners of a cube and just touching each other. (a) Calculate the volume of a cube that will just enclose these spheres. (b) Calculate the percentage of that volume actually occupied by the spheres.

19. An X ray from a copper target of wavelength 1.5374 A is reflected from two parallel planes 2.82 A apart. What is the angle of reflection for $n = 1$?

20. From the ionic radii given in the following table, suggest the most probable structures for the following salts: KBr, RbBr, CsBr, MgO, MgS, MgSe, MgTe, BaS, SrS, and ZnS.

Ionic Radii (A)					
K^+	1.33	Ba^{+2}	1.35	Br^-	1.95
Rb^+	1.48	Sr^{+2}	1.13	O^{-2}	1.40
Cs^+	1.69	Zn^{+2}	0.74	S^{-2}	1.84
		Mg^{+2}	0.65	Se^{-2}	1.98
				Te^{-2}	2.21

SPECIAL PROBLEMS 1. The energy required to separate the molecules of solid oxygen or nitrogen and produce gaseous molecules at an infinite distance from each other is 2.06 and 1.64 kcal/mole, respectively. For solid chlorine, the value is 7.43 kcal/mole. What difference in the nature of the chlorine molecules or atoms as compared to oxygen and nitrogen could account for the higher strength of binding in solid chlorine? In a similar vein, account for the differences in the energies required to separate molecules of methane, ammonia, and water from their positions in the solids (2.70 kcal/mole, 7.07 kcal/mole, and 11.30 kcal/mole, respectively).

2. The energy required to place the ions of solid lithium fluoride at an infinite distance from each other is 244 kcal/mole; for cesium fluoride, it is 173 kcal/mole. A similar difference appears for lithium iodide (177 kcal/mole) and cesium iodide (140 kcal/mole). What differences between lithium and cesium ions might account for this? Why are the values for these four compounds so much larger than the values reported in Question 1 above?

3. Compare and contrast the basic structural units in crystals of sodium, sodium chloride, nitrogen, and carbon, and describe for each what phenomena would be observed as the substance is gradually heated to

eventually reach 3000°C. Estimate the temperature range at which each of these phenomena would be observed.

4. Compute the radius ratio, r_+/r_-, for (a) a square planar crystal structure in which B^- ions are at the corners of the square and an A^+ ion is at the center, and (b) an equilateral triangular crystal structure with B^- at the apices and A^+ at the center.

5. Prove that in the cubic closest-packed structure, the empty space is 25.9%. Estimate the percentage empty space in a body-centered cubic structure.

6. Refer to the lattice structure of sodium chloride (Figure 13.21). The interionic distance is 2.819 A. The density of sodium chloride is 2.165 g/cc. (a) Calculate the molecular weight of sodium chloride to the second decimal place using these data. (b) Compare your calculated value in (a) with the ascribed value, 58.44. It has been proposed that this discrepancy may be due to a defect in the crystal, such that in some positions sodium atoms rather than ions are present. To compensate electrically for this substitution, an equal number of chloride ions are missing completely. On the basis of the discrepancies in molecular weights, what percentage of the anion positions must be vacant in sodium chloride?

7. Compare the crystal systems of the following minerals from the data given.

Mineral	Cell Dimensions in A			Axial Angles		
	a	b	c	α	β	γ
Zircon	6.58	6.58	5.93	90°	90°	90°
Beryl	9.21	9.21	9.17	90°	90°	120°
Topaz	4.64	8.78	8.38	90°	90°	90°

REFERENCES

ADDISON, W. E., *Structural Principles in Inorganic Compounds*, Wiley, New York, 1961.
BRAGG, L., "X-Ray Crystallography," *Sci. Amer.*, **219**, no. 1, 58, (1968).
ETZEL, H. W., "Ionic Crystals," *J. Chem. Educ.*, **38**, 115 (1961).
HOLDEN, A., *The Nature of Solids*, Columbia University Press, New York, 1965.
HUME-ROTHERY, W., *Electrons, Atoms, Metals and Alloys*, Dover, New York, 1963.
MOORE, W. J., *Seven Solid States*, Menlo Park, Calif., 1967.

PART FOUR

FAMILIES
OF
p-TYPE
ELEMENTS

THE OXYGEN FAMILY

The elements oxygen, sulfur, selenium, tellurium, and polonium constitute Group VI of the periodic table. The atoms of these elements all have the valence electron configuration s^2p^4. As a group they are considered nonmetals, although the heavier elements show some properties characteristic of metals. These elements have lower electronegativities than the corresponding halogens but, on the whole, their electronegativities have the relatively high values characteristic of nonmetals and decrease with increasing atomic number.

Oxygen is undoubtedly the most important member of the family. On a weight basis, oxygen atoms make up about one half the earth's crust, about nine-tenths of the oceans, about one-fifth of the air, and about three-fourths of the human body. In terms of the number of atoms, oxygen is the third most abundant element in the universe. In addition, oxygen forms binary compounds with more other elements than any except the halogens. About 180 oxides of the elements are known.

The valence electron configurations and some other important properties of the atoms of the elements in Group VI are given in Table 14.1.

AN OVERVIEW OF
OXYGEN FAMILY CHEMISTRY

Some important characteristics of the oxygen family elements are summarized in the following statements:

1. Because of their high electronegativity, the atoms of these elements, like those of the halogens, show a tendency to react with metallic elements by taking up electrons. To acquire a noble gas configuration, two electrons are needed by each atom; the resulting species is the doubly charged negative ion X^{-2}. The numerous metal oxides, sulfides, selenides, and tellurides are formed in this manner, although many of these compounds contain bonds with some covalent character.

Table 14.1 Some Properties of Oxygen Family Atoms and Ions

Element	Atomic Weight (amu)	Valence Electron Configuration	Atomic Radius (A)	Ionic Radius of X^{-2} (A)	First Ionization Energy (kcal/mole)	Electronegativity
Oxygen	15.9994	$2s^2\ 2p^4$	0.74	1.40	314	3.50
Sulfur	32.064	$3s^2\ 3p^4$	1.04	1.84	239	2.60
Selenium	78.96	$4s^2\ 4p^4$	1.17	1.98	225	2.55
Tellurium	127.60	$5s^2\ 5p^4$	1.37	2.21	208	2.30
Polonium	210	$6s^2\ 6p^4$	1.64			2.1

2. The members of this family exhibit oxidation states varying from $-$II to VI. For example, sulfur, selenium, and tellurium all form compounds having formulas XO_3, XO_2, and H_2X where the oxygen family element X is present in the VI, IV, and $-$II oxidation states, respectively. Some of the most familiar of these are SO_3, SO_2, and H_2S. This pattern of oxidation states is predictable from the structure of the oxygen family atoms. Because of its high electronegativity (second only to fluorine), oxygen shows an oxidation state of $-$II in all oxides except in F_2O, where it has a positive oxidation state of II, and in peroxides (for example, in H_2O_2), where it has an oxidation state of $-$I.

3. Among its compounds in the $-$II oxidation state, the family contains some strong reducing agents, such as hydrogen sulfide, hydrogen selenide, and hydrogen telluride. Some moderate to strong oxidizing agents, such as sulfuric acid and selenium and tellurium trioxides, are found among the compounds of the elements in their higher oxidation states.

4. A variety of oxyacids and the five binary acids (H_2X) containing members of this family are known. The formulas for some of these acids are given in Table 14.2.

5. The elements of this family form a number of compounds in which two or more atoms of the Group VI element are bonded to one another, forming short chains or rings. Examples of such compounds are the peroxides, the peroxyacids, and the thionous and thionic acids.

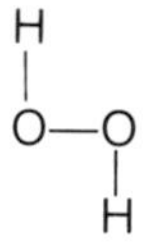

Hydrogen peroxide

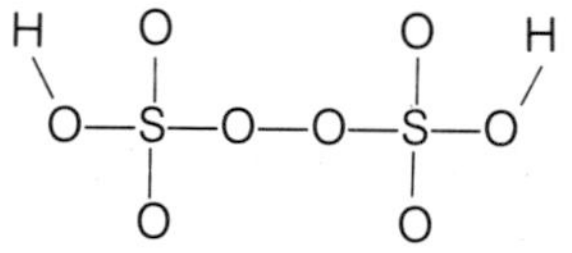

Peroxydisulfuric acid

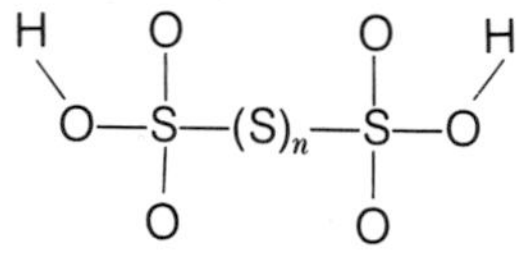

A thionic acid ($n = 2$–6)

Table 14.2 Some Acids Containing Oxygen Family Elements

Oxidation State	Oxygen	Sulfur	Selenium	Tellurium
VI		H_2SO_4	H_2SeO_4	H_6TeO_6
IV		H_2SO_3	H_2SeO_3	
II		H_2SO_2		
$-$II	H_2O	H_2S	H_2Se	H_2Te

This tendency toward catenation also appears in the free elements; for example, ozone, O_3,

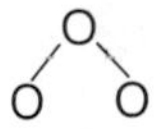

and elemental sulfur, S_8, are formed

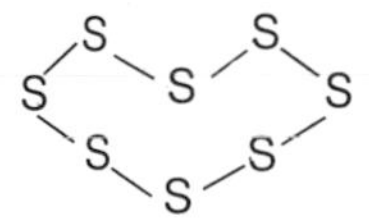

6. Sulfur, selenium, and tellurium form a number of *polymeric* compounds that have a simple repeating unit containing the oxygen family element. Examples are solid sulfur trioxide and selenium dioxide.

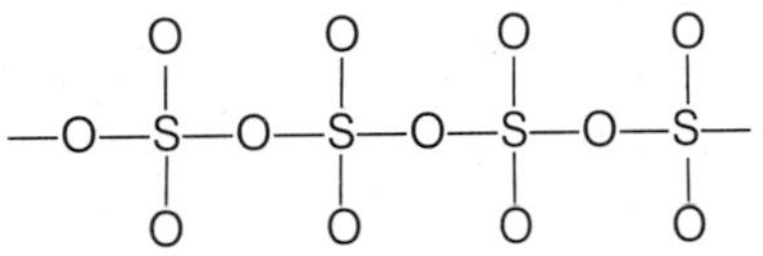

(poly)Sulfur trioxide

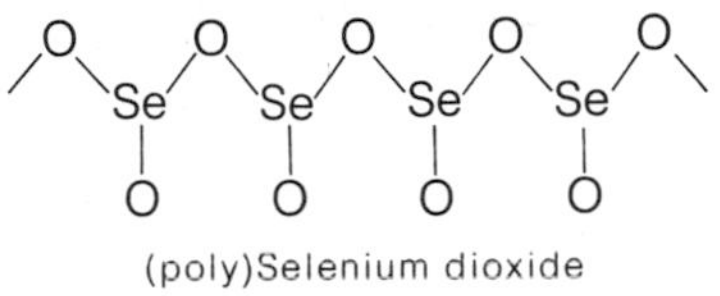

(poly)Selenium dioxide

Bonding

Bonding in species containing oxygen family elements varies from ionic to covalent. In the H_2X compounds, the bonding is predominantly covalent. However, in the MX type of compounds, where M represents a metal, the bonds are often ionic. The melting points of the H_2X and the Na_2X compounds of this family given in Table 14.3 reflect these differences in bonding. As in the halogens, the oxygen family elements show a trend from typically ionic to typically covalent compounds when combined successively with the elements in a horizontal row or period in the periodic table. This is illustrated in Table 14.4, where the melting points of oxides and sulfides of the elements in the third period are recorded. Because of its high electronegativity, oxygen forms ionic bonds with more metals than do the other members of the family.

Table 14.3 Comparison of Melting Points of H_2X and MX Compounds

Compound	Melting Point (°C)	Compound	Melting Point (°C)	Compound	Melting Point (°C)
H_2O	0.00	Na_2O	1193	MgO	3075
H_2S	−85.5	Na_2S	1000	MgS	>2000
H_2Se	−65.73	Na_2Se	>875		
H_2Te	−51	Na_2Te			

Table 14.4 Melting Points of Oxides and Sulfides of Third Period Elements

Oxide	Na_2O	MgO	Al_2O_3	SiO_2	P_4O_{10}	SO_2	Cl_2O
Melting point (°C)	1193	3075	2300	1710	422	−72.5	−116
Sulfide	Na_2S	MgS	Al_2S_3	SiS_2	P_4S_{10}	S_8	S_2Cl_2
Melting point (°C)	1000	>2000	1100	1090	268	116	−76.5

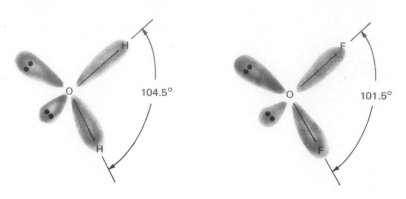

Figure 14.1 Water and fluorine oxide. In these molecules, the four pairs of electrons are believed to exist in sp^3 hybrid orbitals with only two pairs being used to form σ bonds in each molecule. The bond angles are less than the tetrahedral angle of 109°28′ because the lone pair—lone pair repulsion of the electrons in the nonbonding orbitals distorts the regularity of the expected sp^3 tetrahedral orientation.

Bonding in the Oxygen Molecule

A simple electronic formula for the oxygen molecule is sometimes written $:\ddot{O}::\ddot{O}:$, in which each oxygen atom has acquired an octet of electrons and a double covalent bond connects the atoms. However, the high reactivity of oxygen gas and its paramagnetism cannot be explained by such a structure. Only the molecular orbital theory (Chapter 6) can satisfactorily explain the paramagnetism of the O_2 molecule. According to this theory, the electrons of greatest energy are in antibonding $\pi^*_{2p_y}$ and $\pi^*_{2p_z}$ molecular orbitals, thereby giving rise to both paramagnetism and enhanced reactivity.

Stereochemistry in Oxygen Compounds

The spatial arrangements of the atoms in covalent molecules containing oxygen family elements have been determined by various techniques including X-ray diffraction, electron diffraction, and microwave spectroscopy. These arrangements are explicable in terms of the various types of hybridization possible in the atoms of this family.

In most of the simple inorganic covalent oxides, the oxygen atom apparently utilize sp^3 hybrid orbitals in forming bonds; the "lone pairs" affect the geometry of the molecules. Examples are water and fluorine oxide, where the bond angles are 104.5° and 101.5°, respectively (Figure 14.1).

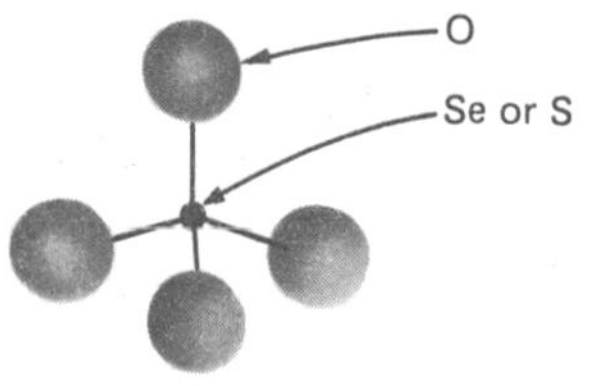

Compounds of sulfur, selenium, and tellurium having formulas of the type XB_2, such as SCl_2 and H_2Se, have bond angles close to 90°. This can be interpreted as a further distortion of the tetrahedral sp^3 hybrid, or as p^2 bonding involving perpendicular p orbitals. Grouping of four similar atoms around the oxygen family atom, as in the ions SO_4^{-2} (Figure 14.2) and SeO_4^{-2}, leads to the tetrahedral sp^3 structure.

Since SO_3 is a trigonal planar molecule, it is reasonable to postulate that in this case the sulfur atom uses sp^2 hybrid orbitals in forming the three σ bonds to the oxygen atoms. The fourth pair of electrons around the sulfur atom forms a delocalized π bond with the oxygen atoms (Figure 14.3).

Table 14.5 gives the shapes of some common covalent species containing oxygen family elements. It also includes the type of hybridization used by the central atom, and the number of σ and π bonds present in the structure.

Figure 14.2 Selenate ion showing the tetrahedral arrangement of oxygen atoms around the selenium atom.

Acids

The acids of this family vary in strength from very strong, such as sulfuric acid, to extremely weak, such as water. The ionization constants for a number of oxygen family acids are given in Table 14.6. All binary acids of this family are weak; water is the weakest of the group.

Table 14.5 Stereochemistry of Oxygen Family Structures

Example	Shape	Probable Hybridization on the Central Atom	σ Bonds in Structure	π Bonds in Structure
H_2O, F_2O	V-shaped	sp^3	2	0
H_2S, Cl_2S	V-shaped	sp^3 (or p^2)	2	0
SO_2	V-shaped	sp^2	2	1
SO_3	Trigonal planar	sp^2	3	1
SO_4^{-2}, SeO_4^{-2}*	Tetrahedral	sp^3	4	2*
SF_4, TeI_4	Distorted tetrahedral	sp^3d	4	0
SF_6, SeF_6	Octahedral	sp^3d^2	6	0
$Te(OH)_6$	Octahedral	sp^3d^2	6	0

*The S—O and Se—O distances in these ions is less than expected for single σ bonds, indicating some π-bond character. The molecules are probably resonance hybrids of the 16 structures which can be written with 8, 10, 12, 14 and 16 electrons on the central atom.

Table 14.6 Acid Strengths of Oxygen Family Elements

Compound	Ionization Constant at 25°C		Compound	Ionization Constant at 25°C	
	K_I	K_{II}		K_I	K_{II}
H_2O	1.3×10^{-16}	$\sim 10^{-24}$	H_2SO_4	large	1.2×10^{-2}
H_2S	8.7×10^{-8}	1×10^{-14}	H_2SO_3	1.0×10^{-2}	1.0×10^{-7}
H_2Se	1.9×10^{-4}	1×10^{-11}	H_2SeO_4	large	1.0×10^{-2}
H_2Te	2.3×10^{-3}	1×10^{-11}	H_6TeO_6	1.6×10^{-9}	

Note: K_I is for first ionization, K_{II} for second ionization: (1) $H_2X + H_2O \rightleftharpoons H_3O^+ + HX^-$; (2) $HX^- + H_2O \rightleftharpoons H_3O^+ + X^{-2}$.

As in all diprotic acids (those having two ionizable protons), the second ionization constant, K_{II}, is invariably much less than the first because the negative charge on the ion HX^- makes removal of the second proton difficult.

The strengths of the oxyacids can be explained in terms of the electronegativity of the central atom and the number of oxygen atoms, as was done for the halogen oxyacids. Thus H_2SO_4 is weaker than $HClO_4$ because chlorine has a higher electronegativity than sulfur; H_2SO_4 is stronger than H_2SO_3 because of the additional oxygen atom on the former. Telluric acid does not appear to follow the pattern set by the oxyacids of simpler structure.

Figure 14.3 A gaseous sulfur trioxide molecule showing the trigonal-planar arrangement of the oxygen and sulfur atoms. Here, the three pairs of electrons forming the σ bonds exist in sp^2 hybrid orbitals on sulfur. The π bond accounts for the fourth pair of electrons. This bond is delocalized so that it includes all three oxygen atoms. This is conventionally indicated by drawing three resonance structures, and by the single orbital diagram at the right.

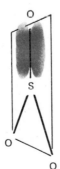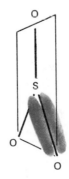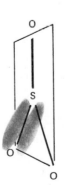

Because of oxygen's small atomic radius and its high electronegativity, the
−II oxidation state is the most stable state for this element. For sulfur,
selenium, and tellurium, the IV state appears to be the most stable, although
this state is only slightly more stable than certain others. As a result, the VI
oxidation-state compounds of sulfur, selenium, and tellurium are all moderate to strong oxidizing agents while the −II and 0 oxidation-state species are
good to strong reducing agents. Thus, when sulfur burns in air, the reaction
may be represented by the equation

$$S_8 + 8O_2 \longrightarrow 8SO_2$$

wherein sulfur acts as the reducing agent, being oxidized to the IV state, and
oxygen acts as the oxidizing agent, being reduced to its stable −II state.
Similarly, the reactions between sulfuric acid and hydrogen sulfide might be
expected to give sulfur dioxide and sulfur only sulfur dioxide as products,
along with water.

$$3H_2SO_4 + H_2S \longrightarrow 4SO_2 + 4H_2O$$

In this case, sulfuric acid acts as the oxidizing agent (sulfur in oxidation state
VI is reduced to the IV state), and hydrogen sulfide acts as the reducing agent
(sulfur in oxidation state −II is oxidized to the IV state).

The relative strength of the H_2X compounds as reducing agents is $H_2Te >
H_2Se > H_2S > H_2O$, as expected from electronegativity considerations.
Among the free elements of the family, oxygen is by far the strongest oxidizing agent. Sulfur is considerably weaker; the other elements are extremely
weak oxidizing agents. This is consistent with the great stability of the −II
oxidation state of oxygen and the decreasing stability of the −II state of the
other members. Interestingly, the oxidizing strengths of the VI oxidation-
state species appear to follow the order $SO_3 < TeO_3 < SeO_3$. Presumably,
this order is related to an interplay among several factors, including an electronegativity factor and an atomic size factor.

Figure 14.4 Plot of standard oxidation potential and oxidation state for
typical oxygen family species. Lowest
points on curves represent more stable
species. Thus the −II state of oxygen
and the 0 to IV states of other members are the most stable states.

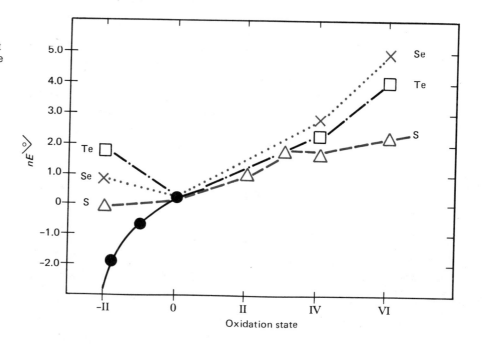

As shown in Figure 14.4, the most stable oxidation states of these elements are the −II state for oxygen and both the −II and II states for the other members.

It has been pointed out that oxygen, in large quantities, is all around us and is a vital part of us. Sulfur is much less abundant but it was one of the few elements known to the ancients. It is the brimstone of the Bible.

Free sulfur occurs in many parts of the world, including Sicily, Japan, Spain, Iceland, and Mexico, as well as in different localities in the U.S., especially in Texas and Louisiana. Some of the most important sulfur compounds in nature are the sulfides galena (PbS), sphalerite (ZnS), chalcopyrite ($CuFeS_2$), and pyrite (FeS_2), and the sulfates gypsum ($CaSO_4 \cdot 2H_2O$), barite ($BaSO_4$), celestite ($SrSO_4$), and Epsom salt ($MgSO_4 \cdot 7H_2O$). The origin of the great deposits of free sulfur is not well understood. A tentative but very probable explanation is that they were formed in the photosynthesis process during the early period of the earth's history, when there was a much higher concentration of H_2S in the atmosphere. Much of the oxygen in the earth's atmosphere is formed by photosynthesis:

$$CO_2 + H_2O \xrightarrow[\text{chlorophyll}]{\text{sunlight}} \text{sugar} + O_2$$

$$CO_2 + H_2S \xrightarrow[\text{chlorophyll}]{\text{sunlight}} \text{sugar} + S_8$$

Sulfur is one of the most important raw materials of the chemical industry. About three-fourths of the total sulfur from all sources is burned and converted into sulfuric acid. Other important uses include those in the vulcanization of rubber, the preparation of wood pulp for paper, and the preparation of medicinals and insecticides.

Tellurium was discovered by Franz Müller (Baron von Reichenstein) and Martin Klaproth toward the end of the 18th century, and Jakob Berzelius isolated selenium in 1817.

Selenium is frequently found in small quantities in natural sulfur. Combined with metals, it also occurs along with some of the sulfides, especially pyrites (FeS_2). Occasionally, compounds of selenium occur in the soil and are absorbed by certain varieties of plants. Thousands of cattle and sheep who have fed on these plants have died from poisoning.

Tellurium is found free and combined with metals, especially gold, silver, lead, and bismuth, forming compounds known as tellurides. No important use has been found for tellurium.

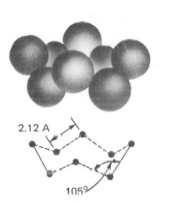

2.12 Å

105°

Figure 14.5 The puckered ring of eight sulfur atoms in solid sulfur.

PROPERTIES OF THE ELEMENTS Under ordinary temperature and pressure, oxygen is a gas consisting of diatomic molecules; sulfur is a yellow solid composed of molecules present as puckered rings of eight atoms (Figure 14.5); selenium is a grey solid containing zigzag chains of atoms; tellurium is a grey metallic solid; and polonium is a metallic substance with the atoms packed in a cubic lattice.

Table 14.7 gives some physical properties of these elements. Among the solid elements, the density, melting point, boiling point, and heats of fusion and vaporization increase systematically from sulfur to tellurium. Polonium has a higher density than tellurium, but both its melting and boiling temperatures are lower.

Table 14.7 Some Physical Properties of Oxygen Family Elements

Element	Density of Solid (g/cc at 25°C)	Melting Point (°C)	Boiling Point (°C)	Heat of Fusion (kcal/mole)	Heat of Vaporization (kcal/mole)
Oxygen		−218.9	−182.96	0.053	0.81
Sulfur	2.06	119.0	440.60	0.34	2.3
Selenium	4.82	220.2	688	1.25	6.29
Tellurium	6.25	450	1390	4.28	12.1
Polonium	9.51	254	962		

The systematic increase in density can be accounted for in terms of the increasing density of the atoms in proceeding from sulfur to polonium, and the changes in molecular and crystal structure. The relatively high and steadily increasing values for melting and boiling points and for the heats of fusion and vaporization reflect increasingly stronger forces of attraction among the atoms of these elements.

Properties of Atoms　　Table 14.1 lists the electronegativities, ionization energies, and the atomic and ionic radii of the atoms of these elements. In electronegativity, oxygen is second only to fluorine.

In spite of the regularities in many properties of the atoms of this family, there are significant differences between the behavior of atoms of oxygen and those of the other members of this family. Three major differences in behavior and their apparent causes are:

1. Oxygen atoms form double bonds that are more stable than those formed by other members of this family. Hence, oxygen forms O_2 molecules rather than chains or rings of atoms as does sulfur. The tendency to form multiple bonds is related to the fact that underlying the valence shell in oxygen is a filled shell of two electrons, while shells of eight or more lie under the valence shells of the other elements. The increased shielding of the nuclei in the larger atoms presumably inhibits π-bond formation, which can occur readily with oxygen atoms.

2. Oxygen atoms only rarely form combinations in which they are in positive oxidation states, whereas the other atoms in the family do so readily. The high electronegativity of the oxygen atom and its small size inhibit the formation of the positive state in most cases. Conversely, the larger, more polarizable, less electronegative atoms of the other elements of the family readily form positive oxidation states.

3. The other members of the family sometimes form compounds by expanding their valence shell or by using d orbitals, but oxygen apparently is unable to do this. Thus, compounds such as SF_6, $Te(OH)_6$, and SeF_6 are stable, presumably because of the ability of the central atom to expand its octet (Figure 14.6).

Chemical Properties　　Oxygen acts almost exclusively as an oxidizing agent, being a reducing agent only in its reactions with fluorine. All other elements in this family may act as both oxidizing and reducing agents.

Figure 14.6 The SF_6 molecule showing (a) the S—F bonds and (b) the octahedral configuration. The sulfur atom uses s, p, and d orbitals in the octahedral sp^3d^2 hybridization.

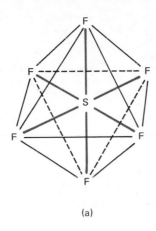

(a)

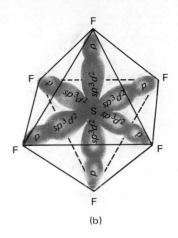

(b)

Reactions with Metals. These elements act as oxidizing agents in their reactions with the metals. In nearly all cases, these reactions give salts with the oxygen family element as an anion in its —II oxidation state.

$$M + X \longrightarrow MX$$

Exceptions to this general reaction are the formation of peroxides and superoxides with alkali metals and the synthesis of polysulfides of the type M_2S_4 and M_2S_6 with active metals.

Reactions with Nonmetals. Most nonmetals form binary covalent compounds in reactions with oxygen family elements. The latter elements may exhibit positive or negative oxidation states, depending upon whether the other element is more or less electronegative. In their binary compounds with hydrogen, oxygen family elements have —II oxidation states. However, positive states are exhibited by the heavier elements of the family in combination with oxygen.

IMPORTANT COMPOUNDS OF OXYGEN FAMILY ELEMENTS

Binary Hydrogen Compounds

All members of the family form covalent binary compounds with hydrogen (Table 14.8). The molecules all have V-shaped structures as illustrated in Figure 14.7. In spite of this similarity in structure, water is markedly different from the others. For example, at room temperature water is a liquid while hydrogen sulfide, hydrogen selenide, and hydrogen telluride are gases. Water has unusually high values for melting and boiling points and for heats

Table 14.8 Some Properties of Binary Hydrogen Compounds of Oxygen Family Elements

Compound	Melting Point (°C)	Boiling Point (°C)	Heat of Fusion (kcal/mole)	Heat of Vaporization (kcal/mole)	Dissociation Energy (kcal/mole)	Dipole Moment (debye units)
H_2O	0	100	1.44	9.72	222	1.84
H_2S	−85.6	−60.8	0.57	4.46	162	0.92
H_2Se	−60.4	−41.5	0.60	4.75	132	
H_2Te	−51	−1.8		5.70	116	

Figure 14.7 H_2X molecules of the oxygen family.

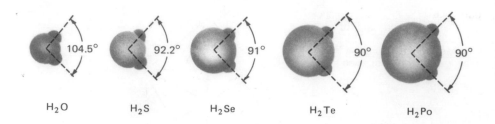

of fusion and vaporization. The high values are a result of the greater polarity of the water molecule and of the importance of hydrogen bonding among water molecules. Even in liquid water, large clusters of molecules appear to be bound together by hydrogen bonds.

Presumably, hydrogen bonding also is responsible for another abnormal property of water—expansion (or decrease in density) as it is cooled from 3.98°C through its freezing point. This results in a lower density for ice (near its freezing point) than for liquid water and is the reason ice floats on the surface of rivers and lakes.

The explanation for the expansion of water at lower temperatures is that cooling results in formation of more and more hydrogen bonds among the molecules. As the number of these bonds approaches a maximum, the water molecules become oriented in a tetrahedral pattern with considerable empty space between neighboring molecules. The rigid bonding prevents other water molecules from entering this space. Therefore, the more "open" or expanded structure of lower density results, as illustrated in Figure 13.26.

Chemical Properties of Water

In addition to its powerful solvent properties, water undergoes a variety of chemical reactions. Four types of reactions are:

1. *Acid-Base Properties.* Water can act as either an acid (proton donor) or a base (proton acceptor) in reactions such as

$$H_2O + CN^- \text{ (from NaCN)} \rightleftharpoons HCN + OH^-$$

and

$$HCl + H_2O \rightleftharpoons H_3O^+ + Cl^-$$

The first ionization constant of water (K_I) is 1.3×10^{-16}; the second (K_{II}) is about 10^{-24} at 25°C.

2. *Oxidation-Reduction Reactions.* Water can act as either a weak oxidizing or a weak reducing agent. In reactions with active metals—for example,

$$2Na + 2H_2O \longrightarrow 2Na^+OH^- + H_2 \uparrow$$

it acts as an oxidizing agent with hydrogen being reduced. When acting as a reducing agent, the oxygen is oxidized, as in

$$2F_2 + 2H_2O \longrightarrow 4HF + O_2$$

3. *Hydrolytic reactions.* Reactions such as

$$PBr_3 + 3H_2O \longrightarrow H_3PO_3 + 3HBr$$

and

$$SO_2Cl_2 + 2H_2O \longrightarrow H_2SO_4 + 2HCl$$

are known as hydrolytic reactions, and the compounds are said to undergo hydrolysis. In such reactions, a water molecule may be imagined to split into

Figure 14.8 Hydration of positive ions: (a) ion-dipole bond; (b) covalent bond.

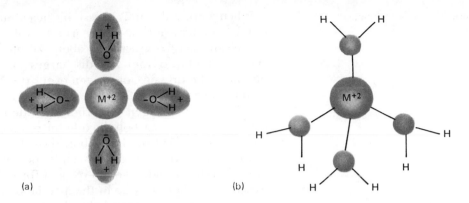

(a)   (b)

a hydrogen ion (H^+) and a hydroxide ion (OH^-) as each of these ions becomes attached to a different portion of the molecule to be hydrolyzed.

4. *Hydration Reactions.* Small cations with large charges, especially transition element ions, form aquo complexes with water. The reaction might be represented by the equation

$$M^{+n} + xH_2O \longrightarrow M(H_2O)_x^{+n}$$

For most ions, x is 6, although for some small ions such as lithium and copper ions, x is 4. For larger ions like barium ion, x is 8.

The water molecules may be bonded to the metal ion by ion-dipole or covalent bonds (as discussed in Chapter 24).

These bonds are so strong in some cases that when the metallic salts are crystallized from water solution, water is carried into the crystals containing the ion. Examples of such hydrates are $Cd(NO_3)_2 \cdot 4H_2O$, $MgCl_2 \cdot 6H_2O$, and $CuSO_4 \cdot 5H_2O$.

Sulfuric Acid, H_2SO_4 Known since the time of alchemy, sulfuric acid is not only one of the most important laboratory reagents but also one of the most important industrial chemicals. It is prepared commercially by the catalytic oxidation of sulfur dioxide either on a catalyst surface (called a heterogeneous catalyst), such as platinum or vanadium pentoxide, in the contact process, or with nitric oxide, NO, in a homogeneous catalytic *process* in the lead chamber process. The sulfur trioxide formed in either process is converted to sulfuric acid by water.

Respiration. The term respiration has a dual meaning. It means breathing in and out of the lungs, and it means the consumption of oxygen and production of carbon dioxide by any metabolic process in an organism. We shall briefly trace the chemical pathway of oxygen in human respiration.

THE PRODUCTION AND USES OF SULFURIC ACID

The annual production of sulfuric acid in the U.S. fluctuates widely, but is more than 35,000,000 tons, calculated as 100% H_2SO_4; about 80% of this is produced by the contact process. The fertilizer industry consumes approximately one-third of the output. Its minor uses are innumerable.

Relative Amounts of Sulfuric Acid Consumed in the U.S.
(Tons of 100% H_2SO_4 on a yearly basis)

Fertilizers	14,500,000	Coal products	798,000
Chemicals	3,940,000	Paints and pigments	2,660,000
Petroleum	980,000	Rayon	798,000
Iron and steel	1,630,000	Miscellaneous	1,920,000

When a normal man is at rest, his heart pumps about 5000 ml of blood per minute into his arteries. His heart rate is about 70 beats per minute; the blood ejected by each contraction is about 70 ml. In the lungs, the blood absorbs about 250 ml of oxygen and discharges about 200 ml of carbon dioxide each minute. The pressure of oxygen over arterial blood is about 100 torr, that of carbon dioxide about 40 torr.

This ratio of oxygen to carbon dioxide in arterial blood is important, as is the exchange of carbon dioxide for oxygen in the lungs. As you know, the body must get rid of the carbon dioxide it is constantly producing and it must obtain oxygen needed by cells to perform their normal functions. If the lungs cannot provide enough oxygen for the O_2–CO_2 exchange, the ratio of oxygen to carbon dioxide in the arterial blood will fall. If this situation continues for even a short time, the blood's acidity will rise. The blood enzymes, as well as those in the tissues receiving the blood, will then either fail to function, or malfunction, producing severe illness and probably death.

The relative acidity or pH (Chapter 19) of human blood is 7.40, which means it is very slightly basic. However, should the pH drop to even 7.00 (that of pure water), irreparable damage may occur. (The difference between a pH of 7.40 and 7.00 is about 3.0×10^{-8} moles of hydroxide ion per liter). But blood is protected from changing in acidity under most normal circumstances by chemical systems known as buffers (Chapter 19). The carbon dioxide dissolved in arterial blood is one of the components of the most important blood buffer. This buffer will function as long as the ratio of O_2 to CO_2 stays in a certain range. The blood level of carbon dioxide also determines the depth and rate of breathing; the higher the concentration of CO_2, the deeper and faster the breathing.

Oxygen diffuses into the blood capillaries of the tiny air sacs of the lungs (the alveoli). In the capillaries, it reacts with the hemoglobin molecule, Hb, to give oxyhemoglobin. A chemical equilibrium

$$Hb + O_2 \rightleftharpoons HbO_2$$

is established in the blood. Over 95 per cent of the oxygen is carried to the body tissues as oxyhemoglobin. As the arterial blood passes through the tissues, it gives up about 250 ml of oxygen per minute (virtually all it absorbed in passing through the lungs). The pressure of oxygen over the now venous blood has dropped to about 40 torr.

In tissue, the oxygen is released from hemoglobin in a process that is facilitated by the higher concentration of carbon dioxide in these regions. The oxygen diffuses into the cell, where it can enter into a series of reactions known as the respiratory chain. In this series, electrons from various products in the decomposition of glucose or other food materials are transferred by a series of steps to molecular oxygen, resulting in its reduction to water and the oxidation of the electron-suppliers (Chapter 29).

Indispensable among the compounds in the respiratory chain are the cytochromes, a group of iron-containing proteins similar to hemoglobin. The iron ions in these structures can receive electrons from a food product and transfer them down the chain eventually to molecular oxygen. In this process, iron (III) is reduced to iron(II) in receiving an electron, and iron(II) is oxidized to iron(III) in transferring an electron. Cytochrome oxidase, the last of at least five cytochromes in the respiratory chain, transfers an electron to oxygen.

A cell such as a muscle cell has relatively low oxygen consumption as long as the muscle is at rest. However, if the muscle is stimulated to a series of contractions, the oxygen consumption is abruptly and dramatically increased. In many cases, this increase is more than 100-fold. When contraction ceases, the oxygen consumption returns quickly to the rest rate.

Studies of respiration have led to improved medical practice in this area. By combining measurement of blood flow through the lungs with data on the concentration of oxygen and carbon dioxide in arterial and venous blood, it became possible to describe quantitatively the whole process of respiratory gas exchange and transport. As a result a physician is able to recognize an abnormal function in a patient.

In emphysema, for example, there is a scarring of the lungs, which leads to their overinflation and a maldistribution of gases. In certain forms of lung cancer, blockage of oxygen diffusion from the lungs to the blood occurs. In many cardiovascular and pulmonary diseases, there are distorted relations between flow of blood through the lungs and ventilation of the lungs.

Measurement of various properties of the lungs and of the gas composition of the blood allows the physician to diagnose the disorder, to determine its extent, and to evaluate the effect of treatment.

SUMMARY Oxygen family elements, because of their relatively high electronegativities, act predominantly as nonmetals, although the heavier members have some metallic properties. Because of their $s^2 p^4$ valence-shell configuration, oxidation states between $-II$ and VI are common. However, oxygen, with its especially high electronegativity, exhibits positive oxidation states only in combination with fluorine, and polonium compounds in the VI state are unknown.

These elements form both ionic and covalent bonds. In reactions with metals, the bonds are predominantly ionic with the oxygen family element forming X^{-2} anions. In reactions with hydrogen or with nonmetals, the bonds usually are predominantly covalent. The stereochemistry of the covalent species varies from angular to trigonal planar and from tetrahedral to octahedral. Complex ions such as sulfate and selenate, SeO_4^{2-}, are common, as are oxyacids such as H_2SO_3, H_2SeO_4, and H_6TeO_6. Acid strengths vary from very weak (H_2O) to very strong (H_2SO_4).

Strong reducing agents such as the H_2X compounds (except H_2O) and strong oxidizing agents such as SeO_3, TeO_3, and SO_3 are found in this family.

Catenation, or the forming of chains or rings of similar atoms, is observed in the oxygen family, particularly with sulfur. Polymerization also is prevalent, as in solid sulfur trioxide or polyselenium dioxide.

IMPORTANT TERMS

Bonding
hydrogen bonding
ion-dipole bond
catenation
polymers

Reactions
photosynthesis
hydrolytic
hydrolysis

hydration
oxidation-reduction

Acid
diprotic acid
acid strength
oxyacid

polarizable atom
heterogeneous catalyst
bleaching agent
stereochemistry

1. Give the electronic configurations of the simple negative ions of the elements of the oxygen family. Write electron-dot structures for the hydrides of these elements.

2. Write electron-dot structures for: (a) SeO, (b) SeO_2, (c) H_2SeO_4, (d) TeO_3^{-2}, (e) $Se_2O_3^{-2}$, (f) HSO_4^-, (g) O_2H^-, and (h) D_2O_2.

3. How many electrons take part in bonding in (a) SO_2? (b) In SO_3? (c) In SO_4^{-2}? (d) In $S_2O_3^{-2}$? Draw electron-dot structures for each of these molecules, labeling the σ and the π bonds.

4. Sulfur dioxide is a V-shaped molecule; carbon dioxide is a linear molecule. Draw the electron-dot structures of the two, and suggest why the structural difference should occur.

5. Draw the electron-dot structure of S_2Cl_2 and predict the molecular shape.

6. Write equations to show reactions you might expect to occur (a) if hydrogen telluride were added to 1-molar sulfuric acid solution; (b) if hydrogen selenide were added to 1-molar selenic acid solution.

7. Mercuric oxide (HgO), potassium chlorate, and water contain, respectively, 7.4, 39.2, and 88.81% of oxygen by weight. (a) What weight of oxygen can be obtained from 10 g of each one of these compounds? (b) What weight of each of the three compounds would yield 100 liters of oxygen at STP?

8. How many grams of sodium peroxide would be required to react with water to produce 48 g of oxygen?

9. Calculate the volume of air at STP required to burn 500 g of glucose, $C_6H_{12}O_6$, converting it to carbon dioxide and water.

10. Table 14.7 shows that the melting point, boiling point, heat of fusion, and heat of vaporization for elements of the oxygen family increase in the order of increasing atomic weight. Is this what would be expected from consideration of the changes in the properties listed in Table 14.1? Discuss.

11. In the complete burning of 16 g of sulfur to produce sulfur dioxide, (a) what weight of oxygen is required, and (b) what weight of sulfur dioxide is produced?

12. (a) One hundred liters of hydrogen sulfide will require how many liters of oxygen for complete combustion? (b) How many liters of sulfur dioxide will be formed (all gases measured under the same conditions)?

13. Write equations for the following reactions: (a) zinc with sulfur; (b) hydrogen with selenium; (c) water with calcium; (d) phosphorus triiodide with water; (e) hydrogen sulfide with excess oxygen; (f) water with selenium dioxide; (g) 1 mole of sulfuric acid with 1 mole of sodium hydroxide, both in water solution; (h) concentrated sulfuric acid with bismuth; (i) dilute sulfuric acid with zinc.

14. Why does selenium have a higher boiling point than sulfur?

15. If 95% of the sulfur is utilized, what is the daily consumption of sulfur in an industrial plant whose daily output is 200 tons of sulfuric acid containing 98.6% of hydrogen sulfate?

16. What weight of hydrogen sulfide is necessary to precipitate all the cadmium from a solution containing 10 g of cadmium nitrate $[Cd(NO_3)_2]$?

17. A compound was analyzed and found to have the following percentage composition: oxygen, 22.25; sulfur, 11.12; zinc, 22.73; water of hydration, 43.88. (a) Calculate the empirical formula of the compound. (b) What is the common name of the compound?

SPECIAL PROBLEMS

1. Consider the trend in the heats of vaporization of the oxygen family hydrides, as shown in Table 14.8. How do you account for the fact that water does not follow the trend set by the other compounds? Discuss.

2. Write equations to show that water can act either as an acid or a base, and to show that pure water is ionized.

3. Discuss and explain the facts that the ionization constants, and the strengths as reducing agents, of the Group VI hydrides increase in the order $H_2Te > H_2Se > H_2S > H_2O$.

4. Why are the bond angles in S_8 slightly smaller than the tetrahedral angle?

5. In the burning of 1 gram-atomic weight of carbon to produce carbon dioxide, explain the fact that 94 kcal of heat is evolved when oxygen is used, but 119.7 kcal of heat is evolved when ozone is used.

6. Calculate the weight of $CuSO_4 \cdot 5\,H_2O$ required to make 1l of (a) 1.5M, Cu^{+2} and (b) 1.5 m, Cu^{+2} (density 1.22 g/cc).

7. For complete reduction, how many moles of electrons must be accepted by 1.8 moles of superoxide ion (O_2^-)?

8. The oxides of a nonmetal, XO_2, and an active metal, MO, are treated with water. Write equations for each reaction, assuming XO_2 gives only a single product.

9. Twenty milliliters of a normal solution of sulfuric acid were added to 30 ml of a normal solution of sodium hydroxide, and the resulting solution evaporated to dryness and heated to expel all the water present. (a) What substances were present in the residue? (b) Calculate the weight of each. (c) What substances and how much of each would have been present if 20 ml of a normal solution of sodium hydroxide had been added to 30 ml of a normal solution of sulfuric acid and the same procedure followed?

10. Calculate from the following data the percentage of hydrogen sulfate in a sample of sulfuric acid: 5 g of the sulfuric acid was dissolved in 100 ml of water and neutralized with a 0.5 N solution of sodium hydroxide; 60 ml of the sodium hydroxide solution was required to effect neutralization.

REFERENCES   BRASTED, R. C., SNEED, M. C., and MAYNARD, J. L., *Comprehensive Inorganic Chemistry*, Vol. 7, *Sulfur, Selenium, Tellurium, Polonium and Oxygen*, Van Nostrand Reinhold, New York, 1961.

GILLESPIE, R. J., "The Electron-Pair Repulsion Model of Molecular Geometry," *J. Chem. Educ.*, **47**, 18 (1970).

JOLLY, W. L., *The Chemistry of the Nonmetals*, Prentice-Hall, Englewood Cliffs, N. J., 1965.

LAIDLER, K. J., and FORD-SMITH, J. H., *The Chemical Elements*, Bogden and Quigley, Tarrytown, N.Y., 1970.

PHILLIPS, C. S. G., and WILLIAMS, R. J. P., *Inorganic Chemistry*, Vol. 1, Oxford, New York, 1965.

THE NITROGEN FAMILY

Die Hochdrucksynthese des Ammoniaks aus seinen Elementen fortan under die Prozesse gerechnet werden darf, auf welche die Landwirtschaft ihre Hoffnungen setzt, wenn sie angesichts der abnehmenden Ergiebigkeit der chilenischen Salpeterlager und der beschränkten Ausdehnungsfähigkeit der Ammoniakgewinnung aus dem gebundenen Stickstoff der Kohle nach neuen Quellen für ihren wichtigsten Bedarfstoff umsieht.*

Fritz Haber (1868–1934)

The chemistry of the Group V elements—nitrogen, phosphorus, arsenic, antimony, and bismuth—is more diverse than that of the groups of elements previously discussed. It illustrates the variations in structure and reactivity resulting from atoms of moderate electronegativity, needing three electrons to complete the valence shell octet. Table 15.1 gives some important properties of the atoms of these elements. All have valence electron configurations of the s^2p^3 type; all are larger and less electronegative than the atoms of the corresponding halogen or oxygen family elements, and all show a distinct tendency to form covalent rather than ionic bonds. As with the halogen and oxygen families, the lighter elements are nonmetals while the heavier elements show characteristics of metals.

AN OVERVIEW OF NITROGEN FAMILY CHEMISTRY

Some important characteristics of the nitrogen family include:

1. Although it might be predicted from a knowledge of halogen and oxygen family chemistry that nitrogen family elements would readily take on three electrons per atom to form an ion of charge −3 according to the relation

$$\cdot \ddot{X} \cdot + 3e^- \longrightarrow : \ddot{X} :^{-3}$$

this type of ion is rare. It is found in a few compounds of nitrogen with active metals and probably in an even smaller number of phosphorus compounds. Both the lower electronegativity of nitrogen family atoms and the high polarizability of the relatively large X^{-3} ions are responsible for the failure of these atoms to form stable simple anions.

2. All members of the family exhibit multiple oxidation states between −III and V, as expected from the electron configurations of their atoms. Perhaps

*"The high-pressure synthesis of ammonia from its elements may henceforth be considered among those processes upon which agriculture sets its hopes, when, in view of the decreasing productivity of the nitrate deposits in Chile and the limited possibility for expanding the production of ammonia from the combined nitrogen of coal, it looks about for new sources of the material it needs most." (The quotation is from *Chemiker-Zeitung*, April 5, 1910, p. 345.)

Table 15.1 Some Properties of Nitrogen Family Atoms

Element	Valence Electron Configuration	Atomic Radius (A)	Electroneg-ativity	First Ionization Energy (kcal/mole)
Nitrogen	$2s^2\,2p^3$	0.74	3.05	336
Phosphorus	$3s^2\,3p^3$	1.10	2.15	254
Arsenic	$4s^2\,4p^3$	1.21	2.10	231
Antimony	$5s^2\,5p^3$	1.41	1.8	199
Bismuth	$6s^2\,6p^3$	1.52	1.7	185

Table 15.2 Some Nitrogen Family Species in Various Oxidation States

Oxidation State	Nitrogen	Phosphorus	Arsenic	Antimony	Bismuth
V	N_2O_5	P_4O_{10}	As_2O_5	Sb_2O_5	Bi_2O_5
IV	NO_2	P_4O_8		Sb_2O_4	
III	N_2O_3	P_4O_6	As_4O_6	Sb_4O_6	Bi_2O_3
II	NO				
I	N_2O				
0	N_2	P_4	As	Sb	Bi
−I	NH_2OH				
−II	NH_2NH_2	PH_2PH_2			
−III	NH_3, NH_4^+	PH_3	AsH_3	SbH_3	BiH_3

the most important of these for the entire family are the −III, 0, III, V states. Nitrogen itself shows all integral oxidation states between −III and V. The oxidation state pattern and some examples of each state are given in Table 15.2.

3. Except for nitrogen, all atoms of this family show a capacity for coordinating five or six other atoms around them. Examples are: PCl_6^-, AsF_6^-, $Sb(OH)_6^-$, and BiF_5. In these compounds, the nitrogen family atoms have expanded their octets and are using d orbitals as well as s and p orbitals in forming bonds.

4. The family contains both strong oxyacids such as HNO_3 (nitric acid) and $H_4P_2O_7$ (pyrophosphoric acid), and weak oxyacids such as HNO_2 (nitrous acid), H_3PO_3 (phosphorous acid), and H_3AsO_4 (arsenic acid). It includes one binary acid, the explosive hydrazoic acid HN_3.

5. The XH_3 compounds are all bases because of the moderate electronegativity and the unshared pair of electrons on X. Thus, the reaction

$$:XH_3 + HY \rightleftharpoons XH_4^+ + Y^-$$

is common, and of the XH_4^+ species, ammonium ion (NH_4^+) and phosphonium ion (PH_4^+) are well known.

6. The III oxidation state compounds that might be represented by the formula HOXO or $(HO)_3X$ are acidic if X is nitrogen or phosphorus, *amphoteric* (acting as both an acid and a base) if X is arsenic or antimony, and basic if X is bismuth. This can be explained by comparing the strength of the X—OH bond with that of the H—O bond. If the X—OH bond is much

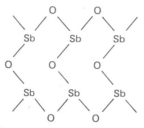

Pyrophosphoric acid

$H_3P_3O_9$
Trimetaphosphoric acid

Portion of Sb$_2$O$_3$ chain in the mineral valentinite

Polymetaphosphoric acid

Figure 15.1 Some polymeric oxygen compounds of phosphorus and antimony.

stronger than the H—O bond, the latter will rupture preferentially, as in the equation

$$H—OXO + H_2O \longrightarrow H_3O^+ + XO_2^-$$

This reaction occurs when X is an element of relatively high electronegativity, such as nitrogen or phosphorus. Conversely, if the H—O bond is the stronger, then the X—OH bond will rupture preferentially, giving

$$HO—XO \longrightarrow HO^- + XO^+$$

This occurs when X is an element of low electronegativity, such as bismuth. The amphoteric compounds are those in which the two bonds are of comparable strengths so that both reactions might occur as in

$$H_2O \;\nearrow\!\!\!\nearrow \quad \overset{HOXO}{} \quad \nwarrow\!\!\!\nwarrow$$
$$H_3O^+ + OXO^- \qquad HO^- + XO^+$$

In the presence of excess acid, the position of equilibrium shifts toward the right and more XO^+ is formed, and we say HOXO acts as a base; in excess base, the position of equilibrium shifts toward the left, more XO_2^- forms, and we say HOXO acts as an acid. This behavior is characteristic of antimony and bismuth, and accounts for species often represented as SbO^+ and BiO^+ in compounds such as $SbOCl$, $(BiO)_2SO_4$, $BiOCl$.

7. The V oxidation state species are strong oxidizing agents with the exception of the phosphorus compounds. Thus HNO_3, Bi_2O_5, As_2O_5, and Sb_2O_5 are readily reduced even by weak reducing agents forming, in most cases, the corresponding III state. In acid solution, the order of oxidizing strength of the V state species is Bi > N > Sb > As ≫ P.

8. Compounds of these elements in negative oxidation states are all good reducing agents; they are oxidized to the 0 or the III state in most cases. The order of reducing strength among −III state compounds is BiH_3 > AsH_3 > SbH_3 > PH_3 ≫ NH_3, with the more metallic hydrides being the better reducing agents.

9. Like the oxygen family, these elements, particularly phosphorus, form polymers and polyacids containing —X—O—X— bonds and —X—X—X— bonds.

The polyphosphoric acids are formed by removing the components of water from adjacent molecules:

$$HO—\overset{O}{\underset{OH}{P}}—OH + H—O—\overset{O}{\underset{OH}{P}}—OH \longrightarrow HO—\overset{O}{\underset{OH}{P}}—O—\overset{O}{\underset{OH}{P}}—OH + H_2O$$

Bonding The elements of the nitrogen family form covalent bonds in nearly all cases. The nitride ion, N^{-3}, is found in the nitrides of active metals, such as Li_3N; hydrated bismuth ions, Bi^{+3}, appear in salts with sulfates and nitrates and some other stable anions. Other common ionic forms include amide ion, NH_2^-; nitrate and nitrate ions, NO_3^- and NO_2^-; phosphate ion, PO_4^{-3}; ammonium ion, NH_4^+; antimonyl ion, SbO^+; and bismuthyl ion, BiO^+, all of which contain covalent bonds. It is important to realize the extremes

Figure 15.2 Structures containing —X—X—X— bonds.

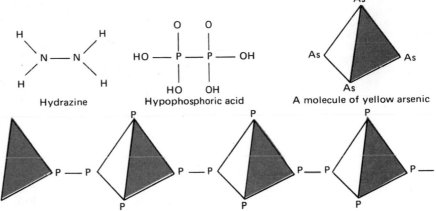

H
H
N — N
H
H
Hydrazine
O
O
HO — P — P — OH
HO
OH
Hypophosphoric acid
As
As
As
As
A molecule of yellow arsenic
P
P — P
P — P
P — P
P —
Suggested arrangement of phosphorus atoms in red phosphorus

in bonding in this family, for the lightest element nitrogen has high enough electronegativity to form a simple anion. The heavier elements antimony and bismuth have such low electronegativities that they form simple cations. Intermediate between these extremes are the vast majority of structures containing these elements in covalently bonded arrays of atoms.

These arrays of covalently bonded atoms may be small molecules (Figure 15.3), or they may be macromolecules—solids in which the entire crystal is one huge molecule and in which the atoms in forming the bonds create a massive three-dimensional network such, as in aluminum nitride, AlN (m.p. 2000°C), or boron nitride, BN (m.p. > 3000°C) (Figure 15.4). Small molecules are usually formed when nitrogen family atoms combine with halogen or oxygen family atoms—those capable of forming only one or two bonds. Macromolecules are formed when nitrogen family atoms combine with other trivalent atoms such as aluminum or boron. In this case, the three-dimensional network is possible.

Bonding in the free elements nitrogen and phosphorus emphasizes the differences in bonding tendencies between an atom having a shell of only two electrons below the valence shell (N) and one having a shell of eight

Figure 15.3 Structure of some small molecules and ions.

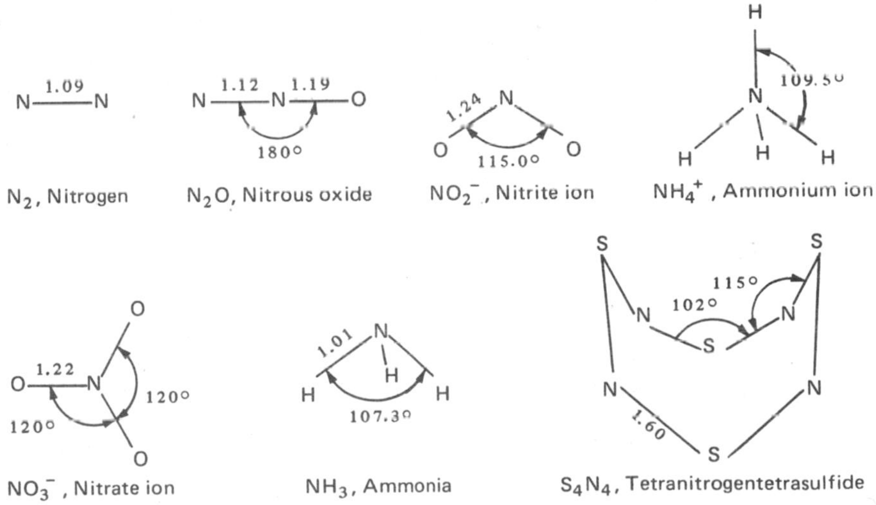

N 1.09 N
N₂, Nitrogen
N 1.12 N 1.19 O
180°
N₂O, Nitrous oxide
1.24
N
O
O
115.0°
NO₂⁻, Nitrite ion
H
N
H
H
H
109.5°
NH₄⁺, Ammonium ion
O
O 1.22 N
120°
120°
O
NO₃⁻, Nitrate ion
1.01
N
H
H
H
107.3°
NH₃, Ammonia
S
S
N
N
S
102°
115°
N
N
S
1.60
S₄N₄, Tetranitrogentetrasulfide

Figure 15.4 The structure of BN
illustrating its macromolecular nature
and similarity to diamond and graphite.
(a) Arrangement of boron and nitrogen
atoms in the diamond form of BN. (b)
Arrangement of the same atoms in the
graphite form of BN.

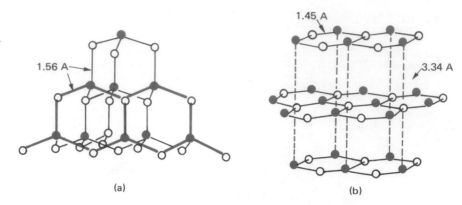

(a)

(b)

electrons below the valence shell (P). The nitrogen atom forms π bonds readily; π bonding in phosphorus is less common. As a result, molecular nitrogen is diatomic. Its molecule, the most stable diatomic molecule known, has one σ and two π bonds between atoms. By contrast, molecular phosphorus consists of tetrahedral molecules where only σ bonds exist between atoms.

Stereochemistry

The shapes of molecules of some nitrogen compounds are shown in Figure 15.3. In N_2O, the shape is presumably due to sp hybridization of the central nitrogen atom; in NO_3^-, the nitrogen atom is using sp^2 hybrid orbitals, while sp^3 hybridization characterizes the bonding orbitals in ammonium ion. Table 15.3 summarizes the stereochemistry of representative structures of the family.

The stereochemistry of phosphorus and arsenic structures is most often tetrahedral, as in PO_4^{-3}, PH_4^+, AsO_4^{-3}, and PCl_4^-. Trigonal bipyramid structures also occur, as in PF_5 and AsF_5. In the tetrahedral structures, phosphorus and arsenic atoms use sp^3 hybrid orbitals to form bonds; in trigonal bipyramid structures, these atoms use sp^3d hybridization. Antimony atoms often exhibit octahedral stereochemistry with accompanying sp^3d^2 hybridization, as in $SbCl_6^-$ or $Sb(OH)_6^-$.

It is interesting to contrast the stereochemistry of the V-state oxyanions in the family. These substances, nitrate ion, NO_3^-; phosphate ion, PO_4^{-3}; arsenate ion, AsO_4^{-3}; and hexahydroxoantimonate ion, $Sb(OH)_6^-$, contain, respectively, three, four, and six oxygen atoms around the central atom. The inability of the nitrogen atom to expand its octet, coupled with its ten-

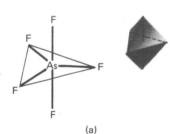

(a)

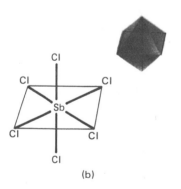

(b)

Figure 15.5 Some halides of arsenic and antimony. (a) AsF_5 molecule showing trigonal bipyramid structure; the valence bonds are shown by heavy lines and the trigonal base outlined by light lines. (b) $SbCl_6^-$ ion showing octahedral structure; the valence bonds are shown by heavy lines and the square base by light lines.

Table 15.3 Stereochemistry of Some Nitrogen Family Structures

Molecule or Ion	Shape	Hybridization of Central Atom	Kinds of Covalent Bonds
N_2O	Linear	sp	$2\sigma, 1\pi$
NO_3^-	Trigonal planar	sp^2	$3\sigma, 1\pi$
NH_4^+, PH_4^+, PCl_4^+	Tetrahedral	sp^3	4σ
NH_3, PH_3, AsH_3, SbH_3	Pyramidal	sp^3	3σ
PO_4^{-3}, AsO_4^{-3}	Tetrahedral	sp^3	$4\sigma, 1\pi$
PF_5, AsF_5	Trigonal-bipyramidal	sp^3d	5σ
$Sb(OH)_6^-, PCl_6^-$	Octahedral	sp^3d^2	6σ

Table 15.4 Some Nitrogen Family Acids and Their Ionization Constants at 25°C

Oxidation State	Nitrogen	Phosphorus	Arsenic	Antimony
V	Nitric acid, HNO_3 $K = \infty$	Orthophosphoric acid, H_3PO_4 $K_I = 7.5 \times 10^{-3}$ $K_{II} = 6.2 \times 10^{-8}$ $K_{III} = 1.2 \times 10^{-12}$ Pyrophosphoric acid, $H_4P_2O_7$ $K_I = 1.4 \times 10^{-1}$	Orthoarsenic acid, H_3AsO_4 $K_I = 2.5 \times 10^{-4}$ $K_{II} = 5.0 \times 10^{-8}$ $K_{III} = 3.0 \times 10^{-13}$	Antimonic acid, $HSb(OH)_6$
III	Nitrous acid, HNO_2 $K = 4.5 \times 10^{-4}$	Orthophosphorous acid, $H_2(HPO_3)$ $K_I = 1.6 \times 10^{-2}$ $K_{II} = 7 \times 10^{-7}$	Arsenious acid, H_3AsO_3 $K_I = 6 \times 10^{-10}$	
I	Hyponitrous acid, $H_2N_2O_2$	Hypophosphorous acid, $H(H_2PO_2)$ $K = 1 \times 10^{-2}$		
$-\frac{1}{3}$	Hydrazoic acid, HN_3 $K = 1.8 \times 10^{-5}$			

dency to form strong π bonds, permits only three oxygen atoms around the nitrogen atom in nitrate ion. The four available orbitals on the nitrogen atom are used in the formation of three σ bonds and one π bond. Phosphorus and arsenic atoms, on the other hand, do not form π bonds readily. Therefore, each of the four bonding orbitals on these atoms forms a σ bond with one oxygen atom, giving rise to the XO_4^{-3} structures. The octahedral symmetry of the V-state antimony ions reflects the ease with which this atom expands its octet. The nearly equivalent $5s$, $5p$, and $5d$ energy levels in this atom readily interact to give six bonding orbitals designated as sp^3d^2 hybrids. Each orbital forms a σ bond with an oxygen atom to produce the $Sb(OH)_6^{-1}$ structure.

Strengths of Acids and Bases

Some important acids of this family and their ionization constants are given in Table 15.4. Nitric acid is the strongest acid in the family, while pyrophosphoric acid, $H_4P_2O_7$, and hypophosphorous acid, $H(H_2PO_2)$, are moderately strong. The remainder are weak acids.

The trends in acid strength within the family can be explained in terms of the electronegativity of the central atom and the number of oxygen atoms in the molecule, as described earlier. Thus orthophosphoric acid, H_3PO_4, is weaker than sulfuric acid, H_2SO_4, which in turn is weaker than perchloric acid, $HClO_4$, because the electronegativity increases in the order $P < S < Cl$. Also, nitric acid is stronger than nitrous acid, HNO_2, because the latter compound has fewer oxygen atoms. As an extrapolation of this principle, one would predict that pyrophosphoric acid,

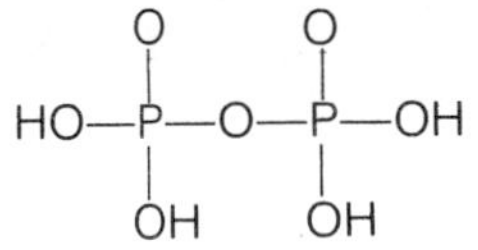

$$
\begin{array}{ccc}
& O & O \\
& \| & \| \\
HO-&P-O-P&-OH \\
& | & | \\
& OH & OH
\end{array}
$$

will be stronger than orthophosphoric acid,

$$
\begin{array}{c}
\quad\;\;\; O \\
\quad\;\;\; \| \\
HO-P-OH \\
\quad\;\;\; | \\
\quad\;\;\; OH
\end{array}
$$

because the former contains the unit

$$
\begin{array}{c}
\quad\;\; O \\
\quad\;\; \| \\
-O-P-OH \\
\quad\;\; | \\
\quad\;\; OH
\end{array}
$$

at the position where the latter contains only an OH. The dihydrogenphosphate unit, being more electronegative than a hydroxyl group, should increase the acid strength of pyrophosphoric acid over that of orthophosphoric acid. This prediction is verified by the measured dissociation constants, which show that the pyro acid is about 19 times stronger than the ortho acid.

Although all the binary compounds of nitrogen family elements with hydrogen are weak bases, ammonia is the strongest and most important base in the series. In water solution, ammonia accepts a proton from water to form ammonium ion, as in

$$: NH_3 + H_2O \rightleftharpoons NH_4^+ + OH^-$$

The ionization constant for this reaction is 1.4×10^{-5} at 25°C. While phosphine, PH_3, undergoes the same reaction, it is a much weaker base than ammonia. Hydrazine, H_2NNH_2, is also a weaker base than ammonia. It is a diprotic base—one capable of accepting two protons—and forms two acids, $H_2NNH_3^+$ and $^+H_3NNH_3^+$. Another important base in the family is bismuthyl hydroxide, $BiO(OH)$, which produces OH^- on ionization, showing the metallic nature of the heavier members.

Occurrence, Preparation, and Uses

Table 15.5 gives a few brief statements about the occurrence, preparation, and uses of the elements.

The occurrence of nitrogen as a constituent of the proteins of living tissues is of interest because humans and animals are not able to synthesize these substances from the free nitrogen of the atmosphere. Animal life therefore depends upon a source of compounds of nitrogen, obtained by eating plant foods containing proteins (especially beans and peas), or other animal flesh or secretions. Plants, especially those belonging to the legume family, act as hosts to certain "nitrogen-fixing" bacteria, which enter the root cells from the soil. The bacteria are parasitic and obtain their food from the plant, assimilating a part of it into their own tissues and oxidizing a part of it. They utilize the energy thus made available in building up nitrates from the free nitrogen absorbed by the soil from the atmosphere. The nitrates formed, as well as other nitrates present in the soil, are converted by the plant into plant protein. Earlier, it was thought that only a few microorganisms were able to fix atmospheric nitrogen in the forms essential to higher plant and animal life, but it is now known that many have this capability, including algae of the sea.

When an animal eats plant protein, it is changed by digestion into nitrogen

Table 15.5 Occurrence, Preparation, and Uses of the Elements of the Nitrogen Family

Element	Occurrence	Preparation	Uses
Nitrogen	Free in air (78%). Nitrates ($NaNO_3$, KNO_3). Proteins in living tissue.	From air by liquefaction.	Source of nitrogen compounds in fertilizers, explosives.
Phosphorus	Phosphates [apatite, $CaCl_2 \cdot 3Ca_3(PO_4)_2$; phosphorite, $Ca(OH)_2 \cdot 3Ca_3(PO_4)_2$]. Bones, teeth, living tissue.	From phosphates by reduction with coke and sand: $$2Ca_3(PO_4)_2 + 6SiO_2 + 10C \longrightarrow 6CaSiO_3 + 10CO + P_4$$	As alloying element in phosphor bronzes; phosphorus compounds are used in fertilizers, matches.
Arsenic	Sulfides (realgar, As_2S_2; orpiment, As_2S_3). Occasionally free.	From the oxide obtained as by-product in metallurgical processes involving ores of various metals. $$As_4O_6 + 6C \longrightarrow As_4 + 6CO$$	In alloys; arsenic compounds in insecticides and medicinals.
Antimony	Sulfides (stibnite, Sb_2S_3). Occasionally free.	From the oxide by reduction with carbon, or from the sulfide: $$Sb_2S_3 + 3Fe \longrightarrow 3FeS + 2Sb$$	In alloys, particularly with lead in lead storage batteries and in type metal.
Bismuth	Often free, also as oxide and sulfide. Not an abundant element.	When free, separated from the ore by heating until bismuth melts ($271°C$) and runs off; from oxide by-product by reducing with carbon.	In low-melting alloys, in sprinkler systems and electrical fuses; compounds in pharmaceutical preparations.

compounds called amino acids, which are rebuilt into animal protein tissues. These in turn are eventually converted by body processes into such animal waste products as urea and ammonia (Figure 15.6).

THE ELEMENTS

Physical Properties Nitrogen is the only element of Group V that is a gas at ordinary conditions. All the others are solids, and most exist in several crystalline modifications. These are so numerous that the physical properties of the elements are difficult to compare directly. Table 15.6 lists some physical properties of these elements.

Figure 15.6 The nitrogen cycle—transformations of nitrogen in nature.

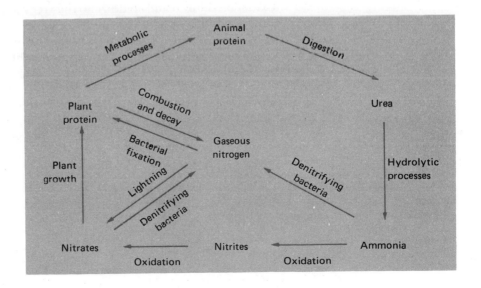

Elementary phosphorus exists in at least three allotropic forms known as white phosphorus, violet or red phosphorus, and black phosphorus. The white form consists of P_4 molecules arranged in a cubic lattice and held in the crystal structure by van der Waals forces. This allotrope is very reactive, low-melting (44.1°C), soluble in a number of solvents, and has a relatively low density (1.82 g/cc) at room temperature. Violet or red phosphorus is less reactive, less soluble, much higher-melting, and has a higher density (2.2 g/cc). It is believed to consist of chains of P_4 molecules arranged as shown in Figure 15.7. Black phosphorus is similar to graphite in appearance and, like graphite, conducts electricity. Its crystals are composed of phosphorus atoms in double layers widely separated from one another. It is stable at room temperature although it is formed by heating white phosphorus to temperatures above 300°C.

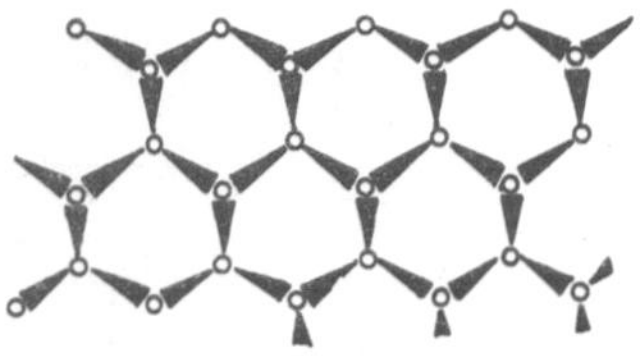

Arsenic and antimony exist in a number of allotropic modifications. A metallic or gray form and a yellow form are common to both. The stable forms at room temperature are the metallic forms; they are soft, brittle, and gray or silver-white in appearance. They are composed of puckered sheets of covalently bonded atoms; each atom in the sheet is bonded to three other atoms, and the different sheets are held together by metallic bonding. The yellow forms of arsenic and antimony resemble white phosphorus in general behavior. However, they are even more reactive. The units in the crystal are As_4 or Sb_4 tetrahedra. Yellow arsenic and antimony are more stable at lower temperatures than are the metallic allotropes of the elements. This suggests that the metallic bonding may change to covalent bonding at lower temperatures.

Bismuth occurs only in a metallic form that is soft, reddish-white, and brittle. The crystal is composed of puckered sheets of covalently bonded atoms; the different sheets are held together by metallic bonds. Bismuth differs from normal metals in having a lower electrical conductivity in the solid than in the liquid.

Figure 15.7 Portion of a layer in elemental bismuth. Note the puckering in the sheets. These sheets of covalently bonded atoms are held together by metallic bonds.

Chemical Properties

In the usual aqueous environment, these elements tend to remain in the uncombined state. This is understandable from the values of the standard oxidation potentials in Table 15.7, which show that compounds in negative oxidation states readily oxidize to the free element, and that these in turn have no great tendencies to oxidize to positive oxidation states. The order of the free elements with respect to oxidizing strength ($N_2 > P_4 > As_4 \sim Sb_4 > Bi$) is the same as the order of decreasing metallic character [increasing ionization energy (Table 15.1)], indicating the correlation between oxidizing strength

Table 15.6 Some Physical Properties of Nitrogen Family Elements

Element	Density (g/cc at 25°C)	Melting Point (°C)	Boiling Point (°C)	Heat of Fusion (kcal/mole)	Heat of Vaporization (kcal/mole)
Nitrogen	—	−209.9	−195.8	0.086	0.67
Phosphorus (white)	1.82	44.1	280	0.15	2.97
Arsenic (gray)	5.727	817 (36 atm)	613(s)	6.62	34.5
Antimony	6.684	630.5	1325	4.74	16.23
Bismuth	9.80	271	1559	2.6	36.2

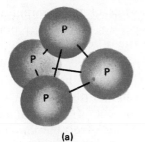

(a)

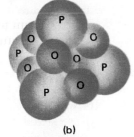

(b)

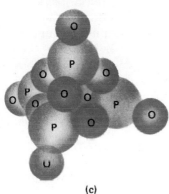

(c)

Figure 15.9 Structures of molecules of phosphorus and its oxides: (a) P_4; (b) P_4O_6; (c) P_4O_{10}.

Table 15.7 Standard Oxidation Potentials for Reactions of the Elements of the Nitrogen Family (in acid solution at 25°C)

Reaction	$E°$ (volts)
$BiH_3 \rightleftharpoons Bi + 3H^+_{(aq)} + 3e^-$	~0.80
$AsH_3 \rightleftharpoons As + 3H^+_{(aq)} + 3e^-$	0.60
$SbH_3 \rightleftharpoons Sb + 3H^+_{(aq)} + 3e^-$	0.51
$P + 3H_2O \rightleftharpoons H_3PO_3 + 3H^+_{(aq)} + 3e^-$	0.50
$PH_3 \rightleftharpoons P + 3H^+_{(aq)} + 3e^-$	−0.06
$Sb + H_2O \rightleftharpoons SbO^+ + 2H^+_{(aq)} + 3e^-$	−0.21
$As + 2H_2O \rightleftharpoons HAsO_2 + 3H^+_{(aq)} + 3e^-$	−0.25
$2NH_4^+ \rightleftharpoons N_2 + 8H^+_{(aq)} + 6e^-$	−0.28
$Bi + H_2O \rightleftharpoons BiO^+ + 2H^+_{(aq)} + 3e^-$	−0.32
$N_2 + 4H_2O \rightleftharpoons 2HNO_2 + 6H^+_{(aq)} + 6e^-$	−1.45

Figure 15.8 Structure of hydrides of the nitrogen family.

and actual ionization of the atoms. Considering the elements as reducing agents, only phosphorus has a positive value of the oxidation potential, and nitrogen has a high negative value. As observed, therefore, we might expect nitrogen to occur in the free state in nature, but not phosphorus.

All of the elements are oxidized by hot concentrated sulfuric acid; nitrogen and arsenic form oxides, phosphorus gives phosphoric acid, and antimony and bismuth form sulfates.

IMPORTANT COMPOUNDS OF NITROGEN FAMILY ELEMENTS

The Binary Hydrides; Ammonia

The simple hydrides of formula XH_3, ammonia (NH_3), phosphine (PH_3), arsine (AsH_3), stibine (SbH_3), and bismuthine (BiH_3) are all colorless gases composed of pyramidal-shaped molecules, as illustrated in Figure 15.8.

In this group, only ammonia is a widely used compound. The others are highly toxic, thermally unstable, spontaneously flammable in air, and used only rarely as reducing agents. There is no close similarity between ammonia and the other hydrides because:

1. Ammonia has a much more polar molecule than the others and forms strong hydrogen bonds between molecules.
2. Nitrogen is more electronegative than hydrogen, but the other elements in the family are less electronegative than hydrogen. This means that nitrogen is at the negative end of the ammonia dipole while hydrogen is the negative portion in the other hydrides.

These factors are reflected in the physical properties of the compounds, which are given in Table 15.8. Thus ammonia has a higher melting point, and higher heats of fusion and vaporization than the others—facts that are consistent with the greater polarity of the ammonia molecule and with the existence of hydrogen bonds between molecules.

Preparation of XH₃ Compounds

Because of the low reactivity of the nitrogen family elements, direct reaction with hydrogen gives very low yields of XH_3 compounds. In spite of this, much commercially available ammonia is prepared by the direct combination of nitrogen and hydrogen in the Haber process. This uses a catalyst at temperatures between 400°C and 600°C and at high pressures (up to 1000 atm). The equation for the reaction is

$$2N_2 + 3H_2 \rightleftharpoons 2NH_3$$

Chemical Properties of Ammonia

Ammonia is a weak base and a moderately strong reducing agent. The products from the reactions in which it acts as a base are, in most cases, salts of the ammonium ion, NH_4^+, such as ammonium chloride, NH_4Cl; ammonium sulfate, $(NH_4)_2SO_4$; and the like. These salts are similar in behavior to the corresponding sodium or potassium salts having the same anion, principally because the ammonium ion has approximately the same radius as the potassium ion (1.45A compared with 1.33A). In acting as a reducing agent, ammonia is usually oxidized to stable molecular nitrogen, as in

$$2NH_3 + 3Cl_2 \longrightarrow N_2 + 6HCl$$

Ammonia undergoes two other kinds of reactions:

1. With active metals it forms hydrogen and the amide salt of the metal, MNH_2, as illustrated by the equation

$$2Na + 2NH_3 \longrightarrow 2NaNH_2 + H_2$$

The metal amides are ionic compounds containing the very strong base amide ion, NH_2^-.

2. With metal ions, especially transition metal ions it forms species of the type $M(NH_3)_x^{+n}$, which are called ammine complex ions. This is illustrated by the equation

$$Co^{+3} + 6NH_3 \longrightarrow Co(NH_3)_6^{+3}$$

The ammonia molecule may be bonded to the metal by ion-dipole bonds or by covalent bonds involving the unshared pair of electrons on the ammonia molecule. Such reactions are discussed in detail in Chapter 24.

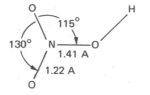

Figure 15.10 Structure of a molecule of hydrogen nitrate, HNO_3.

Table 15.8 Some Physical Properties of Nitrogen Family Hydrides

Compound	Melting Point (°C)	Boiling Point (°C)	Heat of Vaporization (kcal/mole)	Heat of Fusion (kcal/mole)	Dipole Moment of Gas (debyes)
NH_3	−77.74	−33.40	5.58	1.35	1.47
PH_3	−133.75	−87.22	3.49	0.27	0.55
AsH_3	−116.3	−62.5	4.18	0.56	0.16
SbH_3	−88	−17			0.11
BiH_3		22			
NH_2NH_2	1.8	113.5			1.35
PH_2PH_2	−99	51.7			

Ammonia as a Solvent

Liquid ammonia (b.p. −33.4°C), because of the polar nature of its molecules, is a good solvent for a variety of ionic and polar substances. That it is a more basic solvent than water is illustrated by the fact that acetic acid, which is only slightly ionized in water, is almost completely present in the salt form, $CH_3CO_2^- + NH_4^+$, when dissolved in liquid ammonia. Like water, liquid ammonia undergoes self-ionization according to the equation

$$NH_3 + NH_3 \rightleftharpoons NH_4^+ + NH_2^-$$

Acids placed in liquid ammonia react to give ammonium ions:

$$CH_3CO_2H + NH_3 \rightleftharpoons NH_4^+ + CH_3CO_2^-$$

Bases (B) placed in liquid ammonia may form amide ions:

$$B + NH_3 \rightleftharpoons BH^+ + NH_2^-$$

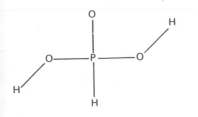

Figure 15.11 The phosphorous acid structure.

As shown by the self-ionization equation above, NH_4^+ and NH_2^- play the same role in liquid ammonia as H_3O^+ and OH^- do in water. Hence compounds containing ammonium ions behave as acids in liquid ammonia while those containing amide ion behave as bases in this solvent. Many reactions which cannot be carried out in less basic solvents can be effected in liquid ammonia.

COMPOUNDS IN POSITIVE OXIDATION STATES

Oxides

The important oxides of the family are listed in Table 15.2. All of the oxides of nitrogen, except dinitrogen pentoxide, N_2O_5, are gases while those of the other elements are solids. Some physical properties and the molecular structures of some representative oxides are given in Table 15.9.

Table 15.9 Properties and Structure of Some Oxides of Nitrogen Family Elements

Compound	Formula	Melting Point (°C)	Boiling Point (°C)	Appearance	Molecular Structure
Nitrous oxide	N_2O	−90.8	−88.5	Colorless gas	
Nitric oxide	NO	−163.6	−151.7	Colorless gas	
Dinitrogen trioxide	N_2O_3	−103	3.5	Brown gas / Blue liquid	
Nitrogen dioxide	NO_2	11.2	21.2	Brown gas	
Dinitrogen pentoxide	N_2O_5	32.4 (sublimes)		White solid	
Phosphorus(V) oxide	P_4O_{10}	422		White solid	Figure 15.9(c)
Phosphorus(III) oxide	P_4O_6	23.8	173	White solid	Figure 15.9(d)
Arsenic(V) oxide	As_4O_{10}	(d 400)		White solid	Similar to P_4O_{10}
Arsenic(III) oxide	As_4O_6	315		White solid	Similar to P_4O_6
Antimony(III) oxide	Sb_4O_6	655			Similar to P_4O_6
Bismuth(III) oxide	Bi_2O_3	817		Yellow solid	Ionic with eight oxygen atoms surrounding each bismuth atom

Some Chemistry of Plant Growth.

While animals get their food in highly concentrated form by eating plants and other animals, plants obtain their food from very dilute sources in the inorganic environment, assimilating what amounts to one ion or one molecule at a time. In doing this, they become the principal accumulators of nutrients from the environment.

Carbon dioxide from the air and mineral nutrients from the soil are the two chief classes of plant food raw materials. The carbon dioxide is converted to organic compounds, chiefly carbohydrates, in photosynthesis.

Mineral nutrients such as nitrate, phosphate, magnesium, and potassium ions are absorbed by the roots from the soil. They are carried in the sap, a water solution, through capillary tubes and other vessels to all parts of the plant. In higher plants, these vessels are of two types: the xylem and the phloem. Both occur in bundles in roots, stems, and leaves. Flow in the xylem is mainly upward, whereas products from leaves and other organs move both upward and downward through the phloem. Many botanists believe that movement of fluids in plants is made possible by the pressure differential created between leaves and roots by evaporation of water from the leaves. Upward of 99 per cent of the water absorbed by the roots is lost in transpiration. However, a significant portion of the movement of fluids appears to be due to a very large number of minuscule osmotic or chemical pumps located throughout the plant. Water and dissolved mineral nutrients are pulled into the roots by an osmotic pump.

The nutrients are unable to pass freely through the cell wall. However, both osmotic and chemical mechanisms—as yet not completely identified, but perhaps similar though not identical to the sodium ion pump described in Chapter 8—presumably aid their passage through the cell membrane and into the cytoplasm of the cell.

Once inside, the mineral nutrients enter into the biochemistry of the cell (see Chapter 29), the nitrate being used in manufacturing amino acids for protein synthesis and the nitrogen bases in RNA and DNA; the phosphate for use in carbohydrate metabolism, in energy storage or release in ATP or ADP, and also in RNA and DNA; the magnesium ion in chlorophyll; and the potassium as the cation present in largest amount in the intracellular fluid.

In transforming nitrate ion to amino acids, the plant first reduces it to nitrite ion (NO_2^-) and then to ammonia. The ammonia is used in the direct synthesis of amino acids. Reduction of nitrate to nitrite is catalyzed by an enzyme *nitrate reductase*, which is widely distributed in plants and fungi. This enzyme contains molybdenum ions, which appear to be oxidized from the V to the VI oxidation state in the process of reducing nitrate. The reduction of nitrite ion to ammonia is catalyzed by an enzyme *nitrite reductase*, which appears to contain the rather strong reducing agent ferredoxin, an iron-containing protein. In ferredoxin, the iron presumably is in the II state and is oxidized to the III state in reducing nitrite ion to ammonia. The iron(II) is regenerated during the photosynthetic reactions in the leaves.

The passage of molecules and ions through the cell membranes of roots is not all one way. Roots also excrete numerous substances, including metabolic products, enzymes and materials that may be toxic to the plant. For example, herbicides, when applied to the leaves of plants, may be excreted by the roots.

Research into the mechanisms of plant growth has resulted in the identification of chemicals that have powerful specific effects on plant development.

Some of these chemicals are naturally occurring, isolated originally from the plant itself; some are synthesized by chemists but with structures closely related to those of natural materials; others are quite unlike known plant materials. The use of chemicals in plant development has become not only helpful but an essential and integral part of agriculture and plant food production.

The herbicide 2,4-D (2,4-dichlorophenoxyacetic acid) has a chemical structure similar to the auxins, the first natural plant growth controllers to be discovered. 2,4-D is selective against broad-leaved plants. It and related compounds are useful in eliminating weeds among grasses, whether in wheat fields or in lawns. Such compounds also are valuable in that they have a low toxicity for man, animals, and soil organisms.

Other growth regulators similar to the auxins are in commercial use for purposes as diverse as promoting flowering and fruiting, increasing fruit set, preventing fruit drop, thinning fruits, and defoliation. The last is especially important in mechanical cotton-picking, where considerable advantage is gained if the leaves of the plant fall off before harvest.

Synthetic compounds related to the gibberellins and the cytokinins, two more recently discovered plant growth regulators, have found important applications. For example, certain of these compounds slow the aging of cut materials by preventing protein destruction. Thus, they strongly increase the keeping qualities of crops such as spinach and broccoli.

Studies on the accurate inhibition of various plant processes by synthetic chemicals have led to applications such as the production of shorter, more robust plants by interfering with stem elongation. Tomatoes and peppers have been treated successfully this way.

We see in the growth and development of plants a host of complicated but understandable and controllable chemical processes. From water, carbon dioxide, the nitrate, phosphate, magnesium, and potassium ions, and a few other simple chemicals come the many varieties of vegetation we see around us and often take for granted, but which not only enrich our lives by providing food, shelter, weather control, and beauty, but without which human existence would be impossible.

SUMMARY   Nitrogen family chemistry is very diverse. Because the atoms are somewhat larger than their corresponding members among the halogen and oxygen families, and because they have only moderate electronegativities, their properties vary from typically nonmetallic (nitrogen and phosphorus) through amphoteric (antimony) to distinctly metallic (bismuth). The s^2p^3 valence electron configuration gives rise to multiple oxidation states between $-$III and V. All elements show a distinct tendency to form covalent bonds. The stereochemistry varies from pyramidal for the MH_3 compounds to trigonal planar for nitrate ion, tetrahedral for phosphate and arsenate ions, trigonal bipyramid for XF_5 structures, and octahedral for hexahydroxo-antimonate(V) ion, $Sb(OH)_6^-$.

Compounds in the $-$III state may act either as bases or as reducing agents. Compounds in the V state, except for those of phosphorus, are good to strong oxidizing agents. In their positive oxidation states, nitrogen, phosphorus, and arsenic form oxyacids. However, the corresponding antimony compounds are amphoteric, and the bismuth compounds are bases.

As in the oxygen family, both catenation and polymerization are observed in the nitrogen family. The polyphosphoric acids and the oxides of arsenic and antimony are examples of polymers.

All elements exist in several crystalline modifications in the zero oxidation state. In white phosphorus, yellow arsenic, and yellow antimony, the lattice sites of the crystal are occupied by molecules consisting of four atoms covalently bonded and tetrahedrally arranged. In violet and black phosphorus, these phosphorus tetrahedra are linked by covalent bonds to make chains or sheets of tetrahedra. This polymerization accounts for the low reactivity of violet and black phosphorus.

Ammonia has many uses, including that as a solvent more basic than water.

IMPORTANT TERMS

Nitrogen fixation
protein
legume
Haber process

Amphoteric substance

Polymeric compound
macromolecule

QUESTIONS AND PROBLEMS

1. Compare the elements of the nitrogen family in the following ways, giving examples when suitable. (a) Tendency to form ionic compounds. (b) Tendency to form multiple bonds. (c) Complete electronic configurations. (d) Covalent radii. (e) Electronegativities.

2. Explain why ions of the type X^{-3} are not commonly found in nature. List the formulas of some compounds of nitrogen in oxidation state $-III$, and discuss the bonding in these compounds.

3. Write formulas for some compounds of the nitrogen family elements in the V and III oxidation states other than those given in Table 15.2.

4. How can gaseous ammonia be obtained from ammonium ion? Write an equation for this conversion. What volume of ammonia at STP would result from the complete conversion of 5 g of ammonium chloride?

5. Write electron-dot structures for: N_2O; NO; NO_2; HNO_2; NO_2^-; HNO_3, and NO_3^-.

6. In terms of the electronic structure, why would you expect ammonia, phosphine, and hydrazine to be basic in character? Are there reactions other than their reactions with water or protons which show their basic character? If so, write equations for the reactions.

7. Write equations to show the acid, basic, and amphoteric (or any one or combination of these) properties of the hydroxyl-containing compounds of the nitrogen family elements in the III oxidation state.

8. Calculate the volume of 1 M nitric acid required to neutralize 5.6 g of calcium hydroxide. Calculate the weight of HNO_3 this volume contains.

9. A certain phosphate rock was found to contain 80% calcium phosphate. (a) What weight of phosphorus can be prepared from 1000 kg of this rock? (b) What weight of orthophosphoric acid?

10. How do you explain the trend in the nitrogen family from a typical anion-forming element (nitrogen) to a typical cation-forming element (bismuth) with increasing atomic weight?

11. Contrast the bonding in N_2 and P_4.

12. Write equations to show the following reactions: (a) bismuth with chlorine at high temperature; (b) hot nitrogen with metallic lithium; (c) hot concentrated sulfuric acid with phosphorus, with bismuth; (d) calcium nitride with water; (e) ammonia with sodium; (f) ammonia with oxygen on a platinum catalyst; (g) phosphoric acid with ammonia; (h) water with phosphorus(V) oxide (i) bismuth(III) oxide with water.

13. In the complete oxidation of 1.8 moles of hypophosphorous acid to orthophosphoric acid, how many moles of electrons are donated?

14. What experimental differences in properties would you expect to find between the gray and yellow forms of arsenic and antimony? How would you interpret these differences in terms of the structures of the two forms?

SPECIAL PROBLEMS

(May require more information)

1. Explain the fact that it is simple to give a molecular weight for nitrogen but not for arsenic.

2. In the Haber process (p. 332), what is the theoretical yield of ammonia obtained from treating 600 tons of nitrogen and 120 tons of hydrogen?

3. What volume of ammonia gas, measured under standard conditions, will be required to form $(NH_4)_3PO_4$ with 500 ml of 1.5 N H_3PO_4?

4. Suggest a reason for the closeness of the melting point (314K) and the boiling point (320K) of nitrogen (V) oxide (molecular formula, N_2O_5).

5. Which of the following anions would have the greatest tendency to give an aqueous alkaline solution, and which the least: $H_2PO_4^-$; PO_4^{-3}; HPO_4^{-2}? Explain.

6. Dilute sodium hydroxide solution is added dropwise to a clear solution of antimony trichloride in dilute hydrochloric acid until the final solution tests basic and is clear. What physical observations may be made during the addition of base? Write appropriate equations for any reactions that have occurred.

7. The first ionization constants of the following acids decrease in the order given: $HClO_4 > HNO_3 > H_2SO_4 > H_4P_2O_7 > H_3PO_4 > H_2 (HPO_3) > HNO_2$. What explanations can you offer for this sequence?

8. Note in Table 15.6 that the heat of vaporization of gray arsenic is nearly 10 times greater than that of white phosphorus. Compare this with the corresponding change in Table 14.7 between sulfur and selenium. What explanations can you offer?

9. Interpret, on the basis of their properties, the fact that nitrogen occurs in the free state in nature but phosphorus does not.

10. What experimental data other than bond angles are recorded in this

chapter that show that the hydrides of the nitrogen family elements are almost surely not planar molecules?

11. How do we know that ammonia forms hydrogen bonds in the liquid state?

12. Write an equation or equations to show how the ion NO_2^+ could be formed in pure hydrogen nitrate. Write the electron-dot structure for this compound. Would you expect it to (a) have a dipole moment? (b) be paramagnetic?

13. Compare the electron-dot structures of nitric acid and nitrate ion. How do you explain the facts that the nitric acid molecule has two N—O bond distances and two O O bond angles, but the nitrate ion has only one of each?

$$\underset{N}{O \quad O}$$

REFERENCES

JOLLY, W. L., *The Inorganic Chemistry of Nitrogen*, Benjamin, Menlo Park, Calif., 1964.

JOLLY, W. L., *The Chemistry of the Nonmetals*, Prentice-Hall, Englewood Cliffs, N. J., 1965.

LAIDLER, K. J., and FORD-SMITH, J. H., *The Chemical Elements*, Bogden and Quigley, Tarrytown, N.Y., 1970.

16 THE CARBON FAMILY

A BROAD OVERVIEW OF THE CARBON FAMILY

The atoms of the five carbon family elements all have the valence electron configuration s^2p^2. The first two members, carbon and silicon, are nonmetals. The remaining elements, germanium, tin, and lead, are metals. But all five elements form predominantly covalent compounds, and all except lead seem to prefer tetrahedral stereochemistry. The electronegativities of the atoms below carbon in the family vary from 2.0 to 1.6—values too high to form cations and too low to form anions readily.

The elements carbon and silicon play a large role in nature. The "organic" world, the world of living and growing things, is largely formed of compounds that have carbon as their central element. The "inorganic" world—the world of the rocks and the inanimate substances of the earth's crust—is in large part composed of compounds of silicon with oxygen and of these two with other elements. The importance of the compounds of carbon and silicon lies in the fact that these compounds often have very large molecules that are chainlike or exist in a network pattern. It is these structural features of carbon and silicon compounds that account for the particular qualities of some rocks, the toughness of wood, the strength of silk, and the elasticity of rubber.

In studying the chemistry of these compounds, therefore, we are attempting to gain an insight into the structural features of the world about us.

Some important features relating to the chemistry of the carbon family are:

1. The atoms of these elements are larger and their electronegativities lower than those of the corresponding members of the nitrogen, oxygen, and halogen families. As a result, they are less nonmetallic than the corresponding elements in these families. And the trend within the carbon family from nonmetals to metals is more pronounced than in the other families of nonmetals. Table 16.1 gives some important properties of carbon family atoms.

Table 16.1 Some Properties of Carbon Family Atoms

Element	Valence Electron Configuration	Atomic Radius (A)	First Ionization Energy (kcal/mole)	Electronegativity	X—X Single-Bond Energy (kcal/mole)
Carbon	$2s^2\ 2p^2$	0.771	260	2.5	82
Silicon	$3s^2\ 3p^2$	1.173	188	1.8	53
Germanium	$4s^2\ 4p^2$	1.223	182	2.0	65
Tin	$5s^2\ 5p^2$	1.412	169	1.8	42
Lead	$6s^2\ 6p^2$	1.538	171	1.6	<23

2. A most important characteristic of the family as a whole is the tendency of all atoms to form four covalent bonds utilizing sp^3 hybrid orbitals, thereby producing tetrahedral structures as in

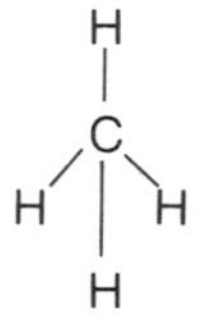

Methane 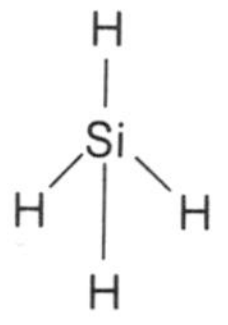   Silane 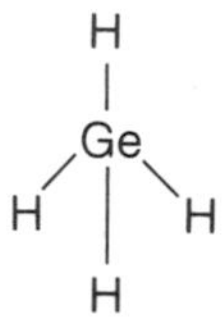   Germane 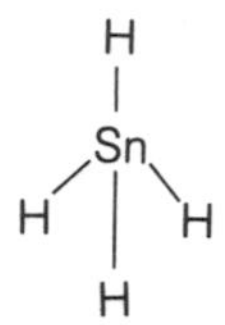   Stannane 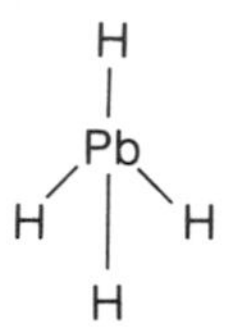   Plumbane

To form the four covalent bonds, the atom must promote an s electron and hybridize the s and p orbitals. Lead has the least tendency to do this.

3. The ability of carbon atoms to combine with each other to form chain or ring structures is well known. Chains are also formed with other members of the family except lead, but the compounds are much more reactive than are those of carbon. Some compounds in this family illustrating catenation (see Chapter 14) are:

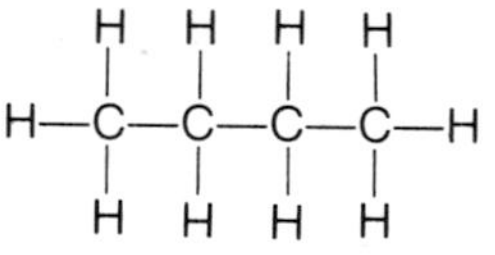

n-Butane 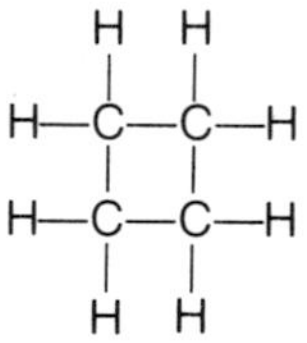   Cyclobutane 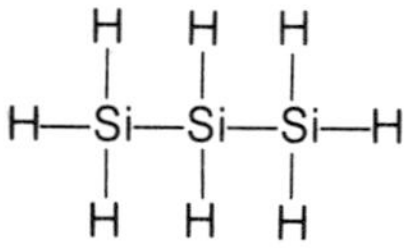   Trisilane

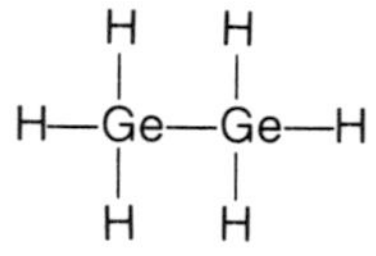

Digermane 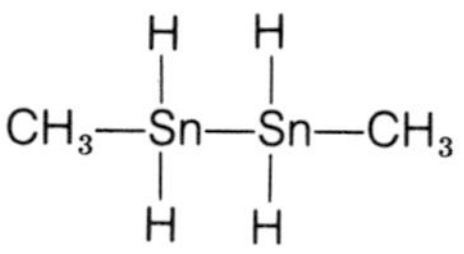   1,2-Dimethyldistannane

Both diamond and graphite are examples of catenation, for these are networks of a great number of covalently bonded carbon atoms (see Figure 16.1). The carbon family atoms maintain tetrahedral stereochemistry in these compounds. Of the family, only carbon forms compounds in which the atoms are held together by double or triple bonds as in ethylene and acetylene.

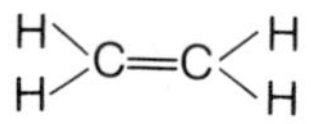

Ethylene 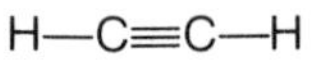   Acetylene

Figure 16.1 The crystal structure of two forms of pure carbon, diamond and graphite. (a) Diamond—the carbon atoms are equidistant and each is bonded to four others. (b) The unit cell of diamond—note the grouping of one carbon atom surrounded by four others. (c) Graphite—the carbon atoms are bonded to three others with strong covalent bonds, but the layers are bonded only weakly to each other. In any one hexagon, such as the heavily outlined one in the center, three of the atoms lie directly over others in adjacent layers, and the other three lie over the centers of hexagons in neighboring planes. Atoms of hexagons in alternate planes lie directly over one another, as shown at the right.

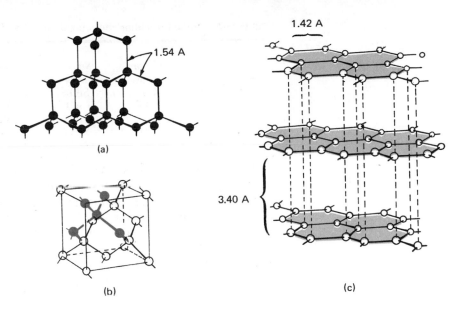

4. Polymeric or macromolecular structures such as those found in polysilicic acid, asbestos, silicones, tin(II) sulfide (SnS), and a host of similar compounds are common in this family. The stability of these polymers testifies to the strong covalent bonds formed between oxygen or sulfur and the carbon family elements. The structural formulas for the substances named above are given below.

Polysilicic acid

Portion of an asbestos structure (see also Figure 16.8)

Portion of a silicone molecule

Portion of tin(II) sulfide structure

5. Because of the many structural variations arising as a result of catenation and polymerization, the concept of oxidation state is not as important in this family as it is in many others. A strict application of the rules for determining oxidation state shows that the carbon family elements appear in the

Table 16.2 Oxidation State Pattern for Carbon Family Elements

Oxidation State	Carbon	Silicon	Germanium	Tin	Lead
IV	CO_2	SiO_2	GeO_2	SnO_2	PbO_2
II	CO	(SiO)	GeO	SnO	PbO
0	C	Si	Ge	Sn	Pb
$-IV$	CH_4	SiH_4	GeH_4	SnH_4	PbH_4

$-IV$, 0, II, and IV states. With the heavier elements, only the 0 and positive states are stable. Table 16.2 gives the oxidation state pattern for the family.

6. The family contains some strong reducing agents and at least one strong oxidizing agent. The reducing agents include the $-IV$ oxidation state compounds, whose reactivity in the gas phase decreases in the order $PbH_4 > SnH_4 > SiH_4 > GeH_4 > CH_4$. In acid solution, methane is a stronger reducing agent than ammonia. The elements act as reducing agents in the 0 oxidation state and, except for lead, in the II states. Since the IV states of all but lead are quite stable, all lower states tend to act as reducing agents, being oxidized in most cases to the stable IV state.

The IV state of lead is very reactive compared with the II state. Consequently, PbO_2, for example, is a potent oxidizing agent, being readily reduced to the II state. The stability of the II state has been ascribed to the tightness with which the high charge on the lead nucleus holds the *s* electrons, making them less available for use in bonding.

7. The hydroxides of these elements are acidic for carbon and silicon, and amphoteric for the remaining members. Carbon apparently forms the very unstable $C(OH)_4$, which easily loses one molecule of water to give weak, unstable carbonic acid, $(HO)_2CO$, or H_2CO_3. Silicon forms silicic acid, H_4SiO_4, and a variety of polysilicic acids formed by removing the components of water from adjacent molecules, as in trisilicic acid, $H_8Si_3O_{10}$.

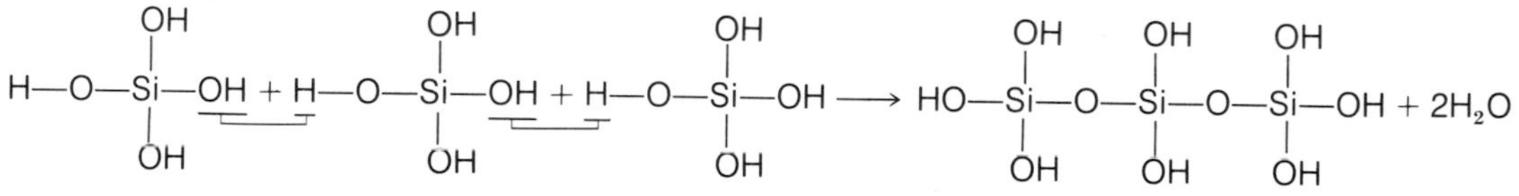

Trisilicic acid, $H_8Si_3O_{10}$

The hydroxides of tin and lead in both the II and IV states are amphoteric, forming salts of both acids and bases. For example, in the presence of bases, these hydroxides form salts in which the carbon family element is present in the anion portion of the salt. Examples are:

$$Sn(OH)_2 + OH^- \longrightarrow Sn(OH)_3^- \quad \text{or} \quad (HSnO_2^- + H_2O)$$

$$Pb(OH)_2 + OH^- \longrightarrow Pb(OH)_3^- \quad \text{or} \quad (HPbO_2^- + H_2O)$$

In the presence of acids, the hydroxides of tin and lead form salts in which the atoms of these metals are present in the cation.

$$Pb(OH)_2 + 2H_3O^+ \longrightarrow Pb^{+2} + 4H_2O$$

Bonding Like those of the nitrogen family elements, the atoms of this family form covalent bonds in nearly all cases. The only stable simple ionic forms of these atoms are:

1. C^{-4} ion, which appears to exist in compounds such as Be_2C and Al_4C_3.

2. Si^{+4} ion, which has been reported to exist in glass in those cases where the silicon atoms are found in a random pattern throughout the structure. For the most part, silicon in glass is in a covalent network of groups of this sort:

$$-O-\underset{\underset{|}{O}}{\overset{\overset{|}{O}}{Si}}-O-$$

3. Pb^{+2} ions, which are recognizable in lead fluoride, PbF_2, and lead sulfide, PbS. Compounds tin(II) chloride, $SnCl_2$; germanium(II) oxide, GeO; and the tetrachlorides of all the elements are predominantly covalent. However, there is an obvious trend toward formation of ionic bonds, or toward covalent bonds with increasing amounts of ionic character, as the electronegativity decreases or as the metallic nature of the elements increases.

The structures resulting from the covalent bonding involving these elements may be relatively small molecules like most of the hydrocarbons and their derivatives which comprise organic chemistry; or they may be macromolecules such as the dioxides of silicon, germanium, and tin; the silicates found in rocks and minerals; or in diamond, graphite, free silicon, germanium, and tin. Simple molecules are usually formed when the carbon family element combines with a monovalent element such as hydrogen or chlorine. Macromolecules result when the combination is with a polyvalent atom. A structural formula for the quartz form of silicon dioxide is given in Figure 13.27(b).

The capacity of carbon for catenation and the ability of silicon to form network Si—O—Si structures have been attributed to the unusually high bond energies of the C—C, C—H, and Si—O bonds, which are 82, 98, and 185 kcal/mole, respectively.

Undoubtedly, silicon forms Si—O—Si bonds in preference to Si—Si bonds because the bond energy of the Si—O bond is 185 kcal/mole, compared with 53 kcal/mole for the Si—Si bond. Conversely, the bond strengths for both the C—C and the C—O bonds are very high and nearly the same, 82 compared with 84 kcal/mole, respectively. This confers great stability on compounds with both C—C and C—O bonds.

There is a marked difference in properties between the anhydrous halides of the heavier members and the hydrated forms of these compounds. For example, anhydrous tin(IV) chloride ($SnCl_4$) is a liquid, while the hydrate $SnCl_4 \cdot 4H_2O$ is an ionic solid. The former consists of $SnCl_4$ molecules while the latter presumably is made up of hydrated tin(IV) ions, $Sn(H_2O)_4^{+4}$, and chloride ions. The more basic water molecule evidently displaces the less basic chloride ion in the covalent bond to tin, as illustrated below:

$$\underset{H}{\overset{H}{\diagdown}}O: + \underset{|}{\overset{\diagup}{Sn}}-Cl \longrightarrow \underset{H}{\overset{H}{\diagdown}}O: \underset{|}{\overset{\diagup}{Sn}}{}^{+} + Cl^{-}$$

Reactions of this type occur generally in compounds in which halogens are covalently bonded to metals, as in Al_2Cl_6, $BiCl_3$, $GeCl_2$, and $SnCl_2$.

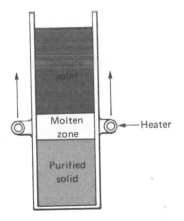

Figure 16.2 Zone refining. A heater encircling the container moves slowly upward. Impurities more soluble in the liquid move toward the top, and purified material crystallizes out below the moving melt. Alternatively, the container may move in the opposite direction past a fixed heater.

Table 16.3 Melting Points of Some Oxides, Sulfides, and Hydrides of the Carbon Family

OXIDES		HYDRIDES.		SULFIDES	
Compound	Melting Point (°C)	Compound	Melting Point (°C)	Compound	Melting Point (°C)
CO_2	−56.5	CH_4	−185.2	CS_2	−111.8
SiO_2	1710	SiH_4	−184.7	SiS_2	1090
GeO_2	1116	GeH_4	−165.9	GeS_2	800
SnO_2	1927	SnH_4	−150	SnS_2	800
PbO_2	752(d)			PbS	1114

Table 16.3 gives the melting points of some oxides, sulfides, and hydrides of carbon family elements and illustrates the variation in bonding possible in the family. All the hydrides and the carbon oxides and sulfides are small molecules, as indicated by the low melting points. The remaining oxides and sulfides, except for lead sulfide, are macromolecules having melting points 1000–2000°C above those of small molecules. Lead sulfide is an ionic solid.

Stereochemistry   As indicated earlier, the atoms of the family form tetrahedral structures using sp^3 hybrid orbitals more than any other type of stereochemistry. However the carbon atom, perhaps because of its capacity to form π bonds, is also able to form the following:

1. sp hybrid orbitals, which result in linear molecules such as carbon dioxide and acetylene, H—C≡C—H.
2. sp^2 hybrid orbitals, which result in trigonal planar structures as in formaldehyde, H_2CO, and ethylene, C_2H_4.

Silicon atoms, with their decreased tendency to form π bonds, only rarely produce structures with other than tetrahedral symmetry. The ability of these atoms to expand the octet accounts for the octahedral hexafluorosilicate ion, SiF_6^{-2}, and its salts.

Germanium, tin, and lead show tetrahedral symmetry in the −IV and IV states, as in GeH_4 and $SnCl_4$. Their dihalides—for example, $SnCl_2$—are angular molecules in the gas phase. In the solid state, they appear to consist of chains of molecules arranged as illustrated below:

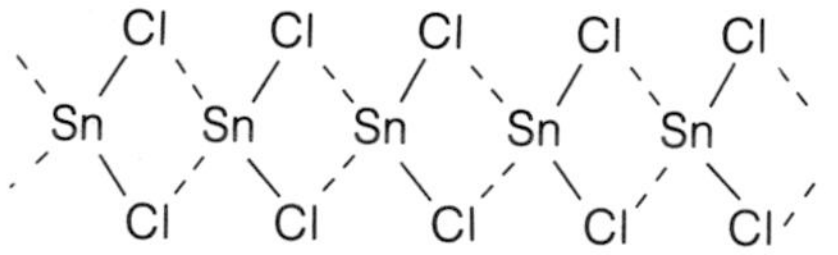

The heavier elements in the family also form stable octahedral complex ions such as hexachlorostannate ion, $SnCl_6^{-2}$,

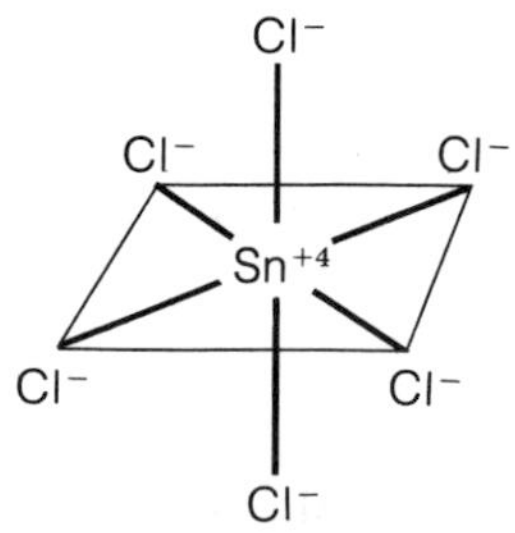

$SnCl_6^{-2}$ complex ion

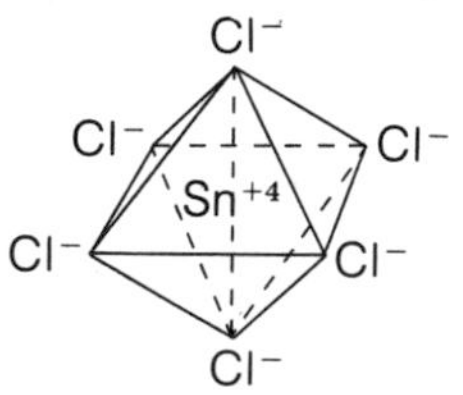

Outline of octahedral structure

and hexafluorogermanate ion, GeF_6^{-2}. A salt, ammonium hexachloroplumbate, $(NH_4)_2PbCl_6$, has been prepared.

Table 16.4 summarizes data on the occurrence and uses of the carbon family elements.

THE ELEMENTS

Physical Properties

Table 16.5 gives some physical properties of the members of this family. All of these elements are solids at room temperature and well above. The first three members are high-melting (3570°C, carbon; 1420°C, silicon; 937°C, germanium); tin and lead melt at considerably lower temperatures (231.8°C and 327.4°C; respectively). The high melting points of the first three are indicative of the macromolecular structure (network of covalently bonded

Table 16.4 Occurrence and Uses of Carbon Family Elements and Compounds

Element	Occurrence	Use
Carbon	Free in diamond, graphite, and the complex mixtures of coal. As gas or liquid in carbon dioxide, natural gas, petroleum. As solid in carbonate rocks and products of living organisms: carbohydrates, fats, proteins.	Important to life in photosynthesis and metabolism; in fuels, foods, drugs, plastics, and others.
Silicon	In the siliceous minerals comprising the overwhelmingly predominant compounds of the earth's crust: granite, sandstone, feldspar, shale, clay, and the like.	In the ceramic industries, in metal refining, and as constituent of alloys.
Germanium	In the mixture of coal, and as the complex ore germanite, which contains the germanium mostly as the ion, GeS_6^{-8}.	A semiconductor, germanium doped with tiny amounts of elements of the boron or nitrogen families is used in transistors.
Tin	As cassiterite, SnO_2. Not found in the U.S.	As corrosion-resistant covering for steel in tin plate; as alloying element in bronzes (with copper), solder (with lead), bearing metals.
Lead	As galena, PbS; cerussite, $PbCO_3$; anglesite, $PbSO_4$. Plentiful in the U.S.	In storage-battery plates; as alloying element in solder, pewter, type metal, shot; compounds as paint pigments.

Table 16.5 Some Physical Properties of Carbon Family Elements

Element	Density (25°C) (g/cc)	Melting Point (°C)	Boiling Point (°C)	Heat of Fusion (kcal/mole)
Carbon	D3.51; G2.0*	3570	4200	
Silicon	2.33	1420	2680	11.1
Germanium	5.35	937.2	2830	7.6
Tin	7.28	231.8	2687	1.72
Lead	11.34	327.4	1751	1.14

*D, Diamond; G, Graphite.

Figure 16.3 Some crystals of quartz found in nature.

atoms) of carbon, silicon, and germanium, and the lower melting points of tin and lead reflect the less rigid metallic binding in their crystals. The heats of fusion of silicon and germanium are seven or more times higher than those of tin and lead, also demonstrating this difference in bonding. Tin and lead have a very wide liquid range, typical of metals, as discussed in Chapter 8. The densities of the solids are relatively high and generally increase from carbon to lead. The higher values reflect the efficient packing in the crystals and the trend from carbon to lead parallels the increasing density of the individual atoms—the lead atom is much more dense than the carbon atom.

Tin, lead, and the graphite form of carbon are moderate electrical conductors.

Allotropes

Carbon exists in two crystalline forms, diamond and graphite. These forms have markedly different properties. For example, diamond is the hardest substance known while graphite is used as a lubricant. The structures of the two forms are shown in Figure 16.3. In the diamond network, the covalently bonded atoms are arranged tetrahedrally and are equidistant. This structure is very hard and high-melting, for it cannot be disrupted without breaking a large number of covalent bonds.

Graphite consists of layers of carbon atoms in which each atom uses three of its electrons in sp^2 hybrid orbitals to form covalent bonds with three other atoms. This produces a network of hexagons, as shown in Figure 16.3. The layers presumably are held to each other by metallic binding made possible by the fourth electron in the unhybridized p orbital on each atom. The distance between layers is 3.40 A, and the distance between bonded atoms within a layer is 1.42 A. The conductivity of graphite is due to this metallic binding. Graphite is thermodynamically more stable than diamond at room temperature, and its density is less than that of diamond.

Silicon, a deep-gray solid, crystallizes in the diamond structure and is very hard and brittle. Since Si—Si bonds are weaker than C—C bonds, it is neither as hard nor as high-melting as diamond. Germanium also crystallizes in the diamond structure but it is not as hard or as high-melting as silicon or carbon.

Tin exists in three solid forms that pass into one another at definite transition temperatures:

$$\text{Gray } (\alpha) \text{ tin} \underset{}{\overset{13.2°C}{\rightleftarrows}} \text{white } (\beta) \text{ tin} \underset{}{\overset{161°C}{\rightleftarrows}} \text{rhombic } (\gamma) \text{ tin} \underset{}{\overset{231.8°C}{\rightleftarrows}} \text{liquid}$$

(diamond structure) (metallic)

The white or metallic form is stable at room temperature, but at temperatures below 13.2°C, the metal loses its luster and often crumbles to a powder, which is the nonmetallic gray form. This transition illustrates the tendency of the electrons in the metallic bonds to become localized at lower temperatures.

Lead crystallizes in the cubic close-packed lattice. The forces holding the large, heavy atoms to particular positions in the crystal are not great, and the crystal is soft and relatively low-melting.

Chemical Properties

Some important chemical properties of the carbon family elements are illustrated by the following general equations, where M represents the element of this group.

1. $M + O_2 \longrightarrow MO_2$. Diamond burns above 800°C; graphite burns at 690°C; silicon burns at 400°C; germanium and tin burn upon heating; lead forms PbO or Pb_3O_4.
2. $M + 2X_2(\text{halogen}) \longrightarrow MX_4$. Occurs with varying degrees of difficulty, depending upon both the halogen and the carbon family element. Lead forms PbX_2 in most cases.
3. $M + 2S \longrightarrow MS_2$. High temperatures are required; diamond requires 1000°C; lead forms PbS.
4. $M + \text{Acids} \longrightarrow M^{+2} + H_2$. Tin and lead form compounds in the II state. Carbon, silicon, and germanium do not react.
5. $3M + 4HNO_3 \longrightarrow 3MO_2 + 4NO + 2H_2O$. Oxidizing acids raise the element to the IV state, except for lead, which goes to the II state.

The strengths of the elements as oxidizing and reducing agents in acids can be seen in Table 16.6. These values show that carbon is a weak oxidizing agent, but it is the strongest of this group. As with all groups of nonmetals, the oxidizing strength parallels the nonmetallic character of the element.

Silicon is the most powerful reducing agent in the group, being stronger than metallic zinc and stronger than all the hydrides in the nitrogen family except for bismuthine, BiH_3. All the elements in the family except carbon are stronger reducing agents than hydrogen.

SOME INORGANIC COMPOUNDS OF THE CARBON FAMILY

Oxides

Table 16.7 lists the known oxides of the family and gives some of their physical properties and their structures.

The Structure of the Natural Silicates

The salts of silicic acid are sometimes quite complex, since silicic acid is found in the form of various condensed acids. The natural silicates, which play such an important role in the inorganic world, are found in a variety of structures: fibrous, as in asbestos; platelike, as in mica; and massive, as in meerschaum.

In mineral silicates, the SiO_4^{-4} grouping of silicon and oxygen atoms plays a dominant role in the crystal lattice. The oxygen atoms are the largest atoms,

Table 16.6 Standard Oxidation Potentials for Carbon Family Elements (in acid solution at 25°C)

Reaction	$E°$ (volts)
$Si + 2H_2O \rightleftharpoons SiO_2 + 4H^+_{(aq)} + 4e^-$	0.86
$GeH_4 \rightleftharpoons Ge + 4H^+_{(aq)} + 4e^-$	0.70
$Sn \rightleftharpoons Sn^{+2} + 2e^-$	0.14
$Pb \rightleftharpoons Pb^{+2} + 2e^-$	0.13
$Ge + 2H_2O \rightleftharpoons GeO_2 + 4H^+_{(aq)} + 4e^-$	0.1
$SiH_4 \rightleftharpoons Si + 4H^+_{(aq)} + 4e^-$	−0.10
$CH_4 \rightleftharpoons C + 4H^+_{(aq)} + 4e^-$	−0.13
$C + 2H_2O \rightleftharpoons CO_2 + 4H^+_{(aq)} + 4e^-$	−0.20

usually much the largest atoms, present in the lattices. They pack together around the silicon atoms as SiO_4^{-4} tetrahedra, and the metallic ions fit between the tetrahedra where they can. X-ray analysis of a large number of silicates shows that there are at least five general types of structures:

1. *Separate SiO_4^{-4} Ions.* These separate ions, SiO_4^{-4} (Figure 16.4) and the metallic ions, arrange themselves in various kinds of three-dimensional checkerboard patterns of general resemblance to the Na^+, Cl^- lattice. Garnet $[Ca_3Al_2(SiO_4)_3$, or $3CaO \cdot Al_2O_3 \cdot 3SiO_2]$ crystallizes in a lattice of this type.

2. *Separate, More Complex Silicon-Oxygen Ions.* The ion $Si_2O_7^{-6}$ is formed when two silicon atoms share the same oxygen atom between them (see Figure 16.5):

$$\begin{matrix} & O & & O & \\ & \| & & \| & \\ (O & - Si & - O - Si & - O)^{-6} \\ & \| & & \| & \\ & O & & O & \end{matrix}$$

Table 16.7 Some Properties of Carbon Family Oxides

Compound	Formula	Melting Point (°C)	Boiling Point (°C)	Appearance	Molecular Structure
Carbon monoxide	CO	−205.1	−190	Colorless gas	
Carbon dioxide	CO_2	−56.5*	−78.5†	Colorless gas	O=C=O
Carbon suboxide	C_3O_2	−107	−6.8	Gas	O=C=C=C=O
Silicon monoxide	SiO			Gas	
Silicon dioxide	SiO_2	1710	2590	White solid	Si—O—Si network
Germanium monoxide	GeO		710†	Black solid	Ge—O—Ge network
Germanium dioxide	GeO_2	1116	1200	White solid	Ge—O—Ge network
Tin monoxide	SnO	Decomp.		Black solid	Sn—O—Sn network
Tin dioxide	SnO_2	1927	1900†	White solid	Sn—O—Sn network
Lead monoxide	PbO			Yellow solid	Pb—O—Pb network
Lead dioxide	PbO_2	Decomp.		Brown solid	
Red lead	Pb_3O_4	830		Orange-red	Pb_2PbO_4

*At 5.11 atm.
†Sublimes.

Figure 16.4 Several methods of representing the silicate ion, SiO_4^{-4}: (a) ionic formula; (b) conventional structural formula; (c) structural model; (d) scale model, showing the tetrahedron produced by connecting the centers of the oxygen atoms.

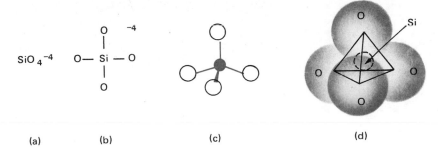

(a) (b) (c) (d)

Figure 16.5 The ion $Si_2O_7^{-6}$. Two tetrahedra, joined at a corner. In some cases the Si—O—Si bond is linear (c); in others the bases of the two tetrahedra lie in the same plane, and the Si—O—Si linkage is bent (e).

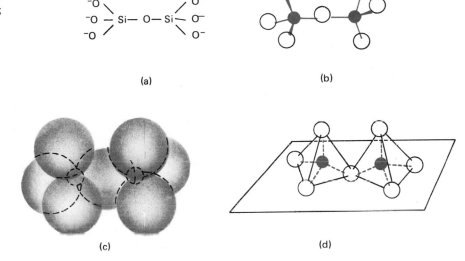

(a) (b)

(c) (d)

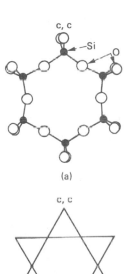

(a)

(b)

Figure 16.6 The ion $Si_6O_{18}^{-12}$. The six silicon atoms and six oxygen atoms shared between tetrahedra lie in the same plane, with the other two oxygen atoms, denoted by c,c, of each tetrahedron lying above and below that plane. (a) A structural formula. (b) Diagrammatic formula.

A ring-shaped ion, $Si_6O_{18}^{-12}$, is also known (Figure 16.6). Such complex negative ions as these, together with metallic ions, make up the crystal lattice. The ring ion is present in the lattice of beryl ($Be_3Al_2Si_6O_{18}$, or $3Be^{+2}$, $2Al^{+3}$, $Si_6O_{18}^{-12}$) (Figure 16.7).

3. *Silicon-Oxygen Strings.* Silicon-oxygen strings are common, occurring in all the fibrous silicates. The strings are extremely long and may run from one end of the crystal fiber to the other. The string may be like those in Figure 16.8, with each SiO_4 tetrahedron attached to its neighbors by two corners, or like those in Figure 16.9, in which alternate tetrahedra are attached by two and by three corners. In both cases, the end-on view of the chain [Figure 16.9(b)] has a trapezoidal shape; the trapezoids are held together by metallic cations. This is shown in the diagram of the diopside structure [$CaMg(SiO_3)_2$] shown in Figure 16.8, where the magnesium ions have a coordination number of 6 and the calcium ions, which are larger, a coordination number of 8. The string of Figure 16.9 is present in tremolite, $Ca_2Mg_2(Si_4O_{11})_2(OH)_2$, and the presence of the OH group in the structure is indicated in the figure.

4. *Silicon-Oxygen Sheets (two-dimensional).* The silicon and oxygen atoms also crystallize in sheets (Figure 13.25), which are held together by positive ions

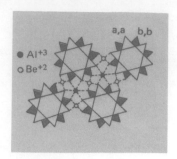

● Al^{+3}
○ Be^{+2}

a,a   b,b

Figure 16.7 Structure of beryl. The ring $Si_6O_{18}^{-12}$ ions of Figure 16.6 are piled in columns one above the other, and the viewer is looking straight down at the columns. At each "point" of the "star" are two oxygen atoms (a,a; b,b) corresponding to those marked c,c in Figure 16.6. The individual $Si_6O_{18}^{-12}$ ions and the columns are held together by Al^{+3} ions (coordination number 6) and Be^{+2} ions (coordination number 4).

Figure 16.8 Strings of SiO_4^{-4} tetrahedra. (a) Conventional structural formula. (b) Side view of a chain. (c) End-on view of four chains in a crystal of diopside, $CaMg(SiO_3)_2$, showing the alternation of the orientation of the chains in the crystal. The large circles between the chains represent calcium ions, and the smaller represent magnesium ions, which bind the chains together. In (c), the tetrahedra at upper left and lower right each present an edge toward the viewer; at upper right and lower left, the viewer sees a face in each case, with the concealed edge represented by a dotted line.

located between the sheets. Each tetrahedron shares three corners with adjacent tetrahedra. All the micas have this style of lattice—for example, muscovite [$KAl_3Si_3O_{10}(OH)_2$].

5. *Silicon-Oxygen Nets (three-dimensional).* Finally, the silicon and oxygen atoms often crystallize in a continuous three-dimensional network, with the metallic ions fitting into the holes. Many examples of this type are found among the feldspars—for instance, albite ($NaAlSi_3O_8$). In these compounds (and also in the micas), some of the SiO_4 tetrahedra are replaced by AlO_4^{-5} tetrahedra. The minerals are aluminosilicates; each corner of a tetrahedron is shared with an adjacent tetrahedron.

Figure 16.9 Double silicon-oxygen strings. (a) Top views, looking down on the SiO_4 tetrahedra. (b) End view of the double chain. (c) End view showing the trapezoidal appearance. (d) Packing pattern of the chains in the end view.

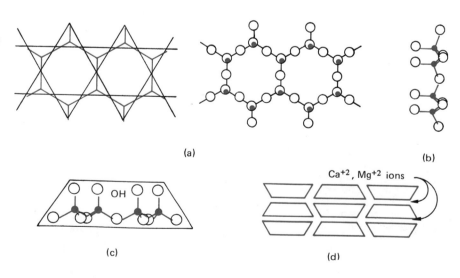

(a)

(b)

OH

(c)

Ca^{+2}, Mg^{+2} ions

(d)

Silicates in Solution

The ions in silicate crystals are usually bound together by very strong forces, with the result that the melting points of the silicates are nearly always rather high. Except for the sodium and potassium silicates, silicates are generally insoluble. Sodium silicate may be prepared by fusing pure silicon with sodium hydroxide or carbonate:

$$4NaOH + SiO_2 \longrightarrow Na_4SiO_4 + 2H_2O$$

$$Na_2CO_3 + SiO_2 \longrightarrow Na_2SiO_3 + CO_2$$

A solution of the products of such fusions in water is called water glass. It consists of a mixture of various silicates of sodium. Sodium silicate has many uses, especially as an ingredient of soap, as a protective coating for porous surfaces (such as those of wood, plaster, or cement), and as a cement or glue, chiefly for pasteboard boxes and cartons.

Fusion of the Silicates

If several different silicates are melted together with an excess of silica, they mix freely to a homogeneous liquid. Crystals may separate from the melt when it is cooled slowly, but usually the liquid solution simply becomes more and more viscous until it is as rigid as a true solid. Such products are called glasses, and they are sometimes regarded as very viscous solutions of one silicate in another or in silica. Actually, they are randomly oriented solids. The high viscosity makes it difficult for the extended structures of the silicates to arrange themselves in the exact positions needed for crystallization, and incomplete or random networks are present (Figure 16.10). The cations appear among the SiO_4 tetrahedra in no particular order, but in random positions.

Modern Ceramics. Many solid oxides, especially those occurring in relatively large amounts in nature, are used in the manufacture of ceramics. Ceramic products include such diverse things as bricks; dinnerware; porcelainware; coatings for refrigerators, washers, and dryers; kitchen stoves and bathroom fixtures; wall plasters; drain and roofing tile; glass; and optical materials.

A tiny piece of ceramic is probably responsible for telling you the temperature of your auto engine. A small splinter of an electronic ceramic mounted in a lightweight phonograph pick-up arm generates from the needle the separate sets of signals necessary to operate the speakers of your stereo phonograph.

Color television depends on the presence of special ceramic materials that glow with the desired light intensity and the required color purity. Ceramic memory cores, often the size of a pinhead, are the very heart of modern computers. Surgical instruments in hospitals and metal plates to be processed can be cleaned more effectively by immersion in a liquid bath containing a ceramic transducer that provides ultrasonic vibrations to more thoroughly remove clinging dirt particles. Ceramic transducers also are used in sonar equipment to find the depth of water, to locate schools of fish, and for detecting the presence of submarines.

A special glass is vital in color television receiver circuits. The television picture is transmitted separately in each of the three primary colors, one color after the other. The receiver collects the color signals in the same sequence. By passing the first and second pictures through special glass ultrasonic delay devices, where the signal travels slower but with virtually no loss

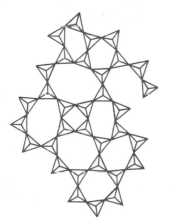

Figure 16.10 Diagram representing the random structure of a glass. Compare this structure with the regular structure of the crystalline material shown in Figure 13.25(b).

or scattering, the first two color signals can be delayed until the third arrives and all three flashed on the screen simultaneously.

These are but a few of the applications of ceramics that help improve or enrich our living. The fact that commonly occurring materials can be fabricated into objects that perform so precisely is an indication of the depth of our understanding of the properties of these materials and a testimony to the value of such knowledge. To illustrate how a small segment of this knowledge is used, we summarize here a few principles of conventional ceramics manufacture. In the following chapter, we illustrate some similar principles for electronic ceramics, those ceramics in electronic circuitry.

Ceramics are mixtures of chemical compounds, chiefly oxides. A large percentage contain silica or silicates. They are made by high-temperature ($\sim$540°C or higher) processing and can generally be used at high temperatures. The conventional steps in preparing a ceramic are preparing the material including mixing the inorganic components, shaping the article, drying, and firing. In the preparation of modern ceramics, each of these steps has become very complex and each must be controlled with extraordinary precision.

The raw materials for ceramics are usually in the form of fine powders. Conventional ceramics may contain large amounts of clay (ideally $Al_2O_3 \cdot 2SiO_2 \cdot 2H_2O$). The ingredients required are weighed out carefully and thoroughly blended and mixed to assure uniformity and reproducibility. This mixture is heated, sometimes to temperatures as high as 1500°C. The solids do not necessarily melt even at these temperatures, but reactions can take place between two or more components. In some cases, such as in china or porcelain, the particles of one of the components melts and forms a liquid that covers the unmelted particles. On cooling, this liquid forms a glass that cements the unmelted particles together and gives the materials a glassy or vitreous appearance.

The process of welding together the solid powders by heating is known as sintering. In this process, the crystal grains may change in size and composition. Usually the large grains grow larger at the expense of smaller grains. There is usually a marked relationship between grain size and properties, including appearance of the material. By controlling grain size, shape, and orientation as well as the sizes and distributions of empty spaces in the material, the ceramic engineer can do much to control the properties of the ceramic.

The effect of changing the mixture of oxides on the properties of a ceramic product can be illustrated as follows. In the manufacture of bricks, clay ($Al_2O_3 \cdot 2SiO_2 \cdot 2H_2O$) is blended with feldspars (potassium aluminum silicates) and sand or quartz. The color of the brick is due to the presence in the clay of oxides such as Fe_2O_3. The colored oxides are excluded in porcelains, and these materials are known for their whiteness. Research showed that replacing feldspar by other metallic oxides such as titanium dioxide, magnesium silicate, zirconium dioxide, or beryllium oxide would impart specific new properties to the ceramic. For example, beryllium oxide imparts a high degree of thermal shock resistance to the porcelain, broadening its use to applications involving extreme temperature conditions and violent thermal shock (as in air- and spacecraft and in nuclear reactors).

Magnesium silicate makes the product machinable; zirconium oxides add mechanical strength and low thermal conductivity, important in refractories for aerospace uses. Increasing the aluminum oxide content to as much as 92% produces products that are especially hard and abrasion resistant. Spark plugs, inert valves, and textile and wire guides (perhaps those in your tape recorder) contain high-alumina ceramics. Ultrapure aluminum oxide ceramics are used as radomes, electromagnetic windows, and cutting tools.

SUMMARY   The atoms of carbon family elements are larger and less electronegative than those of the corresponding members of the nitrogen, oxygen, or halogen families. Thus, carbon family elements are more metallic than their counterparts in the other series of nonmetals. Carbon and silicon are nonmetals; germanium, tin, and lead are metals. However, all five elements form predominantly covalent compounds and all except lead exhibit tetrahedral stereochemistry. The IV oxidation state is stable for all members except lead and, as a consequence, this family has a number of good reducing agents such as carbon, carbon monoxide, and silicon. All of them are oxidized readily to the IV state.

Carbon and silicon form a wide variety of polymeric or macromolecular structures. Many carbon-containing structures are the basic components of plant and animal matter, while polymers containing silicon and oxygen form the base structures of the mineral world.

The chemistry of the elements and of their low- and high-molecular weight compounds is summarized in this chapter.

IMPORTANT TERMS

Inorganic world	**Macromolecular structures**	**Natural silicates**
Organic world	**Network structures**	**Glass**
Chains of atoms	**sp^3, sp^2, sp hybridization**	**Silicon-oxygen strings, sheets and nets**
Rings of atoms	**Allotropes**	

QUESTIONS AND PROBLEMS

1. What differences in properties are observed between the carbon family elements and the oxygen family elements as a result of the smaller electronegativity and larger size of the carbon family elements?

2. Discuss structure and binding in the ion $SnCl_6^{-2}$.

3. Write formulas for some carbon compounds in the IV, II, and −IV oxidation states that are not given in Table 16.2.

4. Write equations to show that tin(II) hydroxide is amphoteric.

5. What difference in the two cases accounts for the fact that C—C bonds are common but Si—Si bonds are unusual?

6. Account for the difference in the melting points of silicon dioxide and carbon dioxide, of germanium disulfide and carbon disulfide, and of lead sulfide and carbon disulfide.

7. The melting points of tin and lead are much lower than those of silicon and germanium, yet the boiling points of all four are equally high. Why?

8. What differences account for the fact that the graphite form of carbon is less dense than the diamond form? That the diamond (gray) form of tin (5.8 g/cc) is less dense than the white form? That the density of

germanium is greater than the density of silicon, but the density of silicon is less than the density of the diamond form of carbon?

9. Draw the electron-dot structures for carbon monoxide, carbon dioxide, and carbon suboxide; indicate whether these are linear or angular molecules; and label the types of bonds (σ or π) and the hybridization of the carbon atom orbitals.

10. Draw electron-dot formulas for *n*-butane and digermane, and indicate the bond angles in these compounds.

11. What volume of carbon monoxide will be required for the synthesis of 40 g of carbonyl chloride ($COCl_2$)?

12. (a) What weight of carbon monoxide must be burned to produce 102,000 cal? (b) What volume will this amount of carbon monoxide occupy under 700 mm pressure and at 26°C?

13. What volume of carbon dioxide under standard conditions can be obtained (a) From 200 g of carbon? (b) By heating gently 500 g of calcium hydrogen carbonate?

14. The fermentation industry is a commercial source of carbon dioxide. If beer, which is 3.2% ethyl alcohol (C_2H_5OH) by weight, is assumed to have a density of 1.0 lb per pint and is essentially a dilute solution of alcohol from fermentation in water, how many pints will have to be made to obtain a pound of carbon dioxide?

15. If we assume that metallic lead costs 10 cents a pound and white arsenic 6 cents, which of the two would be the more expensive material in the manufacture of lead arsenate?

SPECIAL PROBLEMS
(May Require Additional Information)

1. What weight of carbon dioxide is necessary to change the base present in 500 ml of 0.4 *N* NaOH to sodium carbonate?

2. Compare the physical properties of carbon dioxide and nitrous oxide. Account for the many similarities.

3. Silicon monoxide is very unstable (see Table 16.7), but carbon monoxide is stable. Suggest a plausible explanation of this fact.

4. Why is the ratio of sodium to aluminum atoms always the same in minerals that contain only sodium, aluminum, silicon, and oxygen?

5. What is the chemical composition of commercial glasses? Use this information to predict their relative inertness to acids and bases.

6. Account for the sudden, spontaneous cracking of glass objects.

7. Cement is kept cool during the setting process, which is exothermic, but freezing is avoided. Offer an explanation.

8. Explain the following items: (a) Lead has a higher first ionization potential than tin. (b) Solder composed of tin and lead has a lower melting point than either pure metal. (c) An increase in temperature causes the electrical conductivity of lead wire to decrease and that of silicon wire to increase. (d) Silicon dioxide dissolves in hydrofluoric acid. (e) Stannous hydroxide is amphoteric. (f) Carbonic acid is a weaker acid than ortho-

phosphoric acid. (g) Addition of 6 M hydrochloric acid to 1 M sodium silicate results in gel formation.

9. Account for the fact that in minerals, a boron atom is usually located in the center of an equilateral triangle formed by three oxygen atoms, a silicon atom is located in the center of a tetrahedron formed by four oxygen atoms, and a magnesium ion is located in the center of an octahedron formed by six oxygen atoms.

10. How many liters of hydrogen at STP are produced when 23.7 g of tin are dissolved completely in 300 ml of 12.0 M sodium hydroxide? Calculate the final hydroxide ion concentration, assuming all the tin goes to $Sn(OH)_6^{-2}$.

11. Draw sketches of the various ways in which SiO_4 tetrahedra can combine with one another to produce silicate ions having different formulas.

12. Why do silicone coatings make glass water-repellent?

13. Write equations to show the following: (a) The reaction of germanium at high temperatures with oxygen, sulfur, chlorine, and nitric acid. (b) The reaction of sand with coke. (c) The reactions that take place when the ore cassiterite is roasted in air. (d) The combustion of pentane, C_5H_{12}, in excess oxygen. (e) The effect of heating limestone. (f) The reaction of carbon dioxide with sodium hydroxide solution. (g) The preparation of chlorine from hydrochloric acid and lead dioxide. (h) The reaction of carbon dioxide with calcium hydroxide solution.

14. Why are lead nitrate and lead acetate soluble in water, while lead chloride, lead sulfate, and lead chromate are not?

REFERENCES
AMEEN, J. G., and DURFEE, H. F., "The Structure of Metal Carbonyls," *J. Chem. Educ.*, **48**, 372 (1971).

BISSEY, J. E., "Some Aspects of *d*-Orbital Participation in Phosphorus and Silicon Chemistry," *J. Chem. Educ.*, **44**, 95 (1967).

LAIDLER, K., and FORD-SMITH, M. H., *The Chemical Elements*, Bogden and Quigley, Tarrytown, N.Y., 1970.

McCONNELL, D., and VERHOEK, F. H., "Crystals, Minerals and Chemistry," *J. Chem. Educ.*, **40**, 572, (1963).

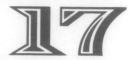

17

THE BORON FAMILY

I claim herein as my invention . . . 2. As an improvement in the art of manufacturing aluminium, the herein-described process, which consists in dissolving alumina in a fused bath composed of the fluorides of aluminum and sodium, and then passing an electric current, by means of a carbonaceous anode, through the fused mass, substantially as set forth.*

Charles Martin Hall (1863–1914)

The elements boron, aluminum, gallium, indium, and thallium constitute Group III of the periodic table. Their atoms all have the valence electron configuration s^2p^1. Boron is a nonmetal but the other elements are metals. Because of its position between families of active metals and families of nonmetals in the periodic table, the boron family might be expected to exhibit bonding and structure characteristics intermediate between those of its neighbors—characteristics that bring out some of the more subtle factors affecting bonding and structure. Many of the similarities and trends in properties found among carbon family elements and compounds appear again in this family, and for the same reasons. Examples are the tendency to form covalent bonds, the formation of macromolecules, and the reluctance of heavier atoms to use valence *s* electrons in bonding. In comparison with the neighboring groups to the left, all members except boron form stable positive ions similar (except in valence) to the alkaline earth ions of Group II.

There are, however, several structural features more or less unique to this family. These include electron-deficient molecules such as boron trifluoride, and bridge bonds such as those in aluminum chloride, Al_2Cl_6, and diborane, B_2H_6:

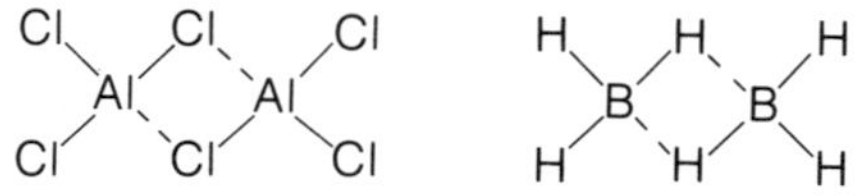

Cl
Cl
Cl
Al
Al
Cl
Cl
Cl
H
H
H
B
B
H
H
H

Aluminum chloride Diborane

A BROAD OVERVIEW OF THE BORON FAMILY

Important aspects of boron family chemistry are:

1. The atoms of these elements are larger and, except for thallium, less electronegative than those of the corresponding members of the carbon family. In addition, they are smaller and more electronegative than those of the corresponding members of Group II. In all cases, however, their atomic

*U.S. Patent 400766, April 2, 1889.

radii and electronegativities are close to those of the Group IV atoms, accounting in part for the family's acting as a whole like a closer relative to the carbon family than to Group II. Table 17.1 gives some properties of boron family atoms, including the ionization energies for removal of first(I), second(II), and third(III) electrons.

The ionization energies for the first three electrons in aluminum are 138, 434, and 656 kcal/mole, respectively, compared with 176 and 346 kcal/mole for the ionization energies for the first two electrons in magnesium and 118 kcal/mole for the ionization energy for sodium. As a result of this high-energy barrier for the formation of the +3 ion, most of the compounds of these elements in the III oxidation state contain predominantly covalent rather than ionic bonds. The M^{+3} ions, when present, are very small; their +3 charge gives them such a large charge density (ratio of charge to radius) that they attract negative ions and other bases very strongly.

2. The stable oxidation states for the family are all zero or positive because of the relatively low electronegativities of these atoms and the fact that they need five electrons to form an anion having an inert gas configuration. The $s^2 p^1$ valence-shell configurations give rise to the I and III oxidation states. Boron and aluminum form stable III states but much less stable I states, while thallium forms a I state that is considerably more stable than the III state. Gallium and indium show a III state and less well-characterized lower oxidation states. The stability of the I state of thallium arises because of the unusual stability of the $6s^2$ energy level, as with lead(II).

3. Bonding in the family is predominantly covalent, but there are numerous examples of ionic compounds. An interesting comparison between some predominantly ionic and some predominantly covalent compounds of this family is given in Table 17.2. The fluorides of all members except boron have

Table 17.1 Some Properties of Boron Family Atoms

Element	Valence Electron Configuration	Atomic Radius (A)	Ionic Radius (A)	Ionization Energies (kcal/mole)			Electro-negativity
				I	II	III	
Boron	$2s^2\ 2p^1$	0.80	0.20	191	580	875	2.0
Aluminum	$3s^2\ 3p^1$	1.25	0.50	138	434	656	1.5
Gallium	$4s^2\ 4p^1$	1.25	0.62	138	473	708	1.6
Indium	$5s^2\ 5p^1$	1.50	0.81	133	435	647	1.5
Thallium	$6s^2\ 6p^1$	1.55	0.95(III), 1.44(I)	141	471	688	1.9

Table 17.2 Melting Points of Some Boron Family Halides (predominantly ionic compounds are below the dashed line)

FLUORIDES		CHLORIDES		BROMIDES	
Compound	Melting Point (°C)	Compound	Melting Point (°C)	Compound	Melting Point (°C)
BF_3	−128.7	BCl_3	−107	BBr_3	−46
AlF_3	1290	Al_2Cl_6	192	Al_2Br_6	97.5
GaF_3	950	$GaCl_3$	77.5	$GaBr_3$	121.5
InF_3	1170	$InCl_3$	586	$InBr_3$	436
TlF_3	550	$TlCl$	429	$TlBr$	456

high melting points and are ionic, while only indium and thallium chlorides and bromides are ionic. As was mentioned in connection with tin halides, many covalent halides form ionic compounds when hydrated. This is especially true of the hydrated halides of this family, as exemplified by hydrated aluminum chloride, $Al(H_2O)_6Cl_3$, which is ionic, being composed of hexa-aquoaluminum(III) ions, $Al(H_2O)_6^{+3}$, and chloride ions.

4. These elements, particularly boron, form a number of macromolecular compounds. Examples are the oxides, X_2O_3, the sulfides, X_2S_3, and the nitrides, XN, of all but thallium. Typical of macromolecular structures is that of boron nitride (Figure 17.4):

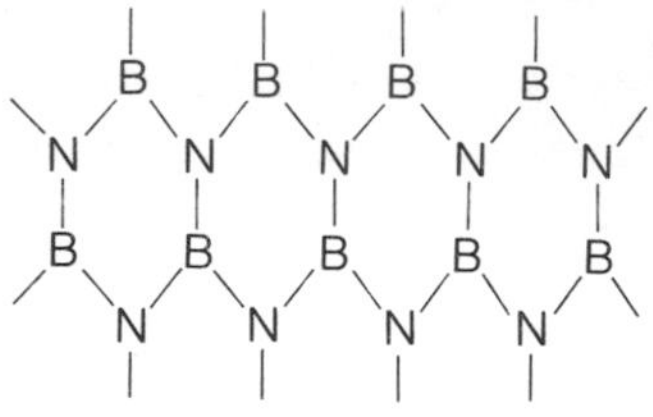

A portion of the boron nitride structure

Since nitrogen atoms supply five and boron atoms supply three valence electrons, this structure is electronically similar to carbon crystals, where each atom supplies four valence electrons. Boron nitride, an unreactive, insoluble, refractory material, has a graphitelike structure consisting of layers of alternating boron and nitrogen atoms arranged as shown above. The bonding within each layer involves sp^2 hybrid orbitals on both kinds of atoms, with the remaining electrons being used to form π bonds between nitrogen and boron atoms. The layers are arranged so that a boron atom lies directly below a nitrogen atom in the layer above. Van der Waals forces are the primary binding forces between layers. At pressures near 70,000 atmospheres and temperatures about 3000°C, this graphite-like structure can be converted to a structure analogous to diamond. This substance, known as borazon, is one of the hardest substances known.

5. Certain of the covalent compounds of these elements in their III oxidation states are electron-deficient molecules—molecules containing atoms with less than a noble gas valence-shell configuration. Examples are boron trifluoride, BF_3, and gallium triiodide, GaI_3, both of which are shown below.

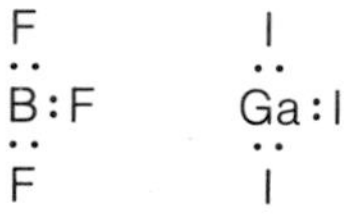

Note the electron-deficient nature of the central atoms. These molecules, especially the chlorides and fluorides, are able to act as Lewis acids by accepting pairs of electrons from various bases, as in the reaction

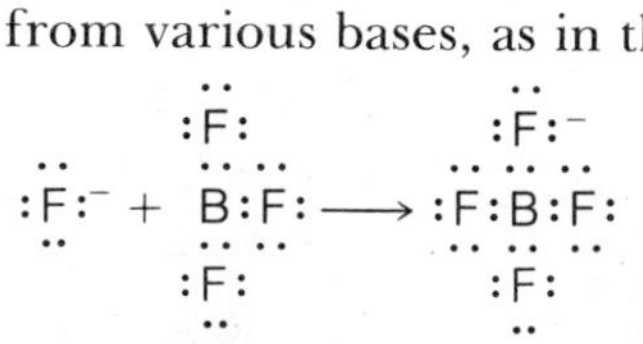

The fluoroborate ion, BF_4^-, is found in numerous salts such as sodium fluoroborate, $NaBF_4$. The bromoaluminate ion, $AlBr_4^-$, and the chlorogallate ion, $GaCl_4^-$, are also known. The ability of aluminum and gallium ions to expand their octets is illustrated by the AlF_6^{-3} and $GaCl_6^{-3}$ ions.

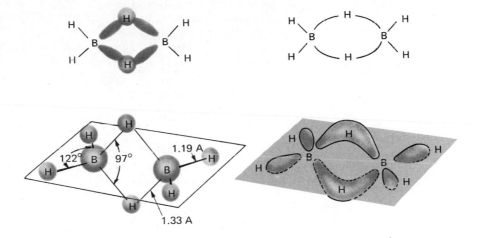

Figure 17.1 Several methods of representing the structure of diborane. The drawings emphasize the overlap of two sp^3 orbitals from boron atoms with the $1s$ orbital of hydrogen to give the bridge-bond, a two-electron bond involving three nuclei. There are two B—H—B bridges in diborane.

6. Some of the electron-deficient compounds form three-center bonds in which *an electron pair interacts with three nuclei.* One of the three nuclei then acts as a "bridge" between the other two. This is shown in Figure 17.1 for diborane, B_2H_6, in which two bridge hydrogen atoms are present. In each of these bonds, the hydrogen $1s$ orbital may be imagined to overlap with the sp^3 hybrid orbitals of both boron atoms, giving rise to a single molecular orbital capable of holding two electrons.

Bridge bonds presumed to involve p orbitals occur in aluminum chloride,

$$\begin{array}{ccc} Cl & Cl & Cl \\ \diagdown & \diagup \cdots \diagdown & \diagup \\ & Al & Al \\ \diagup & \diagdown \cdots \diagup & \diagdown \\ Cl & Cl & Cl \end{array}$$

and the compound

$$\begin{array}{ccc} CH_3 & H & H \\ \diagdown & \diagup \cdots \diagdown & \diagup \\ & Ga & B \\ \diagup & \diagdown \cdots \diagup & \diagdown \\ CH_3 & H & H \end{array}$$

has been reported.

In addition to three-center BHB bonds, the higher boron hydrides show three-center BBB bonds and *multicenter* bonds, as in the five-center bond binding the apical boron atom in B_5H_9 to the other four boron atoms (Figure 17.2). Here, six electrons are used to bind five atoms together.

7. Amphoterism is the outstanding property of the oxides and hydroxides of the family, with the boron compounds showing a stronger tendency to act as acids than as bases and the compounds of the heavier elements showing a stronger tendency to act as bases. An example of amphoteric character among the oxides is given in the following reactions involving boric oxide:

$$\text{Acidic oxide} + \text{basic oxide} \longrightarrow \text{salt}$$

$$B_2O_3 + CuO \longrightarrow Cu(BO_2)_2$$

$$P_4O_{10} + 2B_2O_3 \longrightarrow 4BPO_4$$

In the first reaction, it acts as an acid, being incorporated into the anion of the salt. This reaction is one of a number of reactions involving metal oxides and boric oxide; the products are colored glasses often used to identify the metals in the so-called borax-bead test. In the second reaction, boric oxide acts as a base, the boron atoms forming the cations of the salt.

Aluminum hydroxide exemplifies the amphoteric hydroxides of the family in the following reactions:

$$Al(OH)_3(s) + OH^- \longrightarrow Al(OH)_4^-, \quad \text{or} \quad (AlO_2^- + 2H_2O)$$

$$Al(OH)_3(s) + 3H_3O^+ \longrightarrow Al(H_2O)_6^{+3}$$

The reactions illustrate that either acids or bases will dissolve solid aluminum hydroxide. Bases produce the aluminate ion, AlO_2^-, which is probably present in solution as the hydrated species $Al(OH)_4^-$ or $Al(H_2O)_2(OH)_4^-$. Acids neutralize the hydroxide, giving the hexaaquoaluminum(III) ion, $Al(H_2O)_6^{+3}$.

Two important acids in the family are boric acid, H_3BO_3, which is so weak that in water it acts essentially as a monobasic acid having a K_1 of 6×10^{-10} at 25°C; and hexaaquoaluminum(III) ion, $Al(H_2O)_6^{+3}$, which undergoes the reaction

$$Al(H_2O)_6^{+3} + H_2O \rightleftharpoons Al(H_2O)_5OH^{+2} + H_3O^+$$

for which K_1 is 1.1×10^{-5} at 25°C. This equilibrium accounts for the fact that solutions of most aluminum salts are acidic.

The strongest base in the family is thallium(I) hydroxide, TlOH, which is similar in many ways to potassium hydroxide.

8. The metals of the family are slightly stronger reducing agents than the corresponding members of the carbon family and considerably weaker than the Group II metals. The standard oxidation potentials are given in Table 17.3. All of these metals are stronger reducing agents than would be anticipated from an examination of their high ionization potentials. However, the very high hydration energy of the small M^{+3} ions lowers the overall energy requirements for the reaction in solution, thereby increasing the reducing strength in aqueous media.

The III oxidation state species of this family are all moderate to weak oxidizing agents, with the thallium compounds being the strongest and the boron compounds the weakest in the series.

Figure 17.2 Structures of some boron hydrides: (a) pentaborane-9, B_5H_9; (b) pentaborane-11, B_5H_{11}; (c) decaborane, $B_{10}H_{14}$.

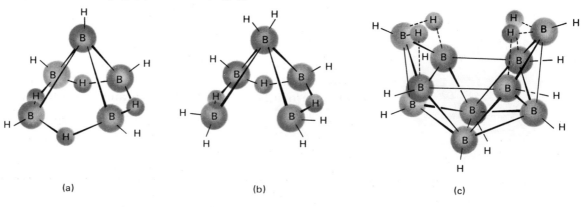

(a)   (b)   (c)

Table 17.3 Standard Oxidation Potentials for Boron Family Elements

Conversion	$E°$ (volts)
$Al + 6H_2O \rightleftharpoons Al(H_2O)_6^{+3} + 3e^-$	1.66
$Ga + 6H_2O \rightleftharpoons Ga(H_2O)_6^{+3} + 3e^-$	0.53
$In + 6H_2O \rightleftharpoons In(H_2O)_6^{+3} + 3e^-$	0.34
$Tl + 3H_2O \rightleftharpoons Tl(H_2O)_3^{+} + e^-$	0.34
$Tl + 6H_2O \rightleftharpoons Tl(H_2O)_6^{+3} + 3e^-$	−0.72

Table 17.4 Some Physical Properties of Boron Family Elements

Element	Density of Solid at 25°C (g/cc)	Melting Point (°C)	Boiling Point (°C)	Heat of Fusion (kcal/mole)	Crystal Structure
Boron	2.4	2300	2500	5.3	Hexagonal
Aluminum	2.7	660	2500	2.6	Face-centered cubic
Gallium	5.93	29.8	2070	1.3	Rhombic
Indium	7.29	156	2100	0.78	Cubic close-packed (distorted)
Thallium	11.85	449	1390	1.0	Cubic close-packed

THE ELEMENTS

Physical Properties

Table 17.4 gives some physical properties of the elements. The densities increase systematically from boron to thallium, with boron and aluminum having low values   The melting point and heat of fusion are much higher for boron than for any of the others, suggesting that the boron crystal is a macromolecular structure in which the boron atoms are covalently bonded. Also, solid boron is a poor electrical conductor while the others are much better conductors, suggesting that the latter form metallic crystals. In support of this are the physical appearance, the extremely wide liquid ranges, the low heats of fusion, and the malleability and ductility of aluminum, gallium, indium, and thallium. All of these are characteristics of metallic crystals.

Boron crystals are dark brown and very hard; aluminum, silvery-white when first cut, acquires a dull luster because of the formation of a thin oxide film on the metal surface; gallium is silvery-white, hard, and brittle; indium is soft, malleable, and ductile with a silver-metal luster; thallium is a soft gray metal that is malleable but with poor tensile strength.

Chemical Properties

Some important chemical properties of boron family elements are illustrated by the following general equations, where M represents the boron family element:

1. $4M + 3O_2 \longrightarrow 2M_2O_3$. High temperature required; gallium is more resistant than others; thallium also gives Tl_2O.
2. $2M + 3X_2 \longrightarrow 2MX_3$. Occurs with all halogens; usually requires higher temperatures; thallium also forms TlX.
3. $2M + N_2 \longrightarrow 2MN$. Occurs with boron and aluminum only.
4. $2M + 3S \longrightarrow M_2S_3$. Occurs at high temperature; boron requires 1200°C; thallium gives Tl_2S.

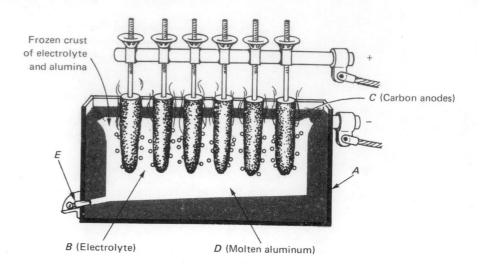

Frozen crust
of electrolyte
and alumina

C (Carbon anodes)

+

−

E

A

B (Electrolyte)

D (Molten aluminum)

Figure 17.3 Electrolytic production of aluminum. The process is carried out in a carbon-lined box (A), which serves as the cathode. Carbon anodes (C) project into the box. Cryolite, Na_3AlF_6, or an artificial mixture of fluorides is used to partially fill the box. The current is then turned on, generating enough heat to melt the cryolite. Purified aluminum oxide is then added; it dissolves in the liquid cryolite. Purified aluminum oxide is then added; it dissolves in the liquid cryolite as an electrolyte (B). The oxide yields aluminum and oxygen upon electrolysis. Temperature is kept above aluminum's melting point. Liquid metal (D) sinks to the bottom of the vessel, from where it is removed from time to time via the taphole (E). Part of the oxygen escapes as gas, and part of it combines with the carbon of the anodes, which are gradually consumed. As the oxide is electrolyzed, more is added, making the process continuous.

5. $2M + 6H_3O^+ \longrightarrow 2M^{+3} + 3H_2 + 6H_2O$. Occurs with all members except boron; thallium forms Tl^+.

6. $2M + 2OH^- + 2H_2O \longrightarrow 2MO_2^- + 3H_2$. Boron forms metal borates on fusion; indium and thallium do not react.

Preparation of the Elements   Boron is best prepared by heating the oxide (B_2O_3) with a large excess of magnesium or aluminum, which act not only as reducing agents but as solvents for the reduced boron:

$$B_2O_3 + 3Mg \longrightarrow 2B + 3MgO$$

By dissolving the excess metal with acids, the crystallized boron is obtained. Prepared in this way, the product is never entirely pure. Pure crystals have been obtained by leading hydrogen and vapor of boron tribromide over a hot (1300°C) wire of tungsten or tantalum.

All the aluminum prepared in the U.S. is obtained by the electrolysis of aluminum oxide (Al_2O_3) dissolved in melted cryolite, as shown in Figure 17.3.

Gallium, indium, and thallium may all be prepared by electrolysis of aqueous (usually basic) solutions of their salts. Obtaining the crude metals from their ores is sometimes difficult.

ARTIFICIAL GEMS

Aluminum oxide can be melted in oxyhydrogen furnaces and obtained in crystalline form. The pure, colorless gem so produced is called white sapphire; by adding the requisite metallic oxide, almost any desired color can be given the gem stone. The rubies and sapphires so produced are identical in almost every respect with the natural stones and are artificial gems, not imitation ones.

Electronic Ceramics—You use them every day. Among the electronic ceramics are conducting and semiconducting glasses; ferrites, a class of non-metallic magnets; ferroelectrics, a class of materials that convert electric signals into mechanical motion such as sound, or mechanical motion into electrical signals; and dielectrics, materials in which electrical energy can be stored. We have space here to discuss only the first two types.

About 90 per cent of all glass produced in the U.S. is made from a mixture of 70 per cent SiO_2 and the oxides of sodium and calcium. This glass is satisfactory for drinking glasses, bottles, and windows, but it is too sensitive to moisture and too low in electrical resistance to be useful in electronic applications. Electronic glasses have lower amounts of sodium oxide than does common glass.

Electrical conductivity is possible in glasses by migration of either sodium ions or electrons. Sodium ion migration, however, is not stable and uniform under long use because of the depletion of these ions by migration to the cathode. By adjusting the glass composition, reducing the amounts of sodium and calcium oxides, and adding vanadium or phosphorus pentoxides to the glass, electronic conduction can be made to predominate over ionic conduction, and the material becomes a semiconducting glass.

Semiconductivity can be further improved by adding large amounts of oxides that are themselves semiconductors, such as FeO, CoO, and MnO. Another variety of semiconducting glass contains the mixture $Na_2O \cdot B_2O_3 \cdot TiO_2$. Semiconducting glasses find numerous applications, including as thermisters, the heat-sensitive resistors that tell you the temperature of the engine in your car and protect your TV receiver and radio or hi-fi set from voltage surges that may damage them. They hold the key to major breakthroughs in computer technology.

A special electronically conducting glass is made by depositing a transparent layer of tin dioxide on a glass surface. Heat is generated on this surface when a voltage is applied to it. The glass is used in windshields, where it prevents fogging and provides a de-icing capability.

Ferrites are ceramics related to magnetite, Fe_3O_4, which is actually a mixture of iron(II) and iron(III) oxides. This oxide is magnetic. Ceramic ferrites contain the iron(III) oxide component, but oxides such as MgO, CuO, MnO, CoO, and BaO have replaced varying amounts of the iron(II) oxide. These materials are made by sintering the oxides together, forming small crystals. Ferrites are classified according to their magnetic properties—from soft or easily demagnetized, such as those containing manganese, copper, or nickel(II) oxides, to hard (more-or-less permanent magnets), such as those containing barium, strontium, or lead oxides.

Soft ferrites are used in Touchtone telephones, in high-frequency fluorescent lighting, in TV, radio, and electronic ignition systems in the recording heads of tape recorders, and in many other applications.

Hard ferrites are used where a strong magnetic attraction is needed—on refrigerator doors; as armatures in electric motors; and to operate windshield wipers, heat blowers, air conditioners, and seat adjusters. Ferrite motors also are used in small appliances and portable electric tools—in cordless shavers, toothbrushes, and electric knives.

A simple explanation for the magnetic behavior of soft ferrites and their modification by added oxides is that the Fe(III) ions in the ferrite crystal are divided equally among the tetrahedral and octahedral holes between the oxide ions. The nuclear spins of the iron ions in these two types of sites are antiparallel to one another and, therefore, cancel each other's contribution to the total magnetization. The resulting magnetization in Fe_3O_4 then is due to the magnetic properties of the Fe(II) ions, which must occupy octahedral sites because of their larger size. Addition of ions such as Zn or Cd, which prefer to occupy the tetrahedral holes, forces Fe(III) ions to concentrate in the octahedral holes, thereby increasing the Fe(III) contribution to the magnetization.

Addition of oxides containing large ions such as barium or lead alters the crystal structure of the ferrite from a cubic (or spinel) structure to a hexagonal structure, a change that stabilizes the magnetization.

SUMMARY   The boron family, located between families of active metals and families of nonmetals, shows the properties of both. In physical properties, boron is a nonmetal; aluminum, gallium, indium, and thallium are metals. The tendency to form covalent bonds is strongest with boron and aluminum; the tendency to form ionic bonds is strongest with gallium and thallium. The oxides of all members are amphoteric.

Common oxidation states of these elements are 0, I, and III, with the III state being the most stable for all except thallium. Many of the simple covalent compounds of these elements are electron-deficient since they have only six valence electrons. Some of these electron-deficient compounds act as Lewis acids, accepting pairs of electrons from bases, as in the formation of BF_4^-. Certain electron-deficient compounds stabilize their structures by forming bridge structures with multicenter bonds.

Members of this family, especially boron, form a number of macromolecular structures. Among these are boron nitride, boric oxide, and aluminum oxide.

IMPORTANT TERMS

Bridge bonds   **Three-center bonds**

Borazon   **Multicenter bonds**

Electron-deficient   acid oxide
molecules   basic oxide
                        glass

QUESTIONS AND PROBLEMS

1. Account for the fact that thallium forms a stable I oxidation state, while the I states of the other elements of the boron family are unstable.

2. Explain why indium bromide is predominantly ionic, but gallium bromide is covalent.

3. In what respects are the following compounds similar to BN: graphite, AlN, and SiC?

4. Draw electron-dot structures for BF_3 and BF^{-4}, and suggest likely shapes for these species.

5. Write equations to show: (a) The ionization of aluminum hydroxide in water solution. (b) The reaction of sodium hydroxide with aluminum hydroxide. (c) The reaction of hydrochloric acid with aluminum hydroxide. (d) The reaction of aluminum chloride with water. (e) The reaction of thallium hydroxide with sulfuric acid. (f) The reaction of

aluminum with oxygen. (g) The reaction of aluminum with chlorine in the absence of water. (h) The reaction of aluminum with hydrochloric acid solution. (i) The reactions of aluminum with sodium hydroxide solution. (j) The electrolysis of bauxite in melted cryolite. (k) The reaction of aluminum sulfate with calcium hydroxide solution.

6. Suggest structures for the following: (a) $Al_2(CH_3)_6$, (b) $Al(BH_4)_3$, (c) B_3H_9.

7. Draw a graph of the energy required to remove three electrons from each of the elements Al, Ga, In, and Tl against the standard oxidation potentials of these elements in water solution, and discuss any deviations you observe from a straight line or a smooth curve.

8. Describe the differences in the properties of boron as compared with those of the other elements of the family that lead to a classification of boron as a nonmetal and the others as metals. How do you account for these differences in terms of such factors as ionization potential, atomic size, and crystal structure?

9. Write an electron-dot structure for the dimer of gallium(III) chloride.

10. Indium is much more electropositive than thallium in going from the zero to the III oxidation state. Explain.

11. What relative weights of sodium, zinc, cadmium, and aluminum will be required to yield the same volume of hydrogen by action on water or on acids?

SPECIAL PROBLEMS

1. Explain the following items: (a) B^{+3} ions are not formed by boron. (b) The classification of boron trifluoride as an acid. (c) The use of borax in laundry products. (d) Indium hydroxide is not amphoteric, but aluminum hydroxide is. (e) Molten aluminum chloride is a poor electrical conductor. (f) Ferric oxide may be reduced to metallic iron with aluminum. (g) Aluminum sulfate gives a strongly acid solution.

2. The unit-cell-edge length is 4.04 A in the face-centered cubic arrangement of aluminum crystals. Calculate the atomic radius and the density of metallic aluminum.

3. Outline the preparation of aluminum from bauxite. How much aluminum can be obtained from 2 tons of bauxite that is 45% Al_2O_3 by weight?

4. Suggest a simple practical separation of sodium chloride (m.p. 800°C) and aluminum chloride from a mixture of these compounds.

REFERENCES

GARRETT, A. B., *The Flash of Genius*, Van Nostrand Reinhold, New York, 1962.
JAMES, B. D., "Structural Studies on Some Complex Species with Bridged Hydrogens," *J. Chem Educ.*, **47**, 176 (1971).
LIPSCOMB, W. M., *Boron Hydrides*, Benjamin, Menlo Park, Calif., 1964.

PART FIVE

ENERGY
AND
CHEMICAL
CHANGE

ENERGY

En poursuivant mes recherches, j'établis entre autres que, quelle que soit la voie par laquelle une combinaison s'accomplisse, la quantité de chaleur dégagée par sa formation était toujours constante, soit que la combinaison ait lieu directement, soit qu'elle ait lieu indirectement et à différentes reprises.*

Germain Henri Hess (1802–1850)

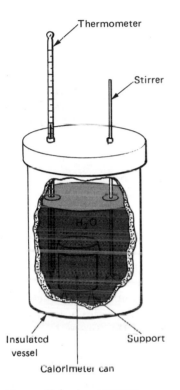

Figure 18.1 A calorimeter.

The chemist is concerned with both of the major entities of the universe— *matter* and *energy*. Having learned something about the particle nature of matter, the structure, combining property, and modes of motion of these particles, chemists then ask such questions as the following about energy:

1. How much energy is involved in chemical change?
2. How is this energy associated with the bonding and motion of chemical species?
3. How can this energy be measured?
4. How can information about energy in chemical change be helpful in predicting whether systems are stable?

The purpose of this chapter is to find the answers to some of these questions and to learn to use energy data to increase our understanding of chemical change. The answers are found in consideration of the following topics:

1. The heat of reaction and its correlation with bond energies.
2. The ways that energy is associated with molecules.
3. Calculation of energies in crystals.
4. The heat capacity of substances.
5. Some applications of thermochemical data.

HEATS OF REACTION

Initial and Final States

In the simple reaction between hydrogen and oxygen to form water,

$$2H_2 + O_2 \longrightarrow 2H_2O \tag{1}$$

---

*"In the course of my investigations I ascertained that by whatever steps a compound may come to be formed, *the quantity of heat developed in the formation is always constant*; it makes no difference whether the compound be formed in a direct or indirect way, all at once, or at different times." (The quotation is from *Bulletin de la classe physico-mathématique publie par l'Académie Impériale des Sciences de St. Petersbourg*, vol. I (1842), p. 150.)

there is a change in which the bonds connecting hydrogen atoms to hydrogen atoms and oxygen atoms to oxygen atoms become bonds connecting hydrogen atoms to oxygen atoms. The *initial state*, which has hydrogen atoms connected to hydrogen atoms and oxygen atoms connected to oxygen atoms, changes to the *final state*, which has hydrogen connected to oxygen. The heat of reaction represents the *energy difference between the initial and final states*. We measure the heat of reaction by arranging for the reaction to take place in a calorimeter.

Factors to be Specified in Going from the Initial to the Final State

For the energy measurements to be meaningful, we must describe the initial and final states exactly. We must know the physical state of the substance: the energy change in the reaction represented by Equation (1) would evidently depend upon whether liquid water or gaseous water was formed. We must know also the average temperature at which the reaction takes place, since the energy change is different at different temperatures (see Figure 18.6). Further, the temperature rise in the calorimeter must be kept small; otherwise the measurements will be imprecise because they refer to a temperature range rather than to a specific temperature. We must know whether the reaction takes place at constant pressure or at constant volume, especially for gaseous substances. We must, of course, know how much material reacts. Since, in chemistry, we may interpret equations in terms of numbers of moles of substances reacting and formed, the measurements are always converted to the reaction of the number of moles specified in the equation. Specifying the necessary quantities for reaction (1),

$$2H_2(g) + O_2(g) \longrightarrow 2H_2O(l); \quad 25°C; \quad V \text{ constant} \tag{2}$$

we state that 2 moles of gaseous hydrogen reacts with 1 mole of gaseous oxygen to form 2 moles of liquid water, all at 25°C in a constant-volume calorimeter.

Internal Energy

Our measurements give us only the difference in energy between the initial and final states, but tell us nothing about the actual energies of the two states. Scientists assign to molecules an internal energy, denoted by E, which will include all the unmeasurable and difficultly measurable energies of formation of the particles in the nucleus, of the nucleus itself, of the internal electron shells, and the like, as well as some energies that can be measured, such as the energies of bonding and the kinetic energies of motion. If one agrees

---

APPARATUS TO MEASURE THE HEAT OF REACTION

A calorimeter consists, typically, of a metal can or thick-walled vessel immersed in a known amount of water in an insulated vessel. Energy liberated in a reaction taking place inside the can will be transferred to the can and water as heat, and raise the temperature. If we measure the temperature rise and know the masses and specific heats of the can and water, we can calculate how many calories have been absorbed by the can and water and therefore, how much energy was liberated in the reaction.

Experiment: A reaction took place in a calorimeter consisting of a steel bomb weighing 4050 g immersed in 1900 g of water and the temperature rise measured 2.80°C. The specific heat of steel is 0.107 cal/g°C and that of water is 1.00 cal/g°C. Calculate the energy in calories liberated by the reaction.

To raise the bomb temperature required 2.80 × 4050 × 0.107 = 1213 cal.
To raise the water temperature required 2.80 × 1900 × 1.00 = 5320 cal.
Total heat liberated = 1213 + 5320 = 6533 cal.

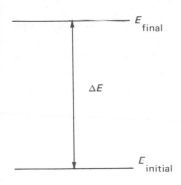

Figure 18.2 Energy difference $\Delta E = E_F - E_I$ between final and initial states.

to this assignment, then the measured energy is the difference in the value of E for the final molecules and E of the initial molecules, which we can write

$$\Delta E = (\text{final state energy}) - (\text{initial state energy}) \qquad (3)$$

Note that the Δ symbol always represents the difference *final* minus *initial* (Figure 18.2). Since the energy for each kind of molecule depends upon the number of molecules present, it is convenient to let the symbol E represent the internal energy per mole of substance. We then have for the change in internal energy in reaction (2):

$$\Delta E - 2E_{H_2O(l)} - (2E_{H_2(g)} + E_{O_2(g)}) \qquad (4)$$

Since energy is evolved in the reaction written, the internal energy $2E_{H_2O(l)}$ is less than the internal energy $2E_{H_2O(g)} + E_{O_2(g)}$, and ΔE is a negative quantity. We write

$$2H_2(g) + O_2(g) \longrightarrow 2H_2O(l); \qquad 25°C; \qquad \Delta E = -134.86 \text{ kcal} \qquad (5)$$

It is no longer necessary to specify that the volume is constant, since the change ΔE is a property of the reactants and products; the quantity we measure in the calorimeter, however, is equal to ΔE only when the measurement is made at constant volume.

Enthalpy

The value of ΔE is obtained in a closed container such as that used in the calorimeter; the experiment is done at *constant volume*. However, most reactions in the laboratory are carried out in a container open to the atmosphere and therefore are done at constant pressure rather than at constant volume. The heat of reaction at constant pressure will be different from that at constant volume.

To illustrate the significance of the last sentence, let us consider what the energy release would be if we were to burn hydrogen in oxygen in an open container (or at least in a cylinder with a weightless piston). We must imagine the atmosphere pressing down on the system and tending to compress it. If compression occurs, then work is being done on the system, and the system will acquire energy because of the work done on it. If, however, the system expands, it will be doing work on the atmosphere and pushing it back; thus, this system will have less energy by the amount used up in doing work on the atmosphere.

When 2 moles of hydrogen burn in 1 mole of oxygen to form liquid water at 25°C in an open calorimeter,

$$2H_2(g) + O_2(g) \longrightarrow 2H_2O(l); \qquad 25°C; \qquad P \text{ constant} \qquad (2')$$

there will be a volume decrease. The volume shrinks as 2 moles of hydrogen, occupying about 48,000 ml at 25°C, and 1 mole of oxygen, occupying 24,000 ml, react to form 36 ml of liquid water. The prevailing atmosphere therefore does work on the system, pushing it down into the smaller volume. As a result, the work energy is added to the system and appears as heat in the calorimeter. The amount of work done is measured by the product of the pressure by the change in volume, $P\Delta V$. (Figure 18.3.)

For our reaction, this becomes

$$P\Delta V = 1 \text{ atm} \times (36 \text{ cc} - 72,000 \text{ ml})$$

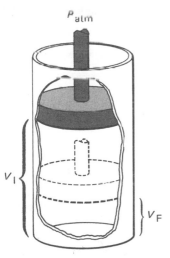

Figure 18.3 Pressure-volume work at constant pressure, $P\Delta V = P(V_F - V_I)$.

which, on conversion of the pressure-volume units to calories (1 l atm = 24.2179 cal), is equal to -1780 cal, and the amount of heat evolved is increased by this amount. Since most chemical reactions are carried out in open containers at constant pressure, pressure-volume work of this sort is almost always obligatory. The amount of $P\Delta V$ work done when only solids and liquids are the reactants and products is small, since they do not show much change in volume on reaction, and for these systems the pressure-volume work may be neglected. Whenever gases are involved, however, and there is a change in the number of moles of gas in passing from the initial to the final state, the pressure-volume work becomes important.

It is convenient to have a name and symbol for the energy that includes the pressure-volume work. It is called the *enthalpy* (en'thal py) and is given the symbol H. The enthalpy change and the change in internal energy are related for constant pressure processes by the equation

$$\Delta H = \Delta E + P\Delta V \tag{7}$$

For the reaction of hydrogen and oxygen at 25°C, ΔH is then $(-134.86 \text{ kcal}) + (-1.78 \text{ kcal})$, and we write

$$2H_2(g) + O_2(g) \longrightarrow 2H_2O(l) \qquad \Delta H_{298} = -136.64 \text{ kcal} \tag{8}$$

Here, the subscript on ΔH indicates that the value is for the reaction at 25°C (298K). Equation (8) implies that each substance has an enthalpy value just as it has a value for the internal energy, and for the reaction written

$$\Delta H = 2H_{H_2O(l)} - [2H_{H_2(g)} + H_{O_2(g)}] = -136.64 \text{ kcal} \tag{9}$$

Since the enthalpy change represents the difference *final* minus *initial*, it is evident that the enthalpy change for the reverse reaction is the negative of the enthalpy change of the forward reaction. Thus, for the reaction

$$2H_2O(l) \longrightarrow 2H_2(g) + O_2(g) \tag{10}$$

$$\Delta H = 2H_{H_2(g)} + H_{O_2(g)} - 2H_{H_2O(l)} = +136.64 \text{ kcal} \tag{11}$$

Enthalpy of Formation   The enthalpy change that occurs when a compound is formed from its elements is a convenient reference quantity.

To make it easy to compare in a meaningful way the ΔH values for different reactions, chemists have agreed to a set of *standard states*. The standard states chosen are the physical forms of the elements and compounds stable at 1 atm pressure and a specified temperature, commonly 25°C. *The enthalpy change when one mole of a compound in its standard state is formed from its elements in their standard states* is known as the *heat of formation* or enthalpy of formation, symbolized by ΔH_f°. Since the stable physical forms of hydrogen and oxygen are the gaseous forms at 1 atm and 25°C, the heat of formation of

---

HEAT AND WORK

The terms *energy, pressure, work,* and *heat* may be defined as follows:

Energy is the capacity to do work.

Pressure is force per unit area, or force/area.

Work is a means of transferring energy that results in a displacement against opposition: Displacement × opposition = distance × force = distance × area × force/area = volume × pressure.

Heat is a means of transferring energy. It may result in a temperature rise, in which case it is measured by the product (heat capacity) × (temperature rise); or it may result in a change of phase such as melting, in which case it is measured by (heat of fusion) × (mass of material).

Table 18.1 Enthalpy of Formation ($\Delta H°_{f298}$ in kcal/mole)

$H_2O(g)$	−57.79	$CO(g)$	−26.41
$H_2O(l)$	−68.32	$CO_2(g)$	−94.05
$HCl(g)$	−22.06	$NaCl(s)$	−98.23
$HBr(g)$	−8.66	$KCl(s)$	−104.18
$HI(g)$	+6.20	$CaCl_2(s)$	−190.0
$SO_2(g)$	−70.96	$CaCO_3(s)$ calcite	−288.45
$H_2S(g)$	−4.81	$AlCl_3(s)$	−166.3
$N_2O(g)$	+19.49	$CuO(s)$	−37.6
$NO(g)$	+21.60	$Cu_2O(s)$	−40.4
$NH_3(g)$	−11.04	$CuSO_4(s)$	−184.00
$Br_2(g)$	+7.34	$CuSO_4 \cdot 5H_2O(s)$	−544.45
$Br_2(l)$	0	$PbCl_2(s)$	−85.85
$CH_4(g)$	−17.89	$C_2H_4(g)$	+12.50
$CH_3Cl(g)$	−19.6	$C_2H_2(g)$	+54.19
$CH_3OH(l)$	−57.02	$C_3H_8(g)$	−24.82
$CHCl_3(l)$	−31.5	$n\text{-}C_4H_{10}(g)$	−29.81
$CCl_4(l)$	−33.3	$n\text{-}C_5H_{12}(g)$	−35.00
$C_2H_6(g)$	−20.24	$iso\text{-}C_5H_{12}(g)$	−36.92
$C_2H_5OH(l)$	−66.35	$neo\text{-}C_5H_{12}(g)$	−39.67
$CH_3CO_2H(l)$	−116.4	$C_6H_6(l)$	+11.72
$H^+(aq)$	0	$OH^-(aq)$	−54.96
$Na^+(aq)$	−57.28	$Cl^-(aq)$	−40.02
$K^+(aq)$	−60.04	$Br^-(aq)$	−28.90
$Ag^+(aq)$	+25.31	$I^-(aq)$	−13.37
$Ca^{+2}(aq)$	−129.77	$S^{-2}(aq)$	+10.0
$Cu^{+2}(aq)$	+15.39	$CO_3^{-2}(aq)$	−161.63
$Zn^{+2}(aq)$	−36.43	$SO_4^{-2}(aq)$	−216.90
$H(g)$	+52.1	$C(g)$	+171.7
$O(g)$	+59.1	$N(g)$	+112.5

$H_2O(l)$ at 25°C is evidently one-half of −136.64, or −68.32 kcal (for 1 mole). Heats of formation of other substances are given in Table 18.1.

Addition of Enthalpy Values A table of heats of formation gives the data for calculating the enthalpy changes of many reactions. These calculations are based upon the law of conservation of energy, as expressed by Hess' law of constant heat summation (1840), which states that *the total change in reaction heat is independent of the number and kind of steps by which the reaction is carried out.* The enthalpy change for a reaction is simply the *sum of the enthalpy changes for each of the steps that may be visualized in a reaction.* So if you wish to know the enthalpy change in the reaction

$$CH_4(g) + 2O_2(g) \longrightarrow CO_2(g) + 2H_2O(l); \qquad 25°C \qquad (12)$$

you may imagine that the reaction occurred as a result of the following steps (Figure 18.4):

$$CH_4(g) \longrightarrow C(s) + 2H_2(g) \qquad (13)$$

$$C(s) + O_2(g) \longrightarrow CO_2(g) \qquad (14)$$

$$2H_2(g) + O_2(g) \longrightarrow 2H_2O(l) \qquad (15)$$

The enthalpy change in each of these reactions is known, since the first reaction is the reverse of the reaction of formation of methane from its elements,

Figure 18.4 Enthalpy changes in the reaction $CH_4(g) + 2O_2(g) \longrightarrow CO_2(g) + 2H_2O(l)$.

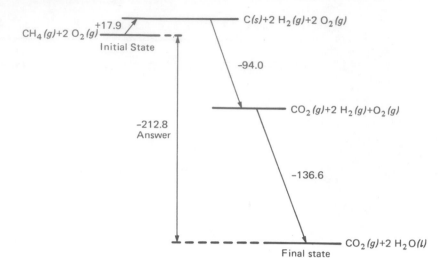

the second represents the formation of carbon dioxide, and the enthalpy change for the third is twice the enthalpy of formation for 1 mole of water, found in the table. Writing the values, you obtain

$$CH_4(g) \longrightarrow C(s) + 2H_2(g) \qquad \Delta H = -\Delta H_f^\circ = +17.9 \text{ kcal} \qquad (13)$$

$$C(s) + O_2(g) \longrightarrow CO_2(g) \qquad \Delta H = \Delta H_f^\circ = -94.0 \text{ kcal} \qquad (14)$$

$$2H_2(g) + O_2(g) \longrightarrow 2H_2O(l) \qquad \Delta H = 2\Delta H_f^\circ = -136.6 \text{ kcal} \qquad (15)$$

Addition of the equations, striking out the quantities appearing on both sides of the arrow, and adding the ΔH values, gives

$$CH_4(g) + C(s) + O_2(g) + 2H_2(g) + O_2(g) \longrightarrow$$

$$C(s) + 2H_2(g) + CO_2(g) + 2H_2O(l) \qquad (16)$$

$$CH_4(g) + 2O_2(g) \longrightarrow CO_2(g) + 2H_2O(l)$$

$$\Delta H_{298} = 17.9 - 94.0 - 136.6 = -212.8 \text{ kcal} \qquad (12)$$

Similar calculations can be made for any reaction in which the enthalpies of formation are known for the compounds appearing in the equation. The step equations to be added are so chosen that addition of them leaves only the over-all equation for the desired reaction, as illustrated above.

Enthalpy Changes in Physical Processes Such as Vaporization

The difference between the enthalpy change in the reaction forming liquid water and that forming gaseous water can be determined by measuring the enthalpy change in a nonchemical process—that of vaporization. The chemist's shorthand for this process is written

$$H_2O(l) \longrightarrow H_2O(g) \qquad (17)$$

This again gives the quantity concerned (1 mole) and specifies the physical state. The heat change in this process, at constant temperature and constant pressure, measures the change in enthalpy. At 100°C, the boiling point of water, the measured value is 9713 cal, and we write

$$H_2O(l) \longrightarrow H_2O(g) \qquad \Delta H_{373} = 9713 \text{ cal} \qquad (18)$$

Figure 18.5 Energies in the Born-Haber cycle for sodium chloride, showing the relative values of the energies.

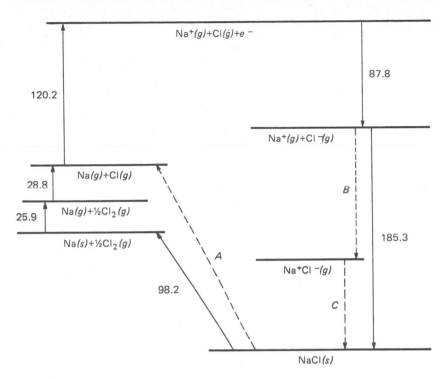

Figure 18.6 Change of heat of reaction with change in temperature.

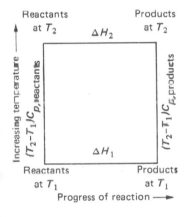

The positive sign indicates that the enthalpy of the final state, $H_{H_2O(g)}$, is larger than that of the initial state, $H_{H_2O(l)}$, and that heat is absorbed in the vaporization process. If, instead of just *one* mole, n moles are vaporized, the equation is written

$$nH_2O(l) \longrightarrow nH_2O(g) \tag{19}$$

and the enthalpy change would be n times as great as that for one mole: $\Delta H = (9713 \times n)$ cal.

If we wish to know the value of the enthalpy change at 25°C, we may calculate it from the value at 100°C by the method given later in this chapter. (See Figure 18.6.)

The heat of solution (Table 18.3) is another useful thermochemical value for a physical process.

CALCULATION OF BOND ENERGIES

Bond Energies One of the important uses of a knowledge of enthalpy changes in reactions is for calculating bond energies. By the bond energy, we mean *the average energy necessary to break all the bonds of a particular type in a mole of gaseous molecules to form gaseous atoms.* Thus the bond energy $E_{H—H}$ of the H—H bond is the enthalpy change in the reaction

$$H_2(g) \longrightarrow 2H(g) \qquad \Delta H = 103 \text{ kcal} \tag{20}$$

$$E_{H—H} = 103 \text{ kcal}$$

The bond energy for the O—H bond in water is one-half the enthalpy change for the reaction

$$H_2O(g) \longrightarrow 2H(g) + O(g) \qquad \Delta H = 221 \text{ kcal} \qquad (21)$$

$$E_{O—H} = 110.5 \text{ kcal}$$

The bond energy of the C—H bond in methane is one-fourth of the enthalpy change for

$$CH_4(g) \longrightarrow C(g) + 4H(g) \qquad \Delta H = 398 \text{ kcal} \qquad (22)$$

$$E_{C—H} = 99.5 \text{ kcal}$$

Bond energy values would be of little use if it were found that their magnitudes were different in different molecules. Fortunately, it is an experimental fact that bond energies remain nearly constant, no matter what other bonds may be present in the molecules. The C—H bond, for example, in all hydrocarbon derivatives has nearly the same value as that in methane. Thus, you can strike an average value for all the compounds that have been investigated, and rely upon the fact that this value will be a good approximation to the bond energy in any particular molecule. There are a few exceptions, especially in very simple molecules; for instance, the C=O bond energy in carbon dioxide, CO_2, is 191 kcal, but the average value for double-bonded C=O bonds in other organic molecules is 173 kcal. Table 18.2 gives average bond energies for a number of common atom combinations.

From the table of bond energies, we can estimate the values of enthalpy changes in reactions. Suppose, for example, we wish to determine the enthalpy change in the reaction for the combustion of ethanol:

$$CH_3CH_2OH(l) + 3O_2(g) \longrightarrow 2CO_2(g) + 3H_2O(l) \qquad (23)$$

Writing out the structural formula for ethanol, we see that it contains five C—H bonds, one C—C bond, one C—O bond, and one O—H bond.

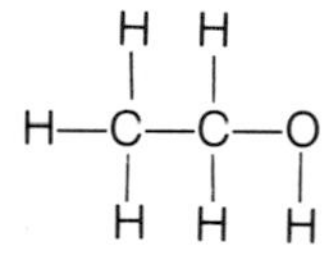

$$
\begin{array}{ccc}
 & H & H \\
 & | & | \\
H- & C- & C-O \\
 & | & | & | \\
 & H & H & H
\end{array}
$$

Table 18.2 Average Bond Energies

Bond	Energy (kcal/mole)	Bond	Energy (kcal/mole)
H—H	103	C—N	72
C—H	98	C—O	84
N—H	93	C—Cl	80
O—H	110	Cl—Cl	57
Cl—H	102	Br—Br	45
Br—H	87	I—I	35
I—H	71	C=C	145
C—C	82	C=O	175
N—N	38	C=O in CO_2	191
O—O	34	C≡C	198
O=O in O_2	117	N≡N	225

Hence the enthalpy change is calculated for the steps

$$CH_3CH_2OH(g) \longrightarrow 2C(g) + 6H(g) + O(g)$$

$$\Delta H \approx 5 \times 98 + 1 \times 82 + 1 \times 84 + 1 \times 110 \quad (24)$$

$$3O_2(g) \longrightarrow 6O(g) \qquad \Delta H \approx 3 \times 117 \quad (25)$$

$$2C(g) + 4O(g) \longrightarrow 2CO_2(g) \qquad \Delta H \approx -4 \times 191 \quad (26)$$

$$6H(g) + 3O(g) \longrightarrow 3H_2O(g) \qquad \Delta H \approx -6 \times 110 \quad (27)$$

Summing of the equations gives, for the gas phase reaction.

$$CH_3CH_2OH(g) + 3O_2(g) \longrightarrow 2CO_2(g) + 3H_2O(g) \qquad \Delta H \approx -307 \text{ kcal} \quad (28)$$

This may now be combined with the vaporization equations (29) and (30)

$$CH_3CH_2OH(g) + 3O_2(g) \longrightarrow 2CO_2(g) + 3H_2O(g) \qquad \Delta H \approx -307 \text{ kcal} \quad (28)$$

$$CH_3CH_2OH(l) \longrightarrow CH_3CH_2OH(g) \qquad \Delta H_{298} \approx 9.4 \quad (29)$$

$$3H_2O(g) \longrightarrow 3H_2O(l) \qquad \Delta H_{298} \approx -3 \times 10.5 \quad (30)$$

adding the three equations

$$CH_3CH_2OH(l) + 3O_2(g) \longrightarrow 2CO_2(g) + 3H_2O(l)$$

$$\Delta H \approx -329 \text{ kcal} \quad (23)$$

The experimental value is -326.7 kcal, for 25°C.

The Ways Molecules Hold Energy

The values for the bond energies in Table 18.2 are calculated from ΔH values converted to 0K. In view of the approximate character of the bond-energy values, and of the approximate constancy of ΔH values with changes of temperature (page 381), we may use the tabulated values for any reasonable temperature. It is of interest, however, to examine the reason for the choice of 0K for the tabulation.

The energy of an isolated molecule at ordinary temperatures may be thought of as being made up of several kinds of energy. There will be the energy of the ultimate particles (neutrons, protons, and electrons) and of the formation of the nuclei (Chapter 23) and of the ingathering of electrons around the nuclei. We may put all these forms together and call them the *atomic energies*. Since chemical reactions do not concern themselves with the breaking apart of atoms themselves (nuclear reactions do), we are not in this chapter interested in these atomic energies, though we may wish to reserve some part of the energies of the electrons in the valence shell for consideration. The atoms combine into molecules, and a second part of the energy we may call the *energy of bonding*. The bonded atoms (new molecules) will normally be in their ground states (Chapter 4), but may, on occasion, have electrons in levels of higher energy than their ground states. This energy of *electronic excitation* constitutes another kind of energy in the molecule; it is relatively unimportant in ordinary chemical reactions, and usually needs to be considered only at very high temperatures or in photochemical processes. The atoms in the molecule may vibrate with respect to each other, and the concept of *vibrational energy* is a useful one. The molecule as a whole may rotate around as many as three mutually perpendicular axes through its center of mass; the energy of this motion is a *rotational energy*. There will be an energy—*translational energy*—associated with the kinetic energy of motion of the molecule in accord with the kinetic theory of gases.

Since we deal with molar quantities, it will be convenient to speak of the energy of a mole of isolated molecules. For the energy of a mole of isolated molecules we write:

$$E_M = \sum_{}^{n} \epsilon_{A_i} + \epsilon_{\text{bonding}} + \epsilon_{\substack{\text{electronic} \\ \text{excitation}}} + \epsilon_{\text{vibration}} + \epsilon_{\text{rotation}} + \epsilon_{\text{translation}} \qquad (31)$$

Here, the symbol $\sum^{n} \epsilon_{A_i}$ represents the *sum* of the atomic energies, for all the n atoms present in the compound. The energy of formation of the mole of isolated molecules is, therefore

$$E_M - E_A = \sum_{}^{n} \epsilon_{A_i} + \epsilon_{\text{bonding}} + \epsilon_{\substack{\text{electronic} \\ \text{excitation}}} + \epsilon_{\text{vibration}} + \epsilon_{\text{rotation}}$$
$$+ \epsilon_{\text{translation}} - \sum_{}^{n} \epsilon_{A_i} \qquad (32)$$

Any energy of electronic excitation that the atoms had before they became molecules is still included in $\epsilon_{\substack{\text{electronic} \\ \text{excitation}}}$, and the energy of translation of the atoms is shared between the three molecule terms $\epsilon_{\text{vibration}}$, $\epsilon_{\text{rotation}}$, and $\epsilon_{\text{translation}}$, all of which are molecular descriptions of the motion of atoms. The energy of binding, $\epsilon_{\text{bonding}}$, represents the additional energy in the molecule resulting from distortion or overlapping of the electron clouds of the atoms when the molecule is formed. If to the term $\Delta E = (E_M - E_A)$ we add the obligatory work $P\Delta V$ representing the volume-change work in the formation of the molecules from atoms, we have

$$\Delta H = \epsilon_{\text{bonding}} + \epsilon_{\substack{\text{electronic} \\ \text{excitation}}} + \epsilon_{\text{vibration}} + \epsilon_{\text{rotation}} + \epsilon_{\text{translation}} + P\Delta V \qquad (33)$$

We are not concerned with electron excitation if the molecules remain in the ground state. The last four terms depend upon the temperature, and become effectively zero at 0K; thus, at 0K, the enthalpy change given, ΔH, closely represents $\epsilon_{\text{bonding}}$. It is for this reason that the ΔH values used in calculating bond energies are those for 0K.

$$\Delta H_{0K} = \epsilon_{\text{bonding}} \qquad (34)$$

Before leaving Equation (33), we note that if we are dealing not with isolated molecules but with matter in bulk where the molecules are close together, it may be convenient to introduce another energy term corresponding to the energy of interaction between molecules, so that the sum of the pertinent energies of a mole of molecules becomes

$$\epsilon_{\text{bonding}} + \epsilon_{\substack{\text{bonds between} \\ \text{molecules}}} + \epsilon_{\substack{\text{electronic} \\ \text{excitation}}} + \epsilon_{\text{vibration}} + \epsilon_{\text{rotation}} + \epsilon_{\text{translation}} \qquad (35)$$

To this we add the $P\Delta V$ term if we wish to consider the enthalpy. The term of largest value, by far, in this expression is the first one, $\epsilon_{\text{bonding}}$.

APPLICATIONS OF THERMO-CHEMICAL MEASUREMENTS
Lattice Energy of a Crystal

The lattice energy of a crystal is the *energy liberated when enough gaseous ions at an infinite distance from each other come together to form a mole of crystalline ionic solid.* For sodium chloride, for example, it would represent the energy change in the reaction

It is convenient to consider the reverse of this process, assumed to take place by the following steps (see Figure 18.5):

$$Na(s) \longrightarrow Na(g) \qquad \text{(energy of sublimation} \qquad +25.9 \text{ kcal)}$$

$$\tfrac{1}{2}Cl_2(g) \longrightarrow Cl(g) \qquad (\tfrac{1}{2}[\text{bond energy of } Cl_2] \qquad +28.5 \text{ kcal)}$$

$$Na(g) \longrightarrow Na^+(g) + e^- \qquad \text{(energy of ionization} \qquad +120.2 \text{ kcal)}$$

$$Cl(g) + e^- \longrightarrow Cl^-(g) \qquad \text{(energy of ionization} \qquad -87.8 \text{ kcal)}$$

$$NaCl(s) \longrightarrow Na(s) + \tfrac{1}{2}Cl_2(g) \qquad (-[\text{energy of formation}] \quad -[-98.2] \text{ kcal)}$$

The first equation represents the sublimation of a mole of sodium. The second equation represents the formation of a mole of chlorine atoms from $\tfrac{1}{2}$ mole of molecules. The third equation represents ionization of a mole of sodium atoms to form isolated sodium ions; the energy required is the ionization energy. The fourth equation represents ionization of chlorine atoms to form chloride ions by absorption of electrons; in this reaction, the energy of the final state is less than the energy of the initial state, and the energy difference between the two is negative. The last equation is the reverse of the equation for the formation of sodium chloride from its elements, the energy of which is recorded in Table 18.1. Adding the several equations, we obtain

$$NaCl(s) \longrightarrow Na^+(g) + Cl^-(g) \qquad \text{(lattice energy} \qquad +185 \text{ kcal/mole)} \qquad (37)$$

This may be compared with the experimental value, determined by subliming $NaCl(s)$ and observing the equilibrium $NaCl(g) \rightleftharpoons Na^+(g) + Cl^-(g)$, equal to 181.3 kcal/mole; the calculated value, 185 kcal/mole, is close to this.

It is instructive to show these values graphically, as in Figure 18.5. The solid arrows in the figure represent the values used in the calculation above; note that they form a closed circuit that may be considered to begin and end with $NaCl(s)$. In deference to this closed-cyclic appearance, the *procedure used for calculating the lattice energy is known as the Born-Haber cycle (1919)*. The steps involved in the experimental determination of lattice energy are represented by the dashed lines B and C, traversed in the reverse direction from the arrows. Here, the symbol $Na^+Cl^-(g)$ represents a gaseous ion pair in the 1 to 1 ratio; note that its energy level is below that of $Na(s) + \tfrac{1}{2}Cl_2(g)$, indicating that this species is stable with respect to the free elements. Note, too, by comparing the vertical height of the arrow A with B plus C, that less energy is needed to dissociate solid sodium chloride into gaseous sodium and chlorine atoms than into gaseous ions. We may thus expect that heating sodium chloride will produce atoms of the element; this is found experimentally to be the case, as evidenced by the yellow light from electronically excited sodium atoms produced when sodium salts are heated in a Bunsen burner flame.

Formation of Ions in Solution

The enthalpy change for the formation of a salt solution is a quantity that can be measured if heats of solution are known. For example, we may add

$$Na(s) + \tfrac{1}{2}Cl_2(g) \longrightarrow NaCl(s) \qquad \Delta H = -98.2 \text{ kcal} \qquad (38)$$

$$NaCl(s) \longrightarrow NaCl(aq, \text{ std. state}) \qquad \Delta H = +0.9 \text{ kcal} \qquad (39)$$

The symbols (*aq*, std. state—aqueous, standard state) in the second equation

Table 18.3 Heats of Solution

[Substance (1 mole, std. state) + water → solution (inf. dil.)]			
Substance	ΔH (kcal/mole)	Substance	ΔH (kcal/mole)
HCl	−17.96	KBr	+4.79
NaCl	+0.93	NaBr	−0.15
NH_4Cl	+3.80	H_2SO_4	−22.99
LiCl	−8.88	$CuSO_4$	−17.51
KCl	+4.12	$CaCl_2$	−19.82
$KClO_4$	+12.1	$ZnCl_2$	−17.48
KNO_3	+8.35	HI	−19.57

have been introduced as a reminder that the properties of a solution, including the enthalpy per mole of solute, depend upon the concentration of the solution (Chapter 12). The numerical value of the enthalpy change for the formation of a solution in which the solute is in its standard state can be measured by measuring the heat of solution of 1 mole of solute in such a large quantity of water that the resulting solution is "infinitely dilute." For sodium chloride, this value is +0.9 kcal, as shown (Table 18.3). Summation of the two equations gives

$$Na(s) + \tfrac{1}{2}Cl_2(g) \longrightarrow NaCl(aq, \text{ std. state}) \qquad \Delta H = -97.3 \text{ kcal} \qquad (40)$$

The sodium chloride solution is completely ionized, so that the above reaction may be considered to be the sum of two reactions:

$$Na(s) \longrightarrow Na^+(aq) + e^-(aq) \qquad (41)$$

$$\tfrac{1}{2}Cl_2(g) + e^-(aq) \longrightarrow Cl^-(aq) \qquad (42)$$

Since sodium ions and chloride ions occur also in other reactions, it would be convenient if we knew ΔH values for each of these reactions separately. The hydrated electron, however, is not a stable species; consequently, the enthalpy change of these reactions has not been measured. The advantage of having a separate value for each reaction of ion formation is so great, however, that chemists have agreed upon a convention that permits them to assign such values. The convention adopted is to assign the value zero for the enthalpy change of the reaction

$$\tfrac{1}{2}H_2(g) \longrightarrow H^+(aq) + e^-(aq) \qquad \Delta H = 0 \qquad (43)$$

On this basis, the enthalpy change of the reaction

$$\tfrac{1}{2}H_2(g) + \tfrac{1}{2}Cl_2(g) \longrightarrow H^+(aq) + Cl^-(aq, \text{ std. state}) \qquad \Delta H = -40.0 \text{ kcal} \quad (44)$$

is all to be assigned to the reaction of chloride ion formation. Subtracting Equation (43) from Equation (44) gives

$$\tfrac{1}{2}Cl_2(g) + e^-(aq) \longrightarrow Cl^-(aq, \text{ std. state}) \qquad \Delta H = -40.0 \text{ kcal} \qquad (45)$$

Subtraction of this equation from Equation (40) gives

$$Na(s) \longrightarrow Na^+(aq, \text{ std. state}) + e^-(aq) \qquad \Delta H = -57.3 \text{ kcal} \qquad (46)$$

The values of enthalpies of formation of ions listed in Table 18.1 are based on this convention.

It is sometimes important to know the value of the enthalpy of reaction at a temperature different from that at which it was measured. This can be calculated by applying the law of conservation of energy. Suppose we know the value ΔH_1 for the reaction at temperature T_1 and wish to find the value ΔH_2 at temperature T_2. We can imagine two different routes for getting from reactants at T_1 to products at T_2, as shown in Figure 18.6. We may heat the reactants from T_1 to T_2 and allow them to react at T_2, or we may allow the reaction to take place at T_1 and heat the products to T_2. The energy change in heating reactants or products from T_1 to T_2 may be calculated if we know the molar heat capacities and the number of moles of each of the substances present and, of course, the temperature change.

The value of the heat capacity of a substance will be different depending upon whether the substance is heated at constant pressure or at constant volume; in the former process, work is done against the opposing pressure as a result of the thermal expansion on heating, and more energy is required than would be needed if the volume were maintained constant. Since the enthalpy changes are determined by measuring heat changes at constant pressure, it is the molar heat capacity at constant pressure, C_p, that is to be used in the calculation of this paragraph.

Returning now to Figure 18.6, we see that the energy needed to go from reaction products at T_1 to reaction products at T_2 will be given by a sum of terms involving C_p for each reaction product times the number of moles of that product times the change in temperature $T_2 - T_1$. Specifically, for the reaction of Equation (12), the energy needed to get from reactants at T_1 to products at T_2 by the horizontal-then-vertical path is

$$\Delta H_1 + C_{p_{CO_2}}(T_2 - T_1) + 2C_{p_{H_2O(l)}}(T_2 - T_1)$$

Alternatively, we may choose the vertical-then-horizontal path, and heat the reactants to T_2 and allow them to react to form products at that temperature. The energy needed is then

$$C_{p_{CH_4}}(T_2 - T_1) + 2C_{p_{O_2}}(T_2 - T_1) + \Delta H_2$$

Since both paths start at the same place—reactants at T_1—and end at the same place—products at T_2—the energy changes in the two paths must be the same, or

$$C_{p_{CH_4}}\Delta T + 2C_{p_{O_2}}\Delta T + \Delta H_2 = \Delta H_1 + C_{p_{CO_2}}\Delta T + 2C_{p_{H_2O(l)}}\Delta T \qquad (47)$$

where the symbol ΔT represents the difference $T_2 - T_1$ between the final and initial temperatures. Rearrangement then gives

$$\Delta H_2 = \Delta H_1 + [(C_{p_{CO_2}} + 2C_{p_{H_2O(l)}}) - (C_{p_{CH_4}} + 2C_{p_{O_2}})]\Delta T \qquad (48)$$

It is important to note, in this equation, that ΔH_2 differs from ΔH_1 only by the product of a difference in heat capacities and ΔT. Since the molar heat capacities of different substances do not differ greatly from each other, and are small in any case, the term in the brackets—representing the difference between small, nearly equal quantities—cannot be large. Therefore, unless the temperature difference ΔT is several hundred degrees, or there is a marked change in the total number of moles as shown by the chemical equation, ΔH_2 and ΔH_1 will not be materially different. Thus as an approximation, we may say that the enthalpy changes in reactions are nearly the same at all temperatures. The only caution needed in making this affirmation is that the physical state of the substances must not change; evidently, a noticeable difference would be observed if, at some higher temperature, gaseous water rather than liquid water were formed in Equation (12).

Applications

1. *Calculation of the Enthalpy of Formation of Gaseous Water at 25°C from the Value for Liquid Water.* The enthalpy of formation of liquid water at 25°C is known [Equation (8)].

$$H_2(g) + \tfrac{1}{2}O_2(g) \longrightarrow H_2O(l) \qquad \Delta H^\circ_{f298} = -68.3 \text{ kcal} \qquad (49)$$

If the heat of vaporization of water at 25°C were known,

$$H_2O(l) \longrightarrow H_2O(g) \qquad \Delta H_{298} = ? \qquad (50)$$

we could obtain the desired quantity by adding the two equations. We can calculate the heat of vaporization at 25°C from the value $+9713$ cal [Equation (18)] measured at 100°C, using the procedure of Figure 18.6, and known values for the molar heat capacities of liquid water and of steam. We have

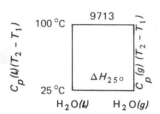

$$\Delta H_{25°C} = 9713 + C_p(l)(100 - 25) - C_p(g)(100 - 25)$$

$$= 9713 + 18 \times 75 - 7.1 \times 75$$

$$= 10.53 \text{ kcal}$$

Using the value above, the enthalpy of formation of gaseous water at 25°C becomes

$$H_2(g) + \tfrac{1}{2}O_2(g) \longrightarrow H_2O(g) \qquad \Delta H = -57.8 \text{ kcal} \qquad (51)$$

2. *Decomposition at High Temperatures.* A problem: Why are substances unstable at high temperatures? Consider, for example, the formation of quicklime by heating limestone (calcium carbonate) in a limekiln:

$$CaCO_3(s) \longrightarrow CaO(s) + CO_2(g) \qquad \Delta H^\circ_{298} = +42.7 \text{ kcal} \qquad (52)$$

According to the equation, 42.7 kcal of heat are necessary at 25°C to decompose 1 mole of solid calcium carbonate. You might guess that the reaction does not take place at this temperature because of the high energy requirement. According to the discussion above, however, the enthalpy change at the high temperature should be about the same as that at the low temperature, and calculation shows that this is indeed the case. In other words, as much energy is needed for decomposition at the high temperature as would be required at the low temperature, yet the reaction proceeds at the high temperature but not at the low. It is evident that the high energy required in the decomposition cannot be the reason—or at least not the only reason—for the stability at 25°C; if it were the reason, limestone should be equally stable at high temperatures. The answer to this problem will be discussed in the next chapter.

The Sun as the Main Source of Energy. Practically all our energy comes directly or indirectly from the sun. Directly, we receive from sunlight an average of about 1 calorie per minute per square centimeter of the earth's surface, or about 4×10^4 kcal per minute per acre. If this could be transformed into electrical or mechanical energy with an efficiency of even 10 per cent, this acre would produce power at the rate of 280 kilowatts while the sun is shining. Effective methods for carrying out this conversion efficiently and cheaply on a large scale and, particularly, methods for storing energy for use at night or during a long period of cloudy days, are still beyond our grasp, and the problem of trapping solar energy is one of mankind's most intriguing problems.

The energy that we obtain indirectly from the sun comes from natural sources of stored energy—namely, from waterfalls, wind, and fuels from recently grown plant materials (especially wood), or from fossil fuels such as coal, oil, and gas (or materials prepared from them).

Two sources of heat not dependent either directly or indirectly on sunshine are the subsurface heat of the earth and nuclear energy.

Renewable and Nonrenewable Sources of Energy. Undoubtedly the sun is fired by nuclear energy, liberated largely by the fusion of hydrogen nuclei to form helium nuclei (p. 162). It is estimated that the sun uses up about 1 per cent of its hydrogen in a billion years. And the expectation is that it will be giving out heat radiation of essentially the present intensity for several billion years to come.

Sunshine, or solar energy, is therefore an inexhaustible source. The sources of energy derived from sunshine, such as waterfalls, wind, and plant materials, and more especially wood and peat recently derived by photosynthesis, are renewable. But the fossil fuels—coal, gas, and oil, also containing stored solar energy—are not renewable; nor are uranium and thorium, from which fissionable materials may be prepared.

Let us now consider, in order, the nature of these several sources of energy and power.

Water Power, Wind, and Tide. The sun's heat causes evaporation of enormous quantities of water every day from the surface of the sea and other bodies of water, and from the leaves of plants. Water vapor, rising in currents of warm air into the atmosphere, is condensed to rain. And rain water, from mountains and hills, gathers in streams and rushes down to the sea. The kinetic energy of flowing water can be transformed into electrical energy by water wheels or turbines coupled to generators. Many industries that require a cheap and readily available supply of energy tend to concentrate in regions near large waterfalls, as at Niagara Falls. Less than 5% of the power consumed annually in the U.S. is generated by water power, but about half of our definitely realizable water power remains undeveloped.

Although it seems likely that wind power could supply as much as 15% of the world's energy requirements, the windmills in the U.S. actually generate each year less than a third as much energy as our electric fans use in creating little winds of their own.

Tide power could, in principle, satisfy half the world's needs, but it cannot presently compete economically with other energy sources.

The future possibilities of making better use of the energy of the sun and waters, winds, and tides are being investigated. Indications are that procedures can be developed for supplying limited amounts of energy locally or regionally from these sources, but that they cannot be expected to make a major timely contribution to our total energy needs.

It has been estimated, for example, that the potential natural geothermal energy available from existing or likely-to-be-developed procedures in the

U.S. is 30,000 megawatts—an amount less than 1/20th our anticipated requirement for 1980. One project now being studied in the Northwest is directed toward locating sources of thermal rock at depths below 3000 feet and to use an atomic device to create deep underground cavities. Water could then be piped into the cavities and returned in a closed-cycle system to generate steam and electric power.

Developments in the capture of solar energy hold promise for generating electric power. The low intensity of solar energy means it must be collected from a huge surface area at a place where there are not too many cloudy days. If collected in space by way of enormous fields of orbiting solar collectors, its capture in large enough amounts may become feasible. Such huge satellites, properly spaced in synchronous orbits, could collect solar energy day and night and without cloud interference. Once collected, this energy could be converted into electrical energy and transmitted to earth by laser or microwaves.

While this concept is far from reality at present, it illustrates the opportunities for applying imagination and ingenuity in attacking important global problems.

SUMMARY

In this chapter, we have considered the energy associated with changes in matter. Energy released or absorbed in such processes is viewed as a difference between the energy contained in the products and in the reactants. This is expressed in terms of changes in internal energy,

$$\Delta E = (E_{\text{products}} - E_{\text{reactants}})$$

or as changes in enthalpy,

$$\Delta H = (H_{\text{products}} - H_{\text{reactants}})$$

Positive values of ΔE or of ΔH for a process mean that the products need more energy than the reactants supply; negative values of these quantities mean reactants supply more energy than products need. For chemical reactions, a major portion of ΔH or ΔE often is associated with bond-making and bond-breaking. Hence, bond energies are considered in some detail.

It is shown that ΔH and ΔE for any process may be regarded as a sum of the ΔH or ΔE values for a series of changes which, added together, give the over-all process. Thermochemical cycles set up this way are useful in explaining why certain processes are energetically more or less feasible than other processes. These cycles also form the basis for estimating the ΔH or ΔE values for processes that are difficult to measure directly.

IMPORTANT TERMS

Molecular energy
energy of electronic
  excitation
energy of rotation
energy of vibration
energy of translation

Calorimeter

Born-Haber cycle

Initial state
final state
standard state

Heat capacity
molar heat capacity

Heat
heat of reaction
heat of vaporization

Total energy
atomic energy
chemical energy
bond energy

lattice energy
internal energy

Enthalpy
enthalpy of formation

Work

1. If we can identify the various energies within a molecule, why do we use the value of the *difference* in energy of the initial and final states rather than the *total* energy of the system?

2. What is the factor that must be added to the value of the internal energy, measured at constant volume, to give the value of ΔH? Show how the mathematical relationship between ΔE and ΔH can be derived.

3. One mole of methane is burned

$$CH_4(g) + 2O_2(g) \longrightarrow CO_2(g) + 2H_2O(l)$$

in a bomb calorimeter (constant volume) at 25°C. The heat of combustion for 0.25 mole of methane measured under these conditions is 52.9 kcal. Assuming that all the gases are ideal, calculate: (a) The heat evolved in the calorimeter. (b) The heat that would have been evolved had the same quantity of methane been burned at a constant pressure of 1 atm. (c) ΔE for the reaction as written above. (d) ΔH for the reaction. (e) In the calorimeter experiment, was work done on the system or by it?

4. Calculate the work done against the atmosphere ($P\Delta V$) in calories when 1 g of zinc dissolves at 25°C in HCl to give H_2.

5. If the reaction

$$2H_2(g) + O_2(g) \longrightarrow 2H_2O(l)$$

goes to completion at room temperature, 25°C, and under 1 atm of pressure, how much work is done on the system?

6. In the following reactions, predict how the heat effect at constant pressure compares with the heat effect at constant volume.
   (a) $\frac{1}{2}H_2(g) + \frac{1}{2}Cl_2(g) \longrightarrow HCl(g)$
   (b) $\frac{1}{2}N_2(g) + \frac{3}{2}H_2(g) \longrightarrow NH_3(g)$
   (c) $H(g) + Cl(g) \longrightarrow HCl(g)$

7. When 16 g of hydrogen gas are burned to water vapor at constant pressure, 462.32×10^3 cal are evolved. What is the molar heat of formation for $H_2O(g)$?

8. Calculate the heat of formation of C_2H_2 gas from the following data:

$$C_2H_2(g) + 2\frac{1}{2}O_2(g) \longrightarrow 2CO_2(g) + H_2O(l) \qquad \Delta H = -311 \text{ kcal}$$

$$C + O_2 \longrightarrow CO_2(g) \qquad \Delta H = -94.1 \text{ kcal/mole}$$

$$H_2 + \frac{1}{2}O_2 \longrightarrow H_2O(l) \qquad \Delta H = 68.1 \text{ kcal/mole}$$

9. Find $\Delta H°$ at 25°C and 1 atm for the following reactions:

   (a) $N_2(g) + 3H_2(g) \longrightarrow 2NH_3(g)$
   (b) $C_2H_4(g) + H_2O(g) \longrightarrow C_2H_5OH(l)$

10. Using Table 18.1, calculate the enthalpy of combustion of propane, C_3H_8, to carbon dioxide and liquid water at 298K and 760 torr.

11. For butane, $C_4H_{10}(g)$, the standard enthalpy of combustion at 25°C is -687.9 kcal/mole. What is $\Delta E°$ for this reaction at 25°C?

12. The enthalpies of formation for gaseous ethane, C_2H_6, ethene, C_2H_4, and ethyne, C_2H_2, are given in Table 18.1. (a) Calculate the *molar* enthalpy of combustion for each. (b) Calculate the heat evolved per

kilogram of each compound on combustion. (c) Which compound would make the most efficient fuel in terms of cal/g?

13. At its normal boiling point, 46.3°C, carbon disulfide vaporizes with the utilization of 84.1 cal/g. Calculate ΔH and ΔE for the vaporization of one mole of carbon disulfide at 46.3°C.

14. The hydrogenation of unsaturated hydrocarbons, such as ethene (C_2H_4) and propene (C_3H_6), results in the formation of the analogous saturated hydrocarbons, ethane (C_2H_2) and propane (C_3H_8). The reactions are:

$$C_2H_4(g) + H_2(g) = C_2H_6(g)$$

$$C_3H_6(g) + H_2(g) = C_3H_8(g)$$

Given the data of Table 18.1 and the value of the enthalpy of formation of propene, 4.88 kcal/mole, calculate ΔH_{298}° for the two reactions given.

15. The internal energy, ΔE, includes what various energies of a molecule?

16. Show the reasoning that leads to the conclusion that ΔH° at 0K is a measure of the energy of bonding.

17. Explain why we can assume that the bond energy for the O—H bond is one-half the value of ΔH for the reaction $H_2O(g) \longrightarrow 2H(g) + O(g)$.

18. From the bond energy values given on page 000, calculate ΔH° for the reaction

$$CH_4 + 2O_2 \longrightarrow CO_2 + 2H_2O$$

19. Using Hess' law, calculate the average Xe—F bond energy from the following data:

$$XeF_4 \longrightarrow Xe^+ + F^- + F + F_2 \qquad \Delta H = 286 \text{ kcal/mole}$$

$$Xe \longrightarrow Xe^+ + e^- \qquad \Delta H = 280 \text{ kcal/mole}$$

$$F + e^- \longrightarrow F^- \qquad \Delta H = -83.5 \text{ kcal/mole}$$

$$F_2 \longrightarrow 2F \qquad \Delta H = 38 \text{ kcal/mole}$$

Assume all species to be in the gaseous state and that all data were obtained at the same temperature.

20. Calculate the NO bond energy from the enthalpies of formation given in Table 18.1.

21. Use the result of problem 20 and data from Table 18.2 to calculate the enthalpy of the reaction

$$4NH_3 + 5O_2 \longrightarrow 4NO + 6H_2O$$

and compare the result with that calculated from the data of Table 18.1.

22. The lattice energy of NaCl, 181 kcal, is a very high value of energy holding the Na^+ and Cl^- ions in the crystal lattice. Nevertheless, NaCl is readily soluble in water, a process in which the Na^+ and Cl^- ions of the crystal are pulled away from each other into solution. What new bonds must be formed in this process and what must be the order of magnitude of these bonds if the enthalpy of solution of NaCl(s) is only 0.9 kcal?

23. Trace the kind of energy and the effect of that kind of energy on a molecule of water as you raise the temperature from solid H_2O at 0K

to a very high temperature such as 1,000,000°C. What may happen at 50,000,000°C?

24. On the basis of what you now know about the different forms of energy that make up the internal energy of a molecule, devise an explanation of why the specific heat of sodium at 25°C is 0.01, but the specific heat of sugar at 25°C is much higher (0.3).

25. (a) Why are the growing seasons usually longer in land areas surrounding a large body of water, such as Lake Erie, than in land 100 miles from the lake?

   (b) Why is the temperature usually lower on a hot day near or in a forest or an area of vegatation than in a city of paved streets?

SPECIAL PROBLEMS

1. Five (5.00) g of carbon were burned to CO_2 in a calorimeter made of copper (sp. heat 0.093) whose mass was 2000 g and which contained 2500 g of water. The initial temperature of the water was 20.0°C and the final temperature 34.5°C. Calculate the heat of combustion of carbon in calories per gram.

2. One mole of an ideal gas undergoes an isothermal expansion from 5 l to 15 l against a pressure of 1 atm. Calculate: (a) The work done in the transformation. (b) The heat absorbed.
   (Hint: $\Delta E = 0$ in an isothermal transformation of an ideal gas.)

3. An ideal gas undergoes a compression under a constant pressure of 2.58 atm from 30.0 l to 7.9 l and evolves 70 calories in so doing. Calculate: (a) The work done. (b) The change in internal energy. Explain the answer to (b).

4. An expanding gas confined in a cylinder forces a piston of radius 5.00 cm out 10.0 cm against an opposing pressure of 720 torr. In so doing, the gas absorbs 80 cal. Calculate: (a) The work done (in calories). (b) The change in internal energy, ΔE. Explain the answer to (b).

5. From the enthalpy of combustion for benzene, $C_6H_6(l)$, −781.0 kcal/mole at 25°C, calculate its enthalpy of formation. Compare your answer with the value given in Table 18.1.

6. For methyl alcohol, CH_3OH, at 25°C, the heat of vaporization is 8950 cal/mole. The enthalpy of formation of liquid methyl alcohol is given in Table 18.1. Considering these two data, what is the enthalpy of formation of gaseous methyl alcohol?

7. Calculate ΔH for the reaction

$$N(g) + 3H(g) \longrightarrow NH_3(g)$$

for the temperature and pressure conditions at which the following thermochemical equations apply:

$$N_2(g) + 3H_2(g) \longrightarrow 2NH_3(g) \qquad \Delta H = -22.0 \text{ kcal}$$

$$H_2(g) \longrightarrow 2H(g) \qquad \Delta H = +104.2 \text{ kcal}$$

$$N_2(g) \longrightarrow 2N(g) \qquad \Delta H = +225.0 \text{ kcal}$$

Explain the difference between the formation of ammonia gas from the atomic forms of hydrogen and nitrogen as opposed to the molecular forms.

8. Find the enthalpy of reaction for the decomposition of glucose according to the following reaction: $C_6H_{12}O_6(s) \longrightarrow 6C$ (graphite) $+ 6H_2O(l)$ from the date given.

$$H_2(g) + \tfrac{1}{2}O_2(g) \longrightarrow H_2O(l) \qquad\qquad \Delta H = -68.3 \text{ kcal}$$

$$C \text{ (graphite)} + O_2(g) \longrightarrow CO_2(g) \qquad\qquad \Delta H = -94.1 \text{ kcal}$$

$$C_6H_{12}O_6(s) + 6O_2(g) \longrightarrow 6CO_2(g) + 6H_2O(l) \qquad \Delta H = -673 \text{ kcal}$$

9. From the data in Tables 8.2, 8.4, and 18.1, the value of ΔH_{f298} for potassium bromide, -93.73 kcal/mole, and the enthalpy change in the reaction

$$Br(g) + e^- \longrightarrow Br^-(g) \qquad \Delta H = -78 \text{ kcal}$$

calculate the lattice energy of KBr by means of the Born-Haber cycle.

10. Use diagrams similar to Figure 18.5, drawn to scale, to discuss the stabilities of the hypothetical compound $CaCl(s)$ and the known compound $CaCl_2(s)$. The necessary data for calcium and chlorine are given in Tables 8.2 and 8.4 and Figure 18.5. Assume the lattice energy for CaCl to be the same as that of NaCl, and use the value 525 kcal/mole as the lattice energy of $CaCl_2$.
    (a) Offer an explanation for the higher lattice energy of $CaCl_2$ as compared to NaCl.
    (b) Will CaCl be a stable compound with respect to the reaction

    $$CaCl(s) \longrightarrow Ca(s) + \tfrac{1}{2}Cl_2(g)$$

    (c) With respect to the reaction

    $$2CaCl(s) \longrightarrow Ca(s) + CaCl_2(s)$$

    (d) Draw a qualitative sketch of Figure 18.5 as it would look for the formation of a possible compound, $CaCl_3(s)$, using approximately the same scale as for CaCl and $CaCl_2$.
    (e) Discuss the relationship between the energy considerations of this problem and the assignment of oxidation state II for element 20 in all compounds on the basis of its electron configuration.

11. Show what data are needed to calculate the value of ΔH at a higher temperature if the value of ΔH is known at a lower temperature. Use as an example the reaction $4NH_3 + 5O_2 \longrightarrow 4NO + 6H_2O$.

REFERENCES   CAMPBELL, J. A., *Why Do Chemical Reactions Occur?* Prentice-Hall, Englewood Cliffs, N.J., 1965.

MAHAN, B., *Elementary Chemical Thermodynamics*, Benjamin, Menlo Park, Calif., 1963.

NASH, L., *Elements of Thermodynamics*, Addison-Wesley, Reading, Mass., 1962.

PIMENTEL, G. C., and SPRATLEY, R. D., *Understanding Chemical Thermodynamics*, Holden-Day, San Francisco, 1969.

CHEMICAL EQUILIBRIA—MEASUREMENT OF EXTENT OF REACTIONS

Soon after the early chemists adopted a set of atomic weights and learned to use them in chemical calculations involving amounts of products formed from a given amount of reactant, they discovered that some reactions do not proceed to completion. Further investigation showed that many reactions are reversible. This led to the conclusion that reactions are incomplete not because reaction stops, but because an equilibrium is established in which the rate of the forward reaction is equal to the rate of the reverse reaction (Chapter 7). Research on mathematical treatment of equilibria has been done in a number of ways on data from a wide variety of sources. This research has been based on:

1. Empirical treatment of measurements of the concentration of reactants and products at equilibrium (described in this chapter).
2. Data on reaction rates (see Chapter 22).
3. Data on the energy changes during reaction (see Chapters 18 and 20).
4. Data on the voltage of cells for oxidation-reduction reactions in solution (see Chapter 21).

THE EQUILIBRIUM CONSTANT
Relations Between Concentrations of Reactants and Products at Equilibrium

One of the most useful methods in the study of science is to search for empirical mathematical relationships that may appear among the data. This kind of approach has been illustrated in developing the gas laws in Chapter 10. Another example of such a search makes use of the data in Table 19.1. The table records the equilibrium concentrations, in moles per liter, of hydrogen, iodine, and hydrogen iodide in the reaction

$$H_2(g) + I_2(g) \rightleftharpoons 2HI(g) \tag{1}$$

---

*"Every system in chemical equilibrium undergoes, as a result of a change in one of the factors of the equilibrium, a transformation in such a direction that, if that transformation alone occurred, it would lead to a change of opposite sign in the factor considered." (The quotation is from *Annales de Mines* [8], vol 13 (1888) p. 157.)

for eight different experiments at the same temperature, starting with different amounts of reactants and products and allowing the reactions to proceed to equilibrium. In the first five experiments, various concentrations of hydrogen and iodine were mixed and allowed to come to equilibrium; in the remaining three experiments, various amounts of hydrogen iodide were allowed to react and come to equilibrium with the hydrogen and iodine formed. The concentration of each component (H_2, I_2, and HI) was determined at equilibrium. Let us examine the data in an effort to find a useful empirical relationship between the molar concentrations of reactants and products at equilibrium.

Two types of mathematical relationships often are sought in the treatment of data: a functional relationship among the observed quantities that remains constant under changing conditions, and a linear-graphic relationship over a wide range of experiments. In this example, we shall search for a treatment of the data that will give a constant value.

Discovering the Equilibrium Constant

The procedure that might be used in trying to discover the form of the functional relation may be summarized as follows.

One convenient way to compare two quantities (equilibrium concentrations of products and reactants, in this case) is to set up a ratio such as:

$$\frac{[\text{Products}]_{eq}}{[\text{Reactants}]_{eq}}$$

Since there are two reactants, hydrogen and iodine, the denominator might be written as a sum, a difference, a product, or a quotient of the equilibrium concentrations of these. Let us try the product, $[H_2]_{eq} \times [I_2]_{eq}$, since this is proportional to the number of collisions in unit time between the two kinds of molecules (Chapter 22).

Let us use the ratio

$$\frac{[\text{HI}]_{eq}}{[H_2]_{eq} \times [I_2]_{eq}}$$

and see if the data of Table 19.1, when substituted into this equation, reveal a simple relationship between the equilibrium concentrations of products and

Table 19.1 Molar Concentrations of H_2 and I_2, and HI at Equilibrium and Ratios of Equilibrium Concentrations at 425.5°C

Trial	$[H_2]_{eq}$	$[I_2]_{eq}$	$[HI]_{eq}$	$\dfrac{[\text{HI}]_{eq}}{[H_2]_{eq}[I_2]_{eq}}$	$\dfrac{[\text{HI}]_{eq}^2}{[H_2]_{eq}[I_2]_{eq}}$
Starting with H_2 and I_2					
1.	4.5647×10^{-3}	0.7378×10^{-3}	13.544×10^{-3}	4.022×10^{3}	54.47
2.	3.5600×10^{-3}	1.2500×10^{-3}	15.588×10^{-3}	3.531×10^{3}	55.04
3.	2.9070×10^{-3}	1.7069×10^{-3}	16.482×10^{-3}	3.322×10^{3}	54.75
4.	2.2423×10^{-3}	2.3360×10^{-3}	16.850×10^{-3}	3.217×10^{3}	54.20
5.	1.8313×10^{-3}	3.1292×10^{-3}	17.671×10^{-3}	3.084×10^{3}	54.49
Starting with HI					
6.	1.1409×10^{-3}	1.1409×10^{-3}	8.410×10^{-3}	6.46×10^{3}	54.34
7.	0.4953×10^{-3}	0.4953×10^{-3}	3.655×10^{-3}	14.90×10^{3}	54.45
8.	0.4789×10^{-3}	0.4789×10^{-3}	3.531×10^{-3}	15.40×10^{3}	54.36

reactants. Column 5 of Table 19.1 shows the results of this calculation; this quotient is *not* constant but varies from 3×10^3 to 15×10^3. Perhaps a different arrangement of terms will prove more successful. Let us try another type of treatment.

If the product $[H_2]_{eq} \times [I_2]_{eq}$ is a measure of the rate of molecular collisions between H_2 and I_2 in the forward reaction, then the product $[HI]_{eq} \times [HI]_{eq}$ should be a measure of the rate of collision between hydrogen iodide molecules in the reverse reaction. Let us try the ratio

$$\frac{[HI]_{eq}^2}{[H_2]_{eq} \times [I_2]_{eq}}$$

Column 6 of Table 19.1 shows that the quotient is a constant (within experimental error) for all eight experiments given in the table. Note that the same constant is obtained in approaching the equilibrium from either side—that is, the ratio is the same whether you start with hydrogen and iodine or with hydrogen iodide. This constant may be called an *equilibrium constant*.

It is evident, then, that a relationship exists between the equilibrium concentrations of products and reactants. The form of this relationship for the reaction

$$H_2 + I_2 \rightleftharpoons 2HI \tag{1}$$

is

$$\frac{[HI]_{eq}^2}{[H_2]_{eq} \times [I_2]_{eq}} = \text{constant} = K \text{ (the equilibrium constant)} \tag{2}$$

It now remains to be seen if such a relation exists for other equilibria.

All Equilibria Can Be Described with an Equilibrium Constant

Extensive experiments with this and many other chemical equilibria reveal that a constant relation between equilibrium concentrations of products and reactants exists for each equilibrium at constant temperature. To obtain this constant, the equilibrium concentrations of products and reactants must be substituted into a ratio-type equation of the proper form. The form needed to give a constant is related to the balanced chemical equation and is illustrated below for a generalized equation.

Consider the chemical equation

$$aA + bB \rightleftharpoons cC + dD \tag{3}$$

The form of the equation shows that the reaction occurs in the ratio *a* moles of A to *b* moles of B to give *c* moles of C and *d* moles of D.

The mathematical form needed to obtain a constant relation between equilibrium concentrations of products and reactants is

$$K = \frac{[C]_{eq}^c \times [D]_{eq}^d}{[A]_{eq}^a \times [B]_{eq}^b} \tag{4}$$

Note that in this form, the equilibrium concentrations are raised to the power of the coefficient of the component in the balanced chemical equation. The constant K is known as the equilibrium constant. It is constant for any reaction at a given temperature. The form of the equilibrium-constant equation for several reactions is given in Table 19.2.

We shall find that the equilibrium constant is a very useful tool in the mathematical treatment of all types of equilibria.

Table 19.2 Form of Equilibrium-Constant Equations for Various Chemical Equilibria

Chemical Equation	Form of Equilibrium-Constant Equation
$2H_2 + O_2 \rightleftharpoons 2H_2O$	$K = \dfrac{[H_2O]_{eq}^2}{[H_2]_{eq}^2 \times [O_2]_{eq}}$
$N_2O_4 \rightleftharpoons 2NO_2$	$K = \dfrac{[NO_2]_{eq}^2}{[N_2O_4]_{eq}}$
$N_2 + 3H_2 \rightleftharpoons 2NH_3$	$K = \dfrac{[NH_3]_{eq}^2}{[N_2]_{eq} \times [H_2]_{eq}^3}$
$4HCl + O_2 \rightleftharpoons 2H_2O + 2Cl_2$	$K = \dfrac{[H_2O]_{eq}^2 \times [Cl_2]_{eq}^2}{[HCl]_{eq}^4 \times [O_2]_{eq}}$

Calculation of Values of K_{eq}

The simplest calculation of an equilibrium constant can be made if all the molar concentrations of reactants and products are known, as they are in Table 19.1. It is also possible to obtain equilibrium concentrations of all components if the *initial* concentration of the starting materials and the *equilibrium* concentration of one component are known; this is based on the fact that the chemical equation relates the amounts of reactants consumed to the amounts of products formed.

This is illustrated by the following problem.

PROBLEM 1 Calculate the value of the equilibrium constant, K_{eq}, for the reaction

$$2NO + O_2 \rightleftharpoons 2NO_2 \tag{5}$$

if, after mixing 1 mole of NO and 1 mole of O_2 in a 2-liter flask at 1 atm and 10°C, it is found that the equilibrium concentration of NO is 0.3 mole/liter.

Solution The formula for the equilibrium constant is

$$K_{eq} = \frac{[NO_2]_{eq}^2}{[NO]_{eq}^2[O_2]_{eq}} \tag{6}$$

(a) *Calculation of equilibrium concentration of NO, O_2, and NO_2.* Values for each of these concentrations are obtained from the above data and from information in the balanced equation, as follows:

$$2NO + O_2 \rightleftharpoons 2NO_2 \tag{5}$$

Concentrations of	NO	O_2	NO_2
At start of experiment	$\dfrac{1 \text{ mole}}{2 \text{ liters}} = 0.5 \dfrac{\text{mole}}{\text{liter}}$	$\dfrac{1 \text{ mole}}{2 \text{ liters}} = 0.5 \dfrac{\text{mole}}{\text{liter}}$	0 moles
At equilibrium	$0.3 \dfrac{\text{mole}}{\text{liter}}$ (given in problem)	$0.4 \dfrac{\text{mole}}{\text{liter}}$ (calculated)	$0.2 \dfrac{\text{mole}}{\text{liter}}$ (calculated)

The calculated values for $[O_2]$ and $[NO_2]$ are obtained by using information given by the chemical equation. We started with 0.5 mole/liter of NO and ended with 0.3 mole/liter; therefore, 0.2 mole/liter reacted. According to the chemical equation, 2 moles of NO will use 1 mole of O_2 in reacting; therefore, 0.2 mole/liter of NO will use 0.1 mole/liter of O_2, leaving $(0.5 - 0.1) = 0.4$ mole/liter. Similarly, the chemical equation tells us that 2 moles of NO will form 2 moles of NO_2; therefore, 0.2 mole/liter of NO will form 0.2 mole/liter of NO_2.

(b) *Calculation of the value of K_{eq}.* Substituting these values of the equilibrium concentrations of NO, O_2, and NO_2 into the formula for the equilibrium constant,

$$K_{eq} = \frac{[NO_2]_{eq}^2}{[NO]_{eq}^2 [O_2]_{eq}} = \frac{(0.2)^2}{(0.3)^2(0.4)} = 1.1 \text{ at } 10°C \tag{7}$$

Variety of Equilibrium Processes   Of the great variety of physical and chemical equilibria, those in which we shall be primarily interested in this chapter are:

1. Equilibria in gaseous systems.
   Example: $H_2(g) + I_2(g) \rightleftharpoons 2HI(g)$
2. Equilibria in ionization of water.
   Example: $H_2O + H_2O \rightleftharpoons H_3O^+ + OH^-$
3. Equilibria in ionization of weak electrolytes.
   Example: $CH_3CO_2H + H_2O \rightleftharpoons H_3O^+ + CH_3CO_2^-$
4. Equilibria in slightly soluble solids.
   Example: $AgCl(s) \rightleftharpoons Ag^+(aq) + Cl^-(aq)$
5. Equilibria involving hydrolysis.
   Example: $H_2O + CH_3CO_2^- \rightleftharpoons CH_3CO_2H + OH^-$

Uses of K_{eq}   The equilibrium constant is an exact measure of the extent of an over-all reaction. The voltage of cells and the standard free-energy changes of reactions are also exact measures of the extent of reaction; they will be discussed in Chapters 20 and 21. *A low value of the equilibrium constant indicates that the over-all reaction proceeds only slightly to the right as the equation is written; a large value of K_{eq} indicates that the over-all reaction proceeds far to the right.*

An example of a system in which the position of equilibrium is far to the right is

$$H_2(g) + Cl_2(g) \rightleftharpoons 2HCl(g) \tag{8}$$

$$K_{eq} = \frac{[HCl]_{eq}^2}{[H_2]_{eq}[Cl_2]_{eq}} = 2.18 \times 10^{33} \text{ at } 25°C \tag{9}$$

An example of a system in which the position of equilibrium is far to the left is

$$N_2(g) + O_2(g) \rightleftharpoons 2NO(g) \tag{10}$$

$$K_{eq} = \frac{[NO]_{eq}^2}{[N_2]_{eq}[O_2]_{eq}} = 4.26 \times 10^{-31} \text{ at } 25°C \tag{11}$$

EXAMPLES OF SEVERAL EQUILIBRIUM PROCESSES

Melting $\rightleftharpoons$ freezing

Dissolving $\rightleftharpoons$ crystallizing

$N_2 + 3H_2 \rightleftharpoons 2NH_3$

$AgCl(s) \rightleftharpoons Ag^+(aq) + Cl^-(aq)$

$H_2O + H_2O \rightleftharpoons H_3O^+ + OH^-$

Vaporizing $\rightleftharpoons$ condensing

A second use of the equilibrium constant is that it makes possible the calculating of the equilibrium concentration of reactants and products at a given temperature.

PROBLEM 2 The value of K_{eq} for the reaction $N_2 + 3H_2 \rightleftharpoons 2NH_3$ is 9.5×10^{-4} at 500°C. Calculate the equilibrium concentrations of each of the components if the starting amounts are 1.000 mole of nitrogen and 1.000 mole of hydrogen in a 2.000-liter flask at 500°C.

Solution   (a) *Calculation of concentrations if x moles of NH_3 are formed per liter.*

$$N_2 + 3H_2 \rightleftharpoons 2NH_3 \tag{12}$$

Concentrations of	N_2	H_2	NH_3
At start of experiment	$\dfrac{1.000 \text{ mole}}{2.000 \text{ liters}} =$ 0.500 mole/liter	$\dfrac{1.000 \text{ mole}}{2.000 \text{ liters}} =$ 0.500 mole/liter	0 moles
At equilibrium	$\left(0.500 - \dfrac{x}{2}\right) \dfrac{\text{mole}}{\text{liter}}$	$\left(0.500 - \dfrac{3}{2}x\right) \dfrac{\text{mole}}{\text{liter}}$	$x\dfrac{\text{mole}}{\text{liter}}$
	since by the chemical equation $\frac{1}{2}$ mole of N_2 is used to form 1 mole of NH_3	since by the chemical equation $\frac{3}{2}$ mole of H_2 is used to form 1 mole of NH_3	

(b) *Calculation of equilibrium concentrations of H_2, N_2, and NH_3.* Substituting the values from (a) in the formula for the equilibrium constant gives

$$K_{eq} = \frac{[NH_3]_{eq}{}^2}{[N_2]_{eq}[H_2]_{eq}{}^3} = \frac{x^2}{\left(0.500 - \dfrac{x}{2}\right)\left(0.500 - \dfrac{3}{2}x\right)^3} = 9.5 \times 10^{-4} \tag{13}$$

Algebraic solution for x gives

$$x = 7.4 \times 10^{-3} \text{ mole/liter} = NH_3 \text{ concentration}$$

Therefore,

$$[N_2] = 0.500 - \frac{x}{2} = 0.500 - 0.004 = 0.496 \text{ mole/liter}$$

$$[H_2] = 0.500 - \frac{3}{2}x = 0.500 - 0.011 = 0.489 \text{ mole/liter}$$

EXACT VALUES OF EQUILIBRIUM CONSTANTS

While we have said that the equilibrium constant is, for constant temperature, a *constant*, it must be recognized that the quotient involving concentrations of reactants and products is not exactly constant but shows a dependence upon the total amount of dissolved substance or, for gases, upon the total pressure. True constancy would be obtained only if the gaseous reactants and products formed ideal mixtures or if the solutions similarly were ideal, showing negligible attractive forces between solutes and between solutes and solvent.

For simplicity, we shall discuss the principles of chemical equilibria in terms of concentrations, and assume for discussion purposes that we are dealing with ideal gas mixtures and ideal solutions. Experimental results indicate that using actual concentrations and pressures of real gases gives values that are good enough approximations for many purposes. But this use will be less reliable at high pressures and for concentrated solutions, where deviations from ideality are greatest. For these systems especially, but for precise values in any case, a corrected concentration called the *activity* must be used (Chapter 20).

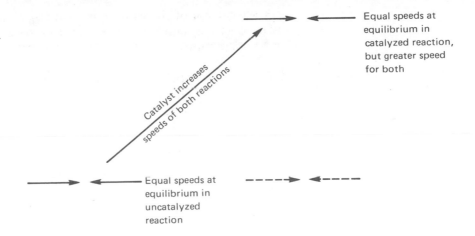

A third use of K_{eq} is to calculate the effect of a change in the concentration of one of the reactants on the position of equilibrium. This effect is called the mass-action effect and will be described later in the chapter.

**SHIFTING THE POSITION
OF EQUILIBRIUM**

We have indicated that the condition for equilibrium *is: speed of forward reaction equals speed of reverse reaction.* We now ask what effect such variables as concentration, catalysts, temperature, or pressure will have on the rates of these reactions and, as a result, on the position of equilibrium. Will any of these change the value of K_{eq} or shift the position of equilibrium? Experimentally, we shall find the following:

1. *A catalyst* increases the speeds of both the forward and reverse reactions, making it possible for the reaction to come to equilibrium more quickly, but does not change K_{eq} or the position of equilibrium (Figure 19.1).

2. *A change of pressure* will affect the position of equilibrium (but not the value of K_{eq}) only for those reactions in which the volume of the products is not the same as the volume of the reactants. For example, the position of equilibrium and the value of K_{eq} for

$$H_2 + I_2 \rightleftharpoons 2HI \qquad (1)$$

are unaffected by pressure changes. However, the position of equilibrium in the reaction

$$\underset{\substack{\text{Total volume of} \\ \text{reactants} = 4}}{N_2 + 3H_2} \rightleftharpoons \underset{\substack{\text{Total volume of} \\ \text{products} = 2}}{2NH_3} \qquad (12)$$

will be shifted by a change in pressure. An increase in pressure will shift the position of equilibrium in the direction of the smaller volume, in this case towards the formation of more ammonia. This is in accord with a sweeping generalization about equilibria first stated by Henri-Louis Le Châtelier in 1887 and quoted at the beginning of this chapter and the bottom of page 396.

For example, for every four moles of reactants that disappear (3 moles of H_2 and 1 mole of N_2), two moles of product (2 moles of NH_3) appear. In a given vessel, two moles will exert less pressure than four moles. If, therefore, the pressure on this equilibrium mixture is increased, at constant temperature, the system can attempt to relieve this stress by shifting the position of equilibrium to the smaller number of moles—in this case, toward the right.

Table 19.3 Percentage of Ammonia Present at Equilibrium in an Original Mixture of 1 Mole of Nitrogen and 3 Moles of Hydrogen at Various Pressures and Temperatures

Temperature (°C)	1 atm	50 atm	100 atm	200 atm	600 atm	1,000 atm
200	15.3	74.4	81.5	85.8	95.4	98.3
400	0.48	15.3	25.1	36.3	65.2	79.8
500	0.13	5.6	10.6	17.6	42.2	57.5
600	0.05	2.25	4.5	8.2	3.1	31.4
800	0.022	0.57	1.19	2.2	—	—
1000	0.004	0.21	0.45	0.9	—	—

If the pressure is decreased, the position of equilibrium will be shifted toward the larger number of moles—in this case, toward the left. These predictions are verified experimentally as shown in each horizontal row of Table 19.3. Each row lists the per cent of ammonia present in equilibrium mixtures at various pressures at a fixed temperature. In each case, the initial molar ratio of nitrogen to hydrogen is 1 to 3.

In general, the position of equilibrium is shifted toward smaller volumes at higher pressures and toward larger volumes at lower pressures. If the number of moles on each side of the equation for a reaction is the same, as in $H_2 + I_2 \rightleftharpoons 2HI$, a change in pressure at constant temperature will not bring about a change in the position of equilibrium, since the system is unable to change in a way that would relieve the stress.

3. *A change in concentration* will affect the position of equilibrium but not the value of K_{eq}. The effect of a change in concentration on an equilibrium can also be used to illustrate the application of Le Châtelier's principle. Let us consider as an example what would happen to the equilibrium systems of Table 7.2,

$$H_2 + I_2 \rightleftharpoons 2HI$$

if, with $[H_2]$ and $[I_2]$ each equal to 0.00224 and $[HI]$ equal to 0.01552, we were to add additional I_2 molecules in an amount equivalent to increasing the concentration of iodine by 0.01000 mole/liter. In terms of Le Châtelier's principle, this addition of iodine would constitute a *stress*. The new equilibrium to be established must relieve the stress; that is, it must decrease the concentration of iodine. The only way it can do this is for iodine to react with hydrogen to form more hydrogen iodide. Thus, in the reestablished equilibrium, there will be more hydrogen iodide and less hydrogen than in the original equilibrium.

Since the quotient

$$\frac{[HI]_{eq}^2}{[H_2]_{eq}[I_2]_{eq}}$$

must remain constant for this reaction at this temperature, we may use the

THE PRINCIPLE OF LE CHATELIER

The effect of forces that tend to disturb a system in equilibrium was studied by Le Châtelier. He made the observation that "if a stress is applied to a system in equilibrium, a new equilibrium will be established in which the position of equilibrium has been shifted in such a direction as to relieve the applied stress."

equilibrium constant to calculate the concentrations of hydrogen and hydrogen iodide in the new equilibrium. In doing this, we find:

$$[H_2]_{eq} = 6.75 \times 10^{-4} \ M \ \text{(down from } 2.24 \times 10^{-3} \ M)$$

and

$$[HI]_{eq} = 1.865 \times 10^{-2} \ M \ \text{(up from } 1.552 \times 10^{-2} \ M)$$

Succinctly stated, the application of stress to the equilibrium, in the form of added iodine, has caused the position of equilibrium to shift toward the right to relieve the applied stress (Figure 19.2).

Conversely, if the concentration of any component is decreased, the position of equilibrium will be shifted so that more of that component will be produced. If, for example, some hydrogen iodide were removed from the equilibrium mixture, the new equilibrium would contain less hydrogen and iodine as a result of a shift of the equilibrium position toward hydrogen iodide. This way, the imposed stress—the removal of some or one of the components—is partially relieved.

4. *A change in temperature* will change the value of the equilibrium constant, and thus the position of equilibrium, in a manner that can be predicted by application of Le Châtelier's principle. Therefore, the temperature must be specified for each equilibrium. One needs only to think of an increase in temperature as resulting from the addition of heat, and of a decrease in temperature as the removal of heat. If heat is absorbed in a chemical reaction, we may consider heat to be a reactant. The attempt to raise the temperature is an attempt to add heat—the *stress*—and the position of equilibrium shifts in a way to relieve the stress—namely, by using heat (the reactant) to produce more of the reaction product. If heat is evolved in a reaction, the attempt to relieve the stress of applied heat (temperature rise) causes the position of equilibrium to shift toward the reactants. The formation of ammonia is an example of the second case:

$$N_2(g) + 3H_2(g) \longrightarrow 2NH_3(g) \qquad \Delta H^\circ_{f298} = -22.08 \ \text{kcal} \qquad (14)$$

Increase in temperature shifts the position of equilibrium to the left, as shown in the figures in any of the vertical columns in Table 21.3.

The application of Le Châtelier's principle may perhaps be more readily obvious if we include the heat of reaction in the chemical equation

$$N_2 + 3H_2 \rightleftharpoons 2NH_3 + \text{heat} \qquad (15)$$

Addition of heat now pushes the position of equilibrium to the left, since displacement in that direction tends to use up heat in the system's attempt to relieve the applied stress. The formation of nitric oxide, however, is favored by a temperature increase, since ΔH° is positive,

$$N_2(g) + O_2(g) \longrightarrow 2NO(g) \qquad \Delta H^\circ = 43.2 \ \text{kcal} \qquad (16)$$

$$\text{Heat} + N_2 + O_2 \rightleftharpoons 2NO \qquad (17)$$

and addition of heat causes the position of equilibrium to shift to the right.

Le Châtelier's principle may be used to predict whether the equilibrium constant for a given reaction increases or decreases when the temperature changes. In endothermic reactions, the forward reaction is favored by an increase in temperature; the reverse reaction is favored by a temperature

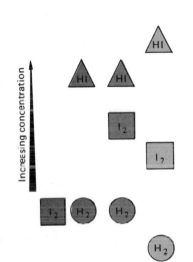

■ Initial equilibrium 1
■ Momentary intermediate state
□ Final equilibrium 2

Figure 19.2 Change of position of equilibrium in $H_2 + I_2 \rightleftharpoons 2HI$ by adding additional iodine. Equilibrium is originally established at the concentration levels indicated at the left (equilibrium 1). Addition of 0.1 mole/liter of iodine momentarily raises the concentration to $[I_2]'$, disturbing the equilibrium. To relieve the added stress, I_2 and H_2 react, lowering their concentrations to the values at the right (equilibrium 2) and forming HI, which raises its concentration to the new position of equilibrium.

decrease. Therefore, the change in K_{eq} for these reactions parallels the temperature change, increasing when temperature is increased and decreasing when the temperature is decreased. In exothermic reactions, the forward reaction is favored by a temperature decrease and the reverse reaction is favored by a temperature increase. So in exothermic reactions, K_{eq} changes in the sense opposite to the change in temperature.

EQUILIBRIA IN IONIC REACTIONS IN AQUEOUS SOLUTION

Reactions involving ionic reactants often proceed at rates too fast to measure by any ordinary method. Nevertheless, such reactions are frequently reversible, and reach the equilibrium state almost instantaneously because of their high rates. The positions of these equilibria can be measured, and the concentrations existing at equilibrium can be fitted into appropriate expressions for the equilibrium constants. In this section, we consider some of the common ionic equilibria.

Ionization of Weak Acids

We have already seen (Chapter 7) that acids react with water to form hydronium ions and the anions of the acid. If the acid is a strong acid, such as hydrochloric acid, this reaction goes so far toward completion that, in the usual dilute solutions of the laboratory, no measurable concentration of hydrogen chloride molecules remains in the solution when equilibrium is established:

$$HCl + H_2O \longrightarrow H_3O^+ + Cl^- \tag{18}$$

With weak acids, however, an equilibrium is established in which undissociated acid molecules remain. For the reaction of acetic acid, for example,

$$CH_3CO_2H + H_2O \rightleftharpoons H_3O^+ + CH_3CO_2^- \tag{19}$$

we write the equilibrium constant

$$K_{eq} = \frac{[H_3O^+]_{eq} \times [CH_3CO_2^-]_{eq}}{[H_2O]_{eq} \times [CH_3CO_2H]_{eq}} \tag{20}$$

However, dilute solutions are virtually all water—the number of moles of solute is negligible when compared with the number of moles of water. Under these conditions, the equilibrium concentration of water is both large and constant while the concentrations of the other components in the equilibrium may vary over a relatively wide range. Because it is both inconvenient and unnecessary to include this large constant value of $[H_2O]_{eq}$ in calculations involving K_{eq}, a modified equilibrium-constant expression, noted here as the ionization constant, K_{ion}, is defined. In K_{ion}, the equilibrium concentration of water is excluded from the right-hand side of the equation as in

$$K_{ion} = \frac{[H_3O^+]_{eq}[CH_3CO_2^-]_{eq}}{[CH_3CO_2H]_{eq}} \tag{21}$$

Such modified equilibrium-constant expressions, from which the concentration of solvent water has been excluded, are used for all ionic equilibria in dilute aqueous solutions.*

The value of the ionization constant may be determined by measuring the concentration of ions in the solution of the weak acid.

---

*The practice of excluding the water concentration from the ionization-constant expression also covers up our ignorance as to the exact formula for the hydronium ion (Chapter 7). If this were $H_9O_4^+$, the water concentration might have to appear in the denominator as $[H_2O]^4$. Actu-

Table 19.4 Ionization Constants for Acids at Room Temperature

Acid	K_{ion}	Acid	K_{ion}
$HClO_4$	$\sim 10^{+9}$	CH_3CO_2H	1.82×10^{-5}
HI	$\sim 10^{+7}$	$CH_3CH_2CO_2H$	1.34×10^{-5}
HBr	$\sim 10^{+6}$	$CH_3CH_2CH_2CO_2H$	1.38×10^{-5}
HCl	$\sim 10^{+5}$	$Al(H_2O)_6^{+3}$	1.1×10^{-5}
H_2SO_4	$\sim 10^{+3}$	$C_5H_5NH^+$	6.0×10^{-6}
HNO_3	$\sim 10^{+2}$	HSO_3^-	5.0×10^{-6}
CCl_3CO_2H	2.2×10^{-1}	H_2CO_3	4×10^{-7}
$CHCl_2CO_2H$	5.7×10^{-2}	H_2S	1×10^{-7}
HO_2CCO_2H	2.1×10^{-2}	$H_3NCH_2CH_2NH_3^{+2}$	1×10^{-7}
H_2SO_3	1.3×10^{-2}	$H_2PO_4^-$	7×10^{-8}
HSO_4^-	1.0×10^{-2}	$HOCl$	1.1×10^{-8}
H_3PO_4	8×10^{-3}	$o\text{-}ClC_6H_4OH$	3.3×10^{-9}
$o\text{-}ClC_6H_4NH_3^+$	2.3×10^{-3}	HCN	2.0×10^{-9}
CH_2ClCO_2H	1.4×10^{-3}	$m\text{-}ClC_6H_4OH$	9.6×10^{-10}
$o\text{-}ClC_6H_4CO_2H$	1.1×10^{-3}	NH_4^+	5.7×10^{-10}
HF	6.7×10^{-4}	$p\text{-}ClC_6H_4OH$	4.2×10^{-10}
$m\text{-}ClC_6H_4NH_3^+$	5.8×10^{-4}	$(CH_3)_3NH^+$	1.6×10^{-10}
HCO_2H	1.77×10^{-4}	$H_2NCH_2CH_2NH_3^+$	8.2×10^{-11}
$m\text{-}ClC_6H_4CO_2H$	1.5×10^{-4}	HCO_3^-	4×10^{-11}
$p\text{-}ClC_6H_4NH_3^+$	1.1×10^{-4}	$CH_3NH_3^+$	2.4×10^{-11}
$p\text{-}ClC_6H_4CO_2H$	1.1×10^{-4}	$CH_3CH_2NH_3^+$	2.4×10^{-11}
$C_6H_5CO_2H$	6.68×10^{-5}	$(CH_3)_2NH_2^+$	1.7×10^{-11}
$HO_2CCO_2^-$	5.42×10^{-5}	HPO_4^-	1.3×10^{-12}
$C_6H_5NH_3^+$	2.6×10^{-5}	HS^-	1×10^{-14}

PROBLEM 3   In a solution made by dissolving 0.10000 mole of acetic acid in enough water to make one liter of solution, the hydronium ion concentration was found to be 0.00134 molar. Calculate the ionization constant of acetic acid.

Solution   By Equation (19), one acetate ion forms for each hydronium ion when acetic acid ionizes. Consequently, neglecting the very small amount of hydronium ion produced by the ionization of water [Equation (27)], the acetate ion concentration is equal to the hydronium ion concentration:

$$[H_3O^+] = [CH_3CO_2^-] = 0.00134 \text{ molar}$$

Similarly, one acetic acid molecule disappears for each hydronium (or acetate) ion formed, and the concentration of acetic acid is equal to the difference between the amount originally present and the amount that ionizes:

$$[CH_3CO_2H] = 0.10000 - 0.00134 = 0.09866 \text{ molar}$$

These values are substituted into the expression for the ionization constant [Equation (21)]:

$$K_{ion} = \frac{(0.00134)(0.00134)}{0.09866} = 1.82 \times 10^{-5} \tag{22}$$

Ionization constants for a number of acids are given in Table 19.4.

---

ally, the symbol $[H_3O^+]$ in the numerator is an approximation for the activity of hydrogen ion (p. 394) and stands for the concentration of hydrogen ion in all its forms, whether H^+, H_3O^+, $H_9O_4^+$, or whatever. Similarly, the term for water in the denominator should represent its activity; since the solutions are dilute, the activity of water is nearly that of the standard state (pure water), namely, 1 (Chapter 20), and the term for water is equal to 1 and disappears.

Common-Ion Effect Since the ionization of a weak acid involves an equilibrium, any change in the concentration of one of the substances present must change the position of equilibrium in the direction predicted by Le Châtelier's principle. If, for example, we add acetate ions to a solution of acetic acid, the position of the equilibrium in reaction (19) must shift toward the left, to relieve the stress of the increased acetate ion concentration. A shift to the left can occur only by the reaction of acetate ions with hydronium ions already present in the solution; this reaction reduces the concentration of hydronium ions. The observed result is thus a decrease in the acidity of the solution because of the shift in the position of equilibrium when acetate ions are added to an acetic acid solution. (Of course, we cannot add a solution containing acetate ions alone; some positive ions must also be present. Sodium acetate solution, for example, will contain sodium ions and acetate ions, since sodium acetate, being a salt and a strong electrolyte, is completely ionized.) Such a displacement of the position of an ionic equilibrium because of the addition of an ion involved in the equilibrium is called the *mass-action effect* or *common-ion effect*.

PROBLEM 4 In Problem 3, the value of the ionization constant for acetic acid was found to be 1.82×10^{-5}. Calculate the $[H_3O^+]$ in a 0.10 molar solution of acetic acid in which the $[CH_3CO_2^-]$ has been increased by adding enough sodium acetate to make the sodium concentration 0.20 M.

Solution Let

$$[H_3O^+] = x$$

further,

$$[CH_3CO_2^-] = 0.20 + x$$

(Acetate ion comes from two sources: the added sodium acetate and the ionized acetic acid.)

and

$$[CH_3CO_2H] = 0.10 - x$$

In setting up these concentrations, we have assumed that all the hydronium ion present has been formed by ionization of acetic acid—that is, that the amount of hydronium ion produced by the reaction of Equation (27) is so much smaller than that produced according to Equation (19) that $[H_3O^+]_{\text{from }H_2O}$ can be neglected. Substituting these values in the formula for the ionization constant for acetic acid, we have

$$K_{\text{ion}} = \frac{(x) \times (0.20 + x)}{0.10 - x} = 1.82 \times 10^{-5} \tag{23}$$

The solution to this problem requires that a quadratic equation be solved. However, an approximate solution may be obtained by observing that even in Problem 3 the hydronium ion concentration was only 0.00134 molar, and that, by the common-ion effect, it will be even less in the present case. Consequently, since x is small, $0.1 - x$ is nearly equal to 0.1 and we may substitute 0.1 for $0.1 - x$; by similar reasoning, we may substitute 0.2 for $0.2 + x$:

$$K_{\text{ion}} = \frac{(x) \times (0.20 + x)}{0.10 - x} \approx \frac{(x) \times 0.20}{0.10} = 1.82 \times 10^{-5} \tag{24}$$

whence

$$x = 0.91 \times 10^{-5} = 0.0000091 \text{ molar}$$

If this value of the hydronium ion concentration is compared with its value in the absence of the added acetate ion (Problem 3), it can be seen that, in this case, the common-ion effect has resulted in decreasing the hydronium ion concentration more than a hundredfold.

Ionization of Weak Bases

What has been said above about the ionization of weak acids can be applied directly to the ionization of weak bases. Here, water acts as the acid in the acid-base reaction (Chapter 7):

$$H_2O + NH_3 \rightleftharpoons NH_4^+ + OH^- \tag{25}$$

The ionization constant for ammonia in dilute water solution is written

$$K_{ion} = \frac{[NH_4^+]_{eq}[OH^-]_{eq}}{[NH_3]_{eq}} \tag{26}$$

The addition of ammonium chloride to the solution will displace the position of equilibrium to the left, since the ammonium ion is common to both ammonium chloride and ammonia solutions.

Ion Equilibrium for Water

Although for many purposes water may be considered an un-ionized compound, it is both an acid [Equation (25)] and a base [Equations (18) and (19)] and undergoes a self-ionization:

$$H_2O + H_2O \rightleftharpoons H_3O^+ + OH^- \tag{27}$$

The equilibrium constant appropriate to this reaction in dilute solution is known as the *ion-product constant*:

$$K_w = [H_3O^+]_{eq}[OH^-]_{eq} \tag{28}$$

Exact measurements have shown that in pure water at 25°C, the numerical value of $[H_3O^+]_{eq}$ is 1×10^{-7} molar; according to the mole-for-mole requirement of the chemical equation, this must also be the concentration of $[OH^-]_{eq}$. Hence, $K_w = 1 \times 10^{-14}$ at 25°C. This value must hold not only for pure water, but for any water solution. If an acid is added to neutral water, $[H_3O^+]$ increases and $[OH^-]$ decreases; if a base is added, $[OH^-]$ increases and $[H_3O^+]$ decreases, but in each case the product, K_w, is 1×10^{-14}. For example, in a 0.01 M hydrochloric acid solution $[H_3O^+] = 1 \times 10^{-2}$, so $[OH^-]$ must be 1×10^{-12}.

pH Value

Very slight changes in acidity are frequently of great significance, and the values of small concentrations of hydronium ions have to be written often. S. P. L. Sørensen, in 1909, suggested that a more convenient measure of acidity would be obtained by adopting the convention of using the negative logarithm of the hydronium ion concentration, for which he used the symbol pH.

$$pH = -\log [H_3O^+] \tag{29}$$

According to this definition, if in a solution $[H_3O^+] = 1 \times 10^{-5}$ molar, the pH of the solution is 5.0, since the logarithm of 1×10^{-5} is -5.00 and the pH is the negative of this. If in a solution $[H_3O^+] = 2 \times 10^{-5}$, the pH is 4.7, since the logarithm of 2×10^{-5} is $\overline{5}.30 = -4.7$. The pH of pure water is 7.0; solutions of pH less than this are acid; those of pH greater than 7.0 are basic.

We now know that Sørensen's definition is not completely satisfactory, especially since modern measurements of the acidity of solutions are commonly made using electrochemical cells (Chapter 21). However, the pH thus measured is approximately equal to $-\log [H_3O^+]$ and, for many purposes, the older definition is still useful.

Buffer Solutions

Many chemical reactions must be carried out at constant and controlled hydronium-ion concentrations. A solution of pH 5.0 can be prepared by making the solution 1×10^{-5} molar in hydrochloric acid, but it cannot readily be maintained at that value, the accidental entrance of the smallest amount of base will neutralize some of the acid and lower the hydronium ion concentration, while addition of a small amount of acid will raise it. This difficulty is avoided by using a buffer solution, *containing relatively high concentrations of a weak acid and a salt of that acid*—for example, acetic acid and sodium acetate. This solution contains in it large reservoirs of acetic acid molecules and of acetate ions from the sodium acetate. When a small amount of base is added to the solution, it is true that some of the hydronium ions are removed to form water. But more of the acetic acid molecules then dissociate to reestablish an equilibrium concentration of hydronium ions only negligibly different from the original concentration. The position of equilibrium in the reaction

$$CH_3CO_2H + H_2O \rightleftharpoons H_3O^+ + CH_3CO_2^- \qquad (19)$$

shifts slightly to the right to compensate for the hydronium ions removed by the base. The addition of a small amount of acid simply causes the hydronium ions of the added acid to react with the acetate ions from the sodium acetate, so that again the $[H_3O^+]$ does not change. The solution is thus protected, or *buffered,* against a change in $[H_3O^+]$ on the addition of either acid or base, and maintains a constant pH regardless of the addition of these substances. Buffer solutions containing weak bases and their salts can also be used.

Human blood is buffered at a pH of 7.2; that is, the hydronium ion concentration is between 10^{-7} and 10^{-8} molar. It is maintained at this value by buffer mixtures, such as the weak acid, carbonic acid, and its salt, sodium bicarbonate ($NaHCO_3$). If, for any reason, the pH of the blood varies from 7.2 by a significant amount, serious consequences, even death, may follow.

Hydrolysis

It is observed that solutions of many salts have a pH different from 7.0. For example, solutions of sodium acetate, sodium cyanide, or sodium sulfide are basic, while solutions of ammonium chloride, aluminum sulfate, and copper sulfate are acidic.

All of these observations can be explained by *assuming that either the cation or the anion (or both) of the salt reacts with water.* Such reactions are known as hydrolysis reactions If the anion reacts with water, as in

$$H_2O + CH_3CO_2^- \rightleftharpoons CH_3CO_2H + OH^- \qquad (30)$$

or

$$H_2O + CN^- \rightleftharpoons HCN + OH^- \qquad (31)$$

the solution will be basic. Anions that react with water this way are the anions of weak acids. Therefore, they are relatively strong Brønsted bases. If the cation reacts with water, as in

$$NH_4^+ + H_2O \rightleftharpoons H_3O^+ + NH_3 \qquad (32)$$

or

$$Al(H_2O)_6^{+3} + H_2O \rightleftharpoons H_3O^+ + Al(H_2O)_5OH^{+2} \qquad (33)$$

the solution will be acidic. Cations that react with water this way are the cations of weak bases. Therefore, they are relatively strong Brønsted acids.

If both the cation and anion react with water, as in ammonium acetate,

$$NH_4^+ + H_2O \rightleftharpoons NH_3 + H_3O^+ \atop H_2O + CH_3CO_2^- \rightleftharpoons CH_3CO_2H + OH^-\Bigg\} \longrightarrow 2H_2O \qquad {(32) \atop (30)}$$

the hydronium and hydroxide ions formed in the separate steps react with one another, causing each of the separate equilibria to be displaced toward the products.

Indicators   Indicators are weak acids or weak bases in which the acid form has a color characteristically different from the color of the basic form:

$$\underset{\text{Color } A}{HInd} + H_2O \rightleftharpoons H_3O^+ + \underset{\text{Color } B}{Ind^-} \qquad (34)$$

It is evident from the equation that if $[H_3O^+]$ is small, the position of equilibrium will be far on the right, so that the solution containing the indicator will show color B.

Increase in $[H_3O^+]$ will shift the position of equilibrium to the left, so that color B will fade and color A will appear.

EXACT SOLUTIONS   In each of the solutions of Problems 3 and 4, we have neglected any possible
FOR IONIZATION   hydronium ion formed by ionization of water, thus ignoring the second term
PROCESSES   on the right in the following equation for the total hydronium ion concentration:

$$[H_3O^+]_{total} = [H_3O^+]_{from\ CH_3CO_2H} + [H_3O^+]_{from\ H_2O}$$

This is almost always a satisfactory approximation in acid solutions, since the ionization of water is repressed by the hydronium ion from the acid. But to prepare for solving more complex ionization problems, it may be useful to examine Problems 3 and 4 when this approximation is not made.

PROBLEM 5   To calculate $[H_3O^+]_{eq}$ in an acetic acid solution when K_{ion} is known.

Solution   In a solution of acetic acid, the following substances are present in addition to water: acetic acid and its ionization products, acetate and hydronium ions; and the additional ionization product of water, hydroxide ion. The four concentrations $[CH_3CO_2H]$, $[CH_3CO_2^-]$, $[H_3O^+]$, and $[OH^-]$ are thus unknown quantities required in the solution to the problem. From algebra, we know that a problem with four unknowns requires four equations connecting these unknowns, which, when solved simultaneously, will evaluate the four unknowns. What are these four equations?

Two of them are the expressions for the ionization constant of acetic acid and the ion-product constant of water:

$$K_{ion} = \frac{[H_3O^+]_{eq}[CH_3CO_2^-]_{eq}}{[CH_3CO_2H]_{eq}} \qquad \text{(a)}$$

$$K_w = [H_3O^+]_{eq}[OH^-]_{eq} \qquad \text{(b)}$$

The other two equations are obtained from the facts that the law of conservation of mass must be obeyed and that the solution as a whole must be electrically neutral, with no excess of positive or negative charge. The law of conservation of mass tells us that all of the acetic acid substance initially added to the solution must still be present after equilibrium has been established. Thus, even though the acetic acid ionizes, the sum of the equilibrium concentrations of acetic acid and acetate ion must equal the initial concentration before reaction. If this initial concentration is denoted by a, then this conservation or *material-balance* relation is

$$a = [CH_3CO_2H]_{eq} + [CH_3CO_2^-]_{eq} \qquad \text{(c)}$$

and this is a third equation for our solution.

The fourth equation, the electroneutrality or *charge balance* relation, simply says that the total of all positive charges must equal the total of all negative charges, or

$$[H_3O^+]_{eq} = [CH_3CO_2^-]_{eq} + [OH^-]_{eq} \qquad \text{(d)}$$

There is only one kind of positive ion present, H_3O^+, but two kinds of negative ion, $CH_3CO_2^-$ and OH^-.

The problem now is to solve these four equations simultaneously for $[H_3O^+]$, knowing K_{ion}, K_w, and a. In what follows, the subscripts, $_{eq}$, will be omitted from the symbols for the convenience of the typesetter; but it must be remembered that all concentrations are equilibrium concentrations.

From (d) $\qquad [CH_3CO_2^-] = [H_3O^+] - [OH^-] \qquad$ (e)

From (c) $\qquad [CH_3CO_2H] = a - [CH_3CO_2^-] = a - [H_3O^+] + [OH^-] \qquad$ (f)

From (b) $\qquad [OH^-] = \dfrac{K_w}{[H_3O^+]} \qquad$ (g)

Substituting (g) into (f) and into (e) gives

$$[CH_3CO_2H] = a - [H_3O^+] + \frac{K_w}{[H_3O^+]} = a - \left([H_3O^+] - \frac{K_w}{[H_3O^+]}\right) \qquad \text{(h)}$$

$$[CH_3CO_2^-] = [H_3O^+] - \frac{K_w}{[H_3O^+]} \qquad \text{(j)}$$

and substituting (h) and (j) into (a) gives

$$K_{ion} = \frac{[H_3O^+]\left([H_3O^+] - \dfrac{K_w}{[H_3O^+]}\right)}{a - \left([H_3O^+] - \dfrac{K_w}{[H_3O^+]}\right)} \qquad \text{(k)}$$

Equation (k) is a single equation in one unknown, $[H_3O^+]$, and can be solved for the hydronium ion concentration, since K_{ion}, K_w, and a are known. The solution would have to be carried out by trial and error, since it is a cubic equation involving $[H_3O^+]^3$. Making use of our knowledge of chemistry and the values of the constants, however, we see that several satisfactory approximations can be made:

In acid solutions $[OH^-] \ll [H_3O^+]$ and, thus,

$$\frac{K_w}{[H_3O^+]} \ll [H_3O^+]$$

Thus $\frac{K_w}{[H_3O^+]}$ in the parentheses in both the numerator and denominator can be ignored, since, in view of the inequality given

$$[H_3O^+] - \frac{K_w}{[H_3O^+]} \simeq [H_3O^+] \qquad \text{(l)}$$

Using this approximation in equation (k)

$$K_{ion} = \frac{[H_3O^+][H_3O^+]}{a - [H_3O^+]} \qquad \text{(m)}$$

which corresponds exactly to Equation (22), and is readily solved as a quadratic equation in $[H_3O^+]$.

Experience shows that if K_{ion}/a is of the order of 10^{-4} or less, a further approximation is valid, since under those conditions

$$[H_3O^+] \ll a$$

and

$$a - [H_3O^+] \simeq a \qquad \text{(n)}$$

Substitution of this approximation in (m) gives

$$K_{ion} = \frac{[H_3O^+]^2}{a} \quad \text{or}$$

$$[H_3O^+] = \sqrt{aK_{ion}}$$

The validity of the approximations used in (l) and (n) can now be tested by substituting the value of $[H_3O^+]$ obtained into these two equations to see if the approximations are satisfactory.

PROBLEM 6   The exact solution to a problem such as Problem 4, to calculate $[H_3O^+]$ in a solution containing both an acid and a salt of the acid.

Solution   Let a be the initial concentration of acetic acid, and s the initial concentration of sodium acetate. The solution now contains, in addition to water: acetic acid, acetate ion, sodium ion, hydronium ion, and hydroxide ion. We appear then to have five unknowns, requiring five equations for the solution. Four of these equations correspond to the four used in Problem 5: the ionization-constant expression (a); the ion-product expression (b); a material-balance expression similar to (c); and a charge-balance equation similar to (d); the fifth equation makes use of the chemical information that sodium acetate is a salt made up of sodium ions and acetate ions, and the formula CH_3CO_2Na shows that one mole of sodium acetate contains one mole of sodium ions. Thus, the simple relation

$$[Na^+]_{eq} = s \qquad \text{(o)}$$

is the fifth equation needed.

Although Equations (a) and (b) are used directly as in Problem 5, the material-balance and charge-balance Equations (c) and (d) become

$$a + s = [CH_3CO_2H]_{eq} + [CH_3CO_2^-]_{eq} \tag{p}$$

$$[Na^+]_{eq} + [H_3O^+]_{eq} = [CH_3CO_2^-]_{eq} + [OH^-]_{eq} \tag{q}$$

since there are now two sources of acetate species, acetic acid and sodium acetate, and two sources of positive charge, sodium ions and hydronium ions. Substituting (o) into (q) and solving for $[CH_3CO_2^-]_{eq}$ gives

$$[CH_3CO_2^-] = s + [H_3O^+] - [OH^-] = s + [H_3O^+] - \frac{K_w}{[H_3O^+]} \tag{r}$$

Solving (p) for $[CH_3CO_2H]_{eq}$ and using (r) gives

$$[CH_3CO_2H] = a + s - [CH_3CO_2^-] = a + s - s - [H_3O^+] + \frac{K_w}{[H_3O^+]} \tag{s}$$

Now (r) and (s) are substituted into (a):

$$K_{ion} = \frac{[H_3O^+]\left(s + [H_3O^+] - \dfrac{K_w}{[H_3O^+]}\right)}{a - \left([H_3O^+] - \dfrac{K_w}{[H_3O^+]}\right)} \tag{t}$$

Equation (t) is again an equation in $[H_3O^+]^3$ and can be solved exactly for the one unknown $[H_3O^+]$.

Again, approximations are in order. Using (l) in the numerator and denominator

$$K_{ion} = \frac{[H_3O^+]\,(s + [H_3O^+])}{a - [H_3O^+]} \tag{u}$$

Equation (u) corresponds to Equation (23). Equation (24) is obtained if a and s are of the same order of magnitude and if the approximation (n) holds; then, also

$$s + [H_3O^+] \simeq s \tag{v}$$

and

$$K_{ion} = \frac{[H_3O^+]\,(s)}{(a)} \tag{w}$$

Again, the solution thus obtained should be tested on Equations (l), (m), and (n) to assure that the approximations are valid.

IONIZATION OF DIPROTIC ACIDS

Procedures similar to those discussed in Problems 5 and 6 can be used to solve exactly problems that involve ionization of acids containing two (or more) ionizable protons. Consider the following problem.

PROBLEM 7

Calculate the hydronium ion concentration in a solution of carbonic acid, H_2CO_3, of concentration a.

Solution

The chemical equations are

$$H_2CO_3 + H_2O \rightleftharpoons H_3O^+ + HCO_3^- \tag{A}$$

$$HCO_3^- + H_2O \rightleftharpoons H_3O^+ + CO_3^{-2} \tag{B}$$

$$H_2O + H_2O \rightleftharpoons H_3O^+ + OH^- \tag{C}$$

When equilibrium is established, there are five unknown concentrations:

$$[H_2CO_3]_{eq}, \; [HCO_3^-]_{eq}, \; [CO_3^{-2}]_{eq}, \; [H_3O^+]_{eq}, \; [OH^-]_{eq}$$

We therefore need five mathematical equations for simultaneous solution. These include the ionization-constant expressions

$$K_I = \frac{[H_3O^+]_{eq}\,[HCO_3^-]_{eq}}{[H_2CO_3]_{eq}} \tag{D}$$

$$K_{II} = \frac{[H_3O^+]_{eq}\,[CO_3^{-2}]_{eq}}{[HCO_3^-]_{eq}} \tag{E}$$

$$K_w = [H_3O^+]_{eq}\,[OH^-]_{eq} \tag{F}$$

and the equations for material balance and charge balance.
Material balance on carbonate species:

$$a = [H_2CO_3]_{eq} + [HCO_3^-]_{eq} + [CO_3^{-2}]_{eq} \tag{G}$$

Charge balance:

$$[H_3O^+]_{eq} = [HCO_3^-]_{eq} + 2[CO_3^{-2}]_{eq} + [OH^-]_{eq} \tag{H}$$

Note that the carbonate ion contributes two units of charge for each ion; in the charge-balance equation, its concentration must be multiplied by 2.

The solution to the problem is now the algebraic one of solving a system of five equations in five unknowns. The solution is left to the student. It will be found convenient to start with Equation (H) and express the quantities on the right-hand side in terms of $[H_3O^+]_{eq}$ and the substance present in largest quantity in the final solution, $[H_2CO_3]_{eq}$, using the ionization equations. The material-balance Equation (G) is then used to eliminate $[H_2CO_3]_{eq}$ in terms of $[H_3O^+]_{eq}$ and a, to give a result that may be cast in the form

$$[H_3O^+]^2 = \frac{aK_I\left(1 + \dfrac{2K_{II}}{[H_3O^+]}\right)}{1 + \dfrac{K_I}{[H_3O^+]} + \dfrac{K_I K_{II}}{[H_3O^+]^2}} + K_w \tag{J}$$

This contains only the single unknown $[H_3O^+]$ and can be solved if a, K_I, K_{II}, and K_w are known.

For carbonic acid, and often in the case of other polyprotic acids, consideration of the values of K_I and K_{II} and a will show that the following are good approximations.

$$1 + \frac{2K_{II}}{[H_3O^+]} \simeq 1 \tag{K}$$

$$1 + \frac{K_I}{[H_3O^+]} + \frac{K_I K_{II}}{[H_3O^+]^2} \simeq 1 \tag{L}$$

$$aK_I + K_w \simeq aK_I \tag{M}$$

Hence

$$[H_3O^+]^2 = aK_I$$

$$[H_3O^+] = \sqrt{aK_I} \tag{N}$$

Figure 19.3 In the saturated solutions silver ions, Ag^+ (small circles), and chloride ions, Cl^- (large circles), are constantly going into the solution and returning to the crystal of AgCl.

This is the result that would have been obtained if the only chemical reaction were (A), ignoring (B) and (C), because if only (A) were involved,

$$[H_3O^+]_{eq} = [HCO_3^-]_{eq} \tag{O}$$

$$[H_2CO_3]_{eq} = a - [H_3O^+] \simeq a \tag{P}$$

$$K_I = \frac{[H_3O^+][H_3O^+]}{a} \tag{Q}$$

Such neglect of Equations (B) and (C) will usually be found to be justified if K_I/a is 10^{-4} or less, K_I/K_{II} is 10^4 or greater, and K_I/K_w is 10^3 or greater.

It is of interest to note that solution of Equations (E) and (D) for the carbonate ion concentration in terms of $[H_3O^+]_{eq}$ gives

$$[CO_3^{-2}]_{eq} = \frac{[H_2CO_3]_{eq}}{[H_3O^+]_{eq}^2} K_I K_{II} \tag{R}$$

so that if (P) and (N) hold,

$$[CO_3^{-2}]_{eq} = \frac{a}{aK_I} K_I K_{II} = K_{II} \tag{S}$$

or, the concentration of the di-anion is numerically equal to the second ionization constant, no matter what the initial concentration of the acid is.

EQUILIBRIA INVOLVING IONIC COMPOUNDS OF LOW SOLUBILITY

In a saturated solution of an ionic compound in which an excess of the solid is present, an equilibrium exists between the ions of the dissolved compound and the ions in the crystals of the solid form (Figure 19.3). If the compound has a low solubility—as do silver chloride, barium sulfate, and calcium fluoride—it is found experimentally that the product of the molar concentrations of the ions in solution is equal to a constant, called the solubility-product constant. Table 19.5 gives data for saturated calcium fluoride solutions, and shows the near constancy of the product $[Ca^{+2}][F^-]^2$. The form of this product, containing the square of the fluoride ion concentration, is suggested by the molar ratios in the equation

$$CaF_2(s) \rightleftharpoons Ca^{+2} + 2F^- \tag{35}$$

The complete equilibrium constant for this reaction would be written according to our previous discussions as

$$K_{eq} = \frac{[Ca^{+2}]_{eq}[F^-]_{eq}^2}{[CaF_2(s)]_{eq}} \tag{36}$$

The experimental data show that a constant is obtained when the concentration of solid calcium fluoride is omitted:

$$K_{sp} = [Ca^{+2}]_{eq}[F^-]_{eq}^2 \tag{37}$$

Table 19.5 Concentration of Calcium Ion in a Saturated Solution of Calcium Fluoride which is M Molar in Fluoride Ion

$M = [F^-]$	$[Ca^{+2}]$	*Product* $[Ca^{+2}] \times [F^-]^2$
0.1	1.8×10^{-8}	1.8×10^{-10}
0.01	1.8×10^{-6}	1.8×10^{-10}
0.001	1.7×10^{-4}	1.7×10^{-10}
0.0001	1.7×10^{-2}	1.7×10^{-10}

Table 19.6 K_{sp} Values for Some Common Salts at 25°C

Silver chloride, AgCl	1.8×10^{-10}
Silver iodide, AgI	8.7×10^{-17}
Lead chloride, $PbCl_2$	1.6×10^{-5}
Calcium fluoride, CaF_2	1.7×10^{-10}
Magnesium hydroxide, $Mg(OH)_2$	8.9×10^{-12}
Iron(III) hydroxide, $Fe(OH)_3$	6×10^{-38}
Barium sulfate, $BaSO_4$	1.0×10^{-10}
Bismuth sulfide, Bi_2S_3	1×10^{-70}

This constant is called the solubility-product constant, or the solubility product.

The solubility product can be defined in the following manner: *In a saturated solution of a slightly soluble electrovalent compound, the product of the molar concentrations of the ions formed, raised to the proper powers as suggested by the chemical equation for the solution process, is constant,* and this constant is called the solubility-product constant.*

Determination of K_{sp}   Values of K_{sp} are calculated from the measured molar solubilities of the salts, as illustrated by the following problem.

PROBLEM 8   The solubility of calcium fluoride is 2.73×10^{-3} g/100 ml at 25°C. Calculate K_{sp}.

Solution

$$\text{Solubility in moles/liter} = 2.73 \times 10^{-3} \frac{g}{100\ ml} \times \frac{1\ mole}{78\ g} \times \frac{1,000\ ml}{1\ liter}$$

$$= 3.5 \times 10^{-4}\ \text{mole/liter}$$

Equilibrium involved:

$$CaF_2 \rightleftharpoons Ca^{+2} + 2F^-$$

$$\text{Solubility} = [Ca^{+2}]_{eq} = 3.5 \times 10^{-4}$$

From the chemical equation, two moles of fluoride ion are formed for each mole of calcium ion; thus

$$[F^-]_{eq} = 2[Ca^{+2}]_{eq} = 2 \times 3.5 \times 10^{-4} = 7.0 \times 10^{-4}$$

$$\therefore K_{sp} = [Ca^{+2}]_{eq} \times [F^-]_{eq}^2 = (3.5 \times 10^{-4})(7.0 \times 10^{-4})^2$$

$$= 1.72 \times 10^{-10}$$

Values for K_{sp} at 25°C for some common salts are given in Table 19.6.

Uses of K_{sp}   The solubility product is often used to: calculate the solubility of a salt; determine whether a precipitate will form if a certain amount of one ion is added to a solution containing the second ion involved in the equilibrium; calculate the concentration of each ion involved in the equilibrium

---

*The naive interpretation for the omission of $CaF_2(s)$ in the solubility-product expression is that, in the solid, the concentration of calcium fluoride is fixed and unchanging, being determined by the density of the solid, which fixes the number of moles per unit volume. On this basis, the fixed value can be incorporated into the equilibrium constant to give a new constant, $K_{sp} = [CaF_2(s)]K_{eq} = [Ca^{+2}][F^-]^2$. The sophisticated explanation is that solid calcium fluoride is in its standard state, where its activity is unity, so that the term for calcium fluoride, being equal to 1, does not appear in the equilibrium-constant expression (cf. footnote, page 399).

when the ions are present in a solution containing the salt in question and a second salt contains one of the ions involved in the equilibrium. Examples of each of these uses of K_{sp} follow.

Calculation of Solubility from K_{sp}. This calculation is illustrated in Problem 9.

PROBLEM 9   The K_{sp} for magnesium hydroxide is 8.9×10^{-12} at 25°C. Calculate the molar solubility of magnesium hydroxide.

Solution   Equilibrium involved:

$$Mg(OH)_2(s) \rightleftharpoons Mg^{+2} + 2OH^-$$

$$K_{sp} = [Mg^{+2}]_{eq}[OH^-]_{eq}^2$$

$$\text{Solubility} = [Mg^{+2}]_{eq} = \tfrac{1}{2}[OH^-]_{eq}$$

Let

$$x = [Mg^{+2}]_{eq}$$

Then, from the chemical equation,

$$2x = [OH^-]_{eq}$$

and

$$K_{sp} = [Mg^{+2}]_{eq}[OH^-]_{eq}^2 = (x)(2x)^2 = 4x^3$$

$$4x^3 = 8.9 \times 10^{-12}$$

$$x^3 = 2.2 \times 10^{-12}$$

$$x = \sqrt[3]{2.2} \times 10^{-4} = 1.3 \times 10^{-4} \text{ mole/liter} = \text{solubility}$$

Will a Precipitate Form?   A precipitate will form in a solution containing the ions of a slightly soluble salt when the product of the ionic concentrations raised to the appropriate power is greater than K_{sp}. When this ion product is less than K_{sp}, no precipitate will form because the solution is not saturated with respect to these ions.

PROBLEM 10   Will a precipitate form when 500 ml of a 4×10^{-4} M $BaCl_2$ solution is mixed with 500 ml of a 4×10^{-4} M Na_2SO_4 solution? K_{sp} for $BaSO_4$ is 1.0×10^{-10}.

Solution   The equilibrium involved is $BaSO_4(s) \rightleftharpoons Ba^{+2} + SO_4^{-2}$ and

$$K_{sp} = [Ba^{+2}]_{eq}[SO_4^{-2}]_{eq} = 1.0 \times 10^{-10}$$

The final volume of the mixed solutions is 1 liter; the actual ionic concentrations upon mixing the solutions is

$$[Ba^{+2}] = 2 \times 10^{-4} \qquad [SO_4^{-2}] = 2 \times 10^{-4}$$

The product of these is

$$[Ba^{+2}] \times [SO_4^{-2}] = 4 \times 10^{-8}$$

This product is greater than K_{sp}, so precipitation should occur. The amount of precipitate formed will be small, and just enough to reduce the product of the concentrations of the ions remaining in the solution to 1.0×10^{-10}.

Solubility and the Common-Ion Effect.   Suppose that barium chloride and sulfuric acid solutions containing exactly equivalent amounts—say, 0.01

mole of each substance—are mixed. Barium sulfate will precipitate as a result of the reaction

$$Ba^{+2} + SO_4^{-2} \rightleftharpoons BaSO_4(s) \qquad (38)$$

Then, after the precipitate has settled, a further slight precipitation from the saturated solution can be obtained by adding an excess of either Ba^{+2} or SO_4^{-2}. This is predicted by Le Châtelier's principle, since increasing the concentration of one of the substances on the left of Equation (38) will shift the position of equilibrium to the right. The amount by which the equilibrium will shift can be calculated from the solubility-product constant and the amount of excess ion added, since an increase in $[Ba^{+2}]$, for example, must be compensated for by a decrease in $[SO_4^{-2}]$ if K_{sp} is to remain constant. The only possibility for decreasing $[SO_4^{-2}]$ is for sulfate ions to react with barium ions to precipitate barium sulfate [Equation (38)]. The use of an excess of one of the precipitating ions to remove all but a negligible quantity of the other from the solution is a common laboratory practice; it is another example of the application of the *common-ion effect*.

PROBLEM 11 Calculate the concentration of the sulfate ion in a saturated solution of barium sulfate.

Solution The K_{sp} for barium sulfate is 1×10^{-10}, or

$$[Ba^{+2}] \times [SO_4^{-2}]_{eq} = 1 \times 10^{-10}$$

Let $[SO_4^{-2}]_{eq} = x$. By Equation (38), the concentration of the sulfate ion must equal the concentration of the barium ion when only barium sulfate is present; so $[Ba^{+2}]_{eq} = [SO_4^{-2}]_{eq} = x$. Substituting in the expression for the solubility-product constant, we get

$$x^2 = 1 \times 10^{-10}$$

Therefore, $x = 1 \times 10^{-5}$, or 0.00001 mole of Ba^{+2} or SO_4^{-2} per liter.

PROBLEM 12 Calculate the concentration of the sulfate ion in a saturated solution of barium sulfate to which barium chloride is added until $[Ba^{+2}] = 0.1\ M$.

Solution Let $[SO_4^{-2}]_{eq} = x$. In this case, $[Ba^{+2}]_{eq}$ is no longer equal to $[SO_4^{-2}]_{eq}$, since additional barium ion is present; in fact, according to the problem, $[Ba^{+2}]_{eq} = 0.1$ molar. Substituting in the expression for K_{sp},

$$[Ba^{+2}]_{eq} \times [SO_4^{-2}]_{eq} = 1 \times 10^{-10}$$

we get

$$0.1 \times x = 1 \times 10^{-10}$$

Therefore, $x = 1 \times 10^{-9}$, or 0.000000001 mole of SO_4^{-2} per liter.

Thus, the addition of more barium ion has decreased the sulfate ion concentration from 0.00001 molar (Problem 11) to 0.000000001 molar, as predicted by the common-ion effect.

Selective Precipitation. Two ions that form salts differing greatly in solubility, on reaction with the same oppositely charged ion can be separated from each other. Thus, it is easy to separate sodium ion from silver ion by adding chloride ion to precipitate insoluble silver chloride, leaving soluble sodium chloride in solution. The silver chloride is simply filtered off.

Also, a separation often can be made even though both salts are insoluble if the solubility-product constants are sufficiently different. A procedure frequently used in qualitative analysis for common cations separates sulfides of those cations that form highly insoluble sulfides from those cations whose sulfides are slightly more soluble by adjusting the sulfide ion concentration to such a value that the solubility product of the less soluble sulfide is exceeded, but that of the more soluble is not. The calculations are as given in the following problem.

PROBLEM 13   A solution is 0.1 M in each of cadmium ion and zinc ion. Can a sulfide ion concentration be chosen to precipitate almost all the cadmium ion without precipitating zinc sulfide?

Solution   The solubility-product constants (Appendix Table E.7) are:

$$K_{spCdS} = 6 \times 10^{-27}$$

$$K_{spZnS} = 1 \times 10^{-20}$$

Zinc sulfide will not precipitate until $[S^{-2}]$ reaches

$$[S^{-2}] = \frac{1 \times 10^{-20}}{0.1} = 1 \times 10^{-19} \; M$$

When $[S^{-2}] = 1 \times 10^{-19} \; M$, the concentration of $[Cd^{+2}]$ cannot be greater than

$$[Cd^{+2}] = \frac{6 \times 10^{-27}}{1 \times 10^{-19}} = 6 \times 10^{-8} \; M$$

Addition of sulfide ion to a concentration of $1 \times 10^{-19} M$ will thus precipitate cadmium sulfide, reducing the cadmium ion concentration in the solution to $6 \times 10^{-8} \; M$, leaving the zinc ion in solution at its original concentration, $0.1 \; M$.

How is the sulfide ion concentration to be maintained at $1 \times 10^{-19} \; M$? This can be accomplished by adding hydrogen sulfide in acid solution. Combining the ionization-constant expressions for hydrogen sulfide ionization

$$K_I = \frac{[H_3O^+]_{eq}[HS^-]_{eq}}{[H_2S]_{eq}}$$

$$K_{II} = \frac{[H_3O^+]_{eq}[S^{-2}]_{eq}}{[HS^-]_{eq}}$$

we obtain

$$K_I K_{II} = \frac{[H_3O^+]_{eq}^2[S^{-2}]_{eq}}{[H_2S]_{eq}}$$

Substituting the values of K_I and K_{II}, the saturation value for the gas hydrogen sulfide at 1 atm pressure (0.1 M), and the desired value for $[S^{-2}]$, namely $1 \times 10^{-19} \; M$, we obtain

$$[H_3O^+]^2 = \frac{K_I K_{II}[H_2S]}{[S^{-2}]} = \frac{1 \times 10^{-7} \times 1 \times 10^{-14} \times 1 \times 10^{-1}}{1 \times 10^{-19}} = 1 \times 10^{-3}$$

$$[H_3O^+] = 3 \times 10^{-2} M$$

Thus, if a solution $0.03M$ in hydrochloric acid is saturated with hydrogen sulfide, the sulfide ion concentration will be $1 \times 10^{-19} \; M$, and the separation of cadmium ion and zinc ion can begin.

It should be noted, however, that the final concentration of sulfide ion after precipitation of cadmium sulfide will not be $1 \times 10^{-19} M$ because of acid generated in the precipitation reaction:

$$Cd^{+2} + H_2S + 2H_2O \longrightarrow CdS(s) + 2H_3O^+$$

The calculation of the final hydronium ion concentration and the final concentration of cadmium ion in the solution is left to the student.

Dissolving Insoluble Salts   If experiments can be arranged so the concentration of silver ion or chloride ion is maintained at such a low value that $[Ag^+][Cl^-]$ is always less than K_{sp}, the usually insoluble salt silver chloride may dissolve completely. In some cases, this can be accomplished by forming *complex ions* (Chapter 24). Thus, if ammonia is added to a solution containing silver ion, the complex ion $Ag(NH_3)_2^+$ is formed. This reduces the concentration of silver ion and, following Le Châtelier's principle, the position of the solubility equilibrium shifts to the right in the attempt to relieve the "stress" of the decrease in $[Ag^+]$. The shift may be represented by the heavy arrows in the equations

$$AgCl(s) \rightleftharpoons Ag^+ + Cl^- \qquad (39)$$
$$+$$
$$2NH_3$$
$$\updownarrow$$
$$Ag(NH_3)_2^+$$

If the vertical reaction of formation of $Ag(NH_3)_2^+$ goes far enough, by adding excess ammonia $[Ag^+]$ may be made so small that the silver chloride dissolves completely.

Salts of weak acids, insoluble in pure water, commonly dissolve in solutions of strong acids as a result of removal of the anions of the salts by the formation of undissociated weak acid molecules. Thus, silver acetate will dissolve in nitric acid as a result of the displacement of the following equilibria when $[H_3O^+]$ is made very large by adding nitric acid:

$$CH_3CO_2Ag(s) \rightleftharpoons CH_3CO_2^- + Ag^+ \qquad (40)$$
$$+$$
$$H_3O^+$$
$$\updownarrow$$
$$CH_3CO_2H$$
$$+$$
$$H_2O$$

The positions of the equilibria are displaced in the direction of the heavy arrows.

Reinforced Buffering in Blood.   The pH of the environment is an important variable in many biological processes. Studies indicate that many such processes occur only at a specified pH, and that this pH must be maintained during the course of the changes that occur. Characteristics such as these are provided by buffer solutions. In living organisms, especially vertebrates, there are complex mechanisms to improve and reinforce the simple buffering mechanism provided by single conjugate acid-base pairs.

Thus, while the major intracellular buffer is the conjugate acid-base pair $H_2PO_4^{-1}/HPO_4^{-2}$ (which holds the pH at 7.2), organic phosphates such as

ATP (Chapter 29) and glucose-6-phosphate also add buffering power in the cell.

The bicarbonate buffer system (H_2CO_3/HCO_3^-) is the major extracellular buffer functioning in the blood and interstitial fluids of vertebrates. However, there are at least four other buffer systems that make important contributions to the overall buffer action in the blood.

The bicarbonate system is in approximate equilibrium with the partial pressure of CO_2 in the lung air. This gives this system some distinctive features. While it functions as a buffer in the same way as other conjugate acid-base pairs, the K_1 of H_2CO_3 is about $10^{-3.8}$, which would suggest that the buffer formed by this system would produce a pH well below the normal range of blood pH. However, because the normal buffer (H_2CO_3/HCO_3^-) is in equilibrium with the CO_2 of the lung air, represented as follows:

$$CO_2 + 2H_2O \rightleftharpoons H_2CO_3 + H_2O \rightleftharpoons H_3O^+ + HCO_3^-,$$

the pH of the buffer becomes especially sensitive to the partial pressure of CO_2 in this air. This can be made clear by considering the relation

$$[H_3O^+] = K_1 \frac{[H_2CO_3]}{[HCO_3^-]}$$

which gives the hydronium ion concentration of the bicarbonate buffer. This equation tells us that increasing H_2CO_3, while keeping other variables constant, will result in an increase in $[H_3O^+]$ and a decrease in the pH. Similarly, decreasing H_2CO_3 will result in an increase in pH. Since the equilibrium $CO_2 + H_2O \rightleftharpoons H_2CO_3$ affects the concentration of H_2CO_3, the indication is that increases in the lung's air concentration of CO_2 will decrease the pH of the bicarbonate buffer, and that decreases in the CO_2 concentration will increase the pH of the buffer.

Calculations show that with normal amounts of CO_2 in the lung air, the bicarbonate system buffers at a pH near 7.0.

It is important to remember also that the CO_2 in the blood and in lung air is the end product of cellular oxidation of foodstuffs such as carbohydrates. Thus, the ratio $[H_2CO_3]/[HCO_3^-]$ in the blood is a reflection of the rate of CO_2 production by cellular oxidation and the rate of CO_2 loss by expiration.

This production of CO_2 in the cells results in a higher $[H_2CO_3]/[HCO_3^-]$ ratio in venous blood than in arterial blood. The resulting effect on the pH of venous blood is largely offset by accompanying changes in the hemoglobin systems, which also exert a buffering action.

The blood hemoglobin behaves as a polyprotic acid with at least five acid groups. As an approximation, we can represent the hemoglobin buffer system by

$$\frac{[\text{hemoglobinate}^{-1}]}{[\text{H-hemoglobin}]}$$

In a similar way, the oxyhemoglobin buffer system can be represented by

$$\frac{[\text{oxyhemoglobinate}^-]}{[\text{H-oxyhemoglobin}]}$$

These two systems differ in that oxyhemoglobin can be regarded as a stronger acid than hemoglobin. In venous blood, where the oxygen content is low, the hemoglobin buffer is more important than the oxyhemoglobin buffer. Thus, in venous blood, the tendency for the pH to fall because of the increased $[H_2CO_3][HCO_3^-]$ ratio is opposed by the effectiveness of the less acidic hemoglobin buffer system. In arterial blood, oxygenation of the hemoglobin in the lungs tends to make the blood more acidic. This opposes the tendency of the blood to increase in pH as CO_2 is liberated in the lungs.

The balance between the bicarbonate and the hemoglobin-oxyhemoglobin systems is such that the actual difference in pH between arterial and venous blood is usually less than 0.03 pH unit.

Other blood buffer systems of some importance are the phosphate system ($HPO_4^{-2}/H_2PO_4^-$) and those from certain plasma proteins. The buffering action of these systems is relatively constant and is not altered by variables such as the CO_2 concentration or the hemoglobin/oxyhemoglobin ratio.

SUMMARY   The principles of chemical equilibria have been presented in this chapter. Starting with the observation that many chemical reactions do not proceed to completion, the concept of chemical equilibrium is developed. This is followed by a study of chemical equilibrium for a single reaction at a given temperature from which a relation between the equilibrium concentrations of products and reactants is discovered. This relation, the equilibrium-constant expression, is shown to have general applicability for all chemical reactions at equilibrium.

The effect of various stresses such as changes in temperature, pressure, or concentration of reactants and products on systems in equilibrium is examined and interpreted in terms of the equilibrium constant, the position of equilibrium, and Le Châtelier's principle.

Application of equilibrium principles to gaseous systems and to reactions occurring in solution is given. This includes equilibria in solutions of weak acids and bases, the water equilibrium, the concept of pH, buffered solutions, hydrolysis, and solubility equilibria.

IMPORTANT TERMS

Equilibrium constant
ionization constant
ion-product constant
hydrolysis constant
solubility-product
    constant

Le Châtelier's principle
common-ion effect
mass-action effect

Brønsted acids and bases

Hydrolysis

Extent of reaction

Position of equilibrium

Indicator

Buffer solution

Activity

pH value

QUESTIONS AND PROBLEMS

1. Why must the temperature be specified when indicating the value of an equilibrium constant?

2. Indicate the effect of the following on (a) the speed of a reaction and (b) the position of equilibrium: catalyst, pressure, temperature, concentration.

3. Show how Le Châtelier's principle can be used in each of the above examples.

4. Explain how an increase in temperature may be expected to affect (a) exothermic reactions, (b) endothermic reactions.

5. Write the expressions for the K_{eq} of the following:
   (a) $H_2(g) + Cl_2(g) \rightleftharpoons 2HCl(g)$
   (b) $C(s) + H_2O(g) \rightleftharpoons CO(g) + H_2(g)$
   (c) $3H_2(g) + N_2(g) \rightleftharpoons 2NH_3(g)$

6. Calculate the value of K_{eq} for the reaction $H_2 + I_2 \rightleftharpoons 2HI$ from the following data:

Equilibrium Concentrations at Approximately 700 K		
mmoles I_2/liter	mmoles H_2/liter	mmoles HI/liter
1.71	2.91	16.5
3.13	1.83	17.7
0.495	0.495	3.66

7. 1.5 moles of nitric oxide (NO), 1.0 mole of chlorine, and 2.5 moles of NOCl were mixed together in a 15-1 container at 230°C. When the reaction

$$2NO(g) + Cl_2(g) \longrightarrow 2NOCl(g)$$

came to equilibrium, 3.06 moles of NOCl were present. Calculate (a) The number of moles of nitric oxide present at equilibrium, (b) the equilibrium constant based on partial pressures. (See Chapter 20.)

8. Consider the reaction $2X + Y = X_2Y$. Initially, 3 moles of X and 3 moles of Y are placed in a 1-liter container, and after the reaction has attained equilibrium, 0.6 mole of X_2Y is present. (a) How many moles of X as such are present? (b) Of Y? (c) Calculate the equilibrium constant.

9. Consider the reaction $A + B = AB$, for which $\Delta H = -15,000$ cal. (a) How much heat evolution accompanies the reaction if 6 moles of A react? (b) If 6 moles of A are in the flask initially and 2 moles are present at equilibrium, how much B reacted? (c) How much AB is formed? (d) Find the value of ΔH and of K_{eq} in this case.

10. Suppose a 0.1-molar solution of a base MOH is 2% ionized at 25°C. Calculate the value of the ionization constant of the base.

11. Calculate the value of K_w for water if, at the neutral point, the pH is found to be 7.0.

12. Calculate the pH value of a solution whose $[OH^-]$ is 10^{-8}.

13. What are the concentrations of H_{aq}^+, ClO_2^- and $HClO_2$ in a 0.10 M solution of chlorous acid?

14. Calculate the concentration of all species present in a 0.10 M solution of nitric acid.

15. How many moles of nitrous acid, HNO_2, must be used to prepare 2.5 liters of a solution that has a pH of 3.0?

16. The dissociation constant for formic acid, HCOOH, is 1.8×10^{-4} at 25°C. Calculate the concentration of H_3O^+, $HCOO^-$, and HCOOH present at

equilibrium in the following solutions: (a) 0.01 M HCOOH; (b) 1.00 M HCOOH

17. Sulfuric acid is a polyprotic acid such that it dissociates completely, forming H_3O^+ and HSO_4^-. The bisulfate ion, however, dissociates further into H_3O^+ and SO_4^{-2}. The dissociation constant of HSO_4^- is 1.20×10^{-2}. Calculate the concentration of all species present in a 0.100 M H_2SO_4 solution.

18. Calculate the value of $[H_3O^+]$ in 500 ml of a 0.05 M solution of acetic acid to which has been added 0.1 mole of sodium acetate. The value of the ionization constant for acetic acid is $K_{ion} = 1.8 \times 10^{-5}$.

19. The ionization constant for a water solution of ammonia is 1.75×10^{-5}. (a) Calculate the hydroxide ion concentration in a 0.05 M solution. (b) Calculate the per cent reaction completion for a 0.05 M solution. (c) Calculate the pH of a 0.05 M solution.

20. Determine the pH in each of the following situations: (a) A solution is 0.02 M in ammonium hydroxide. To this is added enough NH_4Cl to increase the $[NH_4^+]$ to 0.04 M. (b) Twenty ml of a solution of 0.200 M acetic acid is mixed with 40.0 ml of 0.200 M sodium acetate. (c) 0.050 mole of acetic acid and 0.100 mole of sodium acetate are dissolved in enough water to make a liter of solution.

21. Explain how the addition of CH_3CO_2Na to water "buffers" the water and the pH remains almost constant after the first addition of HCl, even though large amounts may be added.

22. If 250 ml of a 1 M acetic acid solution are buffered by the addition of 250 ml of 1 M sodium acetate, what is the $[H^+]$ of the buffered solution?

23. Write the hydrolysis equation for acetate ion and predict whether a 0.1-molar solution will be acidic or basic.

24. (a) Calculate $[H^+]$ in a 0.1-molar solution of NH_4Cl at 25°C. Write the equation for the hydrolysis reaction involved and obtain the necessary data from data tables in this chapter to solve the problem. (b) Calculate the per cent of hydrolysis.

25. Show that an indicator is a weak acid or a weak base and that the indicator action is a shift between the ionized and the un-ionized form.

26. The solubility of SrF_2 at 25°C is 1.22×10^{-2} g/100 ml. What is the value of K_{sp} for SrF_2?

27. The value of K_{sp} of $BaCO_3$ is 8.1×10^{-9} at 25°C. What is the solubility of $BaCO_3$ in moles per liter?

28. The value of K_{sp} of Ag_2CrO_4 at 25°C is 9.0×10^{-12}. Calculate the solubility of silver chromate, Ag_2CrO_4, in moles per liter in a 0.25 M K_2CrO_4 solution.

29. A saturated solution of barium fluoride, BaF_2, at 25°C contains 7.6×10^{-3} mole/liter of barium ion. Calculate (a) the concentration of fluoride ion, (b) the solubility product of barium fluoride.

30. At 25°C, the K_{sp} for $MgCO_3$ is 1.0×10^{-5} and that for $BaCO_3$ is 8.1×10^{-9}. Assuming both magnesium ions and barium ions are present in solution

at 1×10^{-3} M concentrations, what is the concentration of carbonate ion that will precipitate the maximum amount of Ba^{+2} ion without precipitating Mg^{+2}?

31. Calculate the pH of a saturated solution of copper hydroxide. Assume the K_{sp} of Cu $[OH]_2$ to be 1×10^{-20}.

32. (a) Suppose you have a solution of pH 3 and you dilute it a thousandfold; what is the new pH? (b) Suppose you dilute the new solution a thousandfold; what is now the pH?

SPECIAL PROBLEMS

1. Using the data of Table 19.4, determine the equilibrium constant for the process

$$H_2S + 2H_2O \rightleftharpoons 2H_3O^+ + S^{-2}$$

From this, calculate the sulfide-ion concentration in a saturated H_2S solution (0.1 M) to which has been added enough hydrochloric acid to make the hydronium ion concentration 1 M.

2. What are the concentrations of all species present in a solution prepared by mixing 50 ml of 0.6 M sodium formate, $NaHCO_2$, with 50 ml of 0.4 M hydrochloric acid? Assume the volumes are additive.

3. The hydrolysis constant for the reaction with water of a salt of a strong base and a weak acid is defined as

$$K_h = \frac{[HA][OH^-]}{[A^-]}$$

from the chemical equation

$$H_2O + A^- \rightleftharpoons HA + OH^-$$

Derive the relationship

$$K_h = \frac{K_w}{K_{ion}}$$

4. Calculate the pH of a 0.1-molar solution of Na_2CO_3. Obtain the necessary data from Table 19.4.

5. Calculate the concentration of hydronium ion in a solution of sodium bicarbonate of concentration a moles per liter using the exact procedure, followed by approximations. The chemical equations are

$$HCO_3^- + H_2O \rightleftharpoons H_3O^+ + CO_3^{-2}$$

$$H_3O^+ + HCO_3^- \rightleftharpoons H_2CO_3 + H_2O$$

$$H_2O + H_2O \rightleftharpoons H_3O^+ + OH^-$$

6. Calculate the hydronium ion concentration in a solution of concentration a in carbonic acid and s in sodium bicarbonate.

7. Calculate the hydronium ion concentration in a solution of concentration s_1 in sodium bicarbonate and s_2 in sodium carbonate.

8. Apply Le Châtelier's principle to the equilibrium

$$\text{Solid} \rightleftharpoons \text{Liquid}$$

represented by the line OC in Figure 11.9, as is done for the liquid $\rightleftharpoons$

vapor equilibrium on p. 250. For the substance considered in Figure 11.9, the volume occupied by a given mass of the solid substance is smaller than that occupied by the same mass of the liquid substance. (a) Increase in pressure would then cause the position of equilibrium to shift in which direction? (b) To melt a solid requires the addition of heat (heat of fusion). Considering the answer to (a), would you expect the increase of pressure to have the effect of raising, or lowering, the melting point?

9. The substance water is unusual compared to the vast majority of substances in that the solid form, ice, floats on the liquid; for almost all other substances, the solid would sink to the bottom. (a) What does this information tell you about the relative volume occupied by a given mass of ice and of liquid water (recall Archimedes' principle)? (b) From your answer to (a), what would you predict, from Le Châtelier's principle, would be the effect on the ice ⇌ liquid-water equilibrium of the application of pressure? (c) How would line OC, Figure 11.9, look for the substance water? Sketch Figure 11.9 for water, labeling the coordinates of the triple point, O (4.58 mm and 0.0075°C), and those of the critical temperature and pressure (374°C and 218 atm).

10. The substance carbon dioxide has its triple point at −57°C and 5.3 atm. The critical temperature and pressure are 31°C and 73 atm. Sketch a diagram for carbon dioxide using the same letters as shown in Figure 11.9, and labeling the coordinates of points O and A. Describe what would happen as the following experiments are carried out: (a) Liquid carbon dioxide, at 0°C and 60 atm pressure, is gradually heated at constant pressure to 40°C. (b) The pressure on liquid carbon dioxide, originally at 0°C and 60 atm, is gradually reduced at constant temperature to 5 atm. (c) Solid carbon dioxide, originally at −70°C and 15 atm, is heated gradually at constant pressure to 40°C. (d) Solid carbon dioxide, originally at −70°C and 15 atm, is subjected at constant temperature to a gradual pressure reduction to 1 atm. (e) Solid carbon dioxide, originally at −100°C and 1 atm, is allowed to warm to room temperature while exposed to the atmosphere. (f) Solid carbon dioxide, originally at −60°C and 100 atm, is allowed to warm at constant pressure until it reaches 40°C.

11. Compare the solubility of iodine in water and in diethyl ether. Assume the two solvents to be immiscible; that 0.029 g of iodine will dissolve in 100 ml of water at 20°C and 21 g of iodine will dissolve in 100 ml of diethyl ether at the same temperature. (a) What is the ratio of iodine's solubility in water to that in ether (distribution coefficient)? (b) A 100-cc sample of a saturated aqueous solution of iodine was shaken with 40 cc of ether. How much (weight) of iodine remained in the water layer? (c) A 100-cc aqueous solution was treated with two consecutive 20-cc samples of ether. How much iodine remains in the water layer with this method? (d) What can you say about the comparative efficiency of the two extractions based on your calculations?

REFERENCES   BAUMAN, R. P., *An Introduction to Equilibrium Thermodynamics,* Prentice-Hall, Englewood Cliffs, N.J., 1966.
BUTLER, J. N., *Solubility and pH calculations,* Addison-Wesley, Reading, Mass., 1964.
SIENKO, M. J., *Equilibrium,* Benjamin, Menlo Park, Calif., 1964.

20 FREE ENERGY

Early chemists spent much of their time searching for new elements, determining their properties, studying their reactions, and making compounds of them by a variety of reactions. As they learned more about the fraction of the reactants that was changed to products when a particular reaction took place, and about the changes in the position of equilibrium as reaction conditions were altered, they began to seek the underlying principles that govern the stability and reactivity of chemical systems in general.

One avenue of investigation led to the "broad highway" of chemical thermodynamics. Chemical thermodynamics is the branch of physical chemistry concerned with, among other things, predicting where the position of equilibrium will be in chemical reactions. It shows that the position often may be calculated from measurements of heat and energy quantities, thus making it possible to choose the best conditions for the largest yield of products from a given amount of material without performing a large number of experiments to find out, by actual trial of many conditions, which are the best. Equally important, thermodynamics tells the experimenter that some reactions, for which plausible chemical equations can be written, have no chance of success. Thus it saves long hours of experimental trials foredoomed to failure. The importance of chemical thermodynamics lies just here—it offers an answer to the question: How can you tell whether a reaction can take place?

In this chapter, we shall introduce the topic of thermodynamics and discuss the relationship between the thermodynamic quantity called the free energy and the feasibility and yield in chemical reactions.

Heat of Reaction Not the Determining Factor: An Erroneous View

At one time, chemists thought that the requirement for a reaction taking place was a release of energy. In other words, all reactions that took place on the mixing of reactants (called *spontaneous* reactions, as contrasted to reactions

*From *Thermodynamics* (New York: McGraw-Hill, 1923).

Figure 20.1 Mixing of two perfect gases.

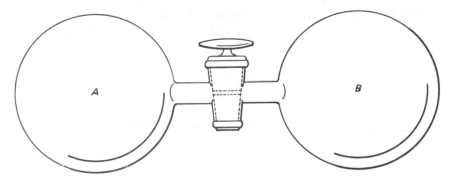

forced to proceed, by electrolysis, photolysis, or the like and therefore *not spontaneous*) must be exothermic reactions, and thus must have a negative value of ΔH. We have seen, however, in Table 18.1, that the formation of nitric oxide is an endothermic reaction in which heat is absorbed (positive value of ΔH); yet this reaction proceeds spontaneously when nitrogen and oxygen, heated separately to 400°C or so, are allowed to mix.* The existence of even this one endothermic spontaneous reaction—there are countless others—is sufficient to show that the hypothesis that only exothermic reactions are spontaneous is incorrect, and must be discarded.

A Spontaneous Process that Is Neither Exothermic Nor Endothermic—Mixing of Gases

To find out what does control spontaneity of reactions, let us consider a process that proceeds spontaneously with no energy change—the mixing of two perfect gases. A quantity of gas A is placed in flask A (Figure 20.1), and flask B is filled to the same pressure with gas B. On opening the stopcock, the two gases will be observed to mix, gradually, so that after a sufficient time each is evenly distributed between the two flasks. The mixing has been a spontaneous process; there has been no temperature change, no energy change, and no outside source of stirring to make the gases mix. There must have been a "natural" tendency causing the system to change from its initial unmixed state to the final mixed state. What is the nature of this driving force?

Examination shows that the final state is a more probable one than the initial state of separated gases. If the mixture were before us, we would say that it was unlikely or improbable that the components would separate of themselves; we could then say that the mixed state had a greater probability of existence than the unmixed state. The driving force in this process is the tendency toward greater "mixed-up-ness," the tendency to go to the less ordered state, to the more random state, to the state of higher probability.

Rule: Spontaneous Processes Go from More Order to Less Order

The mixing of the two gases is a single example of a general and universal rule: *All processes that proceed spontaneously from a given initial state to a given final state do so with an increase in randomness*—with a decrease in order. This is one form of the statement known as the second law of thermodynamics.

---

*The fact that a reaction is spontaneous does not mean it will go to completion. It may go almost to completion or part of the way to completion. Spontaneous simply means that *some* reaction will occur.

System and Surroundings

In applying the above rule, you must understand precisely what is meant by the initial and final states. Chemists often distinguish between a *system* and its *surroundings*. By the *system*, they might mean the material reacting, or a container and its contents, such as a beaker of reactants on a desk top. The beaker is not *isolated*, but can interchange energy with its *surroundings*—by absorbing heat from the laboratory atmosphere, for example. When the second law of thermodynamics is stated as above, the initial state to be considered is that of the system plus the surroundings, and the final state must also include both the system and surroundings. In the example of the gas mixing, we chose a special kind of gas—a perfect gas—so no interchange of energy or stirring between the system (the two connected flasks) and the surroundings took place, and we needed to consider only the system itself. But this would not be the usual case. The involvement of the surroundings, in the usual case, is unfortunate, since it is not always easy to assess changes in randomness in the surroundings. Hence scientists have sought criteria of spontaneity that involve only the system itself. To separate what happens in the system from what happens in the surroundings, let us first consider the transfer of energy between the two.

Energy Transfer to and from Surroundings and Systems

Energy can be transferred into the system from the surroundings either as heat, denoted by the symbol q, or as work, denoted by w. Any heat transferred to the system will increase its energy. Work done on the system will also increase its energy. (As an example, consider the mechanical work done in compressing a spring; the work done is stored as potential energy in the spring.) We may thus write for the energy increase in the system

$$\begin{pmatrix} \text{Energy increase in} \\ \text{the system} \end{pmatrix} = \begin{pmatrix} \text{heat absorbed by} \\ \text{the system} \end{pmatrix} + \begin{pmatrix} \text{work done on} \\ \text{the system} \end{pmatrix} \qquad (1)$$

Conventional choices prefer the same preposition (by) in both terms on the right, and we write

$$\begin{pmatrix} \text{Energy increase in} \\ \text{the system} \end{pmatrix} = \begin{pmatrix} \text{heat absorbed by} \\ \text{the system} \end{pmatrix} - \begin{pmatrix} \text{work done by} \\ \text{the system} \end{pmatrix} \qquad (2)$$

or, using symbols,

$$\Delta E = q - w \qquad (3)$$

The minus sign now appears because we measure work done *by* the system; this uses energy. *The net energy increase, ΔE, is the difference between the energy added to the system as heat and the energy used in performing work.* The italicized statement is a consequence of the law of conservation of energy; the specific application to the transfer of heat energy and work energy is known as the first law of thermodynamics.

Let us now look at this equation from a different point of view. Suppose that a process takes place in which there is a decrease in energy, so that $\Delta E = E_{\text{final}} - E_{\text{initial}}$ is a negative quantity. This energy decrease could be transferred to the surroundings in one of several ways: it could go entirely as heat, with no work done, as in a chemical reaction at constant volume; it could be transferred partly as heat and partly as work; and the ratio of the part transferred as heat to the part transferred as work could be varied. Which one of these methods of transfer actually occurs will depend upon the wishes of the experimenter in deciding how much of the energy decrease he wishes to use as work. But only in certain specialized systems can all of the energy change be converted into work, he will find, however, that there is one method that will produce more work than any other—namely, that method in which the work is done in a "reversible" manner.

Thermodynamic Reversibility   The word "reversible" in the last sentence has a special meaning in thermo-dynamics. *A process is being carried out* reversibly *when an infinitesimal change in the value of a variable can cause the process to proceed in one direction or the other.* Consider, for example, gas at pressure P expanding against a piston. If the expansion were allowed to occur against an opposing pressure $P - P$, where P is an infinitesimally small pressure, then raising the opposing pressure to $P + P$ will cause the piston to move in the opposite direction and compress the gas, thus reversing the expansion. The expansion of gas at pressure P against an opposing pressure $P - P$ is thus an example of a reversible expansion. The pressure-volume work for this case is greater than the work that could be done in any other expansion process that allows the gas to undergo the same volume change. Similarly, the work done in *any reversible process* is the *maximum work* obtainable from that process.

Entropy   It is an experimental fact that when the process in which the energy change is ΔE is arranged to produce the maximum amount of work, the amount of heat transferred becomes a fixed quantity, which we denote by q_{rev}. There-fore, for any process carried out reversibly, we write

$$\Delta E = q_{rev} - w_{max} \tag{4}$$

The heat transferred, q_{rev}, is now a definite quantity determined only by the initial and final states, and, for processes at constant temperature, we find that q_{rev}, divided by the temperature, is a measure of the change in the "mixed-up-ness," or randomness, of the system. The name applied to the randomness is entropy; it is denoted by the symbol S. The change in ran-domness is denoted by ΔS. In symbols, therefore, for constant-temperature processes,

$$\frac{q_{rev}}{T} = \Delta S \tag{5}$$

where the delta symbol again represents the difference between the initial and the final state:

$$\Delta S = S_{final} - S_{initial} \tag{6}$$

Thus, reversible conditions give the maximum work and make $q_{rev} = T\Delta S$, so

$$\Delta E = T\Delta S - w_{max} \tag{7}$$

Free Energy   In dealing with chemical systems in open beakers, we meet the same situation mentioned in Chapter 18—namely, that we are often obliged to perform pressure-volume work on the atmosphere whether we wish to or not. This work is included in w_{max}, but often it is not useful work. For the usual constant-pressure processes, therefore, it is convenient to add $P\Delta V$ to both sides of Equation (7):

$$\Delta E + P\Delta V = T\Delta S - (w_{max} - P\Delta V) \tag{8}$$

We now recognize the left-hand side as ΔH [Equation (7), Chapter 18]; the quantity in parentheses is the maximum *useful* work. Let us conceive that this *maximum useful work* appeared as the result of a decrease in a store of useful work in the system known as a free energy, G, so that

$$G_{initial} - G_{final} = (w_{max} - P\Delta V) \tag{9}$$

As noted earlier, the delta symbol represents *final minus initial*, so

$$G_{initial} - G_{final} = -\Delta G \tag{10}$$

and

$$-\Delta G = (w_{max} - P\Delta V) \tag{11}$$

Substitution gives

$$\Delta H = T\Delta S + \Delta G \tag{12}$$

The Sign of the Free-Energy Change Is the Criterion of Spontaneity for the System

We now have, in ΔG, the criterion of the *system*, sought for on page 422, which determines whether a change from a specified initial state to a specified final state will be a spontaneous one. Since the decrease in G measures the maximum useful work that can be obtained, only when ΔG is negative can work be obtained.

If the process is to be spontaneous, it must take place of itself, with no work being done on the system. However, if ΔG is positive, work must be done on the system to effect the change; thus the change is not spontaneous if ΔG is positive. If ΔG is negative, work can be obtained, and the process is a spontaneous one. If ΔG is zero, there is no tendency to change at all. Since the free-energy criterion has been developed for systems at constant pressure and constant temperature, the result must be stated: *For spontaneous changes at constant pressure and constant temperature, the change in free energy, $(\Delta G)_{P,T}$, must be negative.* The subscripts indicate that pressure and temperature are constant.

The usefulness of a spontaneity criterion for the *system*, mentioned on page 422, may now be made more clear by considering as an example the freezing of supercooled water at $-10°C$ in contact with a heat reservoir at $-10°C$. (We assume the heat reservoir to be so large that its further absorption of heat does not make a measurable change in its temperature.) The freezing of water at a temperature below its freezing point is a spontaneous process, yet the entropy of the water decreases since the random arrangement of water molecules in the liquid changes to the ordered arrangement characteristic of the crystal. These facts seem to contradict the general and universal rule cited on page 421 that in spontaneous processes, the entropy (randomness) must increase. The difficulty arises because we did not include the entropy change of the surroundings (the heat reservoir). In this case, it is possible to calculate the entropy change in the surroundings. It is found that the entropy increase caused by the absorption of the heat of crystallization by the reservoir is greater than the decrease in entropy in the system (the water), so that there *is* a net entropy increase for system *plus* surroundings. No such seeming contradiction arises if we consider the *free-energy change* for the *system* alone; calculation shows that it is negative, as required for a spontaneous process at constant pressure and constant temperature.

Relation of Changes in Enthalpy
and Entropy to Changes in
Free Energy

Rearrangement of Equation (12) shows the effect of changes in enthalpy and entropy on ΔG:

$$\Delta G = \Delta H - T\Delta S \qquad (13)$$

The first term, ΔH, is a measure of the change in energy in the system; the second term, $T\Delta S$, is a measure of the change in randomness in the system. The change in energy (ΔH) is measured by the interchange of heat between system and surroundings when the process is carried out at constant pressure, as in the measurements of heats of reaction in Chapter 18. Change in randomness ($T\Delta S$) is measured by the interchange of heat between the system and surroundings, q_{rev}, when the process is carried out in a reversible manner. The more negative ΔH is, the more likely ΔG is to be negative and the reaction to be spontaneous. A negative value of ΔH means that energy is given off by the system when the process occurs. If ΔS is positive, ΔG will also more likely be negative. A positive ΔS means that the system is changing to a more random state. Both ΔH and ΔS change only little with change in temperature. But $T\Delta S$ is nearly proportional to the temperature, unless there is a change in the physical state of one of the substances, as when the temperature range includes a melting, boiling, or transition point.

Values of ΔH may be directly measured as heats of reaction or estimated from bond energies (Chapter 18). The determination of ΔS from measurements of q_{rev} is not quite so easy, but estimates of ΔS can be made for some cases from the connection of entropy with randomness. Randomness may be considered to be measured by the number of places there are to put molecules without changing the properties of the system. There are more places to put a given number of molecules in a large volume than in a small volume, and the randomness—and thus the entropy—is greater for any one substance, the larger the volume occupied by the substance. This is the explanation for the increase in entropy (randomness) on opening the stopcock between gases A and B, Figure 20.1; a larger volume becomes accessible to both gases. As another example, the change from solid to liquid always represents an entropy increase because the molecules of the liquid may occupy many places and are not restricted to the specific sites of the crystal lattice of the solid. Alternatively, the entropy increase on melting may be looked upon as a change to a more disordered state (the liquid) from the ordered, less random state of the solid. This change in order is still more marked in passing from solid to gas or liquid to gas.

Molecules may not only be put in different places in space, as when the volume changes, but also in different "places" in an energy sense. We have

---

We have established that the change of order-to-disorder is a driving force in making a process occur. However, the usefulness of this measure of driving force is diminished by the fact that its use requires a knowledge of changes in the surroundings as well as in the system. An *improved* measure of the driving force is found in the ability or inability of the system to perform work under reversible conditions. For constant-pressure and constant-temperature processes, the process is spontaneous only in the case that work can be obtained. The magnitude of the reversible useful work that can be performed, $w_{max} = -P\Delta V$, is measured by the change in free energy. These concepts lead directly to a general indicator for predicting the spontaneity of a constant-temperature process: If the process involves a release of free energy (ΔG is negative), the process is spontaneous; if, however, work must be done on the system (ΔG is positive) to make the process take place, the process is not spontaneous.

seen earlier that an isolated molecule may hold energy in many different ways (Chapter 18): as energy of electronic excitation, as vibrational motion of the atoms in the molecule, as rotational motion of the molecule as a whole, and as kinetic energy of translational motion of the molecule as a whole. Each of these energies is quantized, and the values of each are approximately independent of the values of the others. The difference in energy between one energy level and the next, however, varies greatly among the four kinds. This difference is largest for electronic energies and decreases in the order listed, being extremely small for translational energies. Consequently, for electronic and vibrational energies, there may be accessible to the molecules only a few energy "places," whereas for translational energies there will be many places available to a group of molecules with limited energies. Since there are many more places to put the molecules, energy-wise, in translational motion than there are in other types of motion, the entropy for translational motion will be much larger than for the other types.

Applications to Chemical Reactions

Solids Reacting to Form Gases. The reaction of water-gas formation

$$C(s) + H_2O(g) \longrightarrow CO(g) + H_2(g) \qquad (14)$$

proceeds with the absorption of heat, the enthalpy change (ΔH) amounting to +31,400 cal at 25°C when each gas is at 1 atm pressure. The entropy change may be expected to be positive, since a loss of order appears when solid carbon disappears in forming gaseous carbon monoxide. The measured value of ΔS for the reaction written, at 25°C and 1 atm for each gas, is 32.0 cal/deg. Therefore, at 25°C,

$$\Delta G = \Delta H - T\Delta S = +31,400 - 298 \times 32.0 = +21,900 \text{ cal} \qquad (15)$$

Since ΔG is positive, the reaction will not occur spontaneously at this temperature when the gases are at 1 atm pressure. At 1000K, assuming approximate constancy of ΔH and ΔS, the value of ΔG would be

$$31,400 - 32,000 = -600 \text{ cal}$$

ΔG is negative and the reaction as written is spontaneous.

Since heats of reaction are commonly of the order of tens or hundreds of kilocalories, and entropy changes in reactions are commonly of the order of tens of calories per degree, it is seen that at low temperatures the sign of the enthalpy change is the factor that controls the sign of ΔG, and thus the spontaneity of the reaction. An exothermic reaction (ΔH negative) will produce a negative value of ΔG, and the reaction will be spontaneous. It is this fact that makes it reasonable to use the value of ΔH alone as a criterion of stability of compounds and spontaneity of reaction, as has been done several times in the preceding chapters. If ΔH is large and negative (exothermic reaction), ΔG will almost surely be negative at ordinary temperatures, and the reaction will be spontaneous. Similarly, an endothermic reaction (ΔH positive) at low temperatures will produce a positive value of ΔG, and no spontaneous reaction will occur.

The sign and magnitude of ΔS will combine with ΔH and T to determine the actual value of ΔG, through Equation (13). In particular, ΔG may become negative even for an endothermic reaction at high temperatures, if ΔS for the reaction is positive.

The discussion in the last paragraph provides the answer to the question raised in Chapter 18 about the operation of the limekiln [Equation (52),

Chapter 18]. Calcium carbonate is stable at ordinary temperatures but decomposes at the temperature of the kiln. Note that, again, a solid forms a gaseous product,

$$CaCO_3(s) \longrightarrow CaO(s) + CO_2(g) \qquad \Delta H°_{298} = +42.7 \text{ kcal}$$

and there is an entropy increase amounting to 37.4 cal/deg. Even if ΔH did remain the same with change in temperature, therefore, still amounting to 42,700 calories at 1200K, the effect of the entropy increase would be sufficient to make the reaction spontaneous (ΔG negative):

$$\Delta G = 42,700 - 1200 \times 37.4 = -2180 \text{ cal at 1200K} \qquad (16)$$

Gases Formed from Gases. When all reactants are gases, the sign of the free-energy change is often determined by the relative numbers of molecules in reactants and products, as a result of the overwhelming entropy for translational motion. Consider, for example, the dissociation of nitrogen molecules, which, you recall, are very tightly bound together:

$$N_2(g) \longrightarrow 2N(g) \qquad \Delta H = +225.0 \text{ kcal} \qquad (17)$$

From what has been said above, the electronic terms will make little contribution to the entropy effects. Note, however, that there are two particles in the product, and thus two sources of translational entropy, but only one such source in the reactant. The reactant molecules do have, however, vibrational and rotational entropies, absent in the atomic products. When the change $N_2 \longrightarrow 2N$ occurs, molecules are changed to atoms; the entropy terms due to rotation and vibration disappear and there is a change from translational entropy for one particle to translational entropy for two. Because of the closeness of quantum-level spacing for the translational energies, the increase in entropy resulting from the added translational term in the product more than compensates for the loss of the rotational and vibrational terms in the reactant, and the overall result is an increase in entropy (ΔS positive). This result is of general application: An increase in the number of particles produces an increase in entropy, so that dissociation processes in gases proceed with an entropy increase.

Since ΔS is positive for dissociation processes, it means that at some temperature all such processes become spontaneous no matter how much energy the dissociation requires. For at a high enough temperature, $T\Delta S$ must become greater than ΔH, so that the difference $\Delta H - T\Delta S$ becomes negative. Every molecule dissociates at high enough temperature.

FREE ENERGY AND EQUILIBRIUM For a reversible reaction the direction in which the reaction will proceed will always be that which can occur with a decrease in free energy. Consider, for example, the simple gas reaction

$$H_2 + I_2 \rightleftharpoons 2HI \qquad (18)$$

The reaction will proceed from left to right if the free-energy change

$$\Delta G = G_{\text{final}} - G_{\text{initial}} = G_{2HI} - (G_{H_2} + G_{I_2}) \qquad (19)$$

is negative. If the free-energy change for this left-to-right process is positive, then it will be negative for the reverse reaction, by the inequalities

$$G_{2HI} - (G_{H_2} + G_{I_2}) > 0 \qquad (20)$$

$$(G_{H_2} + G_{I_2}) - G_{2HI} < 0 \qquad (21)$$

and the reverse reaction becomes the spontaneous reaction. If the free-energy change is zero—that is, if $G_{products} = G_{reactants}$,

$$G_{2HI} = (G_{H_2} + G_{I_2}) \qquad (22)$$

$$\Delta G = 0 \qquad (23)$$

there is no tendency to go in either direction. This is the situation known as chemical equilibrium, introduced in Chapters 7 and 19 as the equivalence of forward and reverse rates in a reversible reaction. In those discussions, the tendency to react was related to the concentrations of the reactants, and the arrival of the system at an equality of rates signaled the arrival at fixed values of the concentrations. Since the free-energy change has a nonzero value when reaction occurs and a value zero at equilibrium, it is evident that the free-energy change must be related to the concentrations of reactants and products.

Free Energy of the Standard State

Expression of the relation of ΔG to the concentration of reactants and products requires recognition of the fact that we cannot determine absolutely what the free energy of a given initial or final state is, but only the change between the free energy of the initial and final states. The situation here is similar to that for enthalpy and enthalpy changes, which was resolved by adopting a convention about the values of the enthalpy in some chosen standard state. We proceed for free energy in the same way as for enthalpy, and adopt a standard-state convention. We let the free energy of a substance at each temperature have a fixed, although unknown, value of the free energy, denoted by $G°$, when the substance is at unit activity (Chapter 19). It can then be shown that the free energy, G, for the substance in any other state at the known temperature differs from the value in this standard state by the factor $2.303RT \log a$, where $\log a$ is the logarithm of the activity, thus:

$$G = G° + 2.303RT \log a \qquad (24)$$

Here, G and $G°$ represent free-energy values per mole of substance, and R is the molar gas constant, 1.987 cal/mole/deg. The equation is consistent with

---

ACTIVITY AND ITS RELATION TO CONCENTRATION

Warning has been given (page 394) that the setting up of equilibrium quotients using concentrations of products at equilibrium divided by concentrations of reactants at equilibrium does not give a constant value for the calculated equilibrium "constant." It would be useful in thermodynamic calculations if equilibrium "constants" were truly constant. In fact, the advantages to be gained if this were the case would be so great that chemists have invented a kind of effective concentration known as the *activity*, so that when activities instead of concentrations are substituted in the mathematical equations of thermodynamics, all these equations are exact. The advantage gained by this procedure is that the whole of the mathematical apparatus of thermodynamics may now be used with complete confidence, since the deviations from exactness that might have arisen because of the peculiar properties of individual substances have been taken care of at the level of the individual substances themselves, by substituting their activities for their concentrations.

No useful gain will have been made, however, if we do not know how to translate concentrations into activities, or—what amounts to the same thing—how to measure activities as a function of concentrations. Several methods are available for determining this relationship, but most of them share the feature that they require many careful experiments, largely because the activity is a continually changing function of the concentration. Complicating matters further, the activity is not only a changing function of the concentration of a pure substance, as in a gas at varying pressures, or of the concentration of a single solute in solution, but also of the concentration of

the assignment of the value 1 for the activity in the standard state; in the standard state, $\log a = \log 1 = 0$, so $G = G°$. For an ideal gas, the activity is equal to the pressure, and with some approximation we may equate these two quantities for real gases also, so that the standard state for a gas is 1 atm pressure. With considerably less validity, we may, when discussing solution processes, approximate the activity of a substance with its concentration in the solution. Before making either of these approximations, let us calculate the free-energy change in a chemical reaction.

Free-Energy Change in a Reaction

Suppose we have a reaction with the equation

$$M(a_M) + 2N(a_N) \rightleftharpoons P(a_P) \tag{25}$$

in which 1 mole of substance M reacts with 2 moles of substance N to form 1 mole of substance P. The reaction is carried out in solution under conditions such that the activities are a_M, a_N, and a_P, respectively. We calculate the free-energy change in this process:

$$\Delta G = G_{final} - G_{initial} = G_P - G_M - 2G_N \tag{26}$$

The factor 2 appears in front of G_N because 2 moles of N are needed and G_N represents the free energy of 1 mole of N only. Using Equation (24),

$$\Delta G = G_P° + 2.303RT \log a_P - G_M° - 2.303RT \log a_M \\ - 2G_N° - 2(2.303)RT \log a_N \tag{27}$$

Combining terms differently,

$$\Delta G = G_P° - G_M° - 2G_N° + 2.303RT(\log a_P - \log a_M - 2 \log a_N) \tag{28}$$

$$\Delta G = \Delta G° + 2.303RT \log \frac{a_P}{a_M a_N^2} \tag{29}$$

In the last equation, the difference in the standard-state free energies has been expressed in a single term, $\Delta G°$, and the transformation of the terms in the parentheses in the preceding equation comes about by making use of the properties of logarithms.*

---

*$x \log y = \log y^x$; $\log x + \log y = \log xy$; $\log x - \log y = \log \left(\frac{x}{y}\right)$.

---

other solutes that may be in the solutions (gaseous, liquid, or solid solutions), even though these other solutes may have no readily observable chemical effect on the substance whose activity we are trying to measure.

The change in the activity with change in concentration is traced to a change in the interaction between the molecular species, it is thus convenient to establish as a starting point that the activity is equal to the concentration when the molecules do not interact with each other at all. For gases, we already have, in the concept of an ideal gas, such a standard of no interaction, and we may obtain the relation between activity and concentration for a gas in terms of the deviation of its properties from those described by the perfect-gas law (Chapter 10). For solutions, the state of no interaction between solutes can be approached by putting the solute molecules far apart; if we could go to "infinite dilution," there would be an infinite distance between any pair of molecules, and no interaction could occur. While we cannot measure the properties of a solute when its concentration is zero (an infinitely dilute solution), we can make measurements on real solutions of varying concentrations and assess the magnitude of the desired quantity by plotting the measured values against the concentration and extrapolating to zero concentration. This extrapolation gives the standard for no interaction, and comparison of the values for the real solution with those that would be observed if the solutes behaved at all concentrations the way they behave at infinite dilution gives the required relation between activity and concentration.

If $\Delta G°$ and the activities on the right side are such that ΔG on the left side has a negative value, the reaction will proceed from left to right. In fact, the ΔG value calculated will represent the free-energy decrease for a reaction at temperature T when 1 mole of M at activity a_M combines with 2 moles of N at activity a_N to form 1 mole of P at activity a_P, under conditions such that the amounts of substances are so large that this change produces no observable change in these activities.

Equilibrium Constant If ΔG calculated this way were equal to zero, it would mean that there was no tendency for reaction to take place; that is, the system would be at equilibrium, and the activities would be the fixed activities corresponding to the equilibrium conditions. Hence

$$0 = \Delta G° + 2.303RT \log \frac{a_{P,eq}}{a_{M,eq}a_{N,eq}^2} \tag{30}$$

Solving for $\Delta G°$, we have

$$\Delta G° = -2.303RT \log \frac{a_{P,eq}}{a_{M,eq}a_{N,eq}^2} \tag{31}$$

Since $\Delta G°$ is a properly chosen difference between the fixed values of the free energies in the standard states, it has a fixed (that is, constant) value, and the right-hand side of the equation, which is equal to $\Delta G°$, must also have a fixed and constant value. This means that the quotient of the activities at equilibrium is a constant, which we call the equilibrium constant, K:

$$K = \frac{a_{P,eq}}{a_{M,eq}a_{N,eq}^2} \tag{32}$$

Hence,

$$\Delta G° = -2.303RT \log K \tag{33}$$

Equation (33) makes it possible to calculate the equilibrium constant when $\Delta G°$ is known, or to calculate $\Delta G°$ from measurements of equilibrium activities or the equilibrium constant.

For the general reaction between m moles of M and n moles of N to form p moles of P and q moles of Q,

$$mM(a_M) + nN(a_N) \rightleftharpoons pP(a_P) + qQ(a_Q) \tag{34}$$

Equation (29) would have the appearance

$$\Delta G = \Delta G° + 2.303RT \log \frac{a_P{}^p a_Q{}^q}{a_M{}^m a_N{}^n} \tag{35}$$

where $\Delta G°$ has the form

$$\Delta G° = pG_P° + qG_Q° - mG_M° - nG_N° \tag{36}$$

Setting $\Delta G = 0$ gives us the usual form [Equation (4), Chapter 19] of the equilibrium constant for a reaction of this stoichiometry:

$$K = \frac{a_{P,eq}^p a_{Q,eq}^q}{a_{M,eq}^m a_{N,eq}^n} \tag{37}$$

If we make the approximation that the activity is equal to the concentration, the equilibrium-constant expressions, Equations (32) and (37), become

$$K_c = \frac{c_{P,eq}}{c_{M,eq}c_{N,eq}^2} \qquad (38)$$

$$K_c = \frac{c_{P,eq}^p c_{Q,eq}^q}{c_{M,eq}^m c_{N,eq}^n} \qquad (39)$$

These agree with the expressions previously used in Table 19.2. A return to the material on page 429 will show that these are the results which would have been obtained directly if Equation (24) were written

$$G = G° + 2.303RT \log c \qquad (40)$$

This is equivalent to making the approximation that the standard state is that corresponding to unit concentration ($c = 1$). Later (Chapter 21), we shall treat free energies as if that were the case, but it must be remembered that this is an approximation, and often a gross one.

For gaseous substances, the activity can, with good approximation, be set equal to the pressure, so that Equations (32) and (37) become

$$K_p = \frac{P_{P,eq}}{P_{M,eq}P_{N,eq}^2} \qquad (41)$$

$$K_p = \frac{P_{P,eq}^p P_{Q,eq}^8}{P_{M,eq}^m P_{N,eq}^n} \qquad (42)$$

It is important to realize that the standard free-energy change is a constant for each reaction at a specified temperature, and that the equilibrium constant is therefore also a constant. This was found experimentally to be the case in the calculations of Table 19.1, and Equation (31) shows that this is a requirement of the thermodynamic approach. An important feature of the thermodynamic approach is that it distinguishes between the equilibrium concept and the concept of rates of reaction. Here, we are no longer concerned with the specific values for reaction rates, as we were in Table 7.2, but only with the fact that the *free-energy change is zero at equilibrium*. Further, the form of the equilibrium constant quotient is fixed by the over-all stoichiometry of the chemical reaction [Equations (25) and (32); (34) and (37)] and we need not consider the dependence of reaction rates on concentrations or pressures of reactants. This is particularly helpful, because for many reactions this dependence may be very complicated; often it is unknown and, particularly for rapid reactions, difficult to determine.

The value of $\Delta G°$ for a reaction refers to the free-energy change when reactants in their standard states are changed to products in their standard states. While a positive value of $\Delta G°$ means that the reaction is not spontaneous when the several species are in their standard states, it does not mean that the reaction may not be spontaneous under some other conditions. To consider what may happen, let us write Equation (35) for the reaction

$A(g) + B(g) \rightleftharpoons C(g) + D(g)$, using pressure as the measure of activity. We have

$$\Delta G = \Delta G° + 2.303RT \log \frac{P_C P_D}{P_A P_B} \tag{43}$$

Substituting Equation (33) for $\Delta G°$,

$$\Delta G = -2.303RT \log K + 2.303RT \log \frac{P_C P_D}{P_A P_B} \tag{44}$$

In this equation, P_A, P_B, P_C, and P_D represent the pressures of reactants and products under the conditions of the experiment; the quotient

$$\frac{P_C P_D}{P_A P_B}$$

has the same form as the quotient for the equilibrium constant, but the pressures are not equilibrium pressures. For convenience, let us write

$$Q = \frac{P_C P_D}{P_A P_B}$$

ΔG, $\Delta G°$, AND CHEMICAL REACTIONS

Beginning students sometimes say that the value of $\Delta G°$, as calculated for a particular reaction from tables of free-energy values, will tell whether the reaction will take place or nor. More mature students, however, will remember that $\Delta G°$ refers only to the free-energy change under certain specified conditions, namely, when all the products and all the reactants are present in the reaction mixture, and each in its standard state of unit activity. Under all other conditions, the appropriate criterion of spontaneity is ΔG in Equation (29), not $\Delta G°$.

Let us consider a series of chemical situations involving the reaction described by the equation

$$M(a_M) + 2N(a_N) \longrightarrow P(a_P) \tag{A}$$

Case I. All products and reactants are present in the reaction mixture, and $\Delta G°$ and the activities a_M, a_N, and a_P have such values that ΔG is negative. Under these conditions, the reaction will proceed spontaneously to the right as written, so that the activity of P increases while the activities of M and N decrease. The reaction will continue until a_M, a_N, and a_P reach their equilibrium values $a_{M,eq}$, $a_{N,eq}$, and $a_{P,eq}$. After this, no further change in the activities of products and reactants will occur.

Case II. All reactants and products are in the reaction mixture, and the values of $\Delta G°$ and the activities a_M, a_N, and a_P are such that ΔG is positive. Under these conditions, the reaction will not proceed spontaneously as written; that is, it will not proceed from left to right, with an increase in the activity of the product. On the contrary, the reaction will proceed in the reverse direction; P will be transformed into M and N, moving toward equilibrium. The final equilibrium activity of the product [in Equation (A) as written] will be less and the equilibrium activities of the reactants will be greater than the starting activities.

Case III. Only the reactants M and N are present in the reaction mixture. In this case, since no product is present in the reacting mixture, $a_P = 0$, the description of the reaction is

$$M(a_M) + 2N(a_N) \longrightarrow P(a_P = 0) \tag{B}$$

Substitution in the free-energy Equation (29) gives

$$\Delta G = \Delta G° + 2.303RT \log \frac{0}{a_M a_N{}^2} \tag{C}$$

The fraction $\dfrac{0}{a_M a_N{}^2}$ is equal to zero, and the logarithm of zero is minus infinity. Hence, ΔG is

and, using the properties of logarithms, rewrite Equation (44):

$$\Delta G = 2.303RT \log \frac{Q}{K} \qquad (45)$$

Whenever Q/K is less than 1—that is, when the experimental quotient is less than the equilibrium quotient—the logarithm, and hence ΔG, will be negative, and the reaction spontaneous. Ordinarily, under these conditions, the reaction will proceed until the pressures P_A, P_B, P_C, and P_D have adjusted themselves so that the quotient Q equals the equilibrium constant K. Q/K then equals 1, the logarithm is zero, ΔG is zero, and equilibrium has been established. This sequence of events can occur no matter how small K may be; if, initially, the quotient

$$\frac{P_C P_D}{P_A P_B}$$

is less than

$$\frac{P_{C,eq} P_{D,eq}}{P_{A,eq} P_{B,eq}}$$

a certain amount of spontaneous reaction will occur, until equilibrium is established.

---

always negative, no matter whether $\Delta G°$ is positive, negative, large, or small, since $\Delta G = \Delta G° - \infty$. A negative value of ΔG means that the reaction will proceed spontaneously as written until equilibrium is established. Hence, the activity of M and N will decrease while that of P increases, until equilibrium values are obtained and ΔG is zero:

$$0 = \Delta G° + 2.303RT \log \frac{a_{P,eq}}{a_{M,eq} a_{N,eq}^2} \qquad (D)$$

Case IV. All reactants and products are present and their activities are each equal to 1. In this case, $\Delta G = \Delta G°$, since the activity ratio is equal to 1 and the logarithm of 1 is zero. In this case, and in this case only, the direction of the spontaneous reaction is determined by the sign of $\Delta G°$; spontaneous to the right if $\Delta G°$ is negative, spontaneous to the left if $\Delta G°$ is positive. With either sign, reaction will proceed until the equilibrium values of the activities are reached.

Equilibrium-Constant Magnitudes. Although Case IV shows that $\Delta G°$ is of limited usefulness in determining the spontaneous direction of chemical reactions, being applicable only to the special case where all reactants and products are at unit activities, the value of $\Delta G°$ is of great importance as a measure of the magnitude of the equilibrium constant. This may be shown by considering the two equations

$$\Delta G° = -2.303RT \log K \qquad (E)$$

$$\log K = -\frac{\Delta G°}{2.303RT} \qquad (F)$$

Case V. $\Delta G°$ is positive. Here, $\log K$ is negative, corresponding to a value of K less than 1. This means that the experimental result obtained when equilibrium is established is that the position of equilibrium in Equation (A) is to the left as written. Therefore, the reaction of Equation (A) would not be a good choice for converting M and N into P with large yield unless the special procedures for driving the reaction to completion (p. 201) were applied.

Case VI. $\Delta G°$ is negative. Here, $\log K$ is positive, corresponding to a value of K greater than 1. This means that the position of equilibrium in Equation (A) is to the right as written. Therefore, this reaction would be useful for converting M and N into P.

Thus the sign and magnitude of $\Delta G°$ is a useful indicator of the degree of completion of a reaction, but not of its spontaneity in the general case. For the latter purpose, ΔG must be used.

Completion of Reaction	If K is very large—that is, if $\Delta G°$ has a large negative value—the concentration of reaction products at equilibrium will be very large compared to the concentration of reactants at equilibrium, and the reaction has evidently "gone to completion" except for the negligible amounts of reactant left at equilibrium. Even if K were small, however, the reaction could be "driven to completion" by arranging the experimental conditions so that Q always remains less than K. This could be accomplished, for example, by continual removal from the scene of the reaction of one of the reaction products. If the presence of substance D, say, were kept always very small, as by condensing the gas or absorbing it in some inert solvent, it is evident that Q must always remain small, Q/K always less than 1, and the reaction always spontaneous.
Equilibria in Reactions in Which Solids Are Present	In choosing the standard states for free-energy calculations, the standard state of a solid is chosen as the pure solid at the temperature in question, and the activity of the solid is set equal to 1 in accordance with this choice. The standard state of liquids also is chosen as the pure liquid.
Free Energy of Formation; Addition of Free-Energy Values	Just as chemists have found it convenient to speak of an enthalpy of formation, so the concept of free energy of formation is useful. By the free energy of formation, $\Delta G_f°$, is meant the *free-energy change when one mole of a compound is formed in its standard state from its elements in their standard states*; it is usually tabulated for 25°C. Thus the values for $H_2O(g)$ and for $AgCl(s)$ in Table 20.1 represent the free-energy changes in the reactions

$$H_2(\text{gas, 1 atm}) + \tfrac{1}{2}O_2(\text{gas, 1 atm}) \longrightarrow H_2O(\text{gas, 1 atm})$$

$$\Delta G_{f298}° = -54.64 \text{ kcal} \quad (46)$$

$$Ag(s) + \tfrac{1}{2}Cl_2(\text{gas, 1 atm}) \longrightarrow AgCl(s) \qquad \Delta G_{f298}° = -26.22 \text{ kcal} \quad (47)$$

These values may be added and subtracted in the same way as values of enthalpies of formation are handled, as shown in the following sum:

$$CaCO_3(s) \longrightarrow Ca(s) + C(s) + \tfrac{3}{2}O_2(g, \text{ 1 atm}) \quad \Delta G_{f298}° = +269.8 \quad (48)$$

$$Ca(s) + \tfrac{1}{2}O_2(g, \text{ 1 atm}) \longrightarrow CaO(s) \qquad \Delta G_{f298}° = -144.4 \quad (49)$$

$$C(s) + O_2(g, \text{ 1 atm}) \longrightarrow CO_2(g, \text{ 1 atm}) \qquad \Delta G_{f298}° = -94.3 \quad (50)$$

$$CaCO_3(s) \longrightarrow CaO(s) + CO_2(g, \text{ 1 atm}) \qquad \Delta G_{298}° = +31.1 \text{ kcal} \quad (51)$$

Effect of Change in Temperature on the Equilibrium Constant and Equilibrium Concentrations	When Equation (13) connecting enthalpy change, entropy change, and free-energy change is written for the standard states

$$\Delta G° = \Delta H° - T\Delta S° \qquad (52)$$

and the relationship between the standard free-energy change and the equilibrium constant [Equation (33)] substituted, we obtain

$$-2.303RT \log K = \Delta H° - T\Delta S° \qquad (53)$$

Division by $2.303RT$ and change of sign gives

$$\log K = -\frac{\Delta H°}{2.303RT} + \frac{\Delta S°}{2.303R} \qquad (54)$$

Table 20.1 Standard Free Energies of Formation (ΔG°_{f298} in kcal/mole)

Substance	ΔG°_{f298}	Substance	ΔG°_{f298}
$H_2O(g)$	-54.6357	$NaCl(s)$	-91.785
$H_2O(l)$	56.6902	$KCl(s)$	-97.592
$HCl(g)$	-22.769	$CaCl_2(s)$	-179.3
$HBr(g)$	-12.72	$CaCO_3(s)$ (calcite)	-269.78
$HI(g)$	0.31	$AlCl_3(s)$	-152.2
$SO_2(g)$	-71.79	$CaO(s)$	-144.4
$NO(g)$	20.719	$CuO(s)$	-30.4
$NH_3(g)$	-3.976	$Cu_2O(s)$	-34.98
$Br_2(g)$	0.751	$CuSO_4(s)$	-158.2
$Br_2(l)$	0.00	$CuSO_4 \cdot 5H_2O(s)$	-449.3
$CO(g)$	-32.8079	$PbCl_2(s)$	-75.04
$CO_2(g)$	-94.2598	$AgCl(s)$	-26.22
$CH_4(g)$	-12.140	$C_2H_4(g)$	16.282
$CH_3Cl(g)$	-14.0	$C_2H_2(g)$	50.000
$CH_3OH(l)$	-39.73	$C_3H_8(g)$	-5.614
$CHCl_3(l)$	-17.1	$n\text{-}C_4H_{10}(g)$	-4.10
$CCl_4(l)$	-16.4	$n\text{-}C_5H_{12}(g)$	-2.00
$C_2H_6(g)$	-7.860	$iso\text{-}C_5H_{12}(g)$	-3.50
$C_2H_5OH(l)$	-41.77	$neo\text{-}C_5H_{12}(g)$	-3.64
$CH_3CO_2H(l)$	-93.8	$C_6H_6(l)$	29.756
$H^+(aq)$	0.00	$OH^-(aq)$	-37.595
$Na^+(aq)$	-62.589	$Cl^-(aq)$	-31.350
$K^+(aq)$	-67.466	$Br^-(aq)$	-24.574
$Ag^+(aq)$	18.43	$I^-(aq)$	-12.35
$Ca^{+2}(aq)$	-132.18	$S^{-2}(aq)$	20
$Cu^{+2}(aq)$	15.53	$CO_3^{-2}(aq)$	-126.22
$Zn^{+2}(aq)$	-35.184	$SO_4^{-2}(aq)$	-177.34
$H(g)$	48.575	$C(g)$	160.845
$O(g)$	54.994	$N(g)$	81.471

Since ΔH° and ΔS° are usually nearly constant with temperature, Equation (54) shows that a plot of $\log K$ against $1/T$ should be a straight line. The slope of the line will be positive or negative according to whether ΔH° is negative (exothermic reaction) or positive (endothermic reaction). For the formation of hydrogen iodide, ΔH° is positive,

$$H_2(g) + I_2(s) \longrightarrow 2HI(g) \qquad \Delta H^\circ = 12.4 \text{ kcal} \qquad (55)$$

and $\log K$ increases (and K increases) as T increases ($1/T$ decreases). Thus, for a given mixture, the fraction of hydrogen and iodine converted to hydrogen iodide at equilibrium is greater the higher the temperature. For the formation of ammonia, ΔH° is negative,

$$N_2(g) + 3H_2(g) \longrightarrow 2NH_3(g) \qquad \Delta H^\circ = -22.08 \text{ kcal} \qquad (56)$$

and $\log K$ decreases (K decreases) as T increases ($1/T$ decreases). The effect of the decrease in K is to decrease the percentage of ammonia in the equilibrium mixture as the temperature increases, as shown in each column of Table 19.3.

These conclusions are in accord with the qualitative predictions from Le Châtelier's principle already discussed (Chapter 19).

Free Energy and the Oxidation of Glucose. An important application of the free-energy concept is its use to improve our understanding of the energy transfer in the chemical processes of living cells. As discussed in some detail in Chapter 29, the chemical system ATP-ADP (adenosine triphosphate and adenosine diphosphate) functions as a carrier of chemical energy. Thus ADP is able to store energy by accepting a phosphate group in certain energy-producing reactions during the oxidation of foodstuffs, and ATP is able to supply energy by donating its terminal phosphate group in reactions that require energy.

Under the conditions which biological reactions occur—constant temperature and pressure—every chemical reaction has a characteristic standard free-energy change. For biological systems, this is given the symbol $\Delta G^{\circ\prime}$, the standard state being all products and reactants at $1.0M$ concentration and at a pH of 7.0; $\Delta G^{\circ\prime}$ can be calculated from the equilibrium constant for the reaction using an equation analogous to Equation (33) in this chapter. Since ΔG values for various reactions can be added or subtracted, $\Delta G^{\circ\prime}$ values for individual biological reactions can be calculated from equilibrium data of a consecutive series of reactions or from the difference in the standard free energy of formation of reactants and products.

Among the biologically important processes, $\Delta G^{\circ\prime}$ has the largest negative values for oxidation and hydrolysis reactions. These reactions are, of course, the major sources of energy in the cell. For example, in the hydrolysis of ATP to ADP,

$$\text{ATP} + \text{H}_2\text{O} \rightleftharpoons \text{ADP} + \text{phosphate}$$

$\Delta G^{\circ\prime}$ is -7.30 kcal/mole at pH 7.0 and 37°C in the presence of excess Mg^{+2}. For comparison, $\Delta G^{\circ\prime}$ values for other important biological processes are:

Reaction	$\Delta G^{0\prime}_{25^\circ}$(kcal/mole)
Sucrose + H_2O $\rightleftharpoons$ glucose + fructose	-7.0
Glucose + $6O_2$ $\rightleftharpoons$ $6CO_2$ + $6H_2O$	-686
(17 steps required)	
Glucose + phosphate $\rightleftharpoons$ glucose-6-phosphate	$+3.3$
Glucose-6-phosphate $\rightleftharpoons$ Glucose-1-phosphate	$+1.7$

As with other chemical reactions, the $\Delta G^{\circ\prime}$ value for the ATP-ADP hydrolysis varies with temperature, pH, and concentrations of Mg^{+2} and other ions. Under intracellular conditions, it is about -12.5 kcal/mole.

In the sequence of steps by which glucose is oxidized to carbon dioxide and water, at least nine phosphorylated compounds are produced (see Figure 29.4). All of these compounds undergo hydrolysis reactions as part of the over-all oxidation of glucose:

$$\text{Phosphorylated compound} + \text{H}_2\text{O} \rightleftharpoons \text{product} + \text{phosphate}.$$

The $\Delta G^{\circ\prime}$ values for the hydrolysis of some of these compounds, such as those for glucose-6-phosphate and glucose-1-phosphate, are smaller negative numbers than for ATP, while the $\Delta G^{\circ\prime}$ values for the hydrolysis of phosphoenolpyruvate and 1,3-diphosphoglycerate (compounds formed after the glucose molecule is split into fragments) are larger negative numbers than for ATP.

In effect, the $\Delta G^{\circ\prime}$ values give us a measure of the tendency of a phosphorylated molecule to lose or to transfer phosphate. Those compounds with larger negative $\Delta G^{\circ\prime}$ values than ATP transfer phosphate more readily than ATP, and vice versa.

The significance of this knowledge in the over-all oxidation of glucose and, therefore, in the primary energy-producing process of living cells, is that the more reactive phosphorylated compounds must, in effect, transfer their phosphate to ADP, thereby enabling some of the energy liberated in oxidation to be captured and stored. Similarly, the ATP formed by this phosphate transfer process must transfer its phosphate to less reactive compounds such as glucose or glucose-6-phosphate. This transfer results in increasing the energy of the glucose or the glucose-6-phosphate, thereby activating these compounds for oxidation and starting them on their way in the oxidation process.

SUMMARY The concept of free energy, its usefulness in predicting the direction of spontaneous change and its application in determining the position of equilibrium in chemical reactions have been discussed in this chapter. Changes in matter are shown to occur spontaneously when the entropy, the measure of randomness, increases as the change occurs. However, this criterion of spontaneity has limited applicability because both the system and its surroundings must be considered. The free-energy change is developed as a criterion of spontaneity when referring to the system alone. The free-energy change is seen also as a measure of the maximum useful work obtainable from the system. The relationship of free-energy change to changes in enthalpy and entropy makes it possible to evaluate the free-energy change for many systems from measurements of ΔH and ΔS.

The relationship of the free-energy change to the effective concentrations (activities) of reactants and products leads to a thermodynamic understanding of the equilibrium constant and the position of equilibrium. The principle of additivity of free-energy changes is developed and used to calculate free-energy changes and equilibrium constants for processes for which these quantities have not been measured.

IMPORTANT TERMS

Thermodynamics

Entropy
spatial randomness
energy randomness

Enthalpy

System
surroundings

Reversible (thermodynamic) process

Spontaneous reaction
exothermic reaction
endothermic reaction

Activity
activity of standard state

Free energy
driving force
maximum work
maximum useful work
free-energy change
standard free-energy change
standard-state free energy
free energy of formation

Equilibrium
equilibrium constant

Le Châtelier's principle
completion of reaction

QUESTIONS AND PROBLEMS

1. Explain what causes a reaction to proceed in a system in which no energy is gained or lost and there is no energy change in the system.

2. Show how energy can be transferred into a system as heat; as work. Is such a system an isolated system? A closed system?

3. Explain how work done in a reversible manner gives the maximum work that can be done. Must this be an isolated system? A closed system?

4. Under what conditions can we write $q = T\Delta S$?

5. Under what conditions can we use the change in ΔS as an indication of whether a reaction will go? Under what conditions can the value of ΔG be used?

6. What is the relationship between ΔE and ΔH? Between ΔH and ΔG?

7. How does the value of $T\Delta S$ change with temperature? With this information, show that every compound should decompose if the temperature is high enough.

8. A substance undergoes a change in physical state. How can you measure the corresponding free-energy change?

9. Explain in terms of order-to-disorder the tendency of ice to melt to form liquid water. Can you use a similar explanation to explain the tendency of water to evaporate to form water vapor?

10. How do you explain the statement "when you boil water you are contributing to the 'running downness' of the universe"?

11. Consider the dissolution of two springs in acid. One is a coiled spring of iron under tension; the other is an uncoiled one not under tension. Explain in terms of ΔG why you could predict a difference in the tendency of the dissolution reaction to occur.

12. Given the following data:

$$H_2(g) + \tfrac{1}{2}O_2(g) \rightleftharpoons H_2O(l) \qquad\qquad \Delta G° = -56,690 \text{ cal}$$

$$H_2O(g) + Cl_2(g) \rightleftharpoons 2HCl(g) + \tfrac{1}{2}O_2(g) \qquad \Delta G° = 9100 \text{ cal}$$

$$H_2O(l) \rightleftharpoons H_2O(g) \qquad\qquad \Delta G° = 2053 \text{ cal}$$

calculate the standard free energy of formation of hydrogen chloride from its elements at 25°C.

13. The standard free-energy change for the reaction

$$2HI(g) \rightleftharpoons H_2(g) + I_2(s)$$

is −630 cal at 25°C. Calculate the value of the equilibrium constant.

14. At 25°C the value of K_{eq} equals 5.65×10^7 for the reaction

$$H_2(g) + C_2H_4(g) \rightleftharpoons C_2H_6(g)$$

Calculate the value of $\Delta G°$.

15. The heat of vaporization of benzene (C_6H_6) at 80.2°C and 1 atm is 94.4 cal/g. Calculate the value of ΔH, ΔE, and ΔS for the process at that temperature. Note that the vaporization process under the conditions specified is always at equilibrium, since the vapor pressure of benzene at 80.2° is 1 atm.

16. The heat of vaporization of water at its boiling point at atmospheric pressure is 9.72 kcal/mole. Calculate ΔS, ΔH, ΔE, and ΔG for the vaporization process.

$$H_2O(l) \longrightarrow H_2O(g); \; 100°C, 1 \text{ atm}$$

17. Using the data of Table 20.1, calculate the standard free energy change for the process

$$CO_2(g) + H_2(g) \longrightarrow CO(g) + H_2O(g)$$

at 25°C, and the equilibrium constant.

18. Calculate $\Delta G°$, $\Delta H°$, $\Delta S°$, and K at 25°C for the reaction

$$Fe_3O_4(s) + 4H_2(g) \longrightarrow 3Fe(s) + 4H_2O(g)$$

The standard enthalpy of formation of $Fe_3O_4(s)$ at 25°C is -267.0 kcal/mole and its standard free energy of formation at 25°C is -242.4 kcal/mole.

19. (a) Calculate the value of $\Delta G°_{298}$ for the reaction

$$CaCO_3(s) \longrightarrow CaO(s) + CO_2(g)$$

and the pressure of carbon dioxide over a mixture of calcium carbonate and calcium oxide at 25°C.
(b) Calculate ΔG for the reaction at 25°C

$$CaCO_3(s) \longrightarrow CaO(s) + CO_2(g, 1 \times 10^{-8}\ atm)$$

20. Calculate $\Delta G°$, $\Delta H°$, and $\Delta S°$ and the equilibrium constant at 25°C for the reaction

$$\tfrac{1}{2}H_2(g) + \tfrac{1}{2}Br_2(g) \longrightarrow HBr(g)$$

21. (a) Calculate $\Delta G°$, $\Delta H°$, and $\Delta S°$ at 25°C for the reaction

$$\tfrac{1}{2}H_2(g) + \tfrac{1}{2}I_2(g) \longrightarrow HI(g)$$

$\Delta G°_{f298}$ for $I_2(g)$ is 4.63 kcal/mole and $\Delta H°_{f298}$ is 14.88 kcal/mole.
(b) Assuming that $\Delta H°$ and $\Delta S°$ are independent of temperature, calculate $\Delta G°$ and K for the reaction at 1000°C. (c) If initially pure HI is heated in a container to 1000°C, what percentage of it will have decomposed at equilibrium?

22. Calculate $\Delta G°$ and K for the following reaction at 25°C:

$$2C_2H_6(g) + 7O_2(g) \longrightarrow 4CO_2(g) + 6H_2O(l)$$

23. Using data at 1 atm from Table 19.3, calculate the equilibrium constants for the reaction

$$N_2(g) + 3H_2(g) \rightleftharpoons 2NH_3(g)$$

and $\Delta H°$ for that reaction

24. Calculate ΔG for the following processes at 298K
    (a) $H_2(g, 1\ atm) + Cl_2(g, 1\ atm) \longrightarrow 2HCl(g, 1\ atm)$
    (b) $H_2(g, 0.5\ atm) + Cl_2(g, 0.5\ atm) \longrightarrow 2HCl(g, 0.5\ atm)$
    (c) $H_2(g, 0.1\ atm) + Cl_2(g, 0.2\ atm) \longrightarrow 2HCl(g, 0.4\ atm)$
    (d) Which of these are spontaneous processes at 298K?

25. (a) Would you predict that the entropy would increase or decrease in the following processes? Give reasons for your answers.
    (1) $H_2O(s) \longrightarrow H_2O(l)$
    (2) $C(s) + 2H_2(g) \longrightarrow CH_4(g)$
    (3) $2CO_2(g) \longrightarrow 2CO(g) + O_2(g)$

(4) $N_2(g, 1\text{ atm}) \longrightarrow N_2(g, 2\text{ atm})$

(5) $CaCO_3(s) + 2H^+(aq) \longrightarrow Ca^{+2}(aq) + CO_2(g) + H_2O(l)$

(6) $NaCl(s) \longrightarrow Na^+(aq) + Cl^-(aq)$

26. The standard free energy of formation of liquid benzene is positive, yet benzene is a common laboratory solvent. How do you account for this?

27. A refrigerator at a fixed temperature of 5°C is exposed to the air of a room at 25°C. During the course of five minutes, 1000 calories of heat leaked into the refrigerator. Calculate the entropy gained by the refrigerator and that lost by the room. Did the overall entropy of the system plus the surroundings increase or decrease? Was the process

1000 cal in the room at 25° $\longrightarrow$ 1000 cal in the refrigerator at 5°

reversible or irreversible?

SPECIAL PROBLEMS

1. In discussions of atomic structure, a statement is often made that atoms react to form compounds because of the tendency to gain, lose, or share pairs of electrons. How do you rationalize that statement with those made in this chapter: "reactions go in the direction of greater disorder"; or "an indication of whether a reaction will go is the sign of ΔG"?

2. (a) Calculate $\Delta G°$, $\Delta H°$, and $\Delta S°$ for the reaction

$$4HCl(g) + O_2(g) \longrightarrow 2H_2O(g) + 2Cl_2(g)$$

at 25°C, using data from Chapters 18 and 20.

(b) Assuming that $\Delta H°$ and $\Delta S°$ are independent of temperature, calculate the equilibrium constant for this reaction at 600°C.

3. From the following data on the values of $\Delta H°$ and $\Delta S°$, calculate the free energy and equilibrium constant at 600°C for the reaction

$$C(s) + 2H_2(g) \rightleftharpoons CH_4(g)$$

The enthalpy change for this reaction is $-21,000$ cal at 600°C. The entropies of graphite, hydrogen, and methane at 600°C and 1 atm are 4.8, 38.9, and 56.6 cal/mole/deg, respectively.

4. Calculate $\Delta G°$ for the reaction at 25°C:

$$Ag^+(aq) + Cl^-(aq) \longrightarrow AgCl(aq)$$

and from this value, the solubility-product constant for silver chloride at that temperature.

5. Mercuric oxide dissociates at high temperatures according to the equation

$$2HgO(s) \longrightarrow 2Hg(g) + O_2(g)$$

At 450°C, the pressure of the two gases formed is 810 torr, but at 420°C, the dissociation pressure is only 387 torr. (a) Calculate the partial pressures of mercury and of oxygen and, from these, the value of the equilibrium constant, in pressure units, at each temperature. (b) If 10.0 g of HgO are placed in a 1.0 liter evacuated vessel and the temperature raised to 450°C, how many grams of mercuric oxide will remain undecomposed?

6. From the information given in Problem 5 above, estimate the value of $\Delta H°$ for the decomposition of mercuric oxide.

7. Air (assumed to be 1/5 oxygen and 4/5 nitrogen) at 10 atm and nitric

oxide at 1 atm are placed in a container at 25°C. What is the free-energy change in the reaction

$$NO(g, 1 \text{ atm}) \longrightarrow \tfrac{1}{2}N_2(g, \text{ in air at 10 atm}) + \tfrac{1}{2}O_2(g, \text{ in air at 10 atm})$$

8. Determine whether it would be thermodynamically feasible to use acetylene gas, C_2H_2, as the starting material for the preparation of liquid benzene, C_6H_6. If this synthesis were to be attempted, would it be better, from a thermodynamics standpoint, to work at a high temperature or at a low temperature?

9. (a) Calculate the enthalpy change in the solidification of 1 g of supercooled water at −10°C by considering the following steps: (1) heating the supercooled water to 0°, (2) freezing the water at 0°, (3) cooling the ice to −10°. The specific heat of water is 1; the specific heat of ice is 0.5; the heat of fusion of ice at 0° is 80 calories/gram.
(b) The entropy change in the solidification of supercooled water at −10°C can be calculated by considering the same steps as in (a), and comes to −0.275 cal/deg gram. How do you interpret the fact that this change is negative? Compare this value with the entropy change of a heat reservoir at −10° surrounding the container in which 1 g of water supercooled to −10° freezes, and calculate the overall entropy change of system plus surroundings. Is the solidification process spontaneous?
(c) Calculate the free-energy change for the solidification at −10° of water supercooled to −10°. Is the solidification process spontaneous?

10. A common laboratory procedure for the preparation of oxygen is to heat solid potassium chlorate:

$$2KClO_3(s) \longrightarrow 2KCl(s) + 3O_2(g)$$

(a) Calculate the enthalpy change in this reaction, given the data of Table 18.1, and the enthalpy of formation of solid potassium chlorate, −93.50 kcal/mole at 298K. Is heat absorbed or evolved in this reaction? According to Le Châtelier's principle, is heating potassium chlorate to make it decompose a reasonable procedure? (b) Calculate the equilibrium constant for the above reaction at 298K and at 600K using the following standard entropy values $S°$, expressed in calories/degree/mole at 298K

$KClO_3(s)$	34.17
$KCl(s)$	19.76
$O_2(g)$	49.00

and assuming $\Delta H°$ and $\Delta S°$ to be independent of temperature. (c) Why does the laboratory procedure suggest heating potassium chlorate for the preparation of oxygen?

REFERENCES   MACWOOD, G. E., and VERHOEK, F. H., "How Can You Tell Whether a Reaction Will Occur?" *J. Chem. Educ.,* **38,** 334, (1961).
MAHAN, B. H., *Elementary Chemical Thermodynamics,* Benjamin, Menlo Park, Calif., 1964.
STRONG, L. E., and STRATTON, W. A., *Chemical Energy,* Van Nostrand Reinhold, New York, 1965.
RAMAN, V. V., "Evolution of the Second Law of Thermodynamics," *J. Chem. Educ.,* **47,** 331 (1970).
NASH, L. K., "Chemical Equilibrium as a State of Maximal Entropy," *J. Chem. Educ.,* **47,** 353, (1970).

ELECTROCHEMICAL CELLS—THE MEASUREMENT OF FREE ENERGY

Die Vorgänge in einem constanten galvanische Elemente, welche bei verschwindend kleiner Stromintensität vor sich gehen, wobei man die dem Widerstand und dem Quadrat dieser Intensität proportionale Wärmeentwicklung in Schliessungsdrahte als verschwindende Grössen zweiter Ordnung vernachlässigen kann, sind vollkommen reversible Prozesse und müssen den thermodynamischen Gesetzen der reversiblen Prozesse unterliegen.*

Hermann Ludwig Ferdinand von Helmholtz (1821–1894)

An electrochemical cell is a device in which a chemical reaction is caused to take place as the result of the passage of an electric current (electrolysis cell), or in which a chemical reaction takes place and causes an electric current to be produced (galvanic cell). In this chapter we shall discuss both types of cell and refer particularly to those measurements with galvanic cells that provide the data on oxidation potentials. Recall that we have used such data in previous chapters to arrange elements and compounds in the order of their oxidizing or reducing strengths in aqueous solution. The voltages of galvanic cells are direct measures of the free-energy changes in the chemical reactions in the cells. We shall also study variations of cell voltages with concentrations of reactants and products, the use of cell voltage in predicting equilibrium concentrations, and applications of electrochemical cells in chemical analysis.

Let us first examine the nature of electrical conduction in solution, and the reactions that occur in some cells.

Conduction of Electricity in Solution

A solution of an electrovalent compound such as sodium chloride contains sodium ions, positively charged, and chloride ions, negatively charged. When two electrodes are placed in the solution and each is connected to one of the poles of a battery, the positive sodium ions will be drawn toward the negative electrode, and the negative chloride ions will be drawn toward the positive electrode (Figure 21.1). This migration of charged particles through the solution results in a transfer of electric charge; we recognize this as an electric current and say that the solution conducts a current. This type of conduction, which results from the motion of ions through the solution, is called *electrolytic conduction*, distinguishing it from *metallic conduction*, which occurs when electrons flow along a wire.

---

*"The reactions in a given galvanic cell which take place at vanishingly small currents, where one can neglect, as a second-order correction, the heat evolved (proportional to the resistance and the square of this small current), are completely reversible processes, and must obey the thermodynamic laws of reversible processes." (The quotation is from *Monatsberichte der königlichen preussischen Akademie der Wissenschaften* (1877), p. 713.)

Figure 21.1 Migration of ions in electrolytic conduction.

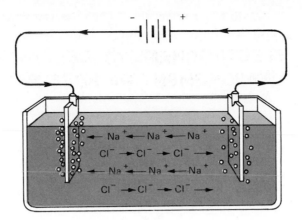

The changeover from metallic conduction along the wires from the battery to electrolytic conduction in the solution occurs at the surface of the electrodes, and always results in chemical reaction, most often forming atoms from ions or ions from atoms or molecules. The exact nature of the electrode reactions depends upon the nature of the dissolved salt and its concentration, the nature of the electrodes, the relative ease of reaction of solute species and solvent at the electrode surfaces, the temperature, and so on. The reactions are those of oxidation and reduction and may be written as *half-reactions*, in which electrons are treated as if they were reactants or products. In the case of sodium chloride solution, the half-reactions are

$$2Cl^-(aq) \longrightarrow Cl_2(g) + 2e^- \quad \text{oxidation, at the anode} \tag{1}$$

$$2H_2O(l) + 2e^- \longrightarrow H_2(g) + 2OH \ (aq) \quad \text{reduction, at the cathode} \tag{2}$$

The first reaction gives electrons to one of the electrodes; the second takes electrons away from the other. The electric circuit is completed by the battery, which serves much as an electron pump, taking electrons away from the first electrode and pumping them along to the other, where they are released (Figure 21.2).

Of course, electrons are not observable products and reactants, as Equations (1) and (2) seem to show; electrons are simply transferred from chloride

Figure 21.2 Electrode reactions in electrolysis of sodium chloride solutions.

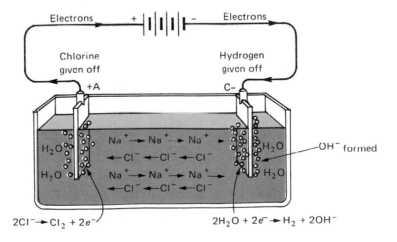

ions to water molecules. The observed reaction, therefore, is the sum of the two half-reactions above:

$$2H_2O(l) + 2Cl^-(aq) \longrightarrow H_2(g) + Cl_2(g) + 2OH^-(aq) \tag{3}$$

The arrangement of solution and electrodes does, however, permit the *oxidation* to take place at one electrode (called the anode) and the reduction to take place at the other (called the cathode). Thus, the chlorine and hydrogen gases appear at different places, and may be collected separately if desired. The sodium ions of the sodium chloride remain unchanged in quantity and may be removed from the solution as sodium hydroxide. The industrial preparation of sodium hydroxide and chlorine by this electrolysis process has already been mentioned (Figure 8.7).

Electrolysis Cell The arrangement of electrodes dipping into a solution of an electrolyte is known as a cell. In the use just described, it is an *electrolysis cell*, in which an oxidation-reduction reaction is caused to take place by passing a current from an external source through the cell. Such cells are widely used in industry in winning certain metals from their ores, as in the Hall process for aluminum, in refining crude metals, as in the refining of copper; in the preparation of elements and compounds such as chlorine and sodium hydroxide, just mentioned; in electroplating, usually for combating corrosion, as in the protection of automobile bright-work with chromium plate; and in the preparation of certain organic compounds.

Faraday's Laws The laws governing the amount of material undergoing reaction in an electrolysis were discovered by Faraday in his early research on electrolytic conduction. Faraday observed that the weight of hydrogen or oxygen or silver or other elements liberated at an electrode depends *solely* on the *quantity of electricity* that flows through the solution undergoing electrolysis. It does not depend at all on the temperature or the concentration of the solution, or the rate at which the charge flows, but only on the quantity of electricity. This fact constitutes Faraday's first law: *The amount of electrochemical change is directly proportional to the quantity of electricity that flows* (Figure 21.3). The quantity of electricity, measured in coulombs, is the product of the current in amperes and the time in seconds.

Faraday's second law has to do with the relation between the relative weights of the *different elements* liberated at the electrodes by the same quantity of electricity, and the gram-equivalent weights of those elements. The *gram-equivalent weight* of an element is its mole weight divided by its valence. Faraday's experiments proved that if *a given quantity of electricity is allowed to flow successively through a number of different solutions in series, the weights of the various elements liberated at the electrodes are in the ratio of their gram-equivalent weights* (Figure 21.4).

The Avogadro Number The explanation of Faraday's laws can now be made in the light of our knowledge that an electric current is a flow of electrons. Since the electrode reactions involve gain or loss of electrons, it is evident that the amount of material deposited depends only upon the number of electrons that pass around the circuit—that is, upon the quantity of electricity that passes around the circuit—and not upon the temperature or other conditions of the electrolysis. Further, since the charge on an ion determines the number of electrons that need to be gained or lost to neutralize it, the number of ions of different elements, such as Ag^+, Ni^{+2}, and Cr^{+3}, that can be neutralized by a given number

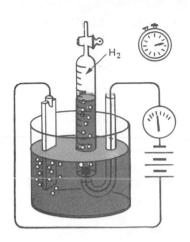

Figure 21.3 The quantity of hydrogen produced in the electrolysis of sodium chloride solution is proportional to the quantity of electricity which flows through the solution—that is, to the product of the current and the time.

of electrons will vary inversely as the number of charges on the ions. Thus for the passage of y electrons, there will be neutralized y atoms of silver, $y/2$ atoms of nickel, and $y/3$ atoms of chromium. The weights of the elements deposited (Figure 21.4) at the electrode will equal the number of atoms deposited times the weight of the atom. Thus, if there are N atoms in a gram-atomic weight, there will be deposited $y \times 108/N$ g of silver, $y/2 \times 59/N$ g of nickel, and $y/3 \times 52/N$ g of chromium. The weights of the elements deposited will therefore be in the ratio of their equivalent weights (atomic weights divided by electrovalence); the actual weights are: $y/N \times 108$ g Ag, $y/N \times 59/2$ g Ni, $y/N \times 52/3$ g Cr, and their ratio is

$$108 : \frac{59}{2} : \frac{52}{3}$$

An electrolysis, in other words, is essentially a counting process. The number of electrons that pass through the circuit is measured by the quantity of material deposited. This fact makes it possible to calculate the Avogadro number from electrolysis experiments. It is found experimentally that a quantity of electricity amounting to 96,487 coul is required to liberate 1 g-equiv. wt. of an element. This quantity (called a faraday) is the same for any element, by Faraday's second law. For a univalent element such as silver, the gram-equivalent weight is equal to the mole weight, and it follows that the passage of 96,487 coul will produce 1 mole of silver. The charge on an electron is 1.6021×10^{-19} coul; thus the passage of 96,487 coul represents the passage of

$$\frac{96,487}{1.602 \times 10^{-19}} = 6.0225 \times 10^{23} \text{ electrons}$$

around the circuit. Each electron neutralizes one silver ion,

$$\text{Ag}^+ + e^- \longrightarrow \text{Ag} \tag{4}$$

so the number of silver atoms in a gram atomic weight is 6.02×10^{23}.

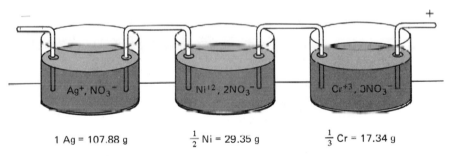

| 1 Ag = 107.88 g | $\frac{1}{2}$ Ni = 29.35 g | $\frac{1}{3}$ Cr = 17.34 g |

Figure 21.4 Illustration of Faraday's second law. To deposit metallic *atoms* silver, nickel, and chromium from the solutions of the salts in which they occur as the ions, Ag^+, Ni^{+2}, and Cr^{+3}, each Ag^+ ion must recover one electron, each Ni^{+2} ion two electrons, and each Cr^{+3} ion three electrons. One faraday of electricity passing through the circuit furnishes enough electrons to deposit 1 mole of silver at the left electrode of the left-hand cell. A corresponding number of electrons is introduced to the circuit at the right electrode but deposits only $\frac{1}{2}$ mole of nickel in the center cell. The like number of electrons released at the right-hand electrode in the center cell will deposit $\frac{1}{3}$ mole of chromium, and the circuit is completed when the faraday of electrons returns to the battery from the right-hand electrode of the chromium nitrate cell. A given current will deposit these metals from solution in the *ratios* of their gram-equivalent weights (107.88 g Ag; 29.35 g Ni; 17.34 g Cr) regardless of whether the quantity of current passing consists of 100 or 100 billion electrons.

Figure 21.5 The zinc-copper
galvanic cell.

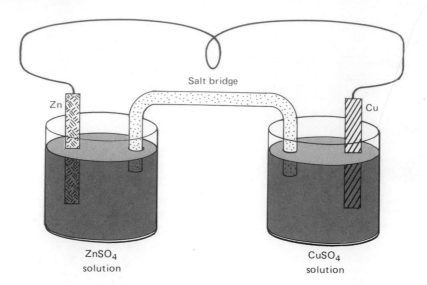

Salt bridge

Zn

Cu

ZnSO₄
solution

CuSO₄
solution

Galvanic Cells

In addition to the use of a cell for electrolysis, it is possible to arrange the electrodes and electrolytes in such a fashion that the electrons transferred, when an oxidation-reduction reaction takes place spontaneously within a cell, will serve as a source of electric current in an external circuit. Such a cell is called a *galvanic* cell.

A typical galvanic cell might consist of a metal rod (such as zinc) dipping into a solution of its ions (zinc ions); this solution is separated mechanically, but not electrically, from a solution of ions of a different metal (such as copper ions) into which dips a rod of the second metal (copper). The connection between the two solutions might be made with a "salt bridge"—typically, a solution of potassium chloride—that conducts a current but keeps the two solutions separate. A diagram of such a cell is shown in Figure 21.5, and the conventional representation is

$$\text{Zn} \mid \text{Zn}^{+2} \parallel \text{Cu}^{+2} \mid \text{Cu}$$

The single vertical lines represent separations of phases between the solid metals and the solutions of their ions, and the double vertical line represents the separation of the two halves of the cell.

Cell Reaction

The reaction taking place in the cell of Figure 21.5 is the same as the one that would occur if a piece of zinc were dropped into copper sulfate solution:

$$\text{Zn}(s) + \text{Cu}^{+2}(aq) \longrightarrow \text{Zn}^{+2}(aq) + \text{Cu}(s) \tag{5}$$

With the salt-bridge arrangement, however, the two half-reactions are separated, so the oxidation process occurs at the left electrode

$$\text{Zn}(s) \longrightarrow \text{Zn}^{+2}(aq) + 2e^- \tag{6}$$

and the reduction at the right

$$\text{Cu}^{+2}(aq) + 2e^- \longrightarrow \text{Cu}(s). \tag{7}$$

Thus the zinc rod dissolves, forming zinc ions, and copper ions deposit on the copper rod. The electrons released by the zinc atoms in forming ions displace electrons in the wire toward the copper rod, where copper ions are neutralized. Since the formation of zinc ions in the left beaker would, if that were the only change there, give that solution a positive charge, electrical neutrality requires the diffusion of some positive ions from the left beaker toward the right, or of some negative ions into the left beaker, or both. These ions are exchanged for ions of similar charge in the salt bridge, which releases any oversupply or makes up any deficiency by diffusion to or from the right-hand beaker, which would otherwise acquire a negative charge as copper ions are removed.

Electrical Work

The net result of using a galvanic cell such as that shown in Figure 21.5 is that the electron-transfer process of Equation (5), which changes zinc atoms to zinc ions while changing copper ions to copper atoms, has been used to transfer these electrons along a wire. If an "engine" is placed in the external circuit, the expenditure of energy in the movement of these electrons through the difference in potential (difference in voltage) between the two electrodes can be made to do work. This work will be a maximum when the whole process occurs reversibly (page 423). Reversibility is easy to arrange with a cell because the cell can be made to buck against another current source whose voltage is adjusted until a further infinitesimal change in voltage will cause the current to flow through the cell in the opposite direction. The work of electron transfer done under these conditions measures the free-energy change in the process, since it is reversible work that does not include pressure-volume work. The voltage needed to produce reversible operation is measured on the bucking source (Figure 21.6).

When an electron moves through a difference in potential, electrical work is done equal to the product of the charge on the electron and the difference in potential, $w = \epsilon E$, where ϵ is the charge and E the potential difference. If a mole of electrons moves under these conditions, the work done will be $N\epsilon E$. The charge carried by a mole of electrons, $N\epsilon$, is the faraday, F, equal to 96,487 coul. Since the reaction of a mole of zinc atoms with a mole of copper ions requires the transfer of two moles of electrons [Equations (6) and (7)], the work done will be $2FE$ or, in general, nFE, where n is the number of moles

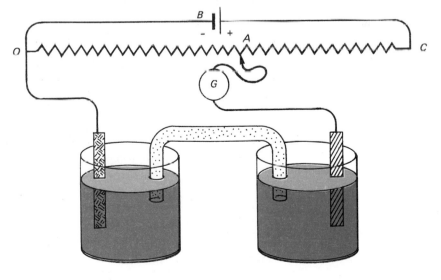

Figure 21.6 Potentiometric measurement of cell voltage. The current from a source of known voltage, B, is passed through a wire of uniform resistance per unit length, OC. The leads from the cell are connected through a galvanometer, G, to the uniform wire at a point A, which is moved along OC until the galvanometer shows no deflection. Slight movement from this setting toward O will cause the galvanometer to deflect in one direction, showing the cell to be a source of current. Movement toward C will cause deflection in the other direction, showing that current is passing through the cell from the battery. When the point of no deflection A has been found, the ratio of length OA to length OC gives the fraction of the voltage of B being produced by the cell.

of electrons transferred when the cell reaction takes place according to the equation written. This is equal to the negative of the free-energy change [Equations (9) and (11), Chapter 20].

$$(w_{max} - P\Delta V) = G_{initial} - G_{final} = -\Delta G$$

and

$$\Delta G = -nFE \qquad (8)$$

If E is the voltage of the cell when the zinc and copper are in their standard states—as they must be, since they are solids—and the zinc and copper ions are at concentrations $c_{Zn^{+2}}$ and $c_{Cu^{+2}}$, ΔG, with $n = 2$, will measure the free-energy change when 1 mole of solid zinc reacts with 1 mole of copper ions at concentration $c_{Cu^{+2}}$ to form 1 mole of zinc ions at concentration $c_{Zn^{+2}}$ and 1 mole of copper metal, according to the equation*

$$Zn(s) + Cu^{+2}(c_{Cu^{+2}}) \longrightarrow Cu(s) + Zn^{+2}(c_{Zn^{+2}}) \qquad (9)$$

If all substances are in their standard states and all ions in solution are at the concentrations chosen as their standard states, we write Equation (8) in the form

$$\Delta G^{\circ} = -nFE^{\circ} \qquad (10)$$

The galvanic cell is thus of particular importance to chemistry because, in contrast to most devices, it can be used easily to measure the work of a chemical reaction, excluding the pressure-volume work, and the value obtained is that for reversible operation. The result of the measurement is therefore a direct measure of the free-energy change in the chemical reaction taking place in the cell. Since ΔG° differs from E° only by the constant term $-nF$, either of the two quantities, E° or ΔG°, can serve as measures of spontaneity. When working mainly with solutions and cells, E° is the more convenient.

Electrode Potentials

Just as the reaction taking place in a cell such as that shown in Figure 21.5 can be separated into two electrode reactions at the cathode and anode [Equations (6) and (7)], so also the voltage of the cell is, by convention, considered to be the sum of a voltage associated with the cathode reaction and a voltage associated with the anode reaction. It is not possible to measure these voltages separately, since every cell must contain both an anode and a cathode. However, numerical values can be assigned to the individual electrode reactions by assigning an arbitrary value to some electrode reaction chosen as a standard. The convention adopted is that the voltage corresponding to the reaction

$$H_2(g,\ 1\ atm) \xrightarrow{\ (H_2O)\ } 2H_3O^+(a = 1) + 2e^- \qquad E^{\circ} = 0 \qquad (11)$$

is assumed to be zero for the reaction proceeding in either direction. Here, we have used the precise expression for the standard state of the hydronium ion—unit activity—to indicate that our usual procedure, in which the free energy or cell potential is expressed in terms of concentrations rather than activities, is an approximation (page 349). In dilute solutions, it is found that

---

*Actually, the measured E for a cell constructed as in Figure 21.5 will not correspond exactly to Equation (9), since there will be "liquid junction potentials" at the two ends of the salt bridge where a solution of potassium and chloride ions comes in contact with solutions of other ions. Often, extended experiments are needed to obtain the precise value, but the simple apparatus gives values precise enough for all but the most exacting work.

both concentrations and activities increase or decrease together and we can, without major loss of validity in our calculations, replace activities with concentrations in our mathematical equations at low concentrations. Because the molality of a solution is temperature independent, it is usually more convenient to use molalities rather than molarities in thermodynamic measurements. We shall often proceed as if the standard state for ions in solutions were a 1-molal solution, while recognizing that the actual concentration of a solution of unit activity probably is not 1 molal.

With the adopted convention that the half-reaction

$$H_2(g, 1 \text{ atm}) \underset{\text{(H}_2\text{O)}}{\rightleftharpoons} 2H_3O^+(a = 1) + 2e^-$$

has zero voltage, we may now find values for other half-reactions by making cells with hydrogen as one of the electrodes. Since hydrogen, being gaseous, cannot of itself serve as a conducting electrode, a chemically inert electrode such as platinum, which absorbs hydrogen readily, is used. While in use, it is kept bathed in an atmosphere of hydrogen.

The electron-transfer reaction between zinc metal and hydronium ions can be used under the above convention to measure the potential for the oxidation of zinc to zinc ions. A cell is constructed in which a zinc rod is immersed in a solution of zinc ions at "1-molal" concentration, and this solution connected through a salt bridge to a standard hydrogen electrode (Figure 21.7). The reaction taking place is known from experiments in dipping zinc into acid solutions:

$$Zn + 2H_3O^+ \longrightarrow Zn^{+2} + H_2 + 2H_2O \tag{12}$$

Thus the half reactions are

$$Zn(s) \longrightarrow Zn^{+2}(a = 1) + 2e^- \tag{13}$$

$$2H_3O^+(a = 1) + 2e^- \xrightarrow{H_2O} H_2(g, 1 \text{ atm}) \tag{14}$$

a fact readily confirmed by electrical measurements showing that electrons flow through the meter M *from* the zinc *to* the hydrogen electrode. The voltage of the cell is measured as 0.762 volt. According to our convention, this is the sum of the voltages of the two half-cells:

$$0.762 \text{ volt} = E°_{cell} = E°_{Zn \rightarrow Zn^{+2}} + E°_{2H_3O^+ \rightarrow H_2} = E°_{Zn \rightarrow Zn^{+2}} + 0 \tag{15}$$

Since the cell has been constructed with all components in their standard states, the measured voltage is a standard voltage, as indicated by the superior zeros on the $E°$ symbols. Further, since by convention the $E°$ value for hydrogen is chosen as zero, the voltage of the cell, 0.762 volt, is the standard potential for zinc. It is called the *standard oxidation potential*, and corresponds to the oxidation half-reaction

$$Zn(s) \longrightarrow Zn^{+2}(a = 1) + 2e \qquad E° = 0.762 \text{ volt} \tag{16}$$

By a series of measurements of this sort, a table (Table 21.1) of standard oxidation potentials can be obtained, not only for the formation of ions from metals but for other electrode reactions as well. All the values in the table are for 25°C and standard states: 1-molal concentrations (more precisely, unit activity) of all reacting ions in water solution, and hydrogen and other gases at 1 atm pressure. The double arrows in the equations indicate that the $E°$ value specifies a cell voltage under reversible conditions of no current flow, and the superior zero indicates that the reactants and products in this determination are at unit activity.

Table 21.1 Standard Oxidation Potentials at 25°C

Half-Reaction	$E°$ (volts)
$Li(s) \rightleftarrows Li^+ + e^-$	+3.05
$K(s) \rightleftarrows K^+ + e^-$	2.93
$Rb(s) \rightleftarrows Rb^+ + e^-$	2.93
$Cs(s) \rightleftarrows Cs^+ + e^-$	2.92
$Ba(s) \rightleftarrows Ba^{+2} + 2e^-$	2.90
$Sr(s) \rightleftarrows Sr^{+2} + 2e^-$	2.89
$Ca(s) \rightleftarrows Ca^{+2} + 2e^-$	2.87
$Na(s) \rightleftarrows Na^+ + e^-$	2.71
$Mg(s) \rightleftarrows Mg^{+2} + 2e^-$	2.37
$Al(s) \rightleftarrows Al^{+3} + 3e^-$	1.66
$Mn(s) \rightleftarrows Mn^{+2} + 2e^-$	1.18
$Zn(s) \rightleftarrows Zn^{+2} + 2e^-$	0.76
$Cr(s) \rightleftarrows Cr^{+3} + 3e^-$	0.74
$Fe(s) \rightleftarrows Fe^{+2} + 2e^-$	0.44
$Cd(s) \rightleftarrows Cd^{+2} + 2e^-$	0.40
$Pb(s) + SO_4^{-2} \rightleftarrows PbSO_4(s) + 2e^-$	0.36
$Co(s) \rightleftarrows Co^{+2} + 2e^-$	0.28
$Ni(s) \rightleftarrows Ni^{+2} + 2e^-$	0.250
$Ag(s) + I^- \rightleftarrows AgI(s) + e^-$	0.151
$Sn(s) \rightleftarrows Sn^{+2} + 2e^-$	0.136
$Pb(s) \rightleftarrows Pb^{+2} + 2e^-$	0.126
$H_2(g, 1 \text{ atm}) \rightleftarrows 2H^+(aq) + 2e^-$	0.000
$Ag(s) + Br^- \rightleftarrows AgBr(s) + e^-$	−0.095
$H_2S(g) \rightleftarrows 2H^+(aq) + S(s) + 2e^-$	−0.14
$Sn^{+2} \rightleftarrows Sn^{+4} + 2e^-$	−0.15
$Cu^+ \rightleftarrows Cu^{+2} + e^-$	−0.153
$Ag(s) + Cl^- \rightleftarrows AgCl(s) + e^-$	−0.222
$Cu(s) \rightleftarrows Cu^{+2} + 2e^-$	−0.337
$S(s) + 3H_2O \rightleftarrows H_2SO_3 + 4H^+(aq) + 4e^-$	−0.45
$2I^- \rightleftarrows I_2(s) + 2e^-$	−0.536
$MnO_4^{-2} \rightleftarrows MnO_4^- + e^-$	−0.56
$H_2O_2 \rightleftarrows O_2(g) + 2H^+(aq) + 2e^-$	−0.68
$Fe^{+2} \rightleftarrows Fe^{+3} + e^-$	−0.77
$2Hg(l) \rightleftarrows Hg_2^{+2} + 2e^-$	−0.79
$Ag(s) \rightleftarrows Ag^+ + e^-$	−0.80
$NO + 2H_2O \rightleftarrows NO_3^- + 4H^+(aq) + 3e^-$	−0.96
$Au(s) + 4Cl^- \rightleftarrows AuCl_4^- + 3e^-$	−1.00
$2Br^- \rightleftarrows Br_2(l) + 2e^-$	−1.09
$2H_2O \rightleftarrows O_2 + 4H^+(aq) + 4e^-$	−1.23
$Mn^{+2} + 2H_2O \rightleftarrows MnO_2(s) + 4H^+(aq) + 2e^-$	−1.23
$2Cr^{+3} + 7H_2O \rightleftarrows Cr_2O_7^{-2} + 14H^+(aq) + 6e^-$	−1.33
$2Cl^- \rightleftarrows Cl_2(g) + 2e^-$	−1.36
$Pb^{+2} + 2H_2O \rightleftarrows PbO_2 + 4H^+(aq) + 2e^-$	−1.455
$Au(s) \rightleftarrows Au^{+3} + 3e^-$	−1.50
$PbSO_4(s) + 2H_2O \rightleftarrows PbO_2(s) + 4H^+(aq) + SO_4^{-2} + 2e^-$	−1.69
$MnO_2(s) + 2H_2O \rightleftarrows MnO_4^- + 4H^+(aq) + 3e^-$	−1.70
$2H_2O \rightleftarrows H_2O_2 + 2H^+(aq) + 2e^-$	−1.77
$2F^- \rightleftarrows F_2 + 2e^-$	−2.87

The electrode reactions are all written as oxidation reactions. In this book, we use the positive sign to indicate that the element is oxidized more readily than hydrogen and the negative sign to indicate that the element is oxidized less readily than hydrogen. Thus, for the cell of Figure 21.7 containing zinc and hydrogen, the oxidation potential of zinc is given a plus sign.

Figure 21.7 The cell $Zn|Zn^{+2}\|H_3O^+|H_2|Pt$.

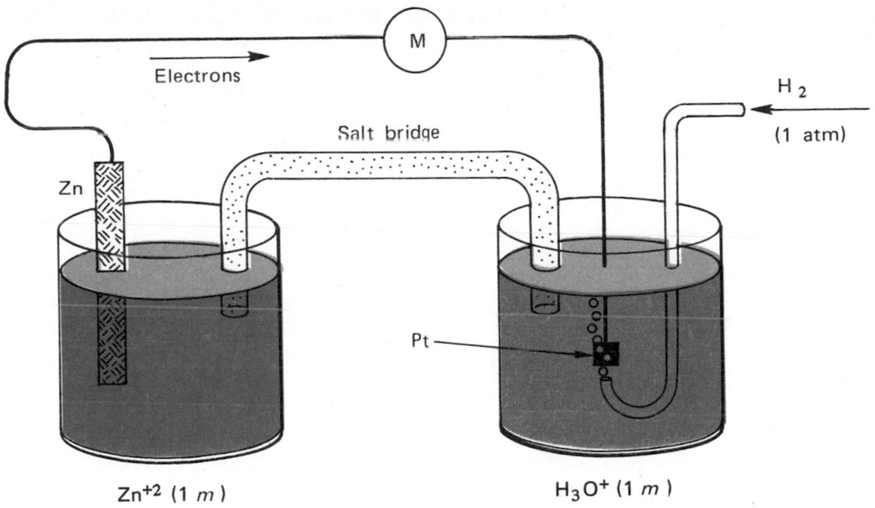

Figure 21.8 The cell $Cu|Cu^{+2}\|H_3O^+|H_2|Pt$.

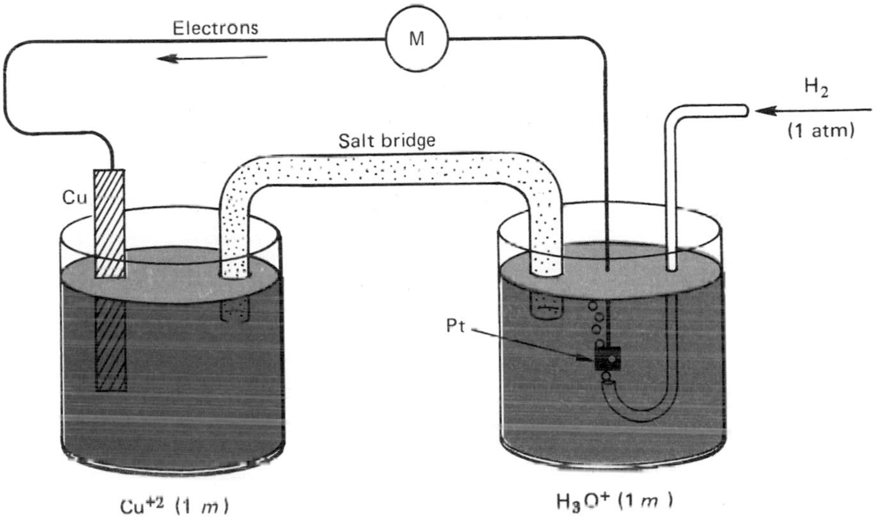

If a cell using copper and copper ions (Figure 21.8) instead of zinc and zinc ions,

$$Cu \mid Cu^{+2}(1\ m) \parallel H_3O^+(1\ m) \mid H_2$$

is constructed for measuring the standard oxidation potential of copper, the direction of flow of current shows that electrons are flowing *toward* the copper *from* the hydrogen electrode. The half-reactions are

$$H_2(g,\ 1\ atm) \xrightarrow{(H_2O)} 2H_3O^+(a = 1) + 2e^- \tag{11}$$

$$Cu^{+2}(a = 1) + 2e^- \longrightarrow Cu(s) \tag{17}$$

The voltage is

$$E^{\circ}_{cell} = E^{\circ}_{H_2 \to 2H_3O^+} + E^{\circ}_{Cu^{+2} \to Cu} \tag{18}$$

Again the hydrogen voltage is assigned the value zero; the total voltage of the cell, 0.345 volt, is assigned to the copper electrode.

$$0.345\ volt = E^{\circ}_{cell} = 0 + E^{\circ}_{Cu^{+2} \to Cu} \tag{19}$$

The half-reaction corresponding to this value is a *reduction* reaction; the oxidation half-reaction is the reverse of this, and gives the negative sign to the standard oxidation potential of copper.

$$E^\circ_{Cu \to Cu^{+2}} = -0.345 \text{ volt} \tag{20}$$

$$Cu(s) \longrightarrow Cu^{+2}(a = 1) + 2e^- \qquad E^\circ = -0.345 \text{ volt} \tag{21}$$

Reducing Agents and Oxidizing Agents

Since $\Delta G^\circ = -nFE^\circ$, and ΔG° is negative for spontaneous reactions, those half-reactions will have the greatest driving force that have the most positive values of E°. Therefore, the strongest reducing agents are the substances appearing as *reactants* in the half-reactions near the top of Table 21.1. At the bottom of the table, the half-reactions will give a negative value of ΔG° when written in the reverse direction. Therefore, the strongest oxidizing agents are the substances appearing as *products* in the half-reactions near the bottom of the table. Of the substances listed, lithium is the best reducing agent, and fluorine the best oxidizing agent. Fluoride ion, on the other hand, is the poorest reducing agent, and lithium ion the poorest oxidizing agent.

Cell Voltage from Oxidation Potentials

From the data in Table 21.1, we can calculate the voltage of a cell containing any pair of electrodes. The voltage of the zinc-copper cell of Figure 21.5, for example, for 1-molal concentrations of zinc and copper ions, would simply be the sum of the potentials corresponding to the half-reactions taking place in the electrode compartments. These are

$$Zn(s) \longrightarrow Zn^{+2}(1\ m) + 2e^- \tag{13}$$

$$Cu^{+2}(1\ m) + 2e^- \longrightarrow Cu(s) \tag{17}$$

The second half-reaction given is the reverse of that given for the copper electrode in the table; to get its value, we must change the sign of the tabulated value:

$$E^\circ_{cell} = E^\circ_{Zn \to Zn^{+2}} + E^\circ_{Cu^{+2} \to Cu} = E^\circ_{Zn \to Zn^{+2}} - E^\circ_{Cu \to Cu^{+2}}$$

$$= 0.762 - (-0.345) = 1.107 \text{ volt} \tag{22}$$

The value 1.107 volt corresponds to the cell voltage when the components are in their standard states.

SIGN CONVENTIONS

The procedure adopted in this book—namely, that potential values shall be strictly related to a half-reaction written as an oxidation reaction and shall be positive for substances more easily oxidized than hydrogen but negative if less easily oxidized—is that used by most physical chemists in the U.S., though not necessarily by most electrochemists. The International Union of Pure and Applied Chemistry focuses attention on the *electrode* rather than on the *electrode reaction* and recommends that the sign of the electrode potential be determined by the electrical behavior of the electrode when in a cell with the hydrogen electrode—negative for negative electrodes and positive for positive electrodes. This has the effect of making the *electrode* potentials exactly opposite in sign to those of the *oxidation* potentials in Table 21.1. One of the easiest ways to make use of either system is to associate the electrode potentials of the IUPAC system with *reduction half-reactions* and use them as such in calculations.

$$Cu^{+2}(a = 1) + 2e^- \longrightarrow Cu(s) \qquad \text{Electrode potential (IUPAC)} = +0.345 \text{ volt}$$

$$Cu(s) \longrightarrow Cu^{+2}(a = 1) + 2e^- \qquad \text{Oxidation potential} = -0.345 \text{ volt}$$

The sign of the oxidation potential does not show whether the electrode is the positive or negative pole of a cell; this depends upon what the other electrode is. Thus, if a zinc electrode is placed in a cell with the copper electrode,

$$Zn \mid Zn^{+2}(1\ m) \parallel Cu^{+2}(1\ m) \mid Cu$$

electrons flow from the zinc to the copper, and the zinc rod must be connected as the negative pole. However, in a cell containing a zinc and a magnesium electrode,

$$Zn \mid Zn^{+2}(1\ m) \parallel Mg^{+2}(1\ m) \mid Mg$$

electrons flow toward the zinc from the magnesium, and the zinc rod must be connected as the positive pole. The half-reactions in the zinc-magnesium cell are

$$Zn^{+2}(1\ m) + 2e^- \longrightarrow Zn(s) \tag{23}$$

$$Mg(s) \longrightarrow Mg^{+2}(1\ m) + 2e^- \tag{24}$$

and the voltage is

$$E^\circ_{cell} = E^\circ_{Zn^{+2} \to Zn} + E^\circ_{Mg \to Mg^{+2}} = -E^\circ_{Zn \to Zn^{+2}} + E^\circ_{Mg \to Mg^{+2}}$$

$$= -(+0.762) + 2.37 = 1.61 \text{ volt} \tag{25}$$

The reaction taking place in the cell will be one that makes the cell voltage positive (ΔG negative).

Relationship Between Cell Voltage and Concentration

Up to this point, all the cells we have dealt with have been made with solutions of ions in their standard states of "1-molal" concentration. The voltages for other concentrations may be readily obtained by making use of Equation (40) of Chapter 20 to obtain an equation similar to Equation (35) of that chapter, but written in terms of the concentration approximation. The result,

$$\Delta G = \Delta G^\circ + 2.3RT \log \frac{c_P{}^p c_Q{}^q}{c_M{}^m c_N{}^n} \tag{26}$$

records the effect of change in concentration on the change in free energy for the general reaction

$$mM(c_M) + nN(c_N) \rightleftharpoons pP(c_P) + qQ(c_Q) \tag{27}$$

The relationship between cell potential and free-energy change is given by Equations (8) and (10) of this chapter:

$$\Delta G = -nFE \tag{8}$$

$$\Delta G^\circ = -nFE^\circ \tag{10}$$

Substitution in Equation (26) gives

$$E = E^\circ - \frac{2.3RT}{nF} \log \frac{c_P{}^p c_Q{}^q}{c_M{}^m c_N{}^n} \tag{28}$$

This equation makes it possible to calculate E° from measurements of the potential E of the cell at concentrations other than 1 molal and to calculate the voltages of cells from tabulated E° values when the ion concentrations are not 1 molal. The following is an example of the second calculation.

Calculate the voltage of the cell of Figure 21.5 at 25°C when the zinc sulfate solution is 0.1 molal and the copper sulfate solution is 0.001 molal.

Substitution of the concentrations in Equation (28) gives

$$E = 1.107 - \frac{RT}{2F}\ 2.303 \log \frac{0.1}{0.001} = 1.048 \text{ volt}$$

The cell voltage is less than the voltage when the concentrations were equal, and the free-energy change is less negative ($\Delta G = -nFE = -48.3$ kcal, compared to -51.1 kcal in the equal-concentrations case). This is to be expected, since the relatively greater decrease in the copper ion concentration compared to the decrease in the zinc ion concentration decreases the driving force in the reaction from left to right.

$$Zn(s) + Cu^{+2}(0.001\ m) \longrightarrow Cu(s) + Zn^{+2}(0.1\ m) \qquad \Delta G = -48.3 \text{ kcal}$$

$$Zn(s) + Cu^{+2}(1\ m) \longrightarrow Cu(s) + Zn^{+2}(1\ m) \qquad \Delta G = -51.1 \text{ kcal}$$

Cell Voltage and Equilibrium Constant

The relations obtained from Equation (10)

$$\Delta G^\circ = -nFE^\circ \tag{10}$$

and from Equation (33) of Chapter 20

$$\Delta G^\circ = -2.303RT \log K$$

permit the immediate calculation of the equilibrium constant for any reaction for which E° is known.

Calculate, from electrode potential data, the equilibrium constant at 25°C for the reaction represented by Equation (A).

$$Br_2(l) + 2Fe^{+2}(1\ m) \rightleftharpoons 2Br^-(1\ m) + 2Fe^{+3}(1\ m) \tag{A}$$

Since the ions are in their standard states (approximation!), the data of Table 21.1 may be used directly. The half-reactions are

$$Br_2(l) + 2e^- \longrightarrow 2Br^-(1\ m) \qquad E^\circ = 1.09$$

$$Fe^{+2}(1\ m.) \longrightarrow Fe^{+3}(1\ m) + e^- \qquad E^\circ = -0.771$$

The E° value for the first reaction is given the positive sign, since the equation is the reverse of that in Table 21.1. The voltage of a cell in which reaction (A) took place would then be

$$E_{cell} = E^\circ_{Br_2 \rightarrow 2Br^-} + E^\circ_{Fe^{+2} \rightarrow Fe^{+3}} = +1.09 - 0.771 = +0.32 \text{ volt}$$

The positive voltage means that the reaction is spontaneous as written. The free-energy change is

$$\Delta G^\circ = -nFE^\circ = -2 \times 96{,}487 \times (+0.32) = -61{,}800 \text{ joules or } -14.8 \text{ kcal}$$

Note that the cell voltage is obtained by direct use of oxidation-potential data obtained from Table 21.1; no alteration of those values is necessary to take into account the fact that the bromine molecule gains two electrons while the iron(II) ion loses only one. Account of this difference *has* been taken in balancing the chemical equation for the oxidation-reduction process. The balanced equation shows that two electrons have been transferred, and hence the value $n = 2$ has been assigned in calculating the standard free-energy change as $-nFE°$.

From the equation relating $\Delta G°$ to the equilibrium constant

$$\log K = \frac{-14{,}800}{2.3 \times 1.987 \times 298}$$

and

$$K = 7.3 \times 10^{10} = \frac{[Br^-]_{eq}{}^2[Fe^{+3}]_{eq}{}^2}{[Fe^{+2}]_{eq}{}^2}$$

The large value of the equilibrium constant in the problem tells us, as you may remember (page 153), that when equilibrium is established in the bromine-bromide, ferrous ion-ferric ion system, the ferrous ion concentration will be low, or, the position of equilibrium is well toward the right.

Galvanic Cells and Hydronium Ion Concentration

An important application of the galvanic cell in the laboratory is its use in measuring the concentration of hydronium ions in a solution. Suppose that a cell is set up as in Figure 21.9. Here, one electrode is a hydrogen electrode and the other a "reference electrode," and the two are connected by a salt bridge. The voltage of the cell can be measured using a potentiometer or, typically, a vacuum-tube voltmeter, V, which draws negligible current and allows the reversible potential difference to be measured. The voltage will be found to depend upon the concentration of hydronium ions and of the metal ions (Me$^+$) of the reference electrode:

$$E = E° - 2.303 \frac{RT}{F} \log \frac{c_{Me^+}}{c_{H_3O^+}} \tag{29}$$

If the concentration of hydronium ion in the left-hand beaker is now changed to $c'_{H_3O^+}$, while leaving the reference half-cell unchanged, the voltage will be

$$E' = E° - 2.303 \frac{RT}{F} \log \frac{c_{Me^+}}{c'_{H_3O^+}} \tag{30}$$

Figure 21.9 Measurement of hydronium-ion concentration.

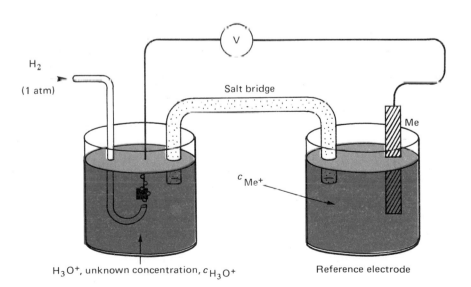

H_2 (1 atm)

Salt bridge

Me

c_{Me^+}

H_3O^+, unknown concentration, $c_{H_3O^+}$

Reference electrode

Thus the magnitude of the voltage, E or E', is a correct measure of the hydronium ion concentration, $c_{H_3O^+}$ or $c'_{H_3O^+}$. The nature of the reference half-cell is unimportant except that it be unchanged from experiment to experiment so it makes a constant contribution to the over-all cell voltage. The arrangement of Figure 21.9 can be calibrated by measuring the voltage for a known hydronium ion concentration. Measured changes in voltage can then be translated into hydronium ion concentrations through the logarithmic term of Equation (30).

In practice, a *glass electrode*, consisting of a thin glass bulb containing, typically, an electrode of silver coated with silver chloride dipping into hydrochloric acid solution (Figure 21.10), is used in place of the hydrogen electrode, for convenience. This device, when combined with a reference electrode, shows differences in potential that depend upon the hydronium ion concentration in the same way that the hydrogen electrode does.

pH Value

We are now in position to explain the present convention defining the quantity pH, for which the historical definition was introduced on page 401.

The pH of a solution, pH(X), is precisely defined in terms of the voltages of two cells, each similar to the one shown in Figure 21.9. One cell contains a carefully chosen solution we may call the standard, S, and the other the unknown solution X. The pH of the unknown is then defined by the equation

$$pH(X) = pH(S) + (E_x - E_s)\frac{F}{2.3RT} \tag{31}$$

where E_x and E_s are the voltages of the two cells and pH(S) is the value assigned to the standard solution. In the U.S., a commonly used standard is a 0.05-molal solution of potassium acid phthalate, to which the National Bureau of Standards has assigned the value pH(S) = 4.008 at 25°C. For measurements in dilute solutions that have a precision no greater than about 0.1 pH unit, we may safely consider that the pH value of a solution is equal to $-\log$ [H_3O^+].

In practice, the pH of a solution is measured on a "pH meter," which measures the voltage of cells consisting of a glass electrode and an electrode of mercury in contact with a saturated solution of mercury(I) chloride, connected through a potassium chloride salt bridge. The meter is first set at a particular value E_s by measurement on a standard solution of known pH(S), and the difference $E_x - E_s$ between that value and the voltage measured for the unknown solution is read off a scale marked directly in pH units.

Potentiometric Titration

Suppose that instead of changing from one solution to another in the left half-cell of Figure 21.9, the experimenter repeatedly adds to the acid solution measured amounts of a solution of a strong base. This will neutralize the acid, changing the hydronium ion concentration and the voltage of the cell. From the form of Equation (29), it can be seen that the change in $c_{H_3O^+}$ from, for example, 0.1 m to 0.01 m, will make the same change in voltage as the change from 0.00001 m to 0.000001 m. It will require, however, only 1/10,000th as much base solution to produce the latter change as would be required for the former. Consequently, a plot of the change in voltage per milliliter of base solution added will show a rise at first, followed by a very rapid change near the neutral point and a slow change on the basic side (Figure 21.11). The volume of base added to reach the most rapidly rising portion of the curve represents the amount needed to neutralize the acid.

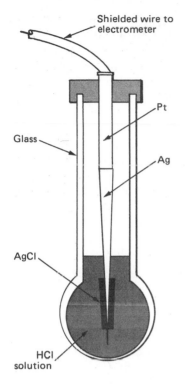

Figure 21.10 Diagram of a glass electrode.

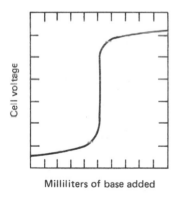

Figure 21.11 Change of cell voltage upon addition of base to an acid solution.

The experiment thus constitutes a *titration* of the acid with the base, using a potentiometric (*potential measurement*) procedure for determining the end-point of the titration. Obviously, for the procedure to be usable, the cation present in the solution of the base used must not react with any compounds in the cell solution in a way that would change the cell voltage.

Curves exactly similar to that of Figure 21.11 will appear in other half-cells when a titrating solution is added that removes the ion in the cell. For example, silver ion in the cell

$$Ag \mid Ag^+ \parallel reference\ electrode$$

may be removed by adding sodium chloride solution to precipitate silver chloride.

Fuel Cells: Infants with Unusual Promise. Current energy research centers on the direct conversion of chemical, nuclear, or solar energy to electrical energy. Fuel cells—devices that generate electricity from chemical fuels—are being designed and fabricated for applications such as auxiliary power sources on spacecraft and submarines or other military vehicles, to replace gasoline- and diesel-powered generators and automobile engines, or to convert household sewage and other wastes to less noxious products. Meanwhile some of the energy released in this process would be transformed to electrical energy. In the future, fuel cells could be used to supply fresh water, heat, and electricity to even the most remote villages of the developing countries.

Fuel cell batteries having high efficiencies (50% or more) with lives of more than 5000 hours and outputs of 10 kilowatts or more (a 3-kilowatt battery can power a golf cart, for example) have been made. Unfortunately, their cost is still too high to make them competitive with prevailing power sources.

We can define a fuel cell as a galvanic cell in which a conventional fuel is supplied to one electrode and an oxidant, usually oxygen or air, to the other. Both electrodes and the electrolyte(s) present should be unaffected by the reaction. At the anode or "fuel electrode," the fuel is oxidized and electrons are released to the external circuit. At the cathode, oxygen is reduced and electrons received from the external circuit. A fuel cell is essentially a low-voltage device, giving an output of about 1 volt or less. To produce a useful output, several cells are connected in series and the collection called a fuel cell battery or fuel battery. A major difference between this type of battery and the common disposable variety of dry cell is that the fuel battery is a continuous feed battery that will, ideally, produce electricity as long as suitable reactants are fed into it. The disposable battery is useful only until the reactants initially put into it have been consumed.

In its simplest form, a fuel cell consists of two electrodes separated by an electrolyte. Provision must be made to supply fuel to one electrode, oxidant to the other, and to remove reaction products and heat from the system. The components must be engineered into a mechanically stable system. Each of these factors must be considered in designing an inexpensive but functional system.

If gaseous fuels such as hydrogen or methane are to be used, the electrode must be porous so the reaction can take place at the gas/liquid/solid interfaces within the pores. If the fuel is soluble in the electrolyte, such as ammonia in a water solution, the electrode need not be porous. Liquid fuels such as methanol or ethylene glycol have the advantage over gases of being easier to transport and store and of providing higher energy per unit volume. However, few liquids meet the many rigorous requirements of a good fuel, such as: a wide liquid range; chemically stable and soluble in but inert toward strong acids or bases; readily ionizable; and rapidly reacting at the electrode to give inoffensive, preferably gaseous products. Methanol and hydrazine are two of the more promising liquids for fuel cells, but both are too expensive at present.

Electrodes also must meet stringent requirements. They must be good conductors and good electron sources or sinks, and must not be consumed or deteriorated by the electrolyte, heat, or electrode reactions. Perhaps most important, they must be excellent catalysts for the reactions that take place on their surfaces. When hydrogen is used as the fuel, electrodes made of graphite impregnated with finely divided platinum, a 75/25 alloy of palladium and silver, or nickel have been used. Nickel, while much cheaper than the other materials, is a far less effective catalyst; cells in which it is used must be designed to compensate for this deficiency. Cathodes, oxygen-reacting electrodes, are made of platinum, nickel, silver oxide, or silver-nickel alloys. The secret of successful, competitive fuel cells probably lies in the development of inexpensive electrodes that are powerful catalysts for the electrode reactions.

Electrolytes used most often are aqueous solutions of potassium hydroxide or sulfuric acid. Potassium thiocyanate dissolved in liquid ammonia is used in a typical low-temperature battery (operating range—54°C to 72°C). One commercial fuel cell uses an ion-exchange resin saturated with water as the electrolyte. This resin functions by exchanging hydronium ions—receiving those produced at the anode and releasing a corresponding number to neutralize the hydroxide ions produced at the cathode. The high mobility of these ions in water improves the effectiveness of the electrolyte.

Removal of reaction products and heat are essential to continuous efficient operation of the cell. In the hydrogen fuel cell, for example, a pint of water is produced for each kilowatt-hour of operation. Fuel cells with power capacities of 1 to 3 kilowatts for 10 to 14 days are components of the service module of the Apollo spacecraft. Water and heat are removed continuously from these cells by cooling the unspent hydrogen gas. Water, carried as a vapor by the hydrogen, condenses in this process and is collected, purified, and used for washing or drinking. It can also be electrolyzed by another source of electricity to regenerate the reactants.

Fuel cells are like infants. They have unrealized potential. If they are fortunate enough to have wise and imaginative parents, they may help enrich the lives of all—even those of the poorest peasants in the most slowly developing countries.

SUMMARY This chapter is an introduction to the topic of electrochemistry of solutions. Herein are discussed the principles underlying the operation of electrolysis cells (cells in which electricity is used to bring about a chemical reaction) and galvanic or voltaic cells (cells in which a chemical reaction is allowed to proceed so that an electric current is produced).

In electrolytic cells, the battery or generator, operating as an electron pump, charges one electrode negatively (cathode) and the other positively (anode) and creates a potential difference across the electrodes. Positive ions are attracted to the cathode, negative ions to the anode, and this movement of ions constitutes the electrical conductivity of the solution. A chemical change occurs at each electrode—an oxidation at the anode and a reduction at the cathode. The over-all chemical reaction of the cell is the sum of the electrode reactions. Faraday's laws establish a quantitative relationship between the quantity of electricity passed through the cell and the amount of reaction occurring at the electrodes.

In galvanic cells, a spontaneous process is allowed to occur in such a way that reactants are not in direct contact but are connected by an external electric circuit. Transfer of electrons takes place from one reactant to another through the wire, creating an electric current as the reaction proceeds. The voltage of the cell depends upon the nature of the reactants, the concentrations (activities) of ions in solution, and on the temperature. The work obtained from a galvanic cell operating under reversible conditions is a precise measure of the free-energy change for the chemical process occurring in the cell. Assuming the cell voltage is the sum of contributions from each electrode, it is possible to obtain a set of relative potentials (oxidation potentials) for each electrode immersed in a solution of its ions. These relative values, based on the arbitrary standard of zero voltage for the hydrogen electrode in a solution of hydronium ions at unit activity, give a quantitative measure of chemical reactivity for these systems.

Applications of standard oxidation potentials to problems of chemical reactivity and equilibrium are given.

IMPORTANT TERMS

Electrochemical cell
electrolysis cell
galvanic cell

Electrolytic conduction
metallic conduction

Anode
cathode

Cell reaction
half-reaction

Reversible potential
electrode potential
oxidation potential
standard potential

Faraday's laws

Hydrogen electrode
reference electrode
glass electrode

Electrical work

Salt bridge

Potentiometric titration

pH value

Fuel cell

QUESTIONS AND PROBLEMS

1. A current of 0.500 amp is passed through a silver nitrate solution for 2.500 minutes. What weight of silver is deposited at the cathode?

2. A current of 1.200 amp is passed in series through a bath of molten sodium chloride and a bath of molten strontium chloride for one hour and 20.00 minutes. (a) What will be the ratio of the weight of sodium deposited to the weight of strontium deposited? (b) What will be the ratio of atoms of sodium to atoms of strontium? (c) What will be the weight of chlorine produced in the sodium chloride cell; in the strontium chloride cell?

3. When a current is passed through a solution of copper sulfate using platinum electrodes, copper and oxygen are produced. (a) Write the

equation for the reaction taking place at the cathode; at the anode. (b) If a quantity of electricity amounting to 1000 coul is used, calculate the weight of copper deposited and the volume of oxygen produced, measured at 25°C and 741 torr.

4. A cell like that of Figure 21.5 is used as a source of current to light for 60.00 minutes a small incandescent bulb requiring 0.100 amp. (a) How many grams of zinc is changed to zinc sulfate? (b) If the working voltage of the cell under the conditions used is 1.100 volt, what is the energy, in calories, expended in lighting the lamp for the time indicated?

5. If a current of 0.2 amp is passed through a copper sulfate solution for 10 minutes, calculate the following: (a) the number of grams of copper deposited; (b) the number of atoms of copper deposited.

6. A weight 3.974 g of copper is deposited when a current of 12,062 coul is passed through a copper sulfate solution using platinum electrodes. Calculate the number of atoms in a gram-atomic weight of copper.

7. Trace the ionic and electron processes in the solution and at the electrodes in the electrolysis of a water solution containing an acid such as HCl; such as H_2SO_4.

8. Trace the ionic and electron processes in solution and at the electrodes in a battery made of copper and zinc electrodes immersed in copper sulfate and zinc sulfate, respectively.

9. Write the equations for the half-reaction occurring at each electrode in Problem 8. Calculate the $E°$ for the cell from the $E°$ for each half-reaction.

10. Write the equations for the reaction at the left-hand electrode; the reaction at the right-hand electrode; and the overall cell reaction for each of the following galvanic cells:
    (a) Mg | Mg^{+2} ‖ Cu^{+2} | Cu
    (b) Mg | Mg^{+2} ‖ Zn^{+2} | Zn
    (c) Pt | H_2 | H_3O^+ ‖ Zn^{+2} | Zn
    (d) Pt | H_2 | H_3O^+ ‖ Cu^{+2} | Cu
    (e) Cd | Cd^{+2} ‖ Hg_2^{+2} | Hg
    (f) Pt | Br_2 | Br^- ‖ Fe^{+2}, Fe^{+3} | Pt

11. Calculate the voltage at 25°C of the cells of Problem 10 if all gases are at atmospheric pressure and all ions are at "1-molal" concentrations.

12. The reactions taking place in the cell

$$Pt \mid Fe^{+2}, Fe^{+3} \parallel Cl^- \mid AgCl \mid Ag$$

are

$$Fe^{+2} \longrightarrow Fe^{+3} + e^-$$

$$AgCl + e^- \longrightarrow Ag + Cl^-$$

$$Fe^{+2} + AgCl \longrightarrow Fe^{+3} + Ag + Cl^-$$

Calculate the voltage of this cell at 25°C when the concentrations of Fe^{+2}, Fe^{+3}, and Cl^- are "1 molal." Does the cell reaction proceed in the direction written above under these concentration conditions?

13. The cell

$$Pt(s) \mid H_2(1 \text{ atm}) \mid H_3O^+(1\ m), Cl^-(1\ m) \mid AgCl(s) \mid Ag(s)$$

at 25°C has a voltage of 0.22 volt; electrons flow outside the cell from the platinum to the silver electrode. (a) Write the cell reactions. (b) Calculate the standard oxidation potential for the reaction

$$Ag(s) + Cl^- \longrightarrow AgCl(s) + e^-$$

14. Using the value for the oxidation potential of lead, calculate the standard free-energy change, in calories at 25°C, for the reaction

$$Pb(s) + 2 H^+(aq) \longrightarrow Pb^{+2} + H_2(g)$$

15. Under what conditions can we write $\Delta G° = -nFE°$?

16. What is the basis for assuming $E° = 0$ for the reaction

$$H_2 \text{ (1 atm)} \xrightleftharpoons{(H_2O)} 2H_3O^+(1\ m) + 2e^-?$$

17. Calculate the electromotive force at 25°C for a zinc-copper cell in which the zinc sulfate concentration is 0.01 molar and the copper sulfate concentration is 0.1 molal.

18. Calculate the value of ΔG for the cell in Problem 17.

19. Calculate the value of the equilibrium constant at 25°C for the reaction

$$Zn + 2H^-(aq) \longrightarrow Zn^{+2} + H_2$$

using the information that $E°$ for the zinc half-cell is -0.76 and for the hydrogen electrode is 0.

20. Calculate the voltage of the cell

$$Zn(s) \mid Zn^{+2}(0.01\ m) \parallel Cu^{+2}(0.1\ m) \mid Cu(s)$$

21. Under what conditions can we write $\Delta G = -2.303RT \log K$?

22. Explain how the pH value of a solution can be determined from voltage measurement.

23. Explain how you might determine the endpoint of a titration of Ag^+ with Cl^- using a silver rod as one electrode in a cell whose voltage is measured as the solution is titrated.

24. Calculate the equilibrium constant for the reaction

$$Cr_2O_7^{-2} + 6Cl^- + 14H^+ \longrightarrow 2Cr^{+3} + 3Cl_2 + 7H_2O$$

from electrode potential data.

25. Look up the oxidation potentials of zinc and iron in Table 21.1. Which metal is more active? Explain why zinc can be used as a protective coating for iron (galvanized iron) even though its $E°$ is higher than the $E°$ of iron.

SPECIAL PROBLEMS 1. Combine the result of (14) with the free-energy change in the reaction

$$Pb^{+2}(aq) + SO_4^{-2}(aq) \longrightarrow PbSO_4(s)$$

to determine (a) the voltage associated with the reaction

$$PbSO_4(s) + H_2(g, 1\ atm) \longrightarrow Pb(s) + 2H_2^+(aq)(1\ m) + SO_4^{-2}(1\ m)$$

(b) the standard oxidation potential of the reaction

$$Pb(s) + SO_4^{-2} \longrightarrow PbSO_4(s) + 2e^-$$

The solubility-product constant of lead sulfate at 25°C is 1.6×10^{-8}. (c) How do you account for the fact that the standard oxidation potential for lead is higher when sulfate ion is present than in its absence?

2. The voltage of a cell constructed as follows

$$Ag(s) \mid Ag^+(x\ m) \parallel Cu^{+2}(1\ m) \mid Cu(s)$$

in which the concentration of silver ion is unknown, is found by experimental measurement to be 0.40 volt; electrons flow from the copper electrode to the silver electrode. What is the concentration of silver ions in the solution?

3. A cell constructed as in Figure 21.9 gave a voltage of 0.216 volt when the hydronium ion concentration in the left-hand beaker was fixed at 0.100 molal and a value 0.096 volt when a solution of unknown hydronium ion concentration was used in the left-hand beaker. Calculate the hydronium ion concentration of the unknown.

4. A cell such as that in Figure 21.9 gave a voltage of 0.412 volt when a 0.05-molal solution of potassium acid phthalate was in the left-hand beaker and 0.427 volt when a buffer solution of unknown hydronium ion concentration was used. Both measurements were at 25°. What was the pH of the buffer solution? What was the hydronium ion concentration?

5. A solution of hydrochloric acid of unknown concentration was titrated potentiometrically by adding measured amounts of 0.100 M NaOH solution to 50 ml of the acid contained in the left-hand beaker of a cell like the one shown in Figure 21.9. The cell voltage was found to change as follows:

ml NaOH added	E(volts)	ml NaOH added	E(volts)
0	0.339	49.50	0.475
10.00	0.350	49.90	0.516
20.00	0.361	50.10	0.870
30.00	0.374	50.50	0.911
40.00	0.395	55.00	0.969
45.00	0.415	60.00	0.986
49.00	0.457	70.00	1.002

Plot these data and calculate the concentration of the hydrochloric acid solution.

6. Can lead dioxide, PbO_2, be used in acid solution to prepare chlorine from chloride ion? Support your answer with appropriate calculations.

7. Calculate the free-energy change in the reaction

$$Sn^{+2}(0.1\ m) + 2Fe^{+3}(0.01\ m) \longrightarrow Sn^{+4}(0.001\ m) + 2Fe^{+2}(0.1\ m)$$

8. What concentrations of dichromate ion, hydronium ion, and chlorate ion, and what pressure of chlorine gas would you recommend to prepare chlorate ion by the following reaction?

$$5Cr_2O_7^{-2} + 3Cl_2 + 34H^+ \longrightarrow 10Cr^{+3} + 6ClO_3^- + 17H_2O$$

The standard oxidation potential for the half-reaction

$$Cl_2 + 6H_2O \longrightarrow 2ClO_3^- + 12H^+ + 10\ e^-$$

is −1.47 volts. On the basis of your calculations, is it reasonable to expect that chlorate ion could be prepared in this way?

9. (a) Combine the equations

$$MnO_2(s) + 2H_2O \longrightarrow MnO_4^- + 4H^+(aq) + 3e^- \qquad E° = -1.70 \text{ volts}$$

$$MnO_4^{-2} \longrightarrow MnO_4^- + e^- \qquad E° = -0.56 \text{ volt}$$

so to obtain the $E°$ value for the reaction

$$MnO_2(s) + 2H_2O \longrightarrow MnO_4^{-2} + 4H^+(aq) + 2e^-$$

(b) Use this result to calculate $\Delta G°$ for the reaction

$$3MnO_4^{-2} + 4H^+(aq) \longrightarrow 2MnO_4^- + MnO_2(s) + 2H_2O$$

and decide if manganate ion, MnO_4^{-2}, is stable in acid solution.

REFERENCES   SANDERSON, R. T., "On the Significance of Electrode Potentials," *J. Chem. Educ.*, **43**, 584 (1966).
WASER, J., *Basic Chemical Thermodynamics*, Benjamin, Menlo Park, Calif., 1966.
LAWRENCE, R. M., and BOWMAN, W. H., "Electrochemical Cells for Space Power," *J. Chem. Educ.*, **48**, 359 (1971).

CHEMICAL KINETICS

Early chemists were primarily interested in the products of reactions. Later they became interested in the amounts of reactants and products. Then their interest was directed toward knowing more about the atoms, ions, and molecules that react. In the early and middle parts of this century, to understand better the industrial and biological processes and especially the many new methods of synthesis, chemists have turned to a study of pathways or mechanisms of reactions—an area of study known as chemical kinetics.

The data required in kinetics are those obtained by studying reaction rates and factors that affect those rates. From such data, information about the probable reaction mechanism can be deduced. Kinetics studies also include determining the effect of several variables—such as temperature, catalyst, concentration of reactants, pressure, and surface area—on the rate.

Such information also has proven valuable in interpreting the mechanisms of catalytic action, particularly in biochemical processes, as well as in the many industrial processes requiring catalysts, such as the syntheses of ammonia and sulfuric acid, the production of plastics and synthetic rubbers, the cracking and reforming of petroleum, the reactions in internal combustion engines, and many others.

Some very useful and surprising results have been obtained about the mechanisms of even the simplest reactions by studying their rates. Such are the findings, for example, in studying the reaction of gaseous bromine with hydrogen

$$H_2 + Br_2 \longrightarrow 2HBr$$

which might be assumed to be the result of a simple collision of a hydrogen molecule with a bromine molecule, resulting in a switch of partners to form two hydrogen bromide molecules. The study of the reaction rate, however,

---

*The quotations are from *J. Chem. Soc.*, **113**, 471 (1918), and *J. Chem. Phys.*, **3**, 502 (1935).

shows that the path from hydrogen and bromine to hydrogen bromide involves several reaction steps in which free atoms of bromine and hydrogen are involved, as well as hydrogen and bromine molecules. In contrast, rate studies of the reaction between hydrogen and gaseous iodine,

$$H_2 + I_2 \longrightarrow 2HI$$

long thought to result from a simple collision and shift of partners, show that the path from reactants to products involves a collision between two iodine atoms and a hydrogen molecule. How does the study of reaction rates enable us to get such information about reaction pathways?

RATE AND MECHANISM OF REACTION

Rate of Reaction

By the rate of a reaction, we mean the quantity of material per unit volume undergoing transformation in a unit of time (second, minute, hour, day). Usually, the rate is a rapidly changing quantity since, as reaction proceeds, the reactants are used up and there remains progressively less and less material to undergo reaction. Such a change in rate with change of time and concentration of reactants is shown in the graphs for the hydrogen iodide reaction in Chapter 7. The rate must, therefore, be looked upon as an instantaneous quantity determined by plotting the quantity of reactant per unit volume (that is, by plotting the concentration) against time, and measuring the small change in concentration Δc in a brief instant Δt. The change per unit time (that is, the rate of reaction) is then $\Delta c/\Delta t$. It is evident from the graph that $\Delta c_1/\Delta t$ is larger than $\Delta c_2/\Delta t$, showing that the rate decreases as the reaction proceeds (Figure 22.1).

Neither $\Delta c_1/\Delta t$ nor $\Delta c_2/\Delta t$ corresponds to any particular instant of time, but rather to the intervals of time covered by the Δt values. The instantaneous

Figure 22.1 Change of concentration of trinitrobenzoic acid as a function of time, showing change in $\Delta c/\Delta t$ as reactant is used up.

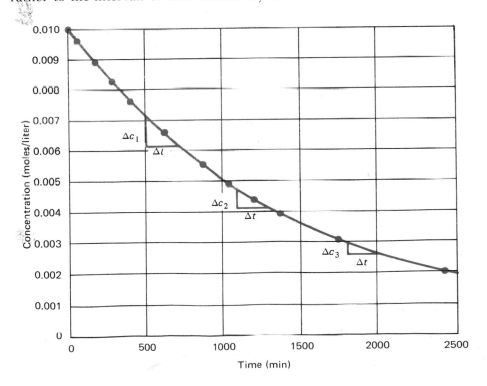

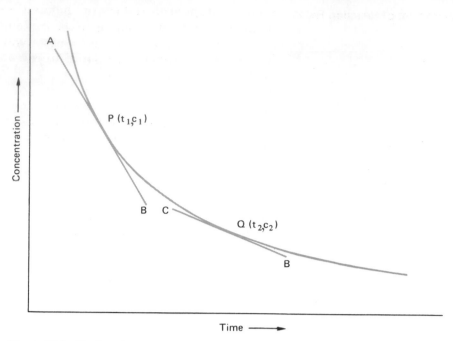

Figure 22.2 Finding the slope of the curve dc/dt at the precise instant t_1. The line *AB* is drawn tangent to the curve at *P*; the line *CD* is drawn tangent to the curve at *Q*. The slope of *AB* is the slope of the curve at *P*; the slope of *CD* is the slope of the curve at *Q*. These slopes give dc/dt at the instants t_1 and t_2, respectively; when multiplied by -1 they become the reaction rates at these times. Concentrations c_1 and c_2 are the concentrations of reactants existing at t_1 and t_2, respectively. One way to determine the rate of a reaction is to first determine experimentally the concentration-time curve (as in Figure 23.1), and then determine the slope of a tangent to that curve at a point that corresponds to a particular concentration and a particular time. In this figure, the rate at *P* is clearly greater than the rate at *Q*, showing that the reaction speed decreases with time and with decreasing concentration of reactants.

quantity, the reaction rate at any given time, is obtained by imagining the time interval, Δt, to become smaller and smaller. As Δt becomes very small and approaches 0, $\Delta c/\Delta t$ becomes equal to the slope of the curve at that particular time (Figure 22.2). It is this value of the slope of the curve at the time in question that is the true value of the reaction rate at that time, and at the concentration of the reactants existing at that instant. Thus, on the graph (Figure 22.1), the reaction rate 172 min after the beginning of the reaction is the slope of the curve at 172 min; this slope has the value 5.89×10^{-6} mole/liter/min, and this is the value of the reaction rate at that time. At 405 min after the start of the reaction, the rate is 5.19×10^{-6} mole/liter/min.

The slope of the curve, expressed as

$$\left(\frac{\Delta c}{\Delta t}\right)_{\Delta t \to 0}$$

is defined in mathematics as the derivative of c with respect to t and is written dc/dt. Because the concentration of reactants decreases with time, dc/dt (the slope) is negative. The rate of reaction is defined as $-dc/dt$, where c is the concentration of reactant.

From what has been said above, the measurement of reaction rate demands the collection of data from which the concentration-time curve of Figure 22.1 may be plotted. The data may be obtained many ways, depending upon the system under investigation. Some procedures frequently used are described in the following paragraphs.

If the reaction involves the removal of an acid or a base, for example, samples may be pipetted from the reaction mixture at measured times (t_1, t_2, t_3, etc.) and the concentrations of acid (c_1, c_2, c_3, etc.) determined in each sample. This was the procedure used in obtaining the data in Table 22.1. Here, the reaction is

Trinitrobenzoic acid → Trinitrobenzene

Samples of the reacting solution were removed at the times recorded, and quickly titrated with standard base to determine the concentration of the acid remaining unchanged at that instant. A graph such as Figure 22.1 was plotted, showing the change of concentration of the acid as a function of time. Tangents to the curve were then drawn; slopes of these tangents represent the instantaneous values of the rates ($-dc/dt$) at the concentrations and times corresponding to the points of tangency, and are recorded in the third column of the table.

For a reaction involving gases, the pressure of the gas is a measure of concentration [from $P = (n/V) \times RT$], and a pressure measurement at known times will give data for the curve.

In another method, the formation or disappearance of a colored substance as a result of the reaction may be observed by watching the changing depth of color as a function of time.

Factors Determining the Rate of Reaction

When data on reaction rates are examined, it is found that the rates depend upon four major quantities: nature of the reactants, presence or absence of a catalyst, temperature, and concentration of the reactants. Let us discuss each of these.

Table 22.1 Decomposition of Trinitrobenzoic Acid in Water at 70°C

Time (min)	Concentration of Trinitrobenzoic Acid (moles/liter)	Rate (moles/liter/min)
0	0.00998	6.69×10^{-6}
58	0.00959	6.42×10^{-6}
172	0.00888	5.89×10^{-6}
275	0.00828	5.55×10^{-6}
405	0.00759	5.19×10^{-6}
618	0.00658	4.51×10^{-6}
870	0.00555	3.72×10^{-6}
1040	0.00495	3.21×10^{-6}
1200	0.00444	2.90×10^{-6}
1360	0.00398	2.69×10^{-6}
1740	0.00308	2.05×10^{-6}
2425	0.00196	1.35×10^{-6}

Nature of Reactants.   Most reactions between simple ions in solution are so fast that their rates cannot be observed by ordinary methods. Many reactions of ions reach equilibrium almost instantaneously. This is not true for most reactions involving covalent bonds such as those in organic molecules, and the rates of reactions involving these substances can often be conveniently measured by methods similar to those described earlier. A useful rule, then, is that reactions involving simple ions are rapid, and reactions of most covalent substances are slow. There are exceptions to both generalizations.

Catalysis.   The rates of many reactions are changed when the reaction mixture is permitted to include a substance known as a catalyst, which itself remains unchanged in quantity and chemical composition at the end of the reaction. For example, the decomposition of hydrogen peroxide in water solution,

$$2H_2O_2 \longrightarrow 2H_2O + O_2 \tag{1}$$

proceeds much faster when bromide ion is present than when the solution contains only hydrogen peroxide. The bromide ion is still in the solution, unchanged in amount, at the end of the reaction; *apparently* it has changed the reaction rate somehow by being present! Chemists do not accept an interpretation that catalysis is due to the mere presence of a catalyzing substance, however, and seek an explanation in terms of a preliminary reaction of the catalyst with the reactant, followed by another reaction that regenerates the catalyst while forming the final product. Thus, the catalyst actually takes some part in the reaction.

The bromide ion catalysis of hydrogen peroxide decomposition is an example of *homogeneous* catalysis, since only one phase (the solution) is present. The reactions in many biological systems involve such homogeneous catalysis by complex substances known as *enzymes*. More usual in industrial chemical processes are cases of *heterogeneous* catalysis, in which a gas or liquid phase containing the reactants is in contact with a solid catalyst. Thus, the ammonia synthesis (Chapter 15 and Table 19.8) uses iron containing small amounts of potassium and aluminum oxides as a catalyst; the oxidation of sulfur dioxide to sulfur trioxide (Chapter 14) uses divanadium pentoxide; and catalytic cracking of hydrocarbons uses silica-alumina mixtures. In each of these cases, the reaction proceeds more rapidly than it would in the absence of the catalyst at the same temperature. This permits the ammonia synthesis, for example, to operate profitably at 500°C instead of going to the higher temperatures that would otherwise be required (see paragraphs on "Temperature," below) to bring the reaction rate to a high enough value for profitable operation.

It has been noted (Figure 19.1) that the presence of a catalyst has no effect on the position of equilibrium, which is determined only by the free-energy change in passing from the initial to the final state. Since the catalyst is a component of both the initial and final states, it makes the same contribution to both, and the terms involving it disappear when the free-energy difference is taken between the two states.

Temperature.   With a few exceptions, the rate of a chemical reaction, under fixed conditions of concentration, increases with increase in temperature. The increase is more than proportional to the increase in temperature;

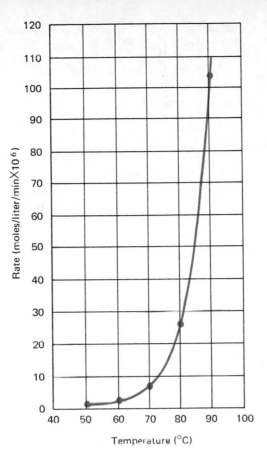

Figure 22.3 The decomposition rate of trinitrobenzoic acid as a function of temperature. The rates are for a concentration of 0.0080 moles/liter.

for many reactions in solution, for example, the rate more than doubles for each 10°C rise in temperature. A reaction proceeding at a given rate at 50°C will go more than twice as fast at 60°C, and more than twice as fast again at 70°C, making the rate at 70°C more than four times as fast as at 50°C. Data for the decomposition of trinitrobenzoic acid at 0.00800 mole/liter are shown in Figure 22.3.

Exact measurements show that the logarithm of the rate is proportional to the reciprocal of the absolute temperature, as expressed by the following relation,

$$\log (\text{rate}) = -\frac{A}{T} + \log B \tag{2}$$

where A and B are constants that differ for each reaction and, for B, for each of the constant concentrations of reactants used in the measurements at the different temperatures (Figure 22.4).

Concentration. It is a general rule that the reaction rate increases with increase in the concentration of reactants. This is what we ought to expect, since a reaction between two species such as a hydrogen molecule and a bromine atom can only occur when they are in contact, or at least, close together, so that a new bond may form between the atoms, while existing bonds are broken

$$H—H + Br \longrightarrow H—Br + H$$

The numbers of such contacts in unit time will depend upon how intensively the molecules are crowded together—that is, upon the concentration—just as the number of collisions between couples on a dance floor depends upon how crowded the dancing area is. The number of contacts between the *two* kinds of molecules, H_2 and Br, will be proportional to the concentration of *each* and, therefore, to the product of their concentrations. We might predict that the rate would be given by an equation of the sort

$$\text{Rate} = kc_{H_2}c_{Br} \tag{3}$$

If this is found to be the case, the proportionality constant, k, is known as the *rate constant for the reaction*. Its value depends upon the temperature at which the rate is measured.

Figure 22.4 The logarithm of the rates of decomposition of trinitrobenzoic acid given in Figure 24.3 plotted according to Equation (2). From Figure 24.3, the rate at 70° C is 5.36×10^{-6} moles/liter/min. The logarithm of this is $\bar{6}.730$ or -5.270 as plotted above for $1/T = 1/(273 + 70) = 0.00292$ (deg K)$^{-1}$.

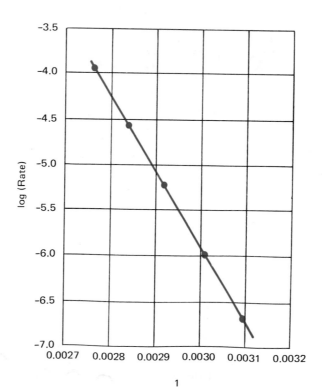

Table 22.2 Rate Constants for the Decomposition of Trinitrobenzoic Acid at 70°C

Concentration (moles/liter)	Rate (moles/liter/min)	Rate Constant, k $\left(\dfrac{\text{rate}}{\text{concentration}}\right)$*
0.00998	6.69×10^{-6}	6.70×10^{-4}
0.00959	6.42×10^{-6}	6.69×10^{-4}
0.00888	5.89×10^{-6}	6.63×10^{-4}
0.00828	5.55×10^{-6}	6.70×10^{-4}
0.00759	5.19×10^{-6}	6.83×10^{-4}
0.00658	4.51×10^{-6}	6.85×10^{-4}
0.00555	3.72×10^{-6}	6.70×10^{-4}
0.00495	3.21×10^{-6}	6.48×10^{-4}
0.00444	2.90×10^{-6}	6.53×10^{-4}
0.00398	2.69×10^{-6}	6.75×10^{-4}
0.00308	2.05×10^{-6}	6.66×10^{-4}
0.00196	1.35×10^{-6}	6.88×10^{-4}

*See Equation (4).

We can also conceive that a reaction might involve only one substance in decomposition, and that the rate of the reaction—that is, the amount decomposing per unit volume per unit time—would depend upon the amount of reactant present in unit volume, so that a rate equation of the form

$$\text{Rate} = kc \tag{4}$$

might appear. This is experimentally the case for the trinitrobenzoic acid decomposition plotted in Figure 22.1 (see Table 22.2).

Order of a Reaction

Reactions that have a rate equation like Equation (4), whose rates depend mathematically upon the concentration of only one substance raised to the first power, are known as *first-order reactions*. Those following an equation like Equation (3) are known as *second-order reactions*, because they involve the product of two concentration terms. An equation corresponding to the reaction of two molecules of the same substance

$$\text{Rate} = kc_A{}^2 \tag{5}$$

is also an equation for a second-order reaction; an example would be that of the reverse of hydrogen iodide formation:

$$2HI \longrightarrow H_2 + 2I \tag{6}$$

$$\text{Rate} = kc_{HI}c_{HI} = kc_{HI}{}^2 \tag{7}$$

A few *third-order reactions* are also known, for example,

$$2NO + O_2 \longrightarrow 2NO_2 \tag{8}$$

with the rate equation

$$\text{Rate} = kc_{NO}{}^2 c_{O_2} \tag{9}$$

There are many additional examples of the dependence of reaction rate on concentrations, giving rate equations similar to those given. A general law, known as the *law of mass action*, can be formulated: *The rate of a chemical reaction is proportional to the concentration of each of the reacting substances.* The *order* of the reaction is equal to the sum of the exponents appearing on the concentration terms in the equation expressing the rate in terms of concentrations.

Complex Rate Equations The rate equations given in the preceding section have the exponents in the rate equation identical to the coefficients in the chemical equation. However, this connection between the chemical equation and the rate equation, which seems to emerge from the results of the preceding paragraphs, is not correct for many reactions. The reason is that reactions often involve several steps, some of which are slow and others fast, and require rate equations that have exponents quite different from the coefficients of the chemical equation. For example, consider the following three reactions and their rate equations:

$$2N_2O_5 \longrightarrow 2N_2O_4 + O_2 \tag{10}$$

$$\text{Rate} = k_{obs} c_{N_2O_5} \tag{11}$$

$$Br^- + OCl^- \longrightarrow OBr^- + Cl^- \tag{12}$$

$$\text{Rate} = k_{obs} \frac{c_{Br^-} c_{OCl^-}}{c_{OH^-}} \tag{13}$$

$$H_2 + Br_2 \longrightarrow 2HBr \tag{14}$$

$$\text{Rate} = k_{obs} \frac{c_{H_2} c_{Br_2}^{1/2}}{1 + k' \dfrac{c_{HBr}}{c_{Br_2}}} \tag{15}$$

In the first case, Equation (10), the chemical equation shows that two molecules are required in the over-all reaction. But the rate equation shows that the rate does not depend upon the number of contacts (collisions) between pairs of molecules, which would require a c^2 term, but that the reaction apparently involves only a single, isolated molecule (c^1). In the case of the oxidation of bromide by hypochlorite, Equation (12), the rate depends upon the concentration of a substance (the hydroxide ion) that does not even appear in the chemical equation for the reaction. In the third case, Equation (14), the rate equation is quite complicated, containing two constants, k_{obs} and k' and a sum and quotient in the denominator.

Mechanism of Reaction Rate equations, such as those given, are established by experiment, by finding what changes in concentration change the rate of reaction, and whether the concentrations should appear as the first power, the square or as some other exponent in the rate equation. The experimental nature of the rate equation has been indicated by labeling the rate constant $k_{observed}$. The chemist then proceeds to suggest a *mechanism* for the reaction, which will explain how a particular rate equation can arise. A complex rate equation such as Equation (13) or (15) indicates that the over-all reaction is taking place with more than one step, so that an intermediate first product formed reacts further in subsequent steps to form the final product. Even a simple rate equation may represent a multistep mechanism, as shown in the next section.

Determination of the Mechanism of a Reaction of Simple Order: The Conversion of Ammonium Cyanate to Urea

Consider, as an example of a mechanism study, the transformation of ammonium cyanate (NH_4OCN) to urea, NH_2CONH_2, a reaction which takes place in water solution

$$NH_4OCN \longrightarrow \begin{array}{c} H_2N \\ \diagdown \\ C=O \\ \diagup \\ H_2N \end{array} \qquad (16)$$

Some data for this reaction are given in Table 22.3.

The Rate Equation. The reaction represented by Equation (16) seems to require only the rearrangement of a single molecule, and suggests that a first-order process is involved. It is evident from column 7 of the table, however, that a rate "constant" calculated for an assumed first-order rate equation shows a definite trend as the concentration decreases, whereas that calculated for a second-order rate equation is constant (column 8). The rate of the reaction is thus expressed by

$$\text{Rate} - k_{\text{obs}} c_{NH_4OCN}^2 \qquad (17)$$

Application of Chemical Information. Equation (17) is simply explained if two molecules of ammonium cyanate are involved in the reaction:

$$2NH_4OCN \longrightarrow 2(H_2N)_2CO \qquad (18)$$

But there seems no compelling chemical reason why the reaction should require two molecules of reactant, and we seek some other explanation of the second-order rate equation. If we remember that ammonium cyanate is a salt and, therefore, completely ionized, we recognize that dissociation into NH_4^+ and OCN^- must occur. When we dissolve 0.1 mole of ammonium cyanate in one liter of solution, therefore, we are really dissolving 0.1 mole of ammonium ions and 0.1 mole of cyanate ions.

If these are the reactants, so that the equation for the reaction is written

$$NH_4^+ + OCN^- \longrightarrow (H_2N)_2CO \qquad (19)$$

then the rate equation would become

$$\text{Rate} = k c_{NH_4^+} c_{OCN^-} \qquad (20)$$

Table 22.3 Rate of Disappearance of Ammonium Cyanate in the Formation of Urea in Water Solution at 50°C*

Time (min)	Concentration (moles/liter)	Δc	Δt	Rate	Average Concentration	$\dfrac{Rate}{Concentration}$	$\dfrac{Rate}{(Concentration)^2}$
0	0.1000	−0.0192	45	0.000427	0.0904	0.00472	0.0523
45	0.0808	−0.0092	27	0.000341	0.0762	0.00448	0.0587
72	0.0716	−0.0078	35	0.000223	0.0677	0.00329	0.0487
107	0.0638	−0.0079	50	0.000158	0.0599	0.00263	0.0440
157	0.0559	−0.0096	73	0.000132	0.0511	0.00258	0.0506
230	0.0463	−0.0083	82	0.000101	0.0422	0.00239	0.0567
312	0.0380	−0.0131	288	0.000045	0.0315	0.00143	0.0455
600	0.0249						

*The rates are approximated as $-(c_2 - c_1)/(t_2 - t_1) = -\Delta c/\Delta t$; these rates are attributed to the average concentration $(c_1 + c_2)/2$ of the change from c_1 to c_2.

But the concentrations of the two ions are equal, since we dissolved them in pairs at the start; they remain equal throughout the reaction, since the chemical equation uses one of each. Thus, $c_{NH_4^+} = c_{OCN^-} = c$, and Equation (20) becomes

$$\text{Rate} = kc \times c = k_{obs}c^2 \qquad (21)$$

as observed in the experimental data of Table 22.3.

Test of the Mechanism. Equation (19) represents a suggested mechanism; it represents a possible explanation for the second-order rate equation found by experiment. In this particular case, the suggestion can be further tested by adding ammonium ion, which is not paired up with cyanate ion. Table 22.4 gives data for such an experiment. Here, ammonium nitrate (NH_4^+, NO_3^-) has been added to an ammonium cyanate solution and the rate equation examined by applying the law of mass action to Equation (19). It is evident that the reaction in the presence of additional ammonium ion is faster than before, and that a constant value of k is obtained when Equation (20) is applied to the data.

Alternative Mechanism. It might be supposed that Equation (19) is now confirmed as the mechanism. Ammonium cyanate, however, is the salt of a weak acid, HOCN, and a weak base, NH_3. Ammonium ions and cyanate ions in water solution therefore will react according to the equation

$$NH_4^+ + OCN^- \rightleftharpoons HOCN + NH_3 \qquad (22)$$

Furthermore, equilibrium will be rapidly established in this reversible acid-base reaction, so every solution containing ammonium ion and cyanate ion contains ammonia and cyanic acid. The reaction forming urea might then be

$$NH_3 + HOCN \longrightarrow (H_2N)_2CO \qquad (23)$$

Table 22.4 Rate of Disappearance of Cyanate in a Solution 0.1 Molar in Ammonium Cyanate and 0.1 Molar in Ammonium Nitrate in Water at 50°C*

Time (min)	Concentration of OCN^- (mole/liter)	Concentration of NH_4^+ (mole/liter)	Δc	Δt	Rate	Average c_{OCN^-}	Average $c_{NH_4^+}$	$\dfrac{\text{Rate}}{c_{OCN^-} \times c_{NH_4^+}}$
0	0.1000	0.2000						
			0.0150	16	0.000912	0.0925	0.1925	0.0526
16	0.0850	0.1850						
			0.0218	34	0.000641	0.0741	0.1741	0.0497
50	0.0632	0.1632						
			0.0131	28	0.000468	0.0566	0.1566	0.0528
78	0.0501	0.1501						
			0.0078	23	0.000342	0.0462	0.1462	0.0502
101	0.0423	0.1423						
			0.0112	46	0.000243	0.0367	0.1367	0.0484
147	0.0311	0.1311						
			0.0078	42	0.000186	0.0272	0.1272	0.0537
189	0.0233	0.1233						

*The rates are approximated as $-(c_2 - c_1)(t_2 - t_1) = -\Delta c/\Delta t$; these rates are attributed to the average concentration $(c_1 + c_2)/2$ of the change from c_1 to c_2.

The rate of this reaction should be expressed by

$$\text{Rate} = k' c_{NH_3} \times c_{HOCN} \tag{24}$$

But the equilibrium in Equation (22) requires that

$$K = \frac{c_{NH_3} \times c_{HOCN}}{c_{NH_4^+} \times c_{OCN^-}} \tag{25}$$

so that

$$c_{NH_3} \times c_{HOCN} = K c_{NH_4^+} \times c_{OCN^-} \tag{26}$$

which, on substitution in Equation (24), gives

$$\text{Rate} = k' K c_{NH_4^+} \times c_{OCN^-} \tag{27}$$

This is of the same form as Equation (20) if $k = k'K$. So the observed rate equation, Equation (20), agrees with both Equations (19) and (23), and we do not know whether the constant multiplying the concentration term $c_{NH_4^+} \times c_{OCN^-}$ is the single constant k or the product of two constants $k'K$. There are thus two explanations for the same data, and kinetic experiments cannot distinguish which one is correct. From other considerations, and by analogy with similar reactions between amines and organic isocyanates, it seems that the mechanism represented by Equation (23) is the more likely.

Rate-Determining Step   The two explanations for the formation of urea are represented by the two mechanisms:

$$NH_4^+ + OCN^- \longrightarrow (H_2N)_2CO \qquad NH_4^+ + OCN^- \rightleftarrows HOCN + NH_3$$
$$HOCN + NH_3 \longrightarrow (H_2N)_2CO \tag{28}$$

Mechanism I   Mechanism II

Each leads to the experimental rate equation. The individual reactions in the mechanisms are known as the *elementary reactions*. In the second mechanism, the first of the elementary reactions is very fast, while the rate of the second reaction is small. The rate of this second reaction determines the over-all rate of urea production, through Equation (24); it is called the *rate-determining step*. It is often true, even in more complex cases, that one of the reactions in a mechanism involving several elementary reactions is much slower than the others, so that the over-all rate is determined by the rate of the slow, rate-determining step.

Although for many reactions only one of the proposed mechanisms seems to offer a reasonable explanation of the experimental facts, the possible existence of other applicable mechanisms cannot be excluded. In fact, in many reactions in nature, several pathways are possible and may be used, depending upon the chemical environment and temperature. The student will not be far wrong if he treats a mechanism as a theory for interpreting the experimental results. As seen earlier in this book, the same experimental data can often be interpreted according to more than one theory. This is also true of theories governing the rates of chemical reaction in general, as discussed in the next sections.

THEORIES OF REACTION RATES   The form of the rate equation, the mechanism of the over-all reaction, the orders of the elementary reactions, and the change of their rates with temperature have been determined from experimental data for many reactions.

It remains to construct a theory to explain, in terms of our knowledge about atoms and molecules, why elementary reactions behave as they do. Actually, two theories have proved useful; the *collision theory* and the *activated-complex theory*. The collision theory suggests that reaction between two molecules takes place on collision of the two, provided that the collision is at least of a certain minimum violence, and that the rate is determined by the number of such violent collisions in unit volume in unit time. The activated-complex theory suggests that an equilibrium is set up in which the reacting molecules form a high-energy association complex containing both of them, and that the rate is determined by the concentration of this complex at equilibrium and by the rate at which it decomposes.

Collision Theory

We have already suggested, in the introductory paragraphs of this chapter, that for reaction to occur between two molecules, they must come together—that is, a *collision* must occur. If this were the only requirement, then a calculation of the number of collisions per unit volume per unit time between molecules A and B should give the rate of reaction between A and B. On the basis of the kinetic theory of gases, it is possible to calculate the collision rate for gaseous molecules; it is found that the observed reaction rates are less than the collision rate by many powers of ten. Further, the increase of the calculated collision rate for a 10°C rise in temperature is only one per cent or less, whereas the observed increase in reaction rate is 100 per cent or more.

The collision theory explains this by suggesting that it is not enough for reaction simply that a collision occur, but that the collision must be a violent one, involving an energy amounting to at least 10–60 kilocalories for a mole of such collisions. Assuming that the effective energy is that obtained from the head-on components of the ordinary kinetic theory velocities of the molecules (Figure 22.5), an equation of the following form is derived:

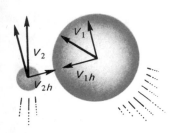

Figure 22.5 In collision theory, the energy available for reaction is assumed to be that associated with the head-on components V_{1h} and V_{2h} of the velocities V_1 and V_2 of the two colliding molecules.

$$\log \begin{bmatrix} \text{number of collisions} \\ \text{per cubic centimeter} \\ \text{per second with} \\ \text{energy in head-on} \\ \text{component} \\ \text{greater than } E_{\min} \end{bmatrix} = -\frac{E_a}{2.3RT} + \log \begin{bmatrix} \text{total number of collisions} \\ \text{per cubic centimeter per} \\ \text{second for the concentration} \\ \text{and temperature of the} \\ \text{experiment} \end{bmatrix}$$

$E_{\min}$ is the minimum energy in the head-on component of collision which will lead to reaction. R is the molar gas constant, 1.987 cal/deg/mole, and E_a is $E_{\min}$ multiplied by Avogadro's number to put the expression on a molar basis

$$E_a = NE_{\min}$$

E_a is known as the *energy of activation*: It is the minimum head-on energy, per mole of collisions, needed for reaction to take place on collision. Equation (29) is of the same form as the experimental equation, Equation (2); if $E_a/2.3R$ is set equal to the experimental constant A in Equation (2), it is found that, for many second-order reactions, B is equal to the calculated logarithmic term on the right of Equation (29) when it is assumed, in accordance with the theory, that the reaction rate is given by the number of collisions with energy greater than E_a. For these reactions, the theory is thus confirmed.

For some second-order reactions, the number of violent collisions calculated on this basis is greater than the observed rate, suggesting that requirements other than a collision of minimum violence are necessary. One such requirement might be that the molecules must collide with a specific orienta-

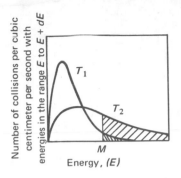

Figure 22.6 **Figure 22.6** Numbers of collisions with specified total energies as a function of the energy of the collision.

tion at the moment of collision; since this orientation would be only one of many possible orientations, the reaction rate would be lessened considerably.

The minimum energy E_a (the activation energy) provides an interpretation of the empirical constant A in Equation (2), as shown by the comparison of Equations (29) and (2). To understand its effect in producing the marked increase in reaction rate as a function of temperature, it is helpful to consider Figure 22.6. This figure shows the number of collisions in which the energies of the collisions lie in the narrow energy range between energy E and $E + dE$, plotted as a function of E. Each curve shows that the number of collisions increases as the energy of the collision increases, and then decreases again. Note that the curves for the two temperatures cross; at the higher temperature T_2, the number with any particular energy is smaller than at the lower temperature T_1 for low energies, but larger than for T_1 at higher energies. Thus, the number of violent collisions, represented by the sum of all collisions with energies greater than that marked by the line M, is markedly increased as the temperature is increased. On the average, half of the collision energy will be available for reaction. The other half, corresponding to the component of relative velocity perpendicular to the line of centers, represents a glancing blow that does not lead to reaction. The increase of the reaction rate with temperature therefore corresponds closely to the ratio of the area to the right of M under curve T_2 to the corresponding area under T_1, where M represents the minimum collision energy necessary to provide the activation energy E_a.

First-Order Reactions

The collision theory in its simple form makes no allowance for the possibility of first-order reactions, since a collision process is of necessity second order, involving two molecules. As a way out of this difficulty, it was suggested that activation by a violent collision does not necessarily lead immediately to decomposition. Instead, decomposition occurs only after a time delay. During this delay, the energy acquired by the molecule in the collision rearranges itself among the various modes of the molecule's vibrational motion until it reaches the one vibration that leads to bond breaking and reaction. Since the rearrangement of energy could take place in an isolated molecule, the collision history of the *acquisition* of the energy would no longer affect the concentration dependence of the actual decomposition (Figure 22.7).

Since the mechanism postulates a time lag between the acquisition of energy and its arrival at the vibration leading to reaction, it would be expected that first-order processes would be more likely the more complex the molecule and, therefore, the more vibrational modes of motion it had. This prediction is verified experimentally, and first-order reactions of gaseous molecules are mainly confined to large and complex species. In the simplest case of a diatomic molecule, there is only one possible vibration and dissociation of diatomic molecules is never observed to be first order.

Figure 22.7 In a first-order reaction in gases, acquisition of energy does not immediately lead to reaction; this is a separate event involving only one energy-rich molecule.

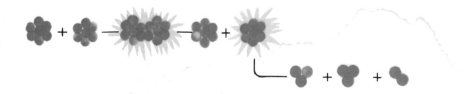

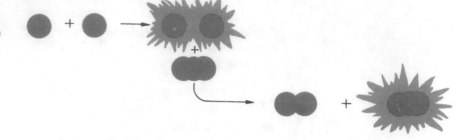

Figure 22.8 Combination of two atoms to form a diatomic molecule requires the presence of a third body to carry off the energy produced when the two atoms combine.

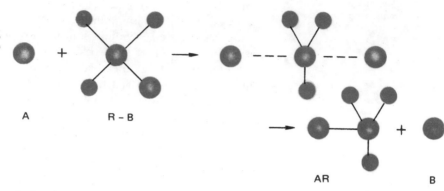

Figure 22.9 Formation and decomposition of an activated complex in the reaction of molecules A_2 and B_2 to form AB molecules.

A   R – B

AR   B

Three-Body Processes

A further consequence of the single vibrational degree of freedom in a diatomic molecule X_2 is that the combination of two atoms to form such a molecule must always be a three-body collision:

$$X + X + M \longrightarrow X_2 + M \tag{30}$$

The "third body," molecule M, must act as a sink to remove excess energy from X_2 (Figure 22.8). If M were not there, all the bond energy of forming X_2 would be stored as vibrational energy in the newborn molecule, and it would dissociate in the first vibration after forming. Different substances have different capacities for removing some or all of this excess energy, and molecules M may be classified according to their efficiencies as third bodies. Larger molecules are more efficient than smaller ones because they are more capable of storing energy among their many degrees of vibrational freedom.

Activated-Complex Theory

This theory presumes that the reacting molecules in an elementary reaction combine to form a transient intermediate known as an activated complex (Figure 22.9). The concentration of the activated complex is assumed to be calculable on the basis of an equilibrium between it and the reactants,

$$A + B \rightleftharpoons M^{\ddagger} \tag{31}$$

The observed reaction rate is proportional to this concentration. An important property of this equilibrium is that the activated complex is presumed to be at a much higher energy level than that of the reactants. Thus, in the passage from the initial state to the final state, which results in the enthalpy change for the reaction, ΔH, the individual molecules must acquire a high energy to put them into the state represented by the activated complex. The difference in energy, per mole of complexes, between the initial state and the state representing the activated complex is about equal to the energy E_a of

Equation (29), and reproduces in the theory the observed temperature dependence of the reaction rate (Figure 22.10).

Application of thermodynamic principles to these concepts predicts that the rate of a second-order reaction between molecules A and B should be given by

$$\text{Rate} = \frac{kT}{h} K^{\ddagger} [A][B] \qquad (32)$$

Here, k is Boltzmann's constant, h is Planck's constant, and $K^{\ddagger}$ is an equilibrium constant for the assumed equilibrium between activated complex and reactants, modified slightly to take account of the fact that the activated complex is unstable and reacts to form products. Comparison of Equation (32) with the experimentally observed expression for the rate of a second-order reaction between A and B

$$\text{Rate} = k_{\text{obs}}[A][B] \qquad (33)$$

shows that $k_{\text{obs}} = (kT/h)K^{\ddagger}$. The dependence of the rate upon the temperature exemplified in Figure 22.3 appears mainly as the dependence of the equilibrium constant $K^{\ddagger}$ on the temperature [Equation (64), Chapter 19]. If molecules A and B were hard spheres without internal structure, as assumed in collision theory, and if the energy-level difference between them and the activated complex were set equal to the activation energy E_a of collision theory, the rate of reaction calculated from Equation (32) would be the same as that calculated from the collision theory. For this special case, therefore, the two theories give identical results.

An important feature of the activated-complex theory is that the term kT/h appears in all rate equations for elementary reactions, no matter what the reactants or order, and that $K^{\ddagger}$ can (in principle) be calculated from the structures and energy levels in the reacting molecules and the activated complex. The parenthetical "in principle" is necessary, for although we can obtain the necessary experimental data on structures and levels in the reactants which, being stable, permit experimental examination of their spectra, dipole moments, and the like—such experimental examination for

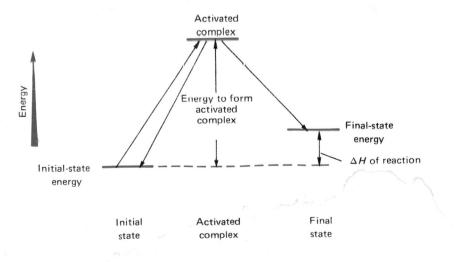

Figure 22.10 Energy relationships in formation of the activated complex and in the over-all reaction.

the transient activated complex is not possible. Various estimates, analogies, semiquantitative calculations, and so forth must be used to guess at its structure and energy levels. These difficulties prevent us from carrying out the precise calculation of the rate that the theory seems to promise except in the simplest cases of hydrogen atoms reacting with hydrogen molecules. In that case, however, the theory does give correct answers, and encourages us to believe that the general postulates are valid.

Enthalpy and Entropy of Activation

The presence of the equilibrium constant, $K^{\ddagger}$, in the expression for the reaction rate on the basis of the activated complex theory suggests the extension to free-energy changes. The change in free energy can, in turn, be expressed in terms of changes in enthalpy and entropy. Recalling the relations between the equilibrium constant and these quantities [Equation (33), Chapter 20], we apply these relations to the postulated equilibrium between activated complex and reactants:

$$-2.303RT \log K^{\ddagger} = \Delta G^{\ddagger} = \Delta H^{\ddagger} - T\Delta S^{\ddagger} \qquad (34)$$

The "enthalpy of activation," $\Delta H^{\ddagger}$, represents the difference in energy between reactants and activated complex. It can be related to the experimentally determined activation energy E_a [Equation (29)], obtained from A in Equation (2), and to the energy level difference between the initial state and the activated complex shown in Figure 22.10. $\Delta S^{\ddagger}$ is the "entropy of activation," which can also be determined from experiment; it is related to B in Equation (2). Thus, in terms of the concepts of enthalpy and entropy, meanings are given to the purely empirical quantities A and B of the earlier equation, and chemists are led to think of variations from reaction to reaction in terms of these concepts. The enthalpy of activation is always positive; the entropy of activation may be positive or negative. The *larger the enthalpy of activation* and the *smaller* or more negative *the entropy of activation*, the *smaller* the reaction rate, and conversely.

In general, the enthalpy of activation will depend upon the strengths of the bonds to be broken and formed in producing the activated complex. We can estimate the changes in rate when a change is made in these bond strengths. A substitution on an organic molecule, for example, which changes the bond strength in the reacting functional group, will be expected to change the reaction rate in the direction corresponding to the change in bond strength. The entropy of activation is a measure of the change in randomness in forming the activated complex. If the activated complex is more loosely joined together than the reactants, there will be an increase in disorder and, consequently a positive value for $\Delta S^{\ddagger}$. If, on the other hand, the complex were more ordered than the reactants, as might happen if a ring structure were produced from chain structures, the entropy of activation would be negative, and the predicted reaction rate less than it would be if $\Delta S^{\ddagger}$ were zero or positive.

The concepts of enthalpy and entropy of activation have been particularly useful to organic chemists in interpreting and predicting reaction rates on the basis of the structures and bonding in the reacting compounds. No such relationship to internal structure is available from collision theory, which treats molecules essentially as if they were hard spheres, colliding like billiard balls with no internal "softness" to absorb or transfer energy.

Chain Reactions

The formation of hydrogen bromide from hydrogen and bromine is an example of one of the classes of reactions known as chain reactions. A mechanism leading to the experimental rate equation

$$\text{Rate} = k_{obs} \frac{c_{H_2} c_{Br_2}^{1/2}}{1 + k' \dfrac{c_{HBr}}{c_{Br_2}}} \tag{15}$$

is

$$
\begin{align*}
\text{(a)} \quad & M + Br_2 \longrightarrow 2Br + M \\
\text{(b)} \quad & Br + H_2 \longrightarrow HBr + H \\
\text{(c)} \quad & H + Br_2 \longrightarrow HBr + Br \\
\text{(d)} \quad & H + HBr \longrightarrow H_2 + Br \\
\text{(e)} \quad & M + Br + Br \longrightarrow Br_2 + M
\end{align*}
\tag{35}
$$

Note that, in this mechanism, a bromine atom is a reactant in reaction (b) of Equation (35) and a product in reaction (c). A combination of elementary reactions such as these two, in which a reactive species serving as reactant in one of the steps is regenerated in a subsequent step, is known as a *chain reaction*. The reactive species—here, the bromine atom—is known as the *chain carrier*. It is evident that *if no other processes intervene*, the chain could continue indefinitely until all the hydrogen or bromine were used up. A single bromine atom, if introduced into a mixture of hydrogen and bromine, could cause all the material to change to hydrogen bromide, by repeating reaction (b), followed by (c), then (b), (c), and so on. Actually such extensive reaction does not occur because reactions such as (e), which remove chain carriers, intervene.

It is characteristic of a chain reaction that there must be a *chain-initiating step* that produces chain carriers [reaction (a)]; *chain-propagating steps*, which bring about the actual transformation to reaction products [reactions (b) and (c)]; and *chain-breaking steps*, which remove chain carriers [reaction (e)]. The chain-breaking steps "shorten the length of the chain" because, whenever a chain carrier is removed, all the reactions of the subsequent chain steps that would have been built up on that chain carrier are prevented from taking place. We can, in fact, prevent a chain reaction from taking place by adding to the reaction mixture a substance that will react with chain carriers as they are formed and keep the chain from propagating. A substance used for this purpose is known as an *inhibitor* for the chain reaction.

Reaction (d) is somewhat special to the hydrogen-bromine reaction, and serves to explain the presence of the concentration of hydrogen bromide in the denominator of the rate equation (15). As the hydrogen bromide concentration increases, hydrogen atoms, which are also chain carriers, disappear wastefully in reaction (d) without producing HBr as the reaction product.

Chain carriers are extremely reactive. Under normal conditions, it can be presumed that their rates of reaction with molecules are so much greater than their rates of formation or removal by reaction with each other, that their concentrations reach a constant value characteristic of a *steady state*, in which the rate of formation of chain carriers is equal to the rate of their disappearance.

Branched-Chain Reactions   In the chain-propagating steps of some chain reactions, the reaction of one chain carrier leads to the formation of more than one chain carrier. This, for example, is the case in the reaction of hydrogen and oxygen to form water. Here, the chain-propagating steps are

$$
\begin{aligned}
H + O_2 &\longrightarrow OH + O \\
O + H_2 &\longrightarrow OH + H \\
OH + H_2 &\longrightarrow H_2O + H \\
\underline{OH + H_2} &\underline{\longrightarrow H_2O + H} \\
H + O_2 + 3H_2 &\longrightarrow 2H_2O + 3H
\end{aligned}
\tag{41}
$$

The summation of these equations shows that from one chain carrier hydrogen atom, three are produced. The chain is said to *branch*. Since each new hydrogen atom can initiate new chains, which can also branch, the net production of water goes on faster and faster. If enough hydrogen and oxygen are present to permit attaining enormously high reaction rates before all the reactants are used up, an explosion will occur.

Thermal Explosions   Not all explosive chemical reactions result from a chain-branching mechanism. In highly exothermic reactions, even of a nonchain type, the heat of reaction may not be carried away fast enough to maintain a constant temperature. Therefore, the temperature of the reaction mixture rises. As a result of the temperature effect on reaction rates, the reaction proceeds more rapidly, liberating heat at a still faster rate with still less chance of its being conducted away. Thus the temperature continues to rise and the reaction rate to increase, until it becomes so great that an explosion occurs.

Are Reaction Rates Important to Life?   Perhaps the most common manifestations of the speeds or dynamics of chemical processes and the versatility of these dynamics are in the response systems of animals. To survive, any organism must be sensitive to stimuli from unusual environmental changes that have essential effects on its activity. Thus all organisms react to considerable changes in pressure, chemical composition, temperature, radiation, and electrical properties of the environment.

Discrimination of particular stimuli and reactions to them differ greatly among organisms. For example, we have the special chemical senses of taste and smell; our sense of sight discriminates among radiations by direction, intensity, pattern, and frequency (color). Studies show that responses to stimuli depend more on the organism than on the stimuli themselves. Beyond doubt, the human response system, with its incredible speed, versatility and sophistication, is one of the great marvels of the world.

The nervous system is the most characteristic part of the body's mechanism for specific responses to stimuli. Special receptor organs—such as those for sight, taste, smell, and heat—transfer stimuli to nerve impulses. The nerve impulse is an electrical signal transmitted along fiberlike extensions of nerve cells that serve as conductors from receptors to effectors. Their pathways and connections determine what the ultimate response will be. Memory, perception, association, and all the richness and misery of our mental lives are the end products of these responses. The nervous system also is involved in homeostatic regulation—that is, in maintaining stability in the internal dynamic chemical environment of the body.

Although the nerve impulse is electrical, it is associated with and set up by very rapid and specific chemical reactions. Different mechanisms are involved for the conduction of the electrical impulse along a nerve fiber, crossing from one nerve to another (at synaptic junctions), and crossing from a nerve into other tissues (for example, neuromuscular junction). Migration of the impulse is accompanied by exchange of Na^+ and K^+ ions across the nerve sheath, the net effect being that the electrical potential across the nerve sheath is increased perhaps by as much as 100 millivolts. This creates the condition for transmission of the electrical impulse.

At the synapse or junction between two nerve cells, the impulse may be transmitted by chemical or electrical mechanisms. The chemical mechanisms may reinforce or inhibit transmission. Arrival of an impulse at a chemical synapse results in release of transmitter molecules, such as acetylcholine, which are thought to be stored in vessels on one side of the synaptic junction. These molecules diffuse across the junction to receptor sites on the other side where they increase the permeability of the membrane of the receptor nerve to ions such as Na^+ or K^+. The presence of these ions changes the electrical potential at this point, creating again the conditions for electrical transmission of the impulse.

If all impulses are transmitted this way, and if the central nervous system can be thought of as a complex, preprogramed, rigidly wired series of electrical circuits, how can human behavior be modified or learning occur? One thought is that this is related to the adaptability of the chemical reactions at the synaptic membranes. Frequent excitation at a certain junction may result in the reaction occurring more readily. Persistent negative feedback at a junction may result in production of inhibitors to the reaction. A considerable research effort currently is focused in this direction.

The nervous system also is involved in maintaining the internal dynamic environment of the body. It performs this function in part through its modulation of autonomic functions such as respiration, heart rate, and blood pressure, and in part by its critical influence on the pituitary, the master gland of the endocrine system.

The endocrine glands synthesize hormones and secrete them into the blood stream. The hormones act as chemical messengers to stimulate or moderate the chemical processes of the body. Physiological effects of hormones may be grouped as follows: preservation of internal stability, control of body growth, and development of sex characteristics.

Because of their potency, hormone concentrations in body tissues must be carefully regulated. However, to protect the organism, hormones must be instantaneously available. This is accomplished by the presence of the pituitary. The master gland, located just under the brain, controls the excretion rates of the other endocrine glands.

We realize, then, that in every bodily response we are aware of, and in the many thousands each minute of which we are unaware, the speed, versatility, and adaptability of chemical reactions play a most vital role.

SUMMARY   Rates of chemical reactions, their measurement, factors that affect them, theories concerning them, and uses of rate data in elucidating mechanisms of reactions are summarized in this chapter.

The rate of a reaction, defined as the quantity of material per unit volume undergoing transformation in a unit time and expressed mathematically as $-dc/dt$, where c is the concentration of a reactant and t is the time, is found by experiment to depend upon the nature of the reacting substances, their concentrations, and temperature. A quantitative relation showing the dependence of the reaction rate on the concentrations of reactants can be obtained by studying the rate at a series of reactant concentrations. From this, the order of the reaction is obtained. The order of the reaction often provides information about the number and kind of species involved in the rate-determining step of the reaction.

Studies of the effect of temperature on rate give data for calculating the activation energy of the reaction—the energy, above the average, needed for reaction.

Two theories of reaction rates, the collision theory and the activated-complex theory, are presented. The collision theory focuses attention on the collisions that result in reaction, on their number, and their energy. The activated-complex theory emphasizes the nature of the chemical intermediate formed in a successful collision, on its geometry, bonding, and structure.

Applications of rate studies to mechanisms of molecular and ionic reactions and to chain reactions including explosions are given.

IMPORTANT TERMS

Mechanism of reaction
complex reaction
rate-determining step

Slope of a curve
negative slope
linear equation
dc/dt

Catalysis
heterogeneous
homogeneous
enzyme

Rate of reaction
law of mass action
first-order reaction
second-order reaction
reaction-rate constant

Collision theory
activated-complex
 theory

Energy of activation
enthalpy of activation
entropy of activation

Chain reaction
chain carrier
chain-initiating step
chain-propagating step
chain-breaking step
steady state
inhibitor
branched-chain reaction

Thermal explosion
branched-chain
 explosion

Three-body process

QUESTIONS AND PROBLEMS

1. Distinguish between (a) chemical kinetics and thermodynamics; (b) unimolecular and bimolecular reactions; (c) first-order and second-order reactions; (d) enthalpy of activation and enthalpy of reaction; (e) rate constant and equilibrium constant; (f) branching chains and nonbranching chains; and (g) activity and concentration.

2. Describe five methods of following the rate of a reaction.

3. Show that the equation, $\log(\text{rate}) = -(A/T) + \log B$, can be cast into the form of an equation for a straight line. What is the interpretation of the quantity A? Of the quantity B?

4. How do you explain the fact that "a catalyst affects the rate of a reaction but it does not affect the position of equilibrium"?

5. In 35 seconds, 0.0013 mole of substance A is produced by chemical reaction in 2.40 liters of solution. What is the rate of the reaction?

6. Substance C is produced by a second-order reaction between substances A and B at a rate equal to 0.0042 mole per minute when 4.0 moles of A and 3.0 moles of B are present in a 1.20-liter volume of solution. Determine the rate equation and calculate the rate constant.

7. The half-life of a reaction is the time required for one-half the original material to be used up. What is the half-life of the reaction of Figure 23.1?

8. Substance P is formed by reaction of substances M and N at constant temperature at the rates given below. Determine the rate equation and calculate the rate constant.

Rate of formation of P(moles/l/sec $\times 10^3$)	Conc. of M (moles/l $\times$ 10)	Conc. of N (moles/l $\times 10^2$)
8.13	1.46	3.92
12.26	2.19	3.92
1.98	1.46	1.95

9. A substance A undergoes reaction to form substances B and C. At a fixed temperature, the concentration of A is observed to change with time according to the measurements recorded in the table. Plot these data and determine the rate of reaction at 50, 100, 200, 400, and 700 min, and the concentration of A at those times. From these corresponding values of rate and concentration, determine the order of the reaction and the value of the rate constant for the fixed temperature of the experiment.

Time (min)	Concentration (moles/liter)	Time (min)	Concentration (moles/liter)
0	0.1000	180	0.0597
20	0.0930	240	0.0526
40	0.0870	300	0.0471
60	0.0817	400	0.0400
80	0.0769	500	0.0348
100	0.0727	700	0.0276
125	0.0696	1000	0.0210
150	0.0640		

10. Substance Q is formed by a second-order reaction between R and S, for which the rate constant is 4.16×10^{-2} liter per mole per minute. Estimate the number of moles of Q that will be formed in 10 minutes in 20 liters of solution containing initially 20 moles of R and 3.0 moles of S.

11. The data plotted in Figure 22.3 are

Temperature (°C)	Rate at c = 0.0080 M (moles/liter/minute $\times 10^6$)
50.0	0.191
60.0	1.10
70.0	5.36
80.0	24.9
90.0	103.2

Plot these data as in Figure 22.4 and calculate the activation energy.

12. Assuming that reaction (b) is reversible, and that there is an equilibrium between iodine atoms and iodine molecules, derive the expression for the equilibrium constant of the reaction of hydrogen iodide formation.

13. The general equation for the reaction of the functional group —Cl with hydroxide ion is

$$R—Cl + OH^- \longrightarrow ROH + Cl^-$$

For some hydrocarbon residues (R—), this reaction follows the rate equation

$$Rate = k[R—Cl][OH^-]$$

For others, the rate equation is

$$Rate = k'[R—Cl]$$

Show that these rate equations agree with the following mechanisms, in which the first reaction is rate determining:

$$R—Cl + OH^- \longrightarrow ROH + Cl^- \qquad R—Cl \longrightarrow R^+ + Cl^-$$

$$R^+ + OH^- \longrightarrow ROH$$

$$\text{Mechanism I} \qquad\qquad \text{Mechanism II}$$

14. Oxides of nitrogen are used as catalysts in the oxidation of sulfur dioxide, for which the reaction is

$$2SO_2 + O_2 \longrightarrow 2SO_3$$

Show that NO and NO_2 will not be used up in the oxidation if the reactions involved are

$$SO_2 + NO_2 \longrightarrow SO_3 + NO$$

$$2NO + O_2 \longrightarrow 2NO_2$$

15. Which of the following molecules would you expect to be most effective, and which least effective, as a third body in the combination of two atoms to form a molecule: Ne, CO_2, C_2F_6?

16. How can the values of $K^{\ddagger}$, $\Delta G^{\ddagger}$, $\Delta H^{\ddagger}$, and $T\Delta S^{\ddagger}$ be determined?

17. Discuss the effect of a temperature change on the values of k, K, ΔG, ΔH, and $T\Delta S$.

18. What is the difference in the assumptions on which the following theories are based: (a) the collision theory; (b) the activated-complex theory?

19. How are (a) the energy of activation of the reaction $A + B \longrightarrow C + D$, (b) the energy of activation of the reaction $C + D \longrightarrow A + B$, and (c) the enthalpy of the reaction $A + B \longrightarrow C + D$ related, if both the reactions written are elementary reactions?

20. How are chain reactions related to explosions? How can explosions be avoided in such reactions?

21. How do reactions in solution differ from those in the gaseous state?

1. A substance T decomposes in water solution, forming substances U and V. From the chemical equation for the reaction, it is known that one mole each of U and V is formed from one mole of T. Substance V is a base, and the decomposition is followed by titrating the base formed, using an acid solution of known concentration. The data obtained in an experiment at 70°C in which 9.98-ml samples of the decomposing solution were titrated with 0.0167 M H_2SO_4 at the time intervals listed are given below. Determine the rate equation for the decomposition, and calculate the rate constant.

Time (hr)	ml H_2SO_4
0.0	0.0
6.0	8.87
12.0	15.24
18.0	19.59
Very long	29.39

Hint: Note that the titration data determine the concentration of reaction *product*. The concentration of *reactant* needed for application of the law of mass action is proportional to the difference between the amount of base present at complete reaction (time = very long) and the amount of base present at any measured time. The rates may be approximated as in the footnotes to Tables 22.3 and 22.4.

2. Show that the rate equation for the reaction

$$Br^- + OCl^- \longrightarrow OBr^- + Cl^-$$

is consistent with the mechanisms

$H_2O + OCl^- \rightleftharpoons HOCl + OH^-$	rapid acid-base equilibrium, constant K_1
$Br^- + HOCl \longrightarrow HOBr + Cl^-$	rate-determining, constant k
$HOBr + OH^- \rightleftharpoons H_2O + OBr^-$	rapid acid-base equilibrium, constant K_2

or

$H_2O + OCl^- \rightleftharpoons HOCl + OH^-$	rapid acid-base equilibrium, constant K_1
$Br^- + HOCl \longrightarrow BrCl + OH^-$	rate-determining, constant k'
$BrCl + 2OH^- \longrightarrow OBr^- + Cl^- + H_2O$	fast, constant k''

and determine the relation between the observed rate constant k_{obs} and the constants given in the mechanisms.

3. Can kinetic measurements distinguish between the rate equations

$$\text{Rate} = k_{obs} \frac{[Br^-][OCl^-]}{[OH^-]}$$

$$\text{Rate} = k'_{obs}[Br^-][OCl^-][H_3O^+]$$

for the reaction of bromide ion with hypochlorite ion?

4. Show that the rate equation for the decomposition of dinitrogen pentoxide is consistent with the mechanism

$$N_2O_5 \rightleftarrows NO_2 + NO_3 \qquad \text{rapid equilibrium, constant } K_1$$

$$NO_2 + NO_3 \longrightarrow NO_2 + O_2 + NO \qquad \text{rate-determining, constant } k$$

$$NO_3 + NO \longrightarrow 2NO_2 \qquad \text{fast, constant } k'$$

$$2NO_2 \rightleftarrows N_2O_4 \qquad \text{rapid equilibrium, constant } K_2$$

and express the observed rate constant, defined by

$$\text{Rate} = k_{obs}(N_2O_5)$$

in terms of k, k', K_1, and K_2.

5. Enzymes catalyze biochemical reactions. The mechanism of many such reactions conform to the following equations (E = enzyme molecule; S = reacting molecule; ES = a molecule formed by combination of enzyme and reactant; P = product molecule):

$$E + S \longrightarrow ES \qquad \text{rate constant } k_1$$

$$ES \longrightarrow E + S \qquad \text{rate constant } k_2$$

$$ES \longrightarrow E + P \qquad \text{rate constant } k_3$$

In most reaction systems, $[E] \ll [S]$ and the amount of reactant S complexed as ES is negligible compared to the total amount of reactant present. Thus, a valid approximation is

$$[S] = S_t - [ES] \simeq S_t$$

where S_t represents the total concentration of reactant. Such an approximation is not valid for $[E]$, which is given by

$$[E] = E_t - [ES]$$

Set up the steady-state approximation (see an advanced text) for $[E]$ on the basis of the mechanism given and make the substitutions for $[E]$ and $[S]$ suggested above. Use the value of $[ES]_{ss}$ thus obtained in the expression for the rate of production of product P, and cast the result in the form

$$\frac{d[P]}{dt} = \frac{k_3 E_t S_t}{S_t + \dfrac{k_2 + k_3}{k_1}}$$

This is known to biochemists as the Michaelis-Menten law, and the quantity $(k_2 + k_3)/k_1$ is called the Michaelis constant for the enzyme; the fraction is often denoted as K_m.

6. The rate of a bimolecular reaction, in moles per liter per minute, when the concentrations of the reactants are $0.100\ M$ and $0.0100\ M$, is given as a function of temperature in degrees Kelvin by the equation

$$\log (\text{rate}) = -\frac{3163}{T} + 8.899$$

Calculate: (a) The activation energy. (b) The rate at 10°C. (c) The rate constant at 10°C. (d) The rate at 10°C when the concentration of each reactant is $0.100\ M$.

7. From values in the Appendix A, calculate the value at 27°C of the quantity kT/h that appears in the rate expression in the activated-complex theory. Be sure to give the units of the answer.

8. A substance Q undergoes a reaction in water solution. Analyses of the products of the reaction indicate that they contain hydrogen and oxygen that must have come from the water. Analyses of the concentration of Q as a function of time gave the following results for an experiment carried out at a constant temperature. Determine, from these data, the order of the reaction with respect to Q. What can you say about the number of molecules involved in the chemical equation for the rate-determining step?

Time (min)	Concentration (moles/liter)	Time (min)	Concentration (moles/liter)
0	0.200	150	0.088
20	0.178	190	0.071
40	0.161	200	0.057
60	0.144	275	0.045
80	0.129	310	0.037
100	0.116	350	0.030
125	0.101		

9. The number of collisions for two gaseous molecules A and B at 27°C and a specified pressure is calculated to be 2.12×10^{28} per cubic centimeter per second. Calculate the number of collisions per cubic centimeter per second in which the head-on component has energy greater than 21,600 cal/mole. If each such collision causes reaction to occur, how many molecules of A react per cubic centimeter per second? What is the activation energy of the reaction?

10. Show, by integration with respect to time, that the concentration of reactant A in an elementary reaction obeying the equation

$$2A \longrightarrow P$$

changes with time according to the equation

$$\frac{1}{c} - \frac{1}{c_0} = kt$$

where c_0 is the concentration at $t = 0$, c the concentration at time t, and k the rate constant.

11. If you assume that rates of chemical reactions are affected by a change of temperature and also assume that atomic nuclei are made of particles that interact with each other to give nuclear reactions, how do you account for the validity of the statement found in many textbooks that "the radiochemical process occurs at the same rate at the temperature of liquid air as it does at the temperature of boiling iron"? If this is true, does it therefore follow that the rate of the nuclear disintegration process is independent of temperature? Defend your answer.

REFERENCES KING, E. L., *How Chemical Reactions Occur*, Benjamin, Menlo Park, Calif., 1963.
LATHAM, J. L., *Elementary Reaction Kinetics*, Butterworth, London, 1962.
LERNER, J., "Kinetic Aspects of Biological Membrane Transport Processes," *J. Chem. Educ.*, **48**, 391 (1971).
PATTON, A. R., *Biochemical Energetics and Kinetics*, Saunders, Philadelphia, 1965.
SHEEHAN, W. R., "Along the Reaction Coordinate," *J. Chem. Educ.*, **47**, 254 (1970).

PART SIX

NUCLEAR CHEMISTRY

NUCLEAR CHEMISTRY

The search for understanding of the atomic nucleus is taking place in chemistry and physics laboratories around the world, where both experimental and theoretical tools of great sophistication often are used. Such experimental tools include the cyclotron, the synchrotron, the linear accelerator, and the Van de Graaff generator—devices for exciting nuclei by bombardment with high-energy particles. By following the nuclear changes that occur or the radiation emitted as an excited nucleus returns to a stable state, much knowledge about properties of the atomic nucleus has been gathered. The theoretical tool is quantum mechanics, for it is found that with nuclear phenomena, as with atomic structure, essential insight into observed behavior is provided by this powerful mathematical approach that relates energy to shape and symmetry.

Yet the nature of the atomic nucleus remains one of the great mysteries of modern science. To date, no comprehensive theory explaining all observed nuclear phenomena has been developed, and we are just beginning to understand the nature of nuclear forces. Several kinds of symmetry factors known as spin, parity, and strangeness have been used successfully to interpret various aspects of nuclear behavior, but even so the picture remains cloudy and somewhat obscure.

Various theories of the nucleus have emerged: a liquid-drop model, a shell theory, a unified theory, a cluster theory. All of these account reasonably well for important nuclear properties such as mass, charge, spin, magnetic moment, binding energy, and energies of nuclear excited states. It appears that neutrons and protons are very similar, and the general term *nucleon* is used for both. Nuclear structure often is described in a manner similar to atomic structure—as an assembly of protons and neutrons existing in definite energy levels analogous to the electron energy levels of the atom. An ideal theory would enable us to calculate the various properties of a given nucleus from quantum mechanical principles and the charge and mass of the nucleons present. To do this, the forces between two nucleons must be known exactly. Unfortunately, the nucleon-nucleon force does not appear to

493

be as simple as the electrostatic force and, at present, it is not known with accuracy. Nevertheless, considerable progress has been made in understanding nuclear structure, and even the incomplete picture developed in this book will enable us to predict reliably certain nuclear properties.

Unstable Nuclei

Unstable or energy-rich nuclei throw off their excess energy and drop to more stable energy states by emitting various high-energy particles and radiations; some nuclei split or fission in this process. These processes of nuclear change have become known as radioactive processes, or radioactivity. The more general term *nuclear processes*, or nuclear chemistry, is now used to refer to the *reactions in the nucleus in which particles or radiations are emitted*, and to the *opposite process, in which the nucleus is bombarded by particles or radiations and captures them*. The new nucleus formed may exist for infinite time, or it may disintegrate instantly or slowly. Examples of these emission and capture processes are given below.

Emission Process. Radium emits helium nuclei (called alpha particles) and changes to radon. This reaction is described by the *nuclear* equation

$$_{88}^{222}\text{Ra} \longrightarrow \ _{2}^{4}\text{He} + \ _{86}^{218}\text{Rn}$$

The masses of the nuclei are indicated with superscripts and the atomic numbers are given in subscripts. A nuclear equation such as this is "balanced" when the sum of the subscripts on the left side of the equation equals the sum of the subscripts on the right side of the equation ($88 = 2 + 86$), and when the same equality holds for the superscripts ($222 = 4 + 218$). This indicates that there is no loss in the number of nucleons or in total charge in a nuclear reaction.

Capture Process. In alpha-particle bombardment, nitrogen nuclei capture helium, immediately emit protons, and are changed to oxygen nuclei, as shown by the equation

$$_{7}^{14}\text{N} + \ _{2}^{4}\text{He} \longrightarrow \ _{1}^{1}\text{H} + \ _{8}^{17}\text{O}$$

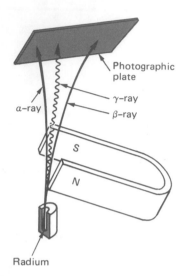

Figure 23.1 Gamma rays are not bent, but alpha and beta rays are bent to the left and right, respectively, in a magentic field with orientation as shown.

THE NATURE AND PROPERTIES OF THE RAYS

Alpha, Beta, and Gamma Rays

In a very simple experiment, Rutherford identified three kinds of rays that may be emitted in the radioactive process. He named them alpha (α), beta (β), and gamma (γ) rays. He placed a small quantity of a material containing radium at the bottom of a hole bored in a piece of lead (Figure 23.1). The intensity of all the rays, except those traveling directly upward, was severely reduced by the shielding power of the lead. He then placed the radium between the poles of a magnet, as shown, with a photographic plate (protected from light) arranged above to receive the rays. The alpha rays were bent to the extent and in the direction expected for relatively heavy, positively charged particles passing through a magnetic field; further study proved them to be *helium nuclei*, He^{+2}. The beta rays, deflected in the opposite direction and to a much greater extent, behaved like negatively charged particles of very small mass; they are *electrons*. The gamma rays, not deflected at all, are *waves of light* similar to X rays, but most of them are of shorter wavelength and therefore of higher energy.

Speed and Energy of the Rays

The speed of the gamma rays is the same as that of light, 3×10^{10} cm/sec, or 186,000 miles/sec. The speed of the alpha particles depends on the kind of nucleus from which they are ejected, but ranges from 9000 to 14,000 miles/sec. Likewise, the beta particles travel at various speeds that can be as fast as 100,000 miles/sec.

The energy of these rays is given in units of electron volts and is usually several million electron volts (Mev). This is about a million times more energy than that required to remove the outside electron from an atom of sodium (5.1 ev). The energies of the alpha and beta particles result from their high velocities and are the kinetic energy ($\frac{1}{2}mv^2$) of the particle; the energy of the gamma rays is calculated by Planck's equation, $E = h\nu$.

The alpha particles from a given isotope either all have the same energy or their energies are distributed in a few energy groups, one of a few specific energy values characteristic of that isotope. This monoenergetic character of the alpha particles is evidence of energy levels in the nucleus.

Beta particles, in contrast to alpha particles, are emitted with a continuous energy distribution extending from near zero energy to a maximum energy that may be as high as 15 Mev (Figure 23.2). If energy levels exist in nuclei, it is puzzling to find that alpha particles from a given isotope are monoenergetic but beta particles appear not to be. To explain this paradox as well as several other problems, the existence of another particle, the neutrino, has been postulated. Neutrinos are assumed to be ejected at the same time as the beta particles and to carry away varying amounts of energy. The resulting energy of the observed beta radiation is thus distributed over a range of values, although the sum, (beta-particle energy) + (neutrino energy), is a constant for any one beta emission.

Gamma rays from a given nucleus are in single or in a few energy groups. The scheme for the decay of ^{60}Co to ^{60}Ni is shown in Figure 23.3, indicating the gamma rays that are observed when the excited nickel nucleus first formed loses energy in dropping to the ground state. The overwhelming fraction (99.9 per cent) of the beta particles have the maximum energy 0.31

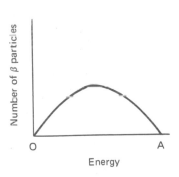

Figure 23.2 A β-decay energy spectrum. Here the energies of the β-rays emitted from a given element are plotted along the abscissa with the number of such emissions having each energy recorded along the ordinate. In contrast to α- and γ-radiations, β-radiation is emitted with a wide range of energies.

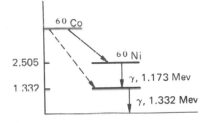

Figure 23.3 An excited ^{60}Co nucleus emits a β-ray and is transformed into an excited nickel nucleus. The excited nickel nucleus (^{60}Ni) emits a 1.173-Mev γ-ray and falls into a lower energy state. Before reaching the ground state, the nickel nucleus emits a second γ-ray (1.332 Mev). It thus seems that two excited states for the nickel nuclei lie at 1.332 and 2.505 Mev above the ground state. There is some evidence for still a third excited state of the ^{60}Ni nucleus.

Mev; very few, as indicated by the dotted line, have an energy maximum of 1.48 Mev, leading to another energy state of the nickel nucleus. Direct beta emission to the ground state of nickel is not observed.

Penetrating Power of the Rays Alpha particles penetrate only the thinnest sheets of metal, and are stopped by even an ordinary piece of paper. Beta particles have greater penetrating power than the alpha particles, but are stopped by relatively thin sheets of metal. Gamma rays can penetrate thick layers of metal, and can be detected after traversing 8 or 10 in of lead. The *relative* penetrating power of the three types of particles, tested toward aluminum, are approximately as follows: alpha particles, 1; beta particles, 100; gamma rays, 10,000.

Ionizing Power of the Rays As they pass through air or other gases, all three kinds of rays have the power to *ionize* gas molecules by knocking off electrons from the outer regions of the atoms of which the gas molecules are composed. Gas or air in the neighborhood of a radioactive substance is thus made electrically conducting. On an average, about 35 ev are required to produce an ion pair (a positive ion and an electron) when the radiations pass through gases, liquids, or solids. Thus, an alpha particle of 3.5 Mev energy will produce at least 100,000 pairs of ions before its speed is reduced to that of a normal gas particle.

SOME PROPERTIES OF THE NUCLEUS

Radius From Rutherford's early experiments in which he bombarded metal foil with alpha particles (see Chapter 2), he calculated the nuclear radius of atoms of light elements to be about 3×10^{-13} cm. The equation now used to calculate nuclear radii is

$$r = A^{1/3} \times 1.4 \times 10^{-13} \text{ cm}$$

where r is the nuclear radius and A is the sum of the protons and neutrons in the nucleus.

Binding Energy For the nucleus to remain intact, the nucleons must be bound together by attractive forces strong enough to overcome the repulsive electrostatic forces between protons. Calculations of the binding energy of the particles in the nucleus indicate that the particles of most nuclei, except deuterium, are held to each other with a binding energy of about 7 Mev per particle; the range is from 2 Mev for deuterium to 8 Mev for the most stable nuclei. This is from 100,000 to 1,000,000 times more than the ionization energy required to remove outside electrons from atoms. The binding energy can be calculated from the mass lost when nuclei of light atoms fuse together to form heavier nuclei.

PROBLEM Calculate the binding energy per particle in a nucleus of helium.

Solution Assume

$$2p + 2n \xrightarrow{\text{fuse}} \text{helium nucleus} + \text{binding energy}$$

The sum of the atomic mass units is $2 \times 1.007825 + 2 \times 1.008665$, or 4.032980. The mass of the helium nucleus is 4.00258 g/mole. This means that 4.032980 − 4.00258 or 0.0304 g/mole of mass is lost on fusion. Hence the binding

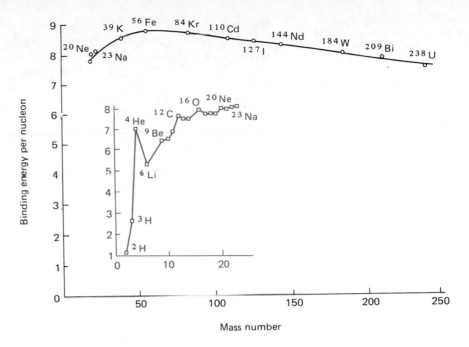

Figure 23.4 Binding energy per nucleon as a function of mass number. The inset shows the light nuclei on an expanded mass number scale. Note the stability of ^{4}He, ^{12}C, ^{16}O, and ^{20}Ne.

energy of the particles of the nucleus is equal to the energy equivalent to this mass lost, or

$$\frac{0.0304 \text{ g/mole}}{6.02 \times 10^{23} \text{ atoms/mole}} = 5.05 \times 10^{-26} \text{ g lost/atom}$$

when two protons and two neutrons fuse to form one helium nucleus.

The energy associated with this mass loss is calculated from the Einstein equation

$$E = mc^2 = (5.05 \times 10^{-26})(3.0 \times 10^{10})^2 = 45.4 \times 10^{-6} \text{ ergs}$$

Since

$$1 \text{ Mev} = 1.6 \times 10^{-6} \text{ erg}$$

$$E = 28.4 \text{ Mev for the nucleus containing four particles, or 7.1 Mev per particle}$$

This means that in the fusion process, the excess energy of 7 Mev per particle is emitted, and the nucleus is *more stable* by 7 Mev per particle than it would be if this energy were not released. This energy is called the *binding energy* of each particle. Figure 23.4 is a plot of binding energy as a function of atomic number illustrating that very light and heavy nuclei are less stable than nuclei of intermediate mass.

Neutron-Proton Ratio   The stability of a nucleus seems to be associated with the ratio of the number of neutrons to the number of protons in it (Figure 23.5). The graph shows the neutron-proton ratio in the stable nuclei that occur naturally. If the n/p ratio is above the curve of the graph, the nucleus tends to stabilize itself by

Figure 23.5 Neutron-proton ratio of naturally occurring elements; dashed line is for a neutron-proton ratio of one.

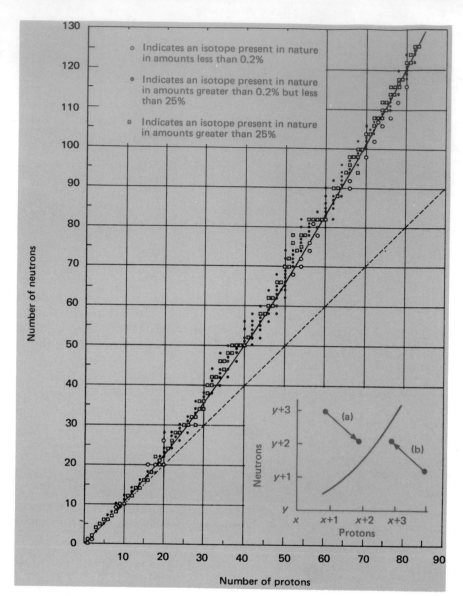

Figure 23.6 Schematic diagram of energy levels in a nucleus. The numbers in parentheses on each level indicate the number of sublevels contained in each main level and hence the number of nucleons of each kind that can occupy that main level. The numbers in the right-hand column are often called "magic numbers". They correspond to a closed shell and are analogous to the completed octet of electrons in the noble-gas structure of atoms. A magic number appears whenever the jump between adjacent levels is particularly large. These numbers represent the number of nucleons required to fill the indicated level and all lower-lying levels.

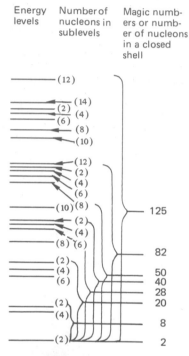

converting a neutron to a proton and a beta particle (this process is called β-decay) [(a) in Figure 23.5].

$$n \longrightarrow p + \beta^-$$

If the n/p ratio is too low, a proton will be converted to a neutron and a positron [(b) in Figure 23.5].

$$p \longrightarrow n + \beta^+$$

Note that the number of neutrons is greater than the number of protons (a straight line at a 45° angle through the origin would represent systems with $n = p$). This indicates that a few "extra" neutrons are required to provide additional binding energy to overcome the electrostatic proton-proton repulsion.

Nucleon-Nucleon Force　An important conclusion concerning nuclear forces follows from a comparison of the binding energies of certain pairs of light nuclei. Consider, for example, the nuclei ^{3}H and ^{3}He. Tritium, $_1^3$H, consists of two neutrons and one proton; helium-3 consists of two protons and one neutron. The binding energies of these nuclei are similar, differing only by an amount comparable to the electrostatic repulsion energy in helium-3. Corresponding observations can be made with other pairs of light nuclei, such as ^{7}Li and ^{7}Be, or ^{11}C and ^{11}B. Such comparisons compel us to conclude that the nucleon-nucleon force remains unchanged when neutrons are replaced by protons or protons by neutrons. The strength of these forces is at least one million times greater than in the chemical bond. Compare H—H bond energy (4.48 ev/molecule) with binding energy (7 Mev/particle).

From studies in which nuclei are bombarded with neutrons or protons, we know that these nucleon-nucleon forces are effective only at very short distances. They fall to zero when the nucleons are separated by as much as 10^{-12} cm.

It thus appears that the nucleon-nucleon force is unusually powerful, that it is effective only at distances shorter than 10^{-12} cm, and that it is equally strong for proton-proton, neutron-neutron, and neutron-proton interactions. Thus if bombarding particles approach within 10^{-12} cm of a nucleus, these powerful nuclear forces draw the particle into the nucleus and cause a fusion reaction to occur.

Energy Levels　Nuclear energy levels assigned on the basis of nuclear radiation studies are summarized in Figure 23.6.

Cross Section　In addition to the physical radius, r, of the nucleus, which is between about 1.4×10^{-13} and 6×10^{-13} cm for atoms from hydrogen to uranium, the term *nuclear cross section*, σ, is also used. The cross section is not a physical dimension but a term used to designate how big in area the nucleus "seems to be" as determined by its ease in capturing neutrons and other bombarding particles. The nuclear cross section is really a measure of the *probability of capture* of a bombarding particle. The unit designating the cross section of a nucleus is the *barn*, defined as 10^{-24} cm^2. Whereas the actual physical radius of a nucleus varies only a small amount from light to heavy atoms, the capture cross sections of the different nuclei vary from 0.0002 barn to 10^5 barns. The cross section of a nucleus for capture of a bombarding particle depends not only upon the kind of nucleus, but also upon the type and the speed of the bombarding particle.

The large cross section of cadmium and boron for capture of neutrons is utilized in their adoption as the important constituents of control rods in nuclear reactors (Figure 23.10).

SOME RADIOCHEMISTRY

The Half-Life of a Radioactive Element　A nuclear reaction such as Ra $\longrightarrow$ He + Rn is known as a nuclear disintegration. Experiments show that the number of nuclei of a given element disintegrating in unit time is proportional to the number of atoms of that element present at the particular moment at which the measurement is made (first-order reaction, Chapter 22). Since the parent nucleus disappears in the disintegration, the number of nuclei of the original element steadily

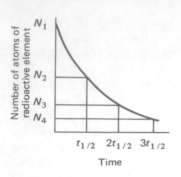

Figure 23.7 Radioactive decay.

decreases. The experimental curve for nuclear disintegrations has a shape similar to that shown in Figure 23.7, which shows the decrease in the number of atoms of a radioactive substance as a function of time. This decrease may also be represented by the logarithmic equation

$$\log A/A_0 = -\frac{\lambda t}{2.303}$$

where A_0 is the number of atoms at the beginning of the measurements, A is the number at time t, and λ is a constant called the disintegration constant that depends upon the radioactive isotope being studied.

The decay curve is such that the number of atoms decrease to zero only after an infinite time. Therefore, it is meaningless to speak of the time of complete decay (full-life) of a radioisotope. However, the term half-life is used; it is defined as the *time required for one-half of the atoms to disintegrate* and thus describes one of the unique properties of radio-isotopes. The equation for the half-life is

$$t_{1/2} = \frac{0.693}{\lambda}$$

where λ is the constant given above. The values for half-lives of several isotopes are given in Table 23.1.

Using ^{14}C, whose half-life is 5760 years, as an example, we would find a decay curve like that shown in Figure 23.7. Inspection of this graph indicates that at the end of every 5760 years, the number of radioactive atoms has decreased to one-half the value at the beginning of that period.

The half-life is an important property with which to identify elements, for each radioactive isotope has its own half-life value. The half-life is unaffected by ordinary temperature or pressure changes and, being a property of the nucleus only, is the same no matter what chemical compound of the radio-active element is used in the measurement. Furthermore, half-life values have become a very useful means of dating geological time periods and archaeological objects.

Radioactivity and Change in Atomic Weight and Number

The loss of one alpha particle, He^{+2}, by the nucleus of an atom results in a lowering of atomic weight by four units; at the same time, it produces a lowering of the positive charge on the nucleus, and a lowering of the atomic number, by two units. With beta-particle emission, there is no change in atomic weight, but there is a gain of +1 in the atomic number. For example,

$$^{238}_{92}U \longrightarrow {}^{4}_{2}He + {}^{234}_{90}Th$$

and

$$^{234}_{90}Th \longrightarrow {}_{-1}\beta + {}^{234}_{91}Pa$$

This information makes it possible to predict the product of a nuclear reaction if the type of radiation emitted is known.

Isotopes—Historical

In the early research on these disintegrations, many more elements were found in the disintegration products than there were places for them in the periodic table. They were given such odd designations as UX_1, UX_2, I_0, RaA, RaC, and the like. In 1913, the English scientist Frederick Soddy found that many of these "different elements" had such similar properties

Table 23.1 The Half-Life of Several Isotopes

Isotope	Half-Life
^{3}H	12.5 yr
^{14}C	5760 yr
^{24}Na	15 hr
^{60}Co	5.2 yr
^{90}Sr	20 yr
^{235}U	4.5×10^9 yr
^{15}O	118 sec
^{11}C	20 min

that they could not be separated by chemical means even though each had its own, unique half-life. This led Soddy to the conclusion that, contrary to Dalton's atomic theory (Chapter 2), atoms of the same element could have different weights; he named such different species of a given element *iso-topes*. Direct proof of the existence of isotopic atoms of the same element was obtained in 1919 by F. W. Aston, who used the mass spectrograph to measure atomic masses. Thus the problem of the larger number of elements found in radioactive disintegration was solved—many of these "new elements" were isotopes of known elements.

Figure 23.8 gives the decay of ^{238}U and its daughter elements, showing the different products formed.

The Radioactive Elements in Nature

The naturally occurring radioactive elements are limited almost exclusively to the heavy elements above lead, with the exceptions of ^{40}K, ^{87}Rb, ^{115}In, ^{138}La, ^{144}Nd, ^{147}Sm, ^{176}Lu, ^{187}Re, and ^{190}Pt. Four disintegration series of the heavy elements are known, in which the parent substances of each series are ^{232}Th, ^{241}Pu, ^{238}U, and ^{235}U, respectively. The end product of the ^{241}Pu series is ^{209}Bi; the end product of ^{232}Th is ^{208}Pb; of ^{238}U, it is ^{206}Pb; and of ^{235}U, it is ^{207}Pb.

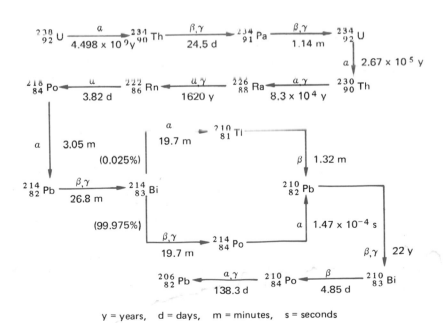

Figure 23.8 The uranium-238 disintegration series.

y = years,   d = days,   m = minutes,   s = seconds

Some radioactive isotopes found in nature are being continually produced by bombardment with cosmic rays. Most likely the carbon-14 in the atmosphere is produced by this process.

Units of Radioactivity

In addition to the energy of radiation and the half-life of isotopes, we are also interested in the number of radiations emitted per unit weight of a radioactive isotope per unit time and the effect of these radiations as they impinge on matter, especially on biological tissue.

The *curie* is the unit now used to indicate the number of radioactive atoms disintegrating per second. It is defined as the quantity of any radioactive substance in which the number of disintegrations per second is 3.7×10^{10}. Since the total effect of these particles as they strike other molecules depends upon the energy of the particle as well as the *number* of particles, another term or unit is required to describe the effect resulting from the bombardment of matter with the high-energy particles.

The destructive power of radiation is difficult to measure, but it can be presumed to be proportional to the number of ions formed in passing through tissue; this, in turn, is presumed to be proportional to the number of ions formed in passing through air. The *roentgen* (r) is defined as the amount of gamma or X radiation that will produce 1.7×10^{12} ion pairs per gram of air, and the number of roentgens to which an object is exposed is known as the *dosage* for that object. One thousand disintegrations producing gamma rays of 3.5 Mev will cause as many ion pairs to form as 1,000,000 disintegrations producing gamma rays of 0.0035 Mev. Thus, the dosage from absorption of radiation from these two sources would be the same (about 6×10^{-5} r). Another unit frequently used as a measure of potential radiation damage in biological tissue is the *rad*, defined as the absorption of 100 ergs of radiation energy per gram of tissue.

Effect of Ionizing Radiation on Tissue—Maximum Permissible Dosage

Alpha, beta, and gamma radiations are called *ionizing* radiations. Alpha particles are the most powerful ionizing particles of the three, but their path in liquids and solids is very short. Gamma rays are deeply penetrating radiations. Neutrons do not cause ionization by knocking out electrons, but they do knock protons off molecules containing hydrogen atoms; they also produce nuclear changes due to neutron capture.

The ions that are formed by these radiations are very reactive in body tissue. The result may be a deep burn in tissue, a reduction in the number of blood cells, or even destruction of genes. Excessive radiation will cause death. Moderate doses of radiation may retard the growth of cancer cells, while heavy doses may produce cancer.

The maximum permissible dosage of full-body radiation that a person may receive with safety is 0.1 roentgen in an eight-hour day, or 0.3 r per week. The lethal dosage is about 600 r of full-body radiation. Great care must be observed when working near radioactive substances to protect against tissue damage from large doses of radiation. Workers should use proper shielding material, such as lead or concrete; they should work as far away from the radioactive material as possible, and should keep the time of exposure to a minimum. They should avoid inhaling or swallowing radioactive material except under supervision of a physician.

Daily Human Exposure to Radiation

We are constantly bombarded by radiations from radioactive substances. The deeply penetrating radiations from interstellar space and the radiations from natural radioactive substances in the dust, air, and building materials are bombarding us at the rate of about 20 radiations per square centimeter per second, representing a dosage of about 0.0003 r per day of full-body radiation. This is called background radiation. Radioactive elements present in the body tissue also produce radiations; for example, a person weighing 150 pounds has enough radioactive carbon, ^{14}C, and potassium, ^{40}K, to give a total of over 300,000 beta particles per minute. Fallout from atomic bomb explosions can cause considerable radiation damage.

Artificially Produced Radioactive Isotopes

In 1919, Ernest Rutherford had showed that hydrogen and oxygen are formed when alpha particles bombard nitrogen atoms:

$$^{14}_{7}N + ^{4}_{2}He \longrightarrow ^{1}_{1}H + ^{17}_{8}O$$

This was the first artificial transmutation, or synthesis, of elements ever reported—a feat tried without success for over 1000 years by the alchemists.

Continuing experiments on the bombardment of elements with alpha particles, Madame Curie's older daughter, Irene Joliot-Curie, and her husband, Frederic, bombarded magnesium and aluminum with alpha particles. In one of these experiments in 1934, they accidently covered up the bombarding source and found that the metal they had been bombarding was radioactive. This was the first radioactive material ever produced artificially. Since that time, all elements have been made radioactive. Today, there are a variety of bombarding particles used to make elements radioactive.

Types of Bombarding Particles

The two general types of bombarding particles are charged particles and neutrons. It is believed that either type of particle will be drawn into the nucleus by the powerful nuclear forces if it comes to within about 10^{-12} cm of the nucleus. With neutrons, capture occurs readily if the velocity of the neutron can be reduced sufficiently for the nuclear forces to be effective. Substances such as graphite, called moderators, are thus used in nuclear reactors to slow down neutrons. Figure 23.9 shows the relative number of low-energy (low-speed) neutrons and high-energy neutrons captured by nuclei.

Positively charged bombarding particles, He^{+2}, H^+, D^+, and so on, are repelled by the positive charge on nuclei. To bring positively charged particles to within 10^{-12} cm of a nucleus so they can be drawn into the nucleus by the nuclear forces, the particle must be traveling at a very high velocity—that is, with a high energy. A high velocity can be obtained by accelerating the particle in an apparatus called a cyclotron, or heating the particle to a high temperature. In the sun, stars, and the hydrogen bomb, where the temperature is 10,000,000K or higher, protons have a sufficiently high velocity to come within 10^{-12} cm of each other and fuse together.

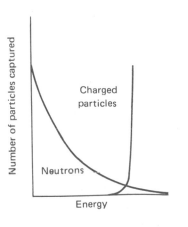

Figure 23.9 Particle capture as a function of energy. These curves indicate that low-energy neutrons are captured more readily than high-energy neutrons but that positively charged bombarding particles must have an energy great enough to overcome electrostatic repulsion with the nucleus to be captured.

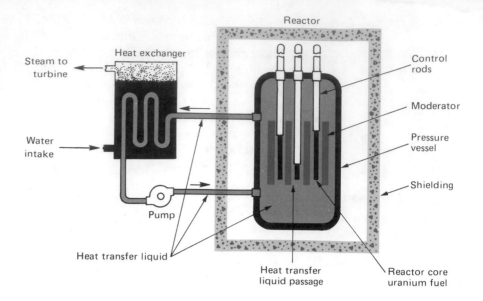

Figure 23.10 The fuel box of a nuclear reactor used for power generation. The control rods regulate the number of neutrons that can cause fission and thus control the rate of the fission process. This, in turn, controls the amount of energy produced. (Adapted from an AEC figure from Argonne National Laboratory.)

Mechanisms of the Process of Induced Radioactivity

When high-energy charged particles are captured by nuclei, they invariably carry a large amount of energy with them into the nuclei. Very likely, this energy is then distributed among the particles of the nucleus and eventually thrown off again by one or more nuclear processes.

When a neutron is captured by a nucleus, an unfavorable balance of protons to neutrons usually results. A common process of emitting the excess energy is for a neutron to change into a proton and a beta particle, which is emitted.

$$ {}_{0}^{1}n \longrightarrow {}_{1}^{1}H + {}_{-1}^{0}e $$

The excited nucleus remaining can drop to a lower energy level and emit energy as a gamma ray.

This process, a very common one, is known as "neutron capture followed by beta decay." A second process by which a nucleus, made unstable by neutron capture, can return to a stable state is fission—fragmenting of the nucleus into two smaller and more stable nuclei.

Neutron Sources— Nuclear Reactors

The nuclear reactor (Figure 23.10) is the main neutron source today. The usual fuel is uranium-235, which undergoes fission when struck with a neutron to produce an average of three neutrons, fission fragments of near but not equal mass, and much energy.

$$ {}_{92}^{235}U + n \longrightarrow \text{fission products} + 3n + 200 \text{ Mev} $$

Note that this is a chain reaction since the neutrons produced can cause fission in other uranium-235 atoms. If uncontrolled, explosion will result. The function of the control rods in the reactor is to remove chain carriers by absorbing the neutrons.

Much research remains to be done to understand fully the mechanism of the fission process. Altogether, about 60 elements have been identified in fission fragments of uranium-235.

Another common source of neutrons is from the bombardment of nuclei of light elements with alpha particles. Beryllium and boron are particularly effective as neutron emitters.

$$_4^9Be + {}_2^4He \longrightarrow {}_0^1n + {}_6^{12}C$$

APPLICATIONS OF NUCLEAR SCIENCE

Manufacture of Elements

A large number (over 1200) artificially radioactive isotopes have now been made. In addition to the artificially radioactive isotopes of elements already known, many new elements, the transuranium elements (lying beyond uranium), have been made. These include the elements neptunium (Np), atomic number 93; plutonium (Pu), 94; americium (Am), 95; curium (Cm), 96; berkelium (Bk), 97; californium (Cf), 98; einsteinium (Es), 99; fermium (Fm), 100; mendelevium (Mv), 101; nobelium (No), 102; and lawrencium (Lr), 103. These elements are placed in the periodic table in the *actinide series*, which contains also the previously known elements actinium, thorium, protactinium, and uranium. They are formed artificially by several processes but chiefly as a result of a sequence of reactions starting with the capture of a neutron by the nucleus of ^{238}U. The ^{239}U thus formed is radioactive (with a half-life of 23 min) and loses a beta particle to form neptunium, which (with a half-life of 2.3 days) in turn loses a beta particle to give an atom of plutonium, which decays with a half-life of 24,000 years, giving off alpha particles. The sequence of results is represented as follows:

$$_{92}^{238}U + {}_0^1n \longrightarrow {}_{92}^{239}U + \text{gamma rays}$$

$$_{92}^{239}U \longrightarrow {}_{93}^{239}Np + {}_{-1}^0e \text{ (beta)}$$

$$_{93}^{239}Np \longrightarrow {}_{94}^{239}Pu + {}_{-1}^0e \text{ (beta)} + \text{gamma rays}$$

By similar nuclear reactions between $_2^4He$ and the successively heavier nuclei of the actinides as they are created, the next higher members of the series have been synthesized. Three elements, *technetium* (43), *astatine* (85), and *promethium* (61), missing from earlier versions of the periodic table, have been made by the particle-bombardment technique.

Atomic Energy by Fission and by Fusion

We know now that atomic energy can be released by the fusion of nuclei of light elements and also by the fission of nuclei of heavy elements. In both processes, the mass of the products is less than the mass of the reactants. The mass "lost" is converted to energy. Such reactions are exothermic (called exoergic). Nuclear reactions also occur in which the mass of the products is *greater* than the mass of the reactants. For such reactions, a large amount of energy must be added; these are endothermic reactions (called endoergic reactions). For atomic-fuel purposes, we are interested in the exothermic reaction; a study of this type of reaction also gives us information about the binding energy of nuclei.

A plot of the binding energy of nuclear particles against the atomic weight is given in Figure 23.4.

Several conclusions can be drawn from this graph:

1. The most stable nuclei—those with greatest binding energy per particle in the nucleus—are those in the atomic weight range of 50 to 60.

2. Exothermic reactions (those with mass loss) can occur by either fusion of light nuclei (such as hydrogen and helium) to heavier nuclei, or fission of heavy nuclei.

3. Since energy is released in fusion of many light elements and in fission of many heavy elements, atomic fuels should be available among the light elements as well as among the heavy ones.

The Fusion Reaction of Hydrogen

At temperatures of 10,000,000 to 50,000,000K, such as might be expected on the surface of the sun, protons are traveling at high enough velocities to approach each other within 10^{-12} cm. At this distance, the powerful nuclear forces overcome the repulsive forces of like electrical charges and cause the nuclei to fuse together.

As protons fuse, helium forms, thus comprising the first step in the synthesis of the elements in the sun and stars. There is a tremendous amount of energy produced when the mass "lost" in converting hydrogen to helium is converted to energy (see page 496). One gram of matter thus converted to energy produces approximately 2×10^{13} cal, or the amount of heat given off when 3000 tons of coal burn. This is probably the major energy-producing reaction in the universe. A similar reaction occurs in the fusion bomb (H-bomb) where a fission bomb (A-bomb) is used as a detonator.

Methods of Synthesis of the Elements in the Sun and Stars

We now have enough information about nuclear reactions to state with considerable confidence a reasonable hypothesis of the formation of the chemical elements. This can be done as follows:

1. We know that hydrogen is a very abundant element in the universe (Figure 23.11), chiefly in the sun and the stars; furthermore, at the high temperatures of the sun and the stars, protons fuse to helium. It is reasonable to assume that the light elements probably are synthesized by a series of successive fusion reactions. As indicated, helium nuclei undoubtedly are the first of these new nuclei to be formed.

2. Laboratory experiments show that the bombardment of light elements, such as beryllium and boron, with alpha particles produces neutrons. Similar reactions probably are a main source of neutrons in the sun and stars.

3. With such neutron sources, the heavier elements can be synthesized by the well-known process of neutron capture followed by beta decay; this process probably constitutes the second step in the synthesis of the elements.

4. According to present theory, these two main processes, fusion and neutron capture followed by beta decay, can account for the formation of most of the elements (Figure 23.12). It is known that several other secondary processes may occur, but fusion and neutron capture are certain to occur to a large extent in the great hot mass of hydrogen in the sun and stars.

The Energy Scale of the Universe

It is significant that our present knowledge of kinetic theory and of nuclear processes now makes it possible for us to gain a fairly complete picture of the types of chemical entities and the reactions they undergo in the various temperature regions of the entire universe (Figure 23.13).

At low temperatures, both atoms and molecules are stable. But as the total energy increases with increasing temperature, the vibrational energy becomes sufficiently large to cause the molecules to dissociate into atoms. Then, as the temperature drops, molecules are formed again. Above 10,000K, few, if any, molecules can exist.

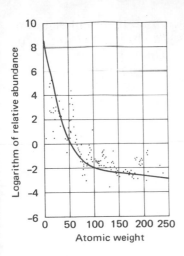

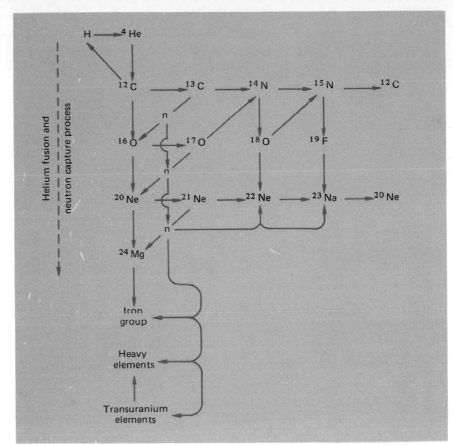

Figure 23.11 The relative abundance of the elements. About 92% of the elements of the universe are hydrogen atoms, and about 7% are helium atoms.

Figure 23.12 This diagram is a simplified explanation of how the elements may have been formed. Hydrogen is believed to be the starting point. The light elements are formed by fusion and neutron capture; the heavier elements are formed mainly by neutron capture. [Adapted from Burbidge, E. M., et al., *Rev. Mod. Physics*, **29**, 547, (1967).]

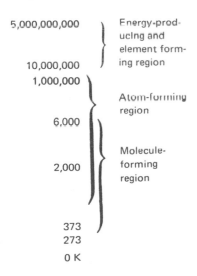

Figure 23.13 The probable temperature scale of the universe, showing the element-forming region.

Above these temperatures, the kinetic energy and electronic energy of atoms becomes greater and greater, causing electrons to move to higher energy levels and finally ionization occurs. At 1,000,000K, probably all atoms are stripped of their electrons; should the temperature drop, the atom systems would be formed again.

As the temperature rises above 1,000,000K, the kinetic energies of the particles increase and finally become so great that they can overcome the repulsion forces between nuclei, and fusion occurs. This is accompanied by massive energy release. Therefore, above 10,000,000K, we identify a third region as the element-forming region of the universe—it is also the energy-producing region, and it is found in the sun and stars.

Nuclear Power: Is It Safe? One of the major problems our society must solve in the next few decades is how to generate enormous amounts of additional electric power without jeopardizing the environment. The Federal Power Commission estimates that our electric-generating capacity will have to be increased at least sevenfold in the next 30 years, and this figure may rise if it becomes necessary to build huge purification plants to clean up the air and water and recycle wastes.

The available energy options for the future have been discussed in Chapters 2 and 18. Considerations such as these make it clear that the prime hope for providing the energy to meet not only U.S. power requirements but those of all other countries in the world resides in the development of safe and economical nuclear power reactors. While first-generation reactors have been developed and are now in limited use, the technology is not yet available for manufacture of reactor systems in either the numbers or with the efficiency required to meet projected power needs. Two types of nuclear reactors now in the development stage hold great promise for meeting these needs.

The first is the breeder-type reactor, which operates so that it produces more nuclear fuel than it consumes. The fuel used is inexpensive low-grade uranium and thorium ores found in large quantities in rocks all over the earth. An efficient breeder reactor would produce enough fuel by nuclear fission to refuel itself and another identical reactor in seven to 10 years. Such commercial-scale breeders are expected to be operative by 1984. Electricity produced in them could cost about 20 per cent less than at present.

Fusion-power reactors constitute a second promising source of energy at low cost. These would be fueled with deuterium, which is abundant and easily extractable from seawater. Until recently, there has been some doubt among experts as to whether such a reactor is feasible—that is, whether its successful development is in fact blocked by the laws of nature. Today there seems to be little doubt that it can be built, and the major problems associated with it center on economic and social aspects of this type of power generation. However, even with the most favorable social climate, a major engineering breakthrough is needed if reactors of this type are to be available for commercial use by the year 2000.

The major question with nuclear power generation is: Is it safe? Following are some facts and factors that must go into any answer to this question:

1. Nuclear reactor fuel requires no burning of the world's oxygen or hydrocarbon resources and thus releases no carbon dioxide or other combustion products to the atmosphere.

2. Breeder reactors produce radioactive wastes. These are the elements in the middle of the periodic table that form when the uranium or thorium atoms fission. Many isotopes of these elements are radioactive. The permanent control of these products is routinely achieved in the reactor fuel cycle.

3. Breeder reactors also produce large amounts of plutonium, which is toxic. Reactors are designed to ensure the confinement of this and all other potentially hazardous substances under all foreseeable conditions including earthquakes and sabotage.

4. Normal operation of a breeder reactor would result in the release of radioactivity in the near vicinity of the reactor that would generally be less than a few percent of the exposure from natural background radiation.

Fusion-power reactors of the type contemplated have no radioactive wastes. The principle reaction products would be nonradioactive neutrons, helium and hydrogen, and radioactive tritium. The tritium is a fuel that could be returned to the reactor to be burned. Precautions for its containment must be provided since tritium has considerable biological-hazard potential.

Breeder reactors must contain a large amount of fissionable material. There is, therefore, the possibility, albeit remote, of a runaway accident. The variety and effectiveness of the safeguards being built into breeder reactors make the probability of an accident that would disperse radioactive material outside the plant virtually nil. Fusion reactors will be incapable of a runaway accident because the amount of "fusioning" material present at any one time would never be enough to produce an incident.

Waste heat dissipation has been a serious problem with breeder reactors, but it can be virtually eliminated in fusion reactors. However, this problem is even now being solved for the breeders by the use of liquid metals (usually sodium) instead of water as the primary coolant. Liquid metal coolants permit reactors to operate at higher temperatures and with a corresponding increase in thermal efficiency. The reduced excess heat is dissipated by transferring it from the liquid sodium to water and cooling the water in giant cooling towers. Under these conditions, thermal pollution is negligible.

SUMMARY   Energy-rich nuclei emit high-energy particles or radiations and become stabilized. This process is known as radioactivity. Natural radioactive materials usually emit high-energy radiations called alpha, beta, or gamma rays. Stable nuclei can be made energy-rich by bombardment with neutrons or high-energy particles.

Nuclear forces are extremely strong forces, but they are effective only at very short distances, dropping to zero at about 10^{-12} cm. Charged particles, when accelerated to high velocities (energies), can overcome the coulomb repulsion of the charged nucleus and approach the nucleus closely enough (10^{-12} cm) to be drawn in and captured by these forces.

Bombarding particles to produce nuclear changes may be neutrons or charged particles of atomic weights ranging from 1 to about 20.

Atomic energy can be produced by mass loss in fusion reactions or in fission reactions.

We can describe with considerable confidence the reactions that, given the proper conditions of nuclear abundance and temperature, must take place to form the various elements of the universe. These nuclear processes have been observed to occur in nuclear reactors, cyclotrons, betatrons, and in the hydrogen bomb. They can be expected to be taking place in the sun and stars—hence producing the chemical elements.

Undoubtedly the hydrogen fusion reaction is the initial step in the element-forming process and is probably the main energy-forming step in the universe.

IMPORTANT TERMS

Alpha particle	**Isotopes**
Barn	**Magic number nuclei**
Beta-decay process	**Maximum permissible dosage**
Beta particle	**Nuclear cross section,** σ
Binding energy of nucleons	**Nuclear process**
Curie	**Nucleon**
Gamma ray	**Radioactivity**
Half-life	**Roentgen**
Ionizing radiation	

1. What is the status of theories about the structure of the nucleus?

2. Write nuclear equations representing an emission process and a capture process.

3. What property of nuclei is indicated by the fact that alpha particles and gamma rays are monoenergetic? Why do beta particles not show monoenergetic properties?

4. How can the magnitude and sign on alpha particles and beta particles be demonstrated?

5. Fill in the proper mass and charge numbers in the nuclear equations:

$$^{238}_{92}U + n \longrightarrow U \longrightarrow Np + \beta$$
$$\hookrightarrow Pu + \beta$$

6. Assume a given radioactive substance has a half-life of 4.0 days and has no radioactive parent. What fraction of the initial amount will be left after (a) 8 days? (b) 32 days?

7. Calculate the number of atoms of uranium-235 disintegrating per minute in 1 mg of ordinary uranium, from the half-life of ^{235}U, $t_{1/2} = 4.51 \times 10^9$ years.

8. $^{239}_{94}Pu$ is an alpha-particle emitter. Calculate its half-life if a 0.100-g sample produces 1.40×10^7 disintegrations per minute.

9. Calculate the half-life of ^{238}U if 1.0 gram of ^{238}U gives 1.3×10^4 alpha particles per second.

10. What is the percentage yield if a quantity of 500 kg of uranium is obtained from 2 tons of pitchblende containing 75% U_3O_8?

11. What is the mass of the nucleus ^{14}C if 0.60 Mev of energy is liberated in the reaction

$$^{14}_{7}N + n \longrightarrow ^{14}_{6}C + p$$

The mass of $^{14}_{7}N$ is 13.9862 amu.

12. Calculate the energy released in the following nuclear reactions:
(a) $^{2}_{1}H + ^{2}_{1}H \longrightarrow ^{3}_{2}He + n$
(b) $^{2}_{1}H + ^{3}_{1}H \longrightarrow ^{4}_{2}He + n$
Assume that the mass of $^{2}_{1}H$ is 2.0147; mass of $^{3}_{1}H$ is 3.0170; mass of $^{4}_{2}He$ is 4.0038; mass of $^{3}_{2}He$ is 3.0160.

13. Calculate the relative size of the nucleus of a helium atom and a uranium atom.

14. Assume that the radius of a nucleus is equal to 1.4×10^{-13} times the number of nuclear particles in the nucleus to the $\frac{1}{3}$ power; calculate the radius of the nucleus in each of the following: helium-4, oxygen-16, arsenic-75, cadmium-112, lead-206, and uranium-238.

15. The energy emitted when nucleons fuse to form an atomic nucleus is called the binding energy of the nucleus. What is the justification for labeling the emitted energy "binding energy"?

16. Calculate the binding energy per nucleon in calcium-40. Calculate the total binding energy for calcium-40.

17. What is the relationship between the neutron-proton ratio and the stability of a nucleus?

18. What types of bombarding particles are used to produce nuclear changes?

19. How does a captured particle "create" an energy-rich nucleus?

20. How do you account for the fact that most particles emitted from the nucleus have very high energies rather than just enough to get out of the nucleus?

21. Cite evidence that indicates that energy levels exist in nuclei.

22. What is the significance of the term "nuclear cross section for capture"?

23. (a) Explain why alpha particles are emitted from the nucleus with energies of several million electron volts rather than low energies covering a range of values. (b) Does the same explanation apply to beta particles?

24. Calculate how much higher the maximum permissible dosage is above the amount of natural daily radiation received by each person.

25. Element 104 has been prepared, where should it be located in the periodic table? What are several nuclear processes that could be used to produce element 104?

26. Assume that a reaction that may be used in a hydrogen bomb is $^6Li + {}^2H \longrightarrow 2\ ^4He$. From the mass differences in the reactants and products, calculate the number of kilocalories produced when 100 g of lithium reacts by this process. Isotopic masses ^6_3Li, 6.015126; ^2_1H, 2.014102 atomic mass units.

27. When Lise Meitner interpreted the fission process, she at once realized that this could be an important source of high energy. Show by means of Figure 23.4 how she could have arrived at this conclusion.

28. How is ^{14}C used as a method of determining the age of a wooden relic?

SPECIAL PROBLEMS

1. What significant contribution did Rutherford and Joliot each make to radiochemistry?

2. Assume an alpha particle of 4.5 Mev passes through a gas. How many positive ions will be formed along its path if approximately 35 ev are required to form an ion pair (a positive ion and an electron)?

3. If the average loss for ion-pair formation in a gas is 35 ev, how many ion pairs are formed in a gas by a passage of a 5.3-Mev alpha particle?

4. Assume that each fission of a fissionable atom can release two neutrons, which, in turn, cause further fissions in a chain reaction. Assume furthermore that 10 steps occur in the fission process. Calculate the number of fissions that will occur in each of the 10 processes; then calculate the total energy in Mev released after the 10 steps if 200 Mev are released per fission in each process.

5. The source of the sun's energy is derived mainly from the conversion of hydrogen into helium. At what rate is hydrogen being consumed, in gram atoms per second, if the radiation from the sun on the earth (93,000,000 miles distant) is 0.0032 calorie per cm^2 per second?

6. Consider the products of disintegration of uranium-to-lead series and assume none of the helium is lost. Calculate the age of a rock found to contain 5×10^{-5} cc of helium at standard temperature and pressure and 3×10^{-7} g of uranium per gram.

REFERENCES   CHOPPIN, G., *Nuclei and Radioactivity*, Benjamin, Menlo Park, Calif., 1964.

FOWLER, W. A., "Theory of the Origin of the Elements," *Sci. Amer.*, **195**, no. 3, 82–91 (1956). *Chem. Eng. News*, **42**, 90–104 (1964).

GARRETT, A. B., *The Flash of Genius*, Van Nostrand Reinhold, New York, 1962. (Useful for stories of the discoveries in radioactivity by Rutherford, Becquerel, Libby, Curie, Joliot, Soddy, Chadwick, Meitner, and de Hevesy.)

GOUGH, W. C., and EASTLAND, B. J., "The Prospects of Fusion Power," *Sci. Amer.*, **224**, no. 2, 50 (1971).

KENDALL, H. W., and PANOFSKY, W., "The Structure of the Proton and the Neutron," *Sci. Amer.*, **224**, no. 6, 60 (1971).

SEABORG, G. T., "Some Recollections of Early Nuclear Age Chemistry," *J. Chem. Educ.*, **45**, 278–289 (1968).

SEABORG, G., and BLOOM, J. T., "Fast Breeder Reactors," *Sci. Amer.*, **223**, no. 5, 13 (1970).

PART SEVEN

THE TRANSITION
ELEMENTS

OVERVIEW OF THE TRANSITION ELEMENTS—
COORDINATION CHEMISTRY

Über die räumliche Lagerung habe ich . . .
die Vorstellung entwickelt, dass die sechs
Gruppen in der relativen Stellung der
Ecken eines Oktaeders um das
Zentralatom angeordnet sind. Diese
Vorstellung führt zu verschiedenen
Folgerungen . . . Einige derselben haben
die experimentelle Prüfung schon
bestanden, so z.B. diejenige, dass
Verbindungen mit komplexen Radikalen
[MeA$_5$B] nur in einer Form auftreten
können, und ebenso die wichtige
Folgerung, dass die Verbindungen mit
komplexen Radikalen [MeA$_4$B$_2$] oder
[MeA$_4$BC] in zwei stereoisomeren Reihen
bestehen können.*

Alfred Werner (1866–1919)

The group of metals in the middle of the periodic table whose inner d or f orbitals are not completely filled are called the transition elements. There are about 55 such elements—a majority of all known elements. The transition elements have certain common properties:

1. All are metals.
2. They are hard, high-melting, brittle, and good conductors of heat and electricity.
3. With few exceptions, their ions and compounds are colored.
4. With few exceptions, they exhibit multiple oxidation states.
5. Many of these elements and their compounds act as catalysts for chemical reactions.
6. Most of their compounds are paramagnetic—they are attracted by a magnetic field.
7. Finally, one of their most general properties is that of forming complex ions, especially for the group known as the d-electron group. Much of the chemistry of the transition elements is associated with the use of d as well as s and p orbitals in forming complex ions.

The d-electron group includes those elements that have partially filled d orbitals (but either completely empty or completely filled f subshells); the second group, called the f-electron group, has partially filled f orbitals. The d-group elements are:

---

*"With regard to the spatial arrangement I have developed the postulate that the six groups are arranged in the relative positions of the corners of an octahedron about the central atom. This postulate leads to various consequences. Several of these have already been confirmed experimentally, for example, the conclusion that compounds with complex radicals [MeA$_5$B] can appear only in one form, and the important conclusion that compounds with complex radicals [MeA$_4$B$_2$] or [MeA$_4$BC] can exist in two stereoisomeric forms." (The quotation is from *Berichte*, **44** (1911) p. 1887.)

First transition series			Sc	Ti	V	Cr	Mn	Fe	Co	Ni	Cu
Electron		$4s$	2	2	2	1	2	2	2	2	1
configuration		$3d$	1	2	3	5	5	6	7	8	10
Second transition series			Y	Zr	Nb	Mo	Tc	Ru	Rh	Pd	Ag
Third transition series			La	Hf	Ta	W	Re	Os	Ir	Pt	Au

The f-group elements are:

Lanthanides

| La | Ce | Pr | Nd | Pm | Sm | Eu | Gd | Tb | Dy | Ho | Er | Tm | Yb | Lu |

Actinides

| Ac | Th | Pa | U | Np | Pu | Am | Cm | Bk | Cf | Es | Fm | Md | No | Lr |

An important distinction between the d and f groups of elements arises because the electron cloud of the partially filled d orbitals apparently projects well out to the periphery of the atoms and ions, so that the electrons in these orbitals introduce or accentuate several important properties of these elements. By contrast, the f orbitals are believed to be buried rather deeply in the atom or ion and thus do not have a profound effect on its properties. As a consequence, d-group elements, while showing the common properties listed above, show a much wider variation in properties than do f-group elements.

PROPERTIES OF TRANSITION ELEMENTS

Metallic Properties

Since the outermost shell of the transition elements contains only one or two electrons, it is not surprising that they are all metals. Moreover, because most of the atoms are relatively small and because some covalent binding between atoms may occur, these metals are high-melting and brittle.

Oxidation States

Most of these elements show several oxidation states because some or all of the electrons in d orbitals may be used along with the valence-shell s electrons in compound formation. For example, vanadium ($3d^3\ 4s^2$) exhibits oxidation states of V, IV, III, and II, corresponding to the use of the two $4s$ as well as 3, 2, 1, or 0 d electrons, respectively.

Color

Color in the ions of these elements arises because the electronic energy levels lie close enough to one another so that electrons can move to higher energy levels by absorbing visible light. The color of the compound is the complement of the color absorbed (Figures 24.1 and 24.2).

Magnetic Properties

The presence of unpaired electrons in the atoms or ions of the transition elements gives rise to paramagnetism. It is possible to measure the paramagnetism of a compound and from this to estimate the number of unpaired electrons present. This is often useful in determining the structure of the compound.

Complex Ions

A complex ion is one that *contains more than one atom*. Complex ions may contain either a metal ion or a nonmetal atom or ion as the central ion. Some examples are given in Table 24.1.

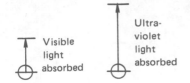

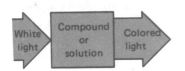

Figure 24.1 Visible and ultraviolet light absorption. Visible light may be absorbed when an electron moves to a slightly higher energy level; ultraviolet light is absorbed when energy levels are more widely separated.

Figure 24.2 Color as a consequence of absorption. White light, a mixture of all colors, passes through a solution and appears as colored light. This is the result of absorption of one or more wave lengths of light from the white light by the compound or its ions.

Table 24.1 Some Complex Ions

Complex Metal Ions	Complex Nonmetal Ions
$Cu(NH_3)_4^{+2}$	ClO_3^-
$Cu(H_2O)_6^{+2}$	SO_4^{-2}
$Co(NO_2)_6^{-3}$	NO_3^-
$Co(NH_3)_6^{+3}$	CO_3^{-2}
$Fe(CN)_6^{-3}$	PO_4^{-3}
$Ag(NH_3)_2^+$	
$Al(H_2O)_6^{+3}$	
$Zn(H_2O)_6^{+2}$	
CrO_4^{-2}	

The deep blue of a solution of copper sulfate containing ammonium hydroxide is caused by the $Cu(NH_3)_4^{+2}$ ions. In nature, we find that the red coloring matter in blood, hemoglobin, is a complex ion of iron; the green coloring matter in plants, chlorophyll, is a complex ion of magnesium; and vitamin B_{12}, a constituent of the vitamin-B group, is a complex ion of cobalt. Because water tends to hydrate ions readily, most ions in aqueous solution may be considered complex ions. Complex ions occur in both the solid state and in solution.

The transition elements are not the only metals whose ions form complex ions, but the tendency is very pronounced among this group of elements. A principal reason why complex ions are formed by the transition elements is the presence of unfilled d orbitals that can be used to form rather strong bonds. We shall therefore discuss the chemistry of complex metal ions with the chemistry of the transition elements in this chapter.

PROPERTIES AND BONDING IN COMPLEX IONS

Ion-Dipole Interaction

Hydration of ions in solution is a familiar example of complex ion formation. In many cases, the bonding is of the ion-dipole type. However, some hydrates may be formed by use of coordinate covalent bonds. Water is a polar covalent molecule; it will "stick" to ions of all kinds in solution because of the attraction of the + or − end of the water molecule to the negative or positive ion. Examples are shown in Figure 24.3.

Coordination Compounds

Hydrates formed by ion-dipole interaction frequently have no definite formulas. Furthermore, the water molecules attached to a particular ion are exchanged frequently for other water molecules in the body of the solution.

Figure 24.3 Hydration of positive ions. The negative end of the water dipole is near the positive charge on the ion. Whether this is purely electrostatic attraction or partially coordinate covalent bond formation is not known in most cases. All of the transition metal ions in solution form complex ions by hydration.

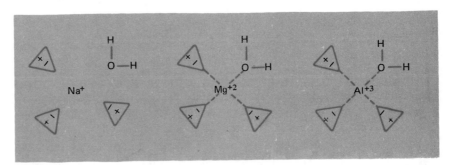

Figure 24.4 (a) Structure of octahedrally shaped $Co(NH_3)_6^{+3}$ ion. (b) and (c) Tetrahedral and planar complex ions, respectively.

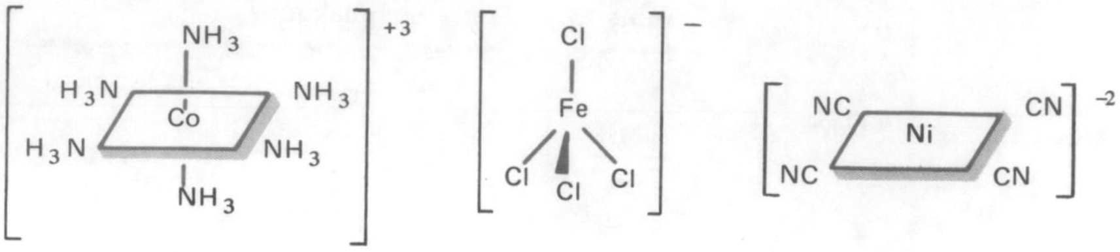

Octahedral Co (III)
complex ion

(a)

A tetrahedral
iron (III)
complex ion

(b)

A square-planar
nickel (II)
complex ion

(c)

The chemically more interesting complex compounds are those of definite formula in which water molecules or other species are attached quite firmly to an ion. These are compounds such as $Co(H_2O)_6Cl_3$, $Cu(NH_3)_4SO_4$, and $K_3Fe(CN)_6$. Experimental data on electrical conductivity in solutions and, for the first two, precipitation of silver chloride and barium sulfate, indicate that these are salts made up of the ions $Co(H_2O)_6^{+3}$, $Cu(NH_3)_4^{+2}$, SO_4^{-2}, K^+, Cl^-, and $Fe(CN)_6^{-3}$.

A German chemist, Alfred Werner, studied compounds of this sort between 1893 and 1918. He was able to show that such groups as H_2O, NH_3, and CN^- assume fixed positions around the central atom in regular geometric patterns. In ions with a *coordination number* of 6—that is, with six groups around the central ion, as in $Co(H_2O)_6^{+3}$—this pattern was shown by Werner to be that of a regular octahedron, as in Figure 24.4. In ions with coordination number 4, the arrangement is either square planar or tetrahedral (Table 24.2).

Coordination numbers of three and five are not common but, when they occur, the shapes are often either planar or pyramidal when three groups surround the central ion, and trigonal bipyramidal when the coordination number is 5.

Table 24.2 Coordination Number and Shape of Complex Ions

Ion	Coordination Number	Shape	Ion	Coordination Number	Shape
$Ag(NH_3)_2^+$	2	Linear	$Fe(CN)_6^{-3}$	6	Octahedral
$Cu(NH_3)_4^{+2}$	4	Square planar	$Co(NO_2)_6^{-3}$	6	Octahedral
$Ni(CN)_4^{-2}$	4	Square planar	$Cr(H_2O)_6^{+3}$	6	Octahedral
$CoCl_4^{-2}$	4	Tetrahedral	$Co(NH_3)_6^{+3}$	6	Octahedral

Table 24.3 Some Simple Ligands

Donor Atom	Ligand	Example
C	CN^-, CO, CNS^-	$Cd(CN)_4^{-2}$, $Co(CO)_4$ $Co(NO_2)_6^{-3}$
N	NH_3, NO, NO_2^-	$Co(NO_2)_6^{-3}$
O	H_2O, OH^-, $C_2O_4^{-2}$	$Al(H_2O)_6^{+3}$
S	SCN^-, $S_2O_3^{-2}$	$Pt(SCN)_6^{-2}$
Halogens	F^-, Cl^-, Br^-, I^-	$CuCl_4^{-3}$, HgI_4^{-2}.

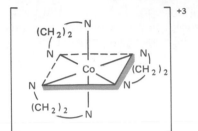

Figure 24.5 A complex ion containing bidentate ligands such as ethylene diamine, $H_2N-CH_2-CH_2-NH_2$.

+3

Table 24.4 Some Polydentate Ligands

$$H_2\ddot{N}-CH_2-CH_2-\ddot{N}H_2$$
Ethylene diamine, a bidentate ligand

$$H_2\ddot{N}-CH_2-CH_2-\ddot{N}H-CH_2-CH_2-\ddot{N}H_2$$
Diethylenetriamine, a tridentate ligand

$$:N\begin{matrix} -CH_2CO_2^- \\ -CH_2-CO_2^- \\ -CH_2-CO_2^- \end{matrix}$$
Anion of nitrilotriacetic acid, a tetradentate ligand

$$\begin{matrix} HO_2C \\ \quad \\ HO_2C \end{matrix}\Big\rangle N-CH_2-CH_2-N\Big\langle\begin{matrix} CO_2H \\ \quad \\ CO_2H \end{matrix}$$
Ethylenediaminetetraacetic acid, a sexadentate ligand

Ligands *The ion or molecule that attaches itself to a central ion or atom to form a complex is called a* ligand. Examples of simple ligands are given in Table 24.3. Ligands frequently are Lewis bases, and carry an unshared pair of electrons, as $\ddot{N}H_3$ or $H_2\ddot{O}:$. Some examples of simple ligands, with only one electron pair for use in bonding, are given in Table 25.2. Some examples of ligands that have an unshared pair of electrons on more than one atom are given in Table 24.4. For example, ethylene diamine, $H_2NCH_2CH_2NH_2$, has an unshared pair of electrons on each of the nitrogen atoms, and thus can attach to an ion at each of them to form an octahedral complex similar to that pictured for cobalt(III) ion in Figure 24.5. Ligands such as ethylene diamine are known as *bidentate* ligands; the general term is *polydentate,* as indicated in Table 24.4.

Chelates The formation of complex ions with bidentate ligands give complex ions that are called *chelates;* the process of forming these ions is called *chelation.* The term chelate means claw, and is derived from the analogy of the attachment of a crab to an object with its claws. The term has been extended to polydentate ligands, so that chlorophyll and hemoglobin, which are tetradentate ligands, are also said to form chelates.

Note that the structure shown in Figure 24.5 contains three rings of five atoms each. Complexes with ring structures such as this, with five or six atoms in the ring, are particularly stable compared to similar complexes involving only monodentate ligands.

Magnetic Data Electrons that are shared in pairs produce molecular species that are *diamagnetic;* the presence of an unpaired electron in a molecule causes the compound to be *paramagnetic.* The experimental observation that classifies substances as diamagnetic or paramagnetic has to do with the behavior of a sample when placed in a magnetic field: diamagnetic substances tend to be pushed out of the field, paramagnetic to be drawn into the field. The direction and intensity of the interaction with the magnetic field is measured by weighing a sample of the substance in the absence and in the presence of a magnetic field (Figure 24.6). The magnetic interaction is described in terms of a magnetic moment; for paramagnetic molecular species, this depends upon the number of unpaired electrons according to the equation

$$M = \sqrt{n(n + 2)}$$

M is the magnetic moment expressed in the units of Bohr magnetons,* and n is the number of unpaired electrons.

Examination of complex ions as in Figure 24.6 shows that many of them are paramagnetic and must, therefore, contain unpaired electrons. For example, hydrated nickel(II) ion has a magnetic moment of 3.2 magnetons, corresponding to two unpaired electrons; hydrated copper(II) ion measures 1.9 magnetons (that is, one unpaired electron); the hexaammine cobalt(III) ion, $Co(NH_3)_6{}^{+3}$, has zero magnetic moment and must therefore contain only paired electrons.

Valence Bond Theory (Coordinate Covalent Bonds)

An early explanation of the bonding in Werner complexes became available in the 1920's. The concept developed by G. N. Lewis of bonding as the result of the sharing of pairs of electrons, as extended to coordinate covalent bonding, was combined with the concept of atomic orbitals to provide the explanation known as the *valence bond theory*. In the theory, we assume that the formation of the complexes depends upon: the orbitals available for coordinate covalent bond formation; the tendency of the ions or groups to share a pair

---

*A magneton is defined by the equation $eh/4\pi mc$, where e is the charge on the electron, h is Planck's constant, m is the mass of the electron, and c is the velocity of light.

NOMENCLATURE FOR COMPLEX IONS

A system for naming complexes is in common use and can be illustrated by the following examples:

Case	Formula	Name
I	$Ni(H_2O)_6{}^{+2}$	Hexaaquonickel(II) ion
II	$Co(NH_3)_6(NO_3)_3$	Hexaamminecobalt(III) nitrate
III	$Cu(NH_3)_4(H_2O)_2{}^{+2}$	Diaquotetraamminecopper(II) ion
IV	$Pt(NH_3)_4ClNO_2{}^{+2}$	Chloronitrotetraammineplatinum(IV) ion
V	$Fe(CN)_6{}^{-4}$	Hexacyanoferrate(II) ion
VI	$VCl_4{}^-$	Tetrachlorovanadate(III) ion

These examples bring out both the general pattern and the more common rules in this system of nomenclature. As is apparent from the examples given, the general pattern for this nomenclature is:

Number and name of ligand—name of metal—oxidation state of metal

In Case I, $Ni(H_2O)_6{}^{+2}$, there are six water molecules surrounding the metal ion. The name given the ligand H_2O is aquo. Thus, hexaaquonickel(II) ion indicates six water molecules surrounding a nickel ion in the (II) oxidation state.

Water and negatively charged ligands are given names ending in o as chloro, cyano (CN^-), and sulfato ($SO_4{}^{-2}$); ammonia, NH_3, is called ammine.

In Case II, the complex salt of cobalt nitrate is named so the cation name comes first—hexaamminecobalt(III)—followed by the name of the anion—nitrate.

Case III illustrates a complex ion with two neutral ligands, $Cu(NH_3)_4(H_2O)_2{}^{+2}$. The least complex ligand is named first—thus, diaquotetraamminecopper(II) ion.

The ion $Pt(NH_3)_4ClNO_2{}^{+2}$ has two negatively charged ligands—chloro, Cl^-, and nitro, $NO_2{}^-$,—and a neutral ligand. The negative ligands are named first in alphabetical order followed by the name of the neutral ligand. Hence the name chloronitrotetraammine platinum(IV) ion.

Anions containing the metal ion end in *ate* as in chromate, permanganate, etc. Thus, the ions $Fe(CN)_6{}^{-4}$ and $VCl_4{}^-$ are hexacyanoferrate(II) ion [where ferrate(II) means iron(II) in an anion] and tetrachlorovanadate(III) ion.

of electrons; and the number of molecules or ions that can be placed around a central ion.

These factors can be illustrated with three complex ions of the transition elements: $MnCl_4^{-2}$, $Fe(CN)_6^{-3}$, and $Co(NH_3)_6^{+3}$.

The atomic orbitals of the *atoms* of manganese, iron and cobalt are:

Atomic Number	Atom	Orbitals Filled	3d	4s	4p
25	Mn	$1s^2\,2s^2\,2p^6\,3s^2\,3p^6$	↓ ↓ ↓ ↓ ↓	⇅	_ _ _
26	Fe	$1s^2\,2s^2\,2p^6\,3s^2\,3p^6$	⇅ ↓ ↓ ↓ ↓	⇅	_ _ _
27	Co	$1s^2\,2s^2\,2p^6\,3s^2\,3p^6$	⇅ ⇅ ↓ ↓ ↓	⇅	_ _ _

The atomic orbitals of the *ions* of Mn^{+2}, Fe^{+3}, and Co^{+3} are:

Atomic Number	Ion	Orbitals Filled	3d	4s	4p
25	Mn^{+2}	$1s^2\,2s^2\,2p^6\,3s^2\,3p^6$	↓ ↓ ↓ ↓ ↓		_ _ _
26	Fe^{+3}	$1s^2\,2s^2\,2p^6\,3s^2\,3p^6$	↓ ↓ ↓ ↓ ↓	_	_ _ _
27	Co^{+3}	$1s^2\,2s^2\,2p^6\,3s^2\,3p^6$	⇅ ↓ ↓ ↓ ↓	_	_ _ _

The ionization process produces ions that can now take part in forming electrovalent bonds as in $MnCl_2$, $FeCl_3$, or $CoCl_3$.

But a second type of bond also can be formed with ligands having a pair of electrons that can enter the unfilled d, s, or p orbitals. According to the valence bond theory, such a bond is of the coordinate covalent type, and the reaction that occurs to form this bond is a Lewis acid-base type of reaction. The number of ligands that will combine this way depends on the relative size of the central ion, the nature of the combining groups, and, to a certain extent, upon the number of orbitals available for bond formation. Let us illustrate these factors with examples.

Example: $Mn^{+2} + 4Cl \longrightarrow MnCl_4^{-2}$

$MnCl_4^{-2}$ Mn^{+2} orbitals   3d   4s   4p

↓ ↓ ↓ ↓ ↓   ⇅ ⇅ ⇅ ⇅
Cl Cl Cl Cl

One pair of electrons on each chloride ion can form a covalent bond with the $4s$ and each of the $4p$ orbitals on the manganese ion. The Mn^{+2} ion is now surrounded by chloride ions. This probably is an sp^3 type of complex. The configuration is tetrahedral.

$$\left[\begin{array}{c} Cl \\ Mn \\ Cl \quad Cl \quad Cl \end{array} \right]^{-2}$$

Example: $Co^{+3} + 6NH_3 \longrightarrow Co(NH_3)_6^{+3}$

$Co(NH_3)_3^{+3}$ Co^{+3} orbitals 3d   4s   4p

⇅ ⇅ ⇅ ⇅ ⇅   ⇅   ⇅ ⇅ ⇅
            NH_3 NH_3 NH_3 NH_3 NH_3 NH_3

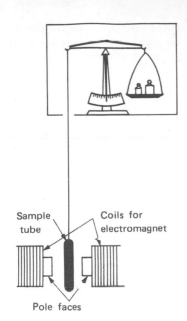

Figure 24.6 Determination of paramagnetic susceptibility. The sample is placed in a tube suspended from one arm of a balance, and the apparent weight of the sample is measured in both the presence and absence of a magnetic field. Paramagnetic material will be pulled into the field and will appear to weigh more in the presence of the field. [From Douglas and McDaniel, *Concepts and Models in Inorganic Chemistry*, Xerox College Publishing, Lexington, Mass., 1965.]

Sample tube

Coils for electromagnet

Pole faces

Here, the extra pairs of electrons from the various ammonia molecules are said to be bonded in two d, one s, and three p orbitals of the metal ion. The Co^{+3} is now surrounded by six ammonia molecules. This is called a d^2sp^3 type of complex. The configuration is octahedral.

$$\left[\begin{array}{c} NH_3 \\ H_3N \diagdown \overset{|}{\underset{Co}{\diagup}} \diagup NH_3 \\ H_3N \diagup \quad \diagdown NH_3 \\ NH_3 \end{array} \right]^{+3}$$

Example: $Fe^{+3} + 6CN^- \longrightarrow Fe(CN)_6^{-3}$

$Fe(CN)_6^{-3}$

Fe⁺³ orbitals	3d				4s	4p	

⇅ ⇅ ↓ ⇅ ⇅ ⇅ ⇅ ⇅
CN CN CN CN CN CN

In this case, the extra pairs of electrons of the cyanide ions are shared with in the d, s, and p orbitals of the metal ion. The Fe^{+3} is now surrounded by six CN^- ions. This type of complex also is the d^2sp^3 type. The configuration is octahedral.

The experimentally determined magnetic moments of $MnCl_4^{-2}$, $Co(NH_3)_6^{+3}$, and $Fe(CN)_6^{-3}$ are 5.9, 0, and 1.7 magnetons, respectively, indicating that these ions contain five, none, and one unpaired electron. In this respect, the experimental data agree with the valence bond theory. A difficulty arises, however, when you attempt to formulate the electron configuration for the ion FeF_6^{-3}, which has a magnetic moment of 5.9 magnetons, indicating the presence of five unpaired electrons.

If we recall that the uncomplexed iron(III) ion has the configuration

3d   4s   4p

↑ ↑ ↑ ↑ ↑   —   — — —

it seems evident that in FeF_6^{-3}, the $3d$ electrons are not rearranged. To account for the large number of unpaired electrons on the basis of valence bond theory, you would have to presume that the orbital occupancy is

3d   4s   4p   4d

↑ ↑ ↑ ↑ ↑   ⇅   ⇅ ⇅ ⇅   ⇅ ⇅ — — —

This assignment also uses two d, three p, and one s orbital, producing the observed octahedral configuration. To reach the required sp^3d^2 hybridization, however, it has been necessary to use the high-energy $4d$ orbitals, and it is difficult to understand why these should be occupied in preference to pairing in the lower energy $3d$ orbitals. There are other cases in which similar questionable assignments to orbitals must be made to obtain agreement with experimental data while adhering to the principles of the valence bond theory. Thus, we conclude that the valence bond theory may not be entirely correct.

WERNER STATEMENT

"Even when, to judge by the valence number, the combining power of certain atoms is exhausted, they still possess in most cases the power of participating further in the construction of complex molecules with the formation of very definite atomic linkages. The possibility of this action is to be traced back to the fact that, besides the affinity bonds designated as *principal valences*, still other bonds, called *auxiliary valences*, may be called into action."

Crystal Field Theory As the research on complex ions progressed, another theory, called the *crystal field theory*, was devised to interpret the colors of the ions and their magnetic properties. The forces between the ligands and the central ion were assumed to be purely electrostatic—an obviously oversimplified point of view, but, nevertheless, one that earlier had made valuable contributions to the interpretation of paramagnetism of ions in crystals, where rigid geometrical patterns of charged bodies also occur (hence the name *crystal field theory*).

The crystal field theory takes into consideration the fact that as an approaching ligand molecule or ion nears a charged particle disturbances, called perturbations, are caused in the electron cloud of both the ion and the ligand. These perturbations in the electric field affect the orientation and energies of the *d*-electron orbitals in the metal ions attached to the ligand and thus produce a new set of unequally spaced energy states in the complex ion. This in turn affects some properties of the ion such as color, bond strength, and reactivity.

In general, we can understand that different arrangements (square planar, tetrahedral, octahedral, and the like) of the ligands around a central ion will change the symmetry of the electrical field around the ion and therefore will have different perturbation effects. This change in symmetry should cause different orientations of the *d* orbitals and produce different energy-state patterns. Let us consider first the case of the free atom.

It is known that the *d* electrons in a free atom in the absence of an electric field are in orbitals all of which have the same energy; they are then said to be fivefold degenerate. (This is represented in Case I, Table 24.5.) In the presence of an electric field, the energies of these orbitals are changed. But, if the field is spherically symmetrical about the ion (that is, if the field is of the same strength at a given distance in any direction from the central ion), the degeneracy of the five orbitals remains. This is shown in Case II. However,

Table 24.5 Relative Energies of Metal Ion *d* Levels in Various External Fields

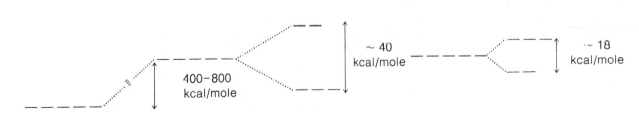

Case I. Free metal ion showing degenerate *d* orbitals all at the same energy—for example, Ti^{+3}.

Case II. Showing change in energy produced by a spherically symmetric electric field. The degeneracy remains.

Case III. Separation of the degenerate levels into a doubly degenerate upper level and a triply degenerate lower level by an octahedral field, as in TiF$_6$$^{-3}$.

Case IV. Separation of the degenerate levels of the spherically symmetric system by a tetrahedral field, as in TiCl$_4$$^-$.

if the field produced by the attached ligands is concentrated in particular directions from the central ion, the degenerate levels are separated into two or more sets of levels. The magnitude of the separation and the number of orbitals in the new levels depend upon the type of arrangement of the ligands around the central ion, and will be different according to whether they are arranged in a square planar, tetrahedral, octahedral, or other manner. The result for an octahedral arrangement is shown in Case III.

Had the ligands been arranged in a tetrahedral pattern around the central ion, the separation in Case III would have been in general less than for the octahedral arrangement, and would have been in a triply degenerate upper level and a doubly degenerate lower level, as illustrated in Case IV.

If the transition metal ion has electrons in the d orbitals, the electrons will enter the split d orbitals so that the lowest levels fill first and will pair only when there is an energy advantage as described below. For example, consider the ions $Ti(H_2O)_6^{+3}$, $Fe(H_2O)_6^{+3}$, and $Co(NH_3)_6^{+3}$. Since all are octahedral complexes, the d orbitals are split as shown in Figure 24.7.

High-Spin and Low-Spin Complexes

The reason the electrons are all placed in the lower levels in $Co(NH_3)_6^{+3}$ is that the ammonia ligands produce a strong field, which causes a rather large separation between the lower and higher d energy levels. Less energy is thus expended in overcoming the repulsion of another electron in two of the lower orbitals than would be required to promote two electrons to the two higher orbitals; therefore, the electrons pair in the lower orbitals. In $Fe(H_2O)_6^{+3}$, by contrast, the water molecules produce a rather weak field around the iron(III) ion, and the separation between the lower and higher d energy levels is less than in the cobalt-ammonia case. The electrons remain unpaired, since the energy expended in promoting two electrons to the higher levels is less than that needed to overcome the repulsion of another electron in two lower levels so that pairing might occur. The $Co(NH_3)_6^{+3}$ ion is thus a "low-spin" complex (few unpaired electrons), while $Fe(H_2O)_6^{+3}$ is a "high-spin" complex (many unpaired electrons).

Whether a low-spin or high-spin complex will form depends upon the relative strength of the field established by the ligands surrounding the central ion and the energies of the d orbitals of the central ion, and the overlap between the orbitals of the central ion and the ligand. In general, water as a ligand produces a smaller difference between the two octahedral d levels than ammonia, and is more prone to produce high-spin complexes. Cyanide ion has the most effect in increasing the separation, and appears at the ex-

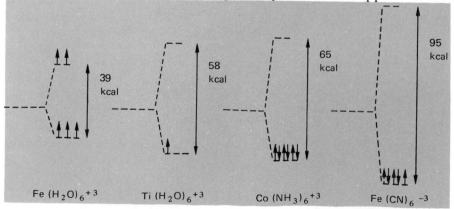

Figure 24.7 Relative energies of d levels in some ions with octahedral fields.

treme of the "spectrochemical series" that *lists ligands in the order in which they increase the separation between the two* d *levels:*

$$I^- < Br^- < Cl^- < F^- < C_2O_4^{-2} < H_2O < NH_3 < NO_2^- \ll CN^-$$

The position of F^- in this series shows that the theory would predict a high-spin complex, FeF_6^{-3}, and thus agrees with the experimentally measured high magnetic moment of FeF_6^{-3} mentioned earlier.

Color of Complex Ions

The color of transition metal complex ions is readily explained in terms of the crystal field theory by assuming that electrons in the lower d levels may absorb photons and move to higher levels. For example, the violet color of the $Ti(H_2O)_6^{+3}$ ion arises when the single d electron absorbs a photon of wavelength 5000 A and moves to one of the higher d levels. The light transmitted by the solution containing many $Ti(H_2O)_6^{+3}$ ions will be white minus the yellow photons of wavelengths 5000–6700 A. Such light is violet to the eye. The difference in energy between the energy levels is obtained in this and other cases by measuring the wavelength of the light absorbed and applying the equation $E_2 - E_1 = h\nu$. Thus the separation of levels as indicated in Figure 24.7 can be determined experimentally by observing the wavelength (or frequency ν) of the light absorbed by the complex.

When the separation between levels is large, so that pairing in the lower levels occurs, the orbital occupancy of electrons is the same as that predicted on the basis of valence bond theory. Thus, for low-spin complexes, the two theories give the same answer.

Ligand Field Theory

The most recent interpretation of the binding in complex compounds is given by a molecular orbital theory known as the *ligand field theory*. In this theory, the energy levels of the various orbitals of the complex are calculated from the atomic orbitals of the central atom and the ligands. It is found that, in addition to the bonding and antibonding orbitals, as depicted for hydrogen in Figure 6.18, there are also *nonbonding orbitals* that remain associated with the central atom and play no role in forming the molecule. The calculations tend to be too complicated to give exact values of energy levels, but the qualitative appearance for an octahedral complex is given in Figure 24.8.

Six orbitals of the central atom (two d, one s, and three p) combine with six orbitals of the six ligands to produce six bonding and six antibonding sigma (σ) orbitals in the molecule. Three of the d-orbitals of the central atom play no bonding role and remain associated with the central atom as nonbonding orbitals. Note the separation of the d orbitals into three lower levels and two upper levels; in this respect, the ligand field theory confirms the crystal field approach. Into this energy-level scheme must now be fitted the d electrons of the central ion and the six electron pairs of the ligands in such a way that the levels of lowest energy fill first.

For a nickel(II) complex ion with six ligands, for example, there will be a total of 20 electrons to be fitted into the scheme (eight d electrons for Ni^{+2} and 12 electrons in the six pairs of the ligands). Of these, 18 will fill up to the two lowest antibonding levels, and the last two will go, one each, into these two. The configuration is thus like that of Figure 24.9, predicting two unpaired electrons as found experimentally in octahedral nickel complexes.

The ligand field theory gives promise of giving the most successful explanation of bonding in complexes.

Figure 24.8 Energies of molecular orbitals in an octahedral complex.

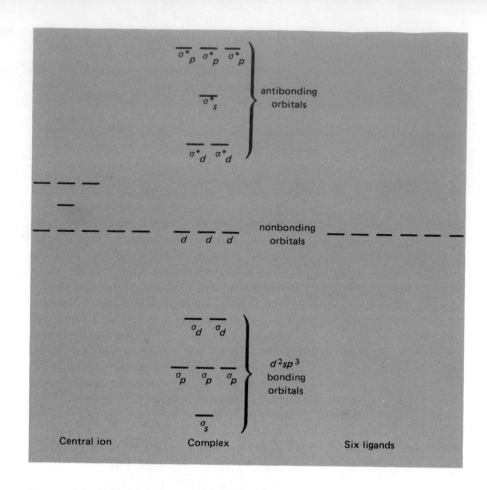

Pi Bonding in Complexes

Up to this point, we have discussed bonding in complex formation as if only σ bonds occurred. Some ligands and some metal ions, however, also have the capability of forming pi (π) bonds. Examination of the compounds indicates that π bonding can be of two sorts: there is metal-to-ligand π bonding in which the bonding electrons that form the π bond come from the metal, and there is ligand-to-metal bonding in which the electrons come from the ligand. The first of these occurs in complexes in which the metal is in oxidation state 0, I, or −I, and the second occurs in complexes in which the metal is in a high positive oxidation state.

Metal-to-Ligand Pi Bonding

Carbon monoxide is the most common ligand in such complexes, and all but two of the transition metals form carbonyl complexes. Examples are $Cr(CO)_6$, $Mn_2(CO)_{10}$, $Co(CO)_4{}^-$, $Ni(CO)_4$, $Mo(CO)_5I^-$, $Co(CO)_3NO$, and $Mn(CO)_5{}^-$. The oxidation states of the metals in the compounds listed are evidently zero or −I, since CO and NO are neutral, and I^- is negative. The neutral compounds are colorless, easily sublimed crystals or low-boiling liquids, pointing to a covalent character; they are highly poisonous. The ionic compounds form salts such as $NaCo(CO)_4$.

The possibility of π bonding in these compounds arises because carbon monoxide, in the resonant structure $:C::\ddot{O}:$, has only six electrons on carbon, so it contains a vacant orbital that an electron pair from the metal can enter. Its unshared electron pair, however, makes it capable of forming bonds with vacant orbitals on the metal. Thus, in $Cr(CO)_6$, for example, the usual six σ bonds are formed, giving a d^2sp^3 octahedral complex. In addition, three π bonds are produced, as six electrons from the chromium are delocalized into the vacant carbon monoxide orbitals. Presumably, these six electrons are those that would be pictured in Figure 24.9 (if that figure were drawn for chromium) as lying in the nonbonding d orbitals. Nickel carbonyl, $Ni(CO)_4$, has a tetrahedral structure, using four σ bonds in the sp^3 configuration, and two π bonds using electrons from the two lower d levels shown in Figure 24.4, Case IV.

Ligand-to-Metal Pi Bonding

This type of bonding is common in oxyanions such as CrO_4^{-2}. This complex is tetrahedral, with four σ bonds in a d^3s configuration; in addition, there are two π bonding pairs. All the bonding electrons are imagined to come from the oxide ions, adding to a chromium ion in oxidation state VI. In the orbital symbolism used in the valence bond theory discussion, we would have

Cr ↑ ↑ ↑ ↑ ↑   ↑   __ __ __

Cr^{+6} __ __ __ __ __ __ __ __ __

CrO_4^{-2} xx xx oo oo oo oo __ __ __

          π σ

Isomeric Complex Ions

Both geometrical and optical (Chapter 28) isomerism are found among complex ions. Geometrical isomerism is of special importance in square and octahedral complexes.

Two isomers of the square-planar complex $Pt(NH_3)_2Cl_2$ are known. These have been assigned the structures

Cl   Cl   NH$_3$ Cl
  \ /   \   /
  Pt   Pt
  / \   /   \
NH$_3$ NH$_3$ Cl   NH$_3$
Cis isomer   Trans isomer

Two such isomers are possible with any square-planar complex of the type MA_2B_2, where M is the metal ion and A and B are ligands.

Geometrical isomers in octahedral complexes are illustrated by the complex $Co(NH_3)_4Cl_2^+$. Here, the isomers have the structures

$$\begin{bmatrix} & Cl & \\ & | & \\ H_3N - | - NH_3 \\ & Co & \\ H_3N - | - NH_3 \\ & | & \\ & Cl & \end{bmatrix}^+$$

Trans isomer

$$\begin{bmatrix} & NH_3 & \\ & | & \\ H_3N - | & Cl \\ & Co & \\ H_3N - & Cl \\ & NH_3 & \end{bmatrix}^+$$

Cis isomer

Optical isomers (structures which are nonsuperimposable mirror images) are possible with polydentate ligands, as illustrated with $Co(en)_3^{+3}$, where en represents ethylenediamine ($NH_2-CH_2-CH_2-NH_2$). The following two

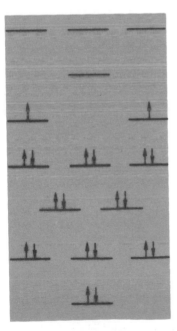

Figure 24.9 Orbital occupancy in an octahedral Ni^{+2} complex oxidation state.

structures are not identical. They are nonsuperimposable mirror images and are known as optical isomers.

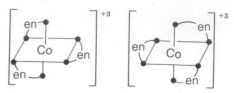

Optical isomers of $Co(en)_3^{+3}$ where $\bullet\frown\bullet \equiv$ en

Stability and Lability in Complex Ions

Complexes in which ligands are rapidly replaced by other ligands are known as *labile complexes*, while those in which such replacement is slow are called *inert*. For example, the exchange

$$NiCl_4^{-2} + Cl^- \longrightarrow NiCl_4^{-2} + Cl^-$$

is very rapid, while the corresponding exchange

$$PtCl_4^{-2} + Cl^- \longrightarrow PtCl_4^{-2} + Cl^-$$

is considerably slower.

A complex is classified as stable or unstable depending on the value of the equilibrium constant for the reaction

$$M^{+x} + n\text{Lig} \rightleftharpoons M(\text{Lig})_n^{+x}$$

If the equilibrium constant is large, the complex is said to be stable; if the equilibrium constant is small, the complex is unstable. The evidence is that formation of the complex occurs in steps, so that in addition to the equilibrium constant for the equation written, constants for the individual steps can be determined. For the tetraammine copper(II) ion, for example, these are

$$Cu^{+2} + NH_3 \rightleftharpoons Cu(NH_3)^{+2} \qquad K = 1.4 \times 10^4$$

$$Cu(NH_3)^{+2} + NH_3 \rightleftharpoons Cu(NH_3)_2^{+2} \qquad K = 3 \times 10^3$$

$$Cu(NH_3)_2^{+2} + NH_3 \rightleftharpoons Cu(NH_3)_3^{+2} \qquad K = 8 \times 10^2$$

$$Cu(NH_3)_3^{+2} + NH_3 \rightleftharpoons Cu(NH_3)_4^{+2} \qquad K = 1.4 \times 10^2$$

The equilibrium constant, K_n, for the over-all reaction

$$Cu^{+2} + 4NH_3 \rightleftharpoons Cu(NH_3)_4^{+2}$$

has the value 4.7×10^{12}. This would be considered a moderately stable complex. Since the copper ion is undoubtedly hydrated in water solution, these equations should probably be more precisely written

$$Cu(H_2O)_6^{+2} + NH_3 \rightleftharpoons Cu(H_2O)_5(NH_3)^{+2} + H_2O$$

and so forth, but the simpler equations are more commonly used. Tetra-cyanonickelate(II) ion, $Ni(CN)_4^{-2}$, is an example of a stable complex, $(K_n = 1 \times 10^{22})$, but the hexaammine cobalt(II) complex is unstable in water solution $(K_n = 1.3 \times 10^5)$.

While the $Ni(CN)_4^{-2}$ ion is stable, it also is labile. Addition of ^{14}C-labeled cyanide ion to an aqueous solution of the complex gives, almost instantaneously, a complex of the same formula containing radioactive carbon.

EFFECT OF LIGANDS ON THE PROPERTIES OF IONS	The attachment of a ligand to an ion may affect several properties of the ion, as illustrated by the following examples.

Solubility The firm attachment of water ligands to an ion presumably is the main reason that ions of solids such as sodium chloride go into solution. Some compounds insoluble in water will dissolve in the presence of other complexing agents. For example, silver chloride is insoluble in water but dissolves in solutions of ammonia because of the formation of the $Ag(NH_3)_2{}^+$ complex ion.

Oxidation-State Stabilization The attachment of six monodentate ligands to a cobalt(III) ion fills the two empty $3d$ orbitals as well as the $4s$ and the $4p$ orbitals; there is, therefore, very little tendency for cobalt(III) to pick up an electron and be reduced to cobalt(II) ions. However, this reduction can be readily accomplished in the following reaction:

$$Co(NH_3)_5Cl^{+2} + Cr(H_2O)_6{}^{+2} + 5H_3O^+ \longrightarrow Co(H_2O)_6{}^{+2} + Cr(H_2O)_5Cl^{+2} + 5NH_4{}^+$$

Here, the electron is believed to be transferred through the chloride ion in a complex of the type

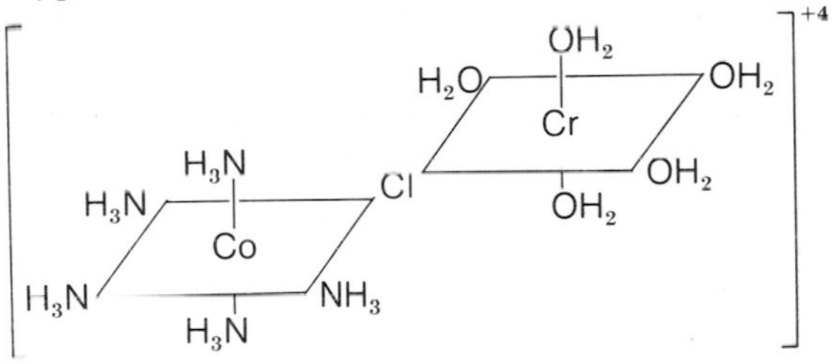

Color The color of $Cu(H_2O)_6{}^{+2}$ is blue-green, $Cu(NH_3)_4{}^{+2}$ is deep blue, and $CuCl_4{}^{-2}$ is green. These are only several of many examples that can be cited to show the effect of the ligand on the color of the ion. The color is a measure of the separation of the d energy levels of Table 24.5.

The much more intense colors shown by some oxyanions, such as $MnO_4{}^-$, arises from a "charge transfer" process and not from the d-d transitions mentioned above. In these cases, absorption of light causes an electron to move from an oxide ligand to the central atom, so that the change may be pictured as

$$[Mn^{+7}(O^{-2})_4] + h\nu \longrightarrow [Mn^{+6}(O^{-2})_3O^-]^-$$

A similar charge transfer absorption is thought to be responsible for the intense color of the thiocyanate iron(III) complex

$$[Fe^{+3}(SCN^-)]^{+2} + h\nu \longrightarrow [Fe^{+2}(SCN)]^{+2}$$

Iron(II) ions react with sulfide ions; iron(II) ions in $Fe(CN)_6{}^{-4}$ complex ions do not. Silver(I) ions precipitate with chloride ions, but $Ag(NH_3)_2{}^+$ complex ions do not form an insoluble compound with chloride ions. The standard oxidation potential of the metal ion also depends on the ligands present, as illustrated in the following examples:

$$Fe(H_2O)_6{}^{+2} \longrightarrow Fe(H_2O)_6{}^{+3} + e^- \qquad E° = -0.77 \text{ volt}$$

$$Fe(CN)_6{}^{-4} \longrightarrow Fe(CN)_6{}^{-3} + e^- \qquad E° = -0.36 \text{ volt}$$

Reactivity   Many other examples can be given to show how the reactivity of an ion may be altered by forming a complex ion.

Importance of Complex Ions   A knowledge of the structure, type of bonding, and properties of complex ions is becoming more and more important to the chemist as he studies enzymes containing metal ions such as zinc, manganese, cobalt, copper and molybdenum; hemoglobin, containing iron (and sometimes copper or vanadium); and chlorophyll, containing magnesium. These are but a few of the many compounds in which complex ions occur.

Transition Elements and Human Health.   Among the ions of transition elements, iron, cobalt, copper, manganese, molybdenum, and zinc play important roles in the biochemistry of the human body. Mercury, present in certain organomercury compounds, has been used for centuries as a pharmaceutical. Excessive amounts of many of these ions can be toxic.

Except for iron, the amounts of transition-element ions needed by the body are minuscule. About four of every million atoms present in an adult are iron atoms, while cobalt, copper, manganese, molybdenum, and zinc are each present in amounts less than one atom per million. The average adult requires only about 100 mg of copper, for example.

Most of these ions possess three chemical properties that are extremely important for carrying out biochemical processes: They form stable coordination complexes with proteins and other biologically active substances; they are versatile catalysts; and they have several oxidation states that are readily formed and interconverted in the environment of the cell, thereby providing a chemical flexibility that is uniquely suited to biochemical functions.

One important function of these metal ions is their role as a cofactor in reactions involving enzymes. While some enzymes depend only on their structures for activity, others require cofactors such as metal ions or coenzymes. The metal ion in such enzyme systems may serve one of two roles. It may function as a coordinating group to bind the reactant species to the enzyme through formation of a coordination complex, or it may act as the catalyst for the reaction itself.

Zinc ion is an essential cofactor in the hydrolysis of proteins to amino acids, which occurs in the small intestine. Manganous ion complexes with ADP or ATP to facilitate the transfer of phosphate, a process that is vital in the energy utilization cycle. Cobalt ion is present in cobamide, a derivative of vitamin B_{12} and one of the most extraordinary of all biologically active substances. Cobamide is important in the *in vivo* (in the body) synthesis of amino acids used to make proteins. Deficiency of cobalt can lead to pernicious anemia. Copper ion is involved in the synthesis of hemoglobin and phospholipids, compounds that are present in cell membranes and in brain and nerve tissue.

The role of copper in human biochemistry is especially interesting. Normally, we are supplied with an abundance of copper in our water and food, and from the copperware used in cooking. However, in rare cases, copper deficiencies or excesses can occur. It has been found that severe copper deficiency leads to degeneration of the sheath around the spinal cord, weakening of the walls of certain blood vessels (including the aorta), decoloration of hair, reduction in the synthesis of hemoglobin and phospholipids, and inhibition of the energy-producing reactions of the cell. It has been suggested that an organism's sensitivity to radiation is correlated with the amount of copper in the tissues—the lower the copper concentration, the greater the sensitivity to radiation damage.

When a radioactive copper salt is injected into the human body, the copper ion first appears in combination with the serum albumin in the blood serum. It then is rapidly absorbed by the liver and later appears in serum as the blue copper-containing protein called ceruloplasmin. This protein serves as a reservoir for copper and as a catalyst for certain reactions, including some that lead to hemoglobin synthesis. Ceruloplasmin also supplies copper ion to certain enzyme systems in the tissues. One of these is tyrosinase, which among other functions, controls hair color and skin pigmentation. Other copper-containing enzyme systems are cytochrome oxidase, the enzyme responsible for one of the last steps in the oxidation of foodstuffs, and hemocyanin, the respiratory blood protein of blue-blooded animals (snails, octopuses, scorpions, and crabs).

Should the ceruloplasmin in the serum be destroyed, or should massive amounts of copper be ingested, the copper ions would diffuse into the tissues where they could accumulate at a high level in the liver and in the brain. This can produce severe mental illness and death. However, this condition can be treated by reducing copper in the diet by eliminating foods such as nuts, mushrooms, liver, and oysters, or by using a chelating drug such as penicillamine, which leaches the copper out of the tissues.

We see in these examples how the chemistry of certain transition elements is utilized by the body, and how that same chemistry can be used to treat diseases caused by malfunctioning of systems involving these elements.

SUMMARY   In this chapter, the composition, bonding, structure, and energies of the complex ions of the metallic elements of the first transition series have been discussed. It is shown that the bonding in these compounds is interpreted as covalent bonding, frequently involving d orbitals. The magnetic properties depend upon the number of unpaired electrons in the complex ions, which, in turn, depends upon the total number of electrons and the distance between energy levels, as determined by the nature of the transition metal and the nature of the ligand. Isomerism, stability, and lability of complex ions are introduced.

IMPORTANT TERMS

Transition elements
first (second, third) transition series
lanthanides
actinides

Complex ions
coordination number
ligand
sp^3 complex
d^2sp^3 complex
bidendate ligand
chelation
"low-spin" complex
"high-spin" complex
π-bonding complex
tetrahedral complex
square-planar complex
octahedral complex
stable complex
labile complex

Isomers
optical
geometrical
substitutional

Magnetic
diamagnetic
paramagnetic
magnetic moment
magneton

Theory
valence bond theory
crystal field theory
molecular orbital theory
ligand field theory

Degenerate level

Spectrochemical series

Oxidation-state stabilization

1. In what ways do the transition elements differ from the representative metals of periodic Groups I and II?

2. How do you explain the fact that transition elements commonly exhibit several oxidation states?

3. List, by name and formula, 15 examples of complex ions.

4. What is the coordination number of the central atom in the following ions and molecules: $Cu(NH_3)_4^{+2}$, $Co(NO_2)_6^{-3}$, SF_6, $Sb(OH)_6^-$, and $Ag(NH_3)_2^+$?

5. Indicate the electronic structures of nickel atoms, Ni, nickel ions, Ni^{+2}, the hydrated ion $Ni(H_2O)_6^{+2}$, the ion $Ni(CN)_4^{-2}$, and the compound $Ni(CO)_4$.

6. The complex $[NiA_4]^{-2}$ is square planar, and the complex $[NiB_4]^{-2}$ is tetrahedral. (a) Draw diagrams for these complexes, according to the valence bond theory. (b) How many unpaired electrons are there in each case?

7. The octahedral complexes of Ni(II) have two unpaired electrons. (a) Does the valence bond theory explain this? (b) Does the ligand field theory? (c) Compare and explain fully.

8. How many d electrons are in the main ion in each of the following complexes? (a) $Pt(NH_3)_4^{+2}$; (b) $Zn,NH_3)_4^{+2}$; (c) ZrF_6^{-3}; and (d) CoF_6^{-3}.

9. (a) Write electron configurations for the elements from manganese to and including zinc, using the s, p, d, and f notations, diagramming the $3d$ and $4s$ electrons as shown on page 521. (b) Show by suitable diagrams the distribution of $3d$ and $4s$ electrons in the ions of these metals having an oxidation state of II. (c) Do the same for any ions exhibiting a III oxidation state.

10. Compare these complexes of cobalt, CoF_6^{-3} and $Co(CN)_6^{-3}$, as to their magnetic types. Explain.

11. Discuss the interpretations of the magnitudes of the magnetic moments of $Fe(H_2O)_6^{+3}$, FeF_6^{-3}, and $Fe(CN)_6^{-3}$. Draw an energy-level diagram to illustrate how $Fe(H_2O)_6^{+3}$ can be a high-spin complex, while $Fe(CN)_6^{-3}$ can be a low-spin complex.

12. Draw the electron configurations for the "high-spin" and "low-spin" forms of the octahedral complexes of Co(III). Explain the d-orbital splitting.

13. Draw the electron configurations for the "high-spin" and "low-spin" forms of the complexes of (a) Fe(II), (b) Fe(III), (c) Cr(II), (d) Mn(II), and (e) Ni(III).

14. Show, by means of an energy-level diagram, what happens when $Ti(H_2O)_6^{+3}$ absorbs a photon in the 5000-A region of the spectrum.

15. The $Cr(H_2O)_6^{+2}$ complex absorbs light of a longer wavelength than $Mn(H_2O)_6^{+3}$. How can this observation be explained in terms of the ligand field theory?

16. How many isomers can there be of the compound $Co(en)_2(NO_2)_2^+$, where en represents ethylene diamine, $H_2NCH_2CH_2NH_2$?

17. Predict the shape and outline the electron configuration for the following

complexes. Distinguish between electrons contributed by the main atom and those contributed by the ligands: (a) $Fe(CN)_6^{-3}$; (b) $Ni(CN)_4^{-2}$, diamagnetic; (c) $Mn(CN)_6^{-4}$; (d) $CuCl_2Br_2^{-2}$.

18. Calculate the concentration of copper ion in a solution prepared by adding 50 ml of 0.4 M copper nitrate solution to 150 ml of 0.4 M ammonia solution.

19. Spots of rust or ink can often be removed from cloth by treatment with oxalic acid solution. Can you suggest what chemical reaction makes this treatment effective?

SPECIAL PROBLEMS

1. Show by diagrams that there can be (a) one and only one complex of the formula $Co(NH_3)_5(NO_2)^{+2}$, (b) two and only two complexes of the formula $Co(NH_3)_4(H_2O)(NO_2)^{+2}$, and (c) two complexes of formula $Co(NH_3)_3(NO_2)_3$. Explain the magnitude of the electrical charges on these species.

2. In terms of the ligand field theory, explain the difference in the standard oxidation potentials of the following reactions:
   (a) $Co(H_2O)_6^{+2} \longrightarrow Co(H_2O)_6^{+3} + e^-$ $\qquad E° = -1.84$ v
   (b) $Co(NH_3)_6^{+2} \longrightarrow Co(NH_3)_6^{+3} + e^-$ $\qquad E° = -0.1$ v
   (c) $Co(CN)_6^{-4} \longrightarrow Co(CN)_6^{-3} + e^-$ $\qquad E° = +0.84$ v

3. Explain, using chemical equations and applying Le Châtelier's principle, why silver chloride dissolves in ammonia solutions.

4. Will the addition of sodium sulfide until the solution becomes 0.1 M in sulfide ion precipitate NiS from a 0.1 M solution of $K_2Ni(CN)_4$? CoS from a 0.1 M solution of $Co(NH_3)_6^{+2}$? The solubility-product constants are 1×10^{-22} for NiS and 5×10^{-22} for CoS.

REFERENCES

Basolo, F., and Johnson, R., *Coordination Chemistry*, Benjamin, Menlo Park, Calif., 1964.

Douglas, B. E., "Stabilization of Oxidation States Through Coordination," *J. Chem. Educ.*, **29**, 119 (1952).

Gray, H. B., "Molecular Orbital Theory for Transition Metal Complexes," *J. Chem. Educ.*, **41**, 2 (1964).

Johnson, O., "Role of *f* Electrons in Chemical Bonding," *J. Chem. Educ.*, **47**, 431 (1970).

Karraker, D. G., "Coordination of Trivalent Lanthanide Ions," *J. Chem. Educ.*, **47**, 424 (1970).

Kirschner, S., "Inorganic Coordination Compounds in General Chemistry," *J. Chem. Educ.*, **35**, 139 (1958).

Liehr, A. D., "Molecular Orbital, Valence Bond, and Ligand Field," *J. Chem. Educ.*, **39**, 135 (1962).

Murmann, R. K., *Inorganic Complex Compounds*, Van Nostrand Reinhold, New York, 1964.

Taube, H., "Mechanisms of Oxidation-Reduction Reactions," *J. Chem. Educ.*, **45**, 452 (1968).

THE TRANSITION ELEMENTS

Jag gienom mine experimenter haft den lycken, att jag är förste upfinnaren af en ny half-metall, neml. cobolt-regulus, som tilförene med wismuth blifwit confunderad.*

Georg Brant (1694–1768)

These elements, occurring in the middle of the periodic table,

$_{21}$Sc	$_{22}$Ti	$_{23}$V	$_{24}$Cr	$_{25}$Mn	$_{26}$Fe	$_{27}$Co	$_{28}$Ni	$_{29}$Cu
$_{39}$Y	$_{40}$Zr	$_{41}$Nb	$_{42}$Mo	$_{43}$Tc	$_{44}$Ru	$_{45}$Rh	$_{46}$Pd	$_{47}$Ag
$_{57}$La	$_{72}$Hf	$_{73}$Ta	$_{74}$W	$_{75}$Re	$_{76}$Os	$_{77}$Ir	$_{78}$Pt	$_{79}$Au

are those in which a d-orbital is filling to its maximum of 10 electrons as the atomic number increases, while maintaining two (or in a few cases one) electrons in the s orbital of next higher principal quantum number. In elements 21 through 29 (first row above), the $3d$ orbitals fill, as shown in Figure 25.1. In the second and third rows the $4d$ and $5d$ orbitals fill in a similar fashion (Table 4.6). At the left-hand end of the group shown above, the $(n - 1)d$ electrons are near enough in energy to the ns electrons that they play an active role in compound formation. In going from left to right, these electrons become progressively lower in energy until in the next succeeding elements, $_{30}$Zn, $_{48}$Cd, $_{80}$Hg, they become a part of the "inner" shells, so that in these three elements only the ns electrons participate effectively in bonding. The "transition" is between the elements of Group IIA, with the valence electron configuration ns^2 and no $(n - 1)d$ electrons, to group IIB, with the same valence electron configuration but with a filled shell $(n - 1)d^{10}$ below it. In all three rows, the atomic sizes remain nearly constant across the row, as shown for the first series in Figure 25.1. This is a reflection of the fact that the valence electron configuration remains effectively constant along the rows, and helps explain why the properties change only gradually with change

*"I had the good fortune, through my experiments, that I am the first discoverer of a new half-metal, namely, cobalt-regulus, which earlier was confused with bismuth." (The quotation is from *Diarium chymicum*, 1741.)

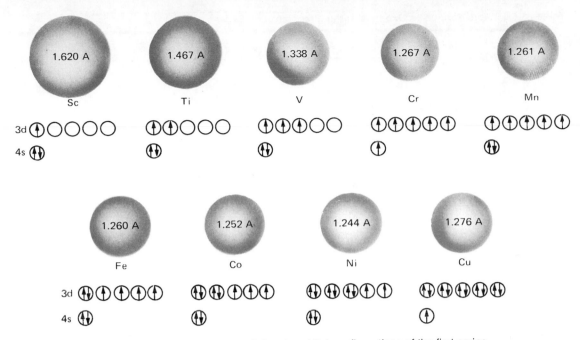

Figure 25.1 Atomic size and *d* and *s* orbital configurations of the first-series transition elements.

in atomic number, in contrast to the great change in properties in passing from Group I to Group VII in the A-group elements.

This chapter discusses the properties of the transition elements, with particular emphasis on those of the first series, 21 through 29.

FIRST-SERIES TRANSITION ELEMENTS
Electron Configuration and Oxidation States

The oxidation states exhibited by these elements in compounds are given in Figure 25.2. The multiplicity of oxidation states arises because the 3*d* as well as 4*s* electrons can be used in compound formation. For the elements scandium to manganese, the highest oxidation state observed is the sum of the electrons in the 4*s* and 3*d* orbitals of the atom. Beyond manganese, the *d*

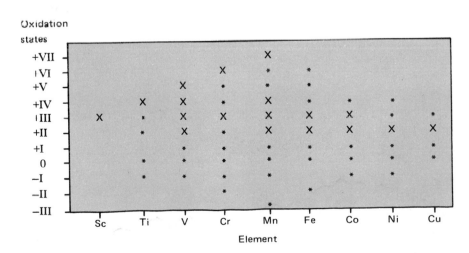

Figure 25.2 Oxidation states of the elements scandium through copper in compounds. X designates the most common states. The * designates a state known to exist but which may not be very stable.

Table 25.1 Colors of Aquo or Oxo Ions of the First Transition Series

Sc	Ti	V	Cr	Mn	Fe	Co	Ni	Cu
III, colorless	II, violet III, violet (green) IV, colorless	II, violet III, green IV, blue V, colorless	II, blue III, gray VI, yellow	II, pink III, red-brown VI, green VII, purple	II, green III, orange	II, pink III, blue	II, green	I, colorless II, blue

NOTE: The aquo ion is the hydrated ion; the oxo ion is the oxygen-bonded type represented by CrO_4^{-2}, MnO_4^-, and so on.

electrons become more tightly bound, and the highest oxidation states, corresponding to (2 + number of unpaired d electrons), become so unstable they are difficult to obtain.

Negative and zero oxidation states are found in certain compounds with π-bonding ligands such as carbon monoxide, as discussed in the preceding chapter. Examples are $Ni(CO)_4$, $V(CO)_6$, $Mn(CO)_5^-$, and $[Fe(CO)_4]^{-2}$.

Colors of the Ions

The transition elements are the only metals that form colored ions. The color, as explained in Chapter 24, is attributed to the absorption of photons from white light to raise the d electrons from the ground state to higher energy levels. The transmitted light will then be colored, the color being the complement of the absorbed color. Figure 25.3 illustrates the electronic transition responsible for the green color of $V(H_2O)_6^{+3}$. The colors of some of the ions of these elements in the various oxidation states in water are given in Table 25.1.

The color of the ions depends upon the ligand or ligands present, as illustrated for Co^{+2} ion in Table 25.2. Different ligands produce different separations of the d-orbital energy levels (spectrochemical series, Chapter 24), so that the hexaaquocobalt(II) ion absorbs in the yellow region of the spectrum but the other two ions absorb in regions of progressively higher energy.

Some Trends in Properties

Before discussing the chemistry of each element in this series, we shall list some general trends in properties that may help in developing a broad view of the chemistry of these elements.

1. The densities of the metals are relatively high, increasing regularly from scandium to copper. Undoubtedly, this is a consequence of decreasing atomic diameter and increasing atomic mass in proceeding across the series. The melting points are very high, reflecting the added stability resulting from participation of d and s electrons in the bonding of the metals. The melting-

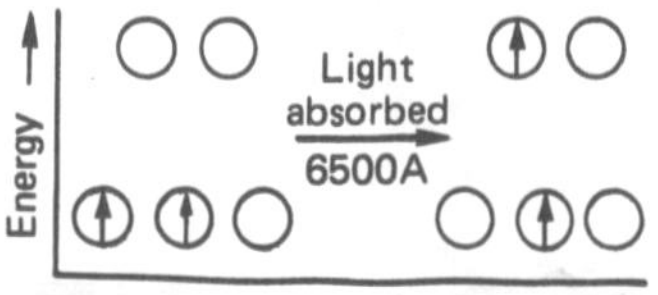

Figure 25.3 Electronic transition responsible for the green color of $V(H_2O)_6^{+3}$. Circles represent d levels as split by the octahedral ligand field.

Table 25.2 Dependence of Color on the Ligand Present

Ion	Color
$Co(H_2O)_6^{+2}$	pink
$Co(NH_3)_6^{+2}$	blue
$Co(en)_3^{+2}$	violet

Element	Density (25°C, g/cc)	Melting Point (°C)	Electrical Conductance (microhms^{-1} cm^{-1} between 0 and 25°C)	Thermal Conductivity (25°C cal/gm/sec/deg)	Tensile Strength of Hard-Drawn Wire (1000 lb/in².)
Sc	2.5	1423			
Ti	4.5	1677	1.3×10^{-2}		
V	5.96	1917	3.8×10^{-2}		
Cr	7.1	1900	7.7×10^{-2}	0.16	
Mn	7.2	1244	5×10^{-3}		
Fe	7.86	1528	1.02×10^{-1}	0.11	80–120
Co	8.9	1490	1.6×10^{-1}	0.16	33
Ni	8.9	1452	1.5×10^{-1}	0.14	155
Cu	8.92	1083	6×10^{-1}	0.99	60–70

point trends appear to be related to the number of unpaired electrons in the atoms, increasing from scandium to vanadium, decreasing from iron to copper.

2. From titanium to manganese, the highest oxidation states are usually found only in the presence of oxygen, fluorine, or chlorine—the most electronegative elements.

3. In the presence of oxyligands, the metal ions usually exhibit tetrahedral coordination in oxidation states IV to VII, as in VO_4^{-3}, CrO_4^{-2}, and MnO_4^{-}. In oxidation states up to IV, the metal ions are usually octahedrally coordinated, as in $Mn(H_2O)_6^{+2}$, ScF_6^{-3}, and $V(CN)_6^{-4}$, or tetrahedrally coordinated, as in $MnCl_4^{-2}$.

4. The oxides of a given element usually produce stronger acids as the oxidation state increases. The halides become more covalent and more susceptible to hydrolysis as the oxidation state increases. For example, Mn_2O_7 is acidic (forming salts in which the manganese appears in the anion, such as $K^+MnO_4^-$) while MnO is basic (forming salts in which the manganese is a cation, such as $Mn^{+2}\ 2Cl^-$); $TiCl_4$ is so sensitive to hydrolysis that it forms white fumes of TiO_2 on exposure to air, while $TiCl_2$ reacts only slowly with water.

5. The standard potentials for oxidation of the metal to the II state (Table 25.4) show that all except copper are reducing agents toward hydrogen ion. The II state is not a stable state in water solution for Ti^{+2} and Cr^{+2}, since these are oxidized by water to the III state; scandium forms only the ion Sc^{+3}.

Table 25.4   Standard Oxidation Potentials of the First-Series Transition Elements

Standard Oxidation Potentials	(volts)
$Sc(s) \longrightarrow Sc^{+3} + 3e^-$	2.1
$Ti(s) \longrightarrow Ti^{+2} + 2e^-$	1.6
$V(s) \longrightarrow V^{+2} + 2e^-$	1.2
$Cr(s) \longrightarrow Cr^{+2} + 2e^-$	0.91
$Mn(s) \longrightarrow Mn^{+2} + 2e^-$	1.18
$Fe(s) \longrightarrow Fe^{+2} + 2e^-$	0.44
$Co(s) \longrightarrow Co^{+2} + 2e^-$	0.28
$Ni(s) \longrightarrow Ni^{+2} + 2e^-$	0.25
$Cu(s) \longrightarrow Cu^{+2} + 2e^-$	−0.34

Scandium

$$1s^2, 2s^2, 2p^6, 3s^2, 3p^6, \text{(↑)} \bigcirc \bigcirc \bigcirc \bigcirc, 4s^2$$

3d orbitals

The element following calcium is scandium, whose atoms have an electron configuration similar to calcium except for an electron in the $3d$ shell. Thus scandium is the first element in the first transition series. The electron levels of the $4s$ and $3d$ electrons are close together; we thus find that scandium, like calcium, reacts with water to form hydrogen and colorless scandium(III) ions. The III oxidation state is the only stable oxidation state of scandium, other than the zero oxidation state, exhibited by the metal. Scandium is similar to aluminum in many ways. However, scandium oxide, Sc_2O_3, is more basic than aluminum oxide, Al_2O_3.

Scandium chemistry was of only little interest until the element was found to be one of the fission products of atomic fuels. Its chemistry then became important in working out methods of separation of scandium from uranium in the purification of atomic fuels. Scandium occurs in nature in the minerals monazite (a phosphate) and gadolinite (a silicate). It is as abundant as arsenic and about twice as abundant as boron.

Titanium

$$1s^2, 2s^2, 2p^6, 3s^2, 3p^6, \text{(↑)} \text{(↑)} \bigcirc \bigcirc \bigcirc, 4s^2$$

3d orbitals

The second transition element is titanium with two $3d$ electrons. The two $3d$ electrons and the two $4s$ electrons are easily removed or used in bonding. Thus we find the IV oxidation state a common and stable one. This is illustrated by such compounds as the dioxide, TiO_2, a white solid, and the tetrachloride, $TiCl_4$, a colorless, fuming liquid that boils at 136.4°C.

TiO_2 is used as a pigment because of its good covering power; $TiCl_4$ reacts with moist air, forming a smoke, and with water to give titanium dioxide. The liquid character of $TiCl_4$ at room temperature indicates that the bonds are predominantly covalent.

$TiCl_4$ and $TiBr_4$ use $3d$ orbitals as electron acceptors and therefore act as Lewis acids toward such compounds as alcohols, ethers, and other oxygen compounds.

$$\begin{array}{c} R \\ \diagdown \\ \diagup \\ R \end{array} O + TiCl_4 \longrightarrow \begin{array}{c} R \\ \diagdown \\ \diagup \\ R \end{array} O{-}TiCl_4$$

There is a sufficient difference in the energy levels of the $4s$ and $3d$ electrons so that titanium forms compounds in the II and III oxidation states, using or losing only two or three electrons, as well as in the IV oxidation state. Titanium compounds in the II and III states are reducing agents. Ti(II) has such a strong tendency to go to Ti(III) that the Ti^{+2} ion reacts rapidly with water to form hydrogen; Ti^{+2} is thus unstable in water solution. Titanium(III) hydroxide, $Ti(OH)_3$, gradually changes to titanium dioxide, TiO_2 (white), on standing.

$$2Ti(OH)_3(s) \rightleftharpoons 2TiO_2(s) + 2H_2O + H_2(g)$$

In the trivalent state, titanium forms salts—for example, $Ti_2(SO_4)_3$ and $TiCl_3$. $Ti(H_2O)_6^{+3}$ is violet and paramagnetic because of the presence of the unpaired *d* electron. At high temperatures, the yellow nitride TiN is formed by the action of nitrogen on titanium.

Titanium has been known for many years but it has not been used as a metal for structural purposes because the common methods of metallurgy produced a brittle product. Recent research has shown that the metal can be worked in a controlled atmosphere (an atmosphere low in oxygen and moisture) to produce a product that is usable for many purposes because it is stronger than iron. The metal is hard, refractory (m.p. 1680°C), a good conductor but of quite low density, and unusually resistant to corrosion (because of a surface coating of oxide and nitride). These properties are responsible for its use in spacecraft and marine equipment.

While resistant to corrosion and unreactive with most elements at ordinary temperatures, titanium reacts at high temperatures with hydrogen, oxygen, sulfur, the halogens, nitrogen, carbon, and boron and with hot steam; it reacts readily with cold sulfuric acid and hot hydrochloric acid to give titanium(III) salts and hydrogen.

Titanium is the 10th most abundant element in the earth's crust.

3d orbitals

Vanadium $1s^2, 2s^2, 2p^6, 3s^2, 3p^6,$ ⬆ ⬆ ⬆ ◯ ◯ $, 4s^2$

Vanadium exists in $-I$, 0, II, III, IV, and V oxidation states. In the V state, V_2O_5 is formed; this is probably the most important compound of vanadium. It is an excellent catalyst for the oxidation of sulfur dioxide to sulfur trioxide.

V_2O_5 is amphoteric. In bases, it forms vanadate ion, VO_4^{-3}; in acids it forms such ions as VO_2^+, $VO_2(OH)_3^{-2}$, $VO(OH)_4^-$, and polymerized forms containing more than one vanadium atom. Vanadium in the V state can be reduced with zinc to form blue vanadyl(IV) ion, VO^{+2}; further reduction gives green V^{+3} and, finally, violet V^{+2}. The latter ion is a fair reducing agent (Table 25-9).

Pure vanadium is difficult to prepare. It is an important alloy metal used to impart toughness and elasticity to steel. Such steels are used to make the jaws of steam shovels, the burrs for rock crushers, and some gears. Vanadium also reacts with oxygen and nitrogen in molten iron. In nature, vanadium occurs as patronite (V_2S_5), vanadinite ($Pb_5V_3O_{12}Cl$), and carnotite ($KUVO_6$).

3d orbitals

Chromium $1s^2, 2s^2, 2p^6, 3s^2, 3p^6,$ ⬆ ⬆ ⬆ ⬆ ⬆ $, 4s^1$

Chromium forms compounds in all states between $-II$ and VI. The III state is probably the most stable, existing as $Cr(H_2O)_6^{+3}$ ions of deep blue-gray color in water solution. The II state is formed by reduction of chromium(III) ion, Cr^{+3}, with zinc, but the II ion is unstable in aqueous solution in the presence of oxygen, with which it reacts rapidly and quantitatively to form the III state. Chromium(II) ion is the strongest useful reducing agent known in aqueous solution.

Chromate ion, CrO_4^{-2}, and dichromate ion, $Cr_2O_7^{-2}$, are the common ions of chromium in the VI state; these can be changed from one to the other by changing the acidity of the solution,

$$2CrO_4^{-2} + 2H_3O^+ \longrightarrow Cr_2O_7^{-2} + 3H_2O$$

$$Cr_2O_7^{-2} + 2OH^- \longrightarrow 2CrO_4^{-2} + H_2O$$

The following formulas show the relationship between these two ions:

$$\left[\begin{array}{c} :\ddot{O}: \\ :\ddot{O}:\underset{..}{\overset{..}{Cr}}:\ddot{O}: \\ :\ddot{O}: \end{array}\right]^{-2} \quad \left[\begin{array}{c} O \\ \| \\ Cr \\ O \diagup \diagdown O \\ O \end{array}\right]^{-2} \qquad \left[\begin{array}{c} :\ddot{O}: \quad :\ddot{O}: \\ :\ddot{O}:\underset{..}{Cr}:\ddot{O}:\underset{..}{Cr}:\ddot{O}: \\ :\ddot{O}: \quad :\ddot{O}: \end{array}\right]^{-2} \quad \left[\begin{array}{c} O \quad O \\ \| \quad \| \\ Cr \quad Cr \\ O \diagup \diagdown O \diagup \diagdown O \\ O \quad O \end{array}\right]^{-2}$$

Chromate ion   Dichromate ion

Chromium is a white, hard, lustrous, and brittle metal that melts at 1890°C. The metal is notorious for being passive, a condition that makes it highly corrosion-resistant and therefore an excellent metal for protecting metal surfaces or for making stainless steel. The cause of passivity is attributed to a thin oxide film that protects the metal from corrosion or electrochemical action, or both. Some chromium alloys are described in Table 25.5.

Chromium reacts slowly in cold dilute hydrochloric and sulfuric acids, and rapidly in hot hydrochloric acid and hot concentrated sulfuric acid. However, it does not react with nitric acid. It also reacts with chlorine or bromine on heating, and with oxygen and water at high temperatures.

Manganese

$$3d \text{ orbitals}$$
$$1s^2,\ 2s^2,\ 2p^6,\ 3s^2,\ 3p^6,\ \uparrow\ \uparrow\ \uparrow\ \uparrow\ \uparrow,\ 4s^2$$

Manganese forms compounds in $-$III, $-$I, 0, I, II, III, IV, V, VI, and VII states. The VII state is the highest exhibited by any of the elements in this series. The I and V states are very unstable. The II state of manganese, in contrast to the II state of chromium, is quite stable in neutral and acid solutions. In alkaline solutions, however, manganese(II) as $Mn(OH)_2$ is quickly oxidized to the III state in the presence of oxygen. In fact, a slurry of $Mn(OH)_2$ is an excellent absorber for oxygen.

The III state of manganese is stable only in the solid form or as a complex ion. Manganese(III) ion, like chromium(III) ion, readily forms complex ions of the high-spin type with fluoride ions and the low-spin type with ligands such as $C_2O_4^{-2}$ and PO_4^{-3}. Unlike chromium(III), manganese(III) is a powerful oxidizing agent; left alone, it disproportionates into manganese(II) and manganese(IV).

$$2Mn^{+3} + 6H_2O \rightleftharpoons Mn^{+2} + MnO_2 + 4H_3O^+$$

$$2MnF_3 + 2H_2O \rightleftharpoons MnF_2 + 4HF + MnO_2$$

Manganese is widely distributed in nature, chiefly as manganese dioxide, MnO_2 (pyrolusite). Metallic manganese is made by electrolysis of manganese sulfate solution or by reduction of MnO_2 with aluminum (Goldschmidt process). It is harder, more brittle, and less refractory than iron. The major use of manganese is as a scavenger of sulfur and oxygen in steel, and as an alloying element in steel.

Iron

$$3d \text{ orbitals}$$
$$1s^2,\ 2s^2,\ 2p^6,\ 3s^2,\ 3p^6,\ \uparrow\downarrow\ \uparrow\ \uparrow\ \uparrow\ \uparrow,\ 4s^2$$

Iron exhibits oxidation states of $-$II, 0, I, II, III, IV, V, and VI. However, the states above III are unstable and oxyanions such as ferrate, FeO_4^{-2},

Table 25.5 Some Alloys of Chromium

Alloy	Composition	Use
Chrome steel	4–10% Cr; Fe	Corrosion-resistant and high-temperature uses
Stainless steel	17–19% Cr; 7–9.5% Ni; Fe	Rollers; furnace parts; process, medical, and restaurant equipment
Nichrome	20% Cr; 80% Ni	Electrical heating elements

rarely form. The II and III states have almost equal stability; only mild oxidizing and reducing agents are required to go from the II to the III or back from the III to the II state. As with chromium, the II state of iron in alkaline solution is readily oxidized to the III state. In both the II and III states, iron forms complex ions of coordination number 6 readily; many of these complexes have high spin.

Reactions of Iron. Iron is a strong enough reducing agent to reduce hydrogen ions to hydrogen; thus, iron reacts with dilute and concentrated hydrochloric and dilute sulfuric acids to give hydrogen. With concentrated sulfuric acid, sulfur dioxide is produced. Concentrated nitric acid causes iron to become passive, probably forming a layer of oxide over the surface. Iron, when heated, reacts with most of the nonmetals; with oxygen, it forms the oxides FeO, Fe_2O_3, and Fe_3O_4.

Complexes of Iron. Iron forms complexes in both the II and III oxidation states. They are octahedral; most are high-spin, but a few are low-spin; $Fe(NH_3)_6^{+2}$ has high spin, but with the stronger-bonding CN^- ion, iron forms $Fe(CN)_6^{-4}$, a diamagnetic complex.

One of the most important complexes of iron is that formed with the porphyrin structure called *heme* in hemoglobin. Heme is an iron(II) complex of porphyrin, as shown in Figure 25.4.

As illustrated, four of the coordination positions on the iron ion are occupied by nitrogen atoms from the porphyrin. A fifth position appears to be occupied by a nitrogen atom from the protein. The sixth coordination

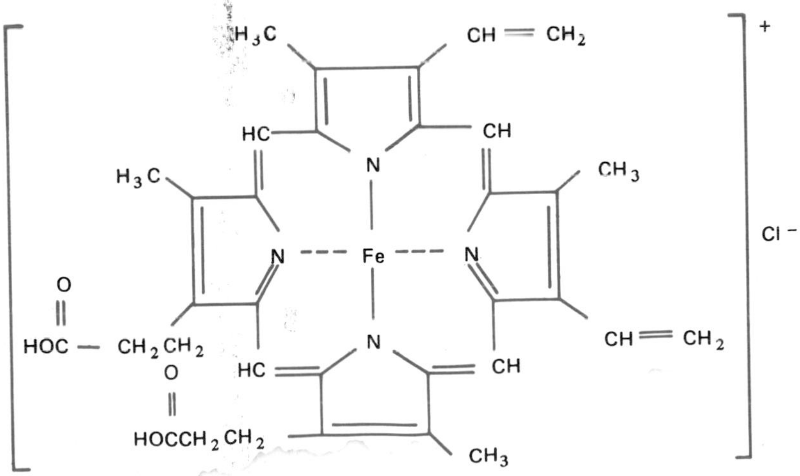

Figure 25.4 Hemin showing the iron(II) complexed to the four nitrogen atoms of the porphyrin ring structure. Oxygen, coordinated to the iron(II) in this structure, is carried throughout the circulatory system.

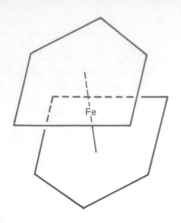

position is undoubtedly occupied by a water molecule but, when the complex is exposed to oxygen, an oxygen molecule displaces the water molecule. The complex has four unpaired electrons when water is a ligand. But when oxygen is present, there are no unpaired electrons. The oxygen-containing complex has four unpaired electrons when water is a ligand, but when oxygen is present, there are no unpaired electrons. The oxygen-containing complex is the substance that carries oxygen throughout the bodies of animals with circulating systems. If the water molecules are displaced by a strongly complexing ligand such as cyanide ion, CN^-, or carbon monoxide, CO, the oxygen molecule cannot enter the complex. This accounts for the poisonous nature of certain substances that contain strongly bonding ligands.

An unusual type of complex is the ferrocene type, made by complexing the unsaturated hydrocarbon, cyclopentadiene, with Fe^{+2} ions to give a complex of the type as shown at left. The complex is diamagnetic and so stable that it is unaffected by boiling sodium hydroxide or hydrochloric acid.

The Metal. Iron, the second most abundant metal in the earth's crust (4.7 per cent), occurs in almost pure oxide forms in many areas, and may be one of the main constituents of the interior of the earth. The solar spectrum is characterized by a large number of iron lines. The indicated high concentration of iron in the earth's interior possibly may be accounted for partially on the basis of the fact that iron atom nuclei are some of the most stable nuclei of the universe; they are in the maximum on the graph of binding energy against atomic masses (Figure 23.4). The cause of the formation of the great deposits of iron ores in various regions of the earth is not known.

Pure iron is a white, lustrous metal melting at 1528°C. It is not a hard metal and it is quite reactive. Finely divided, it is pyrophoric. At temperatures up to 906°C, the metal has a body-centered crystal structure; from 906°C to 1401°C, it has a cubic close-packed structure. Above 1401°C, it again becomes body-centered. It is ferromagnetic up to 768°C.

Iron is obtained from its ore (usually the oxide) chiefly by reduction with hot carbon in the blast furnace. For industrial purposes, it is purified in basic oxygen furnaces, in Bessemer converters, basic open-hearth furnaces, or in electric furnaces. The metallurgy and manufacture of iron and steel is one of the world's greatest industries.

$$3d \text{ orbitals}$$

Cobalt $1s^2, 2s^2, 2p^6, 3s^2, 3p^6,$ (⇅) (⇅) (↑) (↑) (↑), $4s^2$

Although the II and III states in iron have about equal stability, the II state in cobalt is stable in water and in the presence of other weak complexing agents, but the III state is stable only in the presence of strong complexing agents. There are also some important complexes of cobalt in the I state.

Because the uncomplexed cobalt(II) ion is the most stable of the uncomplexed ions, there is an abundance of simple salts and binary compounds of cobalt(II), but those of cobalt(III) are rare. In fact, only the fluoride and sulfate of cobalt(III) are known, while cobalt(II) forms an extensive group of compounds, including the oxide, hydroxide, several sulfides, halides, and others.

Cobalt Complexes. A large number of complexes of cobalt ions are known; a few are illustrated in Figure 25.5 and 25.6. All cobalt(II) complexes

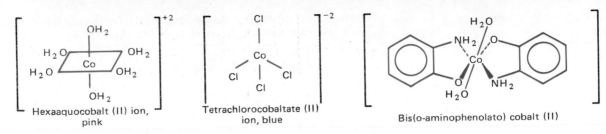

Figure 25.5 Complexes of cobalt(II).

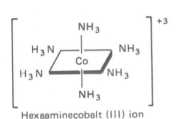

Hexaaminecobalt (III) ion

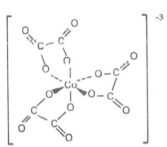

Trisoxalatocobaltate (III) ion

Example of oxygen ligands
bridging two cobalt ions

Figure 25.6 Complexes of cobalt(III).

have high spin—even those formed with strongly bonded ligands. Nearly all cobalt(III) complexes have low spin. An exception is CoF_6^{-3}.

Cobalt(II) ions are oxidized only with extreme difficulty in water solution; the standard oxidation potential for the reaction

$$Co^{+2} \longrightarrow Co^{+3} + e^-$$

in aqueous acid solution is -1.82 volts. However, the standard oxidation potential is greatly increased by the formation of a cobalt(III) complex

$$Co(NH_3)_6^{+2} \longrightarrow Co(NH_3)_6^{+3} + e^- \qquad E° = -0.1 \text{ v}$$

Thus, in the presence of strongly complexing ligands such as ammonia or cyanide ion, cobalt(II) is easily oxidized to cobalt (III) ion by air or by hydrogen peroxide:

$$2Co(NH_3)_6^{+2} + H_2O_2 + 2H_3O^+ \longrightarrow 2Co(NH_3)_6^{+3} + 4H_2O$$

This is a good example of the ability of ligands to alter the properties of the ion with which they are complexed. In this case, the ammonia or cyanide ligands presumably make the III state much more stable by bonding so tightly that a low-spin complex is formed.

The Metal. Cobalt occurs in nature principally as $CoAs_2$ and $CoAsS$. Such minerals are found sparingly in many localities; the richest known deposits are those located at Cobalt, Ontario, and in Africa. These are very rich in silver, and cobalt is worked up as a by-product, together with nickel and arsenic trioxide. In the Congo, cobalt is a by-product of copper refining. The metal is obtained by roasting the ore to the oxide, Co_3O_4, and reducing it with aluminum by the Goldschmidt process. It is a malleable, magnetic, silvery metal that soon takes on a reddish tint upon exposure to the air. It melts at 1490°C and is relatively unreactive. The metal is used chiefly in the form of alloys.

Cobalt Chromium Steels. Collectively known as Stellite, cobalt chromium steels are used extensively in making high-speed cutting tools, since they retain their temper at high temperatures and are important rust-resisting alloys. They are often modified by additions of tungsten, molybdenum, and nickel. An alloy called Konel contains cobalt, nickel, and ferrotitanium, and is harder than steel even when very hot. Metallic cobalt is used as a binding material for crystals of tungsten carbide in making the hard-cutting material Carboloy.

$1s^2, 2s^2, 2p^6, 3s^2, 3p^6,$ ⬆⬇ ⬆⬇ ⬆⬇ ⬆ ⬆, $4s^2$

$3d$ orbitals

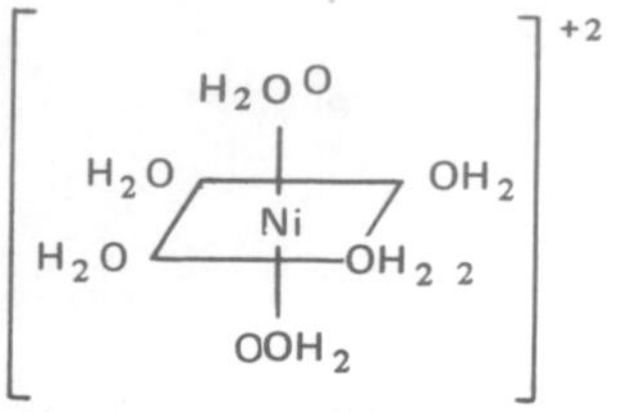

Hexaaquonickel (II) ion, green
(an octahedral complex)

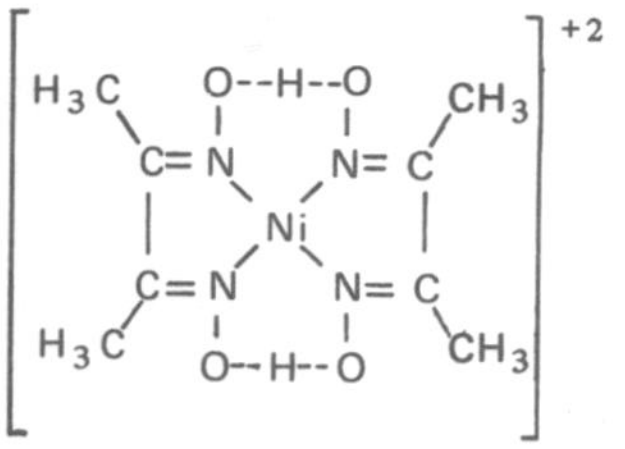

Bis(dimethylglyoximato) nickel
(II) ion, red (a square-planar complex)

Figure 25.7 Octahedral and square planar complexes of nickel(II).

The III oxidation state becomes progressively less stable in the sequence $Fe^{+3} \longrightarrow Co^{+3} \longrightarrow Ni^{+3}$. With iron, the III and II states are of approximately the same stability; with cobalt, the III state is stable only in complexes of strongly bonding ligands; with nickel, the III state occurs only rarely, but nickel can form compounds in both the III and IV states in the solid state and under strongly oxidizing conditions. Nickel in high oxidation states comprises the cathodic material in both the nickel-cadmium battery and the Edison battery. The anode in the latter is metallic iron; the electrolyte is potassium hydroxide, and the cell delivers about 1.3 volts.

Metallic nickel reacts readily in dilute mineral acids to give binary nickel(II) compounds from which a wide variety of both simple and complex salts of the metal in the II state can be made. Thus the oxide, the hydroxide, all the halides, the sulfide, cyanide, sulfate, nitrate, and other salts are well known. A few nonstoichiometric compounds of nickel with the heavier nonmetals, including tellurium, are known.

The complexes of nickel(II) are octahedral, tetrahedral, or square-planar (Figure 25.7). The octahedral complexes have high spin—two of the d electrons are unpaired. The square-planar complexes may be either low- or high-spin depending on the ligand. A very few tetrahedral complexes are known.

Nickel Carbonyl, Ni(CO)$_4$. When carbon monoxide is passed over metallic nickel at a temperature between 30°C and 50°C, the two unite to form a compound of the formula $Ni(CO)_4$, known as nickel carbonyl. It is a colorless liquid boiling at 43.2°C and freezing at −25°C to colorless, needle-shaped crystals. When the vapor of the compound is passed through a tube heated to above 100°C, the compound dissociates into the metal and carbon monoxide. Advantage is taken of this reaction in the Mond process for purifying nickel. Iron and cobalt also form carbonyls.

The Metal. Nickel is almost always associated with cobalt in nature. Like the latter element, it occurs in combination with sulfur and arsenic and is often associated with copper, silver, and iron. Most of its ores are very complex. It was formerly obtained, chiefly as a by-product, in the metallurgy of copper and silver. At present, most of the world's supply of nickel comes from the Sudbury district in Ontario, Canada. There, the nickel occurs in the mineral pyrrhotite (a sulfide of iron) and is associated with chalcopyrite, the nickel-copper content being from 10 to 20 per cent. The Sudbury minerals contain small percentages of platinum, palladium, and iridium, and are a large source of these rare metals. The extraction of nickel from the mineral ores is a complicated process.

Nickel is a silvery metal capable of taking a very high polish. It is very hard, but is quite malleable and melts at 1452°C. It can be welded on iron, and the two can be rolled into sheets. Like iron and cobalt, it is magnetic but less so than iron. It is not attacked by melted alkalies; nickel crucibles are often employed in the laboratory for alkali fusions. Hydrochloric acid with nickel evolves hydrogen very slowly, but dilute nitric acid reacts with nickel readily.

In many countries, pure nickel is used for subsidiary coinage. In the form of a fine powder, it is a most effective catalyst in reactions in which hydrogen is added to unsaturated organic compounds, as in the hydrogenation of oils.

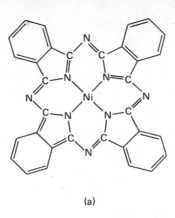

(a)

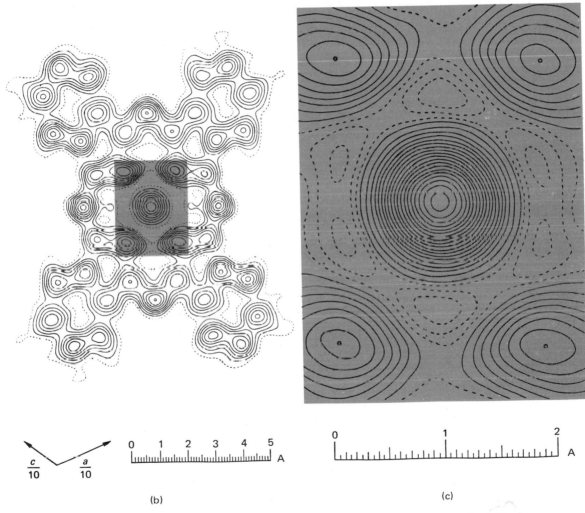

$\dfrac{c}{10}$ $\dfrac{a}{10}$ 0 1 2 3 4 5 A

(b)

0 1 2 A

(c)

Figure 25.8 The nickel phthalocyanine complex. (a) Structural formula showing the coordination bonding with nickel. (b) The electron density profile obtained from X-ray diffraction data, showing the arrangement of the atoms in the complex. Each contour line represents a density increment of one electron per A^2, except on the central nickel atom, where the increment is five electrons per A^2 for each line. (c) Enlargement of the central portion of the electron density profile. [Parts (b) and (c) courtesy of Robertson, J. M., Woodward, Ida, *J. Chem. Soc.*, 220, 222 (1937).

It has been used extensively for electroplating on other metals, such as iron, brass, or copper, to prevent tarnishing. Chromium plating is largely replacing it, but a preliminary plating with nickel usually precedes plating with chromium. For nearly all other purposes, it is used as an alloy metal.

3d orbitals

Copper $1s^2, 2s^2, 2p^6, 3s^2, 3p^6,$ (⇅) (⇅) (⇅) (⇅) (⇅), $4s^1$

Copper atoms have only one 4s electron over a filled 3d subshell. However, in spite of its single valence electron, copper does not resemble the alkali metals in most of its properties. For example, copper is a much more noble metal than any member of the alkali family—it does not liberate hydrogen from water or acids, it oxidizes in air much more slowly, and it reacts with nonmetals at considerably higher temperatures. Moreover, the common stable state of copper in its aqueous chemistry is the II state. Only a few examples of complex compounds containing copper atoms in the I or III states are known.

The nobility of copper is attributed to the very high lattice energy of the copper crystal, where the 3d and 4s electrons are involved in the metallic bonds, and the poor shielding of the 4s electron by the 3d subshell.

The stability of the II state over the I state for copper in water solution is believed to stem from the very high hydration energy in forming the $Cu(H_2O)_6^{+2}$ ion. This is probably the driving force for the very common and rapid disproportionation reaction of copper(I) in water solution:

$$2Cu(H_2O)_6^+ \rightleftharpoons Cu(H_2O)_6^{+2} + Cu + 6H_2O$$

The equilibrium between copper(I) and copper(II) may be displaced in either direction if various ligands are added to the solution. Thus, if cyanide ion or iodide ion is added, the copper(I) species is favored, while perchlorate or sulfate favor copper(II). Both cyanide and iodide ions are capable of forming covalent bonds with the metal ion, while the sulfate and perchlorate ions are not. Here again, the less stable oxidation state may be stabilized by strongly bonding ligands.

Common complexes of copper(II) have a distorted octahedral or tetragonal structure (Figure 25.9). In the distorted octahedral structure, four bonds are of equal lengths, but the remaining two are longer than the others. The long bonds are located opposite one another, as shown in the figure. For ions such as the diaquotetraammine copper(II) ion, the formula $Cu(NH_3)_4^{+2}$ is often used to indicate that the four ammonia molecules are attached by the shorter bonds. The complex may be described as a tetragonal structure.

Occurrence. Metallic copper has been known from the earliest times, and was probably the first metal to come into any considerable use. Well-fashioned copper articles that are at least 6000 years old have been found. The early use of copper is explained by its occurrence as native copper in large pieces and by the ease with which its oxygen compounds are reduced.

The metallurgy of copper is complicated, largely because reduction of the ores (Table 25.6) yields a crude copper containing many other metals such as silver and nickel. This is refined by an electrolysis process.

The Metal. Copper is a heavy metal of characteristic ruddy color. It is rather soft and is very ductile, malleable, and flexible, yet tough and fairly

Figure 25.9 Complexes of copper(II) showing distorted octahedra.

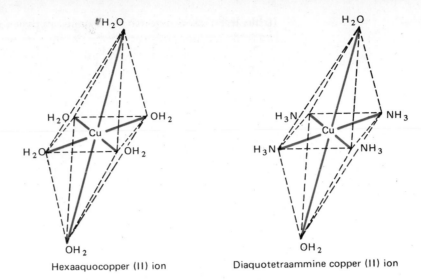

Hexaaquocopper (II) ion · Diaquotetraammine copper (II) ion

strong. It melts at 1083°C and has a density at 25°C of 8.9 g/cc. As an electrical conductor, it is second only to silver. It is not attacked by hydrochloric acid (unless oxygen is present) or by fused alkalies, but strong oxidizing acids convert it into the corresponding salts. In the presence of air, most acids slowly act upon it; even carbon dioxide in moist air gradually covers its surface with a greenish coating of a basic carbonate, patina. When heated in air, the element is oxidized to black copper oxide, CuO. Sulfur and the halogens attack it much more readily than does oxygen.

Table 25.6 Some Ores of Copper

Sulfide Ores	*Oxygen Ores*
Chalcopyrite, $CuFeS_2$	Cuprite, Cu_2O
Chalcocite, Cu_2S	Melaconite, CuO
Bornite, $CuFeS_4$	Malachite, $CuCO_3 \cdot Cu(OH)_2$

Table 25.7 Some Alloys of Copper

Class	*Components*	*Class*	*Components*
Simple brasses	Cu, Zn	Ordinary bronze	Cu, Sn
Leaded brasses	Cu, Zn, Pb	Phosphor bronze	Cu, Sn, P
Tin brasses	Cu, Zn, Sn	Zinc bronze	Cu, Sn, Zn, (P)
Leaded tin brasses	Cu, Zn, Sn, Pb	Leaded zinc bronze	Cu, Sn, Zn, Pb
High-tensile	Cu, Zn, Al	Lead bronze	Cu, Sn, Pb, (Zn)
brasses	Cu, Zn, Fe		Cu, Pb, Sn, (Zn)
Nickel coin	Cu(75%), Ni	Special bronze	Cu, Sn, X
		(X = any metal)	
		Aluminum bronze	Cu, Al, (Fe, Mg, Mn, Ni, Pb, Si)
18-carat gold	Au(75%), Cu, Ag	Sterling silver	Ag(92.5%), Cu
Beryllium copper	Cu, Be	Duraluminum	Cu, Al

NOTE: The elements other than copper are given in the order of their percentages, except where the per cent is given.

Table 25.8 The Second and Third Transition Series

Sc	Ti	V	Cr	Mn	Fe	Co	Ni	Cu
Y	Zr	Nb	Mo	Tc	Ru	Rh	Pd	Ag
La	Hf	Ta	W	Re	Os	Ir	Pt	Au
Ac								

Uses of Copper. About one-fourth of all the copper produced is used for electrical purposes. A second very great use of copper is in making alloys. More than a thousand varieties of these alloys are recognized commercially, and the more familiar names, such as *brass* and *bronze*, now indicate merely the *type* of the alloy, since their composition is subject to wide variation. Some alloys, however, have rather definite compositions.

SECOND- AND THIRD-ROW TRANSITION ELEMENTS

These elements, shown in Table 25.8 below their prototype elements of the first series, have valence electron configurations similar to those of the elements in the first row and, consequently, their chemical properties are similar also. Comparison with the first-row elements shows that one common difference is that the higher oxidation states are more stable than those of the first-series elements, and the lower oxidation states less stable. In fact, many of the lower oxidation states of the second- and third-row elements do not exist as simple ionic species, so that the main chemistry of these elements is in their higher oxidation states. Evidence for the relative stabilities mentioned is shown by comparing oxidation potentials for typical reactions.

$$MnO_2 + 2H_2O \rightleftharpoons MnO_4^- + 4H^+(aq) + 3e^- \qquad E° = -1.70 \text{ v}$$

$$ReO_2 + 2H_2O \rightleftharpoons ReO_4^- + 4H^+(aq) + 3e^- \qquad E° = -0.51 \text{ v}$$

$$Cr \rightleftharpoons Cr^{+3} + 3e^- \qquad\qquad\qquad E° = +0.74 \text{ v}$$

$$Mo \rightleftharpoons Mo^{+3} + 3e^- \qquad\qquad\qquad E° = +0.2 \text{ v}$$

$$Cu \rightleftharpoons Cu^+ + e^- \qquad\qquad\qquad E° = -0.52 \text{ v}$$

$$Ag \rightleftharpoons Ag^+ + e^- \qquad\qquad\qquad E° = -0.80 \text{ v}$$

$$Au \rightleftharpoons Au^+ + e^- \qquad\qquad\qquad E° = -1.68 \text{ v}$$

Comparison of the value for the Au(0) $\longrightarrow$ Au(I) change with that for Au(0) $\longrightarrow$ Au(III), namely −1.50 volts, shows that the Au(I) ion will disproportionate to Au(0) and Au(III) in water solution. Calculation of $\Delta G°$ for each half-reaction (using Equation 8 of Chapter 22) gives

$$3Au^+ + 3e^- \rightleftharpoons 3Au \qquad \Delta G° = -3 \times F \times (+1.68)$$

$$Au \rightleftharpoons Au^{+3} + 3e^- \qquad \Delta G° = -3 \times F \times (-1.50)$$

Addition gives

$$3Au^+ \rightleftharpoons 2Au + Au^{+3} \qquad \Delta G° = -0.54F$$

The negative $\Delta G°$ value indicates that the gold(I) ion is unstable in water.

LANTHANIDES AND ACTINIDES

The two elements below yttrium, $_{57}$La and $_{89}$Ac, are the first members of two series of 14 elements, in each of which an f subshell is filling as the atomic number increases in going to the right in the periodic table. In the *lanthanide* elements, $_{57}$La through $_{71}$Lu, the electrons go into the 4f shell; in the *actinide* elements, $_{89}$Ac through $_{103}$Lr, they enter the 5f shell (Table 25.9).

Lanthanides

The gradual filling of the 4f shell in the lanthanides results in a shrinkage of the atomic radius (Figure 25.10), giving the *lanthanide contraction*. This contraction is believed to arise because the 4f electrons interact more strongly with the nucleus than with one another. Thus, as more electrons are added to the 4f subshell, the nucleus-electron interaction becomes greater, and the atom or ion shrinks. The consequences of the strong interaction of f electrons with the nucleus are twofold. One is that the three valence electrons in all lanthanides—that is, the $5d^16s^2$ electrons—are less tightly held than expected; consequently, the lanthanides are all relatively reactive metals, forming the III state easily. The other consequence is that the chemistry of the lanthanides does not directly involve 4f electrons. Apparently, these electrons are so deep in the atom or so strongly bound to the nucleus that they are not available for bonding. As a result, the lanthanides (which differ only in the number of f electrons and in atomic size) have almost identical chemistry—that related to the III state.

An important aspect of the lanthanide contraction is that the size of *atoms and ions of the elements following lanthanum are smaller than might have been expected.* The usual trend in atomic sizes down a column in the periodic table is that they increase as the number of electron shells increases from one row to the next. However, the lanthanide contraction intervenes between the fifth and sixth periods, so that little or no increase in size occurs between the fifth and sixth periods for the elements following lanthanum, or between the sixth and seventh periods in Groups I and II. The effect of this is that the atoms and ions of, for example, hafnium, are nearly identical to those of zirconium, in spite of the fact that these elements differ by 32 atomic numbers. Since their valence electron configurations are also the same, the chemistry of these elements shows a similarity almost without parallel elsewhere in the periodic table. Similarly, the elements niobium and tantalum are very much alike, as are those of the second- and third-series elements of each of the transition element families.

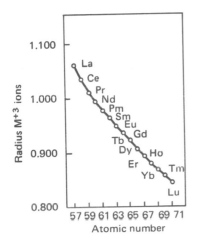

Figure 25.10 Illustrating the lanthanide contraction.

Actinides

These elements are similar to the corresponding elements in the lanthanide series. The actinide series was not identified until several of the transuranium elements were prepared. After a careful study of the properties of elements 90 through 96 and additional evidence from paramagnetic and spectroscopic techniques provided evidence for the presence of f electrons, a suggestion was made by the American chemist Glenn Seaborg that they are all a part of a new transition series similar to the rare earths.

Of this series, only the first four members (actinium through uranium) are found in nature in amounts sufficient to be extracted economically; traces of plutonium are found in uranium minerals. Elements 92 through 103 are

Table 25.9 The Lanthanides and Actinides

$_{57}$La	$_{58}$Ce	$_{59}$Pr	$_{60}$Nd	$_{61}$Pm	$_{62}$Sm	$_{63}$Eu	$_{64}$Gd	$_{65}$Tb	$_{66}$Dy	$_{67}$Ho	$_{68}$Er	$_{69}$Tm	$_{70}$Yb	$_{71}$Lu
$_{89}$Ac	$_{90}$Th	$_{91}$Pa	$_{92}$U	$_{93}$Np	$_{94}$Pu	$_{95}$Am	$_{96}$Cm	$_{97}$Bk	$_{98}$Cf	$_{99}$Es	$_{100}$Fm	$_{101}$Md	$_{102}$No	$_{103}$Lr

Table 25.10 Some Properties of Zinc, Cadmium, and Mercury

Element	Density (g/cc, 25°C)	Melting Point (°C)	Boiling Point (°C)	Heat of Vaporization (kcal/mole)	Standard Oxidation Potential to M^{+2} (volts)	Radius of M^{+2} (A)
Zinc	7.4	419	907	31.2	0.762	0.69
Cadmium	8.65	321	767	26.8	0.402	0.92
Mercury	13.55	−39	357	14.7	−0.854	0.93

made by nuclear transmutation processes that bombard elements with neutrons, protons, alpha particles, and nuclei of heavier elements. All of these elements are radioactive. Most of the isotopes of the heavier members have very short half-lives.

The chemical properties of the actinides are very similar to those of the lanthanides. However, besides the common III oxidation state, they also have states of IV, V, and VI. Like the lanthanides, a number of the actinides form colored ions. The actinide ions are only slightly larger than the analogous lanthanide ions. Similar to the lanthanide contraction is the analogous predicted and observed *actinide* contraction, which shows the same general trends.

ZINC, CADMIUM, AND MERCURY

Zn
Cd
Hg

The family of elements consisting of zinc, cadmium, and mercury has two s electrons beyond filled d subshells. Because the d subshells are filled, these elements are not usually considered transition elements. The common oxidation state is II; there is no evidence for higher states and the I state is known in solution only for mercury, which forms the unique Hg_2^{+2} ion (^+Hg—Hg^+). These metals differ from transition metals in several ways. For example, they are softer, lower melting, and considerably more reactive than their nearest neighbors in the transition series. However, they resemble transition elements in their ability to form complexes with ligands such as ammonia and with cyanide and halide ions.

Zinc and cadmium are very similar in their properties, and their ions resemble the magnesium ion in many ways. Mercury is much less reactive than zinc or cadmium and its complexes are much more stable. All three elements have a greater-than-expected tendency to form covalent compounds, possibly because of the polarizability of the electron cloud around the nucleus. Organometallic compounds of the type R_2Hg, R_2Zn, and R_2Cd are well-known covalent substances.

Some important properties of these elements are given in Table 25.10. Mercury* is the only metal that is a liquid at room temperature; zinc and cadmium are white and shiny but easily tarnished. Zinc and cadmium react with nonoxidizing acids such as hydrochloric acid to liberate hydrogen; mercury is inert to these acids. Zinc reacts with strong bases to form the zincate ion, $Zn(OH)_4^{-2}$. Cadmium and mercury do not react with bases, presumably because their tetrahydroxo complexes are unstable.

---

*Mercury is volatile and toxic. Particular care should be taken not to leave its surface exposed so that evaporation can occur; it should be stored in closed containers, and any spilled droplets should be searched out and cleaned up immediately.

Magnets From Transition Metals. Magnets play a much more important role in our daily lives than most people realize. Each day in the U.S., magnetic materials are responsible for generating at least 3 billion kilowatt-hours of electricity. Most electric and electronic devices contain at least one magnet. Electric motors, generators, and transformers are obvious examples. But telephones, television receivers, speedometers, hearing aids, computers, and thousands of other devices depend for their effectiveness on the presence of magnetic materials. Transition metals, especially iron, cobalt, nickel, and their alloys, are the essential components of most metallic magnets.

These metals have three basic magnetic properties: They are magnetic in the absence of any applied magnetic field; they generally become more magnetic when a relatively weak magnetic field is applied to them; when heated above a certain temperature, their magnetism decreases and only the comparatively weak paramagnetism remains.

An explanation for this behavior in terms of atomic and crystal structure is the following. Transition element magnets have unpaired d electrons. At higher temperatures, only the paramagnetism associated with individual atoms is observable. As the metal cools, the atoms lose thermal energy and neighboring atoms tend to become magnetically aligned. In most transition metals, this alignment is parallel—that is, the atoms orient their "north poles" in the same direction. This results in a greatly increased magnetism known as ferromagnetism. However, this parallel alignment does not extend throughout the entire crystal. Instead, imagine the crystal to be a sintered mass of tiny crystals, known as domains, each containing atoms aligned magnetically parallel but with little or no magnetic alignment between domains; the total magnetic strength of the crystal is simply the sum of the magnetic strengths of the domains, taking into consideration their strengths and orientations. When placed in an external magnetic field, the domains can be reoriented so their magnetic axes become parallel. This accounts for the increase in ferromagnetism produced in many metallic magnets by an applied external field. Thus, even the earth's magnetic field is strong enough to increase the magnetism in a soft iron bar.

Magnetic materials are traditionally divided into "hard" or "soft" categories. Hard materials are difficult to magnetize and demagnetize; soft materials are easily magnetized and demagnetized. Soft materials are better suited to certain electrical devices such as electric motors or transformers, in which the magnetism must be reversed many times each second. Hard materials are used when permanent magnets are needed.

The hard and soft properties of magnetic metals can be explained in terms of the domain model. If the metal contains impurities or a crystal structure such that reorientation of the domains by the external field causes them to be frozen by obstacles (such as impurities or energy barriers) so they cannot easily reorient again, a permanent magnet will form. If the domains are not frozen but are relatively mobile, on reorientation, the material will be magnetically soft.

The hardness or softness of magnetic materials sometimes can be controlled by alloying or by physical treatment such as annealing. A highly successful soft material consists of 79 per cent nickel, 15.7 per cent iron, 5 per cent molybdenum, and 0.3 per cent manganese. One factor that helps make this alloy effective is the tendency of those domains containing mainly iron to align in directions opposite to those containing mainly nickel. In effect, one metal acts as a lubricant for the facile reorientation of the domains of the other metal.

In recent years, a new class of materials for permanent magnets has been developed. These materials are alloys of cobalt and lanthanide elements, notably samarium and cerium. Their resistance to demagnetization is from 20 to 50 times superior to permanent magnets made from conventional materials. The added magnetic hardness undoubtedly is due to the presence in the metal of the compounds $SmCo_5$ or $CeCo_5$ that precipitate in the crystal and form barriers to reorientation of the domains once they are aligned by an external field.

Cobalt magnets are relatively inexpensive. Their availability gives us permanent magnets possessing great strength that can support large loads. This opens vistas for the development of products requiring stronger magnets than have been available in the past. Magnets made of these materials could revolutionize both the mass transportation and the electronics industries.

SUMMARY

This chapter discusses the elements that are filling a d shell after arriving at the electron configuration of calcium, strontium, or barium, and an f shell in the energy level of next lower principal quantum number. The d-shell-filling elements are similar to those discussed in Chapter 24, but the change in nuclear charge has less effect on those elements following lutetium than might have been expected, because of the lanthanide contraction. The usefulness of several of these metals in industry, either as the free metal or as components of alloys, is indicated, since this use is often more important than the use of compounds of these elements.

IMPORTANT TERMS

Ion
aquo ion
oxo ion
oxyanion

Metal-ligand bond

Metal-ligand interaction

Oxyligand

Rare-earth elements
lanthanides
actinides

Fission product

Passive metal

Alloy

High-spin complex

Low-spin complex

Heme

Ferrocene-type complex

Lanthanide contraction

Metallurgy

QUESTIONS AND PROBLEMS

1. Give the electron configuration or electron-dot formulas of the following atoms and ions: Sc, Sc^{+3}, V, Cr, Cr^{+3}, $Cr(NH_3)_6^{+3}$, Co, Co^{+2}, Co^{+3}, $Co(NH_3)_6^{+3}$, MnO_4^-, and $Cr_2O_7^{-2}$.

2. Make a table to summarize information about the first-row transition elements with regard to (a) occurrence in nature, (b) melting point, (c) oxidation states, and (d) formulas of typical compounds in each oxidation state.

3. What would you predict as the limiting value of the freezing-point depression per mole of solute at extremely low concentrations for

the salts $Na_4Fe(CN)_6$, $KMnO_4$, $K_2Cr_2O_7$, $K_3Fe(CN)_6$, $Co(NH_3)_6Cl_3$, $[Co(NH_3)_4(NO_2)_2]_3[Co(NO_2)_6]$?

4. (a) Give reasons for the similar properties of iron, cobalt, and nickel. (b) Does each of these elements form a series of stable compounds in the II and III oxidation states?

5. (a) Explain with diagrams the fact that $K_2[Ni(CN)_4]$ is diamagnetic, whereas $K_2[NiCl_4]$ is paramagnetic. (b) Predict the shapes of the two molecules.

6. From the information in Chapters 19, 20, and 21, and particularly from Table 21.1, compute the standard cell potential and the equilibrium constant for the following reaction.

$$Cu(s) + \frac{1}{2} I_2(s) \rightleftharpoons CuI(s)$$

$$Cu^{+2}(aq) + I^-(aq) + e^- \rightleftharpoons CuI(s); \ E° = 0.88 \ v)$$

7. An electrolytic cell is made with zinc and silver electrodes and a second one with zinc and copper electrodes. Which cell will give the higher voltage? Why?

8. The following transformations can be carried out consecutively in the same beaker:

$$Ag^+(aq) \longrightarrow Ag_2O(s) \longrightarrow AgCl(s) \longrightarrow Ag(NH_3)_2{}^+(aq) \longrightarrow$$

$$AgBr(s) \longrightarrow Ag(S_2O_3)_2{}^{-3}(aq) \longrightarrow AgI(s) \longrightarrow Ag(CN)_2{}^- \longrightarrow Ag_2S(s)$$

(a) At each arrow, list the reagent solution you would add to effect the transformation. (b) Write equations for each of the reactions. (c) Explain, in terms of solubility-product equilibria, why precipitation or solution occurs in each of the reactions.

9. Calculate the percentage of iron in each of the following compounds: (a) $Na_4Fe(CN)_6 \cdot 10 \ H_2O$; (b) $K_4Fe(CN)_6 \cdot 3 \ H_2O$.

10. How many grams of iron(II) chloride can be obtained from the reduction of 500 g of iron(III) chloride by the reducing action of iron?

11. (a) How many grams of hydrogen chloride can be obtained from the reduction of 1 kg iron(II) chloride by hydrogen? (b) Calculate the $E°$ value for this reaction in water solution at 25°C. Could the reduction be carried out in water at room temperature? (c) The standard enthalpy and entropy changes at 25°C for the reduction of one mole of solid iron(II) chloride are +37.4 kcal and 35.9 calories per degree, respectively. Calculate the equilibrium constant for the reaction at 298K and 800K. What conditions would you suggest for carrying out the reduction reaction?

12. Calculate the volume of sulfuric acid, which is 98% hydrogen sulfate, necessary for the conversion of 10 kg of iron(II) sulfate to iron(III) sulfate. The density of the sulfuric acid is 1.84 g/cc.

13. According to the reaction

$$Fe + 4 \ HNO_3 \longrightarrow Fe(NO_3)_3 + NO + 2H_2O$$

concentrated nitric acid oxidizes iron to iron(III). What weight of 67% nitric acid will be required to prepare 1 kg of iron(III) nitrate by this method?

14. How many pounds of pure hematite, Fe_2O_3, and chromite, $Fe(CrO_2)_2$, must be mixed with carbon and reduced to obtain 200 lb of ferrochromium (alloy of iron and chromium) containing 50% chromium and 5% carbon?

SPECIAL PROBLEMS

1. How can you distinguish analytically between (a) Iron(II) ions and nickel(II) ions. (b) Iron(III) ions and cobalt(II) ions. (c) Iron(II) ions and iron(III) ions. (d) Cobalt(II) ions and nickel(II) ions.

2. Using the standard oxidation potentials listed in Table 21.1, in this chapter, and in Appendix E, would you expect that it would be easy to: (a) Oxidize ions of Fe(II) with ions of Cr(VI) in acid solution? (b) Oxidize ions of I^- with ions of Fe(III) in acid solution? (c) Reduce MnO_2 with Cl^- in acid solution? (d) Oxidize ions of Cr(III) with H_2O_2 in acid solution?

3. How do you explain the fact that copper(I) oxide and sulfide are common ores of copper in nature, but that Cu^+ ions are rarely found in solution?

4. (a) Would you expect silver chloride to precipitate from ammonia solution; from sodium cyanide solution? (b) Would you expect silver oxide to change to silver bromide on shaking with potassium bromide solution? (c) Would you expect silver sulfide to change to silver bromide on shaking with potassium bromide solution? Explain your answer in each case.

5. Compare the melting points of zinc, cadmium, and mercury with those of their neighbors in the periodic table, and with the alkaline earth elements. What factors, such as electronic configuration, ion size, ionization energy, and so forth, can be used to help account for the difference, and how do they account for it (if they do)?

6. Write equations for the oxidation of chromite ore with air and the formation of potassium chromate and iron(III) oxide in the process for the production of metallic chromium.

7. In the electrolytic refining of copper, the impure copper is made the anode in a bath of acidified copper sulfate solution with a pure copper cathode. In the process, silver present as impurity in the anode drops off as a solid in the anode compartment, nickel ions accumulate in the electrolyte, and pure copper plates out on the cathode. Show, by consideration of possible reactions at the anode and cathode, that a proper choice of voltage makes this separation of the three metals possible. The standard electrode potentials needed are:

$$Ni \longrightarrow Ni^{+2} + 2e^- \qquad E° = +0.22 \text{ v}$$

$$Cu \longrightarrow Cu^{+2} + 2e^- \qquad E° = -0.34 \text{ v}$$

$$Ag \longrightarrow Ag^+ + e^- \qquad E° = -0.80 \text{ v}$$

8. Calculate the $E°$ value for the disporportionation reaction

$$2Mn^{+3} + 2H_2O \longrightarrow Mn^{+2} + MnO_2 + 4H^+(aq)$$

9. Calculate the $E°$ value for the reaction

$$2Cu^+ \longrightarrow Cu(s) + Cu^{+2}$$

Is this reaction spontaneous in water solution? Copper(II) chloride in solution can be reduced by metallic copper to form insoluble copper(I) chloride, reversing the above reaction. Explain.

REFERENCES BASOLO, F., and R. C. JOHNSON, *Coordination Chemistry*. Benjamin, Menlo Park, Calif., 1964.

FACKLER, J. P., JR., *Transition Elements*. Van Nostrand Reinhold, New York, 1966.

LARSEN, E. M., *The Transitional Elements*. Benjamin, Menlo Park, Calif., 1965.

KARRAKER, D. G., "Coordination of Trivalent Lanthanide Ions," *J. Chem. Educ.,* **47**, 424 (1970).

JOHNSON, O., "Role of f Electrons in Chemical Bonding," *J. Chem. Educ.*, **47**, 431 (1970).

PART EIGHT

ORGANIC CHEMISTRY

ORGANIC CHEMISTRY I: STRUCTURES AND NOMENCLATURE

By showing on the one hand the atomic groups which remain unaffected by certain reactions, and on the other hand those which play a role in frequently recurring metamorphoses, such structural formulae give a picture of the chemical nature of the substance . . .

Kekulé von Stradonitz (1829–1896)

The study of the compounds of carbon is known as organic chemistry. Organic compounds exhibit a remarkable variation in properties. They constitute products which vary widely: plastics and elastomers such as polystyrene and rubber; fibers such as cotton and nylon; gasolines, greases, and lubricating oils; drugs, dyes, insecticides, weed killers, explosives, photographic film, varnishes, lacquers, and many other useful and essential items of our economy. Moreover, the knowledge of organic chemistry has developed to the point where it is now possible in many cases to synthesize compounds meeting the specifications required for a particular use. For example, nylon was synthesized as a substitute for silk, demerol was prepared in a quest for a pain-killer of the caliber of morphine but without the habit forming quality of that drug, and the new rubbers now being used in spaceships are prepared to withstand the extremely wide temperature variation experienced during spaceflight.

Organic compounds and their reactions constitute the foundation of modern biology and are, in fact, the basis of life itself.

We shall begin this series of chapters with an illustration of the age-old technique of the scientist known as simplification by classification. Starting with the more than 2 million known organic compounds, we shall classify them into broad groups and then into smaller families or series of compounds according to their properties and structure. Following this, some important properties, structural features, and a system of nomenclature will be given for 10 series of organic compounds. Physical properties will receive more attention than chemical properties in this chapter; the following chapter is devoted entirely to chemical reactions of organic compounds. In Chapter 28, polymers and some other complex structures will be discussed.

The Diversity of Carbon Compounds

The number and diversity of carbon compounds can be traced to (a) the unusual strength of the carbon—carbon single bond, (b) the four valence electrons on carbon atoms which make it possible for a carbon atom to combine with as many as four other atoms, (c) the ability of the carbon atom to combine with atoms of other nonmetals, (d) the possibility that carbon atoms can combine to form straight or branched chains, or ring structures such as follows:

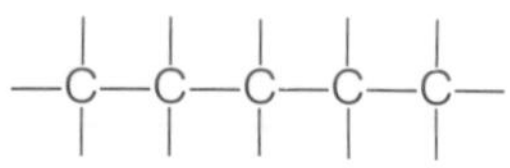

Straight chain 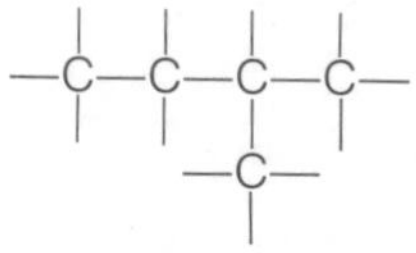   Branched chain 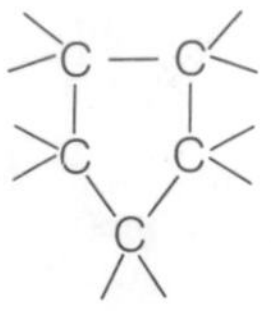   Ring

and (e) that these chain and ring structures may contain only hydrogen atoms bonded at the remaining valence sites on the carbon atoms (in which case the compounds would be known as *hydrocarbons*), or some or all of these hydrogen atoms may be substituted by atoms of other nonmetals, as in

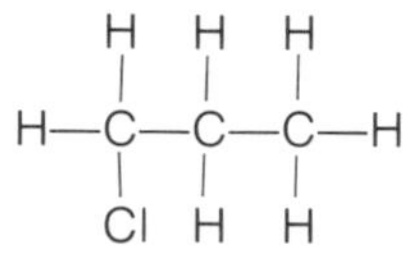

1-chloropropane 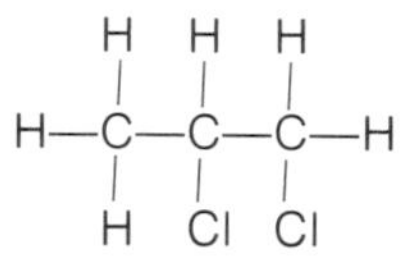   1,2-dichloropropane 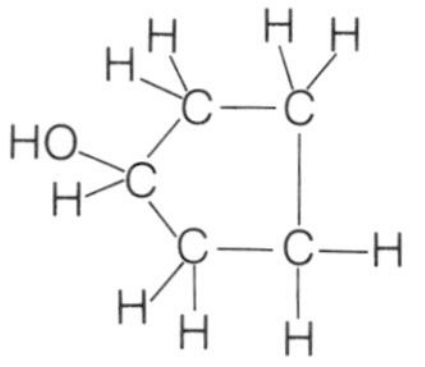   cyclopentanol

CLASSIFICATION OF ORGANIC SUBSTANCES

The early chemists classified organic compounds into two broad divisions: the aliphatic division, made up of fats and waxes, and compounds related to them; and the "aromatic" division, consisting of the essential oils from plants, such as oil of wintergreen and other flavoring compounds. While the names aliphatic and aromatic are still used for the two chief divisions of organic compounds, neither the fatty character nor the aroma of a compound is any longer used as a basis for its classification. Structurally, aromatic compounds are those containing *aromatic rings* such as benzene, naphthalene, and pyridine. Aliphatic compounds contain *chains or nonaromatic rings* of carbon atoms.

Within both the aliphatic and aromatic divisions, further subdivision into various series of compounds, known as homologous series, is made. A homologous series is defined as a group of *compounds containing the same functional group* or groups *in which each member in the series differs from the member of next higher molecular weight by a methylene group,* CH_2. For example, the compounds

$$CH_3—CH_2—CH_3 \qquad CH_3—CH_2—CH_2—CH_3$$
Propane   *n*-Butane

$$CH_3—CH_2—CH_2—CH_2—CH_3$$
n-Pentane

are members of the homologous series known as alkanes; pentane has one more and propane has one less methylene group than butane. Other ex-

*A functional group is an atom or group of atoms that imparts characteristic chemical properties to organic molecules. Examples are the hydroxyl group, —OH, the amino group, —NH_2, and the carboxyl group, —CO_2H.

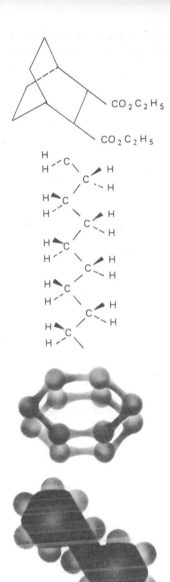

Figure 26.1 Structures of some organic substances.

amples of homologous series in the aliphatic division are given in Table 26.1.

All members of a homologous series contain the same functional group. For example, all alkenes have the carbon-carbon double bond in their molecules; all alkanols (alcohols) have the hydroxyl group in their molecules; all carboxylic acids contain the carboxyl group

$$-\overset{\displaystyle |}{\underset{\displaystyle \|}{C}}-OH$$
$$\quad\;\; O$$

Table 26.2 gives the generalized structural formulas and names for a number of homologous series not given in Table 26.1. In the generalized structural formula, the symbol R is used to represent any aliphatic hydrocarbon group, and Ar is used to represent any aromatic hydrocarbon group.

Table 26.1 Some Homologous Series, Structural Features, and Representative Members

Name of Series	Structural Features	Several Representative Members
Alkanes	Hydrocarbons containing single bonds	CH_3-CH_3, ethane $CH_3-CH_2-CH_3$, propane
Alkenes	Hydrocarbons containing one double bond between carbon atoms	$CH_2=CH_2$, ethene (ethylene) $CH_3-CH=CH_2$, propene
Alkynes	Hydrocarbons containing one triple bond between carbon atoms	$CH\equiv CH$, ethyne (acetylene) $CH_3-C\equiv CH$, propyne
Alkanols (alcohols)	Alkanes with one hydrogen atom replaced by a hydroxyl group	CH_3-CH_2-OH, ethanol $CH_3-CH_2-CH_2-OH$, propanol
Alkanals (aldehydes)	Compounds containing the group $-\overset{\|}{\underset{O}{C}}-H$	$CH_3-\overset{\|}{\underset{O}{C}}-H$, ethanal (acetaldehyde) $CH_3-CH_2-\overset{\|}{\underset{O}{C}}-H$, propanal
Alkanones (ketones)	Compounds containing the group $\overset{\|}{\underset{O}{C}}-C-C$	$CH_3-\overset{\|}{\underset{O}{C}}-CH_3$, propanone (acetone) $CH_3-CH_2-\overset{\|}{\underset{O}{C}}-CH_3$, butanone
Alkanoic acids (carboxylic acids)	Compounds containing the group $-\overset{\|}{\underset{O}{C}}-OH$	$CH_3-\overset{\|}{\underset{O}{C}}-OH$, ethanoic acid (acetic acid) $CH_3-CH_2-\overset{\|}{\underset{O}{C}}-OH$, propanoic acid
Amines	Derivatives of ammonia containing a nitrogen atom bonded to at least one carbon atom	$CH_3-CH_2-NH_2$, ethanamine (ethylamine)
Alkyl halides	Aliphatic hydrocarbons with one hydrogen atom replaced by a halogen atom	$CH_3-CH_2-CH_2-Cl$, 1-chloropropane $CH_3-CH_2-\overset{\|}{\underset{F}{CH}}-CH_2-CH_3$, 3-fluoropentane

Table 26.2 Generalized Structural Formulas for Some Homologous Series

Name of Series	General Formula	Name of Series	General Formula
Phenols	Ar—OH	Aliphatic ethers	R—O—R
Nitroaryls	Ar—NO$_2$	Aromatic ethers	Ar—O—Ar
Nitroalkyls	R—NO$_2$		
Aromatic esters	Ar—C—O—Ar $\parallel$ O	Alkylsulfonic acids	$\underset{\displaystyle O}{\overset{\displaystyle O}{R-\overset{\parallel}{\underset{\parallel}{S}}-OH}}$
Aliphatic esters	R—C—O—R $\parallel$ O		
Mixed esters	Ar—C—O—R $\parallel$ O	Arylsulfonic acids	$\underset{\displaystyle O}{\overset{\displaystyle O}{Ar-\overset{\parallel}{\underset{\parallel}{S}}-OH}}$
	or R—C—O—Ar $\parallel$ O	Alkylthiols (mercaptans)	R—S—H
		Arylthiols	Ar—S—H
Aliphatic amides	R—C—NH$_2$ $\parallel$ O	Alkandioic acids	CO$_2$H $\vert$ (CH$_2$)$_n$ $\vert$ CO$_2$H
Aliphatic anhydrides	R—C—O—C—R $\parallel$ $\parallel$ O O	Alkandiols	OH $\vert$ (CH$_2$)$_n$ $\vert$ OH
Alkanoyl chlorides (acyl chlorides)	R—C—Cl $\parallel$ O	α-Amino acids	R—CH—CO$_2$H $\vert$ NH$_2$

SOME IMPORTANT HOMOLOGOUS SERIES

THEORY: THE ALKANES, RH

The alkanes are *hydrocarbons with single bonds between carbon atoms.* They are also known as *saturated hydrocarbons* because their molecules contain the maximum number of hydrogen atoms possible with tetravalent carbon. The formulas, names, and some of the properties and uses of some members of this series are given in Table 26.3. The alkanes are present in natural gas and petroleum and are among our most important fuels and lubricants.

The alkanes can be thought of as derived from methane by increasing the number of carbon atoms in the chain. This way it is possible to account for both the straight- and branched-chain members of the series. The first alkanes—methane, ethane, and propane—are straight-chain compounds. But for the fourth member of the series, butane, C$_4$H$_{10}$, there are two compounds—a straight-chain compound and a branched-chain compound, shown in Table 26.4.

Isomers. Examination of the molecular formulas of *n*-butane and 2-methylpropane (Table 26.4) shows that they have the same molecular

Table 26.3 Formulas, Names, and Boiling Points of Representative Alkanes

Formula	Chemical Name	Physical State	Boiling Point (°C)	Uses
CH_4	Methane	Gas	−161 ⎫	
C_2H_6	Ethane	Gas	−88 ⎪	Natural gas
C_3H_8	Propane	Gas	−46 ⎬	
C_4H_{10}	Butane	Gas	−1 ⎭	
C_5H_{12}	Pentane	Liquid	36 ⎫	High-grade naphtha
C_6H_{14}	Hexane	Liquid	69 ⎭	
C_7H_{16}	Heptane	Liquid	98 ⎫	
C_8H_{18}	Octane	Liquid	126 ⎪	
C_9H_{20}	Nonane	Liquid	150 ⎬	Gasoline
$C_{10}H_{22}$	Decane	Liquid	174 ⎪	
$C_{11}H_{24}$	Undecane	Liquid	195 ⎭	
$C_{12}H_{26}$	Dodecane	Liquid	215	Kerosine ($C_{12}H_{26}$ to $C_{18}H_{38}$)
$C_{14}H_{30}$	Tetradecane	Liquid	253 ⎭	Lubricating oil (mixtures of higher hydrocarbons)
$C_{20}H_{42}$	Eicosane	Solid		Paraffin ($C_{20}H_{42}$ to $C_{30}H_{62}$)

NOTE: The data given here are for the normal, or straight-chain, hydrocarbons.

formula, namely C_4H_{10}. Similarly, the molecular formulas for the straight and branched-chain alkanes containing five carbon atoms, *n*-pentane, 2-methylbutane, and 2,2-dimethylpropane (Table 26.4) are the same, C_5H_{12}. *Compounds which have the same molecular formula but different structural formulas* are known as isomers. In the examples above, the isomers are known as *position isomers* because they differ only in the position of the carbon atoms in the structure.

As the number of carbon atoms increases, the possibilities for chain branching also increase and the number of isomers increases rapidly. Thus the hydrocarbon with seven carbon atoms has nine possible isomers, that with 10 carbon atoms has 75 possibilities, with 14 carbon atoms, 1858 possibilities, and so on. Table 26.4 gives the formulas for the position isomers of the four-, five-, and six-carbon atom alkanes. In all cases, the correlation between the number of isomers found and the number predicted from our structural theory is exact. Consequently, it is possible to use this theory—which encompasses the covalent bond, the tetravalency of carbon, the univalency of hydrogen, and the idea that carbon-carbon bonds are especially stable—not only to predict the number of isomers corresponding to a given carbon content, but also to write structural formulas for each isomer.

PROBLEM Write the carbon skeletons for the isomeric heptanes.

Solution The first isomer will be the straight-chain compound:

I   C—C—C—C—C—C—C

In the next group of isomers one carbon atom is attached at various places along a chain of six carbon atoms. The two different structures are:

II   C—C—C—C—C—C   III   C—C—C—C—C—C
      | |
      C   C

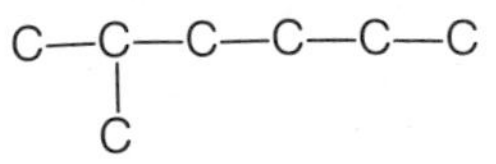

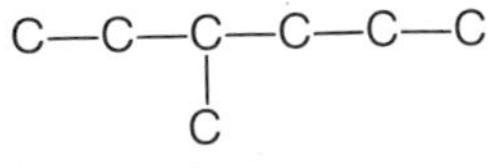

In the third group two carbon atoms are attached at various points along a chain of five carbon atoms. The five possible structures are:

```
IV   C—C—C—C—C        V   C—C—C—C—C       VI             C
         |   |                |   |                        |
         C   C                C   C              C—C—C—C—C
                                                     |
                                                     C
```

```
                 C                                        C
                 |                                        |
VII   C—C—C—C—C         VIII   C—C—C—C—C
             |                         |
             C                         C
                                       |
                                       C
```

The last isomer consists of a chain of four carbon atoms with three carbon atoms attached to it:

```
                    C
                    |
IX   C—C—C—C
         |   |
         C   C
```

Table 26.4 Isomers of Butane, Pentane, and Hexane

Butanes	Pentanes	
CH₃—CH₂—CH₂—CH₃ $\quad$ *n*-Butane $\quad$ (b.p., −0.5°C)	CH₃—CH₂—CH₂—CH₂—CH₃ $\quad$ *n*-Pentane $\quad$ (b.p., 36.1°C)	CH_3 | CH_3-C-CH_3 | CH_3 2,2-Dimethylpropane (b.p., 9.5°C)
CH₃—CH—CH₃ | CH₃ 2-Methylpropane (b.p., −10.2°C)	CH₃—CH—CH₂—CH₃ | CH₃ 2-Methylbutane (b.p., 27.9 C)	

Hexanes		
CH₃—CH₂—CH₂—CH₂—CH₂—CH₃ *n*-Hexane (b.p., 68.7°C)	CH₃—CH₂—CH—CH₂—CH₃ | CH₃ 3-Methylpentane (b.p., 63.3°C)	CH₃ CH₃ | | CH₃—CH—CH—CH₃ 2,3-Dimethylbutane (b.p., 58.0°C)
CH₃—CH₂—CH₂—CH—CH₃ | CH₃ 2-Methylpentane (b.p., 60.3°C)	CH₃ | CH₃—C—CH₂—CH₃ | CH₃ 2,2-Dimethylbutane (b.p., 49.7°C)	

The large number of isomers poses the problem of naming organic compounds. The nomenclature of organic chemistry involves both common and systematic names. The former are usually applied to simple compounds, the latter to more complex substances. While both the simple and systematic names will be used in this text, we shall develop the systematic nomenclature at this point to facilitate communication.

The systematic names are the outgrowth of an international system of nomenclature devised and used by organic chemists throughout the world and recommended by the International Union of Pure and Applied Chemistry (IUPAC). The rules for the IUPAC system are few and simple to use:

1. Straight-chain alkanes are given the names listed in Table 26.3. (It is important that the names of the first 10 alkanes in the table be memorized to use the rules that follow.)

2. For branched-chain hydrocarbons, determine the longest *continuous* chain of carbon atoms in the molecule. Use the name of the straight-chain alkane corresponding to this number of carbon atoms as the basis for the name of the compound. Thus, in the structure

$$\overset{1}{C}H_3 - \overset{2}{C}H - \overset{3}{C}H_2 - \overset{4}{C}H_2 - \overset{5}{C}H_3$$
$$| $$
$$CH_3$$

the longest continuous chain of carbon atoms is five. Therefore, this is known as a *pentane*.

3. Number the carbon atoms of the continuous chain, beginning at the end closest to the branching. In the formula above, numbering starts at the left.

4. Name and number the substituents other than hydrogen atoms attached to the chain. The name of the substituent is taken from the alkane with the same number of carbon atoms, but the *ane* ending is replaced by a *yl* ending. Thus, a CH_3- or

$$\begin{array}{c} H \\ | \\ H-C- \\ | \\ H \end{array}$$

is a *methyl* group because it is derived from methane. Names of other substituents are given in Table 26.5. The number of the substituent is taken from the number of the carbon atom to which it is attached. *Each substituent receives a name and a number.* The name of the compound above is 2-methyl-pentane.

5. Where alternatives exist, numbering of the longest chain is done so that the substituents will have the lowest possible numbers.

PROBLEM Name the following compounds:

1. $$\overset{1}{C}H_3 - \overset{2}{C}H - \overset{3}{C}H - \overset{4}{C}H_2 - \overset{5}{C}H_3$$
$$\qquad\quad | \qquad |$$
$$\qquad\quad CH_3 \quad CH_3$$

Solution (a) Longest continuous chain, five: pentane.
(b) Number from left—closest to branching.
(c) Methyl substituents at positions 2 and 3.

Name: 2,3-dimethylpentane

2.

$$\underset{1}{CH_3}-\underset{\substack{\overset{2}{C}l\\|}}{CH}-\underset{3}{CH}-\underset{4}{CH_2}-\underset{5}{CH_3}$$

Solution

Refer to Table 26.5 for names of substituents.
Name: 2-chloro-3-phenylpentane

Bonding and Structure in Alkanes

The carbon atoms in alkanes exhibit sp^3 hybridization, so each carbon atom may be represented as having its four covalent bonds directed toward the corners of a regular tetrahedron. Carbon-carbon single bonds may be viewed as two tetrahedra that have a corner in common. A chain of carbon atoms might be pictured as a zigzag arrangement of tetrahedra, Figure 26.2.

Rotation Within Molecules

Of considerable importance is the prediction, based on examination of molecular models, that chain-like molecules should have considerable flexibility, this being made possible by rotation of the atoms about the single bonds in the structure. Experimental evidence shows that such rotation occurs freely in most simple chain compounds, with the result that ethane, for example, is a homogeneous mixture of molecules in the two forms shown in Figure 26.3. It is nevertheless true that the stable states are those in which the hydrogen atoms on adjacent carbon atoms are in the "staggered" arrangements, rather than in the "eclipsed" arrangement shown in Figure 26.3. The reason is that the eclipsed arrangement has the hydrogen atoms on neighboring carbon atoms closer together than in the staggered arrangements, so that there is more repulsion between them in the eclipsed structure than in the staggered form. The structures shown in Figure 26.3 are known as different *conformers* of ethane.

Physical Properties of Alkanes

Within a homologous series of organic compounds, the melting and boiling points increase in a regular way as the carbon content increases, as illustrated in Table 26.3. This is consistent with the idea that van der Waals forces increase as the number of electrons in the molecule increase (Chapter 6). Among a series of homologs (members of the same homologous series)

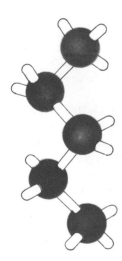

Figure 26.2 Model of a chain of carbon atoms.

Table 26.5 Formulas and Names of Substituents

Formula	Name	Formula	Name
CH_3-	Methyl	$CH_2{=}CH-$	Ethenyl
CH_3-CH_2-	Ethyl	$CH{\equiv}C-$	Ethynyl
$CH_3-CH_2-CH_2-$	n-Propyl	$Cl-$	Chloro
CH_3-CH- $\;\;\;\;\;|$ $\;\;\;\;\;CH_3$	Isopropyl	$F-$	Fluoro
		$HO-$	Hydroxy
		CH_3O-	Methoxy
		CH_3-CH_2-O-	Ethoxy
$\;\;\;\;\;CH_3$ $\;\;\;\;\;|$ CH_3-C- $\;\;\;\;\;|$ $\;\;\;\;\;CH_3$	tert-Butyl	$-CF_3$	Trifluoro-methyl
		$-NH_2$	Amino
		$-NO_2$	Nitro
	Phenyl	$-SH$	Mercapto

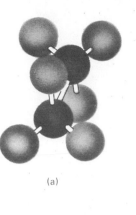

(a)

(b)

Figure 26.3 Forms of the ethane molecules arising from internal rotation. (a) Staggered form. (b) Eclipsed form.

Table 26.6 Uses of Alkanes in the Petroleum and Petrochemical Industries

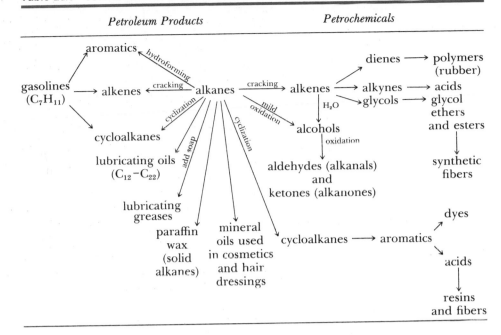

having the same molecular formula, increased branching is accompanied by a decrease in boiling point. This is illustrated in Table 26.4, where the boiling points of the various isomeric compounds are given.

Alkanes containing four carbon atoms or less are gases; the straight-chain homologs containing from five to 16 carbon atoms are liquids, and the higher members of the series are waxlike solids.

The densities of alkanes are always less than 1.0 g/cc; they increase as the carbon content increases, a consequence of the fact that as the chain lengthens, the ratio of molecular weight to molecular volume increases. Because they are nonpolar, the alkanes are not soluble in (or miscible with) water. Liquid alkanes are good solvents for paraffin and other hydrocarbons.

Chemical Properties of Alkanes

The alkanes are relatively inactive. They do not react with acids or bases or with oxidizing agents such as potassium permanganate, $KMnO_4$, or potassium dichromate, $K_2Cr_2O_7$. They do react with oxygen when heated; with halogens, at high temperatures, or in the presence of light; and with concentrated nitric acid, HNO_3.

In the absence of air, alkanes can be caused to "*crack*" (break into smaller fragments) at high temperatures or in the presence of catalysts. The cracking reaction is used extensively in the petroleum industry to convert hydrocarbons of high molecular weight (C_{15} to C_{18}) to gasoline hydrocarbons (molecules having five to 12 carbon atoms).

Catalysts such as aluminum chloride or sulfuric acid can be used to facilitate conversion of straight-chain alkanes to branched-chain or to aromatic hydrocarbons. Both branched-chain and aromatic hydrocarbons have higher octane ratings as gasolines than do their straight-chain derivatives.

Uses of Alkanes

Readily available from petroleum, the alkanes serve as the basis for the petroleum and petrochemical industries. Table 26.6 gives some uses of alkanes.

Figure 26.4 Some cycloalkanes.

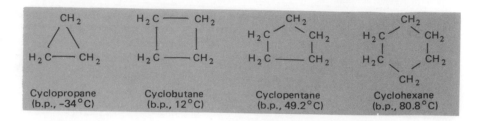

| Cyclopropane (b.p., -34°C) | Cyclobutane (b.p., 12°C) | Cyclopentane (b.p., 49.2°C) | Cyclohexane (b.p., 80.8°C) |

Cycloalkanes

Compounds containing rings of saturated carbon atoms are known as the cyclo-alkanes. Examples are shown in Figure 26.4. These compounds have higher boiling points and densities than the alkanes containing the same number of carbon atoms. Their chemical properties are similar to those of the alkanes, except that the small-ring compounds are more reactive than the corresponding alkanes. This is because the C—C—C bond angles in the small rings are considerably less than the tetrahedral angle, resulting in a *strained* structure—one that reacts by allowing the ring to open, as illustrated by the reaction

$$CH_2\!-\!CH_2 + Br_2 \longrightarrow CH_2\!-\!CH_2\!-\!CH_2$$
$$\underset{\displaystyle CH_2}{\diagdown\ \diagup} \qquad\qquad \underset{\displaystyle Br}{|} \qquad \underset{\displaystyle Br}{|}$$

Note that cyclohexane is not planar, as benzene is. There is less strain and less interference of the hydrogen atoms with each other if the molecule adopts the *chair* form (Figure 26.5).

ALKENES, RCH=CHR

Hydrocarbons containing one double bond between carbon atoms are known as alkenes or olefins. The formulas, names, and some properties of some members of this series are given in Table 26.7. The nature of the double bond between carbon atoms was discussed in Chapter 6. The alkenes are the starting materials for many plastics, synthetic fibers, and numerous alcohols.

Similar to the alkanes in physical properties, the alkenes are considerably more reactive. The characteristic reaction of these compounds involves addition to the double bond, as illustrated by the equations

$$CH_3\!-\!CH\!=\!CH_2 + Br_2 \longrightarrow CH_3\!-\!\underset{\underset{\displaystyle Br}{|}}{C}H\!-\!\underset{\underset{\displaystyle Br}{|}}{C}H_2$$

$$CH_2\!=\!CH_2 + H_2O \xrightarrow{\text{catalyst}} \underset{\underset{\displaystyle H}{|}}{C}H_2\!-\!\underset{\underset{\displaystyle OH}{|}}{C}H_2$$

Perhaps the most important reaction of alkenes is polymerization, or self-addition. In this process, *small molecules unite to form giant molecules* known as *polymers*. The reaction of ethylene molecules to form polyethylene—molecules containing more than 100,000 ethylene molecules—may be illustrated by the equations

$$CH_2\!=\!CH_2 + CH_2\!=\!CH_2 \longrightarrow -CH_2\!-\!CH_2\!-\!CH_2\!-\!CH_2-$$

$$nCH_2\!=\!CH_2 \longrightarrow -CH_2\!-\!CH_2\!-\!(-CH_2\!-\!CH_2\!-)_{n-2}\!-\!CH_2\!-\!CH_2-$$

where $n = 10^5$ and higher.

Figure 26.5 The chair conformation of cyclohexane.

Polyethylene may be waxlike, tough but flexible, or rigid, depending upon how the polymerization reaction is carried out. The uses of polyethylene

Table 26.7 Names, Formulas, and Physical Constants of Some Alkenes

Name	Formula	Melting Point (°C)	Boiling Point (°C)
Ethene (ethylene)	$CH_2{=}CH_2$	−169.4	−103.8
Propene (propylene)	$CH_3{-}CH{=}CH_2$	−185.2	−47
1-Butene	$CH_2{=}CH{-}CH_2{-}CH_3$	—	−6.3
2-Butene	$CH_3{-}CH{=}CH{-}CH_3$	−127*	1.4*
Methylpropene	$CH_3{-}\underset{\underset{CH_3}{\mid}}{C}{=}CH_2$	−140.7	−6.9
1-Pentene	$CH_2{=}CH{-}CH_2{-}CH_2{-}CH_3$	−138	30.1

*Mixture of cis and trans isomers. See page 570.

range from electrical insulation and plastic pipe or tubing to bread wrapping or heavier construction sheeting. Many other alkenes can be polymerized to give plastics, elastomers, or fibers. Various kinds of polymers are discussed in Chapter 28.

Nomenclature of Alkenes

Rules of the IUPAC nomenclature system apply to alkenes with the following additions to the rules for alkanes:

1. The ending -*ene* is used (instead of -*ane*) to indicate that one carbon-to-carbon double bond is present.
2. The *longest continuous chain of carbon atoms containing the double bond* forms the base for the name, and the chain is numbered so that the carbon atoms of the double bond have the lowest possible numbers.
3. The position of the double bond is indicated by the smaller of the numbers of the carbon atoms involved in the double bond. For example: $CH_2{=}CH$ $CH_2{-}CH_3$ is 1-butene. The 1- indicates that the double bond is situated between the first and second carbon atoms.

PROBLEM Name the following alkenes:

1. 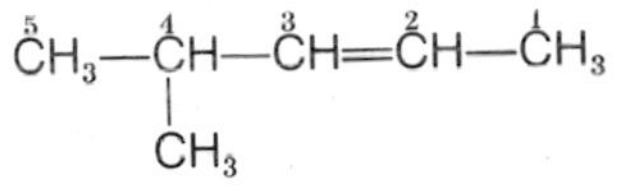
$$\overset{5}{C}H_3{-}\underset{\underset{CH_3}{\mid}}{\overset{4}{C}H}{-}\overset{3}{C}H{=}\overset{2}{C}H{-}\overset{1}{C}H_3$$

Solution 4-methyl-2-pentene

2. 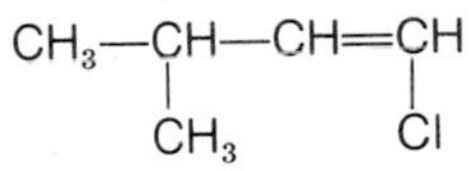
$$CH_3{-}\underset{\underset{CH_3}{\mid}}{CH}{-}CH{=}\underset{\underset{Cl}{\mid}}{CH}$$

Solution 1-chloro-3-methyl-1-butene

3. 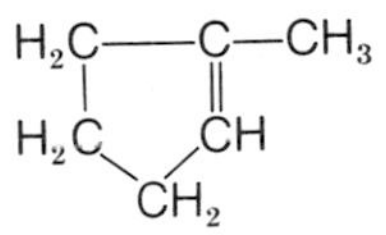
$$\begin{array}{c} H_2C{-}\!\!-\!\!-C{-}CH_3 \\ \mid \quad\quad \parallel \\ H_2C \quad\;\; CH \\ \diagdown\;\;\diagup \\ CH_2 \end{array}$$

Solution 1-methyl-1-cyclopentene

Isomerism Among Alkenes

In addition to the position isomers represented by the formulas illustrating 1-butene, 2-butene, and 2-methylpropene (Table 26.8), the existence of the double bond permits another type of isomerism known as geometrical isomerism An important feature of the double bond is that it does not permit

Table 26.8 Names, Formulas, and Physical Properties of Some Alkynes

Name	Formula	Melting Point (°C)	Boiling Point (°C)
Ethyne (acetylene)	H—C≡C—H	—	−83.6
Propyne	CH₃—C≡C—H	−104.7	−27.5
1-Butyne	HC≡C—CH₂—CH₃	−130	8.6
2-Butyne	CH₃—C≡C—CH₃	−32.2	27.1
1-Pentyne	HC≡C—CH₂—CH₂—CH₃	−95	40
1-Hexyne	H—C≡C—CH₂—CH₂—CH₂—CH₃	−132	71.5

rotation of the group at one end of the bond with respect to the group at the other, in contrast to the behavior of groups connected by a single bond. This means that a compound such as 2-butene can exist in two distinguishable forms, which we can represent as

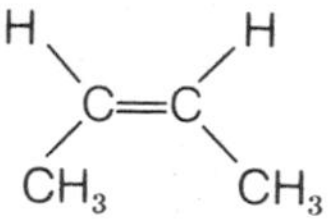

cis-2-Butene 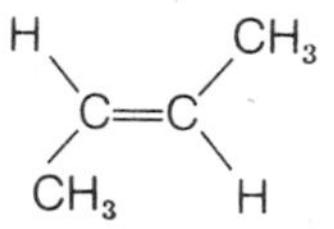   *trans*-2-Butene

Molecules related to each other as *cis*-2-butene* is related to *trans*-2-butene* are known as geometrical isomers. Two geometrical isomers will commonly have different physical properties and *are readily identified as two different compounds*. Chemical reactivity often will show which is the cis and which the trans isomer, but certain physical techniques such as spectroscopy or X-ray diffraction (Chapter 13) are sometimes more convenient.

Alkadienes, Alkapolyenes

Hydrocarbons containing two or more carbon-to-carbon double bonds belong to these homologous series. Examples are:

CH₂=CH—CH=CH₂
1,3-Butadiene

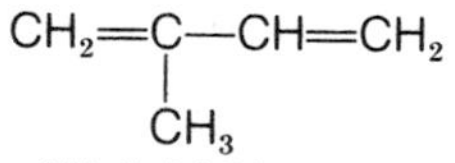

CH₂=C—CH=CH₂
        |
       CH₃
2-Methyl-1,3-butadiene
(isoprene)

CH₂=CH—CH=CH—CH=CH—CH₃
1,3,5-Heptatriene

The two alkadienes given above are raw materials for synthetic rubber. Compounds having more than two double bonds are often found in nature, where they supply the coloring matter for many fruits and vegetables such as tomatoes, carrots, corn, and red peppers. The name for the yellow coloring matter in eggs, carrots, butter, and other yellow-colored vegetables and animal products is β-carotene, which has the structure shown in Figure 26.6.

ALKYNES, R—C≡C—R

Hydrocarbons containing one triple bond between carbon atoms are known as alkynes. These compounds are similar to the alkenes in both their chemical and physi-

---

*The cis- and trans- nomenclature refers to the fact that a plane perpendicular to the paper and passing through the double bond divides the molecules into two parts. In the cis isomer, each part contains two of the same kinds of atoms or groups—in the 2-butenes, these are two hydrogen atoms or two CH₃ groups. In the trans isomer, each part contains one of each kind of atom or group—for example, in *trans*-2-butene, one hydrogen atom and one CH₃ group.

Figure 26.6 β-Carotene.

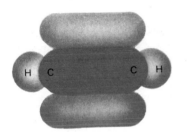

Figure 26.7 Orbital model of ethyne (acetylene).

cal properties, except that the alkynes can add two moles of reagent per mole of hydrocarbon, as illustrated by the reaction

$$H-C\equiv C-H + 2Br_2 \longrightarrow H-\underset{\underset{Br}{|}}{\overset{\overset{Br}{|}}{C}}-\underset{\underset{Br}{|}}{\overset{\overset{Br}{|}}{C}}-H$$

1,1,2,2-Tetrabromoethane

The nomenclature for the alkynes is identical to that of the alkenes, except the ending *-yne* is used to denote the presence of the triple bond. Table 26.9 gives the names, formulas, and some physical properties of representative members of the series.

The nature of the carbon-to-carbon triple bond was discussed in Chapter 6. Ethyne, or acetylene, is a linear molecule in which the carbon atoms use *sp* hybrid orbitals to form σ bonds. The two π bonds between carbon atoms are oriented as shown in Figure 26.7.

Used primarily as intermediates for the synthesis of other organic compounds, including the starting materials for the synthetic fibers Orlon, Acrilan, and Dynel, and in the manufacture of benzene, ethyne is currently the most important member of this family.

ALKYL HALIDES, RX

When a hydrogen atom on an alkane or cycloalkane molecule is replaced by a halogen atom, the resulting compound is known as an alkyl halide, often represented by the formula RX. Since any of the halogens and any straight, branched, or saturated cyclic compound may be used, the number of possible alkyl halides is large. The alkyl halides are especially useful in the preparation of a variety of aliphatic compounds and as solvents. Table 26.10 gives the common names, formulas, and boiling points of some fluorides, chlorides, bromides, and iodides.

The IUPAC names for alkyl halides are derived from the rules given on page 565. The compound is named for the normal alkane having the same number of carbon atoms as are present in the longest continuous chain of carbon atoms in the compound under consideration.

$$CH_3-\underset{\underset{Cl}{|}}{CH}-CH_3 \qquad\qquad CH_3-CH_2-CH_2-I$$

2-Chloropropane   1-Iodopropane

Alkyl halides also may be subclassified into primary, secondary, or tertiary series, depending upon the number of hydrocarbon groups attached to the carbon atom bonded to the halogen atom.

A primary halide
(the carbon atom attached to the halogen atom holds only one hydrocarbon group)

A secondary halide
(two hydrocarbon groups)

A tertiary halide
(three hydrocarbon groups)

Physical Properties of Alkyl Halides

Among the alkyl halides, the alkyl iodides have the highest boiling points, followed by the alkyl bromides, chlorides, and fluorides, in that order. Since the iodine atom is the most polarizable and the fluorine atom the least polarizable of the halogens, the order of boiling points can be explained in terms of van der Waals forces associated with the respective halogen atoms. The observed order is opposite to that expected if dipole forces were of primary importance, for these are greatest in the alkyl fluorides and least in the alkyl iodides. Increasing the carbon content also increases the boiling points of the alkyl halides, as shown in Table 26.9.

Alkyl halides are nearly insoluble in or immiscible with water.

Chemical Properties of Alkyl Halides

An especially important use of alkyl halides is in the synthesis of a large number of organic compounds. The versatility of the halides in such syntheses stems from the ease with which the halogen atom is replaced by a number of other functional groups. The general reaction involves the replacement of the halide ion by some other negative ion:

$$B^- + R{-}X \longrightarrow B{-}R + X^-$$

Examples of such replacements and of the kinds of compounds that can be synthesized from alkyl halides are given in Chapter 27.

Polyhalogen Compounds

A large number of polyhalogen compounds are known, and some of these are of industrial importance. Examples of such compounds and their uses are given in Table 26.10.

Table 26.9 Formulas, Names, and Boiling Points of Some Alkyl Halides

General Formula	Name	BOILING POINT (°C)			
		Fluoride	Chloride	Bromide	Iodide
$CH_3{-}X$	Methyl halide	−78.0	−23.7	4.6	42.6
$CH_3{-}CH_2{-}X$	Ethyl halide	−32	12.2	38.0	72.2
$CH_3{-}CH_2{-}CH_2{-}X$	n-Propyl halide	2	46.6	70.9	102.4
$CH_3{-}CH{-}X$ $\quad\mid$ $\quad CH_3$	Isopropyl halide	−11	36.5	59.6	89.5
$CH_3{-}CH_2{-}CH_2{-}CH_2{-}X$	n-Butyl halide	—	78.0	101.6	127
CHX_3 (an alkyl trihalide)	Haloform	126	61.2	149.5	119 (m.p.)

Table 26.10 Some Polyhalogen Compounds and Their Uses

Name	Formula	Use
Trichloromethane (chloroform)	$CHCl_3$	Anesthetic, sedative, antiseptic
Triiodomethane (iodoform)	CHI_3	Antiseptic
Dichlorodifluoromethane (A Freon)	CCl_2F_2	Refrigerants, aerosol, propellants
1,1,2-Trichloroethene	$CCl_2{=}CHCl$	Dry-cleaning solvent

ALKANOLS OR ALCOHOLS, ROH

Compounds containing a hydroxyl group bonded to a carbon atom are known as alkanols. Alcohols include compounds containing one or more hydroxyl groups such as the alkanediols (two OH groups), the alkanetriols (three OH groups), and so on. Of great industrial importance, the alcohols are used as solvents for shellacs, varnishes, and lacquers; in plastics, explosives, cosmetics, drugs, and special fuels; and for many other purposes.

The formulas, IUPAC names, common names, and physical constants of some alcohols are given in Table 26.11.

A hydroxyl group may be bonded to any carbon atom in the chain. This gives rise to primary, secondary, and tertiary alcohols comparable to the primary, secondary, and tertiary alkyl halides:

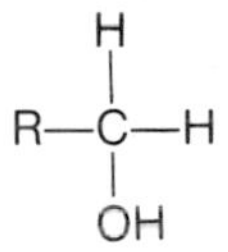

A primary alcohol 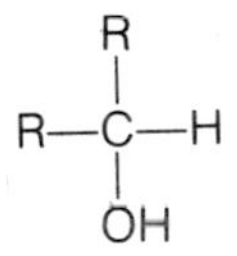   A secondary alcohol 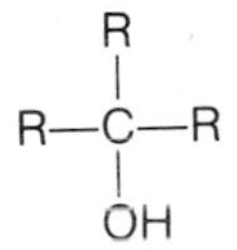   A tertiary alcohol

Physical Properties of Alcohols

The boiling points of the alcohols are higher than those of the corresponding alkyl halides or alkanes, largely because of the hydrogen bonding among alcohol molecules:

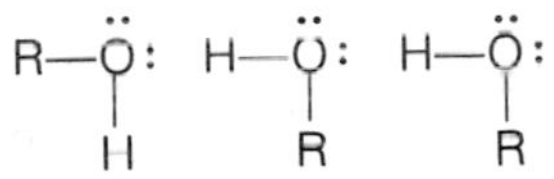

That hydrogen bonding is responsible for the relatively high boiling points of alcohols is suggested by comparison of the boiling point of ethyl alcohol, 78.5°C, with that of its isomer dimethyl ether, $CH_3{-}O{-}CH_3$, which boils at -24°C. In the ether, no hydrogen bonding possibilities exist.

The lower alcohols are all highly soluble in water because, in these compounds, the hydroxyl group comprises an appreciable portion of the molecule, and attraction between the water and alcohol dipoles is important. As the number of carbon atoms in the alcohol molecule increases, the water solubility decreases, because the alcohol is becoming more like a hydrocarbon and less like water. For example, 1-hexanol is only slightly soluble in water but highly soluble in hexane.

Chemical Properties of Alcohols

The chemical properties of alcohols may be summarized under three headings: (1) reactions analogous to those of water; (2) reactions in which the

Table 26.11 Names, Formulas, and Physical Constants of Some Alcohols

Name	Formula	Melting Point (°C)	Boiling Point (°C)
Methanol (methyl alcohol)	$CH_3—OH$	−97.8	64.5
Ethanol (ethyl alcohol)	$CH_3—CH_2—OH$	−117.3	78.5
1-Propanol	$CH_3—CH_2—CH_2—OH$	−127	97.8
2-Propanol (isopropyl alcohol)	$CH_3—\underset{\underset{CH_3}{\mid}}{CH}—OH$	−85.8	82.3
1-Butanol	$CH_3—CH_2—CH_2—CH_2—OH$	−89.8	117.7
2-Butanol	$CH_3—CH_2—\underset{\underset{OH}{\mid}}{CH}—CH_3$	—	99.5
2-Methyl-1-propanol	$CH_3—\underset{\underset{CH_3}{\mid}}{CH}—CH_2—OH$	−108	107.3

hydroxyl group is replaced; and (3) dehydration reactions in which both a hydrogen atom and the hydroxyl group are removed.

1. Reactions analogous to those of water include the following.

Alcohols, like water, show weak acid and base properties:

$$2R—\underset{\underset{H}{\mid}}{O} \rightleftarrows RO—\underset{\underset{H}{\mid}}{H^+} + RO^-$$

Oxonium   Alkoxide
ion   ion

Alcohols, like water, react with active metals to liberate hydrogen:

$$2ROH + 2Na \longrightarrow 2NaOR + H_2$$

Sodium
alkoxide

2. Replacement of the hydroxyl group can be accomplished by several reagents, including phosphorus trihalides and thionyl chloride ($SOCl_2$), and nitric and sulfuric acids.

$$3ROH + PCl_3 \longrightarrow 3RCl + H_3PO_3$$

$$2ROH + SOCl_2 \longrightarrow 2RCl + H_2SO_3$$

$$ROH + H—O—NO_2 \longrightarrow RO—NO_2 + H_2O$$

Alkyl nitrate

When glycerol is allowed to react with nitric acid, nitroglycerine (glyceryl trinitrate) is formed:

$$\begin{array}{l}CH_2—OH \\ | \\ CH—OH + 3HONO_2 \longrightarrow \\ | \\ CH_2—OH \end{array} \quad \begin{array}{l}CH_2ONO_2 \\ | \\ CHONO_2 + 3H_2O \\ | \\ CH_2ONO_2 \end{array}$$

Nitroglycerine

When alcohols react with sulfuric acid, alkyl hydrogen sulfates are formed:

$$ROH + HO-\underset{\underset{O}{\|}}{\overset{\overset{O}{\|}}{S}}-OH \longrightarrow RO-\underset{\underset{O}{\|}}{\overset{\overset{O}{\|}}{S}}-OH + H_2O$$

Alkyl hydrogen sulfate

The sodium salts of alkyl hydrogen sulfates made from long-chain alcohols (C_{12} to C_{14}) are detergents.

3. Dehydration of alcohols results in the formation of alkenes, and may be accomplished using hot, concentrated sulfuric acid:

$$CH_3-\underset{\underset{H}{|}}{CH}-\underset{\underset{OH}{|}}{CH_2} \xrightarrow[\Delta]{H_2SO_4} CH_3-CH=CH_2 + H_2O$$

(H and OH on adjacent
carbon atoms are removed)

ALDEHYDES (ALKANALS),
$$R-\underset{\underset{O}{\|}}{C}-H; \text{ KETONES}$$
$$\text{(ALKANONES), } R-\underset{\underset{O}{\|}}{C}-R$$

Compounds containing the group

$$-\underset{\underset{O}{\|}}{C}-H$$

are known as aldehydes; compounds containing the group

$$C-\underset{\underset{O}{\|}}{C}-C$$

are called ketones   The group

$$-\underset{\underset{O}{\|}}{C}-$$

common to both aldehydes and ketones is known as a carbonyl group. The aldehyde group is always located at the end of a chain of carbon atoms; the carbonyl portion of the ketone group is located on nonterminal carbon atoms. Table 26.12 gives names and formulas for some aldehydes and ketones.

The simplest aldehydes and ketones are known by their common names. The IUPAC nomenclature rules designate that the characteristic ending for aldehydes is -al, and the ending for ketones is -one. The usual rule of selecting the longest continuous chain of carbon atoms containing the aldehyde or ketone group as the basis for the name is followed here as in other homologous series. Thus, the compound

$$CH_3-CH_2-\underset{\underset{O}{\|}}{C}-\underset{\underset{CH_2-CH_3}{|}}{CH}-CH_2-CH_3$$

is named 4-ethyl-3-hexanone, and

$$CH_3-\underset{\underset{H_3C}{|}}{\overset{\overset{CH_3}{|}}{C}}-\underset{\underset{O}{\|}}{C}-H$$

is 2,2-dimethylpropanal.

Table 26.12 Names, Formulas, and Boiling Points for Some Aldehydes and Ketones

Name	Formula	Boiling Point (°C)
Aldehydes		
Formaldehyde (methanal)	$H-\underset{\underset{O}{\|\|}}{C}-H$	−21
Acetaldehyde (ethanal)	$CH_3-\underset{\underset{O}{\|\|}}{C}-H$	20.2
Propionaldehyde (propanal)	$CH_3-CH_2-\underset{\underset{O}{\|\|}}{C}-H$	48.8
Butyraldehyde (butanal)	$CH_3-CH_2-CH_2-\underset{\underset{O}{\|\|}}{C}-H$	61
Benzaldehyde	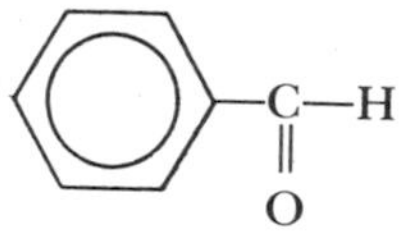 $C_6H_5-\underset{\underset{O}{\|\|}}{C}-H$	179.5
Acrolein (propenal)	$CH_2\!=\!CH-\underset{\underset{O}{\|\|}}{C}-H$	52.5
Ketones		
Acetone (propanone)	$CH_3-\underset{\underset{O}{\|\|}}{C}-CH_3$	56.1
Methyl ethyl ketone (butanone)	$CH_3-\underset{\underset{O}{\|\|}}{C}-CH_2-CH_3$	79.6
Diethyl ketone (3-pentanone)	$CH_3-CH_2-\underset{\underset{O}{\|\|}}{C}-CH_2-CH_3$	102.7
Methyl *n*-propyl ketone (2-pentanone)	$CH_3-\underset{\underset{O}{\|\|}}{C}-CH_2-CH_2-CH_3$	101.7

NOTE: Common names are listed first to emphasize their common usage for these compounds and to point out that the systematic names are not necessarily preferred.

The simpler aldehydes and ketones, except formaldehyde, are liquids, and the C_1 to C_4 compounds are freely soluble in water. All the low molecular weight aldehydes and ketones have a penetrating odor. As the carbon content increases, the odor becomes fragrant. As a result, certain aldehydes and ketones are used in perfumes.

An aldehyde may be distinguished from a ketone by weak oxidizing agents. The aldehyde is very easily oxidized, the

$$-\underset{\underset{O}{\|\|}}{C}-H$$

group being changed to a

$$-\underset{\underset{O}{\|\|}}{C}-OH$$

group, while the ketone is oxidized only by strong oxidizing agents. Two special reagents, Fehling's solution (which contains Cu^{+2} ions) and Tollen's reagent (which contains Ag^+ ions), are weak oxidizing agents that will oxidize aldehydes but not ketones. In Fehling's solution, an aldehyde reduces the copper(II) ion to copper(I) ion, which precipitates as red copper(I) oxide, Cu_2O. An aldehyde also reduces the silver ion, Ag^+, in Tollen's reagent to metallic silver, which forms a silver mirror on the walls of the glass test tube.

Reactions of aldehydes and ketones are of great importance. They will be discussed in Chapter 27.

CARBOXYLIC ACIDS, RC—OH (with O double bonded above the C)

Compounds containing the carboxyl group

$$-\overset{\displaystyle O}{\underset{\displaystyle }{C}}-OH$$

(the name comes from *carb*onyl and hyd*roxyl*) comprise the class of substances known as carboxylic acids. Many of these compounds are found in plants and animals either free or combined with alcohols. The members of this series are weak acids, having ionization constants at 25°C of about 1×10^{-5}. The IUPAC nomenclature for these acids follows the pattern of the previously discussed series, except that the ending *-oic* acid is used as illustrated in Table 26.13.

The lower members of the series have disagreeable odors. Butanoic acid is responsible for the odor in rancid butter. The C_6, C_8, and C_{10} straight-chain acids have the unpleasant "odor of goats." The acids containing one to 10 carbon atoms are liquids at room temperature and the higher members are waxlike solids. All the acids have relatively high boiling points, in part because of hydrogen bonding in the liquid and in part because of relatively strong van der Waals forces. Acids containing one to four carbon atoms are completely miscible with water; those having six or more carbon atoms are almost insoluble in water. This reflects again the interplay between the polar and nonpolar portions of the molecules in determining the properties of the

Table 26.13   Names, Formulas, and Physical Constants of Some Carboxylic Acids

Name	Formula	Melting Point (°C)	Boiling Point (°C)	Ionization Constant at 25°C
Methanoic acid (formic acid)	H—C—OH (with O double bonded to C)	8.4	100.5	2.1×10^{-4}
Ethanoic acid (acetic acid)	CH_3—C—OH (with O double bonded to C)	16.6	118.1	1.8×10^{-5}
Propanoic acid (propionic acid)	CH_3—CH_2—C—OH (with O double bonded to C)	−22	141.1	1.3×10^{-5}
Butanoic acid (butyric acid)	CH_3—CH_2—CH_2—C—OH (with O double bonded to C)	−7.9	163.5	1.5×10^{-5}
2-Methylpropanoic acid (isobutyric acid)	CH_3—CH—C—OH with CH_3 and O below	−47.0	154.4	1.4×10^{-5}

Table 26.14 Reaction Sites in Carboxylic Acids and Possible Reaction Products

Site of Reaction	Description	Products

1. R—C—O—H
   ‖
   O
	Replacement of ionizable hydrogen by reaction with active metals or with base	R—C—O⁻M⁺
		‖
		O
		salts

2. R—C—OH
   ‖
   O
	Replacement of the hydroxyl group—
	(a) by reaction with alcohols,
		R—C—OR′
		‖
		O
		esters

	(b) by reaction with PCl_3 or $SOCl_2$	R—C—X ‖ O acyl halides

3. R—C—OH
   ‖
   O
	Reaction at the carbonyl group—
	with reducing agents
		R—CH₂—OH
		alcohols

4. R—C—OH
   ‖
   O
	Removal of the carboxyl group—
	fusion with sodium or electrolysis of acid
		RH
		alkane

5. R—C—OH
   ‖
   O
	Reactions with the alkyl group—
	Cl_2 will substitute for hydrogen atoms on alkyl portion of molecule
		R′—CH—CO₂H
		    |
		   Cl
		2-chloroalkanoic acid

substance. The short-chain molecules have small nonpolar portions and are soluble in water; the long-chain molecules have large nonpolar portions and are not soluble in water, but are soluble in hydrocarbons.

Chemical Reactions of Acids The chemical reactions of carboxylic acids may be predicted from the structure of the molecule. Five different kinds of reactions are thus predicted, and all are experimentally observed. They are summarized in Table 26.15.

Carboxylic acids and their esters (compounds with alcohols) are found in all living things. Methanoic, or formic, acid is found in bees and in sea nettles, and is responsible for the irritation following the sting from those organisms. Acetic acid and its derivatives are important in both carbohydrate and lipid (fat) metabolism. Esters of this acid have the pleasant odors associated with many fruits and flowers. Long-chain acids such as the C_{16} and C_{18} compounds are found in fats.

Some common polycarboxylic acids are illustrated below:

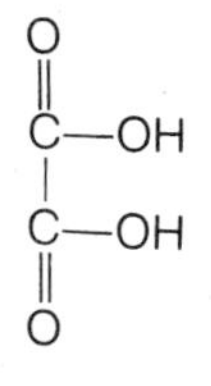

Oxalic acid
(ethanedioic acid)

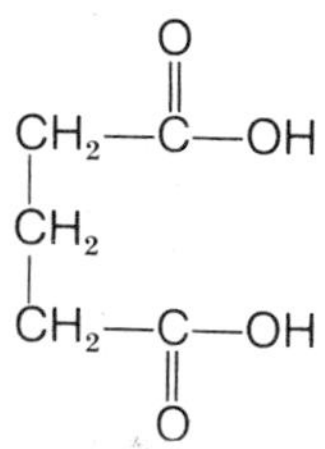

Glutaric acid
(pentanedioic acid)

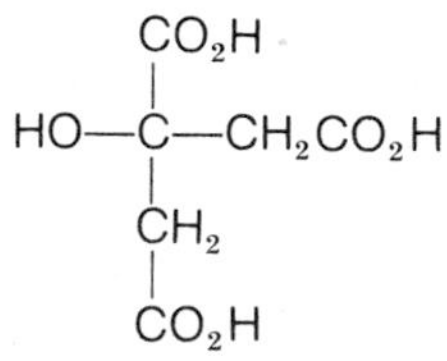

Citric acid
(3-carboxyl-3-hydroxy-
pentanedioic acid)

DERIVATIVES OF CARBOXYLIC ACIDS

Important derivatives of carboxylic acids include esters,

$$R-\underset{\underset{O}{\|}}{C}-OR'$$

metal salts, $RCO_2^-M^+$, acyl halides,

$$R-\underset{\underset{O}{\|}}{C}-Cl$$

acid anhydrides,

$$R-\underset{\underset{O}{\|}}{C}-O-\underset{\underset{O}{\|}}{C}-R$$

and amides,

$$R-\overset{\overset{O}{\|}}{C}-NH_2$$

Esters

These compounds, many of which are liquids and have fragrant odors, *are formed when carboxylic acids react with alcohols.*

$$R-\underset{\underset{O}{\|}}{C}-O-H + HOR' \longrightarrow R-\underset{\underset{O}{\|}}{C}-O-R' + H_2O$$

Esters are named from the alkyl group of the alcohol and the anion of the acid. For example, the ester formed between ethyl alcohol and acetic acid

$$CH_3-CH_2-O-\underset{\underset{O}{\|}}{C}-CH_3$$

is ethyl acetate. Names, formulas, and odors of various esters are given in Table 26.15.

The other derivatives of carboxylic acids will be discussed as needed in subsequent chapters. Nearly all are reactive substances and are used in many syntheses.

Table 26.15 Names, Formulas, and Odors of Various Esters

Name	*Formula*	*Odor*
Isoamyl acetate	$CH_3-\underset{\underset{CH_3}{\vert}}{CH}-CH_2-CH_2-O-\underset{\underset{O}{\|}}{C}-CH_3$	Banana oil
Amyl acetate (*n*-pentyl ethanoate)	$CH_3-CH_2-CH_2-CH_2-CH_2-O-\underset{\underset{O}{\|}}{C}-CH_3$	Apricot, cider
Isoamyl isovalerate	$CH_3-\underset{\underset{CH_3}{\vert}}{CH}-CH_2-CH_2-O-\underset{\underset{O}{\|}}{C}-CH_2-\underset{\underset{CH_3}{\vert}}{CH}-CH_3$	Apple
Ethyl butyrate	$CH_3-CH_2-O-\underset{\underset{O}{\|}}{C}-CH_2-CH_2-CH_3$	Apricot, peach
Octyl acetate	$CH_3-(CH_2)_6-CH_2-O-\underset{\underset{O}{\|}}{C}-CH_3$	Orange

Table 26.16 Names, Formulas, and Boiling Points of Some Aliphatic Amines

Name	Formula	Boiling Point (°C)
Methylamine	CH_3-NH_2	−6.5
Ethylamine	$CH_3-CH_2-NH_2$	16.6
n-Propylamine	$CH_3-CH_2-CH_2-NH_2$	48.7
n-Butylamine	$CH_3-CH_2-CH_2-CH_2-NH_2$	77.8
Dimethylamine	$CH_3\underset{\overset{\displaystyle \vert}{CH_3}}{-N}-H$	7.4
Trimethylamine	$CH_3\underset{\overset{\displaystyle \vert}{CH_3}}{-N}-CH_3$	3.5
Ethylene diamine	$NH_2-CH_2-CH_2-NH_2$	117

AMINES Amines are *derivatives of ammonia in which one or more hydrogen atoms are replaced by alkyl (or aryl) groups,* and they are called primary, secondary, or tertiary amines according to the number of hydrogen atoms replaced.

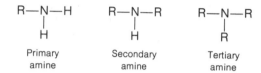

Primary amine   Secondary amine   Tertiary amine

The aliphatic amines resemble ammonia, but are more basic. The lower members of the series are gases and are soluble in water. Volatile amines have a fishlike odor with a hint of the ammonia odor. The aromatic amines are weaker bases than ammonia. Found in both plants and animals, amines are used to make medicinals, including sulfa drugs, local anesthetics, and antimalarials. Table 26.16 gives names, formulas, and boiling points of a number of amines.

AROMATIC COMPOUNDS Compounds having six-membered ring structures in which single and double bonds alternate are somewhat more stable than can be accounted for on the basis of normal σ and π bonding. These substances are known as aromatic compounds. The most common is benzene, C_6H_6. As indicated in Chapter 6, the benzene molecule is highly stabilized by resonance which is the result of the delocalization of the three pairs of π-bonding electrons present in the structure. Two main contributing resonance structures, three of less importance, and a single generalized formula are shown.

The benzene molecule is planar and the C—C—C angles are 120°, as required for sp^2 hybrid bonds. All the C—C bond distances are found to be alike, and equal to 1.39A, in contrast to 1.54A for normal single bonds and 1.30A for normal double bonds; thus the bonds are neither double bonds nor single bonds, but something intermediate between them.

Aromatic Character

The large resonance stabilization of benzene and similar compounds gives to benzene resistance to the characteristic reactions of addition to double bonds that would result in reduction of the stable character of the molecule by loss of resonance. Therefore, the reactions of benzene are not usually those of *addition,* as for other compounds containing double bonds (unsaturated compounds), but rather reactions of substitution, in which *another atom or group is substituted for one of the hydrogen atoms* while the six-membered ring remains intact. This reluctance to undergo the expected addition of reagents is said to be characteristic of the aromatic character of benzene and the compounds shown in Figure 26.8.

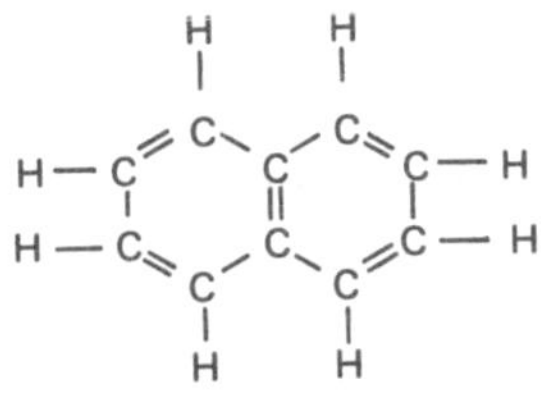

The substitution reactions lead to compounds such as those shown in Figure 26.9. In all of these, the resonance stabilization is maintained; it is enhanced in some compounds as electron pairs of the substituting group play a role in producing additional resonance structures. The reactions of these substituted benzenes are mainly those of the substituted functional group, such as —OH or —NH_2, because of the unusual stability of the aromatic structure.

The usual convention for showing formulas of benzene derivatives is to write the benzene ring as a simple hexagon, or to show a hexagon with a dashed or solid circle inside representing the bonds in the resonance structure,

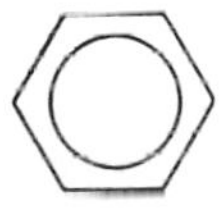

and to show only the attached groups, which have replaced the hydrogen atoms of the parent molecule. Since six hydrogen atoms can be replaced, many isomers can be formed, depending upon whether the substituents are present on adjacent carbon atoms or on more distant ones. Figure 26.10 shows the di- and trisubstituted bromobenzenes. Examples of some aromatic substances and their properties are given in Table 26.17.

Aromatic compounds, many obtained from the distillation of coal tar, are used in pharmaceuticals, dyestuffs, explosives, plastics, fibers, perfumes, insecticides, lacquers, solvents, and many other materials. Benzene is an important fuel and solvent; phenol is a common disinfectant; substituted phenols are widely used to impregnate wood to protect it against decay and insects; naphthalene is used in "moth balls;" aniline is the base of the dye industry; chlorobenzene is used in the manufacture of DDT insecticide; styrene,

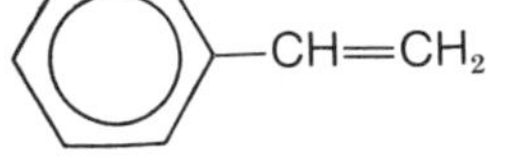

Naphthalene

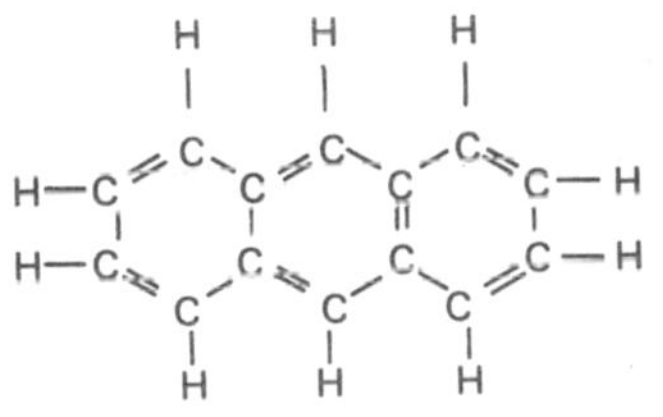

Anthracene

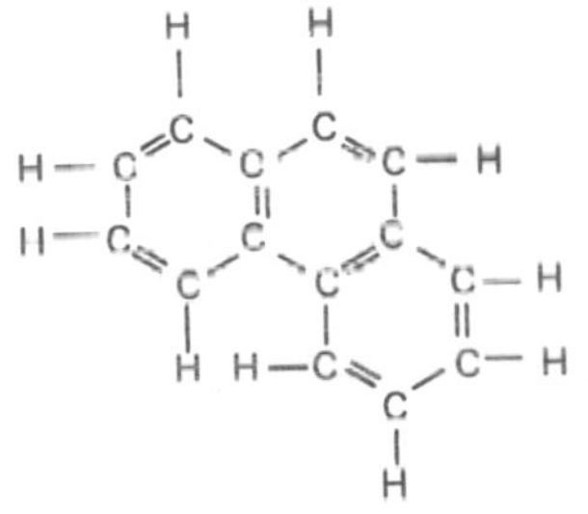

Phenanthrene

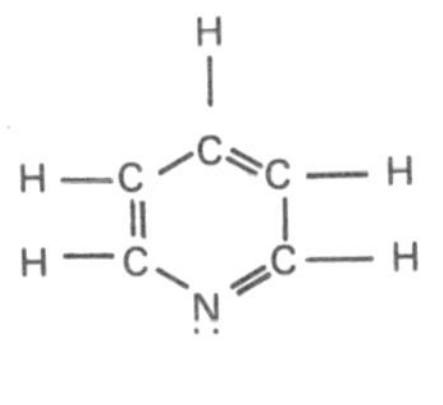

Pyridine

Figure 26.8 Some aromatic hydrocarbons.

Table 26.17 Formulas, Names, and Physical Constants of Some Aromatic Compounds

Formula	Name	Melting Point (°C)	Boiling Point (°C)
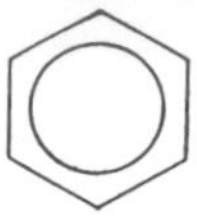	Benzene	5.5	80.1
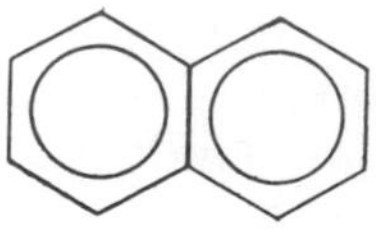	Naphthalene	80.3	218
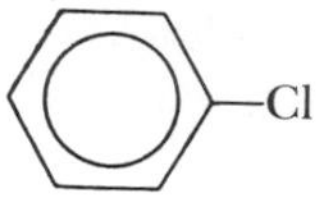 —Cl	Chlorobenzene	−45.2	132.1
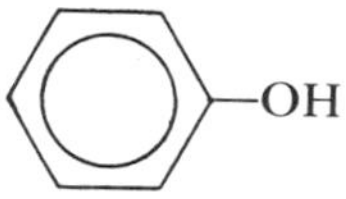 —OH	Phenol	41	182
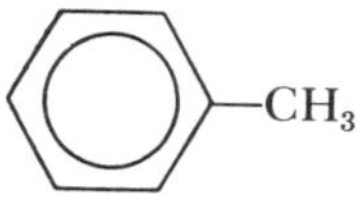 —CH₃	Toluene	−95	110.6
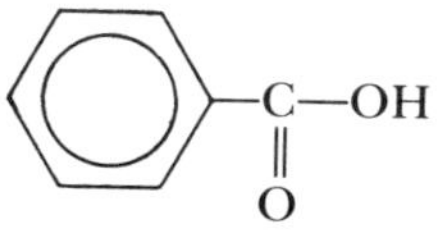 —C—OH ‖ O	Benzoic acid	121.4	—
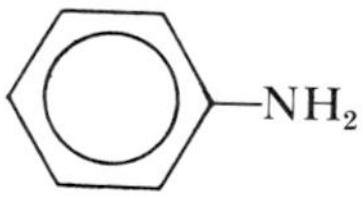 —NH₂	Aniline	−6.2	184.4
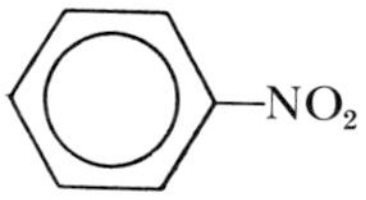 —NO₂	Nitrobenzene	5.8	210

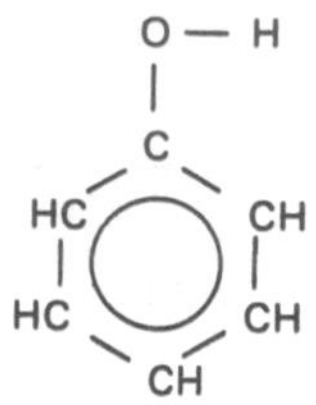

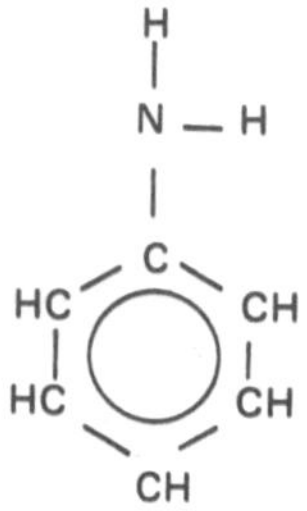

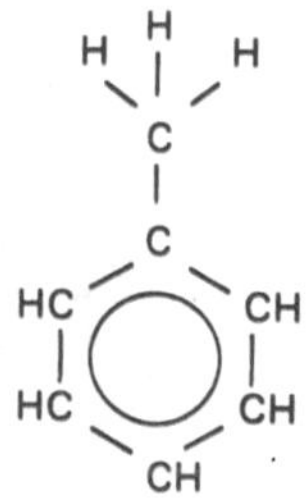

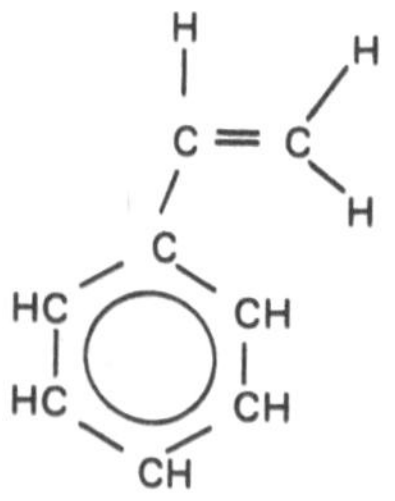

Figure 26.9 Some substituted benzenes.

is a component of synthetic rubber and of plastics; trinitrotoluene and picric acid (trinitrophenol) are high explosives; benzene sulfonic acid,

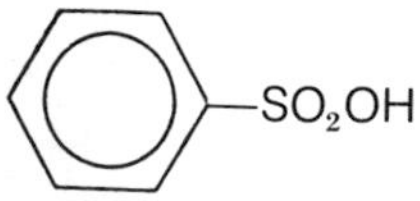

and its derivatives are used as detergents, sulfa drugs, and antiseptics.

The chemistry of aromatic compounds is some of the most interesting and highly developed in all of chemistry.

Figure 26.10 The di- and tri-bromo-
benzenes.

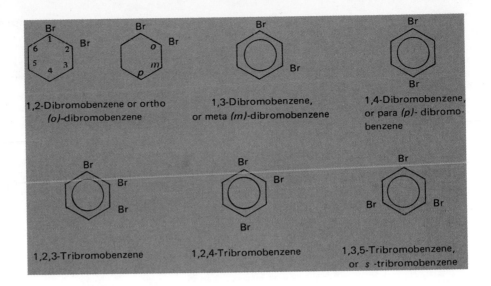

**1,2-Dibromobenzene or ortho
(o)-dibromobenzene**

**1,3-Dibromobenzene,
or meta (m)-dibromobenzene**

**1,4-Dibromobenzene,
or para (p)- dibromo-
benzene**

1,2,3-Tribromobenzene

1,2,4-Tribromobenzene

**1,3,5-Tribromobenzene,
or s -tribromobenzene**

OPTICAL ISOMERISM

We are now ready to consider another type of isomerism known as optical
isomerism.

Certain structures, such as that of lactic acid,

$$CH_3$$
$$HO—C—CO_2H$$
$$H$$

may exist in two forms, differing only in the arrangement of the four groups
around the central carbon atom. These two forms will be nonsuperimposable
mirror images of one another, related to each other as a left hand is to a right.
This situation might be visualized more clearly if you try to build models
of the lactic acid structure using ball and stick models. If several lactic acid
structures are constructed at random and compared, we will discover that
not all of the structures are superimposable and, therefore, all are not identical. In fact, two (and only two) different structures can be constructed. These
are nonsuperimposalbe mirror images of one another, and they are one kind
of optical isomers known as enantiomorphs.

(These are superimposable and identical
The hydrogen atom lies above the plane
of the page (━━━): the OH group lies
below this plane (---).)

(These are nonsuperimposable mirror
images. They are not identical but are
enantiomorphs.)

In general, enantiomorphs are found whenever a molecular structure may
be assembled in either of two forms that are nonsuperimposable mirror
images of one another. Such structures are said to be dissymmetric—wanting
in symmetry.

Dissymmetric organic molecules owe this property to either or both of the following structural features:

1. One or more carbon atoms bonded to four different groups (asymmetrically substituted, or asymmetric carbon atoms), which may be represented by the expression

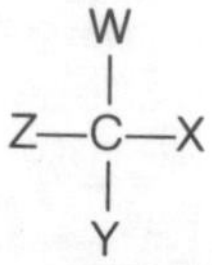

$$Z-\underset{\underset{Y}{|}}{\overset{\overset{W}{|}}{C}}-X$$

2. Rigid or semirigid structural elements, which prevent the substance from assuming a symmetrical spatial arrangement.

Examples of dissymmetry arising from each of these structural features are:

1. Some substances containing asymmetric carbon atoms:

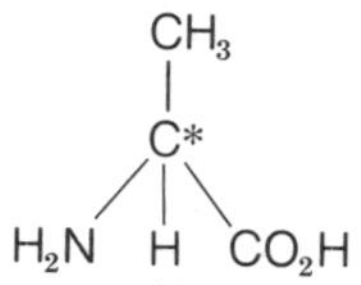

Alanine 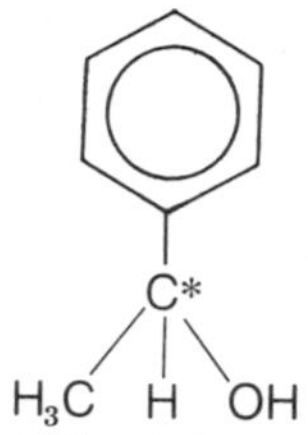   2-Phenylethanol 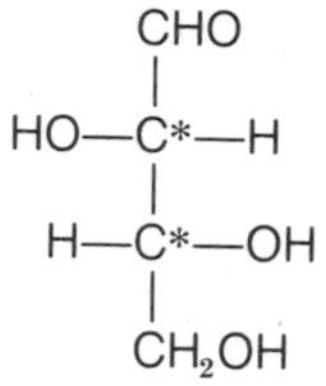   Threose 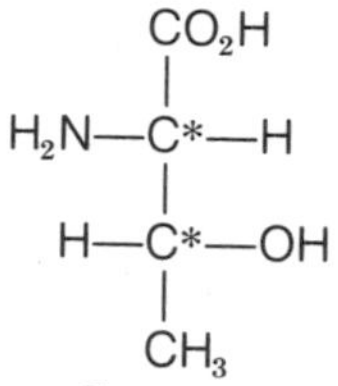   Threonine

2. Some dissymmetric substances containing no asymmetric carbon atoms:

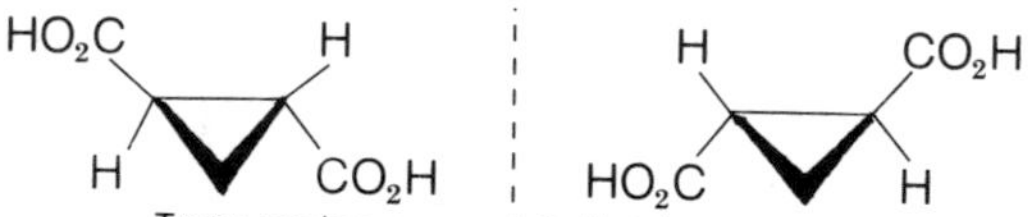

Trans-cyclopropane-1,2-dicarboxylic acid

(The cyclopropane ring is perpendicular to the plane of the page, and the H and CO_2H groups lie in the plane of the page.)

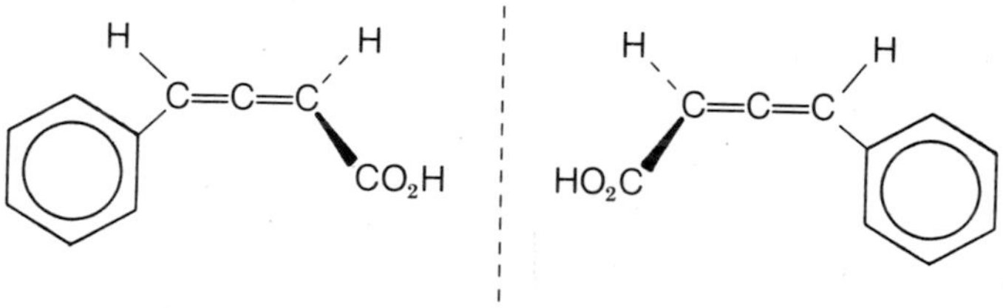

An allene   Mirror image

(The three carbon atoms lie in the plane of the paper, as do the bonds connecting the phenyl group and the hydrogen the first carbon atom.)

Properties of Optical Isomers—Optical Activity

Enantiomorphs are identical in most physical properties—boiling point, solubility, infrared spectra (in solution)—but they rotate the plane of polar-

---

*Asymmetric carbon atoms.

ized light in opposite directions.* It is because of this effect on light that they are said to possess optical activity and are called optical isomers. The enantiomorph that rotates the beam of light to the right is said to be dextrorotatory; the optical isomer that rotates it to the left is levorotatory.

Many naturally occurring substances, including all proteins, carbohydrates, alkaloids, and many vitamins and hormones, are optically active. The various enantiomorphs are often quite different in their biological properties. For example, the active principle secreted by the thyroid gland is one enantiomorph of the amino acid thyroxin; the other enantiomorph of this substance has several times less activity than its mirror image.

To understand more fully the meaning of "optical activity," it will be helpful to examine how it is measured. Optical activity is measured in a polarimeter (Figure 26.11), an instrument that converts ordinary light into plane polarized light, allows this to pass through a sample of the substance under consideration, and provides a technique for measuring the angle through which the original plane of the beam of polarized light is rotated by the sample.

The angle through which the plane of polarized light is rotated depends on the optically active substance, its concentration in solution temperature, wavelength of the polarized light, and length of the light path through the sample. However, at the same concentration, temperature, and so on, a pair of enantiomorphs will rotate the beam by exactly the same extent but in opposite directions. Since one enantiomorph in a pair rotates the plane to the left and the other enantiomorph rotates it to the right, it is important to associate a certain arrangement of groups around an asymmetric carbon atom with the direction of rotation of the light beam.

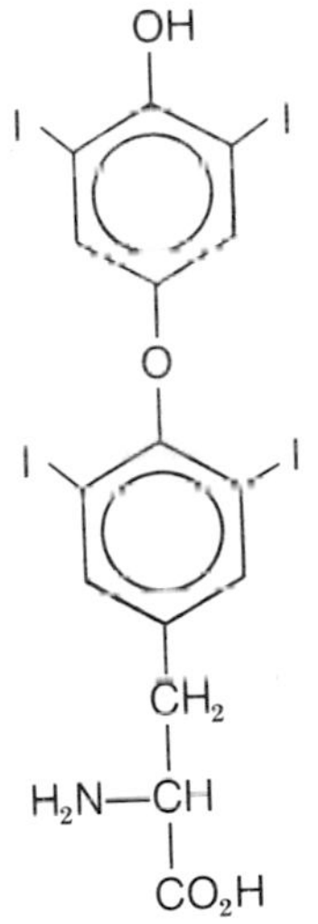

Thyroxin

---

*A beam of ordinary light may be thought of as a series of rays of various wavelengths vibrating in all possible planes perpendicular to the direction of propagation. Monochromatic light consists of a narrow range of wavelengths, with the rays usually vibrating in all possible planes. If a beam of monochromatic light is passed through a substance such as Polaroid, tourmaline crystals, or a Nicol prism, only those rays vibrating in a single plane are transmitted. This transmitted beam is said to consist of plane polarized light. An optically active substance can rotate the plane of vibration of the rays in such a beam.

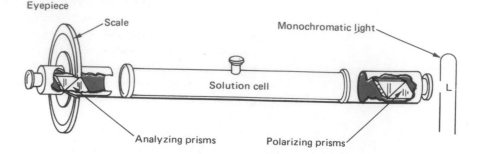

Eyepiece

Scale

Monochromatic light

Figure 26.11 Diagram of a polarimeter.

Solution cell

L

Analyzing prisms

Polarizing prisms

Rotation and Configuration

The particular arrangement of groups around an asymmetric carbon atom is known as the configuration of the structure. For example, in glyceraldehyde

$$
\begin{array}{c}
CHO \\
| \\
H-C-OH \\
| \\
CH_2OH
\end{array}
$$

the groups —CHO, —OH, —CH$_2$OH, and —H may be thought of as being arranged around the asymmetric carbon atom in a clockwise sequence in one enantiomorph, and in a counterclockwise sequence in the other enantiomorph, as illustrated below.

Clockwise sequence
or configuration

Counterclockwise sequence
or configuration

Projections of these formulas in the plane of the paper give the *projection formulas*:

$$
\begin{array}{c}
CHO \\
| \\
H-C-OH \\
| \\
CH_2OH
\end{array}
\qquad
\begin{array}{c}
CHO \\
| \\
HO-C-H \\
| \\
CH_2OH
\end{array}
$$

Clockwise sequence

Counterclockwise sequence

Projection formulas are commonly used and can be interpreted in terms of the tetrahedral arrangement of the groups by constructing an imaginary tetrahedron around the projection formulas, as follows:

Clockwise sequence

Counterclockwise sequence

The problem now is to find out which sequence (the clockwise or the counterclockwise) is associated with rotation of the plane of light to the right (is dextrorotatory) and which sequence rotates the plane to the left (is levorotatory). This has been done using X-ray diffraction, and the result is that the glyceraldehyde structure with the clockwise sequence is dextrorotatory.

In the case of lactic acid, the sequence —CO_2H, —OH, —H, and —CH_3 is similar to that of glyceraldehyde. However, the structure with the clockwise sequence is levorotatory, indicating that *the direction of rotation of the plane of light and the configuration around the asymmetric carbon atom need bear no obvious relationship*. However, such a relationship has been established using an advanced theory known as polarizability theory.

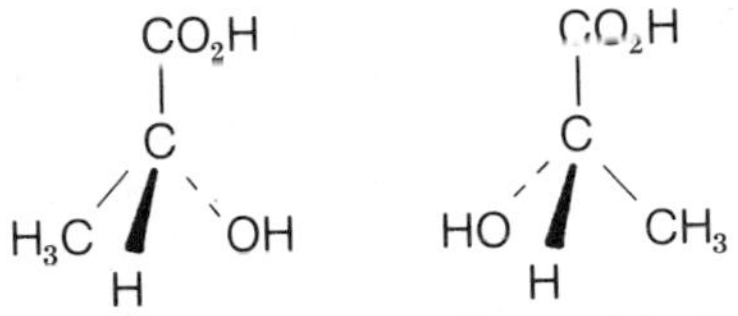

(Levorotatory)   (Dextrorotatory)

Because there is no obvious relation between configuration and direction of rotation, it has proved useful to relate all configurations to a single reference standard, and glyceraldehyde has been chosen. The enantiomorph with the clockwise sequence is called D-glyceraldehyde, and the enantiomorph with the counterclockwise sequence is known as L-glyceraldehyde.

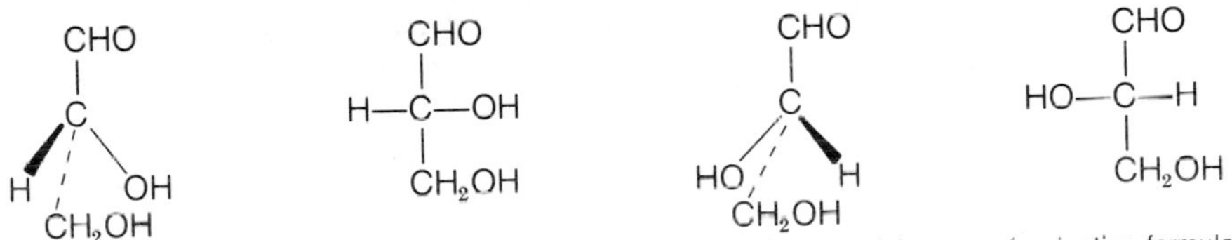

(perspective formula)   (projection formula)   (perspective formula)   (projection formula)
D-Glyceraldehyde   L-Glyceraldehyde

The configuration of the enantiomorphs of other optically active compounds can be specified by relating the configuration of the compound to that of one of the glyceraldehydes. For example, in alanine and lactic acid, the configurations are specified as follows:

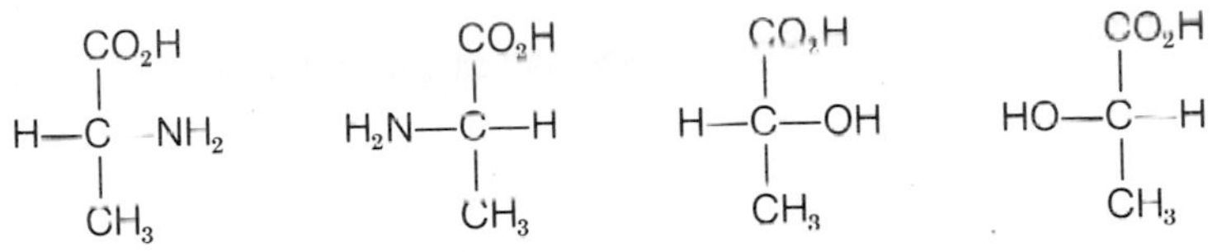

D-Alanine   L-Alanine   D-Lactic acid   L-Lactic acid

It is well to remember that the letters D- and L- designate a configuration related to glyceraldehyde. They do not specify the direction of rotation of polarized light. The convention is to write the formula as a vertical carbon chain with the most oxidized functional group at the top.

To indicate the direction of rotation of light by a given enantiomorph, the signs + for dextrorotatory and − for levorotatory are used. Thus, the nomenclature for optically active compounds often includes D or L to specify configuration and (+) or (−) to indicate the direction of rotation. Examples of this are: D-(+)-glyceraldehyde, L-(+)-lactic acid, and D-(−)-alanine.

Racemic Mixtures   If lactic acid is synthesized in the laboratory, the product has no optical activity; if lactic acid is taken from muscle tissue, dextrorotatory lactic acid is obtained. (Muscle lactic acid is produced under the influence of specific enzymes present in muscles.) The explanation for these differences is that the laboratory preparations give a *mixture containing equal parts of each enantiomorph*. Such a mixture is called a *racemic mixture* or a *racemate*, and is optically inactive since the dextrorotatory effect of the one isomer is exactly balanced by the levorotatory effect of the other. The separation of enantiomorphs from a racemic mixture is known as *resolution*.

How Organic Chemistry Helps Us All.   Early in this chapter, the point was made that organic chemistry has developed to the place where it now is possible in many cases to synthesize compounds having the properties desired for a specific use. Aspirin, for example, first was prepared because its parent drug, sodium salicylate, long known as an excellent antipyretic (fever depressant), caused nausea. Preparation of sodium *acetyl*salicylate, the active component in aspirin, was an attempt to avoid this difficulty by preparing a compound that would pass through the stomach rapidly.

Such a compound would have to be more soluble in the stomach acids than is the parent. Addition of the acetyl group was an inexpensive way of incorporating this feature. However, in making this choice, the chemist had to make use of his understanding of the relation between chemical structure and chemical and physical properties. In this case, for example, he had to know that changing the chemical structure by adding the acetyl group would bring about the desired change in properties—that is, increased solubility.

Learning and capitalizing on the relation between structure and properties is the stock-in-trade of the industrial chemist. Following are a few examples of how this has been used to produce interesting and valuable items of use and importance to us all.

Dyes.   Knowledge of the relation between structure and color has led to the manufacture of inexpensive synthetic dyes that brighten our lives in thousands of uses, including not only in fibers but in paper, soap, detergents, and cosmetics. Dye molecules contain structures having double bonds such as $-N{=}O$, $-N{=}N-$, $-C{=}O$, $-C{=}S$, and $={\bigcirc}=$, which are known as chromophores (color bearers). However, the color of these groups is usually not intense enough, so additional electron-donor groups are built into the molecule. These groups, known as auxochromes (color enhancers), also have much to do with the color of the dye. The following formulas illustrate the change in color with the number and type of auxochrome present in azo dyes:

Benzeneazophenol (yellow)

Benzeneazoresorcinol (orange)

Paranitroaniline Red

Azo dyes and their derivatives are used in cotton, wool, and polyester fibers, in ballpoint pen inks, wood stains, cosmetics, leather, and paper. In fibers, they launder well and are lightfast (stable to light).

New dyes with better fastness and more appealing colors are being developed constantly, and this will continue because the relation between chemical structure and color, and between structure and other important properties such as fixing the dye to the fiber, are well understood.

Drugs. In perhaps no other area of chemistry is precise understanding of the relation between structure and properties more critical than in medicinals. Barbiturates provide an excellent illustration of both the extent and limitation of our knowledge, and our ability to apply it in this vital field. Despite considerable understanding of the relation between structure and major physiological activity of these compounds, we still cannot predict and do not fully understand many of their side effects.

Barbiturates are one of a class of drugs that depress the activity of the central nervous system. They long have been regarded as among the most useful synthetic medicinal substances. Widely used for treatment of insomnia, hysteria, epilepsy, and as preliminary medication before surgical anesthesia, they can cause death if used indiscriminately.

The general formula for the barbiturates is

Potency of these compounds can be altered by varying the hydrocarbon substituents, designated R_1, R_2, R_3, R_4. Duration of their effect can be controlled by placing either an oxygen or sulfur atom at X.

For use as sedatives or as general or intravenous anesthetics, best results are obtained when R_4 is methyl; R_3 is hydrogen; and R_1 and R_2 are alkyl, alkenyl, or cycloalkenyl groups, one of which contains four to six carbon atoms. Intravenous anesthetics must be of short duration. Therefore, a sulfur atom is used at X, or a branched chain group is used at R_1 or R_2.

The presence of one or two phenyl groups at R_1 and R_2 yields structures that are among the most important anticonvulsants. However, the phenyl-substituted compounds have side effects that range from trivial to very serious. Much work has gone into a search for equally potent but safer drugs. While much progress has been made, the phenyl barbiturates remain valuable but potentially very dangerous drugs.

In this chapter, we have sought to classify many of the more than 2 million known organic compounds, first into broad groups known as the aliphatic and aromatic divisions, and then into smaller families or series of compounds called homologous series. The compounds in each series are shown to possess certain similarities in properties and structure. Some important structural features and properties have been summarized for 10 homologous series, and a system of nomenclature has been developed for members of these series. Several types of isomerism have been introduced and illustrated.

IMPORTANT TERMS

Aliphatic	**Alkynes**	**Nitroalkyls**
Aromatic	**Alkanols**	**Esters**
Homologous series	**Alkanals**	**Ethers**
Alkanes	**Alkanones**	**Sulfonic acids**
Alkenes	**Alkanoic acids**	**Thiols**
	Alkanedioic acids	**Diols**
Methyl		
Ethyl	**Phenyl**	**Cycloalkyl**
Propyl	**Ethenyl**	**Acyl**
tert-**Butyl**	**Ethynyl**	
cis and trans isomers	**Methoxy**	
o, m, and *p* **isomers**		
Optical isomerism	**Amines**	
dissymmetric molecules	**Alkyl halides**	
nonsuperimposable mirror images	**Aryl halides**	
optical activity	**Phenols**	
rotation of polarized light	**Nitroaryls**	
dextrorotatory		
levorotatory	**Trifluoromethyl**	
configuration	**Amino**	
projection formula	**Mercapto**	
enantiomorphs	**Halo**	
asymmetric carbon atoms		
racemic mixtures		

QUESTIONS AND PROBLEMS

1. Name each of the following:

(a) 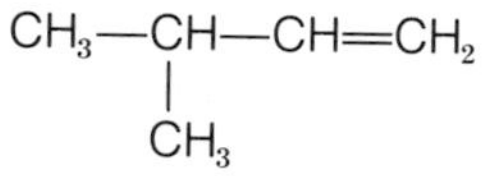 CH_3—CH—CH=CH_2
       |
       CH_3

(b) 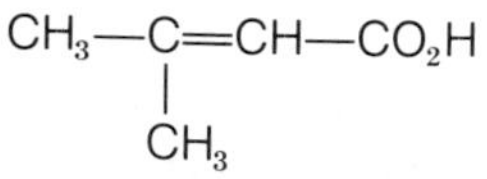 CH_3—C=CH—CO_2H
       |
       CH_3

(c) CH_3—CH—CH_2—CHO
       |
       OH

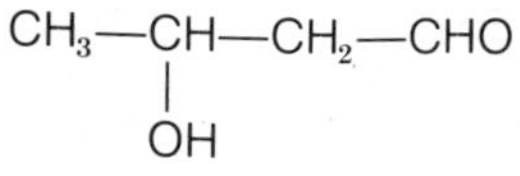

(d) 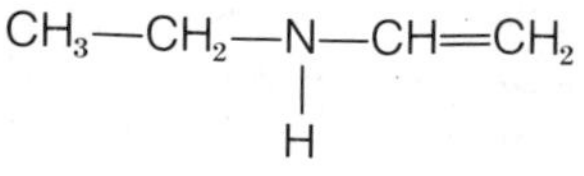 CH_3—CH_2—N—CH=CH_2
            |
            H

(e) 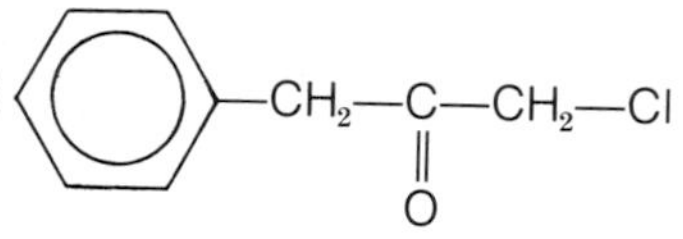 ⟨benzene ring⟩—CH_2—C—CH_2—Cl
                              ‖
                              O

(f) 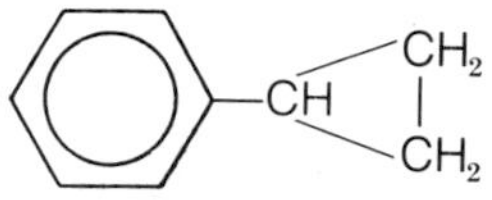 ⟨benzene ring⟩—CH⟨ CH_2 / CH_2 ⟩

2. Name each of the following:

(a) $CH_3—CH—CH—CH—CF_3$
 with substituents CH_3, Br, $CH_2—CH_3$

(b) $CH_2=C—CH_2—CH—CH_3$
 with CH_3 and a phenyl ring

(c) (benzene ring)$—CH—CH_2—CO_2H$ with $CH=CH_2$

(d) $CH_3—C=CH—CH_3$ with C double bonded to CH_2 and CH_3

(e) $CH_3—CH——CH—CHO$ with two benzene rings

(f) (benzene ring with CH_2CH_3 and $—CH_2CH_3$)

3. Name each of the following:

(a) (benzene ring)$—C≡C—$(benzene ring)

(b) $CH_2=CH—CH=CH$ with benzene ring

(c) $HO—$(benzene ring)$—NO_2$

(d) $CH_3O—$(benzene ring)$—CH_2OH$

(e) (benzene ring)$—CH=CH_2$

(f) (benzene ring with $—CHO$ and $—CH_3$)

4. Write a structural formula for one compound in each of the following series: (a) a cycloalkane, (b) a cycloalkene, (c) an amine, (d) an aromatic ketone, (e) an aromatic carboxylic acid, and (f) an alkyl dihalide.

5. Write a structural formula for one compound in each of the following series: (a) a diene, (b) a triol, (c) an ether, (d) an ester, (e) an amine, (f) an aryl bromide, (g) an aromatic ketone, (h) an acetylene, (i) an acid anhydride, (j) an acid halide, (k) a dicarboxylic acid.

6. Write a structural formula for one compound in each of the following series: (a) an ester of butyric acid, (b) a tertiary alcohol, (c) a *cis*-alkene, (d) a *meta* disubstituted aromatic, (e) a cyclic anhydride, and (f) an alkyl phosphate.

7. Write structural formulas for: (a) chloroethanal, (b) ethyl propionate, (c) ethynylcyclopentane, (d) *p*-dimethoxybenzene, (e) 2-mercapto-3-isopropyloctane, and (f) trifluoroacetic acid.

8. Write structural formulas for each of the following: (a) *o*-nitrophenol, (b) 2,4,6-tribromotoluene, (c) *cis*-2-pentene, (d) trichloroacetic acid, (e)

ethylene glycol, (f) *tert*-butyl alcohol, (g) 2-aminopropanoic acid, (h) 3-ethenyl-1,3-pentadiene.

9. Write structural formulas for: (a) 2,3,3-trimethyloctane, (b) phenyldimethylamine, (c) phenyl benzoate, (d) butandioic acid, (e) butendioic acid, and (f) vinylacetylene.

10. Offer an explanation for the fact (a) that the boiling points of the isomeric pentanes decrease in the order

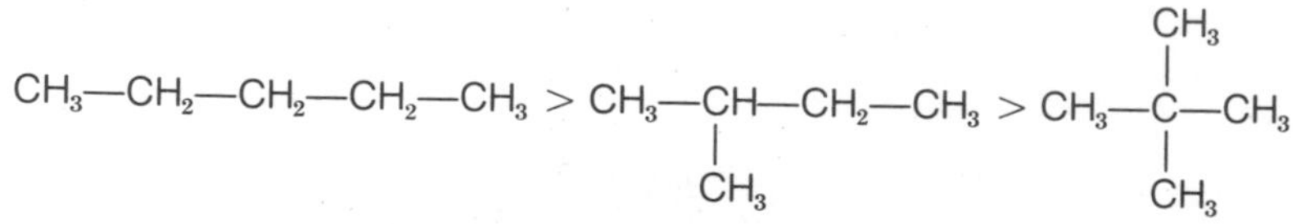

$$CH_3-CH_2-CH_2-CH_2-CH_3 > CH_3-\underset{\underset{CH_3}{|}}{CH}-CH_2-CH_3 > CH_3-\underset{\underset{CH_3}{|}}{\overset{\overset{CH_3}{|}}{C}}-CH_3$$

and (b) that, in general, more highly branched isomers have lower boiling points than their straight-chain analogs.

11. Arrange the following alkanes in order of decreasing boiling point. Write structural formulas for each. (a) 2,2-dimethylpentane, (b) 2-methylbutane, (c) heptane, (d) pentane, and (e) 2,2,3-trimethylbutane.

12. An alkane has a composition of 82.65% carbon and 17.33% hydrogen. What is its empirical formula and its probable molecular formula?

13. Offer an explanation for the observation that the boiling points of alkyl halides generally increase in the order RF < RCl < RBr < RI more or less independently of the particular R group. (Table 26.9.)

14. Give examples of reactions in which alcohols act (a) as acids, (b) as bases.

15. Account for the variation in boiling point among the following two-carbon molecules.

Substance	Boiling Point (°C)
Ethyl chloride	12.2
Ethyl alcohol	78.5
Ethylamine	16.6
Acetic acid	118.1

16. Why should acetic and propionic acids be soluble in water? Would you expect butyric acid to be more or less soluble than acetic acid? Why?

17. Describe the shapes of each of the following: (a) C_2H_2, (b) $C_2H_2Cl_2$, (c) butadiene, (d) acetic acid, (e) acetone, and (f) an aldehyde group.

18. It is said that amines are stronger bases than ammonia. Offer an explanation for (a) the fact that amines are basic, and (b) the fact that, in general, they are stronger bases than ammonia.

19. What common structural feature accounts for both the insolubility of alkanes in water and their relatively low boiling points?

20. Define and give an example of each of the following: (a) an aromatic ether, (b) an aromatic amine, (c) a cycloalkadiene, (d) an alkenone, (e) an alkadienoic acid, and (f) a haloalkanal.

21. Which of the following compounds contain asymmetric carbon atoms? Draw configurational formulas for these compounds.

(a) $CH_3-CH_2-\underset{\underset{CH_3}{|}}{CH}-CH_3$

(b) $CH_3-CH_2-\underset{\underset{CH_3}{|}}{CH}-CH_2-CH_3$

(c) $CH_3-CH_2-\underset{\underset{CH_3}{|}}{CH}-CH_2-CH_2-CH_3$

(d) $CH_3-CHOH-COOH$

22. Define each of the following terms: (a) dissymmetric molecule, (b) asymmetric carbon atom, (c) enantiomorph, (d) dextrorotatory, and (e) racemization.

23. Draw configurational formulas for each of the following:

(a) The L-configuration of

$$CH_3-\underset{\underset{NH_3}{|}}{CH}-CO_2H$$

(b) the L-configuration of

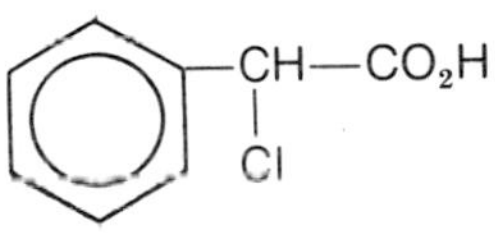

(c) all possible configurations of

$$CH_3-\underset{\underset{Cl}{|}}{CH}-CO_2-CH_3$$

24. Write configurational names for each of the following:

(a)

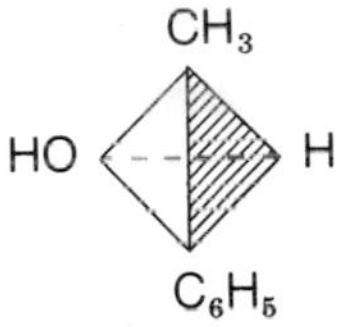

(b)

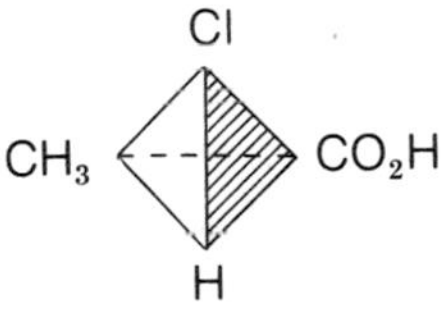

(c) 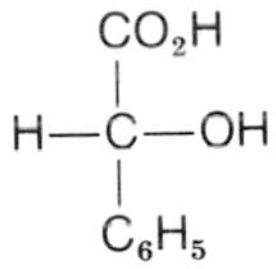 $H-\underset{\underset{C_6H_5}{|}}{\overset{\overset{CO_2H}{|}}{C}}-OH$

(d) 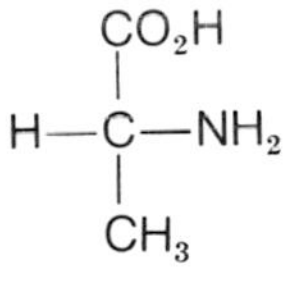 $H-\underset{\underset{CH_3}{|}}{\overset{\overset{CO_2H}{|}}{C}}-NH_2$

SPECIAL PROBLEMS

1. Write structural formulas for: (a) all isomers of C_2H_6O, (b) all isomers of $C_3H_6Cl_2$, (c) all alcohols having the formula C_4H_9OH, (d) all alcohols having the formula $C_4H_8(OH)_2$, (e) all amines having the formula $C_5H_{13}N$, and (f) all possible dichlorobutynes.

2. Write structural formulas for: (a) all isomers of heptane, (b) all alcohols having the formula $C_5H_{11}OH$, (c) all isomers of trichlorobenzene, (d) all possible dichloro-2-butenes.

3. Write structural formulas for: (a) all isomers of C_5H_{10}, (b) all isomers of $C_5H_7Cl_3$, (c) all isomers of methylpropylbenzene, (d) all carboxylic acids having the formula $C_4H_7CO_2H$, (e) all isomers of phenylethanol, and (f) all ketones having the formula C_5H_8O.

4. On analysis a certain secondary alcohol was found to contain 60.0% carbon, 13.3% hydrogen, and 26.6% oxygen. Write a structural formula for this alcohol.

5. Using the bond energies given in Chapter 18, determine the sign of ΔH for each of the following conversions: (a) $CH_4 + Cl_2 \longrightarrow CH_3Cl + HCl$ (b) $CH_3Cl + Cl_2 \longrightarrow CH_2Cl_2 + HCl$ (c) $CH_2Cl_2 + Cl_2 \longrightarrow CHCl_3 + HCl$ (d) $CHCl_3 + Cl_2 \longrightarrow CCl_4 + HCl$

6. Estimate the length of a propane and a benzene molecule.

7. Write structural formulas and names of the isomers of pentyne.

8. Is it possible to distinguish among 1-butanol, 2-butanol, and 2-methyl-2-propanol on the basis of boiling points? Why or why not?

9. A mixture is composed of ethane, ethanol, and acetic acid. How could you separate these by physical methods?

10. From the data of Chapter 18, calculate the heat of combustion at 25°C for each of the following: (a) C_2H_4, (b) C_3H_8, (c) benzene, and (d) n-pentane.

REFERENCES

HENDRICKSON, J. B., CRAM, D. J., and HAMMOND, G. S., *Organic Chemistry*, 3rd ed., McGraw-Hill, New York, 1970.

MORRISON, R. T., and BOYD, R. N., *Organic Chemistry*, 2nd ed., Allyn and Bacon, Boston, 1966.

VAN ORDEN, H. A., and LEE, G. L., *Elementary Organic Chemistry, A Brief Course*, Saunders, Philadelphia, 1969.

OUELLETTE, R. W., *Introductory Organic Chemistry*, Harper & Row, New York, 1970.

27

ORGANIC CHEMISTRY II: REACTIVITY OF SOME ORGANIC STRUCTURES

The preparation of the organomagnesium ethers is, in general, extremely simple. The apparatus is just a round-bottomed flask connected with a good ascending condenser and with a dropping funnel with stopcock, but it is indispensable that everything be absolutely dry.

Victor Grignard (1871–1935)

The basis for the considerable contributions organic chemistry has made to our health and welfare is twofold. The first is an understanding of the physical and chemical properties of organic substances that makes it possible to explain, predict, and, in some cases, control the behavior of systems containing these substances. The second is the very highly developed skill of the organic chemist in synthesizing complex and uniquely structured compounds in high purity and often in large quantities from plentiful and readily available starting materials. This chapter illustrates, in an elementary way, how this skill in synthesis begins.

Each functional group in organic compounds is able to undergo several chemical reactions more or less unique to that group. Nearly all members of the homologous series containing a given functional group will undergo the reactions associated with that group. For example, all alkanes react with chlorine under the proper conditions to give alkyl chlorides, and nearly all aklyl chlorides react in the presence of aqueous sodium hydroxide to give alcohols. The existence of reactions associated with specific functional groups makes it possible to plan a sequence of chemical reactions that will result in the synthesis of a desired chemical structure or substance.

In planning the synthesis of a particular structure, the chemist usually attempts first to construct, through chemical reactions, the desired carbon skeleton and, having done this, to locate the various functional groups at the appropriate points on the skeleton.

Constructing the skeleton and locating the functional groups requires a familiarity with a large number of chemical reactions. At least five classes of reactions useful in organic synthesis are recognized. We shall look at examples of these and attempt to use some of the examples in planning the synthesis of several structures.

CLASSES OF
ORGANIC REACTIONS

1. Oxidation and Reduction Reactions. Oxidation reactions in organic chemistry are *those in which carbon-hydrogen bonds are broken and carbon-oxygen bonds are formed.* An illustration of the scope of oxidation reactions is given by the following sequence, in which an alkane is oxidized first to an alcohol,

then to an aldehyde, then to a carboxylic acid, and finally to carbon dioxide and water.

$$CH_4 \xrightarrow{[O]} CH_3OH \xrightarrow{[O]} HCHO \xrightarrow{[O]} HCO_2H \xrightarrow{[O]} CO_2 + H_2O$$

Alkane   Alcohol   Aldehyde   Carboxylic acid

Reduction reactions *are those in which hydrogen atoms are added to an organic compound*; this is sometimes accompanied by removal of oxygen atoms or atoms of other nonmetals.

Reductions may be illustrated by the equations

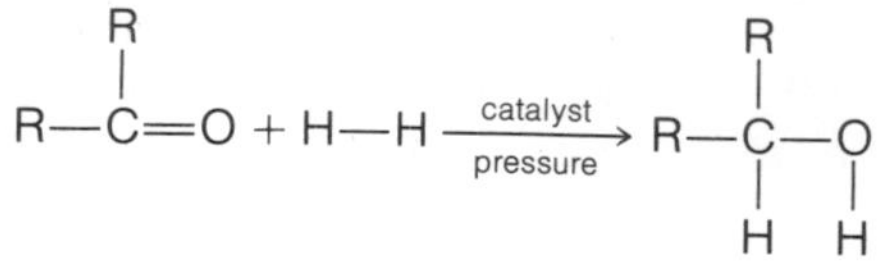

$$\underset{\displaystyle \overset{|}{R}}{R-C}=O + H-H \xrightarrow[\text{pressure}]{\text{catalyst}} R-\underset{\displaystyle \overset{|}{H}}{\overset{\displaystyle \overset{R}{|}}{C}}-\underset{\displaystyle \overset{|}{H}}{O}$$

$$R-CH_2-X + \text{reducing agent} \longrightarrow R-CH_3 + \text{other products}$$
(X is a nonmetal)

2. **Condensation Reactions.** These are *reactions in which two molecules join together* (condense), often *with the elimination of a small molecule* such as water, ammonia, or hydrogen chloride, as illustrated below:

$$X-OH + Y-OH \longrightarrow X-O-Y + H_2O$$

3. **Addition Reactions.** In these reactions a *small molecule*, or molecules, *adds* or joins *to another molecule*. Such reactions occur at double or triple bonds and may be represented by the generalized equation

$$A-A + X=Y \longrightarrow \underset{\displaystyle \overset{|}{A} \quad \overset{|}{A}}{X-Y}$$

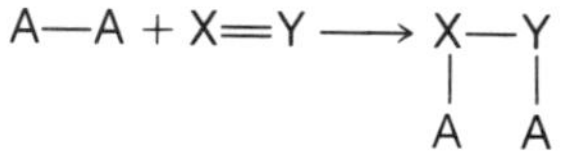

4. **Elimination Reactions.** In these reactions, *a small molecule* such as water or hydrogen chloride *is removed from an organic compound.* Usually the portions of the small molecule are removed from adjacent carbon atoms in the organic compound, as illustrated by the equation

$$R-\underset{\displaystyle \overset{|}{X}}{CH}-\underset{\displaystyle \overset{|}{Y}}{CH}-R' \longrightarrow RCH=CHR' + X-Y$$

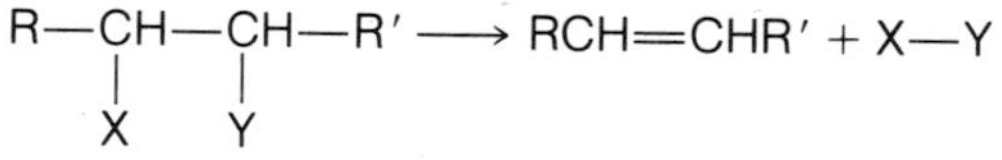

Eliminations usually result in the formation of double or triple bonds.

5. **Substitution Reactions.** In these reactions, *one atom or group displaces another atom or group*, as illustrated by the generalized equation

$$A + R-X \longrightarrow A-R + X$$

The entering and leaving groups include monovalent groups such as hydrogen, halogen, hydroxyl, amino, and many others.

OXIDATION AND REDUCTION REACTIONS

Oxidation and reduction reactions are among the most important and useful reactions of organic compounds. Because of the many possibilities for oxidation or reduction in the variety of organic structures known, these reactions are probably the most diverse in scope, conditions, and required reagents of any class of organic reactions.

Oxidations

The types of reactions involved in the oxidation of a saturated carbon atom appear to involve:

1. Replacement of —H by —OH,

$$R\text{—}H + [O]^* \longrightarrow R\text{—}OH$$

2. Successive oxidation on the carbon atom that is already partially oxidized,

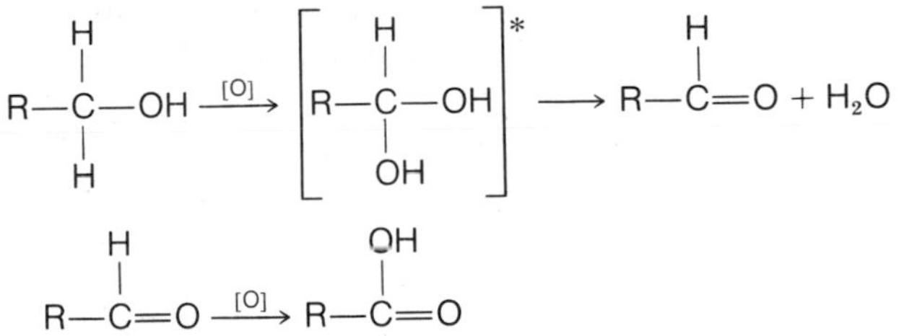

3. Complete oxidation to carbon dioxide and water.

Using these possible reactions, we discuss some typical oxidation reactions.

Oxidation of Alkanes. Alkanes burn in air or oxygen with the formation of carbon dioxide and water, and with liberation of large amounts of heat. As a result, hydrocarbons are among our most important fuels for heat and power.

$$CH_4(g) + 2O_2(g) \longrightarrow CO_2(g) + 2H_2O(l) \qquad \Delta H° = -210{,}800 \text{ cal}$$

$$C_7H_{16}(l) + 11O_2(g) \longrightarrow 7CO_2(g) + 8H_2O(l) \qquad \Delta H° = -1{,}146{,}000 \text{ cal}$$

In the presence of insufficient oxygen, carbon monoxide and carbon may form, resulting in less efficient heat or power production and in a safety hazard.

Oxidation of Alcohols. Primary alcohols are readily oxidized to aldehydes and to carboxylic acids; secondary alcohols are readily oxidized to ketones; tertiary alcohols are difficult to oxidize and give a variety of oxidation products. All of this is in accord with the types of oxidation reaction given above.

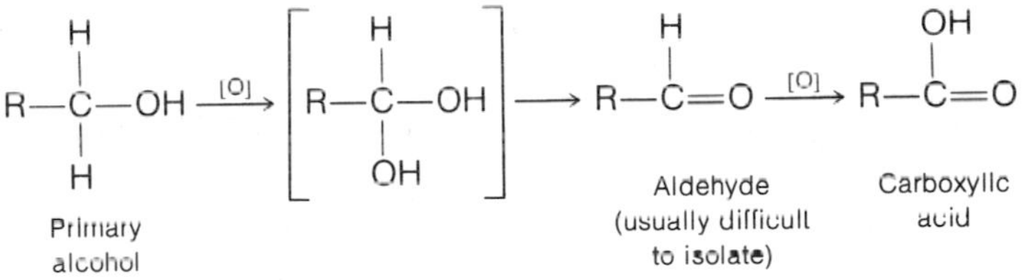

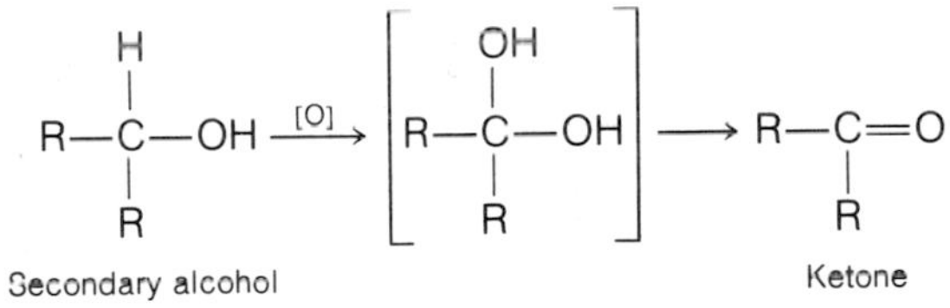

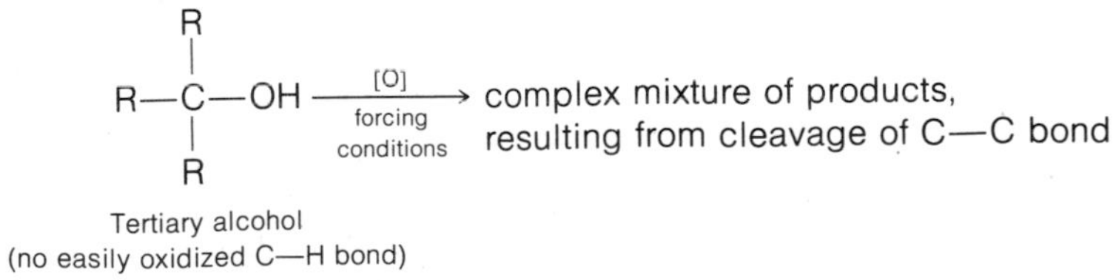

---

*Brackets in these cases represent an unstable or, sometimes, an assumed intermediate species.

597 ORGANIC CHEMISTRY II

Oxidation of alcohols is used as a method of preparing a few aldehydes and a number of ketones, and as a diagnostic test to ascertain if a given alcohol is primary, secondary, or tertiary. Oxidizing agents commonly used to oxidize alcohols are acidified solutions of potassium dichromate, $K_2Cr_2O_7$, or potassium permanganate, $KMnO_4$. Hot copper oxidizes (dehydrogenates) low molecular weight alcohols in the vapor state to aldehydes or ketones.

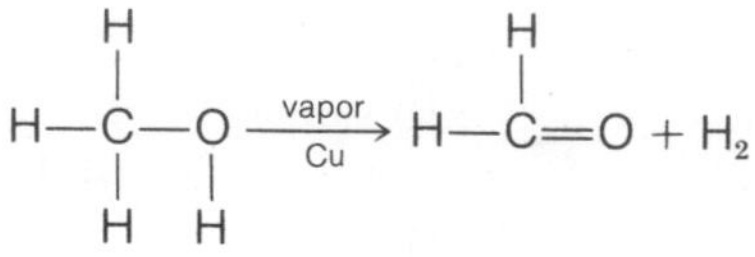

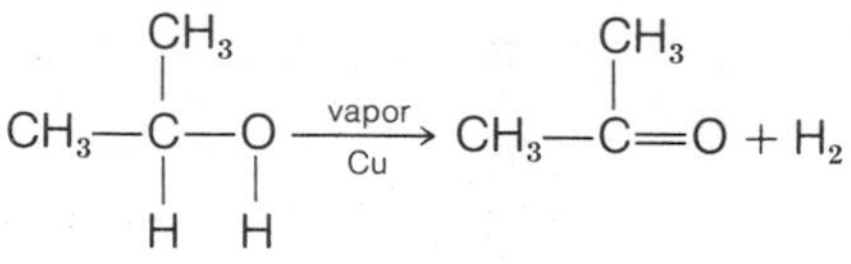

$$H-\underset{\underset{H}{|}}{\overset{\overset{H}{|}}{C}}-\underset{\underset{H}{|}}{O} \xrightarrow[\text{Cu}]{\text{vapor}} H-\overset{\overset{H}{|}}{C}=O + H_2$$

$$CH_3-\underset{\underset{H}{|}}{\overset{\overset{CH_3}{|}}{C}}-\underset{\underset{H}{|}}{O} \xrightarrow[\text{Cu}]{\text{vapor}} CH_3-\overset{\overset{CH_3}{|}}{C}=O + H_2$$

Oxidation of Aldehydes and Ketones. As indicated in Chapter 26, aldehydes are very readily oxidized to carboxylic acids, and ketones are oxidized only with difficulty. The reason for this is the presence of the aldehydic hydrogen

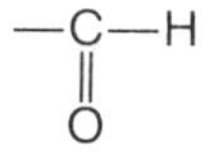

$$-\underset{\underset{O}{\|}}{C}-H$$

in aldehydes and its absence in ketones.

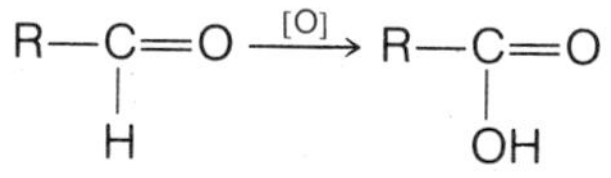

$$R-\underset{\underset{H}{|}}{C}=O \xrightarrow{[O]} R-\underset{\underset{OH}{|}}{C}=O$$

Aldehyde   Carboxylic acid

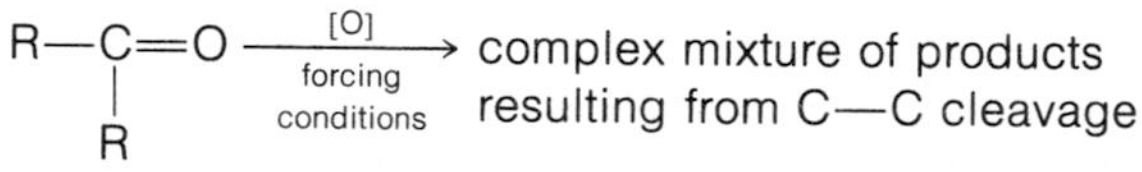

$$R-\underset{\underset{R}{|}}{C}=O \xrightarrow[\substack{\text{forcing}\\\text{conditions}}]{[O]} \text{complex mixture of products resulting from C—C cleavage}$$

Ketone

Oxidation of Alkenes. The π bond often is much more susceptible to oxidation than the σ bond. Consequently, oxidizing agents such as potassium permanganate in solution readily destroy the double bond to give glycols:

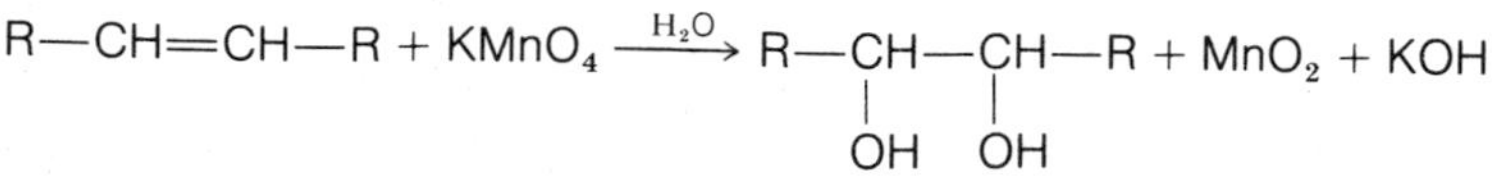

$$R-CH=CH-R + KMnO_4 \xrightarrow{H_2O} R-\underset{\underset{OH}{|}}{CH}-\underset{\underset{OH}{|}}{CH}-R + MnO_2 + KOH$$

A glycol

This reaction may be used to distinguish alkanes from alkenes. Alkanes do not react with potassium permanganate, while alkenes reduce the purple permanganate ion to brown manganese dioxide.

Further oxidation of the glycol brings about cleavage of the carbon-carbon bond between the two hydroxyl groups:

$$\underset{\underset{OH}{|}}{R-CH}-\underset{\underset{OH}{|}}{CH-R} \xrightarrow[-H_2O]{[O]} 2R-\underset{\underset{O}{\|}}{C}-OH$$

The two fragments are isolated as the corresponding acids, as indicated. Oxidation of an alkene is useful for locating the position of the double bond in a molecule. For example, oxidation of 2-pentene, $CH_3-CH=CH-CH_2-CH_3$, with potassium permanganate gives both acetic and propanoic acids, showing that the double bond was between the second and third carbon atoms.

Oxidation of Side Chains on Aromatic Compounds. Potassium permanganate or nitric acid will oxidize alkyl side chains on aromatic rings to give carboxylic acids in which the carboxyl group is bonded to a carbon atom in the ring.

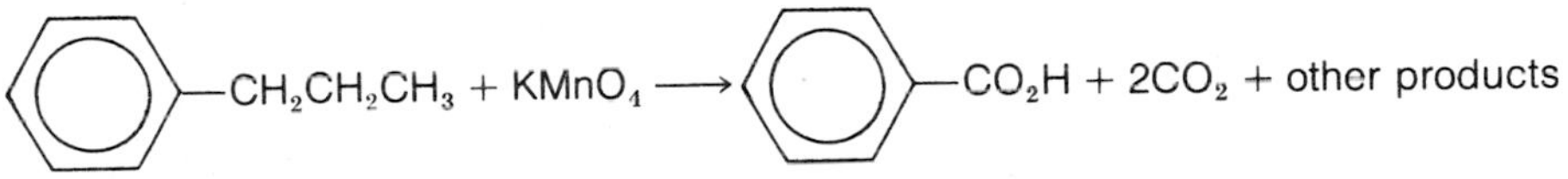

n-Propylbenzene   Benzoic acid

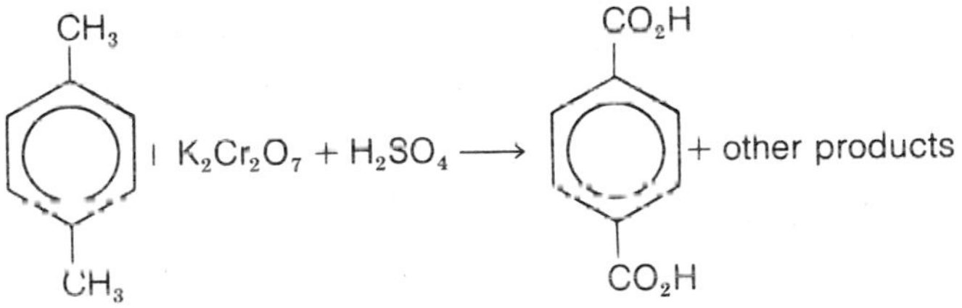

Phthalic acid

Reductions   Nearly all common inorganic reducing agents have been used in organic chemistry. However, the most useful reagents are hydrogen with solid catalysts, metal hydrides such as lithium aluminum hydride ($LiAlH_4$), and active metals such as sodium and zinc.

Reduction of Carbon-Carbon Multiple Bonds. Catalytic hydrogenation is the most generally applicable procedure for reducing alkenes, alkynes, or aromatic compounds. The commonly used catalysts are the finely divided metals nickel, palladium, or platinum, so prepared that the catalyst will present a large surface area. Hydrogen, at low pressures (25 to 40 lb/sq in.) or at high pressures (up to 5000 lb/sq in.), is used. Low-pressure hydrogenations are usually carried out at room temperature, while temperatures of 100°C or 200°C are common for high-pressure hydrogenations. Examples of reductions and their conditions for several types of unsaturated compounds are given below.

Alkenes:

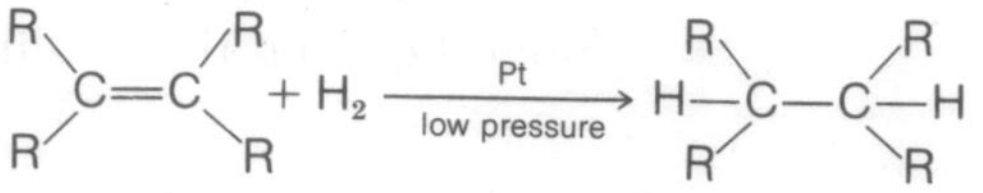

$$R_2C=CR_2 + H_2 \xrightarrow[\text{low pressure}]{\text{Pt}} \text{H—C—C—H}$$

Alkynes:

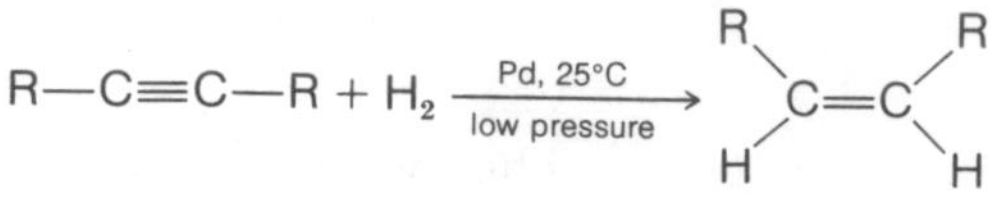

$$R—C\equiv C—R + H_2 \xrightarrow[\text{low pressure}]{\text{Pd, 25°C}}$$

Cis product

Aromatics:

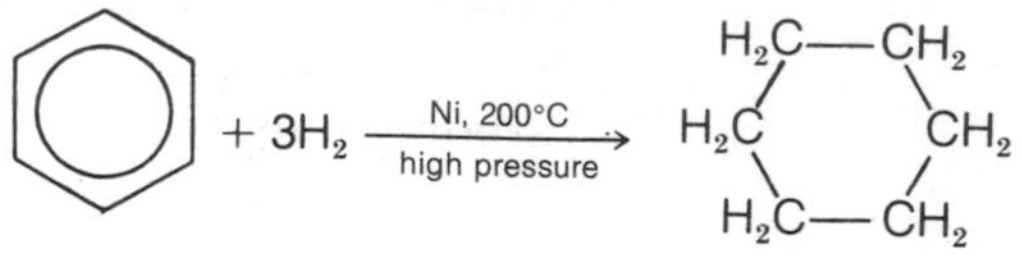

$$\bigcirc + 3H_2 \xrightarrow[\text{high pressure}]{\text{Ni, 200°C}}$$

Cyclohexane

$$+ 2H_2 \xrightarrow[\text{high pressure}]{\text{Ni, 100°C}}$$

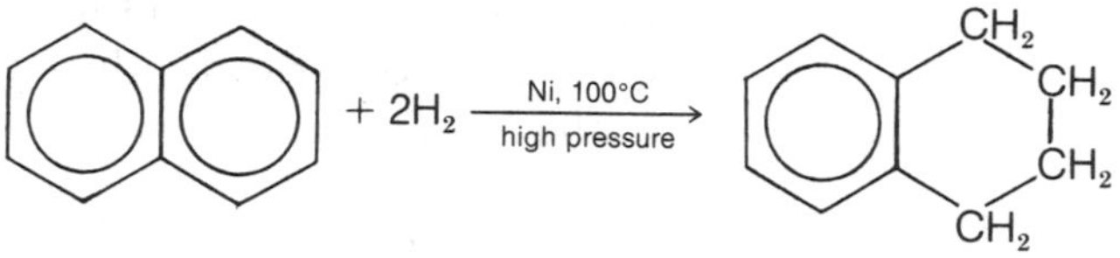

Tetralin (tetrahydronaphthalene)

Reduction of Carbon-Oxygen and Carbon-Nitrogen Multiple Bonds. These may be reduced by catalytic hydrogenation, but the reductions often are accomplished more easily with metal hydrides or active metals. Illustrations of the use of metal hydrides include:

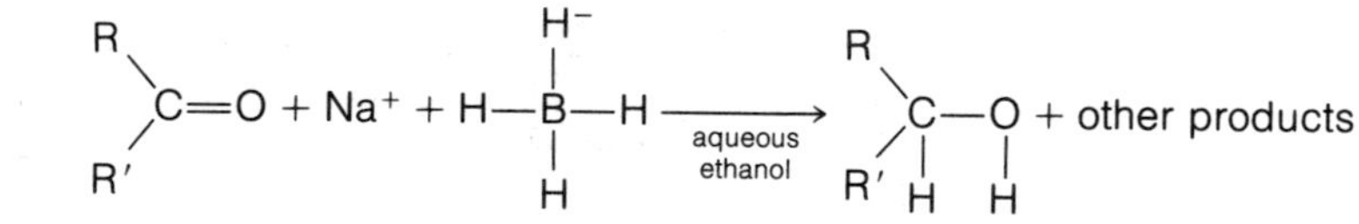

$$\begin{matrix}R\\R'\end{matrix}C=O + Na^+ + \text{H—B—H} \xrightarrow[\text{ethanol}]{\text{aqueous}} \begin{matrix}R\\R'\end{matrix}C\text{—O} + \text{other products}$$

Sodium borohydride

$$\begin{matrix}R\\(R)HO\end{matrix}C=O + Li^+ + \text{H—Al—H} \xrightarrow{\text{dry ether}} R—CH_2OH + \text{other products}$$

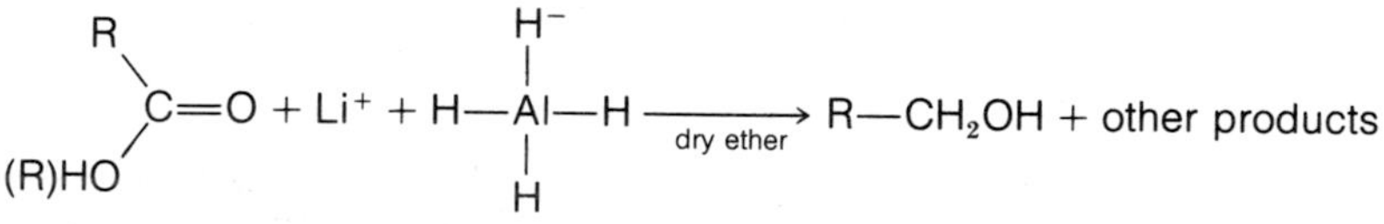

Lithium aluminum hydride

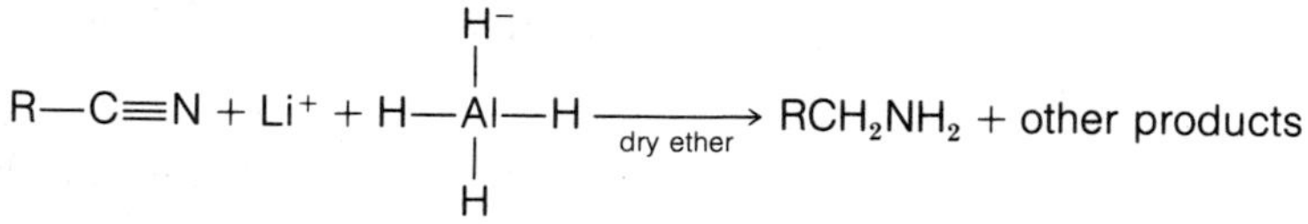

$$R—C\equiv N + Li^+ + \text{H—Al—H} \xrightarrow{\text{dry ether}} RCH_2NH_2 + \text{other products}$$

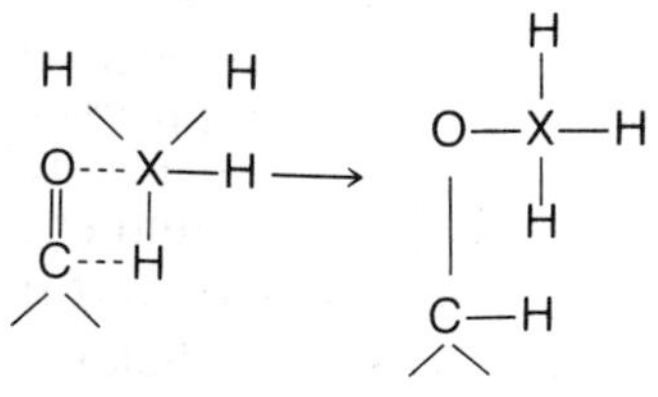

The first step in each of these reactions probably involves a transfer of the type where X is boron or aluminum. This may be followed by reaction of the intermediate with water, illustrated as follows:

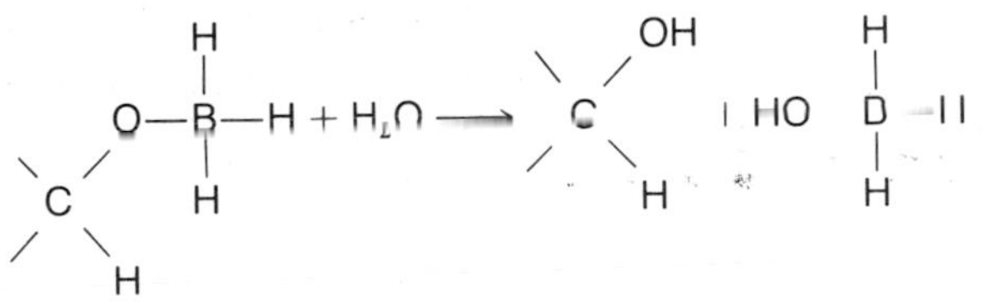

$$\begin{array}{c} H \\ | \\ O-B-H \\ / \quad | \\ C \quad H \\ / \backslash \\ H \end{array} + H_2O \longrightarrow \begin{array}{c} OH \\ \backslash \; | \\ C \\ / \; \backslash \\ H \end{array} + HO \begin{array}{c} H \\ | \\ B-H \\ | \\ H \end{array}$$

Metal hydrides will reduce most aldehydes, ketones, carboxylic acids, and their esters to alcohols. Lithium aluminum hydride may be used to reduce nitriles $(R-C \equiv N)$ and nitro compounds (RNO_2) to amines:

$$RNO_2 + LiAlH_4 \longrightarrow RNH_2 + \text{other products}$$

Hydrogen may be substituted for a halogen by a number of methods, such as treating the halide with zinc in acetic acid, as in the following example:

$$ClCH_2CO_2H \xrightarrow[HO_2CCH_3]{Zn} CH_3CO_2H + HCl$$

CONDENSATION REACTIONS
Condensation Reactions in Which Water Is Evolved

The general equation for condensations of this type is

$$X-OH + HO-Y \longrightarrow X-O-Y + H_2O$$

Reactions of this type are known for all classes of organic compounds containing hydroxyl groups. Some reactions proceed spontaneously; others require elevated temperatures, dehydration catalysts, long reaction times, or any one or combination of these conditions. Examples of condensation reactions of this type include:

1. Alcohols condense to form ethers.

$$ROH + HOR \xrightarrow[240-280°C]{Al_2O_3} ROR + H_2O$$
$$\text{Ether}$$

2. Carboxylic acids condense to form acid anhydrides.

$$R-\underset{\underset{O}{\|}}{C}-OH + HO-\underset{\underset{O}{\|}}{C}-R \xrightarrow[\Delta]{P_4O_{10}} R-\underset{\underset{O}{\|}}{C}-O-\underset{\underset{O}{\|}}{C}-R + H_2O$$
$$\text{Acid anhydride}$$

3. Alcohols condense with acids to give esters.

$$ROH + HO-\underset{\underset{O}{\|}}{C}-R' \longrightarrow R-O-\underset{\underset{O}{\|}}{C}-R' + H_2O$$

$$ROH + HONO_2 \longrightarrow R-O-NO_2 + H_2O$$

$$ROH + HO-\underset{\underset{OH}{|}}{\overset{\overset{O}{\|}}{P}}-OH \longrightarrow RO-\underset{\underset{OH}{|}}{\overset{\overset{O}{\|}}{P}}-OH + H_2O$$

In the reactions of alcohols with carboxylic acids, it is known from oxygen isotope tracer studies that the hydroxyl group of the expelled water molecule comes from the carboxyl group, and the hydrogen atom comes from the alcohol.

By preparing the carboxylic acid with a labeled oxygen atom, it was possible to show that the labeled atom was present in the water formed as a reaction product.

$$\text{R—C—O}^{18}\text{—H} + \text{HOR}' \longrightarrow \text{R—C—OR}' + \text{H}_2\text{O}^{18}$$
$$\quad\;\; \underset{\text{O}}{\|} \qquad\qquad\qquad\quad\; \underset{\text{O}}{\|}$$

Condensation Reactions in Which Hydrogen Chloride Is Evolved

Many reactions of this type are known. Usually, the chlorine atom of the liberated molecule comes from a compound having a reactive chlorine atom, such as the acyl or aroyl halides,

$$\text{R—C—Cl} \quad \text{or} \quad \text{Ar—C—Cl}$$
$$\quad\underset{\text{O}}{\|} \qquad\qquad\qquad \underset{\text{O}}{\|}$$

The hydrogen atom comes from a compound having reactive hydrogen atoms, usually those containing —OH or —NH— groups. Sometimes, a base is added to the reaction mixture to neutralize the hydrogen chloride as it is formed and to prevent reversal of the reaction. Examples of condensation reactions where hydrogen chloride is evolved are:

1. Alcohols and acyl halides give esters.

$$\text{R—OH} + \text{Cl—C—R}' \longrightarrow \text{R—O—C—R}' + \text{HCl}$$
$$\qquad\qquad\quad \underset{\text{O}}{\|} \qquad\qquad\qquad \underset{\text{O}}{\|}$$

2. Ammonia or amines and acyl halides give amides.

$$\underset{\text{H}}{\overset{\text{H}}{\underset{|}{\overset{|}{\text{N}}}}}\text{—H} + \text{Cl—C—R} \longrightarrow \underset{\text{H}}{\overset{\text{H}}{\underset{|}{\overset{|}{\text{N}}}}}\text{—C—R} + \text{HCl}$$

Amide

3. Alkyl halides react with ammonia or amines to give amines.

$$\underset{\text{H}}{\overset{|}{\text{H—N}}}\text{—H} + \text{Cl—R} \xrightarrow[\Delta]{\text{base}} \underset{\text{H}}{\overset{|}{\text{H—N}}}\text{—R} + \text{HCl} \text{ (actually a salt of the base used is formed)}$$

Primary
amine

The product of this reaction may react with additional alkyl halide to produce secondary amines (R_2NH) and tertiary amines (R_3N).

PROBLEM Show by equations how N-ethylacetamide,

$$\text{CH}_3\text{—CH}_2\text{—}\underset{}{\overset{\text{H}}{\underset{|}{\text{N}}}}\text{—C—CH}_3$$
$$\qquad\qquad\qquad\qquad \underset{\text{O}}{\|}$$

can be prepared from ethyl chloride, acetyl chloride, and ammonia, using condensation reactions

Looking at the structure desired, we see it can be made from $CH_3CH_2NH_2$ and $CH_3{-}\overset{\underset{\|}{O}}{C}{-}Cl$ by a condensation reaction such as (II) below. The $CH_3CH_2NH_2$ is not available, and can be synthesized by the condensation reaction (I).

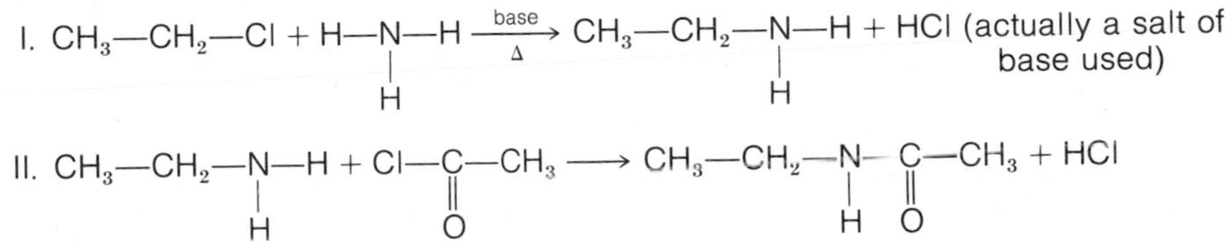

I. $CH_3{-}CH_2{-}Cl + H{-}\underset{\underset{H}{|}}{N}{-}H \xrightarrow[\Delta]{base} CH_3{-}CH_2{-}\underset{\underset{H}{|}}{N}{-}H + HCl$ (actually a salt of base used)

II. $CH_3{-}CH_2{-}\underset{\underset{H}{|}}{N}{-}H + Cl{-}\overset{\underset{\|}{O}}{C}{-}CH_3 \longrightarrow CH_3{-}CH_2{-}\underset{\underset{H}{|}}{N}{-}\overset{\underset{\|}{O}}{C}{-}CH_3 + HCl$

Condensation Reactions in Which a Carbon-Carbon Bond Is Formed

These reactions, which might also be classed as addition reactions, occur with aldehydes, some ketones, and esters that have a hydrogen atom on the carbon atom adjacent to the carbonyl group:

$$\underset{R}{\overset{R}{\diagdown}}\underset{\underset{H}{|}}{C}{-}\overset{\underset{\|}{O}}{C}{-}$$

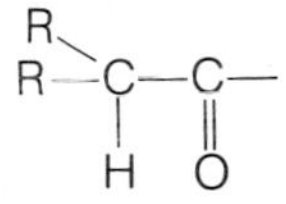

In the presence of base, most compounds having this structural feature undergo self-addition, as illustrated in the generalized reaction:

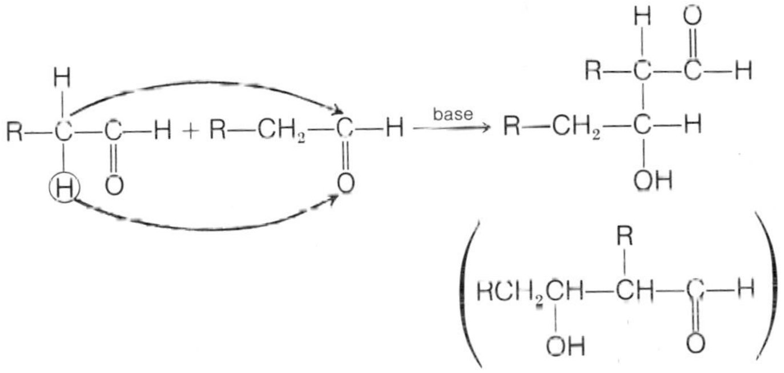

$$R{-}\underset{\underset{(H)}{|}}{\overset{\underset{|}{H}}{C}}{-}\overset{\underset{\|}{O}}{C}{-}H + R{-}CH_2{-}\overset{\underset{\|}{O}}{C}{-}H \xrightarrow{base} R{-}CH_2{-}\underset{\underset{OH}{|}}{C}{-}\overset{\underset{\|}{H\ O}}{\overset{R-\underset{\underset{OH}{|}}{C}-\overset{\underset{\|}{O}}{C}-H}{|}}{C}{-}H$$

$$\left(RCH_2CH{-}\underset{\underset{OH}{|}}{CH}{-}\overset{\underset{\|}{R}}{\underset{\underset{O}{\|}}{C}}{-}H \right)$$

Such reactions, known as *aldol-type condensations*, belong to a very wide class of reactions that can be used to double the chain length and, in some cases, introduce several functional groups into a compound. Sometimes two aldehydes are used, one of them having no hydrogen atom on the carbon atom adjacent to the carbonyl group.

ADDITION REACTIONS

Addition to Alkenes or Alkynes

The characteristic reaction of alkenes and alkynes is addition. Such reactions usually proceed at a very rapid rate, even at low temperatures, whereas most substitution reactions in hydrocarbons proceed slowly, often requiring catalysts and elevated temperatures. Reagents that add to carbon-carbon multiple bonds include: chlorine, bromine, hydrogen; binary acids such as

HCl, HBr, HI; and H_2O, sulfuric acid, and hypochlorous acid. Examples of some addition reactions are:

$$CH_2=CH_2 + H-Br \longrightarrow \underset{\underset{H}{\vert}\quad\underset{Br}{\vert}}{CH_2-CH_2}\ (CH_3-CH_2Br)$$

$$CH\equiv CH + 2Br_2 \longrightarrow \overset{H}{\underset{Br}{}}C=C\overset{H}{\underset{Br}{}} + Br_2 \longrightarrow H-\underset{\underset{Br}{\vert}}{\overset{\overset{Br}{\vert}}{C}}-\underset{\underset{Br}{\vert}}{\overset{\overset{Br}{\vert}}{C}}-H$$

$$CH_3-\underset{\underset{CH_3}{\vert}}{C}=CH_2 + Cl_2 \longrightarrow CH_3-\underset{\underset{CH_3}{\vert}}{\overset{\overset{Cl}{\vert}}{C}}-\underset{}{\overset{\overset{Cl}{\vert}}{C}H_2}$$

$$CH_3-CH=CH_2 + H_2 \xrightarrow{\text{Pt or Ni}} CH_3-\underset{\underset{H}{\vert}}{C}H-\underset{\underset{H}{\vert}}{C}H_2 \quad (CH_3-CH_2-CH_3)$$

The mechanism of addition to carbon-carbon double bonds is believed to involve attack by an electron-seeking species, such as the positive end of the HBr dipole or even a very reactive bromonium ion, Br^+ (formed by a dissociation of the type $:\overset{..}{\underset{..}{Br}}:\overset{..}{\underset{..}{Br}}: \longrightarrow :\overset{..}{\underset{..}{Br}}{}^+ + :\overset{..}{\underset{..}{Br}}:^-$). The electron-seeking reagent attacks (and bonds to) the pair of electrons in the π bond of the alkene, creating a carbonium ion*:

---

*Carbonium ions are reactive intermediates in which one carbon atom is bonded to only three other groups and possesses a positive charge. Examples are:

$$CH_3-\underset{+}{\overset{\overset{CH_3}{\vert}}{C}}-CH_3 \quad (\textit{tert}\text{-butyl carbonium ion}) \qquad H-\underset{\underset{H}{\vert}}{\overset{\overset{H}{\vert}}{C}}-\underset{\underset{H}{\vert}}{\overset{\overset{H}{\vert}}{C}}{}^+ \quad (\text{ethyl carbonium ion})$$

MECHANISM OF AN ALDOL CONDENSATION

The reaction mechanism is believed to involve the following steps:

$$R-\underset{\underset{H}{\vert}}{\overset{\overset{H}{\vert}}{C}}-\underset{\underset{O}{\vert\vert}}{C}-H + OH^- \longrightarrow R-\overset{\overset{H}{\vert}}{\underset{..}{C}}-\underset{\underset{O}{\vert\vert}}{C}-H + H_2O$$

Removal of proton from carbon atom adjacent to C=O, giving reactive anion.

$$R-\underset{\underset{O}{\vert\vert}}{\overset{\overset{H}{\vert}}{C}}-C-H + \underset{\underset{O}{\vert\vert}}{\overset{\overset{CH_2}{\vert}}{C}}\overset{R}{}-H \longrightarrow R-\underset{\underset{H}{\vert}}{\overset{\overset{C=O}{}}{C}}-\underset{\underset{O_-}{\vert}}{\overset{\overset{CH_2}{}}{C}}\overset{R}{}-H$$

Anion attacks C=O in a second molecule, forming C=C bond and producing a new anion and a new structure.

$$R-\underset{\underset{O^-}{\vert}}{\overset{\overset{C=O}{}\atop\overset{H}{}}{CH}}-\underset{}{CH}\,CH_2-R + H_2O \longrightarrow R-\underset{\underset{OH}{\vert}}{\overset{\overset{C=O}{}\atop\overset{H}{}}{CH}}-\underset{}{CH}\,CH_2R + OH^-$$

Reaction of anion with water to give reaction product.

$$X\!-\!X \longrightarrow X^+ + X^-$$

$$X^+ + CH_2\!=\!CH_2 \longrightarrow \underset{\displaystyle X}{CH_2}\!-\!\overset{+}{CH_2}$$

Carbonium ion

The carbonium ion thus formed can then react with an anion or with another molecule of the adding reagent, as in

$$\underset{\displaystyle X}{CH_2}\!-\!\overset{+}{CH_2} + X\!-\!X \longrightarrow \underset{\displaystyle X}{CH_2}\!-\!\underset{\displaystyle X}{CH_2} + X^+$$

or

$$\underset{\displaystyle X}{CH_2}\!-\!\overset{+}{CH_2} + X^- \longrightarrow \underset{\displaystyle X}{CH_2}\!-\!\underset{\displaystyle X}{CH_2}$$

Addition at Primary or Secondary Carbon Atoms

An interesting question arises in considering the addition of an unsymmetrical reagent (HX, as opposed to X—X) to an unsymmetrical alkene (CH_3—CH=CH_2, as opposed to CH_2=CH_2). Here, two isomeric products are possible, depending upon which carbon atom bonds to the X and which bonds to the H of the HX reagent. For example, in the reaction

$$CH_3\!-\!CH\!=\!CH_2 + HBr \longrightarrow \underset{\displaystyle H \quad Br}{CH_3\!-\!CH\!-\!CH_2} \text{ and } \underset{\displaystyle Br \quad H}{CH_3\!-\!CH\!-\!CH_2}$$

hydrogen bromide may add so that bromine is attached to the end carbon atom or to the central carbon atom. The question is: Which isomer is preferred? The answer, determined by measuring the yields of products in many reactions of this type, is: *In general, when an unsymmetrical reagent molecule adds to an unsymmetrical molecule with a double bond, the positive portion of the unsymmetrical reagent molecule bonds to the double bonded carbon atom bearing the greater number of hydrogen atoms.* In the case shown, 2-bromopropane is obtained in high yield.

Other examples of unsymmetrical addition are:

$$CH_3\!-\!CH\!=\!CH_2 + H\!-\!OSO_3H \ (H_2SO_4) \longrightarrow \underset{\displaystyle \underset{\displaystyle SO_3H}{O} \quad H}{CH_3\!-\!CH\!-\!CH_2}$$

$$\underset{\displaystyle CH_3}{CH_3\!-\!C\!=\!CH_2} + H\!-\!OH \xrightarrow{\text{acid}} \underset{\displaystyle CH_3}{CH_3\!-\!\overset{\displaystyle OH \quad H}{C}\!-\!CH_2} \text{ or } \underset{\displaystyle CH_3}{CH_3\!-\!\overset{\displaystyle OH}{C}\!-\!CH_3}$$

An interesting addition reaction has been developed by the petroleum industry to convert low-molecular-weight hydrocarbons (obtained from petroleum and from the cracking process) to high-octane gasolines. An illustration of this process is the reaction

$$\underset{\displaystyle CH_3}{CH_3\!-\!C\!=\!CH_2} + \underset{\displaystyle CH_3}{H\!-\!\overset{\displaystyle CH_3}{C}\!-\!CH_3} \xrightarrow{H_2SO_4} \underset{\displaystyle CH_3}{CH_3\!-\!CH\!-\!CH_2\!-\!\overset{\displaystyle CH_3}{\underset{\displaystyle CH_3}{C}}\!-\!CH_3}$$

2,2,4-Trimethylpentane
(a high-octane gasoline)

The mechanism of this reaction is complex, but the products can be explained if the tertiary butyl group of isobutane acts like a positive group and the hydrogen atom of that molecule acts like a negative group.

PROBLEM Show by equations a method for converting 1-butene to 2-cyanobutane,

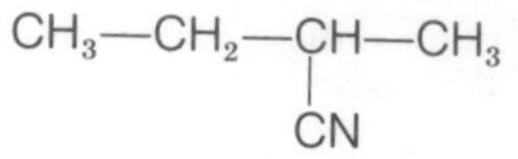

$$CH_3-CH_2-\underset{\underset{CN}{|}}{CH}-CH_3$$

Solution Recalling that HCN is not among the molecules that can be added to alkenes, but that a CN group can be put into a molecule by displacing a halide ion, we might use the following reaction sequence:

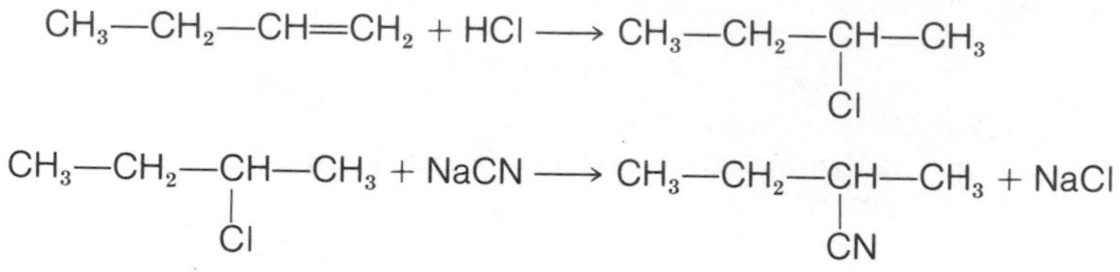

$$CH_3-CH_2-CH=CH_2 + HCl \longrightarrow CH_3-CH_2-\underset{\underset{Cl}{|}}{CH}-CH_3$$

$$CH_3-CH_2-\underset{\underset{Cl}{|}}{CH}-CH_3 + NaCN \longrightarrow CH_3-CH_2-\underset{\underset{CN}{|}}{CH}-CH_3 + NaCl$$

Addition to Carbonyl Groups The π bond in the carbonyl group makes addition reactions possible at this site. However, the greater electronegativity of oxygen over carbon and the possibilities for resonance

$$\overset{}{\underset{}{>}}C\overset{\longrightarrow}{=}O \longleftrightarrow \overset{}{\underset{}{>}}C^{\pm}-O^-$$

make the carbon atom more positive and the oxygen atom more negative than the carbon atoms sharing the π bond in alkenes. The positive character of the carbonyl carbon atom makes this group sensitive to electron-donating reagents, in contrast to the alkene double bond, which is sensitive to electron-seeking reagents. Thus, the type reaction for addition to carbonyl groups is

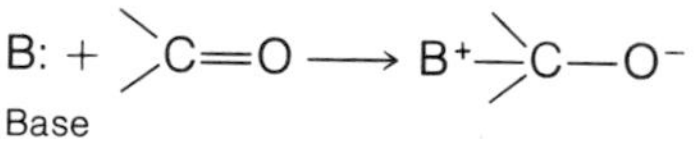

$$\underset{\text{Base}}{B\colon} + \overset{}{\underset{}{>}}C=O \longrightarrow B^+\overset{}{\underset{}{>}}C-O^-$$

The reagents that add to carbonyl groups are generally not the same as those that add to alkenes. Reagents that add to carbonyl groups include hydrogen cyanide (HCN), water and alcohols, ammonia, and many organometallic compounds such as alkyl or aryllithiums (RLi, ArLi, and Grignard reagents, RMgX). Grignard reagents, similar in many ways to the organolithium reagents, are made by adding magnesium to an alkyl or aryl halide dissolved in anhydrous ether. In adding to the carbonyl group, the positive portion of the reagent bonds to the oxygen atom, while the negative portion bonds to the carbon atom. Examples of carbonyl addition reactions include:

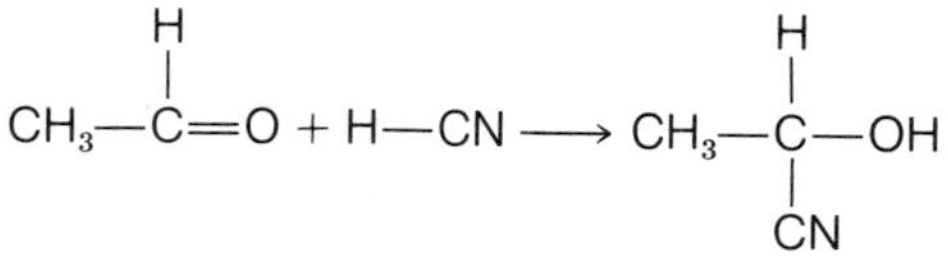

$$CH_3-\overset{\overset{H}{|}}{C}=O + H-CN \longrightarrow CH_3-\overset{\overset{H}{|}}{\underset{\underset{CN}{|}}{C}}-OH$$

Cyanohydrin

$$\underset{H_3C}{\overset{H_3C}{>}}C=O + H-OR \longrightarrow CH_3-\overset{\overset{CH_3}{|}}{\underset{\underset{OR}{|}}{C}}-OH$$

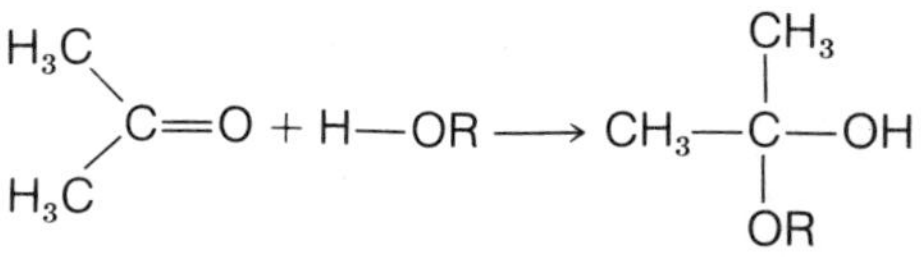

Hemiacetal

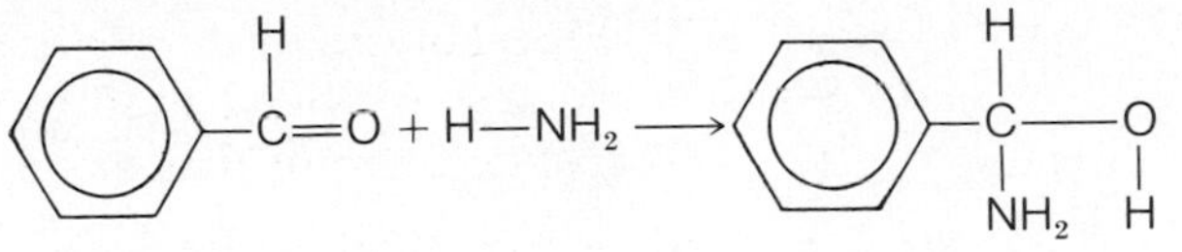

$$\text{C}_6\text{H}_5\underset{\overset{|}{\text{C}}}{\overset{\text{H}}{=}}\text{O} + \text{H}-\text{NH}_2 \longrightarrow \text{C}_6\text{H}_5\underset{\overset{|}{\text{NH}_2}}{\overset{\overset{\text{H}}{|}}{\text{C}}}-\text{O}-\text{H}$$

Aldehyde ammonia

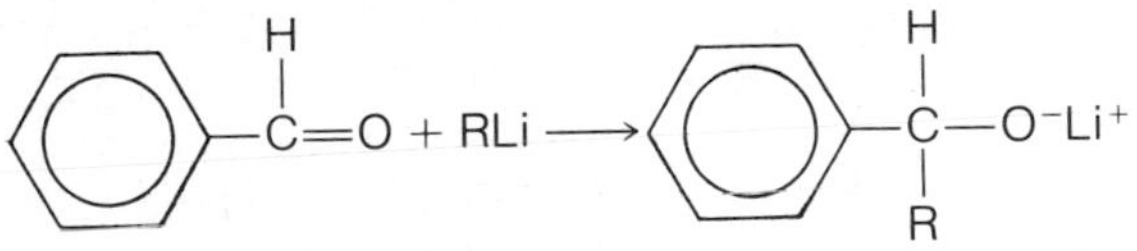

$$\text{C}_6\text{H}_5\underset{\overset{|}{\text{C}}}{\overset{\text{H}}{=}}\text{O} + \text{RLi} \longrightarrow \text{C}_6\text{H}_5\underset{\overset{|}{\text{R}}}{\overset{\overset{\text{H}}{|}}{\text{C}}}-\text{O}^-\text{Li}^+$$

Alcoholate

$$\text{CH}_3-\underset{\overset{||}{\text{O}}}{\text{C}}-\text{CH}_3 + \text{RMgBr} \longrightarrow \text{CH}_3-\underset{\overset{|}{\text{OMgBr}}}{\overset{\overset{\text{R}}{|}}{\text{C}}}-\text{CH}_3$$

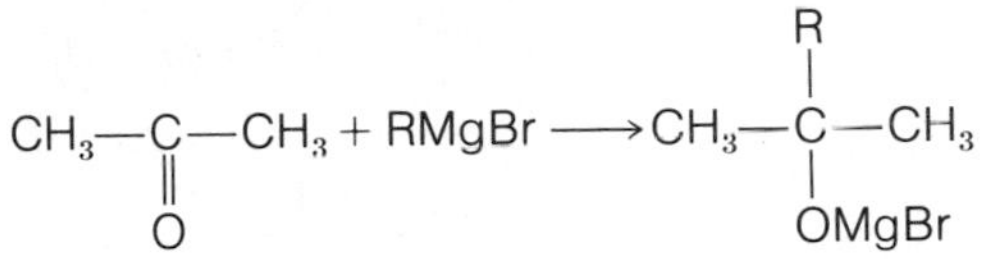

Preparation of Secondary and Tertiary Alcohols from Aldehydes and Ketones. The last two reactions above are particularly useful for preparing secondary and tertiary alcohols, especially those containing different R groups in the same compound, such as

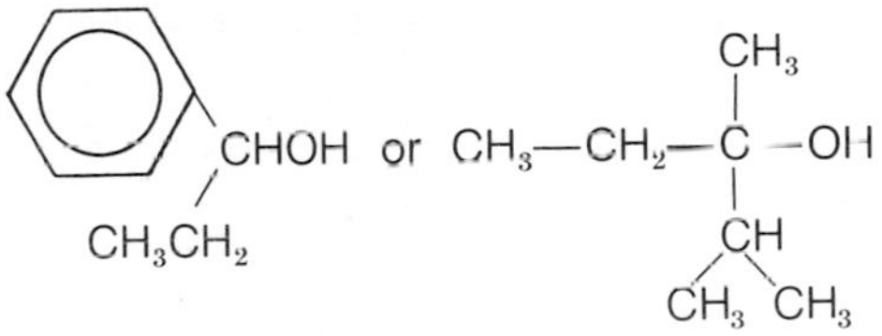

$$\text{C}_6\text{H}_5\underset{\overset{/}{\text{CH}_3\text{CH}_2}}{\text{CHOH}} \quad \text{or} \quad \text{CH}_3-\text{CH}_2-\underset{\overset{|}{\underset{\overset{|}{\text{CH}}}{}}}{\overset{\overset{\text{CH}_3}{|}}{\text{C}}}-\text{OH}$$
$$\underset{\text{CH}_3\;\;\text{CH}_3}{}$$

The aldehyde and the organometallic compound can be selected to prepare a secondary alcohol containing any two desired R groups. For example, the compound 2-butanol,

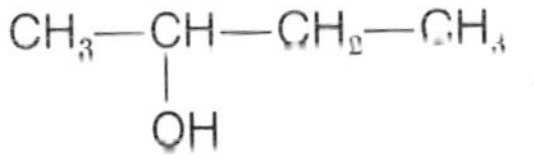

$$\text{CH}_3-\underset{\overset{|}{\text{OH}}}{\text{CH}}-\text{CH}_2-\text{CH}_3$$

can be prepared from ethanal and ethyllithium by the reaction sequence

Step 1 $$\text{CH}_3-\underset{\overset{|}{\text{C}}}{\overset{\text{H}}{=}}\text{O} + \text{CH}_3-\text{CH}_2-\text{Li} \longrightarrow \text{CH}_3-\underset{\overset{|}{\text{CH}_2\text{CH}_3}}{\overset{\overset{\text{H}}{|}}{\text{C}}}-\text{OLi}$$

Step 2 $$\text{CH}_3-\underset{\overset{|}{\text{CH}_2\text{CH}_3}}{\overset{\overset{\text{H}}{|}}{\text{C}}}-\text{OLi} \xrightarrow[\text{H}_2\text{O}]{\text{HCl}} \text{CH}_3-\underset{\overset{|}{\text{CH}_2\text{CH}_3}}{\overset{\overset{\text{H}}{|}}{\text{C}}}-\text{OH} + \text{LiCl}$$

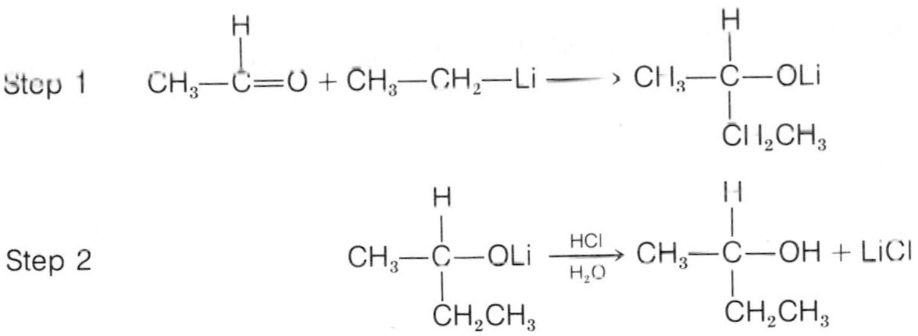

The second step in the process, treatment of the reaction product from step 1 with hydrochloric acid, is the replacement of the lithium ion by a hydrogen ion, thereby producing the desired alcohol.

Tertiary alcohols may be made from ketones and organolithium reagents by similar addition and replacement reactions.

Certain derivatives of ammonia add to the double bond of the carbonyl group. This may be followed by a loss of water from adjacent atoms to form a double bond, as illustrated by the reaction sequence

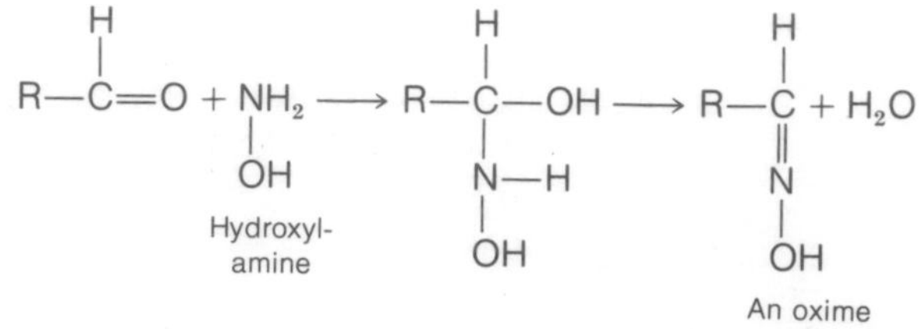

$$R-\overset{\overset{\displaystyle H}{|}}{C}=O + NH_2 \longrightarrow R-\overset{\overset{\displaystyle H}{|}}{\underset{\underset{\displaystyle OH}{|}}{C}}-OH \longrightarrow R-\overset{\overset{\displaystyle H}{|}}{\underset{\underset{\displaystyle OH}{|}}{C}} + H_2O$$

$$\underset{\underset{\displaystyle \substack{Hydroxyl-\\amine}}{|}}{OH} \qquad \underset{\underset{\displaystyle OH}{|}}{N-H} \qquad \underset{\underset{\displaystyle \substack{OH\\An\ oxime}}{|}}{N}$$

Such reactions are often used in the laboratory to identify aldehydes and ketones by preparing oximes or similar derivatives that are crystalline solids and that can, in turn, be identified by their melting points and X-ray diffraction patterns. In this case, examination of the crystalline oxime serves to identify the aldehyde from which it was derived.

Addition of Hydrogen to Carbonyl Groups: Reduction Reactions. While the addition reactions discussed here occur readily in aldehydes and ketones, they do not occur to any appreciable extent in carboxylic acids and their derivatives, even though these substances contain carbonyl groups. All types of carbonyl groups, however, will add hydrogen in the presence of appropriate reducing agents.

Addition to $-C\equiv N$ Nitriles, $R-C\equiv N$, add hydrogen to give amines or water to form carboxylic acids.

The second reaction is especially useful for preparing carboxylic acids having a particular R group. For example, phenylacetic acid

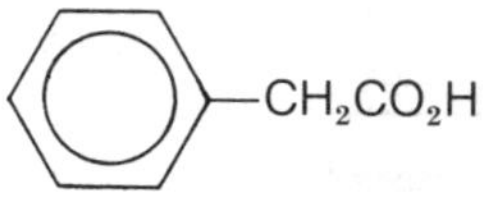

$$\langle\bigcirc\rangle-CH_2CO_2H$$

can be prepared from toluene by the reactions

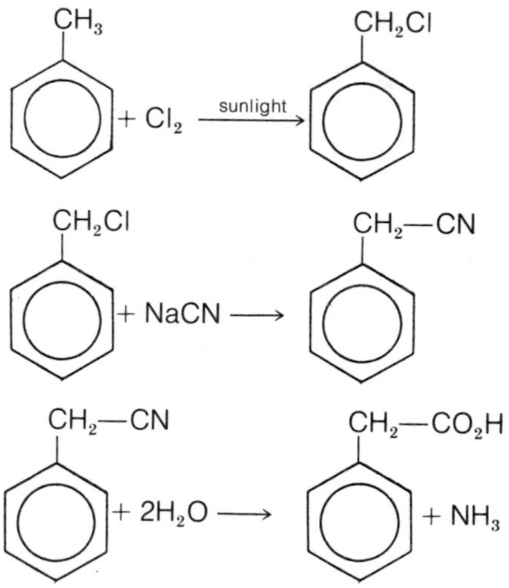

ELIMINATION REACTIONS
Elimination of Water

When water is eliminated from adjacent carbon atoms in an alcohol, as in

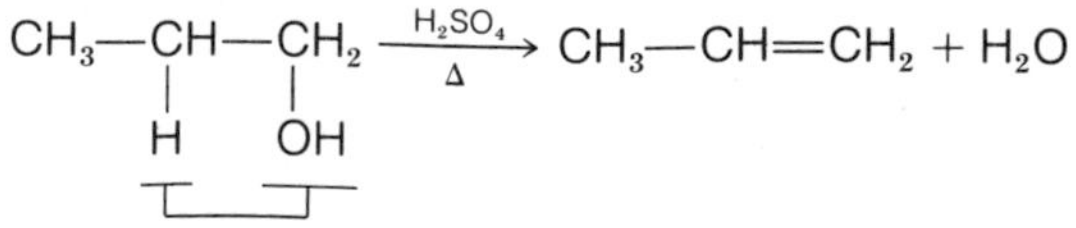

$$CH_3-\overset{\overset{\displaystyle }{|}}{\underset{\underset{\displaystyle H}{|}}{CH}}-\overset{\overset{\displaystyle }{|}}{\underset{\underset{\displaystyle OH}{|}}{CH_2}} \xrightarrow[\Delta]{H_2SO_4} CH_3-CH=CH_2 + H_2O$$

the process is called *dehydration*; the product is an alkene. Dehydrating agents for alcohols are substances that have a great affinity for water, and include concentrated sulfuric or phosphoric acids or activated aluminum oxide, Al_2O_3, at high temperatures. Dehydration of alcohols is a commonly used reaction for the preparation of alkenes. Reaction conditions vary widely, depending on the reactivity of the alcohols. Primary alcohols are usually more difficult to dehydrate than are tertiary alcohols.

Some alcohols yield more than one alkene on dehydration. For example, 2-butanol gives both 1- and 2-butene, as shown below:

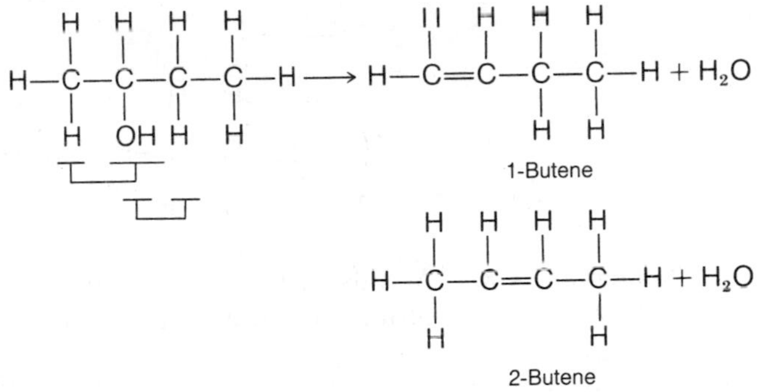

1-Butene

2-Butene

Elimination of HX

When a hydrogen halide molecule is eliminated from adjacent carbon atoms in an alkyl halide, the process is known as *dehydrohalogenation*. It is often accomplished with a solution of potassium hydroxide dissolved in alcohol. Most alkyl halides undergo dehydrohalogenation.

Examples of these reactions are:

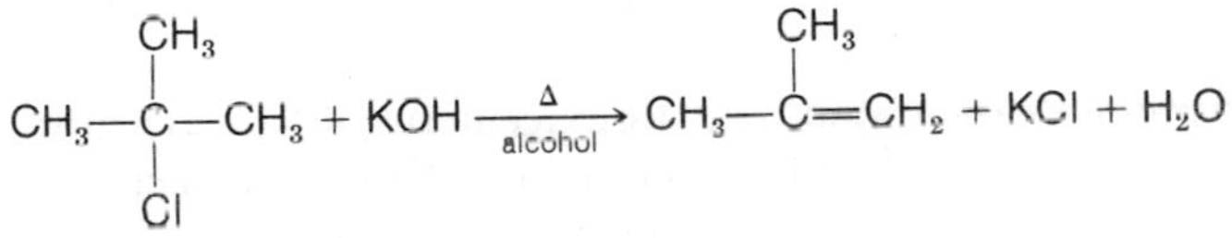

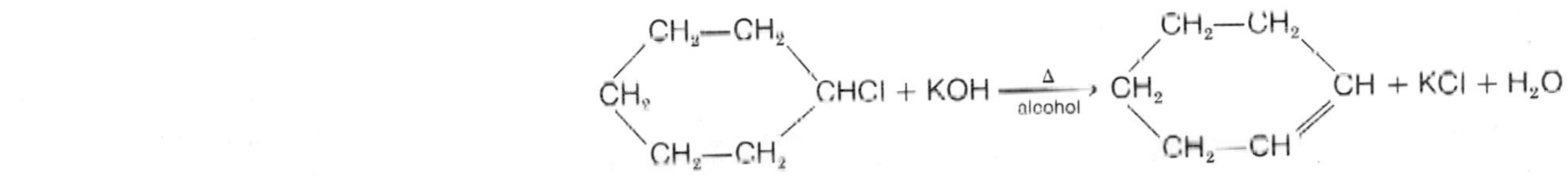

Alkynes can be prepared by the dehydrohalogenation of dihalides, as in the reaction

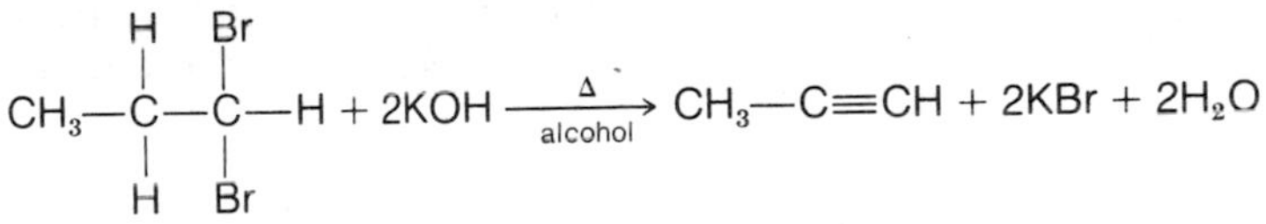

PROBLEM

Show by equations a method of converting phenylethanol,

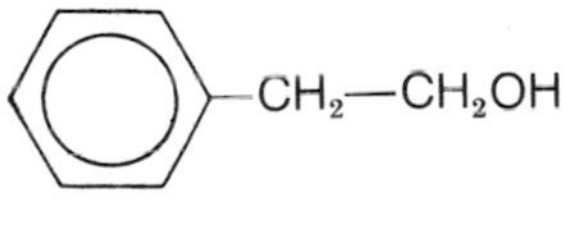

to phenylethyne.

Solution

A method for creating a triple bond is to start with a compound that can eliminate two HBr molecules from adjacent carbon atoms. Such a compound can be made by first dehydrating the phenylethanol to an alkene, then adding Br_2 to the double bond of the alkene.

$$\text{C}_6\text{H}_5-\underset{\underset{\text{H}}{|}}{\overset{\overset{\text{H}}{|}}{\text{C}}}-\underset{\underset{\text{OH}}{|}}{\overset{\overset{\text{H}}{|}}{\text{C}}}-\text{H} \xrightarrow[\Delta]{\text{H}_2\text{SO}_4} \text{C}_6\text{H}_5-\overset{\overset{\text{H}}{|}}{\text{C}}=\overset{\overset{\text{H}}{|}}{\text{C}}-\text{H} + \text{H}_2\text{O}$$

$$\text{C}_6\text{H}_5-\overset{\overset{\text{H}}{|}}{\text{C}}=\overset{\overset{\text{H}}{|}}{\text{C}}-\text{H} + \text{Br}_2 \longrightarrow \text{C}_6\text{H}_5-\underset{\underset{\text{Br}}{|}}{\overset{\overset{\text{H}}{|}}{\text{C}}}-\underset{\underset{\text{Br}}{|}}{\overset{\overset{\text{H}}{|}}{\text{C}}}-\text{H}$$

$$\text{C}_6\text{H}_5-\underset{\underset{\text{Br}}{|}}{\overset{\overset{\text{H}}{|}}{\text{C}}}-\underset{\underset{\text{Br}}{|}}{\overset{\overset{\text{H}}{|}}{\text{C}}}-\text{H} + 2\text{KOH} \xrightarrow[\text{alcohol}]{\Delta} \text{C}_6\text{H}_5-\text{C}\equiv\text{C}-\text{H} + 2\text{KBr} + 2\text{H}_2\text{O}$$

Elimination of X$_2$ Metals such as sodium or zinc will remove halogen atoms from two carbon atoms in a molecule. Examples are:

$$\text{CH}_3-\text{CH}_2-\underset{\underset{\text{Br}}{|}}{\text{CH}}-\underset{\underset{\text{Br}}{|}}{\text{CH}_2} + \text{Zn} \longrightarrow \text{CH}_3-\text{CH}_2-\text{CH}=\text{CH}_2 + \text{ZnBr}_2$$

$$\text{Br}-\text{CH}_2-\text{CH}_2-\text{CH}_2-\text{CH}_2-\text{Br} + 2\text{Na} \longrightarrow \begin{matrix} \text{CH}_2-\text{CH}_2 \\ | \qquad | \\ \text{CH}_2-\text{CH}_2 \end{matrix} + 2\text{NaBr}$$

The last reaction is an example of removal of bromine atoms from non-adjacent carbon atoms, and results in the formation of cycloalkanes.

SUBSTITUTION REACTIONS

Replacement of Hydrogen by Halogen

Replacement of one or more hydrogen atoms bonded to carbon atoms by chlorine or bromine occurs slowly at room temperature, but rapidly at higher temperatures or in the presence of sunlight. An example of this reaction is the chlorination of ethane:

$$\text{Cl}-\text{Cl} + \text{H}-\underset{\underset{\text{H}}{|}}{\overset{\overset{\text{H}}{|}}{\text{C}}}-\underset{\underset{\text{H}}{|}}{\overset{\overset{\text{H}}{|}}{\text{C}}}-\text{H} \longrightarrow \text{H}-\underset{\underset{\text{H}}{|}}{\overset{\overset{\text{H}}{|}}{\text{C}}}-\underset{\underset{\text{H}}{|}}{\overset{\overset{\text{H}}{|}}{\text{C}}}-\text{Cl} + \text{HCl}$$

Almost all carbon-hydrogen bonds can be made to undergo this reaction. Very often, such reactions result in mixtures of products, as illustrated by the reaction of methane with excess chlorine, which gives all four possible chlorinated products:

$$\text{H}-\underset{\underset{\text{H}}{|}}{\overset{\overset{\text{H}}{|}}{\text{C}}}-\text{H} + \text{Cl}-\text{Cl} \longrightarrow \text{H}-\underset{\underset{\text{H}}{|}}{\overset{\overset{\text{H}}{|}}{\text{C}}}-\text{Cl} \ (+\text{HCl}) \xrightarrow{\text{Cl}_2} \text{H}-\underset{\underset{\text{Cl}}{|}}{\overset{\overset{\text{H}}{|}}{\text{C}}}-\text{Cl} \ (+\text{HCl})$$

$$\downarrow \text{Cl}_2$$

$$\text{Cl}-\underset{\underset{\text{Cl}}{|}}{\overset{\overset{\text{Cl}}{|}}{\text{C}}}-\text{Cl} \ (+\text{HCl}) \xleftarrow{\text{Cl}_2} \text{Cl}-\underset{\underset{\text{Cl}}{|}}{\overset{\overset{\text{H}}{|}}{\text{C}}}-\text{Cl} \ (+\text{HCl})$$

However, it is possible to control reaction conditions to get good yields of one of several possible products.

The reaction pathway or *mechanism* for halogenation reactions accelerated by sunlight is known to be a *free-radical chain process*. Halogenation reactions are fast and exothermic, and may lead to explosions.

Reaction at Primary and Secondary Carbon Atoms. An interesting structural question arises when considering the chlorination of propane and higher alkanes. If conditions are fixed so that only one hydrogen atom in each molecule of alkane is replaced, two isomers are possible—1-chloropropane and 2-chloropropane:

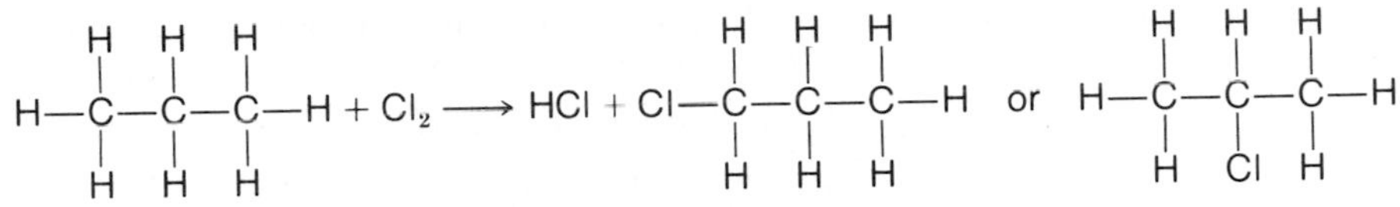

1-Chloropropane   2-Chloropropane

The 1-chloropropane forms when a hydrogen atom on either of the end carbon atoms in propane is replaced; 2-chloropropane forms when a hydrogen atom on the central carbon atom is replaced. The question then is: Do both isomers appear in the reaction products and, if so, what are the relative amounts of each? The answer to this question, found by experiment, is: Both isomers invariably appear. At temperatures of 400°C and above, the ratio of 1-chloropropane to 2-chloropropane is 3 to 1—consistent with statistical considerations. These considerations are based on the fact that there are six hydrogen atoms on terminal carbon atoms and two hydrogen atoms on the central carbon atom in propane. Thus, there are six chances to form 1-chloropropane for every two chances to form 2-chloropropane in this case. As the reaction temperature is lowered, the ratio of the two isomers approaches 1. This suggests that attacking chlorine atoms are more selective in removing hydrogen atoms at lower temperatures, and that they remove hydrogen atoms attached to the central carbon atom in propane (a secondary carbon atom) more readily than those attached to the end carbon atoms (primary carbon atoms).

Reaction at Tertiary Carbon Atoms. A similar situation obtains in the monochlorination of isobutane,

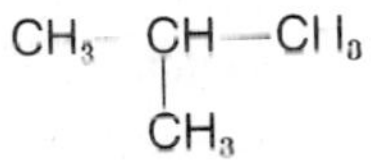

$$CH_3 \quad CH-CH_3$$
$$\underset{CH_3}{|}$$

Here again, two isomers are possible, and statistical considerations predict a 9 to 1 ratio of 1-chloro-2-methylpropane compared to 2-chloro-2-methylpropane. At high temperatures, the product ratios approach this prediction. But at lower temperatures, the substitution occurs preferentially on the tertiary carbon atom, and 2-chloro-2-methylpropane is present in a larger amount.

These results are general for the chlorination of organic compounds, and may be summarized as follows: high-temperature chlorinations give nearly statistical isomer ratios; at lower temperatures, substitution occurs most readily at tertiary carbon atoms and least readily at primary carbon atoms. At 300°C, the relative rates of substitution are: primary, 1, secondary, 3.2, and tertiary, 4.4.

Table 27.1 Replacement Reactions of Alkyl Halides

Reagent	Reaction	Organic Product
1. NaOH	$OH^- + RCl \longrightarrow Cl^- + ROH$	An alcohol
2. NaOR′	$OR'^- + RCl \longrightarrow Cl^- + ROR'$	An ether
3. NaSH	$SH^- + RCl \longrightarrow Cl^- + RSH$	A mercaptan
4. NaNH$_2$	$NH_2^- + RCl \longrightarrow Cl^- + RNH_2$	A primary amine
5. NaCN	$CN^- + RCl \longrightarrow Cl^- + RCN$	A nitrile
6. NaC≡C—R′	$C{\equiv}CR'^- + RCl \longrightarrow Cl^- + R{-}C{\equiv}C{-}R'$	An alkyne

Replacement of Halogen Bonded to Carbon

As indicated in Chapter 26, alkyl halides are used in synthesizing a number of kinds of organic compounds. These syntheses are possible because of the ease with which a halogen atom bonded to carbon is replaced by a base. Examples of such replacements and the kinds of compounds that can be synthesized from alkyl halides are given in Table 27.1.

In each of these reactions, an anion replaces a chloride ion in the organic compound.

PROBLEM Starting with ethane, show by equations a method for preparing ethylamine, CH_3CH_2—NH_2.

Solution The amine can be made from sodium amide, NaNH$_2$, and an alkyl halide (Reaction 4, Table 27.1). In this case, the alkyl group must be an ethyl group, so we need first to prepare ethyl chloride and then allow it to react with sodamide to give the amine. The reaction sequence then becomes

$$CH_3{-}CH_3 + Cl_2 \xrightarrow{\text{sunlight}} CH_3{-}CH_2{-}Cl + HCl$$

$$CH_3{-}CH_2{-}Cl + NaNH_2 \longrightarrow CH_3{-}CH_2{-}NH_2 + NaCl$$

PROBLEM Starting with cyclohexane, show by equations a method for preparing methyl cyclohexyl ether,

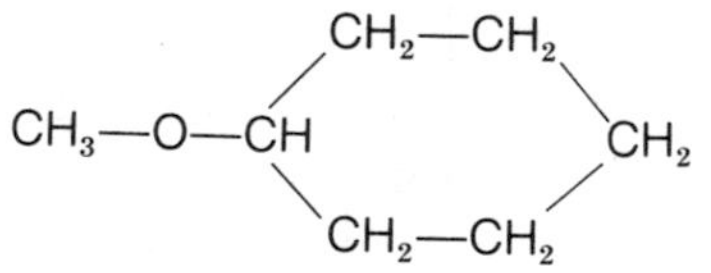

Solution An ether is prepared from an alkyl halide and the sodium salt of an alcohol, NaOR (Reaction 2, Table 27.1). A methyl ether can be prepared from the sodium salt of methyl alcohol, NaOCH$_3$. The alkyl halide needed in this case is cyclohexyl chloride, which can be made by chlorinating cyclohexane. Thus, the sequence in this case becomes

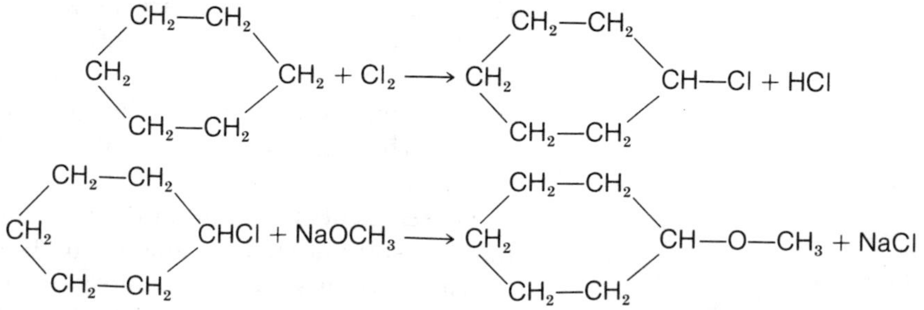

Mechanism of Halogen Replacement. The mechanisms for halogen replacements involve ions as intermediates, in contrast to the free-radical intermediates involved in replacement of hydrogen atoms attached to carbon

atoms. Two important mechanisms for halogen replacement are recognized, and the mechanism dominant in a given case depends on the nature of the alkyl group in the alkyl halide. Primary halides undergo a one-step replacement in which the reaction rate is proportional to the concentration of each reactant. This can be called a *bimolecular mechanism*. Much evidence shows that in this mechanism, the attacking anion approaches the carbon atom to which it will become bonded from a side opposite to that occupied by the halide ion. As the attacking anion approaches, the halide ion leaves, as illustrated by the process

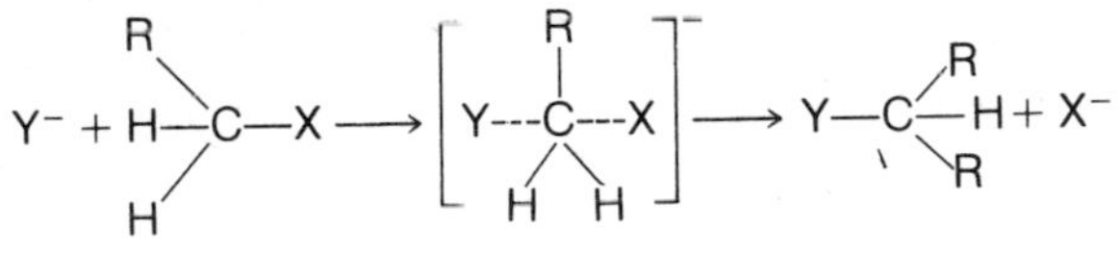

$$Y^- + H-\underset{\underset{H}{|}}{\overset{\overset{R}{|}}{C}}-X \longrightarrow \left[Y\text{---}\underset{\underset{H}{|}\;\underset{H}{|}}{\overset{\overset{R}{|}}{C}}\text{---}X\right]^- \longrightarrow Y-\underset{\overset{\diagdown}{\diagdown R}}{\overset{\diagup R}{C}}H + X^-$$

Tertiary halides probably follow a two-step reaction sequence, as illustrated below:

$$R-X \underset{\text{fast}}{\overset{\text{slow}}{\rightleftarrows}} R^+ + X^-$$

$$R^+ + Y^- \xrightarrow{\text{fast}} RY$$

In the first step, the alkyl halide ionizes slowly to give the very reactive *carbonium ion*, R^+, which reacts very rapidly with the anion or even with the solvent. Since the first step is slow (rate-determining) and involves only the alkyl halide, the kinetics of this process are first-order, and we may call this a *unimolecular mechanism*. This mechanism is favored in polar solvents which assist the ionization step. Secondary halides follow both the first- and the second-order process, with the relative proportion of each depending on conditions and the nature of the alkyl group.

Replacement of Hydroxyl Bonded to Carbon

Direct displacement of an hydroxyl group by another anion,

$$ROH + X^- \longrightarrow RX + OH^-$$

seldom occurs, because the anion X^- often shows a greater stability as an independent species than does the small, poorly polarizable OH^- ion. If the alcohol is placed in acid solution, however, the hydroxyl group is converted to the corresponding oxonium ion, according to the reaction

$$ROH + H_3O^+ \rightleftarrows R-\underset{}{\overset{\overset{H}{|}}{O}}-H^+ + H_2O$$

In oxonium ions, water is the leaving group, and the displacement reaction

$$X^- + R-\underset{}{\overset{\overset{H}{|}}{O}}-H^+ \longrightarrow RX + H_2O$$

proceeds with relative ease. For this reason, the displacement of hydroxyl groups is usually carried out in acid solutions.

Even in the presence of acids, primary alcohols are converted to alkyl chlorides very slowly. However, tertiary alcohols are rapidly converted to the corresponding chlorides under these conditions.

Aromatic Substitution Reactions

Replacement of hydrogen atoms bonded to an aromatic ring can be accomplished by the following reactions.

1. *Nitration:*

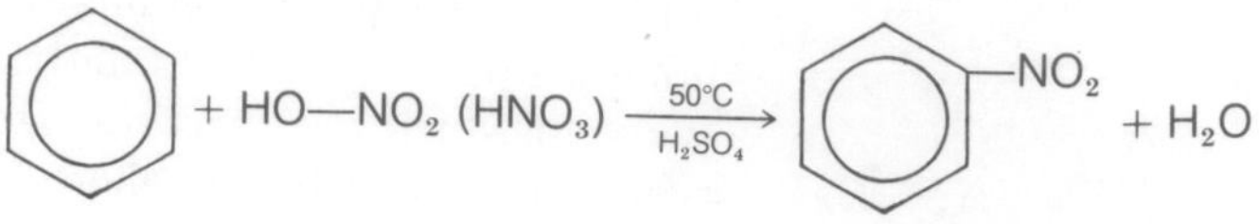

$$\bigcirc + \text{HO—NO}_2 \ (\text{HNO}_3) \xrightarrow[\text{H}_2\text{SO}_4]{50°\text{C}} \bigcirc\text{—NO}_2 + \text{H}_2\text{O}$$

Nitrobenzene

2. *Sulfonation:*

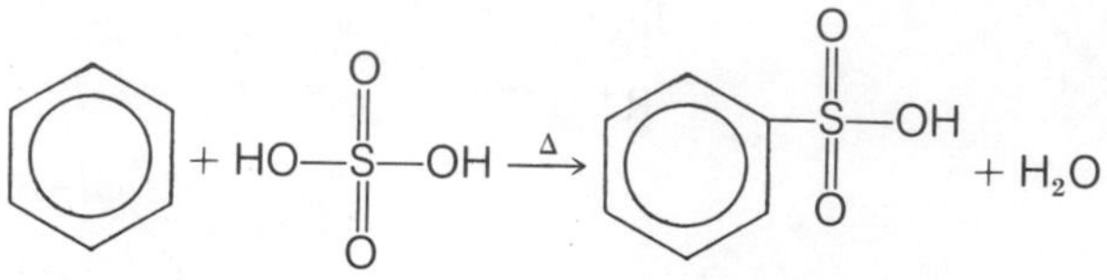

Benzene sulfonic acid

3. *Halogenation* using a Lewis acid catalyst:

$$\bigcirc + \text{Cl}_2 \xrightarrow{\text{FeCl}_3} \bigcirc\text{—Cl} + \text{HCl}$$

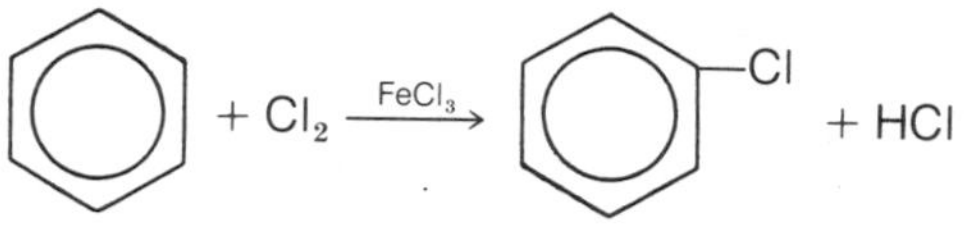

(Sunlight does not catalyze this reaction.)

4. *Alkylation* using a Lewis acid catalyst:

$$\bigcirc + \text{RCl} \xrightarrow{\text{AlCl}_3} \bigcirc\text{—R} + \text{HCl}$$

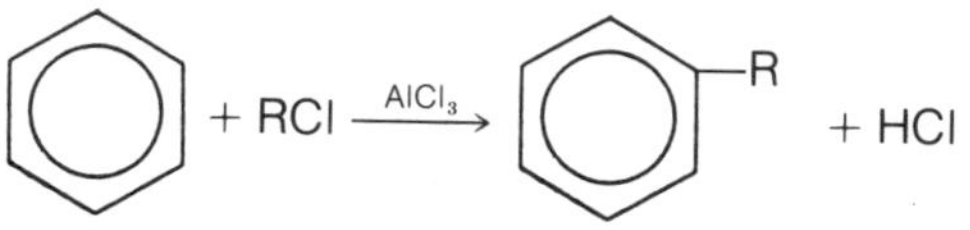

(This reaction, known as the Friedel-Crafts reaction, is a method for obtaining aromatic structures having aliphatic side chains.)

The mechanism of aromatic substitution usually involves attack by a reactive positive ion on the aromatic ring. This is in contrast to aliphatic substitution reactions, which often involve attack by an electron-rich (often negative) species. In the aromatic reactions, the ion, seeking a pair of electrons, adds to the ring. This is followed by expulsion of a proton:

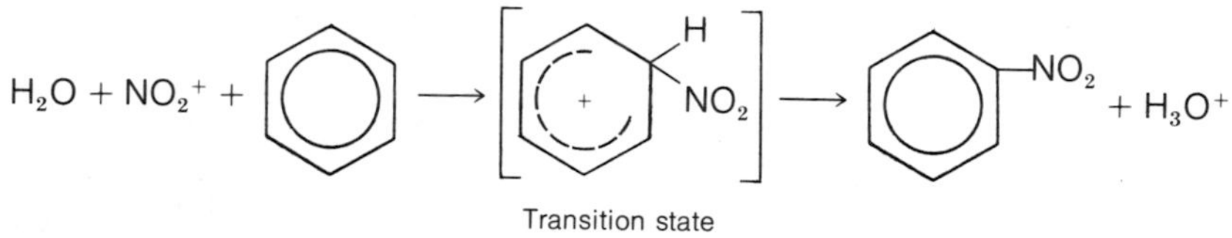

Transition state

Other positive ions that will react are SO_3H^+, Cl^+, and R^+. The reactive positive ions are probably generated as follows:

In nitration, the sulfuric acid catalyst donates a proton to nitric acid, giving a species that may lose water to give NO_2^+ (nitronium ion). In sulfonation, a similar reaction may occur:

$$H_2SO_4 + HO\!-\!SO_3H \longrightarrow \left[H\!-\!\overset{\overset{\displaystyle H}{|}}{O}\!-\!SO_3H \right]^+ + HSO_4^-$$

$$\left[H\!-\!\overset{\overset{\displaystyle H}{|}}{O}\!-\!SO_3H \right]^+ \longrightarrow H_2O + SO_3H^+$$

In alkylation, the aluminum chloride catalyst helps form the carbonium ion, R^+, according to the reaction

$$RCl + AlCl_3 \rightleftharpoons R^+ + AlCl_4^-$$

It is possible to replace more than one hydrogen atom in a benzene ring by using an excess of the substituting reagent or by carrying out successive substitutions using different reagents. Examples of these possibilities are:

1. Use of excess reagent.

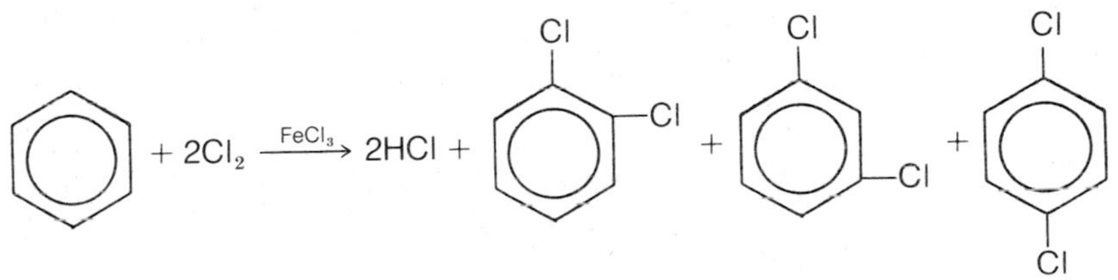

o-Dichloro-
benzene
(11%
of product)

m-Dichloro-
benzene
(1.6%
of product)

p-Dichloro-
benzene
(87%
of product)

Here, a mixture of all three disubstituted benzenes is obtained, but the yield of the 1,3-dichlorobenzene (or m-dichlorobenzene) is very poor—only 1.6% of the product.

2. Successive substitutions.

Step 1

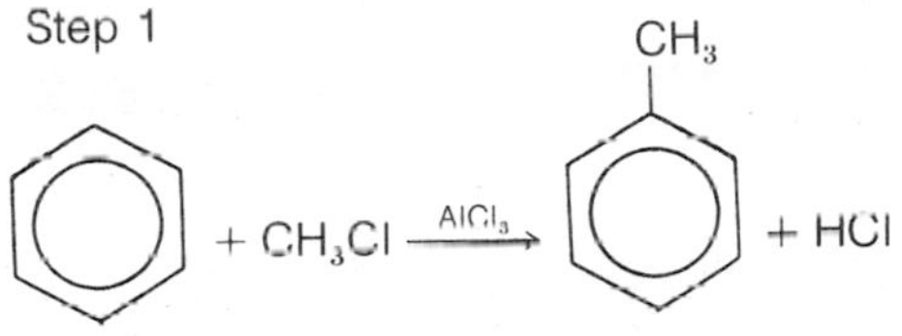

Toluene

Step 2

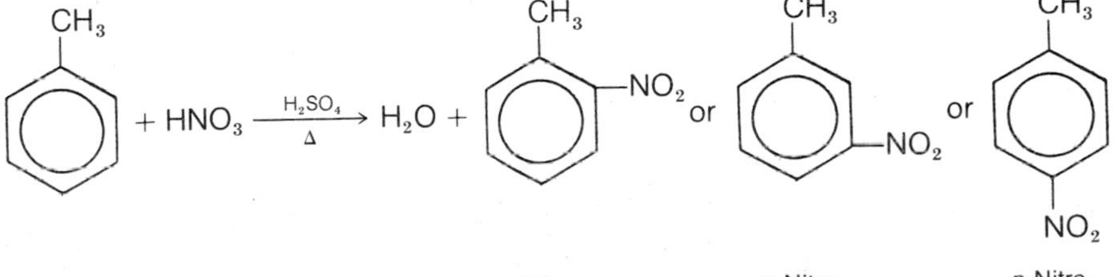

o-Nitro-
toluene
(56%
of product)

m-Nitro-
toluene
(3%
of product)

p-Nitro-
toluene
(41%
of product)

In the second step, a mixture of all three disubstituted benzenes is obtained, but here again the yield of the *m*-nitrotoluene is very low.

When a second substituent is introduced into a monosubstituted benzene, a mixture of the three isomeric disubstituted benzenes is usually obtained. However, the yields of the three isomers vary greatly. In the two examples given above, good yields of the *o*- and *p*- isomers were obtained, but the yield of the *m*- isomer was very low. In the reaction

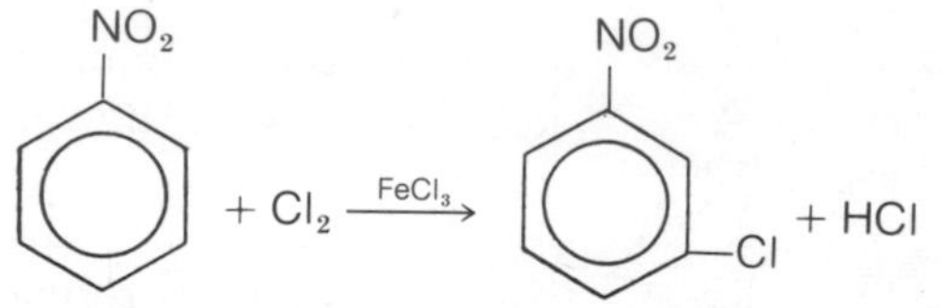

the *m*- isomer is obtained in highest yield. From these and numerous similar results, chemists conclude that *the position taken by an entering substituent depends upon the substituent already present in the ring.* For example, in the first two examples given, the first substituents in the ring were a chloro and a methyl group respectively. These groups control the position taken by the second substituent (a chloro and a nitro group in the two cases in point). The yield of products clearly shows that both the methyl and chloro groups direct second substituents to the *o*- and *p*- positions, but not to the *m*- position. On the other hand, the nitro group in nitrobenzene directs the second substituent to the *m*- position but not to the *o*- or *p*- positions. This behavior is rather general, and it is possible to identify groups as being *ortho-para directing* or *meta-directing*.

Some ortho-para directing groups:

$$-NH_2, \quad -OH, \quad -N(CH_3)_2, \quad -OCH_3, \quad -CH_3, \quad -Cl, \quad -Br, \quad -I$$

Some meta-directing groups (note the presence of multiple bonds in these groups):

Usually, meta-directing groups decrease the activity of the ring toward further substitution. The ordering influence of substituents can be explained in terms of electronic structures. The entering group, a positive ion, seeks out the carbon atom that has the highest electron density. Substituents already present affect the electron density in the ring. Ortho-para directing groups tend to increase the electron density in the ring, and meta-directing groups tend to decrease the electron density in the ring.

PROBLEM   Show by equations a method for converting benzene (a) to *p*-bromonitrobenzene, and (b) to *m*-bromonitrobenzene.

Solution   (a) To get the *p*- isomer, we must first put an ortho-para directing group into the molecule, and then substitute the second group. Of the two, the bromo group is ortho-para directing, so the sequence must be:

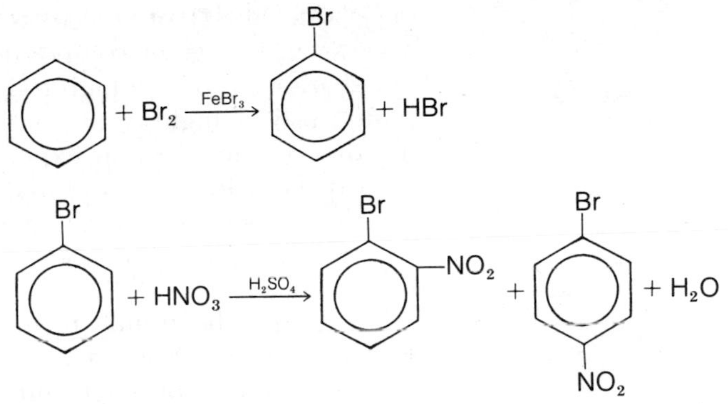

$$\text{C}_6\text{H}_6 + \text{Br}_2 \xrightarrow{\text{FeBr}_3} \text{C}_6\text{H}_5\text{Br} + \text{HBr}$$

$$\text{C}_6\text{H}_5\text{Br} + \text{HNO}_3 \xrightarrow{\text{H}_2\text{SO}_4} o\text{-Br-C}_6\text{H}_4\text{-NO}_2 + p\text{-Br-C}_6\text{H}_4\text{-NO}_2 + \text{H}_2\text{O}$$

(b) To get the *m-* isomer, the meta-directing nitro group must be put in first; therefore, the sequence becomes:

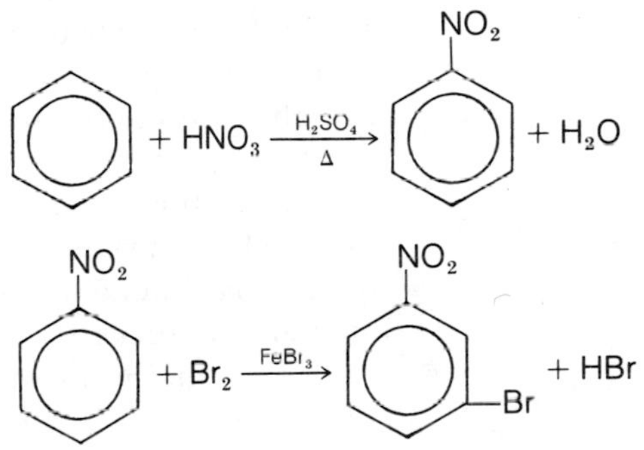

$$\text{C}_6\text{H}_6 + \text{HNO}_3 \xrightarrow[\Delta]{\text{H}_2\text{SO}_4} \text{C}_6\text{H}_5\text{NO}_2 + \text{H}_2\text{O}$$

$$\text{C}_6\text{H}_5\text{NO}_2 + \text{Br}_2 \xrightarrow{\text{FeBr}_3} m\text{-Br-C}_6\text{H}_4\text{-NO}_2 + \text{HBr}$$

PROBLEM   Devise a method for converting benzene to benzoic acid,

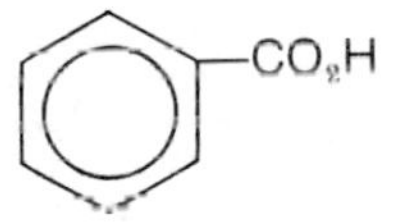

Solution

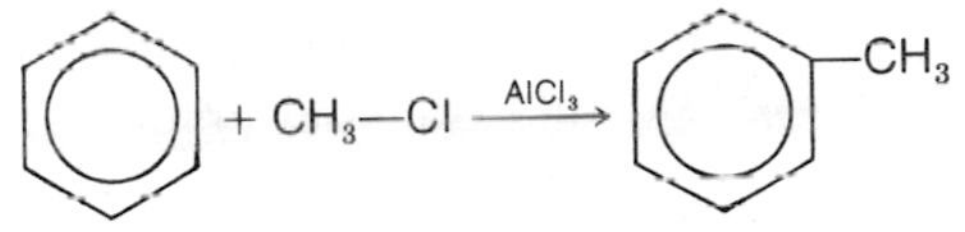

$$\text{C}_6\text{H}_6 + \text{CH}_3\text{-Cl} \xrightarrow{\text{AlCl}_3} \text{C}_6\text{H}_5\text{-CH}_3 \quad \text{Toluene}$$

$$\text{C}_6\text{H}_5\text{-CH}_3 \xrightarrow{\text{KMnO}_4} \text{C}_6\text{H}_5\text{-CO}_2\text{H}$$

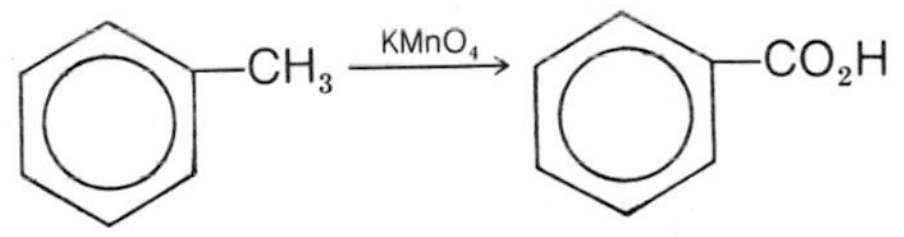

The Need and Basis for Safe Insecticides.   It has been estimated with some justification that if we were to eliminate chemicals in food production, about one-third of the world's people would not eat at all, and that if pesticides were outlawed in the U.S., the food supply would have to be rationed. Calculations have put world food losses to weeds and insects at 70 billion dollars per year, with insects and plant diseases each destroying about 12 per cent, and weeds about 10 per cent of potential production.

More than 200 chemical insecticides are in use. Some of these, such as inorganic arsenicals and those in the nicotine class, are effective against a very wide variety of insects as well as being toxic to animals. Conversely, a great many organic compounds observed to possess insecticidal properties are characterized by a high degree of specificity, so that any individual compound may be highly toxic to some insects or animals and relatively harmless to others. The synthesis expertise of the organic chemist has given us both the capability to make compounds to specification and the tool for discovering the properties materials must have to make them not only effective in controlling insects, but also safe for humans and compatible with the environment.

We are just beginning to understand the biochemical mechanisms of insecticidal action. A summary of some chemistry of insecticides will illustrate the state of our knowledge and how we have applied it in this important area.

Insecticides can be classified according to their mechanical mode of application as contact insecticides, stomach poisons, and fumigants. Contact insecticides penetrate to the vital organs either by penetration of the insect cuticle (hard cover) or through the spiracles (respiratory apertures). Stomach poisons are ingested by mouth. Fumigants enter mostly through the spiracles, though some penetration of the cuticle undoubtedly occurs also.

Many insecticides are mixed with oil or another substance that will dissolve in hydrocarbons or fats. This material will adhere to the cuticle, giving the insecticide it carries a better chance to penetrate. This effect is illustrated by an experiment in which blowfly larvae immersed for an hour in either ethyl alcohol or kerosene were unaffected, whereas a mixture of the two solvents gave highly toxic reactions in seconds. Ideally, an insecticide should contain both a toxic and a fat-soluble component in the same molecule.

One theory that summarizes the properties a good contact insecticide must possess and the role these properties play in its functioning is the following: It must possess a degree of solubility in fats, enabling it to penetrate the cuticle. It also must possess a group that will combine with a vital constituent inside the cell. This is the toxic group; for maximum potency, this group must be neither too reactive nor too inert. If it is too reactive, it can react with less critical cellular components and be wasted; if it is too inert, the reaction will be inefficient.

DDT and its derivatives, whose structures are given below, can be used to illustrate this point.

DDT I
Very reactive

II
Less reactive

III
Inactive

Compounds I and III differ in toxicity, as determined by independent methods, but have about the same fat solubility. Compounds I and II have about the same toxicity, but compound II has less fat solubility. From this, it is suggested that the toxic group is —CCl_3, and that the fat solubility portion

is Cl—⟨benzene ring⟩—

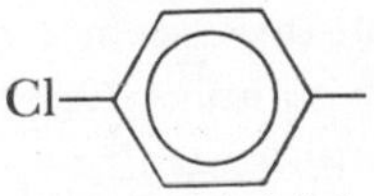

Compound I has the best combination of fat solubility and toxicity and is, therefore, the most active. This kind of thinking has led to the synthesis and testing of many derivatives of DDT and other insecticides. Test results have been correlated with chemical structure and an effort made to relate this information to the biochemistry of insects and mammals. From this is coming information on toxicity differences in insects and mammals, which, in turn, serves as the basis for making safer insecticides.

An important series of powerful insecticides in common use are the organophosphorus compounds, represented by the formulas:

Malathion

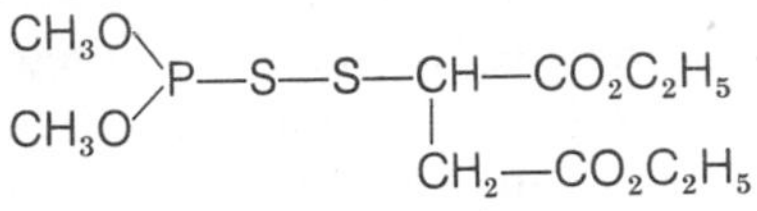

$$\begin{array}{c}CH_3O \\ \\ CH_3O\end{array}\!\!\diagdown P{-}S{-}S{-}\underset{\underset{CH_2{-}CO_2C_2H_5}{|}}{CH}{-}CO_2C_2H_5$$

Parathion

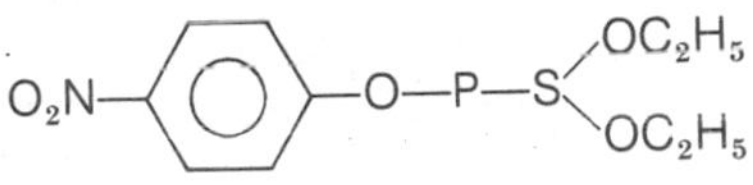

$$O_2N{-}\langle\!\bigcirc\!\rangle{-}O{-}P{-}S\!\!\diagup^{OC_2H_5}_{\diagdown OC_2H_5}$$

Chlorthion

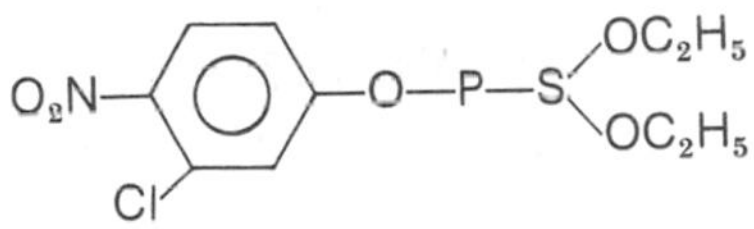

$$O_2N{-}\underset{\underset{Cl}{|}}{\langle\!\bigcirc\!\rangle}{-}O{-}P{-}S\!\!\diagup^{OC_2H_5}_{\diagdown OC_2H_5}$$

Demeton O

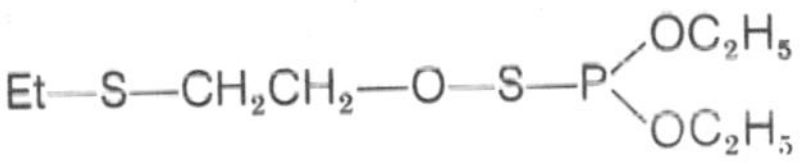

$$Et{-}S{-}CH_2CH_2{-}O{-}S{-}P\!\!\diagup^{OC_2H_5}_{\diagdown OC_2H_5}$$

Demeton S

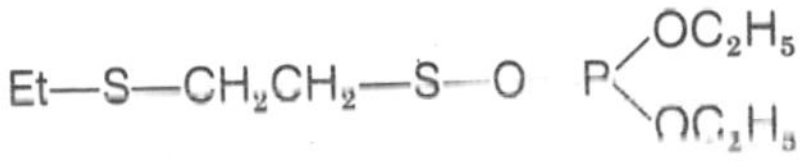

$$Et{-}S{-}CH_2CH_2{-}S{-}O\quad P\!\!\diagup^{OC_2H_5}_{\diagdown OC_2H_5}$$

Menazon

$$\underset{H_2N}{\overset{H_2N}{}}\!\!\text{(triazine ring)}{-}CH_2{-}S{-}P{-}S\!\!\diagup^{OCH_3}_{\diagdown OCH_3}$$

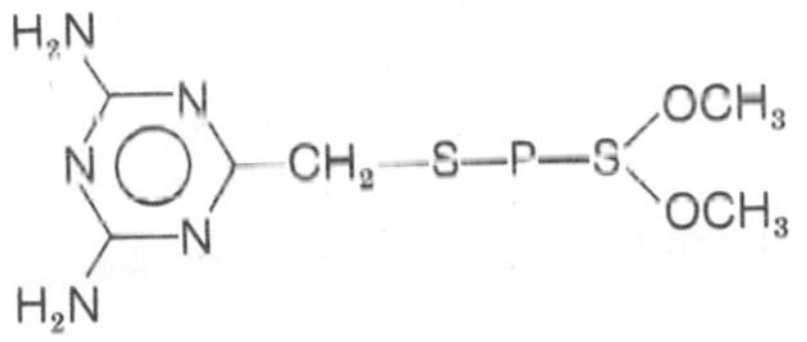

The first compounds in this series tested were excellent insecticides but showed high toxicity to animals. Much research has been devoted to improving this deficiency, leading not only to the development of effective but safer materials, such as the commonly used malathion, but also to some insight into the probable mechanism of toxicity.

The conversion of parathion to chlorthion by adding a chlorine atom reduces mammalian toxicity remarkably. The heterocyclic ring in compounds such as menazon not only reduces the mammalian toxicity, but introduces specificity of action.

These differential effects probably result from a combination of causes. For example, variations in physical properties of the compounds will affect factors such as absorption and membrane permeability. In addition, the compounds can undergo a number of chemical reactions during their metabolism in the insect or host plant, giving products having different toxicities. It is generally believed that the insect toxicity of these materials is caused by the production of compounds that act to inhibit the important cellular reaction that produces acetylcholine, the chemical transmitter of nervous impulses. With its nerve-impulse transmission mechanism blocked or severely hampered, the insect is, of course, in serious difficulty.

SUMMARY   Reactivity of organic structures has been discussed in terms of six types of reactions: (a) oxidations, (b) reductions, (c) condensations, (d) additions, (e) eliminations, and (f) substitutions. Emphasis has been placed on the relation between the structure and bonding in a given functional group and the type of reaction under consideration. Use of the various kinds of reactions in single- and multistep syntheses of organic structures has been illustrated.

IMPORTANT TERMS

Oxidation reaction   **Grignard reagents**   **Halogenation**
Reduction reaction   **Aldol condensations**   **Alkylation**
Condensation reaction   **Dehydrohalogenations**   **Sulfonation**
Addition reaction   **Free-radical reactions**   **Friedel-Crafts reaction**
Elimination reaction   **Mechanism of reaction**   **Ortho-para directing groups**
Substitution reaction   **Carbonium ions**   **Meta-directing groups**
Organolithium reagents   **Nitration**

QUESTIONS AND PROBLEMS

1. Name some oxidizing agents commonly used in organic chemistry and illustrate the use of each with a specific equation.

2. Name some reducing agents commonly used in organic chemistry and illustrate the use of each with a specific equation.

3. Write reactions to show how you could distinguish chemically between: (a) hexane and 3-hexene, and (b) 1-hexene and 2-hexene.

4. What alkenes can be reduced with hydrogen to give 3-methylpentane?

5. Using structural formulas and indicating the reagents needed, show how you could make each of the following conversions:

(a) 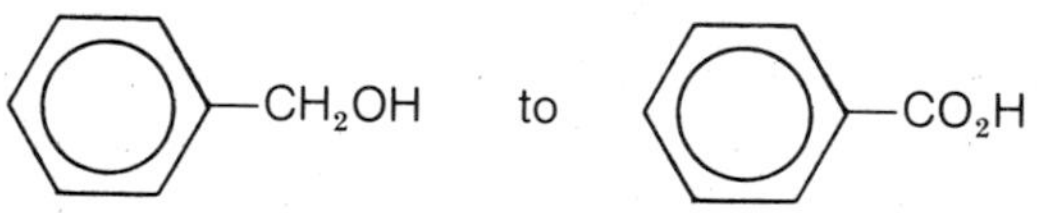 ⬡—CH_2OH to   ⬡—CO_2H

(b) 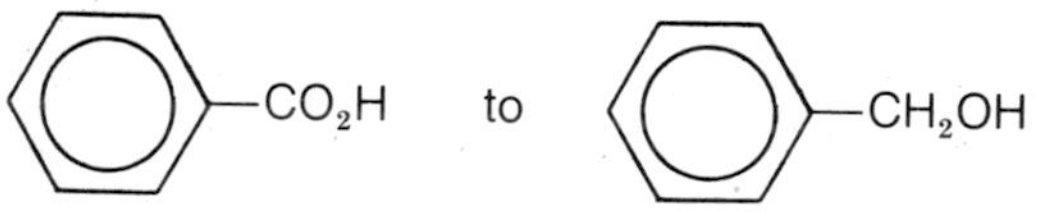 ⬡—CO_2H to   ⬡—CH_2OH

(c) $CH_3C{\equiv}N$ to $CH_3CH_2NH_2$

(d) 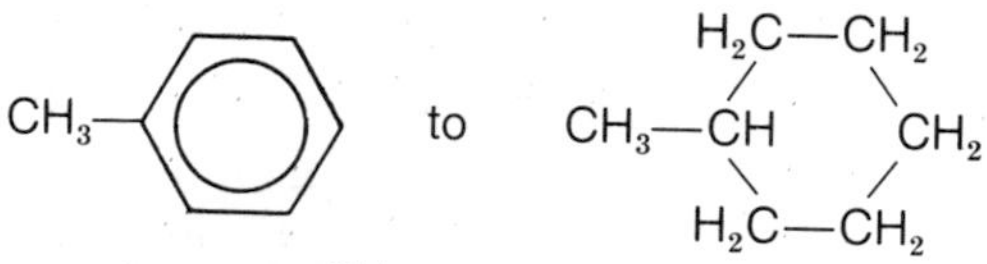 CH_3—⬡   to CH_3—CH cyclopentane ring

(e) 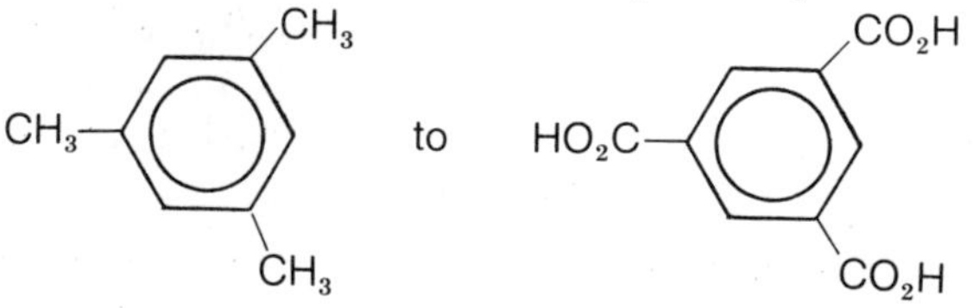 CH_3—⬡ with CH_3 (2,4-dimethyl)   to HO_2C—⬡ with CO_2H and CO_2H

6. Give a simple chemical test that will distinguish: (a) an aldehyde from a ketone, (b) an alcohol from an aldehyde, (c) a primary alcohol from a sec-

ondary alcohol, (d) an alkane from an alkene, (e) a carboxylic acid from an alcohol, and (f) an alcohol from an amine.

7. Which heptanol can be dehydrated to give only 3-heptene?

8. Using structural formulas and indicating the reagents and conditions needed, show how you could make each of the following conversions:

(a) CH_3—$\underset{\underset{CH_3}{|}}{CH}$—$CH_2OH$ to CH_3—$\underset{\underset{CH_3}{|}}{CH}$—$CH_2$—$O$—$CH_2$—$\underset{\underset{CH_3}{|}}{CH}$—$CH_3$

(b) CH_3—$\underset{\underset{CH_3}{|}}{CH}$—$CH_2OH$ to CH_3—$\underset{\underset{CH_3}{|}}{CH}$—$\underset{\underset{O}{||}}{C}$—$O$—$\underset{\underset{O}{||}}{C}$—$\underset{\underset{CH_3}{|}}{CH_3}$—$CH_3$

(c) —CH_2OH to —$\underset{\underset{O}{||}}{C}$—$O$—$CH_2$—

(d)

(e) —CHO + CH_3—CH_2—CHO to —$\underset{\underset{HO}{|}}{C}$—$\overset{\overset{H}{|}}{\underset{}{}}$$\underset{\underset{CH_3}{|}}{CH}$—CHO

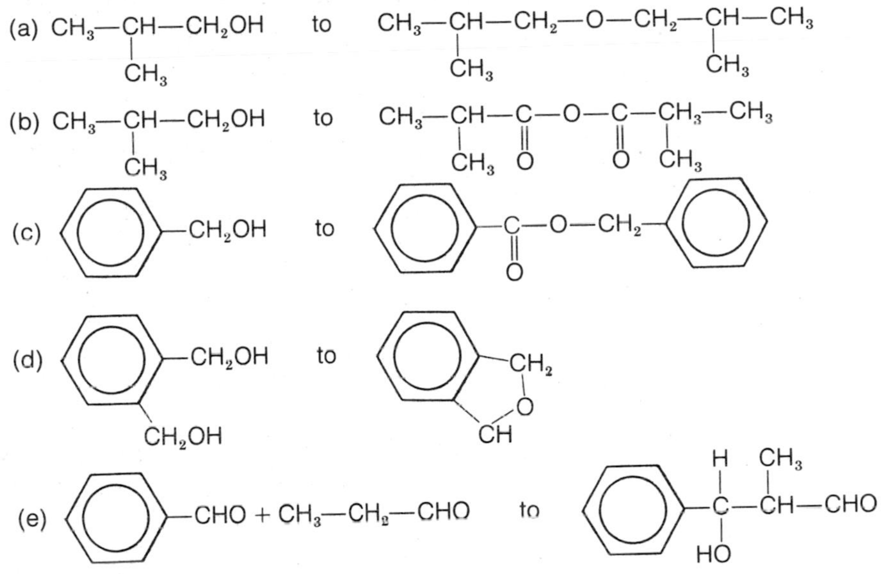

9. What chemical tests can be used to distinguish among propanal, 2-pentanone, and 3-pentanone?

10. Write equations to show the aldol condensation between: (a) ethanal and 2-methylpropanal, and (b) propanal and benzaldehyde.

11. Using structural formulas and indicating the reagents and conditions needed, show how you could make each of the following conversions:

(a) CH_3—$\underset{\underset{CH_3}{|}}{C}$=$CH_2$ to CH_3—$\overset{\overset{O-SO_3H}{|}}{\underset{\underset{CH_3}{|}}{C}}$—$CH_3$

(b) CH_3—CH—CH—CH_3 to CH_3—$\underset{\underset{O}{||}}{C}$—$CH_2$—$CH_3$

(c) CH_3—CH=CH—CH_3 to $\underset{\underset{CH_3}{}}{\overset{\overset{CH_3-CH_2}{}}{CH}}$—$O$—$\overset{\overset{CH_2-CH_3}{}}{\underset{\underset{CH_3}{}}{CH}}$

(d) —Li and CH_3—$\underset{\underset{O}{||}}{C}$—$CH_3$ to —$\overset{\overset{CH_3}{|}}{\underset{\underset{CH_3}{|}}{C}}$—OH

(e) $CH_3\underset{\underset{O}{||}}{C}$—H to CH_3—$\overset{\overset{H}{|}}{\underset{\underset{OH}{|}}{C}}$—

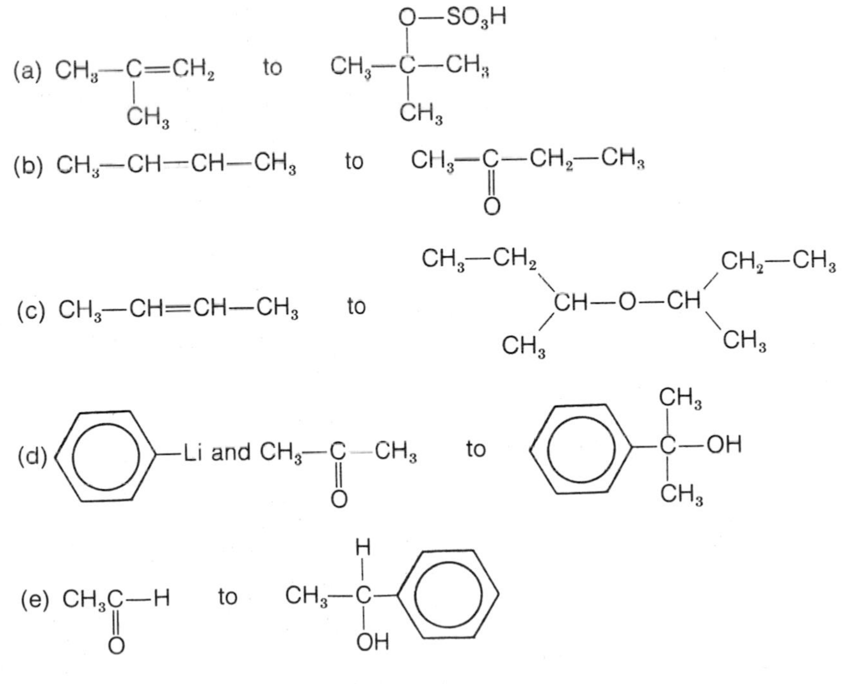

12. Using structural formulas and indicating the reagents and conditions needed, show how you could make the following conversions:

(a)  [benzene ring]—CH—CH$_3$ with OH   to   [benzene ring]—CH=CH$_2$

(b) 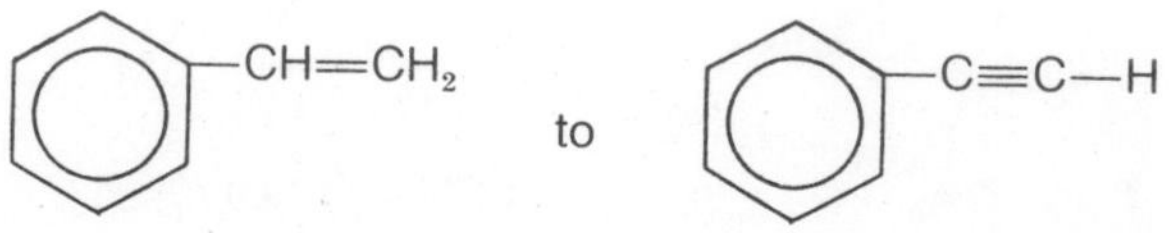 [benzene ring]—CH=CH$_2$ to   [benzene ring]—C≡C—H

(c) CH_2=CH—CH_2—CH=CH_2 to CH_3—CH⟨CH$_2$⟩CH—CH$_3$

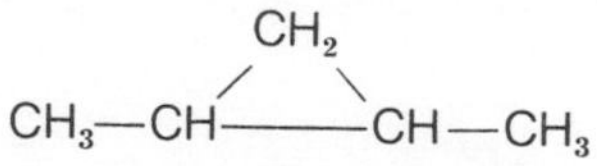

(d) CH_2=CH—CH_2—CH=CH_2 to CH_3—C(=O)—CH_2—C(=O)—CH_3

(e) CH_3—C(=O)—H   to CH_2=CH—CH_2—CHO

13. Using structural formulas and indicating the reagents and conditions needed, show how you could make the following conversions:

(a) CH_3—CH(CH$_3$)—CH_3 to CH_3—C(Cl)(CH$_3$)—CH_3

(b) CH_3—CH(CH$_3$)—Cl   to CH_3—CH(CH$_3$)—C≡CH

(c) CH_3—C(CH$_3$)=CH_2 to CH_3—C(CH$_3$)(NH$_2$)—CH_3

(d) CH_3—[benzene ring]   to CH_3—[benzene ring]—CH(CH$_3$)(CH$_3$)

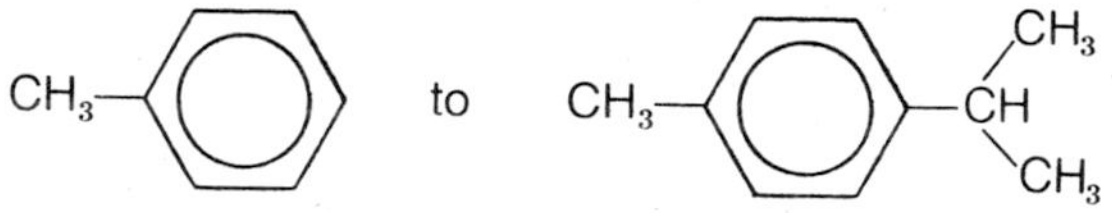

(e)  [benzene ring]   to   [benzene ring]—C(cyclohexene ring: HC—CH_2, CH_2, H_2C—CH_2)

14. Predict the products in the following reactions:

(a) CH_3—CH(CH$_3$)—CH_3 + Cl_2 $\xrightarrow{\text{light}}$

(b) 2 [benzene ring]—Li + CH_3—C(=O)—CH_2—CH_2—C(=O)—CH_3 ⟶

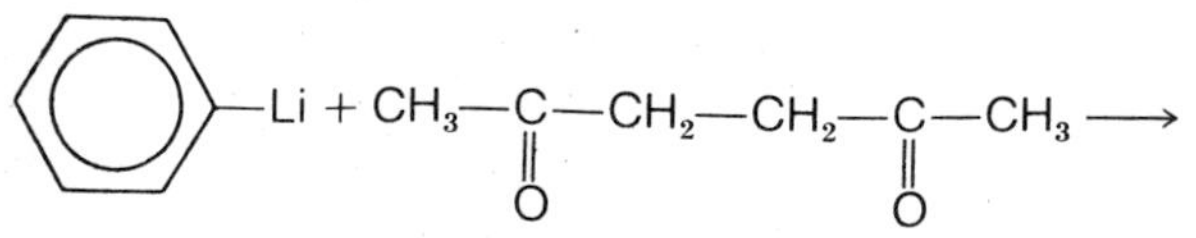

(c) 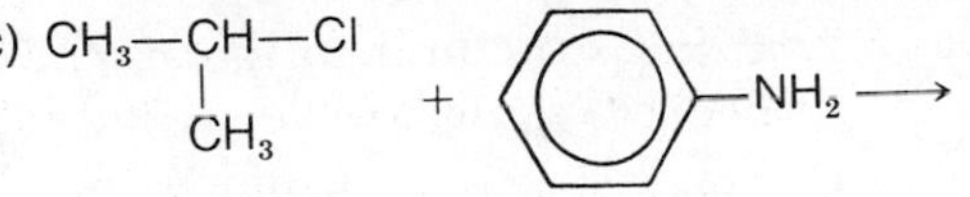 $CH_3-\underset{\underset{CH_3}{|}}{CH}-Cl$ + —NH$_2$ ⟶

(d) $2\ CH_3-\underset{\underset{CH_3}{|}}{CH}-Cl$ + $NH_2-\underset{\underset{O}{\|}}{C}-NH_2$ ⟶

(e) 2 —OH + $\underset{\underset{Cl}{|}}{\overset{\overset{O}{\|}}{C}}-\underset{\underset{Cl}{|}}{\overset{\overset{O}{\|}}{C}}$ ⟶

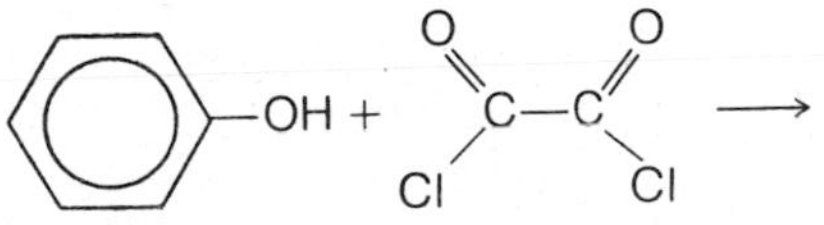

(f) 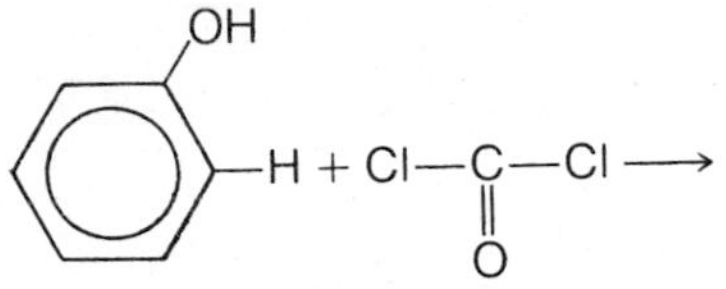 —H + $Cl-\underset{\underset{O}{\|}}{C}-Cl$ ⟶

15. A given alkene reacts with sulfuric acid to form an alkyl hydrogensulfate. This, in turn, hydrolyzes to form tertiary butyl alcohol. What is the structural formula of the original alkene?

16. The carboxylic acids have relatively high boiling points and, when reacted with carbonyl agents such as hydroxylamine, fail to yield derivatives. How can these observations be explained?

SPECIAL PROBLEMS

1. Using structural formulas and indicating the reagents and conditions needed, show how you could make the following conversions: (a) 1-bromopropane from 1-propanol; (b) 2-bromopropane from 1-propanol; (c) 1,2-dibromopropane from 1-bromopropane; (d) propyne from 2-propanol; (e) 2-propanol from propyne; and (f) 1-propanol to 1,2-dibromopropane.

2. Using structural formulas and indicating the reagents and conditions needed, show how you could make the following conversions: (a) 1-butanal from 1-butanol; (b) 2-butanone from 2-butanol; (c) 2,3-dibromo-1-propanol from 1-propanal; (d) phenyl acetate from phenol; (e) propanoic acid from 1-chloropropane; and (f) cyclopentane from 1,5-pentanediol.

3. Upon heating with potassium permanganate, an alkene having the formula C_6H_{12} gave propanoic acid as the only organic product. Write the structural formula for the alkene.

4. Upon treatment with lithium aluminum hydride, a dione having the formula $C_6H_{10}O_2$ gave 3-methyl-2,4-pentanediol. Write the structural formula for the dione.

5. Using structural formulas and indicating the reagents and conditions needed, show how you could make the following conversions: (a) diisopropyl ether from propene; (b) 2-bromopropane from propanone; (c) 1,2-dibromopropane from propanone; (d) 1,2,3-propanetriol from propenal; (e) propene from propanal; and (f) propanone from propanal.

6. Using structural formulas and indicating the reagents and conditions needed, show how you could make the following conversions: (a) propanoic acid from CH_3CH_2CN; (b) propanoic acid from 1-bromopropane; (c) butanoic acid from 1-bromopropane; (d) 1-propanol from propanoic

acid; (e) acetic acid from 2-butene; and (f) 2-chloropropanoic acid from propene.

7. Using structural formulas and indicating the reagents and conditions needed, show how you could make the following conversions:

(a) ethanal from $CH_3—C=O$
$\qquad\qquad\qquad\qquad\quad |$
$\qquad\qquad\qquad\qquad\quad Cl$

(b)

(c) 1,1,2,2-tetrachloroethane from ethanol
(d) 2-cyanopropane from 1-propanol
(e) ethyl acetate from ethyl chloride
(f) $CH_2—CO_2H$
$\quad |$ from 1,2-dichloroethane
$\quad CH_2—CO_2H$

8. Using structural formulas and indicating the reagents and conditions needed, show how you could make the following conversions:
(a) phenylacetylene from 2-phenylethanol
(b) 1-phenylethane-1,2-diol from 2-phenylethanol
(c) propylbenzene from benzene

(d) 1,1-diphenylethene from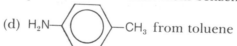

(e) triphenylmethane from benzene and chloroform

(f) dicyclohexylmethane from benzene and $Cl—CH_2—$⬡

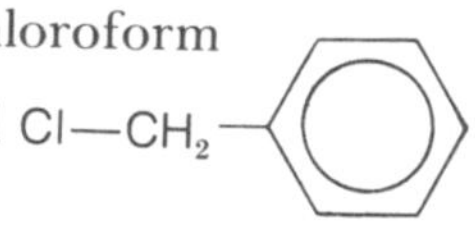

9. Using structural formulas and indicating the reagents and conditions needed, show how you could make the following conversions:
(a) phenylenediamine from m-dinitrobenzene
(b) m-bromoaniline from benzene
(c) p-nitrochlorobenzene from benzene

(d) $H_2N—$⬡$—CH_3$ from toluene

(e) diphenylamine from aniline

10. Three monochloroalkanes are produced by the chlorination of a hydrocarbon. The three products are then hydrolyzed with sodium hydroxide, and the resulting products are oxidized. The compounds, at this point, are identified as 2-pentanone, 3-pentanone, and pentanoic acid. Write the formula for the original hydrocarbon.

11. A compound composed of 71.65% chlorine reacts with potassium cyanide. The compound produced is hydrolyzed and produces an acid with a neutralization equivalent of 59.04. If the acid is treated with phos-

phorus oxychloride, a solid compound of carbon, hydrogen, and oxygen is formed, having a melting point of 119.6°C and a boiling point of 261°C. Write the structural formulas for the original compound and each compound produced along the way.

12. A hydrocarbon reacts with hydrogen iodide to produce compound *X*. Compound *X* reacts with aqueous potassium hydroxide to form compound *Y*. Oxidation of compound *Y* forms propanone. Write the structural formulas for the original compound, compound *X*, compound *Y*, and the final compound.

13. A compound consisting only of carbon, hydrogen, and nitrogen was vaporized. In the gaseous state, a 1-l sample of it weighs 2.68 g at standard conditions. A 1-g sample of the compound treated with nitrous acid gives 746.5 cc of nitrogen, also at standard conditions. What is the structural formula of the original compound?

14. An alkylated acetoacetic ester was hydrolyzed with concentrated potassium hydroxide. The products of the reaction were ethanol and the potassium salts of ethanoic and pentanoic acids. Write the structural formula of the original compound.

REFERENCES   HART, H., and SCHUETZ, R. A., *A Short Course in Organic Chemistry*, 3rd ed., Houghton Mifflin, Boston, 1966.

HENDRICKSON, J. B., CRAM, D. J., and HAMMOND, G. S., *Organic Chemistry*, 3rd ed., McGraw-Hill, New York, 1970.

VAN ORDEN, H. A., and LEE, G. L., *Elementary Organic Chemistry, A Brief Course*, Saunders, Philadelphia, 1969.

OUELLETTE, R. W., *Introductory Organic Chemistry*, Harper & Row, New York, 1970.

POLYMERS AND OTHER COMPLEX MOLECULES

The hypothesis that high polymers are composed of covalent structures many times greater in extent than those occurring in simple compounds, and that this feature alone accounts for the characteristic properties which set them apart from other forms of matter, is in large measure responsible for the rapid advances in the chemistry and physics of these substances.

Paul J. Flory (1910—)

Life depends fundamentally on organic substances. They provide not only food, clothing, shelter, transportation, and medication, but also a means for producing more and higher quality varieties of the same materials. Thus, studies of the chemistry of food have led to more and better foods; studies of the chemistry of fibers, building materials, fuels, medicines, and many other vital materials have led to more and better products to meet the same needs. This chapter deals with the structure and properties of several classes of organic compounds that are essential in meeting some of the primary biological needs of man, materials that by themselves or in some improved form must be counted on to feed, clothe, house, and maintain the health of future generations.

POLYMERS

Polymers are *large molecules made up of many smaller molecules bonded together*. The smaller molecules are called monomers. Naturally occurring polymers include rubber, cellulose (the structural material of plants), and the proteins of muscle and living cells. Synthetic polymers include many familiar products: fibers such as nylon, Dacron, and Acrilan; plastics for uses varying from brush handles to brush bristles and food wrappings, such as Bakelite, Styron, Plexiglas, Saran, Mylar, and polypropylene; elastomers such as synthetic rubber and Neoprene; and resins, such as Glyptal, Duralon, or Lucite, for finishes on automobiles, boats, and other objects. More than half of the products of the chemical process industries of the U.S. are based, to some degree, on the manufacture or use of polymers.

Chemically, polymers are classified according to the way in which the bonds are formed, as addition polymers and condensation polymers. In addition polymers, the small molecules contain double bonds, and polymerization occurs by opening of the double bonds. Polymerized tetrafluoroethylene (commercial Teflon) is an example:

$$nCF_2{=}CF_2 \longrightarrow n[{-}CF_2{-}CF_2{-}] \text{ or } {-}CF_2{-}CF_2({-}CF_2{-}CF_2{-})_{n-2}CF_2{-}CF_2{-}$$

In condensation polymers, a small molecule such as water is eliminated between functional groups on two molecules. Polymerized terephthalic acid and ethylene glycol (commercial Dacron polyester), in which bonding occurs through ester linkages, is an example:

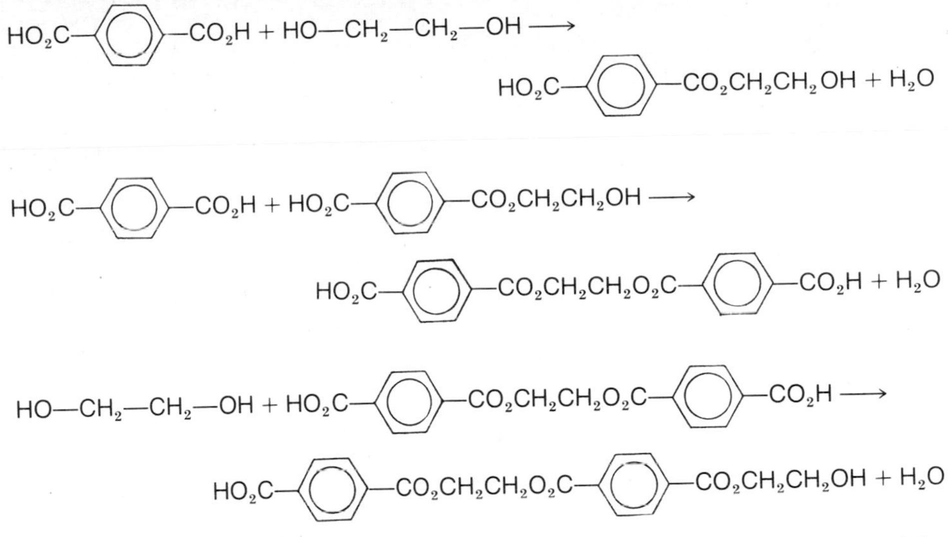

$$HO_2C-\langle\bigcirc\rangle-CO_2H + HO-CH_2-CH_2-OH \longrightarrow$$

$$HO_2C-\langle\bigcirc\rangle-CO_2CH_2CH_2OH + H_2O$$

$$HO_2C-\langle\bigcirc\rangle-CO_2H + HO_2C-\langle\bigcirc\rangle-CO_2CH_2CH_2OH \longrightarrow$$

$$HO_2C-\langle\bigcirc\rangle-CO_2CH_2CH_2O_2C-\langle\bigcirc\rangle-CO_2H + H_2O$$

$$HO-CH_2-CH_2-OH + HO_2C-\langle\bigcirc\rangle-CO_2CH_2CH_2O_2C-\langle\bigcirc\rangle-CO_2H \longrightarrow$$

$$HO_2C-\langle\bigcirc\rangle-CO_2CH_2CH_2O_2C-\langle\bigcirc\rangle-CO_2CH_2CH_2OH + H_2O$$

The size of the polymer molecules depends upon the conditions used in polymerization, and a sample of a given polymer will contain huge molecules of many different sizes. The average molecular weight of synthetic polymers may be as low as 5000 or as high as several million. Of the naturally occurring polymers, rubber has molecular weights ranging from 60,000 to 350,000; cellulose has an average molecular weight of 300,000 to 500,000; some proteins have molecular weights as high as 15,000,000.

The Structure of Synthetic Polymers

Most polymers can be described in terms of their structural unit, or units. This unit usually is the monomer residue as it exists in the backbone of the polymer. For example, in polyvinyl chloride, the polymer made from vinyl chloride,

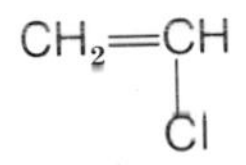

$$CH_2{=}CH$$
$$\overset{|}{C}l$$

the structural unit is the group

$$-CH_2-\overset{|}{C}H-$$
$$\overset{|}{C}l$$

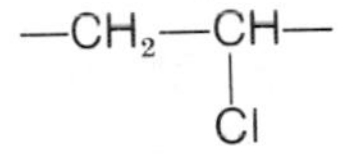

In polystyrene, in which the monomer is styrene,

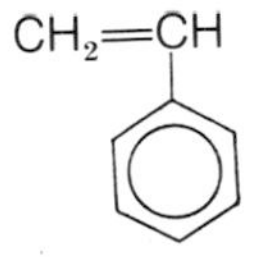

$$CH_2{=}CH$$
$$\bigcirc$$

the structural unit is the group

$$-CH_2-CH-$$
$$\bigcirc$$

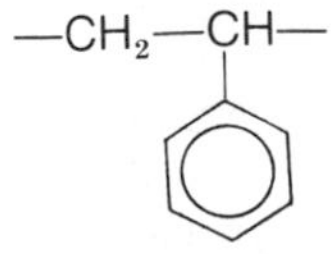

Table 28.1 Monomers and Structural Units in Some Addition Polymers

Polymer	Monomer	Structural Unit
Polystyrene	$CH_2{=}CH$— (phenyl)	—CH_2—CH— (phenyl)
Polyacrylonitrile	$CH_2{=}CH$—CN	—CH_2—CH—CN
Polyvinylidine chloride	$CH_2{=}CCl_2$	—CH_2—$C(Cl)(Cl)$—
Polymethylmethacrylate	$CH_2{=}C(CH_3)CO_2CH_3$	—CH_2—$C(CH_3)CO_2CH_3$—
Polyisobutylene	$CH_2{=}C(CH_3)_2$	—CH_2—$C(CH_3)_2$—

In the polyesters Dacron and Mylar, it is the

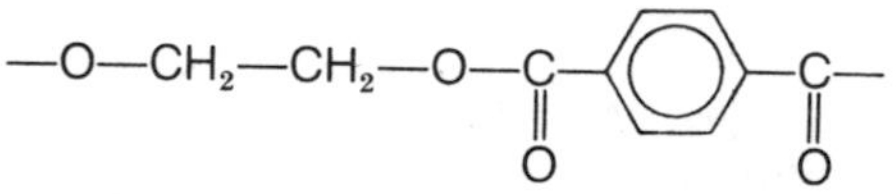

$$-\text{O}-\text{CH}_2-\text{CH}_2-\text{O}-\overset{\text{O}}{\underset{}{\text{C}}}-\langle\text{C}_6\text{H}_4\rangle-\overset{\text{O}}{\underset{}{\text{C}}}-$$

unit, the residue of the two monomers used to make these polymers.

In many common addition polymers, the structural unit is

$$-\text{CH}_2-\overset{}{\underset{\text{X}}{\text{CH}}}-$$

(or —CH_2—CX_2—), where X may be any of numerous groups as illustrated in Table 28.1. The nature of X and the molecular weight of the polymer determines, to a great extent, the properties of these polymers. For example, polyethylene (X is H) is used as tough, water-impermeable films; polyacrylonitrile (X is CN) is used as fibers; polystyrene is used as a moldable plastic (combs, brush handles) and as an insulating plastic foam; polyisobutylene,

$$\left(-\text{CH}_2-\overset{\text{CH}_3}{\underset{\text{CH}_3}{\text{C}}}-\right)_n$$

is a synthetic rubber. Table 28.2 lists some commercial addition polymers and their uses. We shall discuss the relation between structure and properties in the next section.

Table 28.2 Some Commercial Addition Polymers

Name	Monomer	Polymer	Uses
Polyperfluoroethylene (Teflon)	 Tetrafluoro-ethylene	$-\text{C}-\text{C}-\text{C}-\text{C}-$ with F substituents	Chemically resistant; electrical insulation, liner for laboratory ware, pumps, frying pans
Polystyrene (Styron, Lustron)	$\text{CH}=\text{CH}_2$ (phenyl) Styrene	$-\text{CH}-\text{CH}_2-\text{CH}-\text{CH}_2-$ (phenyl groups)	Plastic foams, wrapping material, handles
Polyacrylonitrile (Orlon, Acrilan)	$\text{CH}=\text{CH}_2$ / CN Acrylonitrile	$-\text{CH}-\text{CH}_2-\text{CH}-\text{CH}_2-$ / CN CN	Fibers
Polyvinylchloride (Koroseal, Geon)	$\text{CH}=\text{CH}_2$ / Cl Vinyl chloride	$-\text{CH}-\text{CH}_2-\text{CH}-\text{CH}_2-$ / Cl Cl	Raincoats, tank lining
Polypropylene	$\text{CH}=\text{CH}_2$ / CH_3 Propylene	$-\text{CH}-\text{CH}_2-\text{CH}-\text{CH}_2-$ / CH_3 CH_3	Wool substitute, transparent films
Polymethylmethacrylate (Lucite, Plexiglas)	CH_3 / $\text{C}=\text{CH}_2$ / CO_2CH_3 Methyl methacrylate	CH_3 CH_3 / $-\text{C}-\text{CH}_2-\text{C}-\text{CH}_2-$ / CO_2CH_3 CO_2CH_3	Plastic windows, handles
Polyvinylacetate (Gelva, Vinylite)	$\text{CH}=\text{CH}_2$ / O_2CCH_3 Vinylacetate	$-\text{CH}-\text{CH}_2-\text{CH}-\text{CH}_2-$ / O_2CCH_3 O_2CCH_3	Latex paints, adhesives
Poly-N-vinylpyrrolidone (Periston)	$\text{CH}=\text{CH}_2$ / N / CH_2 $\text{C}=\text{O}$ / CH_2-CH_2 N-vinylpyrrolidone	$-\text{CH}-\text{CH}_2-\text{CH}-\text{CH}_2-$ / N N / CH_2 $\text{C}=\text{O}$ CH_2 $\text{C}=\text{O}$ / CH_2-CH_2 CH_2-CH_2	Blood plasma substitute, hair sprays
Vinyl chloride, vinyl acetate copolymer (Tygon)	—	$-\text{CH}-\text{CH}_2-\text{CH}-\text{CH}_2-$ / Cl O_2CCH_3	Flexible tubing, sheets
Vinylidene chloride, acrylonitrile copolymer	Cl / $\text{C}=\text{CH}_2$ / Cl Vinylidene chloride	Cl / $-\text{C}-\text{CH}_2-\text{CH}-\text{CH}_2-$ / Cl CN	Transparent film

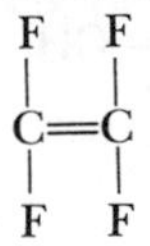

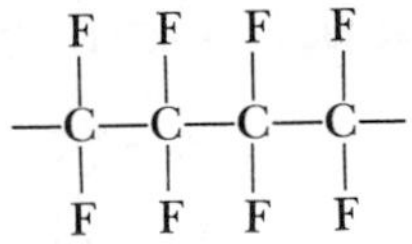

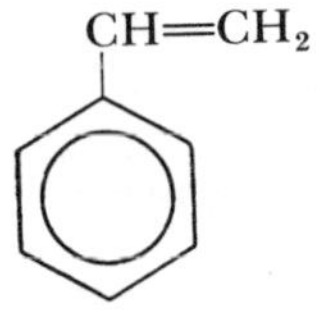

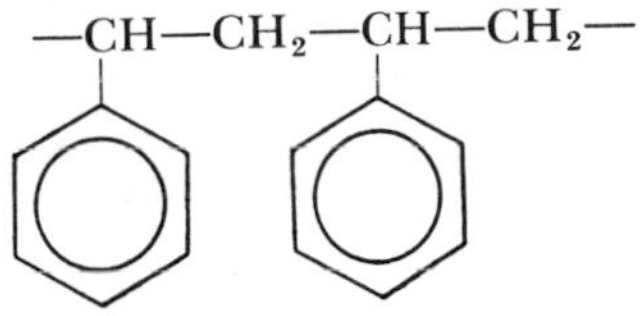

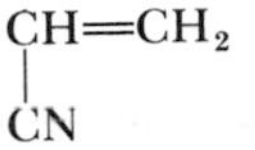

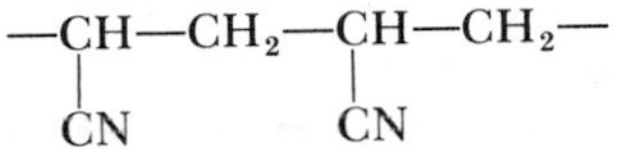

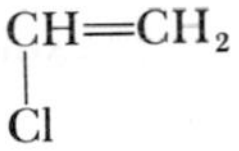

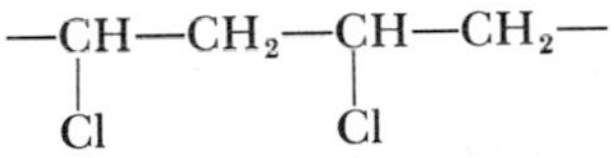

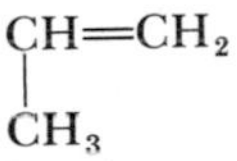

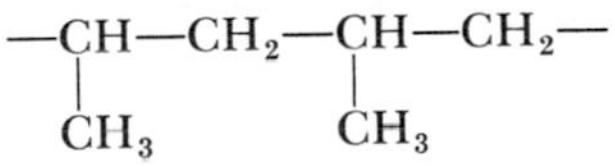

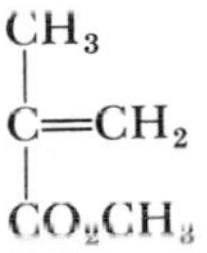

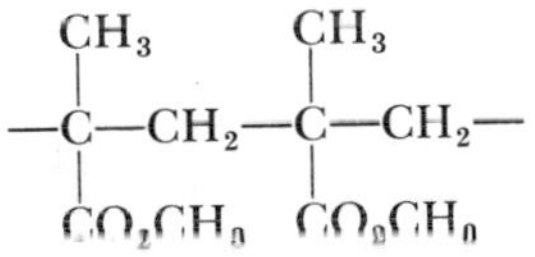

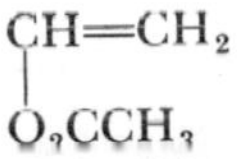

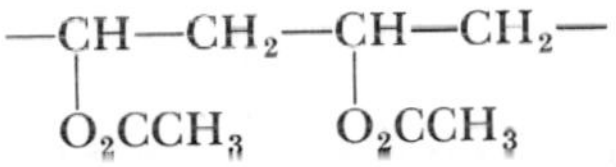

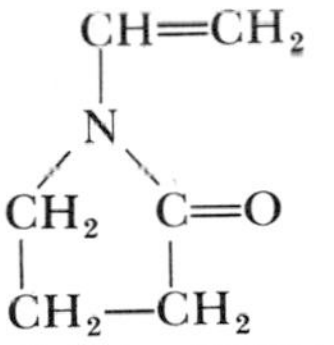

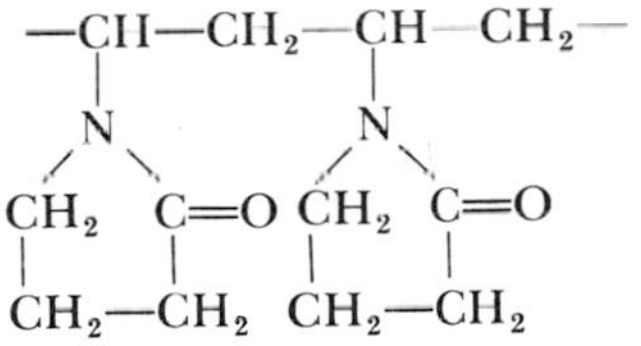

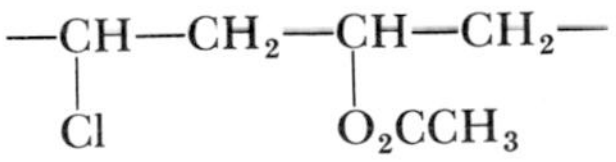

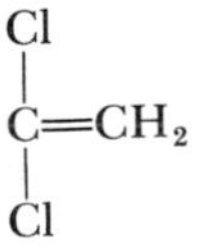

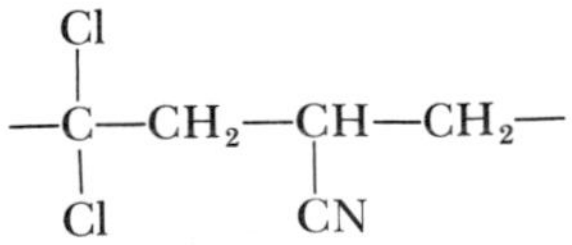

Table 28.2 (Continued)

Name	Monomer	Polymer	Uses
Styrene, butadiene copolymer	$CH_2{=}CH{-}CH{=}CH_2$ Butadiene	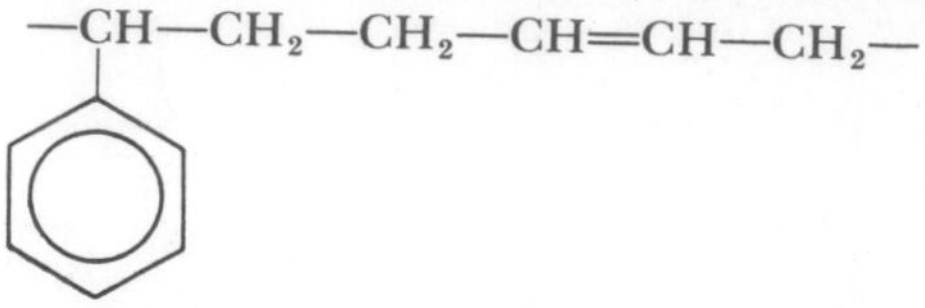 $-CH{-}CH_2{-}CH_2{-}CH{=}CH{-}CH_2-$ (with phenyl group attached to first CH)	Synthetic rubber
Polyisoprene	$CH_2{=}C{-}CH{=}CH_2$ $\quad\ \ \underset{\textstyle CH_3}{\vert}$ Isoprene	$-CH_2{-}C{=}CH{-}CH_2-$ $\qquad\ \ \underset{\textstyle CH_3}{\vert}$	Synthetic rubber

Table 28.3 Some Commercial Condensation Polymers

Name	Monomer	Structural Unit	Uses
Dacron, Terylene	 CO_2H (on benzene ring) CO_2H Terephthalic acid $HOCH_2CH_2OH$ Ethylene glycol	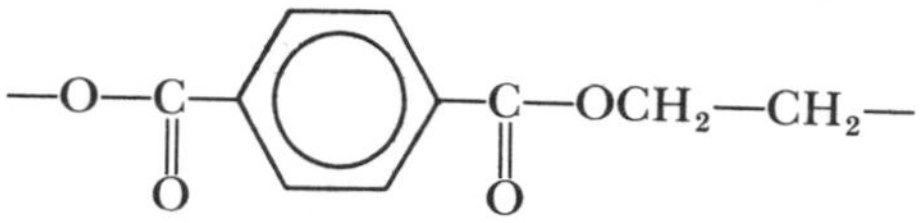 $-O{-}C({=}O){-}$ (benzene ring) $-C({=}O){-}OCH_2{-}CH_2-$	Textile fibers
Nylon 6-6	$HO_2C(CH_2)_4CO_2H$ Adipic acid $H_2N(CH_2)_6NH_2$ Hexamethylene diamine	$-C({=}O){-}(CH_2)_4{-}C({=}O){-}NH(CH_2)_6NH-$	Textile fibers, bristles, sheets
Bakelite Durez	$H_2C{=}O$ Formaldehyde 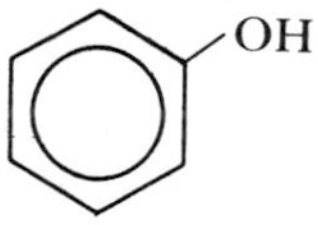 Phenol (with OH)	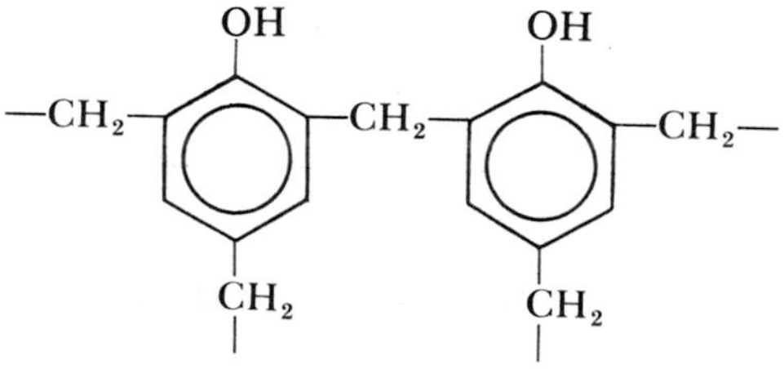 $-CH_2-$ (benzene ring, OH) $-CH_2-$ (benzene ring, OH) $-CH_2-$ with CH_2 chains below	Records, telephone receivers
Beetle	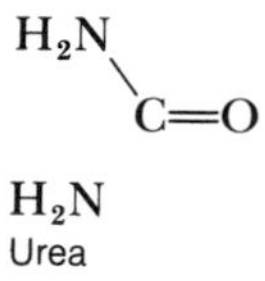 $\underset{\textstyle H_2N}{}{C{=}O}$ H_2N Urea $H_2C{=}O$ Formaldehyde	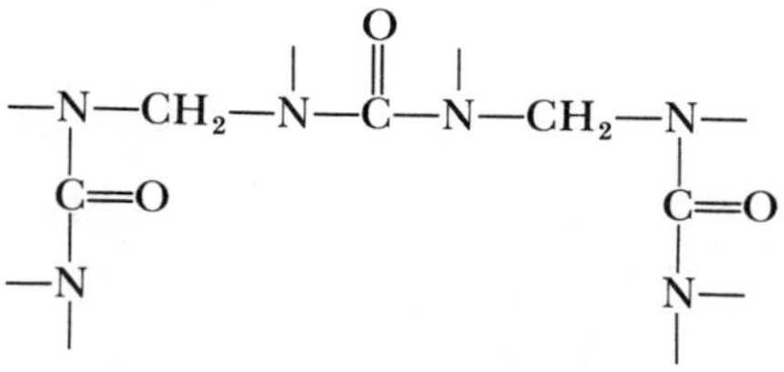 $-N{-}CH_2{-}N{-}C({=}O){-}N{-}CH_2{-}N-$ with $C{=}O$ and N branches	Molding powders, buttons, bottle caps

In most addition polymers the monomers are aligned in the polymer in a head-to-tail arrangement—that is,

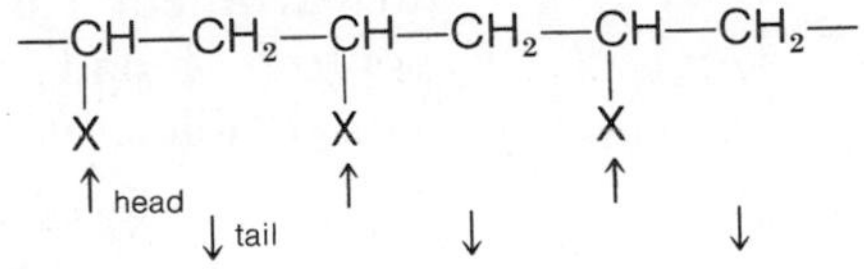

$$-CH-CH_2-CH-CH_2-CH-CH_2-$$

with X groups indicated: ↑ head ↓ tail ↑ ↓ ↑ ↓

Homopolymers are those containing a single repeating unit; copolymers contain two repeating units. Polypropylene is a homopolymer. It contains only the propylene structural unit. Saran is a copolymer. It contains two structural units, $-CCl_2-CH_2-$ and $-CHCl-CH_2-$.

The structural units in the polymer may be arranged or connected several ways. In linear polymers, the units are connected to one another in a chain arrangement, as X—M—M—M—M—M · · · Y, where M is a bivalent structural unit and X and Y are the end groups that prevent the polymer chain from growing longer. Branched polymers have the general structure

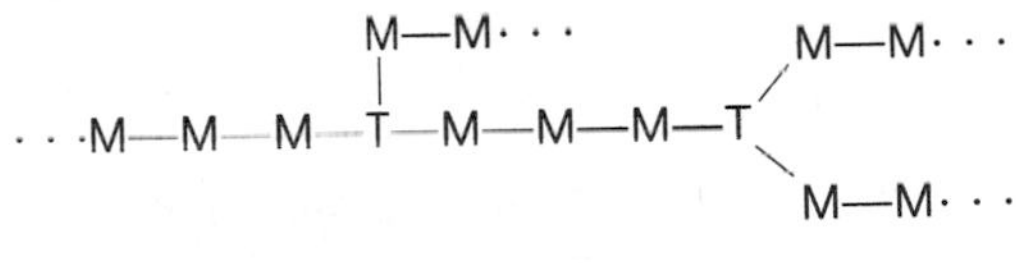

$$\cdots M-M-M-T-M-M-M-T\begin{matrix} M-M\cdots \\ M-M\cdots \end{matrix}$$

with branch M—M· · · above the first T.

which arises when a monomer having three bonding sites is incorporated into the polymer.

Example:

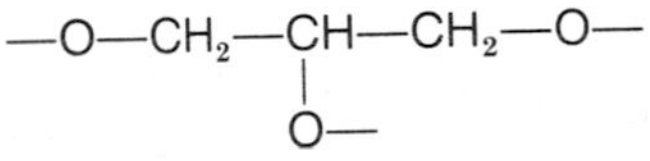

$$-O-CH_2-CH-CH_2-O-$$
$$\qquad\qquad\quad | $$
$$\qquad\qquad\quad O-$$

Network polymers are possible when a tetrafunctional monomer such as divinyl benzene,

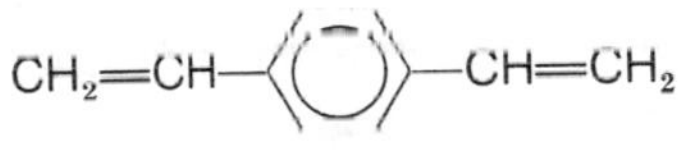

$$CH_2=CH-\langle \bigcirc \rangle-CH=CH_2$$

is used with a comonomer such as styrene. This is illustrated in Figure 28.1.

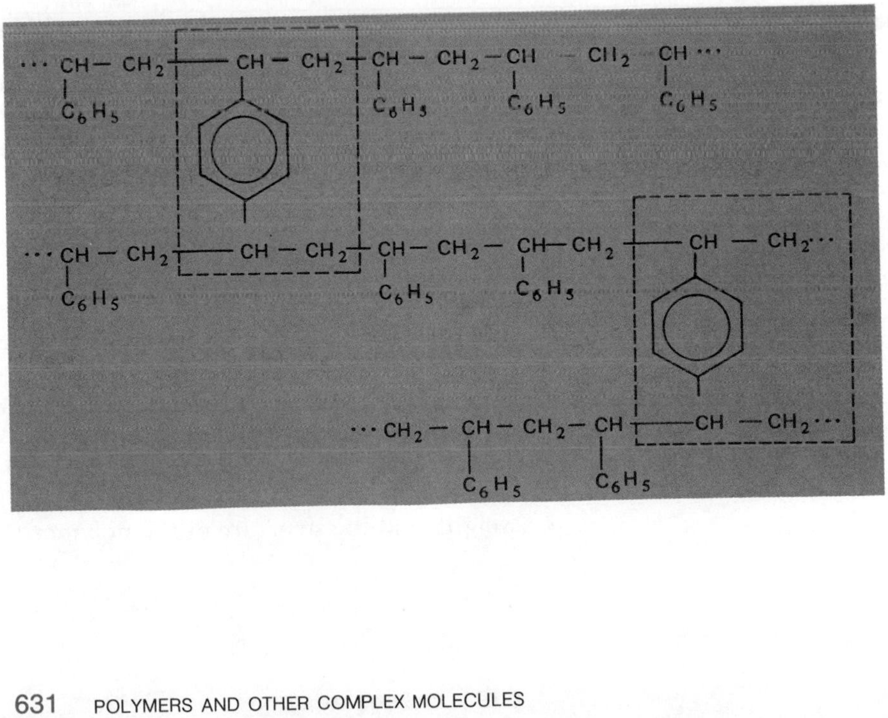

Figure 28.1 Portion of a network copolymer of styrene and divinyl benzene. The divinyl benzene residents are indicated by the dashed lines.

The common phenol-formaldehyde resin known as Bakelite, one of the first synthetic polymers made, is a network polymer.

Much synthetic rubber is a copolymer of about 25 per cent styrene and 75 per cent butadiene (CH_2=CH—CH=CH_2). The polymer backbone contains double bonds arising from 1–4 addition to butadiene:

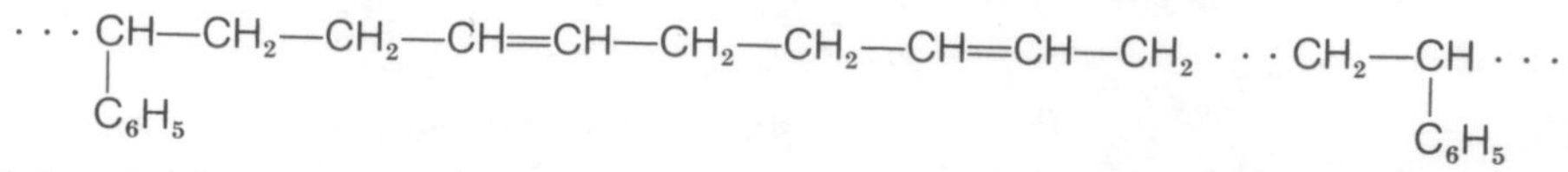

$$\cdots CH-CH_2-CH_2-CH=CH-CH_2-CH_2-CH=CH-CH_2 \cdots CH_2-CH \cdots$$
$$\quad\; | \qquad\qquad\qquad\qquad\qquad\qquad\qquad\qquad\qquad\qquad\qquad | $$
$$\quad\; C_6H_5 \qquad\qquad\qquad\qquad\qquad\qquad\qquad\qquad\qquad\qquad\quad C_6H_5$$

Vulcanization usually involves treating the polymer with sulfur, which reacts with the double bonds to form cross-links between polymer chains, illustrated as follows:

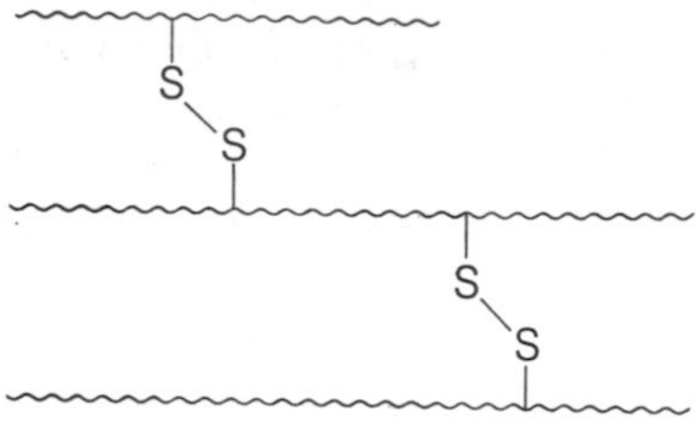

This cross-linking destroys some or nearly all of the double bonds. Soft rubber contains 1 to 2 per cent sulfur and still has double bonds. Hard rubber may contain as much as 35 per cent sulfur. It has very few double bonds and very little elasticity.

Relation of Polymer Structure to Properties

In this section, we shall attempt to relate the structure of some linear polymers to the properties they exhibit in some of their commercial applications. We shall consider four types of properties: those suitable for plastics, rubber, films, and fibers.

The molecular weight and the molecular-weight distribution of a polymer have profound effects on its physical properties. All synthetic polymers and many natural polymers consist of mixtures of molecules of various molecular weights. The softening point, the toughness, and the degree of crystallinity of polymers are related to this distribution. In general, a narrow range of molecular weights in a sample gives more useful properties. Above a certain molecular weight for each substance, small changes in molecular weight have only a negligible effect on the properties of the polymer. It is in this molecular-weight range that the following considerations apply.

Let us then consider the nature of the solid state in linear polymers. Here is a system composed of very long, threadlike structures that have lost nearly all of their translational energy. As in other solids, these molecules have a tendency to arrange themselves in an orderly way. However, because the threads are so long and so flexible, it is difficult to attain a high degree of order. Instead, with few exceptions, the situation may be described as a generally amorphous assembly of threads, perhaps like cooked spaghetti. There may be small highly organized crystallites of about 100 A in length embedded in this assembly. The interior of these crystallites resembles that of ordinary crystals. In effect, then, we view the solid polymer as composed of a matrix of threads that may contain stiff rodlike or disklike crystallites embedded in it. The degree of crystallinity or the per cent of the total matrix composed of crystallites varies with the temperature, the molecular weight, and the structure of the polymer (Figure 28.2). Very high molecular

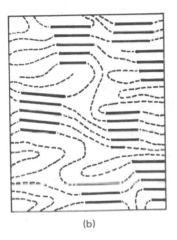

Figure 28.2 Crystallites in a linear polymer: (a) unoriented; (b) oriented after stretching.

weight polyethylene is about 95 per cent crystalline; most polymers have much less crystallinity.

Considering now the energy and structure within the amorphous matrix of threads, the evidence—both experimental and theoretical—shows that this region consists of randomly arranged, coiled, uncoiled, knotted, and entangled threads. At lower temperatures, the atoms or groups in the threads have vibrational energy, but neither the thread nor its parts undergo appreciable rotational motion. Rotational motion of threads involves twisting and writhing like a very long, wounded snake. Such rotations of the groups in the backbones of linear polymers almost always begin sharply at some temperature from −100°C to +200°C. This temperature is known to polymer chemists as the glass transition temperature, and it marks the separation between plastic properties and rubber properties.

Polymers with glass transition temperatures well below room temperature will be rubbers (elastomers) at room temperature; those with glass transition temperatures well above room temperature are likely to find use as plastics. Some adhesives are polymers that make the transition near room temperature. Mechanical properties important for plastics include tensile strength at break (the resistance to longitudinal stress); impact strength; and flexural strength. For rubbers, the tensile strength, the per cent elongation, and the per cent rebound often are important.

Just as rubber must be vulcanized for many uses, so many plastics require a plasticizer, often called an internal lubricant. For example, the addition of about 40 per cent of dioctyl phthalate to polyvinylchloride lowers the glass transition temperature to about 0°C, thus turning it into the form used to make transparent raincoats.

The question now arises: What structural features tend to make a polymer act like a rubber or like a plastic at room temperature? Stating the question another way, we are asking why rotational motion (and ease of mechanical distortion) sets in at lower temperatures for some polymers than for others. The answer seems to be related to the attractive forces between chains or to the bulk of the groups attached to the backbone, which must move as the polymer portions rotate.

The three polymers on the left in Table 28.4 all are plastics at room temperature. Polystyrene and polymethylmethacrylate both have large groups attached to the backbone. These groups should retard rotation, thus increasing the glass transition temperature. In polyvinylchloride, the polar chlorine atoms probably form hydrogen bonds with hydrogen atoms in adjacent molecules, again causing the glass transition temperature to be relatively high. By contrast, styrene-butadiene rubber, which contains 75 per cent butadiene and only 25 per cent styrene, has a glass transition temperature of −55°C. Since butadiene has only hydrogen atoms, these portions of the thread are expected to start rotating at relatively low temperatures.

Table 28.4 Glass Transition Temperatures for Some Well-Known Polymers

	Polymer	T_G (°C)	Polymer	T_G (°C)	
Plastics	Polystyrene	100	Styrene-butadiene polymer	−55	*Rubbers*
	Polymethylmeth-acrylate	110	Polyisobutylene	−65	
	Polyvinylchloride	80	*trans*-Polyisoprene	−53	

Returning now to crystallinity in polymers, we ask: What properties are characteristic of polymers having a relatively high degree of crystallinity? Experience shows that such polymers are characterized by high tensile strength, stiffness, hardness, and low solubility. Further, if the crystallites can be oriented in a single direction within the matrix, the material increases considerably in tensile strength and in toughness. Such properties are highly desirable for fibers and for films.

Most of the polyethylene manufactured is made into films for packaging. These films are permeable to atmospheric gases but impermeable to water. They have low density, high flexibility, and high tear strength. They can be sterilized without losing their shapes. Polyethylene also can be molded into pipe, containers, and many other objects which are inert to most inorganic materials.

At first thought you might expect polyethylene to be a rubber at room temperature; it has no polar forces between threads and no bulky groups to retard their rotation. Instead of forming a matrix of randomly oriented, freely rotating threads, it organizes into a solid having a large per cent of rodlike crystallites. Evidently, it is easier for polyethylene molecules than for those of many other polymers to lie side by side in a well-organized crystallite. The suggested reason for this is illustrated in Figure 28.3. Here, we see that the polyethylene threads can pack very efficiently in a crystal, because the small hydrogen atoms do not disrupt the symmetry of the packing. By contrast, polyvinylchloride has very little crystallinity because the bulky chlorine atoms prevent efficient packing in the crystallite.

Polyacrylonitrile also shows a relatively high degree of crystallinity. This is believed to be due to the formation of a large number of hydrogen bonds between chains, as illustrated below. Polyacrylonitrile is one of the best synthetic fibers.

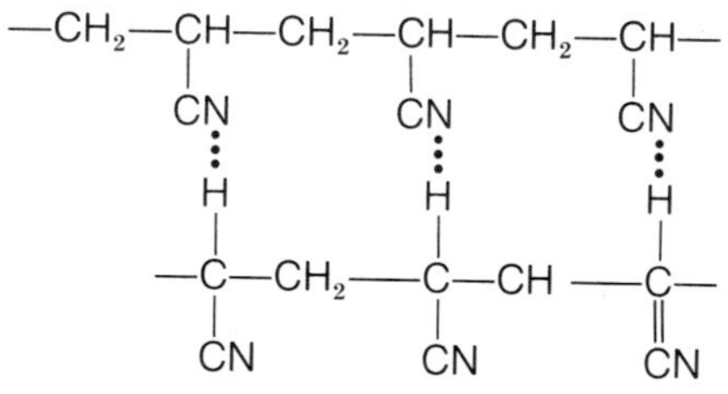

$$
\begin{array}{c}
-CH_2-CH-CH_2-CH-CH_2-CH- \\
\quad\quad\ \ |\quad\quad\quad\ \ |\quad\quad\quad\ \ | \\
\quad\quad\ CN\quad\quad\ CN\quad\quad\ CN \\
\quad\quad\ \ \vdots\quad\quad\quad\ \vdots\quad\quad\quad\ \vdots \\
\quad\quad\ \ H\quad\quad\quad H\quad\quad\quad H \\
\quad\quad\ \ |\quad\quad\quad\ \ |\quad\quad\quad\ \ | \\
-C-CH_2-C-CH-\ C- \\
\ |\quad\quad\quad\ |\quad\quad\quad\ \| \\
CN\quad\quad\ CN\quad\quad\ CN
\end{array}
$$

When fiber-forming polymers crystallize in the absence of external stresses, the crystallites orient at random. If an external stress is applied to the polymer, the crystallites tend to orient in the direction of the external stress. When this is done at temperatures below the melting point of the crystals, it is called cold drawing. Orientation greatly increases the hardness and tensile strength of the fiber.

Natural Polymers The important naturally occurring polymers include rubber, cellulose, starch, glycogen, and proteins. The structural stability of plants depends to a great extent on cellulose. Starch and glycogen are essential sources of carbohydrates for plants and animals, and proteins seem to be the substances most intimately concerned with life itself. In short, natural polymers occur in all living things and play an essential role in life processes.

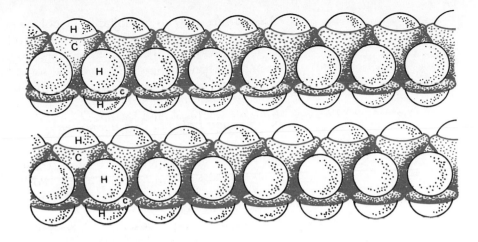

Figure 28.3 Illustrating the packing of polyethylene in a crystallite.

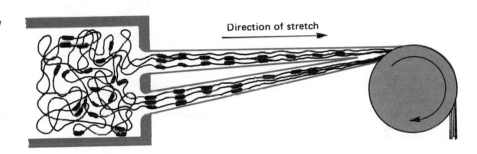

Direction of stretch

Figure 28.4 The drawing shows how the molecules of a high polymer are aligned to form a filament when the matrix is stretched during drying or coagulation.

Polymers of Glucose: Cellulose, Starch, Glycogen. One formula for glucose* is

which can be simplified for our purposes to

The presence of five hydroxyl groups in the molecule suggests immediately the possibility of condensation polymers of glucose—both linear and branched polymers—as illustrated below.

---

*The heavy lines are present to create the impression that the ring is tipped out of the plane of the paper. Glucose has five asymmetric carbon atoms (Chapter 26). There are 32 optical isomers having this structure. Glucose is distinguished from the others by the configurations of the groups around each of the five carbon atoms in the ring structure. Thus, the arrangement of the H, OH, and CH_2OH groups in this structure has special and important significance.

A linear polymer of glucose:

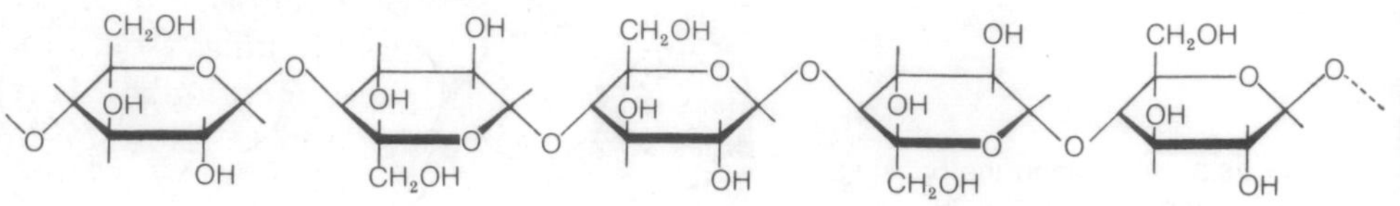

A branched polymer of glucose:

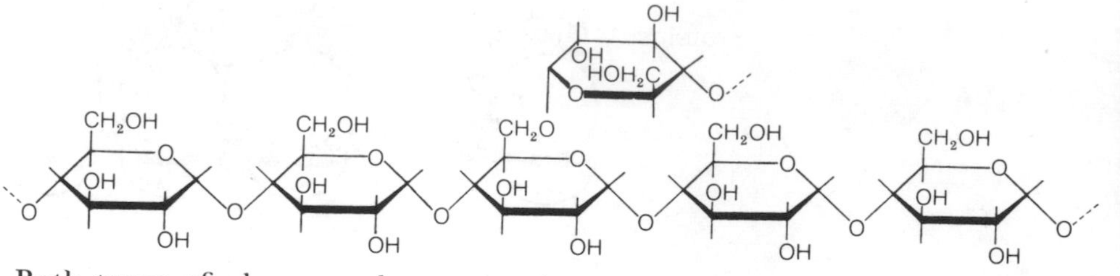

Both types of glucose polymers are known. Cellulose is a linear polymer, while starch and glycogen have considerable branching and a helical structure. In both starch and cellulose, the backbone of the polymer is formed by elimination of water between the first carbon atom on one glucose molecule and the fourth carbon atom on the second molecule, as illustrated below,

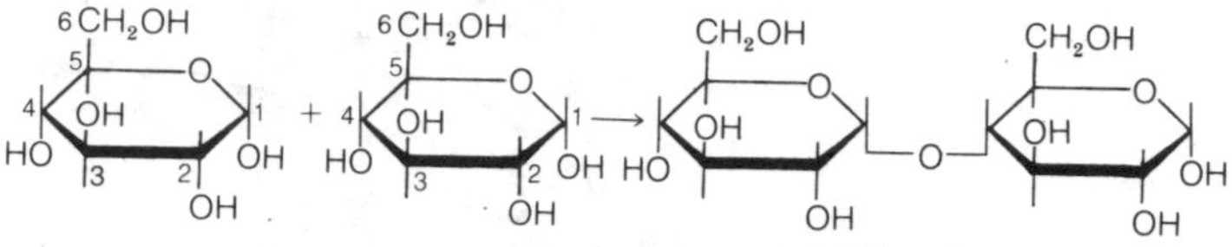

1,4 linkage in a
glucose polymer

The branch points in starch and glycogen appear to involve the first carbon atom on one glucose molecule and the sixth carbon atom in a glucose molecule in the backbone, as illustrated in the branched polymer structure above. For reasons that will be discussed later, two spatial arrangements are possible for the distribution of glucose units around the oxygen atoms that connect them. One of these arrangements is found in cellulose, the other in starch and glycogen.

Starch. This is the reserve carbohydrate of plants; it makes up large fractions of cereals, potatoes, and rice. Starch granules from different sources vary in appearance both in shape and size. The chemical content of all starches is similar. All are branched polymers of glucose, and the differences in properties are related to the chain length and degree of branching. Starch contains a soluble component, amylose (10 to 20 per cent of sample), and an insoluble component, amylopectin (80 to 90 per cent). Amylose contains from 60 to 300 glucose units/molecule; amylopectin contains from 300 to 6000 glucose units/molecule. Amylopectin is very highly branched; amylose is a linear polymer.

Glycogen. This is the energy-reserve carbohydrate of animals, found mainly in muscles and liver. Although it resembles starch in appearance, it has a lower molecular weight.

Cellulose. The structural material of plants, the principal component of cell walls, the chief constituent of cotton fiber (90%), wood (50%), and paper, cellulose is a linear polymer containing between 1800 and 3000 glucose units/

molecule. X-ray studies show that the polymer chains lie approximately parallel to the axis of a cellulose fiber (Figure 28.5). This accounts for the strength of such fibers. Cellulose, because of its abundance, has found many uses in addition to those mentioned. It reacts with nitric acid to form explosives such as *guncotton* and cellulose nitrate. With acetic anhydride, it gives cellulose acetate, which is used as a textile fiber (rayon) and in motion picture film.

Proteins These substances may be thought of as copolymers of α-amino acids,

$$R-CH-C-OH$$
$$\overset{|}{NH_2}\ \overset{\|}{O}$$

and a given protein molecule may contain as many as 20 or more different amino acids. The molecular weight of proteins ranges from several thousand to several million. Proteins occur in all cells of all animals and plants.

The amino acids are linked in the polymer by peptide (amide) linkages resulting from the elimination of water between an amino group in one molecule and a carboxyl group in a second molecule:

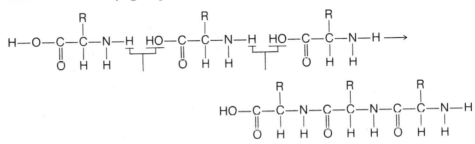

The general formula for a polypeptide (a portion of a protein chain) is

A polypeptide

A protein molecule consists of one or more polypeptide chains that are often coiled and bonded to one another in a variety of ways.

The 20 or so amino acids found in proteins differ in the structure of the R group in the general α-amino acid formula

$$H-N-C-C-OH$$
$$\overset{|}{H}\ \overset{|}{H}\ \overset{\|}{O}$$

Examples of some common amino acids are given in Table 28.5. The distribution of amino acids in proteins varies widely. For example, about half of silk fibroin is glycine, whereas egg albumin is composed of 18 amino acids in amounts varying from 1.5 to 16.5 per cent.

The exact sequence of amino acids in a protein chain is very important for it determines, to a great extent, the properties of the protein molecule. Thus, in the polypeptide illustrated below, the recurring sequence of amino acids

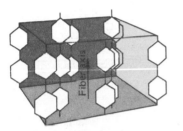

Figure 28.5 Diagram of crystalline cellulose, showing that the polymer chains lie approximately parallel to the axis of the fiber.

Table 28.5 Some Amino Acids Found in Proteins

Name	Formula	Name	Formula
Glycine	$H-N-\overset{\overset{H}{|}}{\underset{\underset{H}{|}}{C}}-CO_2H$	Serine	$H-N-\overset{\overset{CH_2OH}{|}}{\underset{\underset{H}{|}}{C}}-CO_2H$
Alanine	$H-N-\overset{\overset{CH_3}{|}}{\underset{\underset{H}{|}}{C}}-CO_2H$	Threonine	$H-N-\overset{\overset{CH\,\big\langle {}^{OH}_{CH_3}}{|}}{\underset{\underset{H}{|}}{C}}-CO_2H$
Phenylalanine	$H-N-\overset{\overset{H_2C-\bigcirc}{|}}{\underset{\underset{H}{|}}{C}}-CO_2H$	Tyrosine	$H-N-\overset{\overset{H_2C-\bigcirc-OH}{|}}{\underset{\underset{H}{|}}{C}}-CO_2H$
Leucine	$H-N-\overset{\overset{CH_2-CH\,\big\langle {}^{CH_3}_{CH_3}}{|}}{\underset{\underset{H}{|}}{C}}-CO_2H$	Histidine	$H-N-\overset{\overset{CH_2}{|}}{\underset{\underset{H}{|}}{C}}-CO_2H$ (imidazole ring)

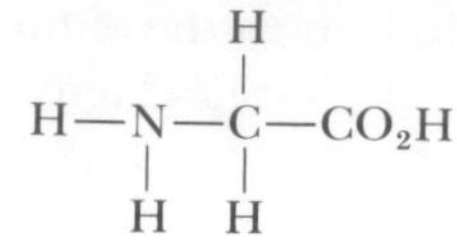

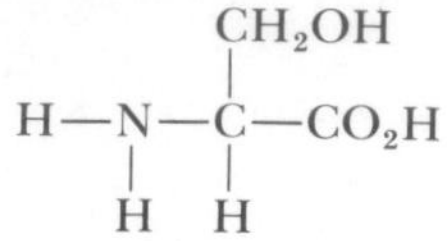

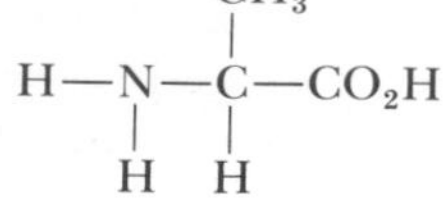

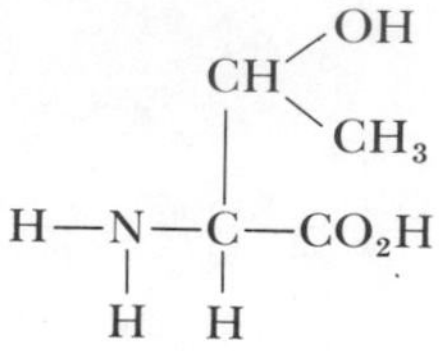

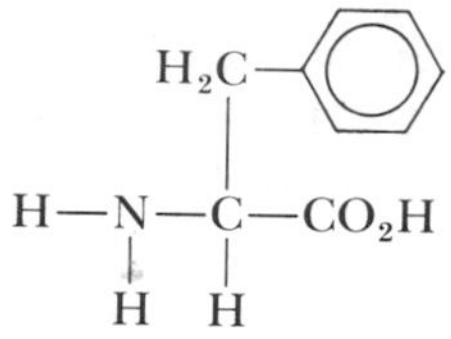

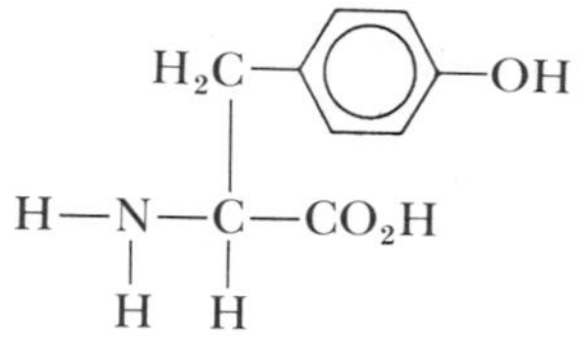

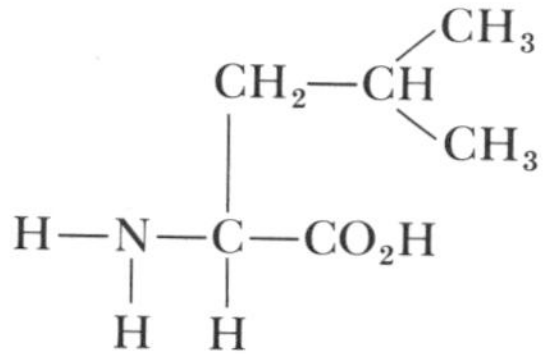

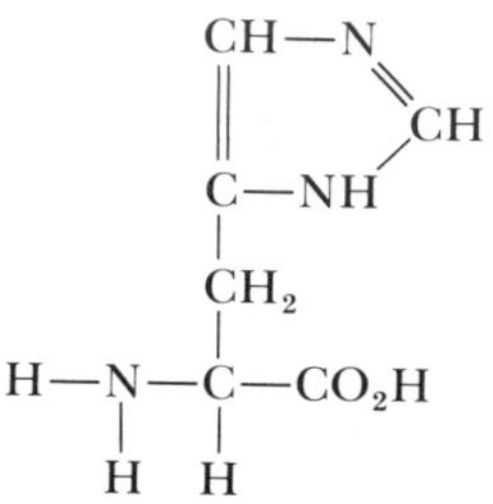

is alanine-glycine-phenylalanine-alanine-glycine-phenylalanine-alanine, and so on.

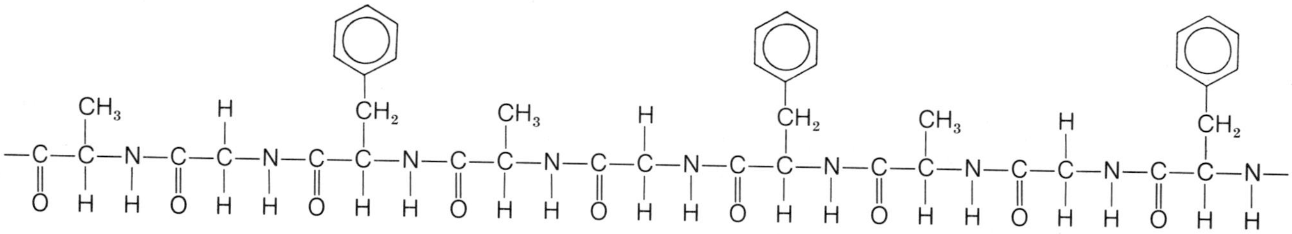

The sequence of amino acids in the protein may determine the shape of the coiled molecule. In the case of enzymes (one kind of protein molecule), the molecular geometry is believed to be intimately associated with the chemical activity (Chapter 29).

The possible sequences of amino acids in proteins is enormous. For example, how many ways can you arrange three amino acids, A, B, and C, in a chain? Some possibilities are

A—B—C—A—B—C—A—B—C, etc.
A—C—B—A—C—B—A—C—B, etc.
A—A—B—B—C—C—A—A—B—B—C—C, etc.
A—B—B—C—A—B—B—C—A—B—B—C, etc.

Since many proteins contain 20 or more amino acids, it is clear that the number of different kinds of proteins is almost unlimited. Fortunately, the living cell knows how to make the particular proteins needed to carry out the life processes. Chapter 29 summarizes some of these processes and indicates how proteins might be synthesized in the cell.

OTHER IMPORTANT COMPLEX MOLECULES
Alkaloids

Alkaloids are a large group of nitrogen-containing bases having marked physiological activity. They are usually obtained from plants by extraction with dilute acids. The acid reacts with the basic alkaloid to form a soluble salt of the base. Some common alkaloids, their formulas, and sources or uses are given in Table 28.6.

Table 28.6 Some Common Alkaloids, Their Formulas, Sources, and Uses

Name	Formula	Source	Uses
Nicotine	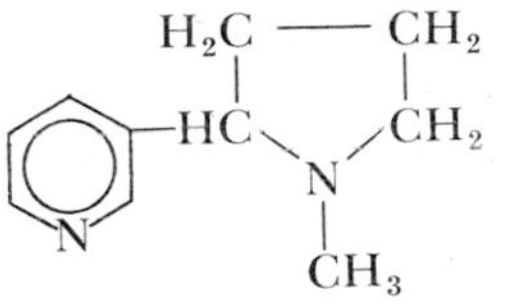	Tobacco (4–6% of dry leaf)	Insecticide
Quinine	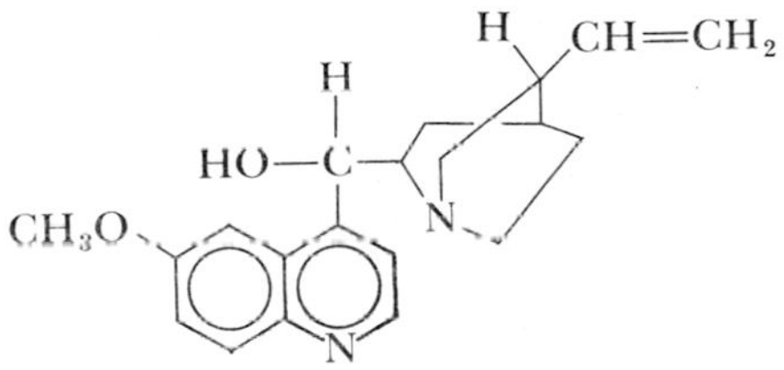	Bark of cinchona tree	Malaria cure
Morphine	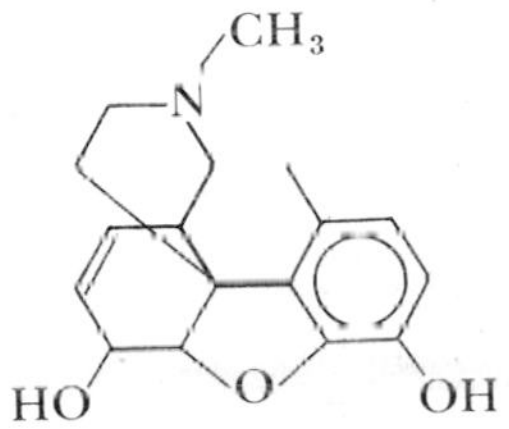	Opium poppy	Powerful analgesic
Strychnine	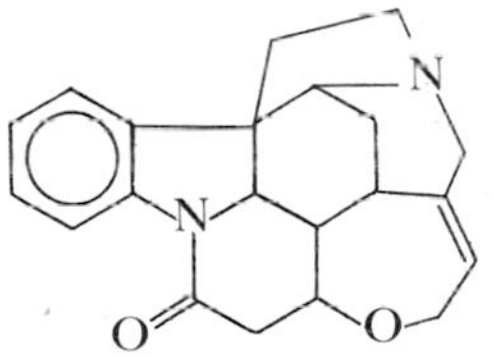	Seeds of *strychnos nux vomica*	Deadly poison, stimulant for nervous system
Reserpine (newest group)	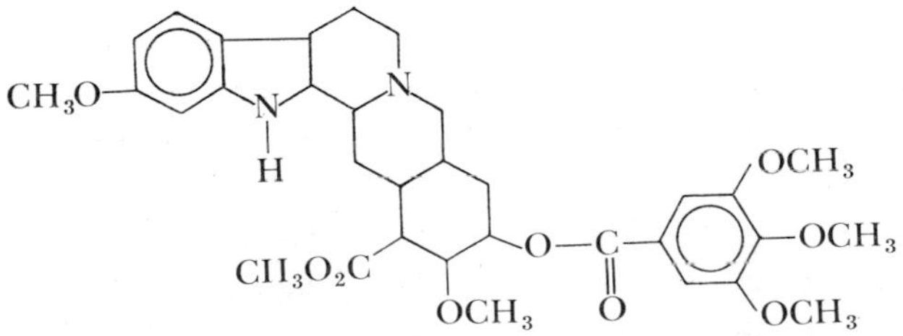	*Rauwolfia serpentina* (Indian snake root)	Tranquilizers, lower blood pressure

Table 28.7 Formulas for Some Steroids

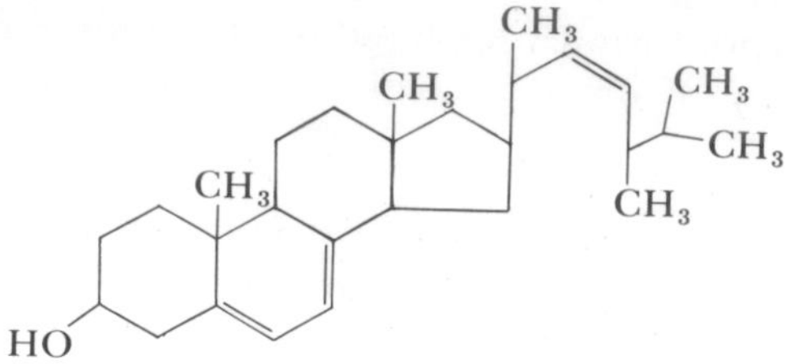

Ergosterol (precursor of vitamin D-2)

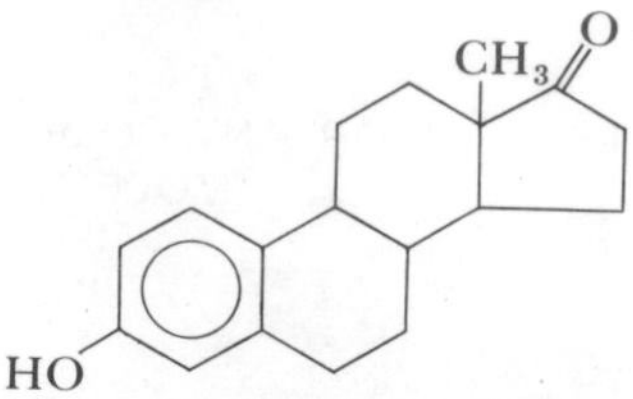

Estrone (a female hormone)

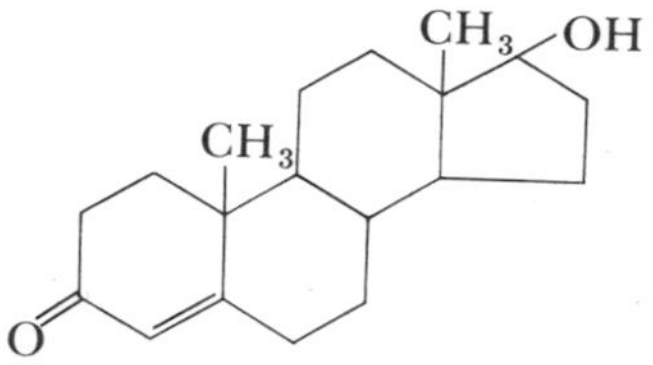

Testosterone (a male sex hormone)

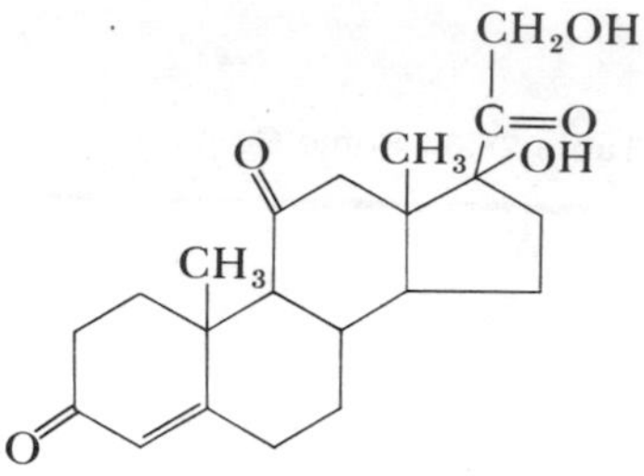

Cortisone (effective in treatment
of rheumatoid arthritis)

Steroids   Steroids are a group of compounds containing the cyclopentanophenan-
threne ring structure, illustrated below:

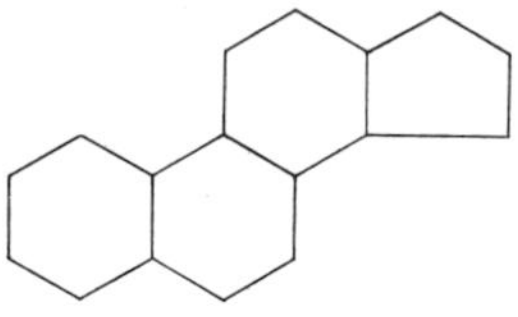

The steroids are widely distributed among plants and animals and are
essential constituents of brain and spinal tissue, bile, sex hormones, the
cardiac aglycones, and the hormones of the adrenal cortex.

A commonly occurring steroid is cholesterol, which is present in relatively
large amounts in blood plasma and in the brain, and in smaller amounts in
all animal cells. The structure of cholesterol is

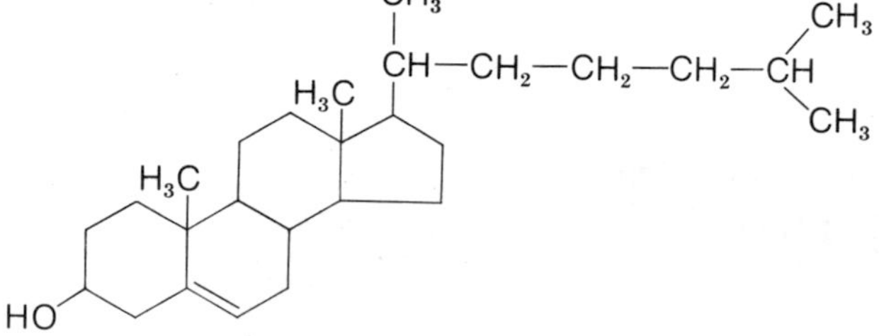

Cholesterol

It has been established that cholesterol can be synthesized in the cells of animals starting from acetic acid. Formulas for other steroids are given in
Table 28.7.

Fats and Oils These are the organic substances occurring in plant and animal tissue that are soluble in nonpolar or weakly polar solvents. Fats and oils are *triesters of glycerol and long-chain carboxylic acids.* They are known as triglycerides. The general formula for a fat or oil is

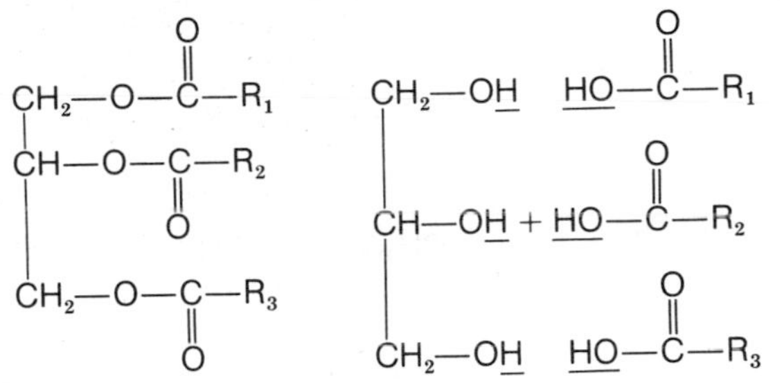

Triglyceride   Glycerol   Carboxylic acid

Components of a fat

The distinction between fats and oils is not sharp, but triglycerides that are solids at room temperature are usually called fats while those that are liquids at room temperature are called oils. If the R groups in the triglyceride contain few or no double bonds, the substance will probably be a solid at 25°C; if double bonds are present in the R groups, the substance will probably be a liquid at room temperature.

The R groups found in triglyceride molecules vary widely, depending on the species of plant or animal. However, the most common R groups found contain 15 or 17 carbon atoms. Not only are the R groups in a given triglyceride molecule likely to be different (for example, C_7, C_{15}, and C_{17}), but a group of triglyceride molecules from a single source will contain a variety of R groups. Consequently, the composition of fats is usually given in terms of the per cent of each kind of R group found when the fat is analyzed. For example, an analysis of human fat was reported to contain the per cents of various R groups given in Table 28.8.

The hydrolysis of fats with base is known as *saponification* (soap-making), because one product of the reaction is soap.

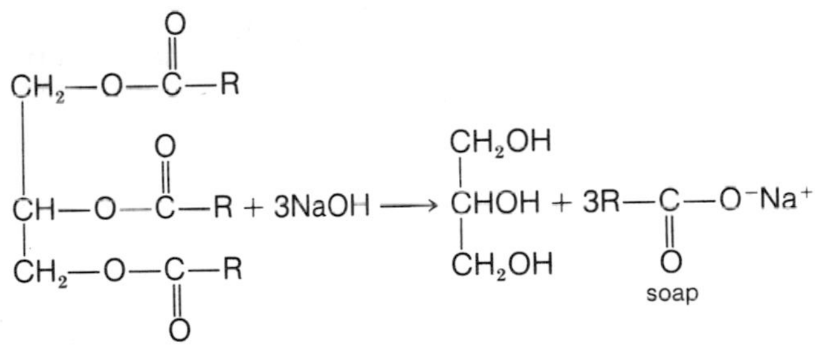

Table 28.8 Analysis of Human Fat (per cent)

Saturated R				Alkenyl R			Alkadienyl R	Alkatetraenyl R
C_{11}	C_{13}	C_{15}	C_{17}	C_{13}	C_{15}	C_{17}	C_{17}	C_{19}
0.5	3.3	25.0	8.4	0.4	6.2	45.9	9.8	0.6

Most soaps are sodium or potassium salts of long-chain (C_{12}–C_{18}) carboxylic acids. The carboxylate end of the soap anion is water soluble while the hydrocarbon end is oil soluble. Thus, one end of the soap anion acts to dissolve grease deposits while the other end "sticks" to the water. The result is that the soap carries with it, into the water, the grease and dirt that associates with the hydrocarbon end of the soap anion.

Vegetable oils may be converted to fats by hydrogenation of the double bonds in the R groups of the triglyceride, as illustrated below:

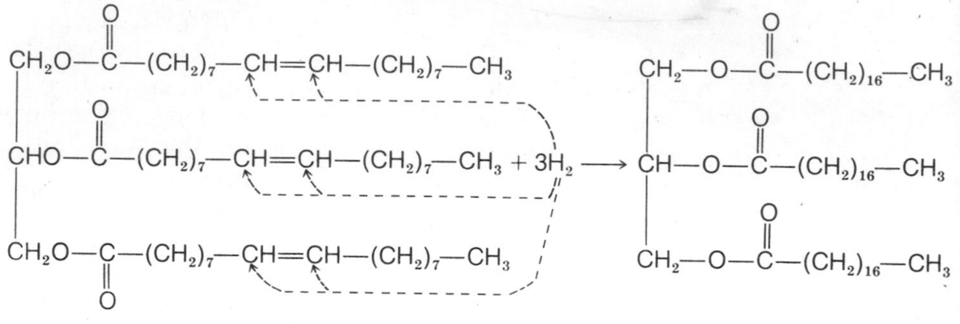

$$CH_2O-\overset{\overset{\displaystyle O}{\|}}{C}-(CH_2)_7-CH=CH-(CH_2)_7-CH_3$$
$$CHO-\overset{\overset{\displaystyle O}{\|}}{C}-(CH_2)_7-CH=CH-(CH_2)_7-CH_3 + 3H_2 \longrightarrow$$
$$CH_2O-\underset{\underset{\displaystyle O}{\|}}{C}-(CH_2)_7-CH=CH-(CH_2)_7-CH_3$$

oil

$$CH_2-O-\overset{\overset{\displaystyle O}{\|}}{C}-(CH_2)_{16}-CH_3$$
$$CH-O-\overset{\overset{\displaystyle O}{\|}}{C}-(CH_2)_{16}-CH_3$$
$$CH_2-O-\overset{\overset{\displaystyle O}{\|}}{C}-(CH_2)_{16}-CH_3$$

fat

Fats prepared this way from vegetable oils are used as shortening or as butter substitutes.

Highly unsaturated oils such as linseed oil and tung oil polymerize to resins under the influence of oxygen from air and in the presence of transition metal soaps. The formation of such resins in oil paints firmly binds the pigment to the surface, and serves as a protective coating for the surface. Oils containing R groups in which the double bonds are conjugated polymerize more rapidly than others, and paints containing these oils are fast-drying. Highly unsaturated, resin-forming oils are known as *drying oils*.

Simple Carbohydrates   The simple carbohydrates are glucose and fructose, their derivatives, and certain related compounds. Formed in plants by photosynthesis from carbon dioxide and water, the carbohydrates are not only the probable original sources of all our body fuel and energy, but also the starting materials for the synthesis of many substances found in plants and animals.

The simple six-carbon sugars have the formula $C_6H_{12}O_6$. A number of them contain five asymmetric carbon atoms, as indicated by the structural formula

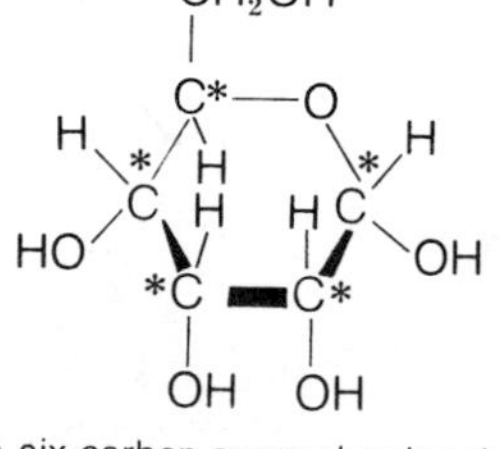

A six-carbon sugar showing the
five asymmetric carbon atoms

Consequently, there are 32 optical isomers of the structure given above. These isomers differ in the configuration on one or more of the asymmetric carbon atoms. The most common naturally occurring isomers among the 32 are α-D-glucose and β-D-glucose, whose formulas are given below.

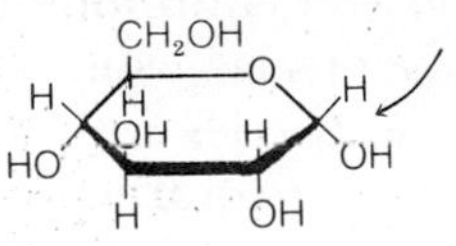

α-D-Glucose β-D-Glucose

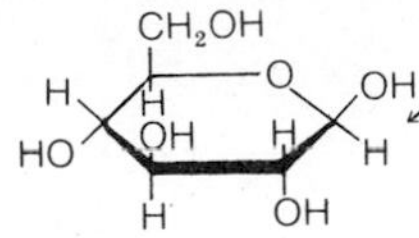

The configurations on the various asymmetric carbon atoms in these two glucose structures are identical except for those on the carbon atom marked by an arrow. This is known as the anomeric (or first) carbon atom. It is derived from an aldehyde group that has undergone a cyclization reaction with one of the hydroxyl groups on the glucose chain. Two forms are possible, depending on how the ring closes. The first is called the α-D-anomer; the second, the β-D-anomer.

Units of D-glucose are condensed into polymer chains in the biological macromolecules starch and cellulose. In starch, the glucose units have the α-configuration:

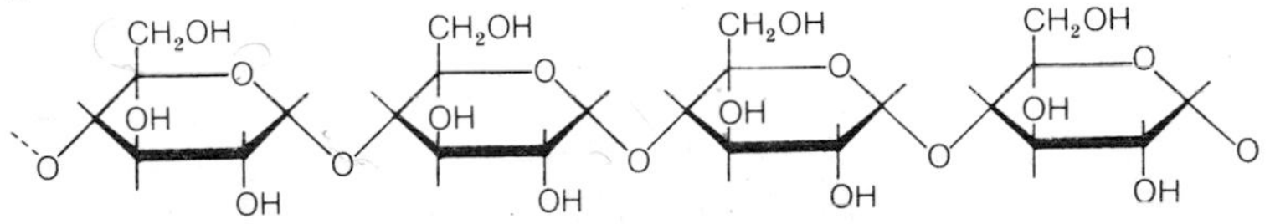

Because of the orientation of the bonds in the oxygen atoms connecting the glucose units, the starch chain is a helical structure.

In cellulose, the glucose units have the β configuration:

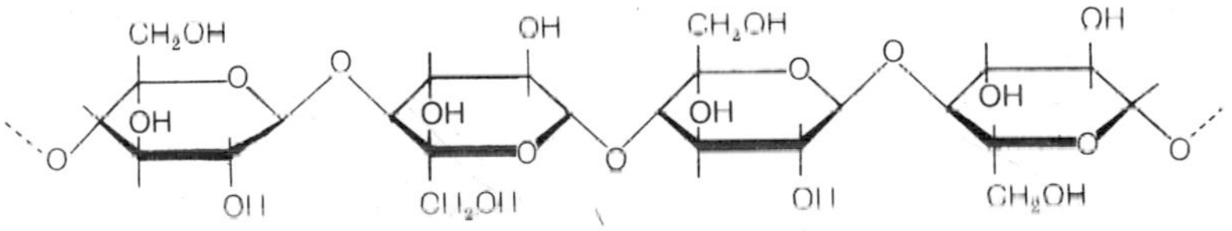

The geometry of the cellulose molecule is linear, appropriate for the structural material in plants.

Pure glucose is a white crystalline solid that resembles cane sugar in its properties, but is only about half as sweet. It is an inexpensive and healthful food. A small percentage (about 0.1 per cent) is present in the blood of persons in good health, and a much larger percentage in the blood of persons afflicted with diabetes. Glucose will ferment in the presence of certain enzymes, especially in the zymase of yeast, to form alcohol:

$$C_6H_{12}O_6 \xrightarrow{enzyme} 2C_2H_5OH + 2CO_2$$

Fructose (Fruit Sugar, Levulose) $(C_6H_{12}O_6)$. Fructose is a white solid that occurs, along with glucose, in fruits and honey.

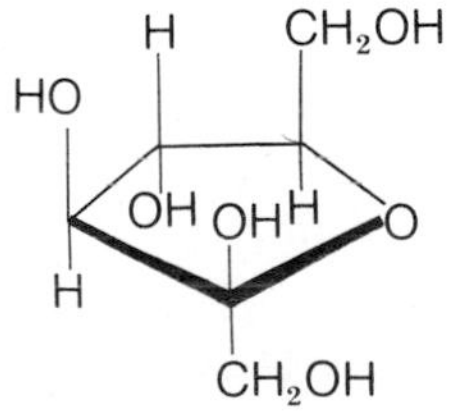

Fructose (a common form)

Sucrose (C₁₂H₂₂O₁₁). Sucrose, a compound containing one glucose and one fructose molecule condensed together, is an example of the compounds containing two simple sugar molecules, known as disaccharides. Sucrose is obtained from sugar cane and sugar beets, and is used principally for food.

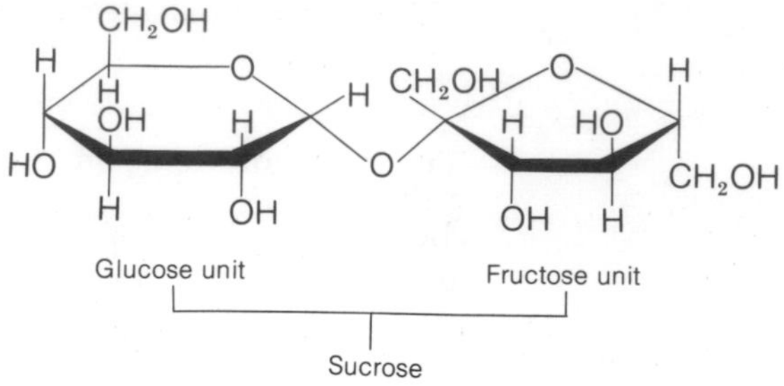

Glucose unit   Fructose unit

Sucrose

Lactose (*Milk Sugar*) *(C₁₂H₂₂O₁₁).* The disaccharide lactose occurs in the milk of all mammals. The average composition of cow's milk is as follows:

Component	Per Cent	Lactose
Water	87.0	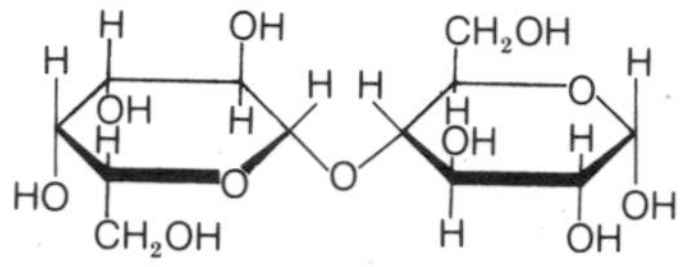
Casein		
(protein)	3.3	
Butterfat	4.0	
Lactose	5.0	
Mineral matter	0.7	

Lactose resembles sucrose in appearance, but is not as soluble and is only about one-fourth as sweet. The souring of milk is the result of the conversion of milk sugar into lactic acid, a compound whose formula is

$$CH_3—CH—CO_2H$$
$$\qquad\ \ |$$
$$\qquad\ \ OH$$

This change is known as lactic fermentation, and is brought about by certain bacteria. These bacteria (or their spores) are present everywhere in the air and are associated especially with dust and dirt.

Maltose (C₁₂H₂₂O₁₁). This sugar, also a disaccharide, is only about half as sweet as sucrose. It is prepared by the action of malt upon starch, accounting for the name maltose. It is an isomer of lactose.

Malt is the name applied to barley that has been moistened, kept in a warm place until it has germinated, and heated until the vitality of the grain has been destroyed. In the process of germination, an enzyme is formed known as diastase, and it is this substance that gives malt its property of changing starch into maltose. Amylase, an enzyme in saliva, brings about a similar change in the starch present in our foods.

Vitamins Systematic experimental studies have shown that the health of humans and other animals cannot be maintained on a diet consisting of only proteins, fats, carbohydrates, and minerals. Very small amounts of vitamins also are necessary for the proper functioning of the body. The body cannot synthesize vitamins; it must depend upon plants and other animals for them.

Because vitamins function in such small amounts, their role in life chemistry must be that of a catalyst. Generally, a vitamin is the main or sole component of a coenzyme (Chapter 29), and many function in the transfer of hydrogen atoms or larger groups from one molecule to another in the living cell. For example, pantothenic acid, part of the vitamin-B complex, is the major component of coenzyme A. This coenzyme is of vital importance in the tricarboxylic acid cycle, the sequence of reactions by which foodstuffs are converted to energy, carbon dioxide, and water. In this process, coenzyme A is responsible for transferring an acetyl group, CH_3CO-, from pyruvic acid to oxalacetic acid. The connection between the biochemical role of the coenzyme and the clinically observable manifestations of vitamin deficiency—arrested growth, deficiency diseases—is now being established.

Some vitamins have simple formulas, such as nicotinamide:

Others, such as vitamin A, are more complex:

Table 28.9 summarizes the vitamins, their roles as coenzymes, and the deficiency diseases they prevent in humans.

Table 28.9 Vitamins

Vitamin	Name	Deficiency Disease of Man	Daily Requirement*
Fat-Soluble Vitamins			
A	(Axerophthol)	Night blindness, xerophthalmia	1.5–2.0 mg
D	Calciferol	Rachitis	0.025 mg
E	Tocopherol	Unknown (muscular dystrophy?)	[5 mg]
K	Phylloquinone	Delayed blood clotting	[0.001 mg]
Q	Ubiquinone	Unknown	
F	Essential fatty acids	Debated	Unknown
	Thioctic acid	Unknown	Unknown
Water-Soluble Vitamins			
B_1	Thiamine	Beriberi (polyneuritis)	0.5–1.0 mg
	Riboflavin	"Pellagra sine pellagra"	1 mg
B_2	Nicotinamide	Pellagra	
complex	Folic acid	Megaloblastic anemia	[1–2 mg]
	Pantothenic acid	Burning foot syndrome	[3–5 mg]
B_6	Pyridoxine	Unknown	[1.5 mg]
B_{12}	Cobalamin	Pernicious anemia	0.001 mg
C	Ascorbic acid	Scurvy	75 mg
H	Biotin	Very rare ("egg white injury," dermatitis)	[0.25 mg]

*Values in brackets indicate estimated amounts.

Sulfa Drugs These were found to be particularly useful in combating bacteria that produce pneumonia, gas gangrene, meningitis, and blood poisoning. The first of these compounds prepared was sulfanilamide (below), but it caused serious reactions in some patients. Chemists modified the properties of the sulfa drugs by replacing one of the hydrogen atoms on the $-SO_2-NH_2$ group of sulfanilamide with other groups, thus producing compounds that are more effective in combating bacteria, but less harmful to the patient. Some of the more common sulfa drugs are sulfadiazine, sulfathiazole, sulfaguanidine, and sulfapyridine.

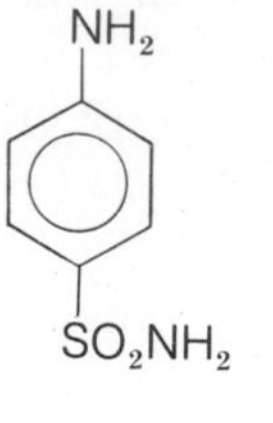

Sulfanilamide 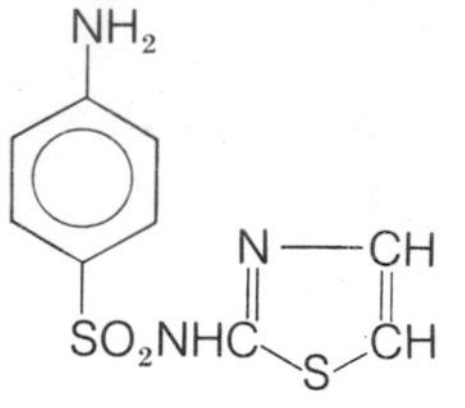   Sulfathiazole 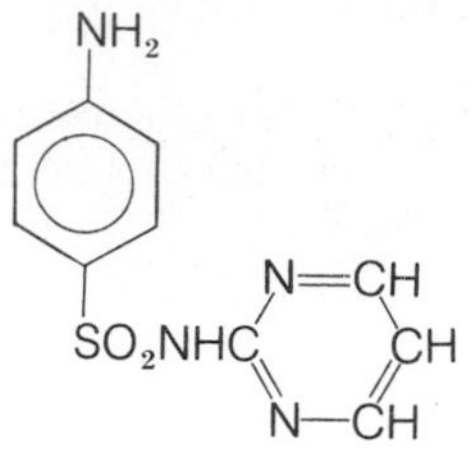   Sulfadiazine

Antibiotics Close upon the development of the sulfa drugs came the discovery of penicillin, which has proved to be extremely effective in the treatment of gonorrhea, pneumonia, meningitis, osteomyelitis, gas gangrene, and streptococcus and staphylococcus infections. Penicillin is called an antibiotic agent; it inhibits or kills microorganisms. It is produced by various strains of mold called *Penicillium notatum*, and is now prepared on a large scale. Penicillin has been synthesized. The general formula for penicillin is

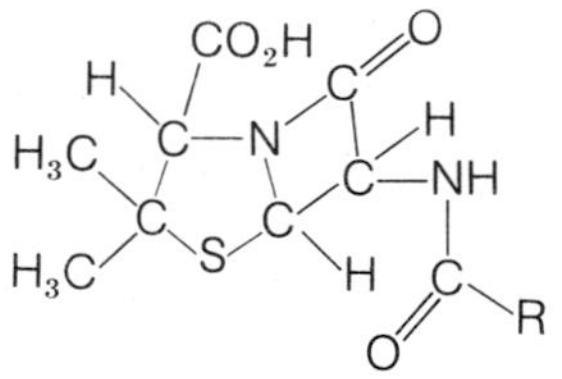

R may be
$-CH_2-CH=CH-CH_2-CH_3$
$-CH_2-C_6H_5$
$-CH_2-C_6H_4-OH(p)$
$-CH_2-(CH_2)_5-CH_3$

Some other antibiotics now used are streptomycin, chloromycetin, aureomycin, and terramycin.

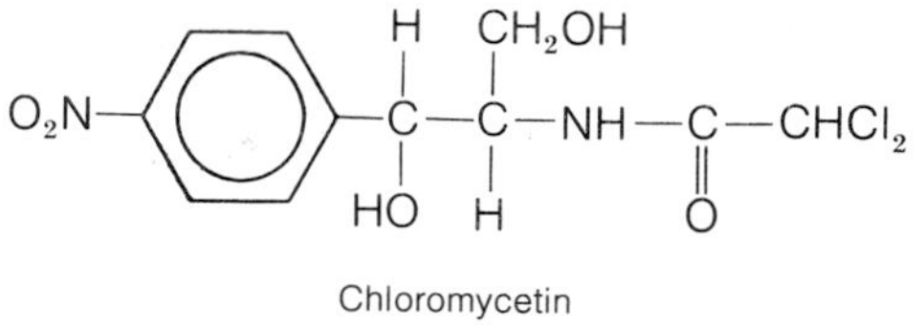

Chloromycetin

Table 28.10 Uses of Some Antibiotics

Name	Effective Against
Penicillin	Pneumonia, peritonitis, diphtheria, gas gangrene, meningitis, syphilis, gonorrhea
Streptomycin	Dysentery, undulant fever, urinary tract infections, typhoid fever, human tuberculosis, certain gastric infections
Chloromycetin	Influenza, whooping cough, typhoid fever, typhus
Aureomycin⎱ Terramycin⎰	Viral pneumonia, human tularemia, psittacosis, typhus, some urinary infections, peritonitis, skin lesions, syphilis

Aspirin
(acetylsalicylic acid
analgesic and antipyretic)

Merthiolate

Picric acid (burns)

Phenobarbital
(sedative and hypnotic)

A Challenge for the Plastics Industry.   The synthetic plastics industry in the U.S. produces in the neighborhood of 20 million pounds of polymeric materials each year. Its principal products are fibers, packaging materials, synthetic rubbers, coatings, adhesives, furniture, building materials, and a host of products known as plastics. Perhaps the most creative of all industries, it has given us an incredible variety of inexpensive items for personal comfort and convenience.

From this industry comes a continuous flow of new and novel products, everything from spandex fibers, wigs, and heart valves to plastic bathtubs and carpets that serve as road beds or artificial ski slopes. Among the latest contributions are: corrosion-resistant bearings, adhesives 30 times stronger than cement mortar, elastomeric building materials for air-supported structures, and gill-like silicone sheets that are watertight but, when used underwater, allow oxygen from the water to pass through them. Its current challenge is to develop stronger, lighter, cheaper, more flexible construction materials, not only for homes and other buildings but also for airplanes, automobiles, trains, and ships.

The challenge here centers on ways to greatly improve the strength and durability of organic polymers. To qualify as structural materials, a product must possess these characteristics: a rigidity sufficient to bear 700,000 pounds/square inch; a tensile strength of at least 100,000 pounds/square inch; an elasticity of at least 10 per cent (to resist tearing or breaking); a softening temperature above 500°C; and high resistance to damage by solvents, corrosive chemicals, radiation, or heat.

Most organic polymers do not even approach these specifications. Polyethylene, for example, which forms a strong, tough, abrasion-resistant film, melts at 130°C. Polystyrene, as another example, melts at 230°C but is easily damaged by solvents. Recently, however, a fiber produced by the decomposition of rayon or acrylonitrile polymers under carefully controlled conditions was found to have the highest rigidity and tensile strength of any known substance. This material, described as having one-fourth the weight and four times the strength of steel, has been used in spacecraft nosecones. While its present cost is prohibitive for large-volume applications, its very existence gives promise that organic polymers having properties suitable for use as structural materials can be produced economically.

There are at least four ways to improve the strength of a polymeric material: The backbone of the molecule can be stiffened; the crystallinity of the matrix can be increased; cross-linking between chains can be increased; and a second material can be added to the matrix. Combinations of some or all of these are being exploited in the search for improved polymer strength.

Stiffening the backbone is accomplished by adding groups to the chain that restrict its motion. Bulky or heavy groups, such as chlorine atoms or phenyl groups, replacing hydrogen atoms on the backbone can do this. The presence of rings of atoms in the backbone, such as the glucose residues in cellulose, also restrict its motion. This concept has been exploited in the synthesis of "ladder" polymers, in which the backbone consists of a series of

condensed rings of atoms This material is very

hard, insoluble, and unmeltable.

Crystallinity of the polymer matrix can be increased by orienting the chains so they can line up in parallel and form crystallites. This usually is done by mechanically stretching the matrix. Crystallization also is facilitated by the presence of small polar groups on the backbone, as was illustrated with polyacrylonitrile on page 634. Highly stretched nylon fibers actually have a higher tensile strength than steel. Highly oriented polyester film (Mylar) is so strong that full-size automobiles wrapped in sheets of it can be lifted and transported by helicopters whose cables are attached to the film.

Cross-linking has been discussed in connection with rubber; in general, the more cross-linking, the harder the rubber product. This too has been exploited, most successfully perhaps by adding a monomer such as styrene to an existing polymer matrix and polymerizing (or grafting) the styrene onto the chains of the original polymer. In this process, many cross-links are formed. Materials such as this can reach a rigidity strength of about 450,000 pounds/square inch and a melting temperature of 350°C. A combination of increasing crystallinity (using polyester) and cross-linking (grafting) can increase the strength only slightly above this.

Adding a second material such as glass fibers to the polymer has produced materials strong enough to be used in furniture, boat hulls, and bathtubs. This is known as reinforced plastic. Fiber reinforcement produces remarkable increases in strength because the fibers can carry appreciable portions of the load and are themselves supported by the matrix. Glass fibers are the most common reinforcing material used at present. The resulting products have many advantages but, because the fibers can stretch, the material tends to bend under load. For this reason, it cannot be used for such things as bridges or airplane wings. However, the cabs of the lead cars in the San Francisco rapid transit trains are made of chopped glass fiber in a flame-retardant polyester polymer.

The plastics industry has not yet met the challenge of developing inexpensive, high-quality construction materials, but who can doubt that this challenge will be met?

SUMMARY   In this chapter we have surveyed very briefly some of the large organic molecules that have importance in our lives. Starting with man-made polymers, we have examined the molecular structures and properties of important plastics, fibers, films, and elastomers. We have attempted to relate properties to molecular structure and motion within the polymer. Naturally occurring polymers—starch, cellulose, proteins—were described briefly. This was followed by descriptions of the structures of alkaloids, steroids, fats and oils, simple carbohydrates, vitamins, sulfa drugs, and antibiotics.

IMPORTANT TERMS

Polymers
addition polymers
condensation polymers
monomers
plastics
fibers
films
elastomers
linear polymers
cross-linked polymers
homopolymers
copolymers
glass-transition temperature
crystallinity
polymer backbone
cellulose
starch
polypeptides
amino acids
glucose

Alkaloids
nicotine
quinine
morphine
reserpine

Vitamins
A, B_1, B_2, B_6, B_{12}
C, D, E, F, H, K, Q

Steroids
cholesterol
ergosterol
estrone
cortisone

Sulfa drugs
sulfanilamide
sulfathiazole
sulfadiazine

Fats and oils
triglycerides

Simple carbohydrates
fructose
lactose
sucrose
maltose

Antibiotics
penicillin
streptomycin
chloromycetin

QUESTIONS AND PROBLEMS

1. Distinguish between addition and condensation polymers.

2. Give structural formulas for the monomers of six commercially important polymers and illustrate the reactions by which each of the monomers is converted into the polymer.

3. Draw structural units for specific polymers to illustrate each of the following: (a) a rigid plastic, (b) a flexible plastic, (c) a film, (d) a fiber, (e) an elastomer (rubber).

4. Explain why the head-to-tail orientation predominates in vinyl polymerizations.

5. Show, by structural diagram, a typical segment of each of the following vinyl polymers (see Table 28.2): (a) Orlon, (b) Koroseal, (c) poly-*N*-vinylpyrrolidone, (d) Teflon, (e) Plexiglas, and (f) Saran.

6. Discuss the relation of polymer structure to properties. Account for the fact that a given polymer may be used in several applications, for example, as a rubber in one application, as a flexible plastic in another.

7. Contrast the starch and cellulose polymer structures.

8. Explain in terms of molecular structure why there are so many kinds of proteins.

9. Show how the molecules glycine, threonine, and serine can react to form a polypeptide chain.

10. What features do alkaloids have in common?

11. Name the heterocyclic rings found in the following compounds: (a) uric acid, (b) nicotine, (c) reserpine, and (d) morphine.

12. Strong oxidation of nicotine yields 3-pyridinecarboxylic acid. Explain this observation.

13. What structural unit is common to most steroids? What is the shape of this structure?

14. Distinguish structurally between a fat and an oil.

15. Linseed oil is commonly used in paint, linoleum, and similar products. Explain its function in such uses.

16. How are vegetable oils converted into solid fats?

17. What types of products are formed when the reactants waxes and sodium hydroxide are boiled?

18. Compare the structures and properties of sucrose, lactose, and maltose.

19. How is starch related from a structural point of view to: (a) cellulose, (b) glycogen, and (c) glucose?

20. What features—structurally or physiologically—do vitamins have in common?

21. Can you find any obvious structural similarities between the sulfa drugs and the antibiotics such as penicillin?

SPECIAL PROBLEMS

1. The preparation of silicone may be accomplished either by the controlled hydrolysis of dimethyldichlorosilane

or by the acid-catalyzed ring-opening polymerization of the cyclic tetramer

Is the silicone polymer that forms an addition polymer or a condensation polymer?

2. (a) Is one of the following polypeptides identical with either or both of the other two? (b) Will their products on hydrolysis be related? (c) Write their structural formulas and the necessary equations to substantiate your answers: alanylglycyltyrosine; glycyltyrosylalanine; and tyrosylglycylalanine.

3. Why does D(−)-fructose, which is levoratatory, have the "D" prefix?

4. When glucose is oxidized mildly, a monocarboxylic acid and then a dicarboxylic acid (each with six carbon atoms) are formed. Write a formula for each of these acids.

REFERENCES HENDRICKSON, J. B., *The Molecules of Nature*, Benjamin, Menlo Park, Calif., 1965.

KATRITZSKY, A. R., and LAGOWSKI, J. M., *Heterocyclic Chemistry*, Wiley, New York, 1960.

O'DRISCOLL, K. F., *The Nature and Chemistry of High Polymers*, Van Nostrand Reinhold, New York, 1964.

MARK, H. F., "The Nature of Polymeric Materials," *Sci. Amer.*, **217**, No. 3, 149 (1967).

SOME CHEMISTRY OF LIVING CELLS

Now our model for deoxyribonucleic acid is, in effect, a pair of templates each of which is complementary to the other. We imagine that prior to duplication the hydrogen bonds are broken and the two chains unwind and separate. Each chain then acts as a template for the formation on to itself of a new companion chain so that eventually we shall have two pairs of chains where we only had one before.*

James D. Watson (1928–) and
F. H. C. Crick (1916–)

Biochemistry is the study of the chemistry of life processes, and we have seen numerous examples of this chemistry in previous chapters. The major research effort in this science today is the study of chemical reactions that are taking place in living cells. Previously, the technique and knowledge in this area limited research workers to the isolation and identification of substances present in the cell. From such studies, they were able to learn a tremendous amount about nutrition, blood chemistry, digestion, and many other processes that keep the organism alive and healthy. The persistent efforts of scientists to look into the cell and identify the chemical reactions occurring under life conditions led to the techniques and the know-how which, in the last half of this century, have revealed to us the secret of heredity and many other features of life processes.

The experiments that led to these revelations are themselves monuments to the intellect. Much of the work involved the use of radioactive tracers. Bacteria, because of their speed in reproduction, were the organisms used most often.

While the chemistry of the cell is extremely complicated, the underlying principles are not difficult to grasp. The chemical reactions of the cell are believed to be determined and controlled by template molecules that adsorb the reactants on their surfaces, catalyze the reactions, and release the products to the cell; energy conversion and utilization revolves largely around one molecule that absorbs energy in its formation and releases energy when it decomposes; and heredity is controlled by threadlike molecules keyed to a four-unit code.

This chapter gives an overview of the chemistry of the living cell. While the picture presented is a simplified one, it represents the current view held by most biochemists.

---

*The quotation is from "Genetical Implications of the Structure of Deoxyribonucleic Acid," *Nature,* **171** (1953), p. 964.

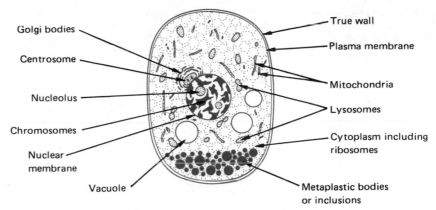

Figure 29.1 Diagram of a living cell.

Golgi bodies

Centrosome

Nucleolus

Chromosomes

Nuclear membrane

Vacuole

True wall

Plasma membrane

Mitochondria

Lysosomes

Cytoplasm including ribosomes

Metaplastic bodies or inclusions

The Living Cell

Biology students are accustomed to thinking of a cell as a drop of viscous liquid surrounded by a membrane. Inside the membrane is the cytoplasm—mostly water but containing numerous heterogeneous regions identified as mitochondria, ribosomes, lysosomes, and other materials. Toward the center of the cell is the nucleus, where the genetic information is stored. While this is a familiar picture to biology students, the biochemist has a somewhat more penetrating view of the cell. He sees it as a highly organized chemical factory, capable of carrying out a myriad of very specific chemical reactions rapidly and efficiently, able to control these reactions to meet the needs of the cell and the organism, and so skillful in regulating its energy requirements that it captures 60% of the energy of its fuel—a feat only a few man-made machines can claim.

Two of the most important questions concerning this complex chemical factory are:
1. How is it possible for the cell to carry out only certain chemical reactions when the materials needed for carrying out so many other reactions are present?
2. What is the chemical secret associated with efficient energy utilization of the cell?

We shall use these two questions as the basis for studying some chemistry of the cell.

Enzymes—The Reaction Controllers

Experiments have shown that nearly every reaction in the cell proceeds readily only in the presence of a specific enzyme. In the absence of appropriate enzymes, most reactions would proceed so slowly that the cell would die. Enzymes are protein molecules often having molecular weights in the 50,000 range, though many are larger than this (Figure 29.2). Almost all of these huge molecules are specific catalysts. This means that a given enzyme will speed up one chemical reaction but not another, even though the second reaction may be very similar to but not exactly the same as the first. For example, the enzyme sucrase will catalyze the hydrolysis of sucrose into glucose and fructose, but it will not catalyze the hydrolysis of maltose into two glucose molecules. All of this means that the cell can carry out only those reactions for which an appropriate enzyme is present, and very few others. This also means the cell must contain a very large number of different types of enzyme molecules. It is estimated that there are about 1700 enzyme units in one mitochondrion of a beef heart cell.

Not only is a specific enzyme required for nearly every reaction in the cell,

Figure 29.2 Model of an enzyme: bovine ribonuclease. Each circle represents an amino acid.

but cellular reactions usually proceed through a number of steps, each one of which requires an enzyme (Figure 29.3). For example, in the oxidation of glucose to carbon dioxide and water—a reaction every student can carry out in one operation using a test tube and a burner—the cell uses 17 steps and 17 enzymes. Eleven of these steps take place in the cytoplasm; six occur in the mitochondrion.

Perhaps the next logical question to ask is, Why are enzymes specific? What is there about these huge protein molecules that enables them to catalyze only one of many possible chemical reactions? Many biochemists believe that the shape of the enzyme is the key here. They believe that the shape of a given enzyme is such that only those molecules involved in the

Figure 29.3 Enzyme specificity; enzyme catalysts bring reactants in close contact and proper orientation for the reaction to proceed. (a) Enzyme and substrate approach each other. (b) Enzyme-substrate complex. (c) Enzyme and products separate.

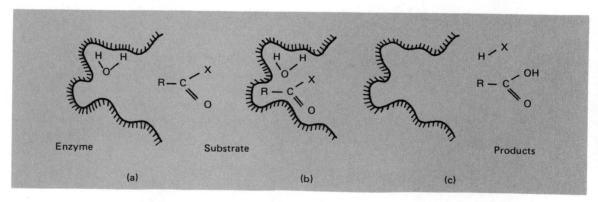

reaction it controls will fit at the active site on the enzyme surface. At the risk of oversimplification, imagine that in terms of this theory, the enzyme sucrase, for example, might have a crater in its surface into which a sucrose molecule could exactly fit. No other molecule in the cell could fit into that crater. Adjacent to the sucrose position on the enzyme surface might be another position into which only a water molecule could fit. The enzyme may then bring sucrose and water together at the active site, catalyze the hydrolysis, release the products from its surface, proceed to pick up a second pair of reactants, and so on. This process will be repeated until the cell has sufficient glucose and fructose, at which time these products may remain on the enzyme surface, thereby temporarily inactivating it.

Other important characteristics of enzymes are:

1. Most enzymes have names ending with the suffix -ase—for example, amylase, maltase, lipase, nuclease—although some, such as renin, pepsin, and trypsin, retain an earlier nomenclature.
2. To be effective, some enzymes require the presence of metal ions, such as zinc, or magnesium, or other molecules known as coenzymes. Often these coenzymes are vitamins or are related to vitamins.
3. The action of enzymes is influenced by pH and by temperature. Each enzyme has an optimum pH and an optimum temperature range where its activity is greatest. High temperatures will permanently destroy enzyme activity.
4. A very few enzymes catalyze more than one reaction.

Examples of some enzymes and the reactions they catalyze are:

1. *Carbohydrases*—enzymes that catalyze the hydrolysis of starch, starch fragments (dextrins), or disaccharides into simple sugars.

   Sucrase—hydrolysis of sucrose to glucose and fructose.
   Lactase—hydrolysis of lactose to glucose and galactose.

2. *Proteases*—enzymes that catalyze the conversion of proteins or protein fragments (polypeptides) to simple polypeptides and amino acids.

   Exopeptidases—hydrolyze amino acids from the end of a polypeptide chain. Carboxypeptidases attack the end having a free carboxyl group; amino peptidases attack the end having a free amino group. *Endopeptidases*—hydrolyze linkages between amino acids within the protein or polypeptide chain. Pepsin, trypsin, and chymotrypsin each catalyze the hydrolysis of only certain bonds in polypeptide chains. For example, pepsin functions at the bond connecting aromatic amino acids to acidic amino acids.

3. *Esterases*—enzymes that catalyze the conversion of esters to alcohols and acids.

   Lipases—hydrolysis of fats to glycerol and fatty acids.
   Phosphatases—hydrolysis of phosphate esters

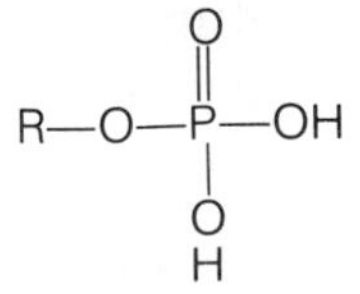

$$R-O-\overset{\overset{\displaystyle O}{\|}}{\underset{\underset{\displaystyle H}{O}}{P}}-OH$$

to phosphoric acid and alcohol, a key reaction in energy storage and transfer, especially in glucose metabolism.

4. *Transferases*—enzymes that catalyze the transfer of a group from one molecule to another. For example, the body is able to make certain amino acids by transferring an amino group from an amino acid in large supply to a structure capable of being converted to a desired amino acid, as illustrated below.

Transaminases—transfer of amino group—for example,

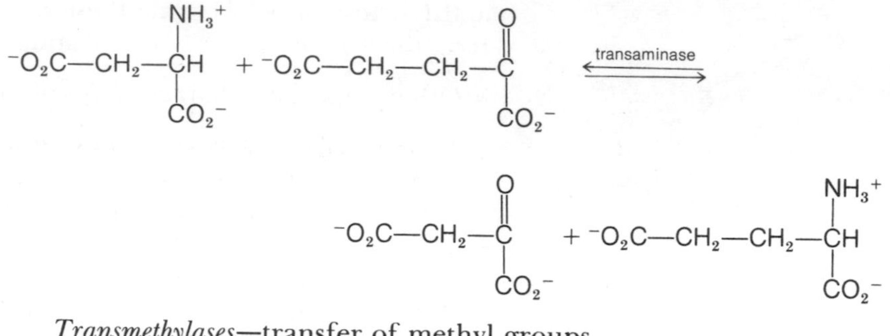

$$^-O_2C—CH_2—\overset{\overset{\displaystyle NH_3^+}{|}}{CH}—\underset{\underset{\displaystyle CO_2^-}{|}}{\ } \quad + \quad ^-O_2C—CH_2—CH_2—\overset{\overset{\displaystyle O}{\|}}{C}—\underset{\underset{\displaystyle CO_2^-}{|}}{\ } \quad \underset{\xrightarrow{\hspace{1cm}}}{\overset{transaminase}{\xleftarrow{\hspace{1cm}}}}$$

$$^-O_2C—CH_2—\overset{\overset{\displaystyle O}{\|}}{C}—\underset{\underset{\displaystyle CO_2^-}{|}}{\ } \quad + \quad ^-O_2C—CH_2—CH_2—\overset{\overset{\displaystyle NH_3^+}{|}}{CH}—\underset{\underset{\displaystyle CO_2^-}{|}}{\ }$$

Transmethylases—transfer of methyl groups.
Transphosphorylases—transfer of phosphate groups.
Oxidases—transfer of hydrogen atoms.

5. *Nucleases*—enzymes that catalyze the hydrolysis of nucleic acids ultimately to pentoses, phosphoric acid, pyrimidines, and purines.

DNAases—degradation of DNA.
RNAases—degradation of RNA.

ATP-ADP—The Energy Converters

Like most other chemical factories, the cell requires relatively large quantities of fuel to sustain its operation. In normal operation, a large fraction of the cell's reactions are endothermic or, more important, they are nonspontaneous. This is evidenced by the spontaneous decomposition of components of cells upon death. Living cells use glucose (and decomposition products from both fats and proteins) as a fuel source. While glucose, in being oxidized to carbon dioxide and water, can supply an adequate quantity of energy, the cell cannot always use that energy at the time the glucose is being oxidized. Neither can the cell allow too much energy to be released to the surroundings. If even half of the energy from oxidizing glucose were released to the cell, the temperature of the cell would rise so high that the enzymes would be inactivated. Thus the cell needs a mechanism for capturing and storing the energy from its fuel, and another mechanism for supplying this energy when and where it is needed.

The basis for these two mechanisms lies in some very elementary chemistry. When a compound *A* is converted to a compound *B*, energy may be released or absorbed, and when *B* is converted back to *A*, the reverse will be true. Suppose, then, that as glucose goes through its 17-step degradation to carbon dioxide and water, it releases energy a bit at a time in many of these steps. Suppose this energy is used to convert compound *A* to compound *B*. All along the way, *B* molecules are being formed and energy is being stored. Now, suppose the cell needs energy. Can it not obtain this by converting some *B* molecules back to *A* molecules?

This is believed to be the energy-conversion mechanism of the cell. The molecules *A* and *B* are adenosine diphosphate, ADP, and adenosine triphosphate, ATP, although there are several other pairs of molecules that may serve the same function.

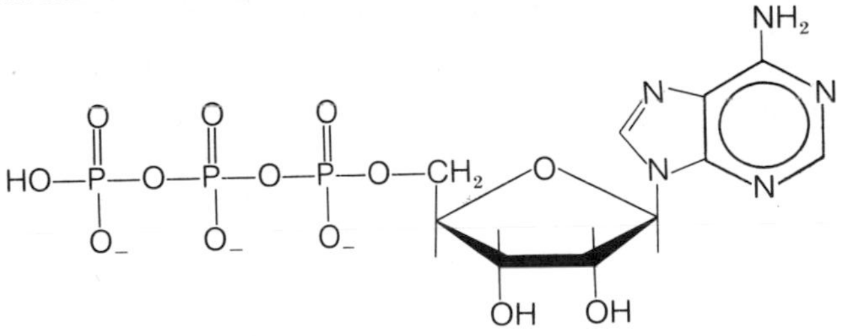

Adenosine triphosphate, ATP

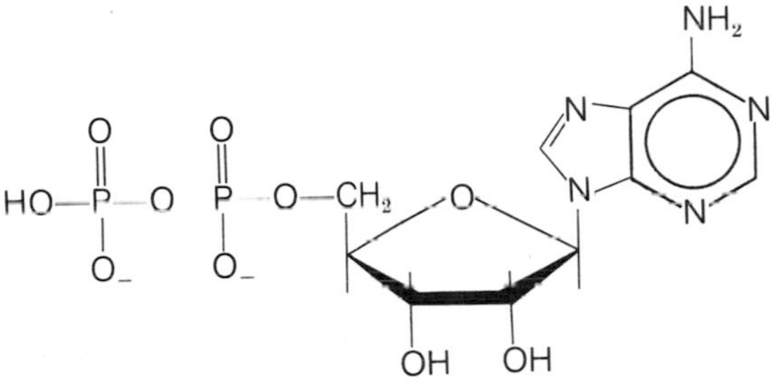

Adenosine diphosphate, ADP

In spite of the complicated structures of ATP and ADP, their energy conversion function lies in the phosphate portion of the molecule. There are two phosphate groups in ADP; in ATP, there are three such groups. Addition of the third phosphate group to ADP requires considerable energy— four to five times as much as is needed to link a phosphate group to an alcohol, for example. Therefore, when ADP is converted to ATP, energy is required and is stored in the ATP formed:

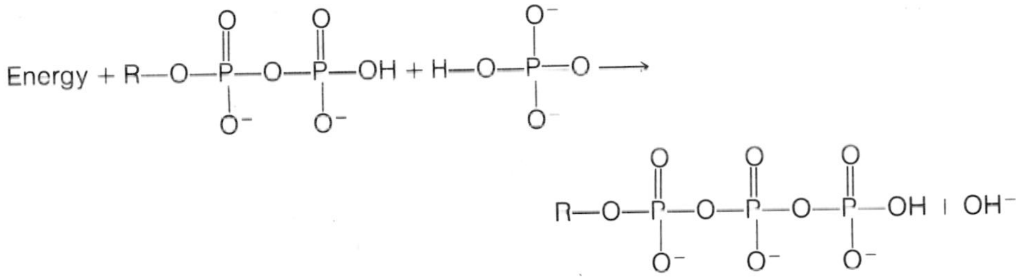

When energy is needed, ATP reacts with water to give ADP and energy.

$$ATP + H_2O \longrightarrow ADP + H_2PO_4^- + energy$$

The question often asked is: Why does it take so much energy to form the triphosphate bond? The answer seems to be related to the fact that in forming a covalent bond to the ADP molecule, the third phosphate unit must localize some of its delocalized (resonance-involved) electrons. Evidently, the energy required to do this is appreciable.

ATP is the key compound in the cell as far as energy utilization is concerned. When the cell needs to carry out an endergonic (energy-absorbing) reaction, that energy is supplied by hydrolyzing ATP back to ADP. When the supply of ATP runs low, the cell produces more by oxidizing glucose or other energy-releasing molecules.

More than 90 per cent of the ATP is made in the mitochondria of the cell. Many cells contain from 50 to 5000 mitochondria. Cells having high-energy requirements have more mitochondria than those having low-energy requirements.

Oxidation of Glucose

The actual sequence of reactions in which a glucose molecule in the cell is oxidized to carbon dioxide and water is given in Figure 29.4. The sequence may be divided into two main portions. The first consists of the breakdown of the glucose molecule into two three-carbon fragments—pyruvic acid or lactic acid, or both. This portion of the oxidation sequence is called glycolysis. The second portion of the sequence is the conversion of the pyruvic (or lactic) acid to carbon dioxide and water, which occurs via the tricarboxylic acid, or Krebs, cycle.

During glycolysis, a glucose molecule first reacts with a molecule of ATP, resulting in a transfer of a phosphate unit from ATP to carbon atom 6 on the glucose molecule (Figure 29.4). The glucose-6-phosphate then is transformed to an isomer, fructose-6-phosphate, which reacts with a second ATP molecule to produce fructose-1,6-diphosphate. The diphosphate decomposes to two molecules of glyceraldehyde-3-phosphate, which go through a series of steps, some involving oxidation or reduction of the three-carbon fragments; the final result of this series of reactions is the formation of pyruvic or lactic acid. Two oxidation steps in this series each produce two ATP molecules (one for each three-carbon fragment oxidized), with the result that the over-all glycolysis sequence gives a net production of two ATP molecules (two ATP molecules were used in transferring phosphate units to glucose or fructose molecules, and four were produced in the oxidations). The ATP molecules, of course, have stored energy.

Glycolysis is important in the muscular activity of the body. When the body calls on its muscles to perform continuous labor, the muscles must produce energy without the presence of enough oxygen for the complete oxidation of carbohydrate to carbon dioxide and water. In glycolysis, the over-all reaction

$$C_6H_{12}O_6 \rightleftharpoons 2CH_3CHOHCO_2H + 2ATP$$
Lactic acid

involves no oxygen. Under strenuous activity, muscles show a decrease in their glycogen (reserve carbohydrate) content and an increase in lactic acid content. No oxygen need be consumed. However, at some stage (perhaps after muscular activity is over), lactic acid must be reconverted to glycogen. The energy required to do this must come from the oxidation of some of the lactic acid, a process that does require oxygen (see discussion of the Krebs cycle). Panting and gasping for breath after strenuous muscle activity is an indication of the body's demand for oxygen to do this job.

The Tricarboxylic Acid Cycle (Krebs Cycle)

The second portion of the glucose oxidation sequence involves the oxidation of pyruvic (or lactic) acid to carbon dioxide and water by a series of reactions that form a cyclic process (Figure 29.4). The intermediates formed in the first steps of this cycle are tricarboxylic acids such as citric acid,

$$HO-\underset{\underset{\displaystyle CH_2CO_2H}{|}}{\overset{\overset{\displaystyle CH_2CO_2H}{|}}{C}}-CO_2H$$

hence the name tricarboxylic acid cycle.

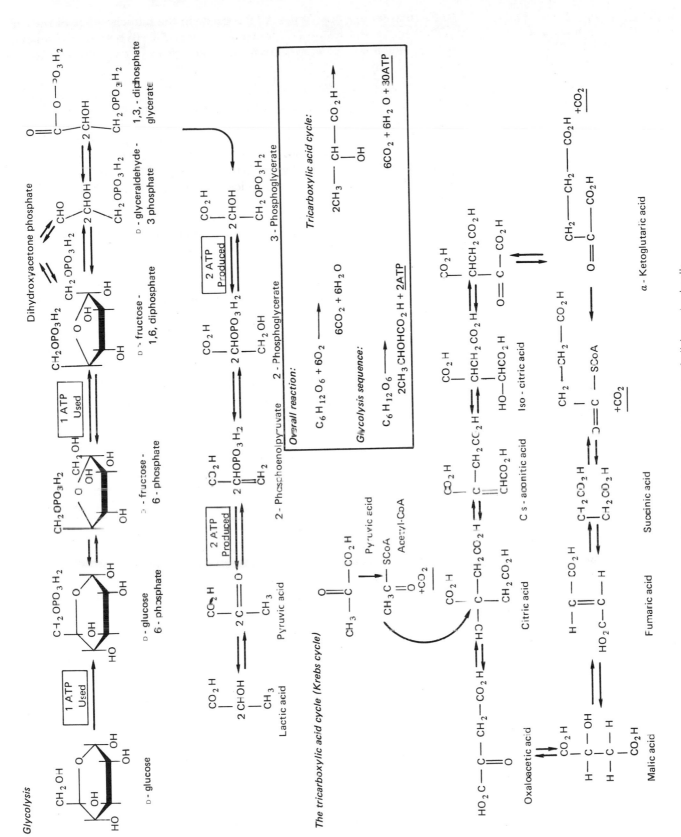

Figure 29.4 Sequence of steps by which glucose is oxidized to carbon dioxide and water in living animal cells.

Figure 29.5 Coenzyme A. The portion on the left is from vitamin B_4 (pantothenic acid).

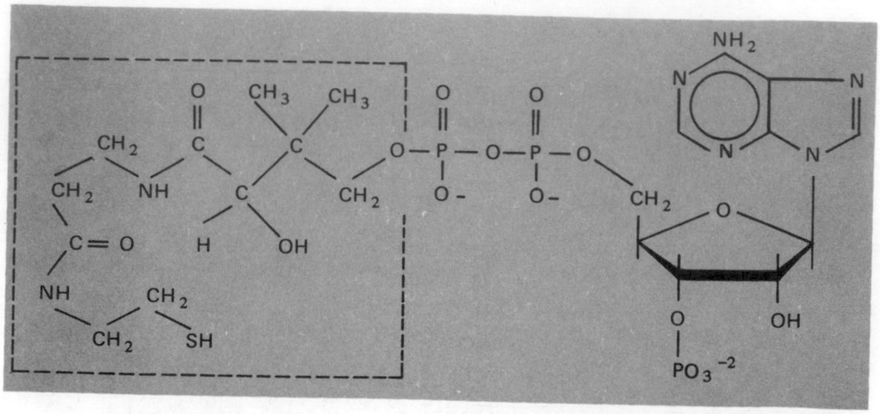

This cycle is the only known reaction sequence by which a foodstuff can be completely oxidized in the body. Foodstuffs other than carbohydrates must be converted first to pyruvic acid or to some other intermediate in the cycle before they can be oxidized to carbon dioxide and water.

Pyruvic acid itself is not an intermediate in the tricarboxylic acid cycle. It is first converted to a substance known as acetyl coenzyme A (CoA),

$$CH_3\overset{\displaystyle O}{\underset{\displaystyle \|}{C}}\!-\!SCoA$$

in one of the most complex and surely one of the most important reactions of the cell. This reaction may be written

$$CH_3-\underset{\underset{\text{Pyruvic acid}}{\overset{\|}{O}}}{C}-CO_2H + CoA-SH \xrightarrow{[-2H]} CH_3-\underset{\underset{\text{Acetyl coenzyme A}}{\overset{\|}{O}}}{C}-SCoA + CO_2$$

It involves both oxidation and loss of carbon dioxide. Moreover, five cofactors are needed in addition to the enzyme. These cofactors include coenzyme A, magnesium ion, and compounds related to other vitamins in the B group.

Acetyl CoA enters the tricarboxylic acid cycle by reacting with oxalacetic acid

$$O{=}C-CO_2H$$
$$CH_2-CO_2H$$

to give citric acid, which is transformed by the series of steps shown in Figure 29.4 again to oxalacetic acid with a loss of two molecules of carbon dioxide along the way. In effect, then, the tricarboxylic cycle takes the two-carbon fragment left from the original glucose, attaches it to an oxalacetic acid molecule, and subjects this to a series of steps that results in the liberation of two molecules of carbon dioxide and regeneration of the starting material, oxalacetic acid. In addition, 15 molecules of ATP are produced as each molecule of lactic acid from glycolysis is converted first into pyruvic acid, then into carbon dioxide and water as the pyruvic acid passes through the tricarboxylic acid cycle.

The Fate of Food   When food enters the organism, it must be transformed into useful form. This may mean breaking down molecules such as fats, starch, and proteins into smaller segments. In our bodies, such reactions ordinarily take place in the alimentary tract under the influence of digestive enzymes secreted into various regions of the tract by digestive glands such as the pancreas. Before entering the bloodstream, proteins are hydrolyzed to amino acids, and carbohydrates to glucose or other simple sugars able to pass through the cell membrane.

Inside the cell, there also are large molecules that must be broken down into smaller segments before these parts can be utilized effectively. An example is glycogen, a polymer of glucose similar to starch, which the cell uses to store glucose until it is needed. Breakdown of large molecules in the cell usually takes place in the lysosomes, and a different team of enzymes is needed for each type of substance broken down.

If the food is to be used as fuel, it will be oxidized with the accompanying formation of ATP. As mentioned, most of the ATP is formed in the mitochondria during the tricarboxylic acid cycle.

Much of the food material entering the cell is used to build new cell material—replacements for worn or damaged parts, or components for new cells. Since most of this material is protein in nature, the cell must first assemble the amino acids, making some if necessary, and then synthesize the extremely complicated proteins. A large fraction of protein synthesis occurs in the ribosomes. To illustrate the complexity of protein synthesis, we shall now take a closer look at the structure of enzymes.

What Gives Enzymes Their Shape?

We have seen that the specificity of enzymes is associated with their shape or structure. One wonders, then, how these huge protein molecules acquire a definite shape. Why are they not jellylike shapeless molecules? The current answer to this question is summarized in the abstract "Shapes of Proteins."

We see from Figure 29.6 that each protein, a polymer of amino acids, has those amino acids arranged in a definite order or sequence (Chapter 28). The number of ways the 20 or so amino acids found in proteins can be arranged in a structure is extremely large. This explains why there are so many different proteins. Once formed, the protein may acquire a helical structure stabilized by a large number of hydrogen bonds formed between portions of the same huge molecule, or it may form fibers or sheets where the hydrogen bonds are formed between parallel chains. The shape of the protein structure is maintained by the large number of hydrogen bonds and, in the case of enzymes, also by other attractive forces (van der Waals, dipole interaction).

Having established why enzymes have definite shapes or structures, we are now ready to ask: How does the cell get the amino acids in the correct order* in each of the many enzymes it must synthesize?

RNA—Template for Protein Synthesis

The compounds responsible for getting the amino acids in the correct order during protein synthesis are known as RNA (ribonucleic acids). These are polymers of ribose (a five-carbon sugar), phosphate, and four organic bases containing nitrogen, two of which are purines and two pyrimidines. The structure of a segment of a RNA molecule is given in Figure 29.7.

---

*As an illustration of how important it is that this sequence of amino acids be correct, it is known that if one amino acid in the 574 that make up a hemoglobin molecule is placed incorrectly, an individual becomes sick and may die of sickle-cell anemia.

Figure 29.6 Structure of a poly-peptide, a portion of a protein mole-cule. (a) The structural formula. (b) A formula emphasizing the α-helix of the protein molecule.

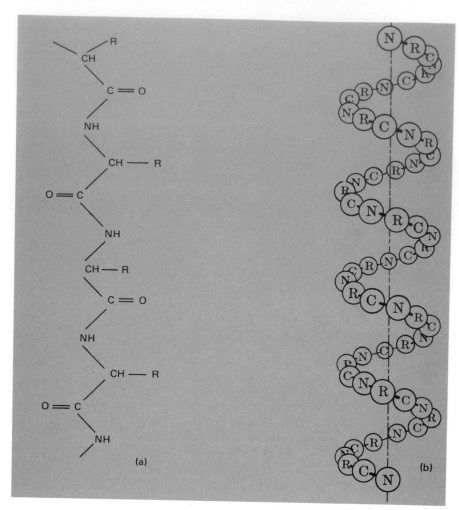

(a)

(b)

Figure 29.7 Representation of a segment of a RNA chain.

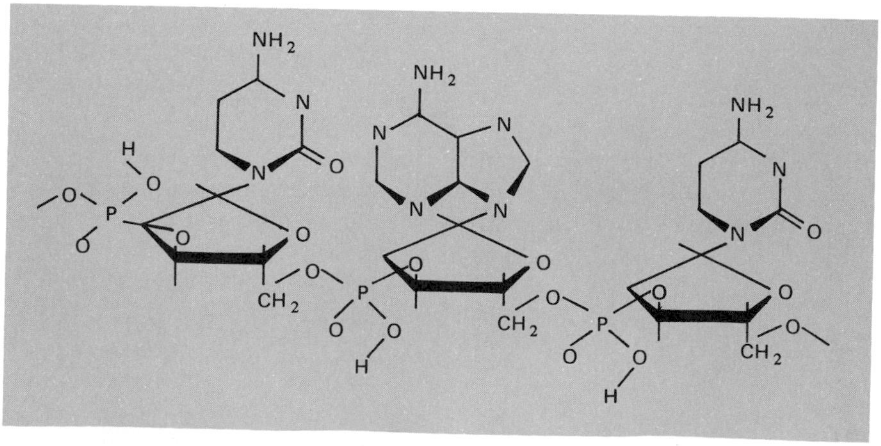

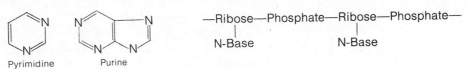

Pyrimidine   Purine

There are many different kinds of RNA molecules, the differences arising in the particular order or sequence of the four nitrogen bases along the polymer chain. This is similar to the situation in proteins and can be illustrated as follows. If the nitrogen bases are designated as I, II, III, and IV, some possible RNA structures include

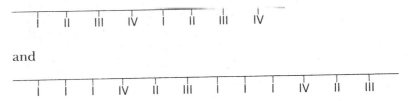

I   II   III   IV   I   II   III   IV

and

I   I   I   IV   II   III   I   I   I   IV   II   III

Two forms of RNA are needed in a protein synthesis. The first, known as *messenger-RNA*, has imprinted on it all the directions needed to make one

---

SHAPES OF PROTEINS

The structure of a protein may be described in terms of four basic structural levels:

Primary structure—related to an amino acid sequence, R_1, R_2, R_3, and so on, and illustrated as follows:

$$HN-CH(R_1)-C(=O)-NH-CH(R_2)-C(=O)-NH-CH(R_3)-C(=O)-$$

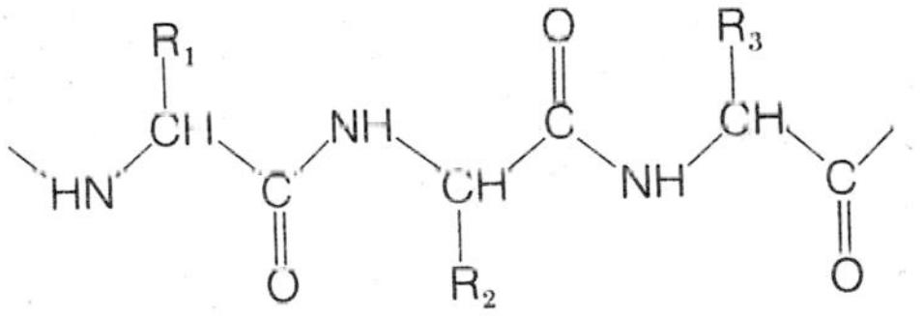

Secondary structure—resulting from hydrogen bonding. Long protein chains have thousands of H-bonds between various portions of the same chain or between adjacent chains.
a) Internal hydrogen bonding gives helical structure in globular proteins (enzymes).
b) Hydrogen bonding between molecules gives sheet or fibrous structures (hair, skin).

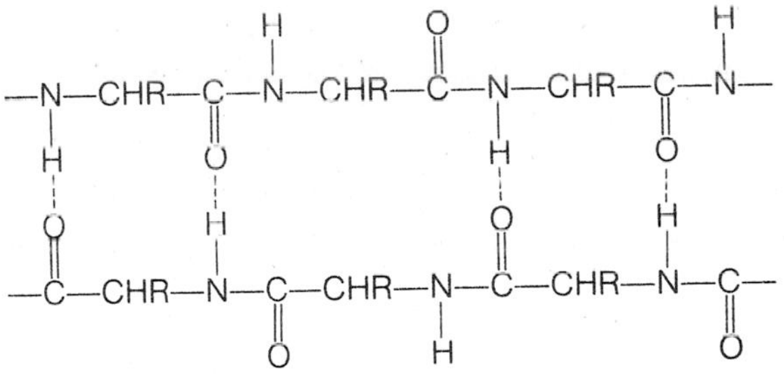

$$-N-CHR-C-N-CHR-C-N-CHR-C-N-$$
$$-C-CHR-N-C-CHR-N-C-CHR-N-C-$$

Tertiary structure—folding or coiling of helix on itself. This is the result of all types of inter-molecular attractive forces: H-bonds, dipole interaction, van der Waals forces—even S—S bridges between segments of the chain form in some cases. The tertiary structure of a protein is related also to the sequence of amino acids, the size and properties of the R groups in the chain being especially important. The unique catalytic properties of enzymes presumably are closely related to the tertiary structure.

Quaternary structure—grouping of subunits to give a large, active structure. Some active proteins are believed to be clusters of several subunits, each of which is inactive separately but if put together in appropriate order, become physiologically active.

kind of protein. There are at least as many different messenger-RNA molecules as there are different proteins. The second form of RNA is known as *transfer-RNA*. Its job is to pick up one particular amino acid and bring it to messenger-RNA. There are at least as many kinds of transfer-RNA molecules as there are amino acids.

Now let us consider an oversimplified but not unreasonable picture of protein synthesis as controlled by the two forms of RNA by using the following schematic diagrams.

1. This represents messenger-RNA. The four nitrogen bases are indicated by line formulas.

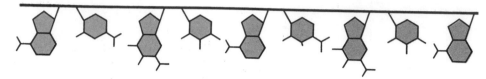

2. These represent three kinds of transfer-RNA.

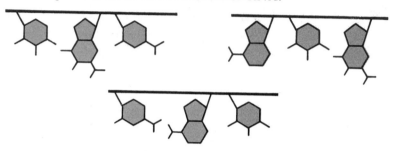

3. These represent three different amino acids.

NUCLEIC ACIDS AND THEIR DERIVATIVES

Nucleic acids are macromolecules present in all living cells. They exist either free or combined with proteins as *nucleoproteins*. Two types of nucleic acids are recognized: DNA, *deoxyribonucleic* acids, and RNA, *ribonucleic* acids. The molecular weights of isolated DNA molecules range from 6 million to above 120 million. RNA molecules have molecular weights from 20,000 to 2 million or more.

DNA contains the purines *adenine* and *guanine* and pyrimidines *cytosine* and *thymine* suspended from *deoxyribose* (a sugar derivative) which, along with *phosphate*, forms the backbone of the DNA polymer.

RNA contains the purines *adenine* and *guanine* and pyrimidines *cytosine* and *uracil* suspended from *ribose* (a five-carbon sugar) which, along with *phosphate*, forms the backbone of the RNA polymer.

Nucleic acids may be considered to be repeating sequences of *nucleotides*. A nucleotide unit consists of a *base (purine or pyrimidine)-sugar-phosphate* complex. A *nucleoside* is a nucleotide without the phosphate member.

4. This illustrates the pick-up of amino acids by transfer-RNA.

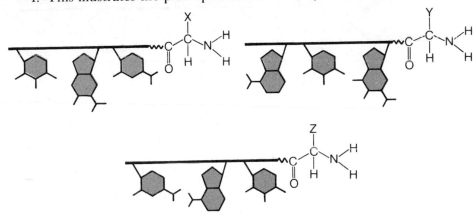

5. This illustrates the attachment of transfer-RNA to messenger-RNA.

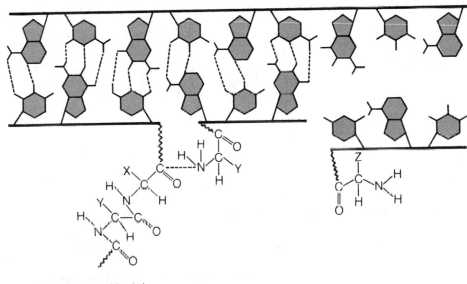

Growing polypeptide chain

Uracil
(pyrimidine)   Adenine
                    (purine)

Guanine   Cytosine
(purine)   (pyrimidine)

Figure 29.8 The matching pairs of nitrogen bases in RNA, showing the alignment through hydrogen bonding.

Note that the transfer-RNA can enter only at certain points. Once in position the amino acids can enter the polypeptide or protein chain.

The two new features brought out by the schematic representations are:

1. There is a complementary relation between pairs of the nitrogen bases. This is brought about by hydrogen bonding. In RNA the complementary pairs are adenine and uracil, guanine and cytosine, as shown in Figure 29.8. When transfer-RNA brings an amino acid to messenger-RNA, a union between the two RNA molecules occurs only when cytosine is opposite guanine and adenine is opposite uracil at points of contact between the two RNA chains.

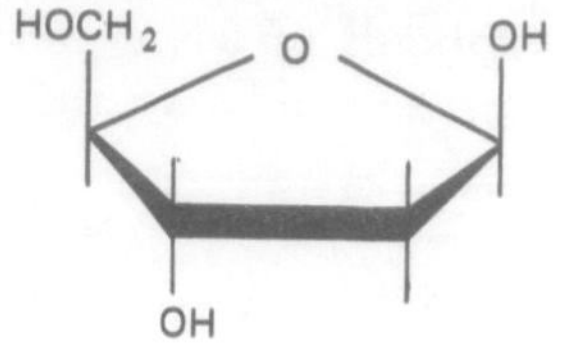

Deoxyribose

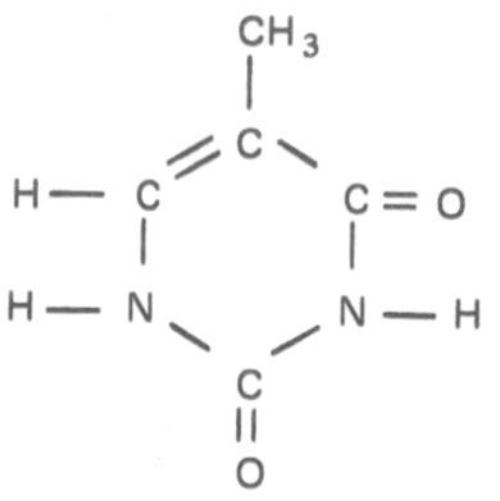

Thymine

Figure 29.9 Two components of DNA not present in RNA.

Table 29.1 RNA Code Triplets Associated with Various Amino Acids

Amino Acids	*Code Triplets* (C, cytosine; G, guanine; U, uracil; A, adenine)
Alanine, CH_3—CH—CO_2H	GCU, GCC, GCA, GCG
Serine, $HOCH_2$—CH—CO_2H	AGU, AGC, UCC, UCU
Leucine, CH_3—CH—CH_2—CH—CO_2H	CUU, CUC, CUA, CUG, UUA, UUG
Tyrosine, 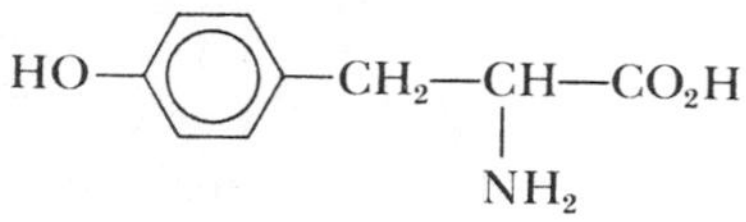 HO—⟨ ⟩—CH_2—CH—CO_2H	UAU, UAC
Proline, 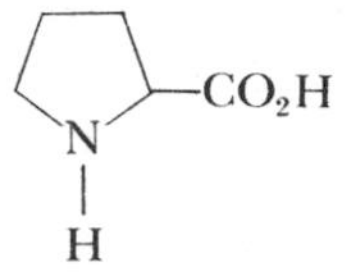	CCU, CCC, CCA, CCG

NOTE: These code triplets are associated with messenger-RNA and are called *codons*. Transfer-RNA for a given amino acid contains the corresponding anticodons.

2. Transfer-RNA needs three nitrogen bases (known as code triplets) to fix an amino acid. For example, transfer-RNA with three adenine groups will pick up the amino acid phenylalanine,

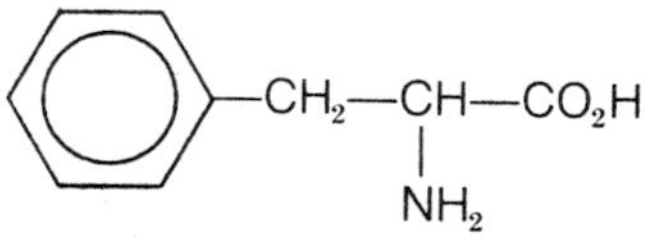

Code triplets for some other amino acids are given in Table 29.1. Most amino acids respond to more than one code triplet, as indicated in the table. This means several different transfer-RNA molecules may be able to pick up a given amino acid.

Where Does RNA Get Its Information?

Most messenger-RNA is thought to be synthesized in the nucleus of the cell in a manner analogous to the synthesis of proteins just described. However, in this case the template is the gene, a material known as DNA—deoxyribonucleic acid. Chemically, DNA is similar to RNA with two exceptions: DNA contains the sugar deoxyribose (similar to ribose but having one less oxygen atom) instead of ribose, and it contains the pyrimidine thymine instead of uracil (Figure 29.9). A portion of a DNA chain can be represented as

—Deoxyribose—Phosphate—Deoxyribose—Phosphate—Deoxyribose—Phosphate
　　|　　　　　　　　　　　　　|　　　　　　　　　　　　　|
　Thymine　　　　　　　　　Adenine　　　　　　　　　Cytosine

Apparently DNA is able to line up the appropriate segments of RNA (the so-called nucleotides) along its chain, perhaps in a manner similar to that represented in the scheme

Schematic nucleotide

Once in position, the nucleotides are linked together by enzymes known as RNA-DNA polymerases to give messenger-RNA. Messenger-RNA leaves the nucleus and finds its way to the ribosomes, where it supervises the synthesis of proteins.

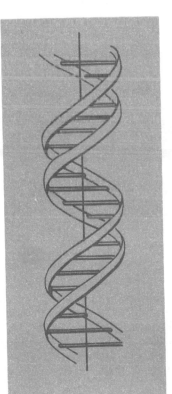

DNA—Key to All Life

Molecules of DNA not only carry the chemical code needed to manufacture the proteins that sustain the cell, but they also have the ability to split into two and make exact copies of themselves—a quality essential for cellular reproduction. In essence, this means that they make the cell what it is and see to it that new cells are produced in the image of the parent. Evidently, DNA molecules are the molecules of life—they define the life processes of the cell, and they go on faithfully reproducing themselves in exactly the same form and generating exactly the same enzymes time after time in immortal splendor.

The secret of DNA's ability to make exact copies of itself over and over again lies, of course, in its structure. Each DNA molecule is a helix composed of two strands of polynucleotides. The two strands are held together by hydrogen bonding between the nitrogen bases. Every adenine unit in one strand is hydrogen-bonded to a thymine unit in the other strand, and every cytosine unit in one strand is bonded to a guanine unit in the other. The DNA molecule appears then like a spiral rope ladder, as illustrated in Figure 29.10.

When ready to replicate, the two strands unwind and separate. Each strand rebuilds its missing partner from the nucleotides present in the nucleus of the cell. In the rebuilding process, the four nitrogen bases line up in the complementary relation described for RNA. When the rebuilding process is complete, there are two identical DNA molecules where one existed before. On cell division, one of these remains with each daughter cell to pass on the genetic information.

DNA molecules are present in every living cell. Apparently, these molecules remain in the nucleus, well protected from potential dangers and shocks

Figure 29.10 The Watson-Crick model of a molecule of DNA. The two ribbons denote the two complementary chains, and the horizontal bars represent the pairs of bases holding the chains together.

more likely in the cytoplasm. The molecular weight of DNA is about 6 million; one molecule may contain 20,000 nucleotides.

One DNA molecule may contain several genes—that is, several coded sequences, each able to make a different RNA molecule. In a complex organism such as man, all the genetic information (DNA) needed to make the full-grown adult is packed into the fertilized human egg and weighs approximately 6×10^{-12} g. The DNA in every human cell contains all the genetic information the fertilized egg contained. One problem now being solved is: Why are cells in a complex organism so different if all contain the same DNA?

Blue Eyes and DNA. The color of eyes, hair, and skin is determined by the amount of pigment called *melanin* (black) present in the pupil of the eye, the hair, or the skin. Melanin is formed from the amino acid tyrosine in a

$$HO \!-\! \bigcirc \!-\! CH_2 \!-\! \underset{\underset{NH_2}{|}}{CH}CO_2H$$

series of steps. Each step requires an enzyme, and each enzyme is produced from a specific RNA template. The amount of melanin produced will depend upon the amount of one of the messenger-RNA molecules present. If much of this particular RNA is present, much melanin will be produced; if only a small amount is present, only a small quantity of melanin will result. A DNA molecule in the nucleus of the cell apparently controls the production of this RNA and thus controls the color of the eyes, hair, and skin of the subject. Since the DNA molecule is passed on to the progeny, they will resemble their parents in coloring. Should something happen to the gene responsible for melanin production, none of this pigment will be made and the individual will be an albino.

Viruses and Heredity

A virus is simply a DNA (or RNA) molecule surrounded by a protective protein. When injected into a cell, the virus DNA directs the formation of enzymes not previously present in the cell. The cell can then perform chemical reactions not possible previously. Upon cell division, the virus DNA will be transmitted to the daughter cells and will then become part of the cell's genetic characteristics. If the chemical process initiated by the virus DNA is harmful to the cell, it could destroy the cell or possibly produce a series of self-destroying cells. This may do great harm to the organism if the infection is not stopped.

If artificial and beneficial viruses could be made, they might be injected into cells and certain genetic defects might be corrected or memory might be improved. In effect, this is a mechanism for controlling heredity. Scientists everywhere hope that if and when such heredity control is possible, it will be used for the benefit of mankind and not for his derogation. Viruses have very recently been synthesized in the biochemical laboratory.

Evidence for the Role of DNA and RNA

The preceding picture of the chemistry of the cell was presented with no experimental evidence. Actually, it would take several volumes to summarize the evidence for what has been condensed in this chapter. However, as an indication of the kind of experiments and reasoning employed by biochemists and microbiologists in this work, the following is presented.

In 1946, Wendell Stanley received a Nobel prize for demonstrating that a chemical substance he had crystallized could be stored indefinitely and, on

Table 29.2   Analysis of Individual Purines and Pyrimidines in DNA From Several Sources

| | MOLAR RATIOS | | |
Source	Adenine: Guanine	Thymine: Cytosine	Purines: Pyrimidines
Man (liver)	1.42	1.80	1.09
Ox (thymus)	1.29	1.43	1.1
Sheep (spleen)	1.26	1.36	1.01
Hen (crythrocytes)	1.45	1.29	0.99
Trout (sperm)	1.32	1.36	1.1

contact with a tobacco plant, would produce a viral disease; indeed, it would actually multiply in the plant and produce more virus. In effect, he showed that a crystalline substance was *infective*, was able to reproduce or to stimulate reproduction of itself. The crystalline substance was found to contain a protein and a nucleic acid.

About 1952, A. D. Hershey and M. Chase grew bacteriophage (bacterial virus) in the presence of radioactive sulfur and radioactive phosphorus. They knew sulfur was present in some amino acids and would therefore be incorporated in the protein portion of the virus. Similarly, phosphorus is a component of nucleic acids, so the radioactive phosphorus would be incorporated in the nucleic acid portion of the freshly made virus. They then infected the host cells with the doubly labeled virus and looked for the radioactive sulfur and phosphorus within the cells. They reasoned that if only the radioactive phosphorus were there, then only the nucleic acid penetrated the cell; if only the radioactive sulfur were found, then only the protein penetrated. They found much radioactive phosphorus and only a trace of radioactive sulfur in the contents of the cell; this showed that the nucleic acid DNA alone is the infective material.

Analysis of the purines and pyrimidines of the DNA from various sources, reported about 1952, gave some interesting results. A small sample of these is given in Table 29.2. While the ratios of the two purines or the two pyrimidines varied from source to source, the ratio of the purines to the pyrimidines remained constant at about 1.0.

This, of course, helped J. D. Watson and F. H. C. Crick, in 1953, make their famous postulate concerning the pairing of purines and pyrimidines and enabled them to construct the double-strand model for DNA given in Figure 29.10.

M. Meselson and F. W. Stahl, in 1958, supported the Watson-Crick postulate by some most ingenious experiments. They grew bacteria on a medium (food) containing a heavy isotope of nitrogen so that all proteins and nucleic acids in the cells became labeled with heavy nitrogen. Then, they grew two generations of these bacteria on ordinary nitrogen and isolated the DNA. Using centrifugation, they determined how the heavy nitrogen was distributed among the DNA molecules. Results indicated that half of the DNA molecules had some heavy nitrogen; the other half had only natural nitrogen. This is exactly what Watson and Crick predicted. For as the bacteria with heavy nitrogen multiplied on the natural nitrogen medium, the DNA strands with heavy nitrogen separated and synthesized their complementary strands using natural nitrogen. On analysis, their DNA contained strands of heavy nitrogen isotopes and strands containing natural nitrogen isotopes; the two kinds of strands could be separated in the ultracentrifuge.

Prior to 1957, A. Kornberg and his group found an enzyme that catalyzes the formation of DNA molecules *in vitro* starting from a broth of nucleotides. The synthesis needs two things to proceed: all four deoxyribonucleotides must be present, and a preexisting molecule of DNA must be present. With these available, new DNA molecules similar to the template DNA are synthesized. Even though the enzyme is isolated from bacteria, it will direct the synthesis of human or other DNA molecules, provided only that the appropriate DNA template is present. Evidently, it is the DNA molecule that carries the information; the enzyme merely directs the linking together of the nucleotides after they are in the proper order. In 1961, Marshall Nirenberg performed a classic experiment in which he used what amounted to a messenger-RNA molecule containing three nucleotide residues arranged in a simple repeating order (e.g., I-II-III-I-II-III-etc.). To this he added transfer-RNA and radioactive amino acids. He found that the radioactive protein synthesized contained only one amino acid residue, indicating that the code for incorporating a particular amino acid in a protein consisted of three nucleotide residues arranged in a specific order. In 1967, Kornberg reported making the first synthetic virus using this information.

While these are but a few of the many thousands of researches directed toward the chemistry of living cells, they provide some insight into the nature of this very exciting field of science.

The Chemistry of Muscle Activity. The biochemical mechanism for contraction and relaxation of muscles has fascinated physiologists, biochemists, and biophysicists for years. Among the many interesting questions posed by the smooth and highly efficient action of muscles in higher animals are: How is muscle action turned on so quickly? What is the energy source for muscle contraction and for the development of force by muscle? What is the chemistry involved in shortening of muscle and in its ease of relaxation?

Electron microscope studies have revealed that a typical skeletal or voluntary muscle fiber consists of a matrix of long slender rods called myofibrils, each of which is made up of a large number of ribbonlike filaments. The filaments are arranged in parallel bundles within the myofibril, and are surrounded and bathed by sarcoplasm, the intracellular fluid of muscle. The sarcoplasm contains glycogen, enzymes to hydrolyze glycogen, ATP, other energy-rich substances, and inorganic electrolytes, including phosphate, calcium, magnesium, sodium, and potassum ions. Active muscles contain large numbers of mitochondria, the cell component with the capability of producing energy by oxidation of foodstuffs.

Muscle myofibrils possess along their length a structural pattern (alternating dark and light stripes or bands) that repeat every 2.5×10^4 A or so. The bands arise from an orderly arrangement of thick and thin filaments within the fibril, as illustrated below:

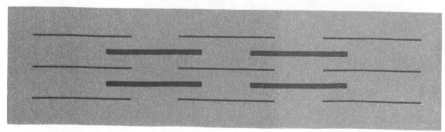

The thick filaments are composed mainly of the protein myosin; the thin filaments contain the protein actin. Muscle contraction apparently involves sliding of the actin filaments over the myosin filaments, somewhat as illustrated:

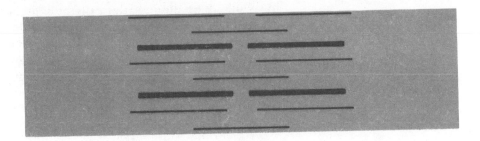

Close examination of myosin reveals it to be a long (about 1600 A) filament made of two polypeptide chains coiled together like the strands in a rope. At one end of the molecule, the two chains are folded into a globular structure, or head, that seems to have enzyme activity.

In the myofibril, the bundle of myosin molecules is arranged so the filaments lie parallel to the long axis of the fiber, with the heads extending sideways out of the bundle in a regular helical pattern. Thus, the thick filament of the fibril gives the appearance of a very long strand with barblike globular projections emerging from its sides. The projections are themselves arranged around the filament to give it a spiral ladder appearance.

These projecting globular heads are believed to form the chemical cross-links between myosin molecules in the thick filament and actin molecules in the thin filament of the myofibril. During muscular contraction or relaxation, these cross-links are constantly being formed and broken. The spiral arrangement of the heads along the thick filament apparently assists the movement of the chains relative to one another. The cross-links are believed to be stable in the absence of ATP, but can be both hydrolyzed and reformed in its presence.

The chemical trigger that sets off a muscle contraction is the release of Ca^{+2} ions into the sarcoplasm. This stimulates the breakdown of ATP, which releases energy to be used in the breaking of cross-links between myosin and actin. Once broken (or hydrolyzed), new cross-links are formed. It is said that one ATP molecule must be converted to ADP for each cross-link broken.

Estimates are that the mammalian muscle requires hydrolysis of very large amounts of ATP during activity—about 10^{-3} moles per gram of muscle per minute. The amount of ATP actually present in muscle is enough for only about one-half second of activity. Clearly, the energy reserve in muscle is not in the ATP present. Instead, it is found in the high-energy compound phosphocreatine, which is present in muscle cells in about five times the concentration of ATP. Phosphocreatine readily transfers its phosphate (and energy) to ADP to produce ATP. As muscle activity uses ATP and produces ADP, phosphocreatine converts ADP back to ATP. Phosphocreatine is produced in the mitochondria as part of the respiration process, and its supply is sufficient to sustain normal muscle activity. Only when muscle is stimulated for long periods of time, in the absence of respiration or glycolysis, will its supply be depleted.

The great speed with which the chemical machinery of our muscles responds to stimulus is explained as follows. A muscle fiber at rest is electrically polarized so that the outside is about 60 millivolts more positive than the inside. Each fiber also has small tubules running perpendicular to its long axis, which also carry a positive charge relative to the fibrils. Stored in small sacs on the surface of the muscle fiber along the walls of the tubules are relatively high concentrations of calcium ions.

When the muscle is excited by the motor nerve (see Chapter 22), the charge differential between the inside and outside of the muscle fiber is reversed. This immediately releases Ca^{+2} ions throughout the entire fiber, which then catalyze ATP hydrolysis, giving energy to break cross-links between myosin and actin, and thereby initiating the biochemical machinery of muscle contraction. The change in Ca^{+2} concentration needed to activate muscle is from $5 \times 10^{-7}\ M$ to $5 \times 10^{-6}\ M$.

Relaxation of muscle occurs when the polarity at the membranes is restored to its original value, and the calcium ions are complexed and pumped back into their storage sacs. This process also involves expenditure of energy that comes from hydrolysis of ATP.

Great amounts of energy are needed in muscle activity; the processes used by the body to obtain and utilize this energy are far more efficient than any devised by man.

SUMMARY

This chapter gives an overview of the chemistry of the living cell. It discusses the views held by most biochemists concerning the role of enzymes, the chemistry of energy conversion within the cell, the function and chemistry of nucleic acids, and the chemical nature of heredity. An indication of the kind of experimental evidence and reasoning employed in developing this picture is also given.

IMPORTANT TERMS

Biochemistry

Living cells
cytoplasm
mitochondria
ribosomes
lysosomes
nucleus

Enzymes
proteins
carbohydrases
proteases
esterases
transferases
transaminases
nucleases

ADP-ATP
glycolysis
tricarboxylic
  acid cycle
coenzyme A

Nucleic acids
nucleotides
messenger-RNA
transfer-RNA
DNA
ribose
pyrimidine
purine
viruses

QUESTIONS AND PROBLEMS

1. Why are enzymes more specific than many other catalysts?

2. What is the substrate of an enzyme?

3. List the five groups of enzymes and summarize the chemistry of each group.

4. Explain why all enzymes are not equally specific.

5. Describe the molecular structure of an enzyme.

6. How does the primary structure of a protein differ from its secondary and tertiary structures? Explain.

7. What is a coenzyme?

8. What is the importance of ATP in the cell? How is it formed?

9. What is glycolysis? What essential role does it play in the chemistry of living things? Write an over-all chemical reaction for glycolysis.

10. Compare the amount of ATP produced in glycolysis with that produced in the Krebs Cycle.

11. For every molecule of glucose oxidized to carbon dioxide and water, how many ATP molecules have been used? How many have been formed?

12. Explain in terms of chemical reactions why we pant for breath after strenuous exercise.

13. Why is the transformation of pyruvic acid to carbon dioxide and water known as a "cycle"?

14. How are proteins synthesized in the cell?

15. Distinguish between the roles of messenger-RNA and transfer-RNA. In what ways does DNA differ from RNA?

16. Distinguish between a nucleoside and a nucleotide, using structural formulas.

17. What is a gene?

18. Describe the Watson-Crick model for DNA replication.

19. How might the injection of a virus into a living cell change the heredity characteristics of the cell?

20. Why must food be digested before it can be used by the body?

21. Can the cells of the human body synthesize all the amino acids it needs? Suggest several nutritional implications in your answer.

22. What is the chemical role of vitamins?

23. Suggest explanations for the fact that the catalytic activity of enzymes is sensitive to both temperature and pH.

SPECIAL PROBLEMS

1. When lactic acid is reconverted to glucose, why must some lactic acid be oxidized to carbon dioxide and water?

2. Which are the energy-producing steps of the tricarboxylic acid cycle?

3. Which steps involve reduction?

4. Which vitamins are involved in the tricarboxylic acid cycle?

5. Write formulas to illustrate each of the types of components obtained by complete hydrolysis of a nucleic acid.

6. Design an experimental approach that would establish the RNA code triplets for several amino acids.

7. Saliva contains a ropy liquid protein called mucin. It is said that mucin makes saliva more viscous and serves to lubricate food, thereby aiding swallowing. Describe in terms of molecular interactions how the protein might both increase the viscosity of water and act as a lubricant for food.

8. How is energy supplied for the activity of muscle cells and those of other motor organs?

9. Amino acids may exist as the dipolar ion, represented as

$$H_3N^+\!-\!CH\!-\!CO_2^-.$$
$$\underset{R}{|}$$

This ion forms as a result of proton donation by the carboxyl group and proton acceptance by the amine group. In general, the pK_I of the carboxyl group is between 2 and 3, while the pK_I for the amine group is between 9 and 10. If an amino acid is present in a solution having a pH between 2.5 and 9.5, will the predominant form of the amino acid be the dipolar ion? Why or why not?

10. Referring to Problem 9, what is the predominant form of an amino acid in a solution of pH > 10? pH < 2?

11. The isoelectric point for an amino acid is the pH at which the solution contains equal numbers of $-CO_2^-$ ions and $-NH_3^+$ ions. Show that for glycine, where $pK_{acid} \simeq 2.5$ and $pK_{Base} \simeq 9.5$, the isoelectric point is 6.0.

12. The pH of the body is $\simeq 7.4$. Does this mean that all amino acids in the body are predominantly in the form of dipolar ions? Why or why not?

REFERENCES

BENZER, S., "The Fine Structures of the Gene," *Sci. Amer.*, **206**, no. 1, 70–84 (1962).

CHANGEUX, J. P., "The Control of Biochemical Reactions," *Sci. Amer.*, **212**, no. 4, 36–54 (1965).

CONN, E. E., and STUMPF, P. K., *Outlines of Biochemistry*, 2nd ed., Wiley, New York, 1967.

CRICK, F. H. C., "The Genetic Code," *Sci. Amer.*, **207**, no. 4, 66–74 (1962).

HURWITZ, J., and FURTH, J. J., "Messenger RNA," *Sci. Amer.*, **206**, no. 2, 41–49 (1962).

KIRSCHBAUM, J., "Biological Excitations and Energy Conversion," *J. Chem. Educ.*, **45**, 28 (1968).

OLBY, R., "Origins of Molecular Biology," *J. Chem. Educ.*, **47**, 168 (1970).

APPENDIX A

Table A.1 General Physical Constants

Constant	Symbol	VALUE Units. Système International (SI)	Units: Centimeter-gram-second (cgs)
Atomic mass unit	amu	$1.660\ 531 \times 10^{-27}$ kg	$1.660\ 531 \times 10^{-24}$ g
Avogadro constant	N_A	$6.022\ 169 \times 10^{23}$ mol^{-1}	$6.022\ 169 \times 10^{23}$ mol^{-1}
Boltzmann constant	k	$1.380\ 622 \times 10^{-20}$ J/K	$1.380\ 022 \times 10^{-16}$ erg/K
Charge on the electron	e	$1.602\ 191\ 7 \times 10^{-19}$ C	$1.602\ 191\ 7 \times 10^{-20}$ cm$^{1/2}$g$^{1/2}$
			$4.803\ 250 \times 10^{-10}$ cm$^{1/2}$g$^{1/2}$s^{-1}
Electron rest mass	m_e	$9\ 109\ 558 \times 10^{-31}$ kg	$9.109\ 558 \times 10^{-28}$ g
Faraday constant	F	$9.648\ 670 \times 10^4$ C/mol	$9.648\ 670 \times 10^3$ cm$^{1/2}$g$^{1/2}$mol^{-1}
Gas constant	R	$8.314\ 34 \times 10^0$ J $\cdot$ K^{-1}mol^{-1}	$8.314\ 34 \times 10^7$ erg $\cdot$ K^{-1}mol^{-1}
			$0.082\ 054$ l $\cdot$ atm $\cdot$ K$^{-1} \cdot$ mol^{-1}
			1.9872 cal $\cdot$ K$^{-1} \cdot$ mol^{-1}
Planck constant	h	$6.626\ 196 \times 10^{-34}$ J $\cdot$ s	$6.626\ 196 \times 10^{-27}$ erg $\cdot$ s
Proton rest mass	m_p	$1.672\ 614 \times 10^{-27}$ kg	$1.672\ 614 \times 10^{-24}$ g
Rydberg constant	R_∞	$1.097\ 373\ 12 \times 10^7$ m^{-1}	$1.097\ 373\ 12 \times 10^5$ cm^{-1}
Speed of light in vacuum	c	$2.997\ 925\ 0 \times 10^8$ m/s	$2.997\ 925\ 0 \times 10^{10}$ cm/s

Table A.2 SI Base Units

Physical quantity	Name of unit	Symbol
Length	meter	m
Mass	kilogram	kg
Time	second	s
Electric current	ampere	amp
Thermodynamic temperature	kelvin	K
Luminous intensity	candela	cd
Amount of substance	mole	mol

Table A.3 Special Names and Symbols for Certain SI Derived Units

Physical quantity	Name of SI unit	Symbol for SI unit	Definition of SI unit
force	newton	N	$kg \cdot m \cdot s^{-2}$
pressure	pascal	Pa	$kg \cdot m^{-1} \cdot s^{-2} (= N \cdot m^{-2})$
energy	joule	J	$kg \cdot m^2 \cdot s^{-2}$
power	watt	W	$kg \cdot m^2 \cdot s^{-3} (= J \cdot s^{-1})$
electric charge	coulomb	C	$amp \cdot s$
electric potential difference	volt	V	$kg \cdot m^2 \cdot s^{-3} \cdot amp^{-1} (= J \cdot amp^{-1} \cdot s^{-1})$
electric resistance	ohm	Ω	$kg \cdot m^2 \cdot s^{-3} \cdot amp^{-2} (= V \cdot amp^{-1})$
electric conductance	siemens	S	$kg^{-1} \cdot m^{-2} \cdot s^3 \cdot amp^2 (= amp \cdot V^{-1} = \Omega^{-1})$
electric capacitance	farad	F	$amp^2 \cdot s^4 \cdot kg^{-1} \cdot m^{-2} (= amp \cdot s \cdot V^{-1})$

Table A.4 Some Conversion Relationships

Electric Charge	
One coulomb	$= 2.778 \times 10^{-4}$ amp $\cdot$ h
	$= 1.036 \times 10^{-6}$ F
Electric Dipole Moment	$= 2.998 \times 10^9$ statcoul
One debye (D)	$= 1 \times 10^{-18}$ statcoulomb $\cdot$ cm
	$= 3.336 \times 10^{-20}$ C $\cdot$ A
	$= 0.21$ electron $\cdot$ A
Energy and Work	(Mass units are included as energy equivalents.)
One erg	$= 10^{-7}$ J
	$= 2.389 \times 10^{-8}$ cal
	$= 6.242 \times 10^{11}$ eV
	$= 1.113 \times 10^{-24}$ kg
	$= 670.5$ amu
One calorie (cal)	$= 4.1840 \times 10^7$ erg
	$= 4.184$ J
	$= 2.613 \times 10^{19}$ eV
	$= 4.659 \times 10^{-17}$ kg
	$= 2.807 \times 10^{10}$ amu
One electron volt (eV)	$= 1.602 \times 10^{-12}$ erg
	$= 1.602 \times 10^{-19}$ J
	$= 3.827 \times 10^{-20}$ cal
	$= 1.783 \times 10^{-36}$ kg
	$= 1.074 \times 10^{-19}$ amu
One kilogram (kg)	$= 8.987 \times 10^{23}$ erg
	$= 8.987 \times 10^{-16}$ J
	$= 2.142 \times 10^{16}$ cal
	$= 5.610 \times 10^{35}$ eV
	$= 6.025 \times 10^{26}$ amu
One atomic mass unit (amu)	$= 1.492 \times 10^{-3}$ erg
	$= 1.492 \times 10^{-10}$ J
	$= 3.564 \times 10^{-11}$ cal
	$= 9.31 \times 10^8$ eV
	$= 1.660 \times 10^{-27}$ kg

Force

One newton (N)	$= 10^5$ dyn $= 0.2248$ lb
One pound (lb)	$= 4.448 \times 10^5$ dyn $= 4.448$ N

Length

One meter (m)	$= 39.37$ in. $= 3.281$ ft $= 6.214 \times 10^{-4}$ mi
One inch (in.)	$= 2.540$ cm
One Angstrom (A) One micron	$= 10^{-10}$ m $= 10^{-6}$ m
One light-year	$= 9.4600 \times 10^{12}$ km

Mass and Weight	(Mass-weight equivalents are valid for terrestrial use only.)
One gram (g)	$= 6.852 \times 10^{-5}$ slug $= 6.024 \times 10^{23}$ amu $= 3.27 \times 10^{-2}$ oz $= 2.205 \times 10^{-3}$ lb
One atomic mass unit (amu)	$= 1.6602 \times 10^{-24}$ g
One pound (lb)	$= 453.6$ g
One ton	$= 2000$ lb $= 907.2$ kg

Pressure

One atmosphere (atm)	$= 1.013 \times 10^6$ dyn/cm^2 $= 76.0$ cm Hg $= 14.70$ lb/in.^{2} $= 2116$ lb/ft^2 $= 760$ torr
One centimeter mercury (cm Hg)	$= 1.316 \times 10^{-2}$ atm $= 1.333 \times 10^4$ dyn/cm^2 $= 5.353$ in. H$_2$O $= 0.1934$ lb/in.^{2} $= 27.85$ lb/ft^2

APPENDIX B

[American Version—abstracted from Report by Committee of the International Union of Pure and Applied Chemistry published in the *Journal of the American Chemical Society*, **82**, 5525 (1960), and in the *Handbook of Chemistry and Physics* (1970–71)].

The adoption of rules for naming chemical compounds is complicated for several reasons:

1. Many common compounds were named before official bodies established rules of nomenclature, and these "trivial" names are ingrained in the science through long usage.
2. Many compounds have complicated structures.
3. Certain forms or names seem to be more adaptable to certain languages —therefore, several versions are agreed upon.

An exhaustive list of rules covers many pages in the official report—some 22 pages for inorganic chemistry and 74 pages for organic chemistry published in the *Handbook* reference above.

We shall list here several of the more common rules gleaned from this international report, which should be adequate for a first course in chemistry. We shall divide the rules into inorganic and organic nomenclature.

I. Inorganic Nomenclature

1. When specifying an isotope, the mass number should be placed at the upper left, and the atomic number at the lower left; for example ^{36}Cl or $^{36}_{17}Cl$. In speaking, these are referred to as "chlorine-36." The symbols D and T for ^2H and ^3H, respectively, are permissible.

2. In formulas, the electropositive element shall be placed first. Example: NaCl.

3. (a) In formulas of complexes, if the complexing group (ligand) is tightly bonded, it shall carry a *subscript* to indicate the number of ligands. Example: $Co(NH_3)_6Cl_3$. (b) In formulas of complexes, if the complexing group (ligand)

is loosely bonded (such as in many hydrates), it shall carry a *coefficient* to indicate the number of ligands. Example: $AlCl_3 \cdot 6H_2O$.

4. In a salt, if the negative constituent is monatomic, the name carries the ending "ide." Example: NaCl is sodium chloride. If the negative constituent is polyatomic, it should be designated as "ate." Examples: Na_2SO_4 is sodium sulfate, $NaClO_3$ is sodium chlorate.

In many cases, the ending "ite" and "hypo . . . ite" have commonly been used for lower oxidation states of the central element, and may still continue to be used. Example: Na_2SO_3 is sodium sulfite, $NaClO_2$ is sodium chlorite (See page 204 on names of halogen salts). For less common salts, it is recommended that the ending "ate" be used, and the number of other atoms be designated in the name.

Examples:

$NaSO_3F$	Sodium trioxofluorosulfate
Na_3PS_4	Trisodium tetrathiophosphate
$K_4Fe(CN)_6$	Tetrapotassium hexacyanoferrate

The Greek numerial prefixes such as mono, di, and tetra, rather than the Latin uni, bi, and quadri, are to be used.

5. Use Roman numerals for oxidation states if English names are used. Example: iron(II) chloride, $FeCl_2$.

6. Radicals (groups with no charge) carry the "yl" ending. Examples: OH is hydroxyl, CO is carbonyl.

7. For common acids such as H_2SO_4, the trivial name sulfuric acid is retained. For other acids, as $HAuCl_4$, the systematic name, hydrogen tetrachloraurate(III) is preferred, but tetrachloroauric(III) acid is acceptable.

8. Oxyacids (a) The "ous--ic" notation to distinguish between different oxidation states is applied in many cases. The "ous" acid names are restricted to certain "ite" anions (see p. 194 on names of oxyhalogens). (b) The terms "hypo" and "per" may be used for lower and higher oxidation states (see names of oxyhalogens). (c) The prefixes "ortho" and "meta" may be used for acids differing in "water" content.

Example:

H_3BO_3	Orthoboric acid
HBO_2	Metaboric acid

9. "Acid" salts should be named as follows:

Examples:

$NaHCO_3$	Sodium hydrogen carbonate
NaH_2PO_4	Sodium dihydrogen phosphate

10. Double and triple salts can be named as follows:

Examples:

$KMgF_3$	Potassium magnesium fluoride
$TlNa(NO_3)_2$	Thallium(I) sodium nitrate
$KNaCO_3$	Potassium sodium carbonate

11. Polyatomic cations formed from monatomic cations, for example $[Al(H_2O)_6]^{+3}$, are named by the rules for complex ions. See page 000 of text.

Examples:

$[Al(H_2O)_6]^{+3}$	Hexaaquoaluminum ion
$[CoCl(NH_3)_5]^{+2}$	Chloropentaamminecobalt(III) ion

12. Names for polyatomic cations formed by additions of protons to monatomic anions are formed by adding the ending "onium" to the root name of the anion element.
Examples:

PH_4^+ Phosphonium ion
H_3O^+ Hydronium ion

13. (a) The names of monatomic anions shall consist of the name (sometimes abbreviated) of the element, with the ending "ide."
Examples:

Cl^- Chloride ion
H^- Hydride ion

(b) Certain polyatomic anions also carry the ending "ide."
Examples:

OH^- Hydroxide ion
S_2^{-2} Disulfide ion
CN^- Cyanide ion

For other rules, see references named in the introduction to this Appendix.

II. Organic Nomenclature

See these specific pages in Chapter 26 and references named in the Introduction to this Appendix.

Alkanes	562
Alkenes	568
Alkanols	573
Alkanals	575
Alkanones	575
Acids	577

APPENDIX C

BALANCING OXIDATION-REDUCTION EQUATIONS

The Oxidation-State Method

Having discovered what element is oxidized and what element reduced, we may write and balance the equation for an oxidation-reduction reaction. The rule is that *the total change in the oxidation state of one element must be equal and opposite to the total change in the oxidation state of another.* This corresponds to the statement that for any oxidation, there must be an equal amount of reduction, and to the statement that the number of electrons gained in the reduction must equal the number lost in the oxidation. The same rule is followed in balancing net ionic equations.

Example. Balance the equation for the reaction

$$\text{change of } -V$$

$$KMnO_4 + NaCl + H_2SO_4 \longrightarrow Cl_2 + K_2SO_4 + MnSO_4 + H_2O + Na_2SO_4 \qquad (1)$$

$$\text{change of I}$$

The changes in oxidation state of manganese and chlorine are indicated. The total change for chlorine must be V to equal the total change for manganese; this will require five chlorine atoms, or 5NaCl. However, chlorine appears as Cl_2 in the products, so that 5NaCl could form only $2\frac{1}{2}$ Cl_2. To balance the equation, therefore, we can multiply by 2, so that we have 10NaCl and $5Cl_2$. This makes a total oxidation-state change for chlorine of 10; so we must also multiply $KMnO_4$ by 2 to give $2KMnO_4 \longrightarrow 2MnSO_4$, with a total oxidation-state change of -10. With these figures, the equation is

$$2KMnO_4 + 10NaCl + H_2SO_4 \longrightarrow$$
$$5Cl_2 + K_2SO_4 + 2MnSO_4 + H_2O + Na_2SO_4 \qquad (2)$$

The remainder of the equation may now be balanced by inspection. There are eight oxygen atoms in the $2KMnO_4$, which evidently appear as H_2O, so there must be $8H_2O$ in the products. To supply the 16 hydrogen atoms (in

$8H_2O$), we need $8H_2SO_4$. When three of the eight sulfate ions have been used for $K_2SO_4 + 2MnSO_4$, the other five appear in $5Na_2SO_4$, which includes the 10 sodium atoms from $10NaCl$, and the equation is balanced.

$$2KMnO_4 + 10NaCl + 8H_2SO_4 \longrightarrow$$
$$5Cl_2 + K_2SO_4 + 2MnSO_4 + 8H_2O + 5Na_2SO_4 \quad (3)$$

Balancing Net Ionic Equations

It was pointed out (Chapter 3) that there is a great advantage in writing equations in ionic form where ions are involved. For example, the equation just shown is equally—in fact, more nearly—correct when written

$$2MnO_4^- + 10Cl^- + 16H_3O^+ \longrightarrow 5Cl_2 + 2Mn^{+2} + 24H_2O \quad (4)$$

since the K^+, Na^+, and SO_4^{-2} ions take no actual part in the reaction. Can we write this equation directly without going through the molecular form? The answer is yes, and the process we use is less complicated than the other form, provided we keep track of the charges on the ions. The steps are as follows:

1. Write down the correct symbols or formulas for the reactants and products (without regard to water or its ions). For this reaction, we would have

$$MnO_4^- + Cl^- \longrightarrow Mn^{+2} + Cl_2 \quad (5)$$

This much must be known: MnO_4^- ion, when it acts as an oxidizing agent in acid solution, yields Mn^{+2}, and that the chloride ion, when oxidized, yields Cl_2. From this point, the equation may be balanced by a systematic procedure.

2. Check oxidation states. Manganese is VII in MnO_4^-, II in Mn^{+2}; chlorine is $-I$ in Cl^-, 0 in Cl_2; there is a change of $-V$ for manganese and I for chlorine.

3. Balance oxidation against reduction. This requires a ratio of $5Cl^-$ to $1MnO_4^-$. So we write

$$2MnO_4^- + 10Cl^- \longrightarrow 2Mn^{+2} + 5Cl_2 \quad (6)$$

4. Balance charges. An ionic equation must balance electrically as well as chemically. Since this reaction takes place in aqueous solution, we may assume that water and its ions are involved. Note that as the equation now stands, there are 12 minus charges on the left and four plus charges on the right. We can balance charges either by adding $16H_3O^+$ on the left or $16OH^-$ on the right. Since the reaction takes place in the presence of acid, the former balancing is preferred, and the equation becomes

$$2MnO_4^- + 10Cl^- + 16H_3O^+ \longrightarrow 2Mn^{+2} + 5Cl_2 \quad (7)$$

5. Complete the material balance. A final examination now shows an excess of 48 hydrogen atoms and 24 oxygen atoms on the left. Presumably, 24 molecules of water are formed, and the equation is complete.

$$2MnO_4^- + 10Cl^- + 16H_3O^+ \longrightarrow 2Mn^{+2} + 5Cl_2 + 24H_2O \quad (8)$$

The Half-Reaction Method

A useful plan to follow is first to write the half-reaction equations, showing the gain or loss of electrons, and then adjust the number of molecules or ions involved so that the number of electrons gained by the oxidizing agent is equal to the number lost by the reducing agent. In writing the half-reaction equations for reactions that take place in solution, water and its ions (the hydronium ion and the hydroxide ion) may be used at will as reagents, since they are always present. Of course, as with any reaction, the products of the reaction must be known before the equation can be written.

Example. When hydrochloric acid is added to potassium permanganate, chlorine is formed.

Step 1. The products of the reaction are the manganese ion (Mn^{+2}), chlorine, and perhaps water; the reactants are the permanganate ion (MnO_4^-), the hydronium ion, and the chloride ion.

Step 2. The half-reaction equation for the oxidation is simply

$$2Cl^- \longrightarrow Cl_2 + 2e^- \tag{9}$$

The reduction involves the addition of electrons to permanganate ion

$$5e^- + MnO_4^- \longrightarrow Mn^{+2} \tag{10}$$

This is not balanced with respect to the atoms present, since no oxygen-containing substance appears on the right. To correct this, add hydronium ions to the left and water to the right, and balance materially:

$$5e^- + MnO_4^- + 8H_3O^+ \longrightarrow Mn^{+2} + 12H_2O \tag{11}$$

Step 3. The balanced half-reaction equations must now be made to correspond with each other. Since two electrons appear as products and five as reactants, it is evident that a total of 10 electrons (least common multiple) must be transferred, or

$$10Cl^- \longrightarrow 5Cl_2 + 10e^- \tag{12}$$

$$10e^- + 2MnO_4^- + 16H_3O^+ \longrightarrow 2Mn^{+2} + 24H_2O \tag{13}$$

and adding,

$$10Cl^- + 2MnO_4^- + 16H_3O^+ \longrightarrow 5Cl_2 + 2Mn^{+2} + 24H_2O \tag{14}$$

Equation (14) is the balanced ionic equation for the oxidation of chloride ions by permanganate ions in an acid solution.

APPENDIX D

A string stretched between two points can undergo vibrations about its straight position. The motion is periodic in time, and also periodic with distance along the string.

Parts (a), (b), (c), and (d) of Figure D.1 represent one type of motion—at the beginning of a cycle, one-fourth of the way through, one-half of the way through, and three-fourths of the way through. At $t = T$, the string will have completed a full cycle, and will look like (a) again. Part (e) shows the maximum displacements [from parts (a) and (c)] and illustrates how individual particles of the string undergo a simple straight-line vibratory motion between these two extremes. The distance these particles travel in their up-and-down motion is known as the amplitude of their vibrations; the figure shows that the amplitude of a particle depends upon its position along the string; particles at the *nodes o, p, q, r*, and *l* do not move at all. Since the string is fixed at each end, points *o* and *l* must always be nodes; thus the vibrations must always be such that the wavelength λ is equal to $2l/n$, where n is an integer. Parts (a) to (e) correspond to $n = 4$; parts (f) to (h) show some other patterns.

The assumption of de Broglie that electrons can be represented as waves means that they can be described in terms of electromagnetic waves. The amplitudes corresponding to the particle motion of parts (e) to (h) correspond to values of the electric and magnetic components of the electron waves (see Figure 4.3). The forces of tension in the vibrating string are analogous, for electrons in atoms, to the forces between electrons and nucleus. The constraints exerted on the motion by tying the string at both ends are analogous to the requirement that these waves be standing waves, with nodes fixed in space and unchanging in time.

The mathematics requires the solution of a complex differential equation known as the Schrödinger equation, which relates the amplitude of the electric and magnetic field to the distance and direction from the nucleus, just as the amplitudes of the string particles' motion are a function of the distance along the string. For a single electron, the Schrodinger equation is

$$\frac{\partial^2 \psi}{\partial x^2} + \frac{\partial^2 \psi}{\partial y^2} + \frac{\partial^2 \psi}{\partial z^2} + \frac{8\pi^2 m}{h^2}(E - V)\,\psi = 0$$

Here m is the mass of the electron, E its total energy and V its potential energy, h is Planck's constant and x, y, and z represent Cartesian coordinates specifying distance and direction. The wave function, denoted by ψ, is the solution to the Schrödinger equation, and is itself a mathematical equation describing how the amplitude of the wave varies with distance (and direction). Just as the string can vibrate only for integral values of n, so the wave equation has solutions, for electrons as waves in atoms, only for integral values of three quantum numbers—three because the amplitudes vary in three dimensions, rather than in the one dimension from o to l.

The results of the calculations are simplest for the case of the hydrogen atom, with its single electron, and, for this atom, the simplest wave functions are those for quantum numbers l and m equal to zero. Then ψ is a function only of r, the distance from the nucleus, and of the principal quantum number, n. We find for

$$n = 1, l = 0, m = 0: \qquad \psi_{1s} = \left(\frac{1}{\sqrt{\pi}}\right)\left(\frac{1}{a_o}\right)^{3/2} e^{-r/a_o}$$

For

$$n = 2, l = 0, m = 0: \qquad \psi_{2s} = \left(\frac{1}{4\sqrt{2\pi}}\right)\left(\frac{1}{a_o}\right)^{3/2}\left(2 - \frac{r}{a_o}\right) e^{-r/2a_o}$$

where a_o is the constant representing the radius of the first Bohr orbit. These two functions are plotted in Figure D.2 (a) and (b); ψ_{1s} shows no node (except at $r = \infty$); ψ_{2s} shows one node, at $r = 2a_o$.

The general expression for the wave function for the s states ($l = 0$) of hydrogen is

$$\psi_{ns}(r) = -\frac{1}{2\sqrt{\pi}}\sqrt{\left(\frac{2}{na_o}\right)^3 \frac{(n-1)!}{2n[n!]^3}}\; e^{-r/na_o} L_n^1\left(\frac{2r}{na_o}\right)$$

where

$$L_n^1\left(\frac{2r}{na_o}\right)$$

is the abbreviation for a polynomial of the form

$$L_n^1\left(\frac{2r}{na_o}\right) = (-1)^n n \left\{ \left(\frac{2r}{na_o}\right)^{n-1} - n(n-1)\left(\frac{2r}{na_o}\right)^{n-2} \right.$$

$$+ \frac{n(n-1)(n-1)(n-2)}{2!}\left(\frac{2r}{na_o}\right)^{n-3}$$

$$\left. - \frac{n(n-1)(n-2)(n-1)(n-2)(n-3)}{3!}\left(\frac{2r}{na_o}\right)^{n-4} + \cdots \right\}$$

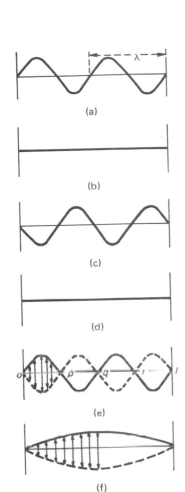

Figure D.1 (a) At $t = 0$; (b) at $t = \frac{1}{4}T$; (c) at $t = \frac{1}{2}T$; (d) at $t = \frac{3}{4}T$; (e) vibratory motion of particles in the string; (f) $n = 1$; (g) $n = 2$; (h) $n = 3$.

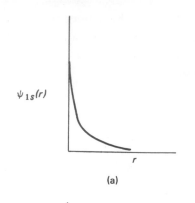

$\psi_{1s}(r)$

(a)

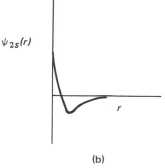

$\psi_{2s}(r)$

(b)

Figure D.2 (a) $\psi_{1s}(r)$ as a function of r (schematic). (b) $\psi_{2s}(r)$ as a function of r (schematic).

Neither of these expressions is important for the student except to show the specific dependence of the hydrogen wave function upon the quantum number n. The energies for these states are given by

$$E = -\frac{2\pi^2 m e^4}{n^2 h^2}$$

The quantum numbers n (and l and m) thus appear naturally in the solutions of the wave equation, and determine the energy levels and the forms of the wave function appropriate to those levels.

APPENDIX E

Table E.1 Densities and Melting Points of Some of the More Common Elements

Name	Density*	Melting Point (°C)	Name	Density*	Melting Point (°C)
Aluminum	2.702	660	Mercury (liquid)	13.546	−38.87
Antimony	6.684	630.5	Molybdenum	10.2	2625
Arsenic (metallic)	5.7	814[36 atm]	Nickel	8.90	1452
Barium	3.5	850	Palladium	12.0	1555
Bismuth	9.80	271	Phosphorus (red)	2.20	590[43 atm]
Boron	2.5	2300	Phosphorus (yellow)	1.82	44.1
Bromine (liquid)	3.119	−7.2	Platinum	21.45	1755
Cadmium	8.6	320.9	Potassium	0.86	62.3
Calcium	1.55	810	Radium	5(?)	960(?)
Carbon (diamond)	3.51		Selenium (gray)	4.80	220
Carbon (graphite)	2.26	3500	Silicon	2.4	1420
Chromium	7.1	1615	Silver	10.5	960.5
Cobalt	8.9	1480	Sodium	0.97	97.5
Copper	8.92	1083	Strontium	2.6	800
Gallium	5.91	29.75	Sulfur (monoclinic)	1.96	119.0
Gold	19.3	1063	Sulfur (rhombic)	2.07	112.8
Iodine	4.93	113.5	Tantalum	16.6	2850
Iridium	22.4	2350	Tellurium (metallic)	6.24	452
Iron	7.86	1535	Tin (white, tetragonal)	7.31	231.85
Lead	11.34	327.5	Titanium	4.5	1800
Lithium	0.53	186	Tungsten	19.3	3370
Magnesium	1.74	651	Vanadium	5.96	1710
Manganese	7.2	1260	Zinc	7.14	419.43

*In g/cc at room temperature.

Table E.2 Weight in Grams of 1 Liter of Various Gases Under Standard Conditions; Boiling Points Under Pressure of 760 Torr

Name	Weight of 1 Liter	Boiling Point (°C)	Name	Weight of 1 Liter	Boiling Point (°C)
Acetylene	1.1621	−83.6	Hydrogen chloride	1.6398	−85.0
Air	1.2930		Hydrogen sulfide	1.5392	−59.6
Ammonia	0.7708	−33.3	Methane	0.7168	−161.4
Argon	1.7824	−185.7	Nitric oxide	1.3402	−151.0
Carbon dioxide	1.9768	sublimes	Nitrogen	1.2506	−195.8
Carbon monoxide	1.2504	−192.0	Nitrous oxide	1.9777	−89.5
Chlorine	3.214	−34.6	Oxygen	1.4290	−183.0
Helium	0.1785	−268.9	Propane	2.0200	−45.0
Hydrogen	0.08987	−252.7	Sulfur dioxide	2.9266	−10.0

Table E.3 Vapor Pressure of Water Expressed in Torr (Millimeters of Mercury) (International Critical Tables)

Temperature (°C)	Pressure	Temperature (°C)	Pressure	Temperature (°C)	Pressure
10	9.209	21	18.650	32	35.663
11	9.844	22	19.827	33	37.729
12	10.518	23	21.068	34	39.898
13	11.231	24	22.377	35	42.175
14	11.987	25	23.756	36	44.563
15	12.788	26	25.209	37	47.067
16	13.634	27	26.739	38	49.692
17	14.530	28	28.349	39	52.442
18	15.477	29	30.043	40	55.324
19	16.477	30	31.824	50	92.51
20	17.535	31	33.695	100	760.00

Table E.4 The Natural Isotopes of the Elements

Atomic Number	Symbol	Natural Isotopes (Mass Numbers)	Atomic Number	Symbol	Natural Isotopes (Mass Numbers)
1	H	1, 2	21	Sc	45
2	He	4, 3	22	Ti	48, 46, 47, 49, 50
3	Li	7, 6	23	V	51, 50
4	Be	9	24	Cr	52, 53, 50, 54
5	B	11, 10	25	Mn	55
6	C	12, 13			
7	N	14, 15	26	Fe	56, 54, 57, 58
8	O	16, 18, 17	27	Co	59
9	F	19	28	Ni	58, 60, 62, 61, 64
10	Ne	20, 22, 21	29	Cu	63, 65
11	Na	23	30	Zn	64, 66, 68, 67, 70
12	Mg	24, 25, 26	31	Ga	69, 71
13	Al	27	32	Ge	74, 72, 70, 73, 76
14	Si	28, 29, 30	33	As	75
15	P	31	34	Se	80, 78, 76, 82, 77, 74
16	S	32, 34, 33, 36	35	Br	79, 81
17	Cl	35, 37	36	Kr	84, 86, 82, 83, 80, 78
18	Ar	40, 36, 38	37	Rb	85, 87
19	K	39, 41, 40	38	Sr	88, 86, 87, 84
20	Ca	40, 44, 42, 48, 43, 46	39	Y	89

Table E.4 (Continued)

Atomic Number	Symbol	Natural Isotopes (Mass Numbers)	Atomic Number	Symbol	Natural Isotopes (Mass Numbers)
40	Zr	90, 94, 92, 91, 96	71	Lu	175, 176
41	Nb	93	72	Hf	180, 178, 177, 179, 176, 174
42	Mo	98, 96, 95, 92, 94, 97, 100	73	Ta	181
43	Tc	—	74	W	184, 186, 182, 180
44	Ru	102, 104, 101, 99, 100, 96, 98	75	Re	187, 185
45	Rh	103	76	Os	192, 190, 189, 188, 187, 186, 184
46	Pd	106, 108, 105, 110, 104, 102			
47	Ag	107, 109	77	Ir	193, 191
48	Cd	114, 112, 111, 110, 113, 116, 106, 108	78	Pt	195, 194, 196, 198, 192, 190
			79	Au	197
49	In	115, 113	80	Hg	202, 200, 199, 201, 198, 204, 196
50	Sn	120, 118, 116, 119, 117, 124, 122, 112, 114, 115	81	Tl	205, 203, 206, 207, 208, 210
51	Sb	121, 123	82	Pb	208, 206, 207, 204, 210, 211, 212, 214
52	Te	130, 128, 126, 125, 124, 122, 123, 120	83	Bi	209, 210, 211, 212, 214
53	I	127	84	Po	210, 211, 212, 214, 215, 216, 218
54	Xe	132, 129, 131, 134, 136, 130, 128, 124, 126	85	At	218, 215
55	Cs	133	86	Rn	222, 220, 219
56	Ba	138, 137, 136, 135, 134, 130, 132	87	Fr	223
57	La	139, 138	88	Ra	223, 224, 226, 228
58	Ce	140, 142, 138, 136	89	Ac	227, 228
59	Pr	141	90	Th	227, 228, 230, 231, 232, 234
60	Nd	142, 144, 146, 143, 145, 148, 150	91	Pa	231, 234
			92	U	238, 235, 234
61	Pm	—	93	Np	
62	Sm	152, 154, 147, 149, 148, 150, 144	94	Pu	239
			95	Am	—
63	Eu	153, 151			
64	Gd	158, 160, 156, 157, 155, 154, 152	96	Cm	—
			97	Bk	—
65	Tb	159	98	Cf	—
66	Dy	164, 162, 163, 161, 158, 160	99	E	—
67	Ho	165	100	Fm	—
			101	Mv	—
68	Er	166, 168, 167, 170, 164, 162	102	No	—
69	Tm	169	103	Lr	—
70	Yb	174, 172, 173, 171, 176, 170, 168	104	—	—
			105	Hn	—

NOTE: The isotopes of each element are listed in the order of abundance (the most abundant being given first). A dash (—) signifies that the element has no natural isotopes. The table lists only natural isotopes. Approximately 292 natural isotopes and 1376 artificial isotopes are known today.

Table E.5 Some Standard Oxidation Potentials in Basic Solution (Values in Acid Solution Are Given in Table 21.1)

Half-Reactions	$E°$ (Volts)
$Ca(s) + 2OH^- \rightleftharpoons Ca(OH)_2(s) + 2e^-$	3.03
$K(s) \rightleftharpoons K^+ + e^-$	2.93
$Al(s) + 4OH^- \rightleftharpoons Al(OH)_4^- + 3e^-$	2.35
$P(s) + 2OH^- \rightleftharpoons H_2PO_2^- + e^-$	1.82
$H_2PO_2^- + 3OH^- \rightleftharpoons HPO_3^{-2} + 2H_2O + 2e^-$	1.65
$Mn(s) + 2OH^- \rightleftharpoons Mn(OH)_2(s) + 2e^-$	1.47

Table E.5 (Continued)

Half-Reactions	$E°$ (Volts)
$Zn(s) + 4CN^- \rightleftharpoons Zn(CN)_4^{-2} + 2e^-$	1.26
$Zn(s) + 4OH^- \rightleftharpoons Zn(OH)_4^{-2} + 2e^-$	1.216
$HPO_3^{-2} + 3OH^- \rightleftharpoons PO_4^{-3} + 2H_2O + 2e^-$	1.05
$S_2O_4^{-2} + 4OH^- \rightleftharpoons 2SO_3^{-2} + 2H_2O + 2e^-$	1.12
$Zn(s) + 4NH_3 \rightleftharpoons Zn(NH_3)_4^{+2} + 2e^-$	1.03
$SO_3^{-2} + 2OH^- \rightleftharpoons SO_4^{-2} + H_2O + 2e^-$	0.93
$Sn(OH)_4^{-2} + 2OH^- \rightleftharpoons Sn(OH)_6^{-2} + 2e^-$	0.90
$PH_3 + 3OH^- \rightleftharpoons P(s) + 3H_2O + 3e^-$	0.89
$H_2 + 2OH^- \rightleftharpoons 2H_2O + 2e^-$	0.828
$Ni(s) + 2OH^- \rightleftharpoons Ni(OH)_2(s) + 2e^-$	0.72
$AsO_2^- + 4OH^- \rightleftharpoons AsO_4^{-3} + 2H_2O + 2e^-$	0.67
$Fe(OH)_2(s) + OH^- \rightleftharpoons Fe(OH)_3(s) + e^-$	0.56
$S^{-2} \rightleftharpoons S + 2e^-$	0.48
$Cr(OH)_4^- + 4OH^- \rightleftharpoons CrO_4^{-2} + 4H_2O + 3e^-$	0.13
$H_2O_2 + 2OH^- \rightleftharpoons O_2 + 2H_2O + 2e^-$	0.076
$Mn(OH)_2(s) + 2OH^- \rightleftharpoons MnO_2(s) + 2H_2O + 2e^-$	0.05
$Cu(NH_3)_2^+ + 2NH_3 \rightleftharpoons Cu(NH_3)_4^{+2} + e^-$	0.0
$Mn(OH)_2(s) + OH^- \rightleftharpoons Mn(OH)_3(s) + e^-$	−0.40
$Co(OH)_2(s) + OH^- \rightleftharpoons Co(OH)_3(s) + e^-$	−0.17
$I^- + 6OH^- \rightleftharpoons IO_3^- + 3H_2O + 6e^-$	−0.26
$ClO_2^- + 2OH^- \rightleftharpoons ClO_3^- + H_2O + 2e^-$	−0.33
$ClO_3^- + 2OH^- \rightleftharpoons ClO_4^- + H_2O + 2e^-$	−0.36
$4OH^- \rightleftharpoons O_2 + 2H_2O + 4e^-$	−0.401
$I^- + 2OH^- \rightleftharpoons IO^- + H_2O + 2e^-$	−0.49
$Ni(OH)_2(s) + 2OH^- \rightleftharpoons NiO_2(s) + 2H_2O + 2e^-$	−0.49
$MnO_4^{-2} \rightleftharpoons MnO_4^- + e^-$	−0.564
$MnO_2(s) + 4OH^- \rightleftharpoons MnO_4^- + 2H_2O + 3e^-$	−0.588
$MnO_2(s) + 4OH^- \rightleftharpoons MnO_4^{-2} + 2H_2O + 2e^-$	−0.60
$ClO^- + 2OH^- \rightleftharpoons ClO_2^- + H_2O + 2e^-$	−0.66
$Br^- + 2OH^- \rightleftharpoons BrO^- + H_2O + 2e^-$	−0.76
$2OH^- \rightleftharpoons H_2O_2 + 2e^-$	−0.88
$Cl^- + 2OH^- \rightleftharpoons ClO^- + H_2O + 2e^-$	−0.89

Table E.6 Instability Constants of Complex Ions

Ligand	Dissociation Equilibrium	K
NH_3	$Ag(NH_3)_2^+ \rightleftharpoons Ag^+ + 2NH_3$	6×10^{-8}
	$Cd(NH_3)_4^{+2} \rightleftharpoons Cd^{+2} + 4NH_3$	2.5×10^{-7}
	$Co(NH_3)_6^{+2} \rightleftharpoons Co^{+2} + 6NH_3$	1.3×10^{-5}
	$Co(NH_3)_6^{+3} \rightleftharpoons Co^{+3} + 6NH_3$	2.2×10^{-34}
	$Cu(NH_3)_4^{+2} \rightleftharpoons Cu^{+2} + 4NH_3$	4.6×10^{-14}
	$Ni(NH_3)_6^{+2} \rightleftharpoons Ni^{+2} + 6NH_3$	4.8×10^{-8}
	$Zn(NH_3)_4^{+2} \rightleftharpoons Zn^{+2} + 4NH_3$	2.6×10^{-10}
F^-	$AlF_6^{-3} \rightleftharpoons Al^{+3} + 6F^-$	1.5×10^{-20}
	$FeF_6^{-3} \rightleftharpoons Fe^{+3} + 6F^-$	5×10^{-16}
Cl^-	$HgCl_4^- \rightleftharpoons Hg^{+2} + 4Cl^-$	1.1×10^{-16}
	$SnCl_4^{-2} \rightleftharpoons Sn^{+2} + 4Cl^-$	3.2×10^{-2}
I^-	$HgI_4^- \rightleftharpoons Hg^{+2} + 4I^-$	5.3×10^{-21}
	$PbI_3^- \rightleftharpoons Pb^{+2} + 3I^-$	3.6×10^{-6}
CN^-	$Ag(CN)_2^- \rightleftharpoons Ag^+ + 2CN^-$	1×10^{-21}
	$Cd(CN)_4^{-2} \rightleftharpoons Cd^{+2} + 4CN^-$	1.4×10^{-19}
	$Co(CN)_4^{-2} \rightleftharpoons Co^{+2} + 4CN^-$	7.9×10^{-20}
	$Cu(CN)_2^- \rightleftharpoons Cu^+ + 2CN^-$	5×10^{-28}
	$Fe(CN)_6^{-4} \rightleftharpoons Fe^{+2} + 6CN^-$	1×10^{-35}

Table E.6 (Continued)

Ligand	Dissociation Equilibrium	K
	$Fe(CN)_6^{-3} \rightleftharpoons Fe^{+3} + 6CN^-$	1×10^{-42}
	$Hg(CN)_4^{-2} \rightleftharpoons Hg^{+2} + 4CN^-$	4×10^{-42}
	$Pb(CN)_4^{-2} \rightleftharpoons Pb^{+2} + 4CN^-$	5×10^{-11}
	$Ni(CN)_4^{-2} \rightleftharpoons Ni^{+2} + 4CN^-$	1×10^{-22}
	$Zn(CN)_4^{-2} \rightleftharpoons Zn^{+2} + 4CN^-$	1.3×10^{-17}
SCN^-	$FeSCN^{+2} \rightleftharpoons Fe^{+3} + SCN^-$	1×10^{-3}
SO_3^{-2}	$Ag(SO_3)_2^{-3} \rightleftharpoons Ag^+ + 2SO_3^{-2}$	3×10^{-9}
$S_2O_3^{-2}$	$Ag(S_2O_3)_2^{-3} \rightleftharpoons Ag^+ + 2S_2O_3^{-2}$	6×10^{-14}

Table E.7 Some Solubility Products at 20°C

Compound	Solubility Product	K_{sp}
Aluminum hydroxide	$[Al^{+3}] \times [OH^-]^3$	5×10^{-33}
Barium carbonate	$[Ba^{+2}] \times [CO_3^{-2}]$	1.6×10^{-9}
Barium chromate	$[Ba^{+2}] \times [CrO_4^{-2}]$	8.5×10^{-11}
Barium sulfate	$[Ba^{+2}] \times [SO_4^{-2}]$	1.0×10^{-10}
Barium oxalate	$[Ba^{+2}] \times [C_2O_4^{-2}]$	1.5×10^{-8}
Bismuth sulfide	$[Bi^{+3}]^2 \times [S^{-2}]^3$	1×10^{-70}
Cadmium hydroxide	$[Cd^{+2}] \times [OH^-]^2$	2×10^{-14}
Cadmium sulfide	$[Cd^{+2}] \times [S^{-2}]$	6×10^{-27}
Calcium carbonate	$[Ca^{+2}] \times [CO_3^{-2}]$	6.9×10^{-9}
Calcium oxalate	$[Ca^{+2}] \times [C_2O_4^{-2}]$	1.3×10^{-9}
Calcium sulfate	$[Ca^{+2}] \times [SO_4^{-2}]$	2.4×10^{-5}
Chromium hydroxide	$[Cr^{+3}] \times [OH^-]^3$	7×10^{-31}
Cobalt sulfide	$[Co^{+2}] \times [S^{-2}]$	5×10^{-22}
Cupric hydroxide	$[Cu^{+2}] \times [OH^-]^2$	1.6×10^{-19}
Cupric sulfide	$[Cu^{+2}] \times [S^{-2}]$	4×10^{-36}
Ferric hydroxide	$[Fe^{+3}] \times [OH^-]^3$	6×10^{-38}
Ferrous hydroxide	$[Fe^{+2}] \times [OH^-]^2$	2×10^{-15}
Ferrous sulfide	$[Fe^{+2}] \times [S^{-2}]$	4×10^{-17}
Lead carbonate	$[Pb^{+2}] \times [CO_3^{-2}]$	1.5×10^{-13}
Lead chromate	$[Pb^{+2}] \times [CrO_4^{-2}]$	2×10^{-16}
Lead iodide	$[Pb^{+2}] \times [I^-]^2$	8.3×10^{-9}
Lead sulfate	$[Pb^{+2}] \times [SO_4^{-2}]$	1.3×10^{-8}
Lead sulfide	$[Pb^{+2}] \times [S^{-2}]$	4×10^{-26}
Magnesium carbonate	$[Mg^{+2}] \times [CO_3^{-2}]$	4×10^{-5}
Magnesium hydroxide	$[Mg^{+2}] \times [OH^-]^2$	8.9×10^{-12}
Magnesium oxalate	$[Mg^{+2}] \times [C_2O_4^{-2}]$	8.6×10^{-5}
Manganese hydroxide	$[Mn^{+2}] \times [OH^-]^2$	2×10^{-13}
Manganese sulfide	$[Mn^{+2}] \times [S^{-2}]$	8×10^{-14}
Mercurous chloride	$[Hg_2^{+2}] \times [Cl^-]^2$	1.1×10^{-18}
Mercuric sulfide	$[Hg^{+2}] \times [S^{-2}]$	1×10^{-50}
Nickel hydroxide	$[Ni^{+2}] \times [OH^-]^2$	1.6×10^{-16}
Nickel sulfide	$[Ni^{+2}] \times [S^{-2}]$	1×10^{-22}
Silver arsenate	$[Ag^+]^3 \times [AsO_4^{-3}]$	1×10^{-23}
Silver bromide	$[Ag^+] \times [Br^-]$	5×10^{-13}
Silver carbonate	$[Ag^+]^2 \times [CO_3^{-2}]$	8.2×10^{-12}
Silver chloride	$[Ag^+] \times [Cl^-]$	2.8×10^{-10}
Silver chromate	$[Ag^+]^2 \times [CrO_4^{-2}]$	1.9×10^{-12}
Silver iodate	$[Ag^+] \times [IO_3^-]$	3×10^{-8}
Silver iodide	$[Ag^+] \times [I^-]$	8.5×10^{-17}
Silver phosphate	$[Ag^+]^3 \times [PO_4^{-3}]$	1.8×10^{-18}
Silver sulfide	$[Ag^+]^2 \times [S^{-2}]$	1×10^{-50}
Stannous sulfide	$[Sn^{+2}] \times [S^{-2}]$	1×10^{-24}
Zinc hydroxide	$[Zn^{+2}] \times [OH^-]^2$	5×10^{-17}
Zinc sulfide	$[Zn^{+2}] \times [S^{-2}]$	1×10^{-20}

INDEX

Absolute temperature, 218
Acetic acid, 578
Acetone, t576
Acetylcholine, 483
Acetylene, 125, 340
Acid anhydrides, 579
Acid-base balance, in body, 188
Acid-base indicators, 403
Acids, 148
  carboxylic, 577
  conjugate, 150
  diprotic, 311
    ionization of, 406
  hydroxy compounds as, 323
  ionization of, 398
  nucleic, 664
  pH, 401
  poly, 324, 342
  strength of, 149, t151, 152, t153
    nitrogen family, t327
    oxyacids, 205, t206
Actin, 670
Actinides, 80, 505, 549
Activated complex, 478
Activity, 394, 399, 409, 428
Acyl halides, 579
Addison, W. E., 304
Addison's disease, 187
Addition polymers, 626, t629
Addition reactions, 603
  unsymmetrical, 605
Adenosine diphosphate, *see* ADP
Adenosine triphosphate, *see* ATP
ADP, 334, 656
Adsorption, 159

Air, 235
  quality, 235, 236
Albite, 350
Alchemy, 3
Alcohols, 573
  names, formulas and physical constants of, t574
Aldehydes, 575
  and ketones, names, formulas, and boiling points for, t576
Aldol condensations, 603
Aliphatic compounds, 560
Alkali and alkaline earth elements,
  crystal structure, t287
  distribution in nature, t180
  hydration energy of ions, t260
  properties of, t172, 174, 175
Alkali metals, 169
Alkaline earths, 169
Alkaloids, 639
Alkanals, 575
Alkanes, 562
  isomers of, t564
  properties of, t563
Alkanols, 573
Alkanones, 575
Alkenes, 568
  names, formulas, and physical constants of, t569
Alkoxides, 574
Alkyl halides, 571
  formulas, names, and boiling points of, t572
Alkylation, 614
Alkynes, 570

names, formulas, and physical properties of, t570
Allotropic forms, 297, 330, 346
Alloys, 288, t289
　of chromium, t541
　of copper, t547
Alpha particles, t23, 494
　bombardment with, 25, 27
　mass of, 27
　penetration of matter by, 27
Aluminum, 356
　production of, 356, 362
Ameen, J. G., 355
Americium, 505
Amides, 579
　metal, 332
Amines, 580
　names, formulas, and boiling points of, t580
Amino acids, 637, t638
　sequence in enzyme, 654
Ammonia, 120, 134, 332
　equilibrium, t396
　freezing point depression by, 269
　Haber process, 322, 332
　heat of vaporization, 248
　solubility in water, 268
Ammonium cyanate, 473
Amorphous solids, 249
Amphoterism, 323, 359
Amylopectin, 636
Amylose, 636
Analysis, chemical, 44
Anglesite, t345
Anode, 444
Anomer, 643
Antibiotics, 646
Anticodons, 666
Antimony, 322
Antiseptics, 208
Apatite, t329
Aq, (aqueous), 47
Aragonite, 183
Argon, 134
Aromatic character, 581
Aromatic compounds, 560, 580
　formulas, names, and physical constants of, t582
Arrhenius, Svante, 148, 269
Arrhenius theory, 149
Arsenic, 322
Arsenic pentafluoride, 326
Asbestos, 341, 347
Aspirin, 647
Astatine, 193, 505
Asthma, 208
Aston, F. W., 29, 30, 39, 501
Asymmetric carbon atoms, 584, 642
Atmosphere, 159
Atmospheric stability, 235

Atomic mass unit, 31
Atomic nucleus, 25, 493
　building blocks of, 26
Atomic numbers, 26
Atomic orbitals, 129
Atomic stability, 98
Atomic theory, 8
Atomic weight, 29, 31
Atoms
　electron configuration of, 76
　electrons in, 20
　light emission by, 63
　masses of, t30, t31
　structure of, 29
ATP, 334, 656
Attraction, interionic, 270
Attractive forces, 133, 243
　between halogen molecules, 197
Austin, A. V., 16
Autonomic functions, 483
Auxins, 335
Auxochromes, 588
Avogadro, Amedeo, 54, 61, 220
Avogadro number, 41, 444
Avogadro's law, 54, 220

Babbitt metal, 289
Bacon, Francis, 4
Bakelite, t630
Balance, Gouy, 521
Balancing equations, vii
Band theory of metals, 290
Barbiturates, 589
Barium, 169
Barometer, 216
Barrow, G. M., 241
Barton, A. F. M., 257
Bases, 148
　conjugate, 150
　ionization of, 401
　pH, 401
Basolo, F., 533, 555
Battery, 457
　solar, 292
Becquerel, Henri, 22
Benzene, resonance in, 128
Benzer, S., 674
Berkelium, 505
Bernal, J. D., 257
Beryl, 349, 350
Beryllium, 169
Beryllium compounds, hybridization in, 122
Beryllium-copper alloys, 181
Berzelius, Jakob, 313
Beta particles, t23, 494
　penetration of matter by, 27
Bicarbonate buffer system, 414
Bickel, C. L., 16

Bile, 274
Binding energy, 496, 505
Biochemistry, 652
Biological reactions, ΔG values for, 436
Biosphere, 164
Bismuth, 322
Bissey, J. E., 355
Blood, 274, 402, 413
　acidity, 318
Bloom, J. T., 512
Boat structure, 568
Body
　fluids of, 273
　ions in, 187, 530
Body-centered cubic structure, 286
Bohr, Niels, 20, 67
Bohr magneton, 520
Boiling point, 248
　of carbon family elements, t345
　elevation of, 263
　of hydrocarbons, t134
　of hydrogen halides, t203
　maximum and minimum, 267
　of nitrogen family hydrides, t332
　of nonmetal-family hydrides, 135
　of water at various elevations, 249
Bond
　bridge, 356
　coordinate covalent, 101
　covalent, 100, 116
　electrovalent, 102
　hydrogen, 135
　ionic, 102
　multicenter, 359
　multiple, 124, 340
　polar and nonpolar, 102
　sigma and pi, 124
　three-center, 359
Bond distances, t117
　in hydrogen halides, t203
Bond energy, t117, 234, 375, t376
　of halogens, t198
　in hydrogen halides, t203
Bond order, 132
Bonding
　comparison of types, 109
　in halogen family, 194
Bordass, W. T., 91
Born-Haber cycle, 375, 379
Boron family, 356
　properties of, t357, t361
Boron hydrides, 360
Boron nitride, 299, 326, 358
Boron trichloride, 101, 122
Bosch process, 163
Boyd, R. N., 594
Bowman, W. H., 463
Boyle, Robert, 215
Boyle's law, 215, 221, 227
Bragg equation, 283

Bragg, L., 304
Bragg, William Henry, 279
Bragg, William Lawrence, 279
Branched-chain reactions, 482
Branched polymers, 631
Brant, Georg, 534
Brass, t547
Brasted, R. C., 321
Bravais, Auguste, 279
Brecia, F., 39
Breeder reactor, 508
Brick, 352
Bridge bonds, 359
Brimstone, 313
Bromine, 193
    electron clouds in, 102
Bromine fluoride, electron clouds in, 102
Brønsted, J. N., 148, 154
Brønsted-Lowry theory, 149, 403
Bronze, 345, t547
Buffer solutions, 402
Buffers, 414
    bicarbonate, 414
    in blood, 318, 414

Cadmium, 550
Calcite, 183, 297
Calcium, 169
Calcium carbonate, 183
Calcium fluoride, solubility of, 409
Calcium fluoride structure, 296
Californium, 505
Calorimeter, 370
Calorie, 11
Campbell, J. A., 388
Campbell, R. J., 241
Canagaratna, S. G., 168
Capture, nuclear, 494
Carbohydrates, 642
Carboloy, 543
Carbon dioxide, 95
    in blood, 318
    bond energy in, 376
    crystal structure of, 298
    properties of, 348
    rotation and vibration in, 232
    solubility in water, 268
Carbon family, 339
    compounds, t344
    oxidation state pattern, t342
    oxides, t348
    properties of, t340, t345
Carbon monoxide, 134
Carbonium ion, 604, 613
Carbonyl group, 575
Carboxylic acids, 577
    names, formulas, and physical con-
        stants of, t577
Carnotite, 539

Carotene, 570
Cassiterite, t345
Catalysis, 468
    enzyme, 653
Catalyst, 49, 395
    heterogeneous, 317
    homogeneous, 317
Catalytic hydrogenation, 599
Catenation, 309, 340
Cathode, 444
Cathode ray, 20, 21, 22
Cell
    chemistry of, 652
    diagram of, 653
    electrolysis, 444
    fuel, 457
    galvanic, 446
    voltage of, 452, 448
        and equilibrium constant, 454
Cell reaction, 446
Cellulose, 636, 643
Cellulose acetate, 637
Celsius, 11, 218
Cementite, 300
Ceramics, 351
Ceruloplasmin, 531
Cerussite, t345
Cesium, 169
Cesium chloride structure, 295
Chadwick, James, 27, 28
Chain reaction, 481, 504
Chair structure, 568
Chalcopyrite, 544
Changeux, J. P., 674
Charge balance, 404, 406
Charge transfer process, 529
Charles' law, 217, 221, 227
Chase, M., 669
Chave, K. E., 278
Chelates, 519
Chemical changes, 8
Chemical cycles, in the biosphere, 164
Chemical equilibrium, 146, 389
Chemical properties, 6, t8
Chemical senses, 136
Chemicals
    in crops, 57
    in food, 57
    in food preparation, 617
Chemistry, 2, 4
Chemotherapeutic agents, 208
China, 352
Chlorate ion, 101
Chlorine, 193
Chloromycetin, 646
Cholesterol, 640
Choppin, G., 512
Chrom steel, t541
Chromium, 539
Chromophores, 588

Cis-trans isomerism, 570
Classification, simplification by, 4, 559
Clay, 352
Clement, Nicolas, 193
Close-packed structures, 285
Clouds, 252
Cobalt, 542
Cobalt chromium steels, 543
Cobalt ions, in body, 530
Cobalt magnets, 552
Codons, 666
Coenzyme, 655
Coenzyme A, 660
Collision theory, 476
Collision, three-body, 478
Color, 86
    of complex ions, 525
    due to charge-transfer, 529
    and structure, 588
    of transition element ions, t536
    and wavelength, t63
Color television, 351
Combining forces, nature of, 92
Combining weights, 9
Common-ion effect, 400, 410
Companion, A. L., 141
Completion of reaction, 434
Complex ions, 413, 516
    bonding, 517
    color of, 525
    configuration of, 521
    coordination number and shape, t518
    isomerism in, 527
    nomenclature, 520
    stability and lability in, 528
Complexes, pi-bonding in, 526
Composition, percent, 42
Compounds, 6
    atomic ratios in, 96
    intermetallic, 289
Compressibility factor, 242
Concentration, 262
    and cell voltage, 453
Condensation, 601
Condensation polymers, 626, t630
Conduction
    electrolytic, 271, 442
    metallic, 271, 442
Conductivity, heat and electricity, t175,
    287
Configuration, of optical isomers, 586
Conformers, 566
Conn, E. E., 674
Constant, Michaelis, 488
Conversion Relationships, tii
Coordinate covalent bond, 101, 520
Coordinates, spherical and Cartesian
        compared, 75
Coordination compounds, 515, 517
Coordination number, 285, 294, 518

Copolymers, 631
Copper, 546
    ions of, in body, 530
    and radiation damage, 530
Cortisone, t640
Coulomb's law, 92
Covalent bond, 99, 116, 118
Covalent bond distance, 118
Covalent substances, 93, 106
Cracking, 567
Cram, D. J., 594, 625
Crick, F. H. C., 652, 669, 674
Critical pressure, 245
Critical temperature, 245
Cromer, D. T., 91
Crookes, William, 20
Cross-linking, 632, 648
Cross section, nuclear, 499
Cryolite, 196, 362
Crystal field theory, 523
Crystal systems, 280
Crystallinity, in polymers, 632, 634, 648
Crystals
    arrangement of ions in, 294
    ionic, coordination numbers in, 295
    molecular, 297
Cubic close-packing, 285
Cubic crystals, 280
Curie, 503
Curie, Marie Sklodowska, 22, 493
Curie, Pierre, 22
Curium, 505
Cycloalkanes, 568
Cyclopentadiene complex with Fe II, 542
Cyclotron, 493, 503
Cytochromes, 318
Cytoplasm, 653

Dacron, 627
Dalton, John, 8, 9, 10, 19, 20, 223, 501
Dalton's law, 223, 228, 268
Dampier, Margaret, 39, 61
Dampier, W. C., 39, 61
Davisson, C. J., 72
Davy, Humphry, 169, 182
de Broglie, Louis, 71
DDT, 618
Decaborane, 360
Dehydration, 609
Dehydrohalogenation, 609
d-Electron group, 515
Delocalization, 126
Democritus, 8
Density, t172
    of elements, txiii
    of gases, txiv
Deoxyribonucleic acid, see DNA
Deoxyribose, 666
Deuterium, 160

Dew, 252
Dextrorotatory, 585
Diamagnetism, 100, 519
Diamond, 299, 341, 346
Diborane, 356, 359
Dielectric constant, 259, 270
Dielectrics, 363
Dienes, 570
Diffusion, 224
Diopside, 349
Dipolar forces, 134
Dipole, 102
Dipole moment, 102
    of hydrogen halides, t203
    of nitrogen family hydrides, t332
Directive influence of substituent groups, 616
Disaccharides, 644
Dissolution, 258
Dissymmetric structures, 583
Distillation, 250
    fractional, 266
Distribution of molecular speeds, 228
Diuretic, 208
DNA, 334, 664, 667
    composition, t669
Doped semiconductors, 291
Dosage, radiation, 502
Douglas, B. E., 141, 212, 533, 555
Dreisbach, D., 257, 278
Drugs, 589
    sulfa, 646
Drying oils, 642
Ductility, 287
Duralumin, 289, 547
Durfee, H. F., 355
Duriron, 289
Dyes, 588

Eastland, B. J., 512
Eclipsed arrangement, 566
Edema, 187
Edison battery, 544
18-carat gold, t547
Einstein, Albert, 65
Einsteinium, 505
Eisch, J. J., 192
Elastomers, 633
Electrical conductivity, in glasses, 363
Electrochemical cells, 442
Electrochemical series, 178
Electrode
    glass, 456
    reference, 456
Electrode potentials, 448
    sign of, 452
Electrolysis, 164, 179, 201, 207, 362, 444, 445
Electrolyte, 269

Electrolytic conduction, 271
Electron cloud, 74, 100, 108, 125, 516
Electron configurations, t81
    alkali and alkaline earth, t170
    noble gas, 98, 100
    of halogen elements, t194
    of transition elements, 535
Electron transfer, 103
Electron-deficient molecules, 356, 358
Electron-density maps, 283, 545
Electron-dot formulas, 99
Electronegativity, 104, t105, 307
    and acid strength, 206, 327
    of halogens, t198
Electroneutrality relation, 404
Electronic ceramics, 363
Electronic glasses, 363
Electrons, 20, 34
    in the atom, 62
    bond pairs, 123
    charge on, 23
    community sharing, 108
    lone pairs, 123
    mass of, 24
    pairing of, 100
    probability representation of, 73
    spin of, 70, 74
    unpaired, 70
    wave nature, 71
Electrovalent binding, 102
Electrovalent substances, 93, 107
Elements, 6
    classification of, 6
    manufacture of, 505
    transition, 515
    transuranium, 505
Elimination reactions, 608
e/m
    alpha particle, t23, 27
    beta particle, t23
    determination of, 23
    electron, 20, 22, 24
    positive rays, 22
    proton, 26
Emission, nuclear, 494
Emphysema, 319
Enantiomorphs, 583
Endocrine glands, 483
Endothermic reaction, 397
Energy, 4, 369, 372
    atomic, 377
    binding, of nucleus, 496
    bond, 375
    bond-breaking, 234
    electronic, 230, 377
    from excited nuclei, 495
    heat, 11
    internal, t232, 370
        of molecules, 230
    kinetic, 11

lattice, 299, 378
  of metal ion d levels, 523
  in molecules, 377
  potential, 11
  radiation, 11
  rotational, 230, 377
  from the sun, 382
  translational, 230, 378
  vibrational, 230, 377
Energy of activation, 476
Energy bands, in metals, 290
Energy of bonding, 377
Energy converter, in the cell, 656
Energy of formation, of ions in solution, 379
Energy of hydration, 260, t178
Energy levels
  and atomic number, 77
  in hydrogen atom, 66
  in nucleus, 499
  rotational and vibrational, 231
  and spectra, t70
  subshell, 69
Energy reserve, in muscle, 671
Energy scale of the universe, 506
Energy transfer, 422
Enthalpy, 371
  of activation, 480
  of formation, 372, t373
    of ions, 380
  of vaporization, 374
Entropy, 423
  of activation, 480
  free energy and, 425
  internal, 426, 427
  of a solid, 426
  translational, 426, 427
Enzyme, 468, 653
  kinetics, 488
  properites of, 655
  types of, 655
Equation, 43, 46
  and Avogadro's law, 55
  balancing oxidation-reduction, vii
  calculations from, 50–56
  information shown in, 49
  interpretation of, 52
  limitations of, 50
  net, 47
  over-all, 47
  writing, 47–50
Equilibrium, 146, 389
  in acid-base reactions, 150
  buffers, 402
  free energy and, 427
  hydrolysis, 402
  ionic, 398
  in oxidation-reduction reactions, 158
  pH, 401
  position of, 147, 395

solid, liquid, and vapor, 250
  solubility, 408
  types of, 393
Equilibrium constant, 390
  calculation of, 392
  and cell voltage, 454
  exact values of, 394
  free energy and, 430
  magnitude, 433
  temperature and, 434
Equivalent weight, 445
Esters, 579, 641
Ethers, 562
Ethyl alcohol, t574
Ethylene, 340
Etzel, H. W., 304
Eutectic, t265
Evans, R. B., III, 241
Evaporation, 246
Exclusion principle, 74
Exothermic reaction, 398
Explosion, 482
Eyring, Henry, 464

Face-centered cubic structure, 286
Fahrenheit, 12, 218
Faraday, Michael, 20, 268, 444, 447
Faraday's laws, 444
Fats, 641
Fehling's solution, 577
Feldspars, 350, 352
f-Electron group, 515
Fermentation, 643
Fermium, 505
Fern, W. O., 192
Ferrites, 363
Ferrocene, 542
Ferroelectrics, 363
Ferromagnetism, 551
Fibers, structure, 634
Films, structure, 634
Fission, 504, 505
Fitzgerel, R. K., 241
Flanagin, D., 115
Flory, Paul J., 626
Fluorescence, 22
Fluorine, 193
Fluorine oxide, 119
  bond angle, 310
Fluoroborate ion, 358
Fluorocarbon, 196
Fluorspar, 196
Food, 660
Forbidden zone, 291
Forces
  between molecules, 133
  nucleon-nucleon, 499
Fogs, 252
Ford-Smith, J. H., 321, 338, 355

Formaldehyde, 124, t576
Formic acid, 578
Formula, 42
  determination of, 44
  electron-dot, 99
  empirical, 42
  molecular, 42, 45
  projection, 586
  simplest, 44
  structural, 42
Fowler, W. A., 512
Francium, 169
Free energy, 420, 423
  and electrical work, 448
  measurement of, 442
Free energy of formation, 434, t435
Free radical, 611
Freezing point, 249
Freezing-point depression, 264, t269
  abnormal, 268
  molecular weight from, 45
Frenkel, J., 257
Friedel-Crafts reaction, 614
Friedrich, Walter, 282
Fructose, 642, 643
Fuel
  fossil, 33
  nuclear, 33
Fuel cells, 457
Functional group, 560, t561
Furth, J. J., 674
Fusion, nuclear, 34, 497, 505
Fusion-power reactor, 508

Gadolinite, 538
Galena, t345
Galileo, 1, 216
Gallium, 356
Galvanic cell, 446
  pH and, 455
Gamma rays, t23, 494
  penetration of matter by, 27
Gamow, G., 91
Garnet, 348
Garrett, A. B., 39, 91, 365, 512
Gas constant, values of, 222
Gas law, general, 220
Gas laws, 215
  deviations from perfect, 242
Gases
  compressible space, 243
  liquefaction of, 245
  mixing of perfect, 421
  solutions of, 267
Gasoline, 567
  high-octane, 605
Gastric juice, 274
Gay-Lussac's law, 221, 227
Geber, 145

Gehman, W. G., 115
Geiger, Hans, 24
Gems, 362
General gas law, 220
Geometrical isomerism, 569
Germanite, t345
Germanium, 339
Germer, L. H., 72
Gershenson, D. E., 39, 91
Gillespie, R. J., 141, 321
Gilman, H., 192
Glass, 351, 363
Glass transition temperatures, 633
Glucose, 635, 642
    oxidation of, 435, 654, 658
Glycerol, 574
Glycogen, 636
Glycolysis, 188, 658
Goldish, D. M., 168
Goldschmidt process, 543
Goldwater, L. J., 555
Goodstein, M. P., 168
Gough, W. C., 512
Graham, Thomas, 224
Graham's law, 223, 228
Gram-atomic weight, 41
Gram-formula weight, 43
Gram-molecular weight, 41, 43
Graphite, 299, 341, 346
Gray, H. B., 533
Greek philosophers, 2
Greenberg, D. A., 39, 91
Grignard, Victor, 595
Grignard reagents, 606
Growth regulators, 335
Guncotton, 637

Haber, Fritz, 322
Half-life, 499, t501
Half-reactions, 443, 449
Hall, Charles Martin, 356
Halogen family, 193
    compounds with boron family
        elements, t357
    hydration energy of ions, t260
    properties of, t197
Halogen replacement, mechanism of,
    612
Hammond, G. S., 594, 625
Heat, 11, 372, 422
    of fusion, t174, 250
        of hydrogen halides, t203
    of reaction, 369, 420
        and temperature, 381
    of solution, 260, t380
    of vaporization, 174, t248, 382
        of hydrogen halides, t203
Heat capacity, 11, 235, 381
Heilbrunn, L. V., 192

Heisenberg, Werner, 72
Helium, properties of, 245
Helmholtz, Hermann Ludwig
    Ferdinand von, 442
Hemocyanin, 531
Hemoglobin, 318, 414, 541
Hendrickson, J. B., 594, 625, 651
Henry's law, 268
Heredity, 668
Hershey, A. D., 669
Hess, Germain Henri, 369
Hess' law, 373
Hexagonal close-packing, 285
Hexagonal crystals, 280
High explosives, 582
High-spin complex, 525
Hildebrand, J. H., 241, 257, 278
Histamine, 208
Hogg, J. C., 16
Holden, A., 278, 304
Homeostatic regulation, 482
Homologous series, 560
Homopolymers, 631
Hormones, 483
Hoyle, G., 192
Human fat, analysis of, t641
Hume-Rothery, W., 304
Hund's rule, 78
Hurwitz, J., 674
Hybrid orbitals, 310, 344
Hybridization, 120
Hydration, 178, 259, 317
Hydration energy, t260
Hydrazine, 328
Hydrocarbons, 340
    boiling points of, t134
    saturated, 562
    Van der Waals constants, t245
Hydrogen, 158
    chemistry of, 158
    occurrence, 158
    preparation, 163
    properties, t159, 245
    reactions, 161
    solubility in water, 268
    uses, 160
Hydrogen bond, 135, 298, 316
    in HF, 203
Hydrogen chloride, solubility in water,
    268
Hydrogen halides, properties of, t134,
    t203
Hydrogen iodide equilibrium, 146, t390
Hydrogen selenide, 119
Hydrogen sulfide, 118
    properties of, t315
    solubility in water, 268
Hydrogen telluride, 119
Hydrogenation, 162, 599, 642
Hydrohalic acids, 203

Hydrolysis, 199, 316, 402
    of ATP, 436
Hydronium ion, 148, 398
Hydronium ion concentration, and
    galvanic cells, 455

Incendiaries, 181
Indicators, 403
Indium, 356
Inert complex, 528
Infinite dilution, 429
Infrared spectra, 233
Inhibitor, 481
Inorganic world, 339
Insecticides, 617
Instability constants, txvi
Insulators, 291
Interhalogen compounds, 202
Interionic attraction, 270
Intermetallic compounds, 289
Internal energy, 230
Internuclear distance, 118
Interstitial solution, 288
Intracellular fluid, 274
Iodine, 193
Iodine heptafluoride, 124
Iodine pentafluoride, 124
Ion-dipole interaction, 517
Ionic equilibria, 403
Ionic reactions, speed of, 271
Ionic solutions, properties of, 272
Ionization, theory of, 268
Ionization constant, 152, t153, t399
    calculations, 400, 403
    of nitrogen family acids, t327
Ionization energy, 83, t84, t85, 184
    alkali and alkaline earth elements,
        t171
    of halogens, t198
Ionization potential, 83
Ionizing radiation
    effect on tissue, 502
    power of, 496
Ion-product constant, 401
Ions, 103, 269
    in body chemistry, 187
    complex, 413, 516
    formulas of, 32
    hydrated, 259
    separation of, by precipitation, 411
Iron, 540
Isomers
    of complex ions, 527
    geometrical, 569
    optical, 583
    position, 562
    properties of, in chlorination reactions,
        611
        of substituted benzenes, 583, 615

Isotope effects, 160
Isotopes, 29, 500
  argon, 32
  copper, t36
  of elements, txiv
  hydrogen. 160
  lithium, t37
  neon, 30
  potassium, 32
  silicon, t37
IUPAC nomenclature, 565, 569, 571, 573, 574, 575, 577

James, B. D., 365
Jancso, G., 257
Jeremiah, 339
Johnson, O., 533, 555
Johnson, R., 533
Johnson, R. C., 212, 555
Joliot-Curie, Frederic, 503
Joliot-Curie, Irene, 503
Jolly, W. L., 321, 338
Joule-Thomson effect, 245

Karraker, D. G., 533
Katritzsky, A. R., 651
Kay, W. A., 39
Kekule, 559
Kelvin, 12, 40, 218
Kendall, H. W., 512
Ketones, 575
Kieffer, W. F., 61
Kinetic energy, of molecules, 226
Kinetic-molecular theory, 225
Kinetics, 464
King, E. L., 489
Kirschbaum, J., 674
Kirschner, S., 533
Klaproth, Martin, 313
Klein, M. J., 91
Knipping, Paul C. M., 282
Kornberg, A.. 670
Krebs cycle, 658

Labile complex, 528
Lactose, 644
Lagowski, J. M., 651
Laidler, K. L., 321, 338, 355
Lanthanide contraction, 549
Lanthanides, 80, 549
Larsen, E. M., 555
Latham, J. L., 489
Lattice energy, 299, 378
Lattice, space, 280
Lavoisier, Antoine, 4
Law
  Avogadro's, 220

Boyle's, 215
Charles', 217
Coulomb's, 67, 92
  of conservation of energy, 9, 179
  of conservation of mass, 9
Dalton's, 223
  of definite composition, 9
Faraday's, 444
Gay-Lussac's, 221
Graham's, 223
Henry's, 268
Hess', 373
  of mass action, 471
  Michaelis-Menten, 488
  of multiple proportions, 10
  periodic, 80
  Raoult's, 263
Lawrence, R. M., 463
Lawrencium, 505
LCAO, 130
Lead, 339
Le Chatelier, Henri-Louis, 389, 395
Le Chatelier's principle, 252, 396, 413, 435
Leddy, J. J., 192
Lee, G. L., 594, 625
Leicester, H. M., 16
Lerner, J., 489
Levorotatory, 585
Lewis, G. N., 92, 99, 420
Lewis acids, 149, 538
Lewis formulas, 99
Lewis, W. C. McC., 464
Liehr, A. D., 533
Ligand field theory, 525
Ligands, 519
  and properties of complex ions, 529, 543
Light, 64
  polarized, 585
  reflection, 86
  scattering, 86
Lime, 426
Limekiln, 184, 382
Limestone, 183
Linear accelerator, 493
Linear polymers, 631
Linnett, J. W., 91
Lippincott, W. T., 16
Lipscomb, W. M., 365
Liquefaction, of air, 246
Liquids, 246
  solutions of, in liquids, 266
  undercooled, 249
Lithium, 169
Lithium aluminum hydride, 601
Logarithms, properties of, 429, 433
Low-spin complexes, 524
Lowry, M., 148
Lucretius, 20

Lymph, 274

Macromolecules, 44, 107, 325
MacWood, G. E., 441
Magnesium, 169
Magnets, 551
Magnetic moment, 520
Mahan, B. H., 388, 441
Malathion, 619
Malleability, 287
Maltose, 644
Manganese, 540
Manganese dioxide, 201
Manometer, 247
Marble, 183
Marcasite, 10
Mark, H. F., 651
Markovnikov's rule, 605
Marsden, Ernest, 24
Mason, E. A., 241
Mass defect, 35
Mass number, 31
Mass spectrometer, 29, 30
Material-balance relation, 404, 406
Matter, 4
Maynard, J. L., 321
McConnell, D., 355
McDaniel, D. H., 141, 212, 555
Measurement, standards of, 56
Mechanism
  of aromatic substitution, 614
  of enzyme action, 654
  of reaction, 472, 604
Meerschaum, 347
Melanin, 668
Melting point, 249
  factors influencing, 173
  of elements, txiii
  of fluorine and chlorine compounds, t195
  of hydrogen halides, t203
  of oxides and sulfides, t309
Mendeleev, Dmitri, 80
Mendelevium, 505
Mercury, 550
Meselson, M., 669
Meta-directing groups, 616
Metal foil experiment, 24
Metallic binding, 107
Metallic luster, 287
Metals, 5, 93
  band theory of, 290
  crystals of, 95, 283
  crystal structures of, t287
  ions of, in body, 530
  and nonmetals contrasted, 186
  properties of, 287
  as reducing agents, 177
  reserves of, 111

Methane, 120
    combustion of, 374
    heat of vaporization, 248
Mica, 297, 347
Michaelis-Menten law, 488
Milk, composition of, 644
Millikan, Robert, 23, 24, 26
Mixtures, 5
    homogeneous, 258
Model, 225
Moderators, 503
Moeller, T., 555
Molal solution, 263
Molar solution, 262
Mole, 41
    in calculations, 43
    of molecules, 43
Mole fraction, 263
Molecular crystals, 297
Molecular orbitals, 129, 310
    antibonding, 130
    bonding, 130
    energy levels of, 131
    in hydrogen-like systems, 130
Molecular structure, 116
Molecular weight
    determination of, 45, 220, 265
    of polymers, 627
Molecules, 116
    shape of, 125
Monazite, 538
Monel, 289
Monoclinic crystals, 280
Monomers, 626
Moore, W. J., 304
Morphine, t639
Morrison, R. T., 594
Morrow, B. A., 39
Moseley, H. G. J., 26, 80
Murmann, R. K., 533
Muscle, 670
Muscle contraction, 188
Muscovite, 297, 350
Myers, J. N., 257
Mylar, 628, 648
Myofibrils, 670
Myosin, 670

Naphthalene, electron-density map, 284
Nash, L. K., 61, 388, 441
Natural silicates, 347
Neon, isotopes of, t32
Nernst equation, 453
Nerve impulses, 188
    chemistry of, 483
Network polymers, 631
Neutrino, 495
Neutron, 27, 34
    mass of, 27, 31

size of, 28
Neutron-proton ratio, 497
Nickel, 544
Nickel carbonyl, 544
Nickel coin, t547
Nicholson, M., 16
Nichrome, t541
Nicotine, t639
Nirenberg, Marshall, 670
Nitrate reductase, 334
Nitration, 614
Nitric acid, 99, 323
Nitriles, 608
Nitrite reductase, 334
Nitro compounds, 562
Nitrogen, 322
    atmospheric, fixing of, 328
    crystal structure of, 298
    properties of, 245
Nitrogen compounds, structure of molecules of, 325
Nitrogen cycle, 329
Nitrogen family, 322
    hydrides, 331, t332
    oxidation states, t323, 333
    properties of, t323, t330
    stereochemistry of, t326
Nitrogen trifluoride, 119
Nitrous oxide, 128
Nobelium, 505
Nobel gases
    boiling points of, t134
    Van der Waals constants, t245
Nomenclature
    inorganic, iv
    organic, see IUPAC
    of oxyacids, t204
    rules for, iv
Nonmetals, 5
    reactive, 193
Normal boiling point, 248
Normal solution, 272
Nosecones, 647
n-Type semiconductor, 292
Nuclear chemistry, 493
Nuclear power, 33, 508
Nuclear reactor, 27, 33, 504, 507
Nuclear science, applications of, 505
Nucleic acids, 661, 664
Nucleon, 493
Nucleon-nucleon force, 499
Nucleoproteins, 664
Nucleoside, 664
Nucleotides, 664
Nucleus, 25
    of the cell, 653
    charge on, 26
    of isotopes, 31
    mass of helium, 28
    particles in, 28

properties of, 496
    structure of, 29, 34
Nylon 6-6, t630

Octet, 98, 100, 314, 326
Odor, 136, 578
O'Driscoll, K. F., 651
Oils, 641
Olby, R., 674
Olefins, 568
Optical activity, 584, 585
Optical isomerism, 583
Orbital, 69, 75
    molecular, 129
    and valence bond theory, 521
Orbital shapes, 75, 121
Order, entropy and, 421
Order of a reaction, 471
Organic chemistry, 559, 595
    nomenclature of, see IUPAC
Organic reactions, 595
Organic world, 339
Organo-phosphorus compounds, 619
Orpiment, t329
Ortho-para directing groups, 616
Orthorhombic crystals, 280
Osmosis, 265
Osmotic pressure, 265
Ostwald, Wilhelm, 258
Ouellette, R. W., 594, 625
Oxidation, 154, 452, 595, 596
Oxidation potential, 177, 178, 449, t450
    alkali and alkaline earth elements, t177
    boron family, t361
    carbon family, t348
    halogens, t200
    in basic solution, txv
    nitrogen family, t331
    oxygen family, 312
    transition elements, 548, t537
Oxidation-reduction, 154, 312
    at electrodes, 443
Oxidation state, 156, t157, 186
    of halogens, t194
    of transition elements, 535
Oxidizing agents, 157
Oxime, 608
Oxyacids
    of the halogens, 204
    of nitrogen family elements, 327
    of oxygen family, t308
    strength of, 311
Oxygen
    in blood, 318
    properties of, 245
    solubility in water, 268
Oxygen-carbon dioxide exchange, in lungs, 318

Oxygen consumption, in muscle, 319
Oxygen family, 307
  compounds of
    acid strengths, t311
      properties, t309
      stereochemistry, t311
    hydrogen compounds, properties of,
      135, t315
    oxyacids, t308
    properties of, t308, t314
Oxyhemoglobin, 318
Ozone, 237, 309

Packing
  of dissimilar spheres, 294
  of similar spheres, 283
Paints, 642
Pancreatic juices, 274
Panofsky, W., 512
Paramagnetism, 100, 310, 519
Particle-waves, 72
Passivity, 540
Patronite, 539
Patton, A. R., 489
Pauli exclusion principle, 74, 78, 99, 108,
  290
Pauling, Linus, 115, 116
Penicillin, 646
  potassium benzyl, 285
Pentaborane, 360
Peptide linkages, 637
Perchloric acid, 99
Periodic law, 80
Periodic table, 7
  electronegativity and, 105
  energy levels and, 76
  group number and chemical similari-
    ties, 186
  and group properties, 184
  hydrogen bonding and, 135
  and ionization energies, 85
  and oxidation states, t187
Peroxides, 308
Peroxyacids, 308
Peterson, Q. R., 141
Petrochemicals, t567
Petroleum, 567
Petroleum products, t567
Pewter, 345
pH, 401, 456
Phase, 5
Phenobarbital, 647
Phillips, C. S. G., 212, 321
Phosphate transfer, in biological reac-
  tions, 436
Phosphine, 119, 328
Phosphocreatine, 671
Phosphorite, 329
Phosphorus, 322

Phosphorus pentachloride, 124
Phosphorus trichloride, 99
Photoelectric cells, 181
Photoelectric effect, 65
Photons, 66
Photosynthesis, 313
Physical changes, 8
Physical constants, ti
Physical properties, 6, t8
Pi bond, 125, 571
  in C = O, 233
  in complexes, 526
Pimental, G. C., 388
Pitt, G. J., 285
Pituitary, 483
Planck, Max, 65
Planck equation, 65, 67
Planck's constant, 65
Plant growth, 334
Plasma, 235, 274
Plasticizer, 633
Plastics, 647
  properties of, 633
  structure of, 633
Plexiglas, t629
Plutonium, 505
Polarimeter, 586
Polarization, 133
Pollutants, 236
Polonium, 307
Polydentate ligands, t519
Polyester, 627
Polyethylene, 568, 628
Polyhalide anions, 200
Polyhalogen, organic compounds, 572
Polymerization, 568
Polymers, 309, 324, 626
  natural, 634
  structure and properties, 632
  structures of, 341
Polypeptide, 637
Polystyrene, t628
Porcelain, 352
Porphyrin, 541
Positive rays, 22
Potassium, 169
Potassium superoxide, 176
Potential
  electrode, see oxidation potential
  liquid junction, 448
Potentiometric titration, 456
Power
  electric, water, wind, and tide, 383
  nuclear, 33
Precipitation, 410
  of sulfides, 412
Pressure, 372
  critical, t245
  osmotic, 265
  units of, 217

Priestley, Joseph, 4, 307
Principle of Le Chatelier, see Le Chate-
  lier's principle
Promethium, 505
Properties
  chemical, 6, t8
  physical, 6, t8
Propylene, 126
Proteins, 188, 637
  shapes of, 661, 663
  structure of, 663
Protein helix, 662
Protein synthesis, 661
Proton, 26, 34
  mass of, 26, 31
p-Type semiconductor, 292
Pupezin, J., 257
Purines, 661
Pyrimidines, 661
Pyrolusite, 540
Pyrrhotite, 544

Quantum, 65
Quantum number, 69
  magnetic, 71
  principle, 71
  relationships, t71, 74
  spin, 71
  subshell, 71
Quantum theory, 65
Quartz, 299

R, values of, 222, t223
Racemic mixtures, 588
Rad, 502
Radiation
  exposure to, 503
  radioactive, 495
Radioactivity, 22, 501
  artificial, 503
  polonium, 22
  radium, 22
  types of, t23
  units of, 502
  uranium, 22
Radiochemistry, 499
Radium, 169
Radius
  alkali and alkaline earth, 171
  of halogens, t198
  ionic, t303
  of nucleus, 496
Radius ratio, 294, t295
Raman, V. V., 441
Randall, Merle, 420
Randomness, 425
  entropy and, 421
Rankine, 218

Raoult's laws, 263, 269
Rate constant, 470
Rate-determining step, 475
Rate equations, 472
Rate of reaction, 465
 concentration and, 469
 temperature and, 468
Raw materials, supplies of, 110
Rayon, 637
Reaction
 addition, 596
 branched-chain, 482
 cell, 446
 chain, 481
 completion of, 434
 condensation, 596
 driving force of, 425
 electrode, 443
 elementary, 475
 elimination, 596
 endothermic, 421
 exothermic, 421
 first-order, 477
 free-energy change, 429
 hydrolytic, 316
 incomplete, 50
 mechanism of, 472
 order of, 471
 oxidation, 595
 products of, 50
 reduction, 596
 spontaneous, 420
 substitution, 596
 weight relationships in, 50–54
 yield in, 61
Reaction rate, 465
 and response system in animals, 482
 theories of, 475
Reactor, nuclear, 27, 504, 507
Realgar, t329
Reducing agents, 157
Reduction, 155, 452, 596, 599
Reinforced plastic, 648
Remy, H., 212
Replacement reactions, t612
Reserpine, t639
Resonance, 126, 580, 606
 conditions for, 127
Respiration, 317
Reversibility, thermodynamic, 423
Rhombohedral crystals, 280
Ribonucleic acid, see RNA
Rigidity, 95
RNA, 334, 661
 code triplets, t666
 messenger, 663
 transfer, 664
Roberts, M. S., 168
Robertson, J. M., 285
Robinson, B. J., 168

Roderick, W. R., 141
Roentgen, 26, 502
Roller, D. H. D., 39
Rotation, within molecules, 566
Rowlinson, J. S., 257
Rubber
 properties of, 633
 structure, 633
Rubidium, 169
Rule of eight, 97
 exceptions to, 100
Rutherford, Ernest, 24, 25, 26, 27, 28,
 62, 67

Saliva, 274
Saltlike compounds, of metals, 289
Salts, dissolving of, 413
Sanderson, R. T., 463
Saponification, 641
Sapphire, 362
Saran, t629
Sarcoplasm, 670
Saturated solution, 261
Scandium, 538
Scavenger, 300, 540
Schneer, C. J., 16
Schrödinger, Erwin, 62, 73, 91
Schrödinger equation, 72, 130, xi
Schubert, J., 192
Science, 1, 12
 meaning of laws in, 225
 power and limitations of, 12
Scientific method, 2, 229
Seaborg, G. T., 512, 549, 555
Sebera, D. K., 141
Seeing nonluminous objects, 85
Selenium, 307
Selvaratam, M., 168
Semiconductors, 291, 363
Semipermeable membrane, 266
Series
 radioactive disintegration, 501
 spectrochemical, 525
Sex hormones, t640
Shapes
 of halogen oxyanions, 204
 of molecules, t119
  predicting, 125
  symmetry and, 122
 of orbitals, 123
Sheehan, W. R., 489
SI Units, ti
Sigma bond, 125
Silicates, 347
Silicic acid, 342
Silicon, 339
Silicone, 341
Silver iodide, 196
Simplest formula units, 44

Singer, P., 278
Size
 of atoms and ions, 185
 of halogen atoms and ions, t198
 of HX molecules, 202
Slip planes, 287
Slope of a curve, 466
Sneed, M. C., 321
Snyder, A. E., 212
Soaps, 642
Soddy, Frederick, 29, 500
Sodium, 169
Sodium borohydride, 600
Sodium carbonate, 181
Sodium chloride, 95, 103
 crystal structure, 294
 freezing point depression by, 269
 heat of solution, 380
 lattice energy, 378
 radius ratio, 295
 structure, 295
Sodium compounds
 preparation of, 183
 production, t183
Sodium hydroxide, 182
 freezing point depression by, 269
Sodium peroxide, 176
Sodium and potassium balance, in body,
 188
Sodium silicate, 351
Solar battery, 292
 diagram of, 293
Solar energy, 383
Solder, 289, 345
Solid state, 279
Solids, deformation of, 110
Solubility, 261
 calculation of, 410
 of gases, 267, t268
 and temperature, 262
Solubility-product constant, 408, t409,
 txvii
Solute, 6, 260
Solutions, 6, 258
 buffer, 402
 constant-boiling, 266
 equilibrium in, 261
 of gases in solids, 268
 solid, 288
 of solids in solids, 268
 varieties of, 260
Solvation, 259
Solvay, Ernest, 181
Solvents, 6, 260
 acid-base, 260
 polar, 260
Sørensen, S. P. L., 401
Sources of energy, 383
Space lattice, 280
Spectrum

absorption, 230
  electromagnetic, 65
  emission, 63, t64
    of hydrogen, 66
  infrared, 233
Spectro-chemical series, 525
Spectrophotometry, infrared, 233
Spectroscope, 63
Speeds, molecular, 228
Spratley, R. D., 388
Spring steel, 289
Staggered arrangement, 566
Stahl, F. W., 669
Stainless steel, 289, 541
Standard conditions, 222
Standard state, 428, 431, 434
Stanley, Wendell, 668
Starch, 636, 643
State, initial and final, 370
States of matter, 5
Steady state, 481
Steels, 299
Stellite, 543
Stereochemistry, 118
  of nitrogen family structures, 326
Sterling silver, 289, 547
Stern-Gerlach experiment, 70
Steroids, t640
Stibine, 119
Stibnite, t329
Stoney, George, 20, 22
Strain, 568
Stratton, W. A., 441
Streptomycin, 646
Strong electrolytes, 270
Strong, L. E., 441
Strontium, 169
Structural materials, 647
Structure
  of atoms, 19
  close packed, 285
  crystal, 280
  of molecules, 118
  molecular, spectra and, 231
  and properties, 94, 106
Strychnine, t639
Stumpf, P. K., 674
Sublimation, 250
Subshell, 69
Substances
  classification of, 93
  compared, t94
  heterogeneous, 4
  homogeneous, 4
Substituents, organic, t566
Substitution reactions, 610
  aromatic, 613
Substitutional solutions, 288
Sucrose, 644
Sulfanilamide, 646

Sulfates, 313
Sulfides, 313
Sulfonation, 614
Sulfur, 307
  solid, 313
Sulfur dioxide, 127
Sulfur hexafluoride, 315
Sulfur trioxide, 128
  resonance in, 311
Sulfuric acid, 317
Superlattice compounds, 289
Supersaturated solutions, 261
Surface tension, 245
Surroundings, 422
Symbols, chemical, 9, 41
Symmetry, and molecular shape, 122
Synapse, 483
Synchrotron, 493
Synthesis, of elements, 506
Synthetic rubber, 628, 632
System, 422

Taube, H., 533
Technetium, 505
Technology, 12
Teflon, 626
Tellurium, 307
Temperature, 11
  absolute, 218
  critical, t245
Temperature scales, 11, 218
Terramycin, 646
Tetragonal crystals, 280
Thallium, 356
Theory, 19
  of acids and bases, 149
  activated complex, 478
  atomic, 8
  band, of metals, 290
  collision, of reaction rates, 476
  crystal field, 523
  of ionization, 268
  kinetic-molecular, 225
  ligand field, 525
  quantum, 65
Thermal explosions, 482
Thermionic emissions, 287
Thermisters, 363
Thermochemical measurements, 378
Thermodynamics, 420
  first law, 422
Thiols, 562
Thionic acids, 308
Thionous acids, 308
Thomson, Joseph J., 20, 21, 23, 24
Tin, 339
Tin chloride, 343
Titanium, 538
Titration, potentiometric, 456

Tollen's reagent, 577
Torricelli, Evangelista, 216
Trans-cis isomerism, 570
Transition elements, 80, 515, 534
  crystal structure, t287
  and human health, 530
  trends in properties, 536
Transmutation, 3
Tremolite, 349
Tricarboxylic acid cycle, 658
Triclinic crystals, 280
Trinitrobenzoic acid, decomposition of,
    t467, t471
Triple point, 252
Tritium, 161
2, 4-D, 335
Type metal, 289

Unit cell, 280
Units
  defining basic, 56
  of measurement, cgs, 11
  of measurement, mks, 11
Unsaturated oils, 642
Unstable nuclei, 494
Urea, from ammonium cyanate, 473

Valence bond theory, 520
Valence electrons, 97, 104
Valence shell, 98
Vanadinite, 539
Vanadium, 539
Vanadium steel, 289
Van de Graaff generator, 493
Van der Waals, J. D., 242
Van der Waals constants, t245
Van der Waals distance, 118
Van der Waals equation, 244
Van der Waals forces, 106, 133, t134,
    298, 358, 566, 572, 577
Van der Waals radii, t140
Van der Werf, C. A., 168
Van Hook, W. A., 257
Van Orden, H. A., 594, 625
Vapor density, molecular weight from,
    45
Vapor pressure, 247
  of water, txiv
  lowering of, 263
Vegetable oils, 642
Velocity, escape, 236
Verhoek, F. H., 115, 241, 355, 441
Viruses, 668
Vitamins, 644, t645
Von Hippel, A. R., 115
Von Laue, Max, 279, 282
Von Reichenstein, Baron, 313
Vulcanization, 632

Waser, J., 463
Water, 119, 134, 316
    in the body, 273
    bond angle, 310
    freezing point constant of, t264
    heat of vaporization, 248, 382
    of hydration, 262
    ion equilibrium, 401
    properties of, 245, t315
    quality of drinking, 57
    as solvent, 259
    structure of solid, 298
    vapor pressure of, txiv
Water glass, 351
Watson, James D., 652, 669

Wave equation, 72
Wave function, 72, x
    angular part, 75
    radial part, 75
Weak electrolytes, 270
Weeks, M. E., 16
Wells, A. F., 212
Werner, Alfred, 515, 518, 522
Williams, L. P., 192, 212
Williams, R. J. P., 212, 321
Wolfenden, J. H., 241
Work, 372, 422
    electrical, 447
    maximum, 423
    pressure-volume, 371

Work hardening, 288
Wurtzite structure, 296
Wynne-Jones, W. F. K., 464

X rays, 26
    diffraction of, 279, 282

Yellow brass, 289

Zinc, 550
Zinc blende, structure, 296
Zinc ions, in body, 530
Zone refining, 343

A B C D E F G H I J   9 8 7 6 5 4 3 2